GW01164000

physica status solidi c
www.pss-c.com
conferences and critical reviews

Editor-in-Chief

Martin Stutzmann, Garching

Regional Editors

Martin S. Brandt, Garching
Peter Deák, Budapest
José Roberto Leite, Saõ Paulo
John I. B. Wilson, Edinburgh

Managing Editor

Stefan Hildebrandt, Berlin

Proceedings

E-MRS 2003 Fall Meeting
Symposia A and C

Warsaw, Poland
15–19 September 2003

Guest Editors

Marek Godlewski and Jacek Kossut

1 · 2 · 2004

WILEY-VCH

physica status solidi (c) – conferences and critical reviews

Editor-in-Chief:	Martin Stutzmann
Managing Editor:	Stefan Hildebrandt
Production Editors:	André Danelius, Heike Höpcke, Irina Juschak
Editorial Assistance:	Katharina Fröhlich, Margit Schütz
Editorial Office:	physica status solidi Bühringstr. 10, 13086 Berlin, Germany Telephone: +49 (0) 30/47 03 13 31, Fax +49 (0) 30/47 03 13 34 e-mail: pss@wiley-vch.de
Publishers:	WILEY-VCH Verlag GmbH & Co. KGaA
Postal Address:	Bühringstr. 10, 13086 Berlin, Germany
Publishing Director:	Alexander Grossmann
Ordering:	Subscription Service, WILEY-VCH Verlag GmbH & Co. KGaA Postfach 10 11 61, 69451 Weinheim, Germany Telephone +49 (0) 62 01/60 64 00, Fax +49 (0) 62 01/60 61 84 e-mail: subservice@wiley-vch.de or through a bookseller
Printing House:	Druckhaus Thomas Müntzer GmbH, Bad Langensalza, Germany Printed on chlorine- and acid free paper.

physica status solidi (c) – conferences and critical reviews is published twelve times per year by WILEY-VCH Verlag GmbH & Co. KGaA.

Annual subscription rates 2004 for pss (a) and (c) *or* pss (b) and pss (c):

		Institutional*	Personal
Europe	Euro	4244/3858	298
Switzerland	SFr	7269/6608	444
All other areas	US$	5406/4914	394

Annual subscription rates 2004 for pss (a), pss (b) and pss (c):

		Institutional*	Personal
Europe	Euro	8488/7716	596
Switzerland	SFr	14538/13216	888
All other areas	US$	10812/9828	788

* print **and** electronic/print only **or** electronic only
pss (a) and/or pss (b) subscriptions: Now including pss (c) – conferences and critical reviews

Postage and handling charges included. All WILEY-VCH prices are exclusive of VAT.
Prices are subject to change.

Single print issues may be ordered by ISBN at www.wiley-vch.de or through your local bookseller.

ISBN 3-527-40504-6

© 2004 WILEY-VCH Verlag GmbH & Co. KGaA, Weinheim

All rights reserved (including those of translations into foreign language). No part of this issue may be reproduced in any form, by photoprint, microfilm or any other means, nor transmitted into a machine language, without written permission from the publisher.

For our American customers:
physica status solidi (c) – conferences and critical reviews (ISSN 1610-1634) is published twelve times per year by WILEY-VCH Verlag GmbH & Co. KGaA, Boschstr. 12, 69469 Weinheim, Germany. Periodicals postage paid at Jamaica, NY 11431. Air freight and mailing in the USA by Publications Expediting Service Inc., 200 Meacham Ave., Elmont, NY 11003. US Postmaster: send address changes to: physica status solidi (c), c/o WILEY-VCH, 111 River Street, Hoboken, NJ 07030.

Valid for users in the USA:
The appearance of the code at the bottom of the first page of an article (serial) indicates the copyright owner's consent that copies of the article may be made for personal or internal use, or for the personal or internal use of specific clients. This consent is given on the condition, however, that the copier pay the stated per-copy fee through the Copyright Clearance Center, Inc. (CCC) for copying beyond that permitted by Sections 107 or 108 of the U.S. Copyright Law. This consent does not extend to other kinds of copying, such as copying for general distribution, for advertising or promotional purposes, for creating new collective works, or for resale. For copying from back volumes of this journal see 'Permissions to Photo Copy: Publishers Fee List' of the CCC.

Contents

Full text on our homepage at: http://www.pss-c.com

This Table of Contents is organized according to the topics presented at the conference. Articles with page numbers marked (a) are reprinted from phys. stat. sol. (b) **201**, No. 2, 181–372 (2004). You may find papers with phys. stat. sol. (a) and phys. stat. sol. (c) citations in the two sections of this volume, separated by coloured sheets for easy orientation.
Note from the Publisher: This issue of physica status solidi (c) has been produced from publication-ready manuscript files, written by the authors using the provided Word or LaTeX templates.

Preface A 189

Committees 190

Preface C 191

Committees 192

Symposium A
Novel Wide Bandgap Materials for Optoelectronic and Electronic Applications

Electronic structure of shallow impurities in GaN studied via bound exciton magnetooptics *(invited)*
    R. Stępniewski, A. Wysmołek, and M. Potemski . . . . (a) 181–189

Optical detection of 2DEG in GaN/AlGaN structures – High magnetic field studies
    B. Chwalis, A. Wysmołek, R. Stępniewski, M. Potemski, W. Knap, J. M. Baranowski, N. Grandjean, J. Massies, P. Prystawko, and I. Grzegory . . . . (c) 193–197

Magnetotransport studies of Ga(Mn,Fe)N bulk crystals
    C. Jastrzebski, W. Gebicki, M. Zdrojek, M. Bockowski, B. Strojek, T. Szyszko, M. Kaminski, and S. Podsiadlo . . . . (c) 198–201

Sub-micron InGaN ring structures for high-efficiency LEDs
    H. W. Choi, P. R. Edwards, C. Liu, C. W. Jeon, R. W. Martin, I. M. Watson, M. D. Dawson, S. Tripathy, and S. J. Chua . . . . (c) 202–205

Surprisingly low built-in electric fields in quaternary AlInGaN heterostructures
    S. Anceau, P. Lefebvre, T. Suski, S. P. Łepkowski, H. Teisseyre, L. H. Dmowski, L. Konczewicz, A. Kaminska, A. Suchocki, H. Hirayama, and Y. Aoyagi . . . . . . (a) 190–194

Multi-phonon processes in C_{6V}^4 ($P6_3mc$) hexagonal semiconductors: GaN, ZnO, ZnS, BeO, CdS, and Al_2O_3 – sapphire: D_{3d}^6 (R3c)
    H. W. Kunert . . . . (c) 206–212

Site symmetry of erbium centers in GaN
    V. Glukhanyuk, H. Przybylińska, A. Kozanecki, and W. Jantsch . . . . (a) 195–198

Luminescent properties of wide bandgap materials at room temperature
    M. Godlewski, J. Szmidt, A. Olszyna, A. Werbowy, E. Łusakowska, M. R. Phillips, E. M. Goldys, and A. Sokołowska . . . . (c) 213–218

Luminescence transients in highly excited GaN grown by hydride vapor-phase epitaxy
    S. Juršėnas, S. Miasojedovas, G. Kurilčik, A. Žukauskas, and P. R. Hageman . . . (a) 199–202

Self-heating and microwave noise in AlGaN/GaN
    L. Ardaravičius, J. Liberis, A. Matulionis, L. F. Eastman, J. R. Shealy, and A. Vertiatchikh . . (a) 203–206

In-depth and in-plane profiling of light emission properties of InGaN-based laser diode
    M. Godlewski, E. M. Goldys, M. R. Phillips, T. Böttcher, S. Figge, D. Hommel, R. Czernecki,
    P. Prystawko, M. Leszczynski, P. Perlin, I. Grzegory, and S. Porowski . . . . (a) 207–211

Cathodoluminescence and atomic force microscopy study of n-type doped GaN epilayers
    M. Godlewski, E. Łusakowska, R. Bożek, E. M. Goldys, M. R. Phillips, T. Böttcher, S. Figge,
    and D. Hommel . . . . (a) 212–215

Compensation mechanisms in magnesium doped GaN
    M. Godlewski, H. Przybylińska, R. Bożek, E. M. Goldys, J. P. Bergman, B. Monemar,
    I. Grzegory, and S. Porowski . . . . (a) 216–220

Study of long-term stability of ohmic contacts to GaN
    E. Kaminska, K. Golaszewska, A. Piotrowska, A. Kuchuk, R. Kruszka, E. Papis, R. Szeloch,
    P. Janus, T. Gotszalk, and A. Barcz. . . . . (c) 219–222

Impact of post-growth thermal annealing on emission of InGaN/GaN multiple quantum wells
    Yung-Chen Cheng, S. Juršėnas, Shih-Wei Feng, C. C. Yang, Cheng-Ta Kuo,
    and Jian-Shihn Tsang. . . . . (a) 221–224

Optical characterization of GaN doping superlattices: as grown, hydrogen implanted, and annealed
    H. W. Kunert, D. J. Brink, M. Hayes, J. Malherbe, L. Prinsloo, J. Barnas, A. G. I. Machatine,
    and M.W. Diale . . . . (c) 223–228

Energy transfer to Er^{3+} ions in silicon-rich-silicon oxide: efficiency limitations
    D. Kuritsyn, A. Kozanecki, H. Przybylińska, and W. Jantsch. . . . . (c) 229–232

Gap state absorption in AlGaN photoconductors and solar-blind photodetectors
    V. Lebedev, I. Cimalla, U. Kaiser, and O. Ambacher . . . . (c) 233–237

GaN on Si substrates for LED and LD applications
    Suzuka Nishimura, Satoru Matsumoto, and Kazutaka Terashima . . . . (c) 238–241

Sensitive $In_{0.53}Ga_{0.47}As$/InP (SI) magnetic field sensors
    T. Przesławski, A. Wolkenberg, K. Regiński, and J. Kaniewski. . . . . (c) 242–246

Thermoelectric and thermal properties of AlInN thin films prepared by reactive radio-frequency sputtering
    Shigeo Yamaguchi, Ryohei Izaki, Yasuo Iwamura, and Atsushi Yamamoto . . . . (a) 225–228

Kinetics of radiative recombination of the "pseudo-donor–pseudo-acceptor" pairs in InGaN/GaN
    A. J. Zakrzewski. . . . . (c) 247–249

Cathodoluminescence properties of zinc oxide nanoparticles *(invited)*
    M. R. Phillips, O. Gelhausen, and E. M. Goldys . . . . (a) 229–234

Optical and magnetic resonance investigations of ZnO crystals doped with TM ions
    V. Yu. Ivanov, M. Godlewski, S. Yatsunenko, A. Khachapuridze, Z. Gołacki, M. Sawicki,
    A. Omel'chuk, M. Bulany, and A. Gorban' . . . . (c) 250–253

Influence of the substrate on the structural properties of sputter-deposited ZnO films
    Hyoun Woo Kim and Nam Ho Kim . . . . (a) 235–238

Optical and RBS studies in Tm implanted ZnO samples
    T. Monteiro, M. J. Soares, A. Neves, M. Oliveira, E. Rita, U. Wahl, and E. Alves. . . . . (c) 254–256

Ion beam fabricated silicon light emitting diodes *(invited)*
M. A. Lourenço, M. S. A. Siddiqui, G. Shao, R. M. Gwilliam, and K. P. Homewood . . . . . (a) 239–244

Defect dynamics in P$^+$ implanted 6H-SiC studied by positron annihilation spectroscopy
G. P. Karwasz, R. Rurali, G. Consolati, and P. Godignon . . . . (c) 257–260

The Si(001)/C$_2$H$_2$ interaction to form a buffer layer for 3C-SiC growth
A. Goryachko, Y. Yeromenko, K. Henkel, J. Wollweber, and D. Schmeißer . . . . (a) 245–248

Temporal response of CVD diamond detectors to modulated low energy X-ray beams
G. Conte, M. C. Rossi, S. Salvatori, F. Spaziani, Y. Avigal, R. Kalish, P. Ascarelli,
and D. Trucchi . . . . (a) 249–252

Deep UV detection by CVD diamond position sensitive devices
G. Mazzeo, G. Conte, M. C. Rossi, S. Salvatori, and V. Ralchenko . . . . (c) 261–264

The interaction of Pr$_2$O$_3$ with 4H-SiC(0001) surface
A. Goryachko, I. Paloumpa, G. Beuckert, Y. Burkov, and D. Schmeißer . . . . (c) 265–268

Hydrogenated nanocrystalline silicon carbide: fabrication, properties and heterostructure device application
H. Colder, P. Marie, L. Pichon, and R. Rizk . . . . (c) 269–273

Self-passivation mechanisms in clusters of N dopants in SiC
R. Rurali, E. Hernández, P. Godignon, J. Rebollo, and P. Ordejòn . . . . (c) 274–277

Micro-Raman study of laser damage in CdTe
M. J. Soares, J. C. Lopes, M. C. Carmo, and A. Neves . . . . (c) 278–280

Photo-ESR and optical studies of Cr photoionization transition in CdZnSe:Cr crystals
K. Swiatek, M. Godlewski, and T.P. Surkova . . . . (c) 281–284

1.54 µm luminescence quenching of erbium-doped hydrogeated amorphous silicon deposited by D.C. magnetron sputtering
M. Kechouane, D. Biggemen, and L. R. Tessler . . . . (c) 285–289

Review Article

Photonic applications of lithium niobate crystals *(invited)*
L. Arizmendi . . . . (a) 253–283

Phase relations in the growth of stoichiometric lithium niobate *(invited)*
K. Polgár, Á. Péter, and M. Ferriol . . . . (a) 284–288

Optical performance of Yb^{3+} in LiNbO$_3$ laser crystal *(invited)*
L. E. Bausá, M.O Ramírez, and E. Montoya . . . . (a) 289–297

Cr^{3+} ions in hydrogenated and proton exchanged lithium niobate crystals
A. Kamińska, L. Arizmendi, A. Barcz, E. Łusakowska, and A. Suchocki . . . . (a) 298–303

Lasers and medicine *(invited)*
Przemysław J. Deren . . . . (c) 290–294

Elementary energy bands in band structure calculations of some wide-bandgap crystals
M. Sznajder, D. M. Bercha, and K. Z. Rushchanskii . . . . (a) 304–307

UV-written channel waveguides in Er^{3+}-doped Bi_2O_3-based glass
A. Favre, E. Lee, V. Apostolopoulos, C.B.E. Gawith, C.Y. Tai, E. Taylor, Y. Kondo,
and F. Koizumi . . . . (c) 295–298

Electrical properties of the five-component chalcogenides of silver
O. L. Kheifets-Kobeleva, V. B. Zlokazov, N. V. Melnikova, L. L. Nugaeva, L. Ya. Kobelev,
and Ya. L. Kobelev . . . . (c) 299–302

Growth and characterization of $AgIn_5S_8$ and $CuIn_5S_8$ thin films
Liudmila V. Makhova, I. Konovalov, and R. Szargan . . . . (a) 308–311

Light scattering topography of excimer grade CaF_2 crystal
Yasunao Oyama, Suzuka Nishimura, and Kazutaka Terashima . . . . (c) 303–306

Effect of deposition conditions and annealing on residual stress of ITO films magnetron sputtered on silica
A. Užupis, S. Tamulevičius, R. Butkutė, B. Vengalis, and V. Lisauskas . . . . (c) 307–311

Thermoluminescence of doped $YAlO_3$ crystals
Ya. Zhydachevskii, A. Durygin, A. Suchocki, A. Matkovskii, D. Sugak, and Z. Frukacz . . . . (c) 312–316

Schottky barrier and ageing effect studies in Au(Cu)/p-CdTe contacts
Ye. O. Bilevych, A. V. Sukach, and V. V. Tetyorkin . . . . (c) 317–320

Symposium C
5th International Workshop on Molecular Beam Epitaxy and Vapour Phase Epitaxy Growth Physics and Technology

MOVPE growth and in situ characterization of GaN layers on sapphire substrates *(invited)*
H. Hardtdegen, N. Kaluza, R. Schmidt, R. Steins, E. V. Yakovlev, R. A. Talalaev,
Yu. N. Makarov, and J.-T. Zettler . . . . (a) 312–319

Growth optimisation of the GaN layers and GaN/AlGaN heterojunctions on bulk GaN substrates using plasma-assisted molecular beam epitaxy
C. Skierbiszewski, Z. Wasilewski, M. Siekacz, A. Feduniewicz, B. Pastuszka, I. Grzegory,
M. Leszczynski, and S. Porowski . . . . (a) 320–323

Comparative analysis of (0001)GaN and (001)GaAs growth kinetics under Ga-rich conditions
V.G. Mansurov, Yu. G. Galitsyn, and K. S. Zhuravlev . . . . (c) 321–324

Growth kinetics of (0001)GaN from Ga and NH_3 fluxes
V. G. Mansurov, Yu. G. Galitsyn, V. V. Preobrazhenskii, and K. S. Zhuravlev . . . . . . (c) 325–328

Influence of TBAs flow in MOCVD growth on nitrogen incorporation in GaAsN alloy and its optical quality detected by photoreflectance
W. Rudno-Rudzinski, R. Kudrawiec, J. Misiewicz, J. Derluyn, and I. Moerman. . . . . . . (c) 329–332

Kinetic and strain-driven growth phenomena on Si(001) *(invited)*
C. Schelling, J. Myslivecek, M. Mühlberger, H. Lichtenberger, Z. Zhong, B. Voigtländer,
G. Bauer, and F. Schäffler. . . . . (a) 324–328

Kinetics of the heteroepitaxial growth of Ge layer at low temperature on Si(001) in UHV-CVD
M. Halbwax, V. Yam, C. Clerc, Y. Zheng, D. Debarre, Lam. H. Nguyen, and D. Bouchier . . . (a) 329–332

Investigation of Si–Ge whisker growth by CVD
A. A. Druzhinin and I. P. Ostrovskii . . . . (c) 333–336

Lateral alignment of SiC dots on Si
   V. Cimalla, J. Pezoldt, Th. Stauden, A. A. Schmidt, K. Zekentes, and O. Ambacher . . . . . . (c) 337–340

Kinetic Monte Carlo simulation of SiC nucleation on Si(111)
   A. A. Schmidt, K. L. Safonov, Yu. V. Trushin, V. Cimalla, O. Ambacher, and J. Pezoldt. . . . (a) 333–337

The role of Ge predeposition temperature in the MBE epitaxy of SiC on silicon
   F. M. Morales, Ch. Zgheib, S. I. Molina, D. Araújo, R. García, C. Fernández, A. Sanz-Hervás,
   P. Masri, P. Weih, Th. Stauden, V. Cimalla, O. Ambacher, and J. Pezoldt . . . . (c) 341–346

3C-SiC:Ge alloys grown on Si (111) substrates by SSMBE
   P. Weih, V. Cimalla, Th. Stauden, R. Kosiba, G. Ecke, L. Spiess, H. Romanus, M. Gubisch,
   W. Bock, Th. Freitag, P. Fricke, O. Ambacher, and J. Pezoldt . . . . (c) 347–350

Growth of InSb thin films on GaAs(100) substrates by flash evaporation epitaxy
   M. Oszwaldowski, T. Berus, A. Borowska, R. Czajka, and M. Zimniak . . . . (c) 351–354

Properties of MBE $Cd_xHg_{1-x}Te$/GaAs structures modified by ion-beam milling
   V. V. Bogoboyashchyy, S. A. Dvoretsky, I. I. Izhnin, N. N. Mikhailov, Yu. G. Sidorov,
   F. F. Sizov, V. S. Varavin, and V. A. Yudenkov . . . . (c) 355–359

Modeling carrier dynamics in quantum-dot lasers *(invited)*
   A. Markus and A. Fiore. . . . . (a) 338–344

Quantum dot semiconductor lasers with optical feedback *(invited)*
   G. Huyet, D. O'Brien, S. P. Hegarty, J. G. McInerney, A. V. Uskov, D. Bimberg, C. Ribbat,
   V. M. Ustinov, A. E. Zhukov, S. S. Mikhrin, A. R. Kovsh, J. K. White, K. Hinzer,
   and A. J. SpringThorpe . . . . (a) 345–352

Formation and optical properties of Ge quantum dots selectively grown on patterned Si(001) substrates
   Lam H. Nguyen, V. Le Thanh, V. Yam, D. Débarre, M. Halbwax, and D. Bouchier . . . . . . (a) 353–356

Formation of Ge nanoislands on pure and oxidized Si surfaces by MBE
   A. I. Nikiforov, V. V. Ulyanov, A. G. Milekhin, O. P. Pchelyakov, S. A. Teys, S. Schulze,
   and D. R. T. Zahn . . . . (c) 360–363

TEM evaluation of strain and stress in III–V semiconductor epitaxial structures *(invited)*
   A. Rocher, M. Cabié, A. Ponchet, A. Arnoult, and E. Bedel-Pereira . . . . (a) 357–363

Optical properties of GaInNAs/GaAs quantum wells: character of optical transitions and carrier localisation effect
   R. Kudrawiec, J. Misiewicz, M. Fisher, and A. Forchel . . . . (a) 364–367

Light induced contrast in Kelvin Force Microscopy of GaN epilayers
   R. Bozek, K. Pakula, and J. M. Baranowski . . . . (c) 364–367

Raman scattering study of InGaAs/AlAsSb and InGaAs/AlAs/AlAsSb heterostructures
   T. Mozume, N. Georgiev, and J. Kasai . . . . (a) 368–371

Optical quality improvement of InGaAs/AlAs/AlAsSb coupled double quantum wells grown by molecular beam epitaxy
   J. Kasai, T. Mozume, H. Yoshida, T. Simoyama, A. V. Gopal, and H. Ishikawa. . . . . (c) 368–371

X-ray photoemission study of manganese thin films deposited on a layered semiconductor
   F. Mirabella, B. A. Parkinson, and J. Ghijsen . . . . (c) 372–377

Optical investigations of two dimensional electron gas in the AlGaN/GaN heterostructures
   R. Kudrawiec, M. Syperek, J. Misiewicz, R. Paszkiewicz, B. Paszkiewicz, M. Tłaczała,
   and W. Strupinski . . . . (c) 378–381

Resistivity and mobility in ordered InGaP grown by MOVPE
   S. Hasenöhrl, J. Betko, M. Morvic, J. Novák, and Fedor. . . . . (c) 382–387

Structural characterization of original 3D gallium structures grown by LP-MOCVD
    L. Imhoff, M. Sacilotti, C. Josse Courty, M. Mesnier, M. C. Marco de Lucas, and S. Bourgeois   (c) 388–391

MBE growth and characterization of InAs/GaAs for infrared detectors
    K. Regiński, J. Kaniewski, K. Kosiel, T. Przesławski, and J. Bąk-Misiuk . . . . (c) 392–395

Investigation of Indium Tin Oxide (ITO) films for the VCSEL laser with dielectric Bragg reflectors
    H. Wrzesińska, L. Ilka, D. Wawer, K. Hejduk, A. Kudła, M. Bugajski, and E. Łusakowska. . . (c) 396–400

Revealing of threading and misfit dislocations in partially relaxed InGaAs/GaAs heterostructures
    Oksana Yastrubchak, Elżbieta Łusakowska, Andrzej Morawski, Oleg Demchuk,
    and Tadeusz Wosiński . . . . (c) 401–404

Magnetic properties of ultra-thin epitaxial V/Gd bilayers
    P. Pankowski, L. T. Baczewski, T. Story, A. Wawro, K. Mergia, and S. Messoloras . . . . . . (c) 405–408

Author Index 409

DOI: The fastest way to find an article online is the *Digital Object Identifier* (DOI).
Starting in Volume 198, issue 2 (January 2003), DOIs have been printed in the header of the first page of every article in physica status solidi (b). On the WWW, one can find an article for example with a DOI of 10.1002/pssa.200306608 at **http://dx.doi.org/**10.1002/pssa.200306608.

Please use the DOI of the article to link from your home page to the articles in Wiley Interscience.

The DOI is a result of a cross-publisher initiative to create a system for the persistent identification of documents on digital networks. More information is available from **www.doi.org**.

Preface

E-MRS 2003 Fall Meeting Symposium A
Novel Wide Bandgap Materials
for Optoelectronic and Electronic Applications

In 2003 a new initiative was started – the Fall Meeting of the European Materials Research Society was organized for the first time in Warsaw, Poland, on 15–19 September 2003 in a venue of the Warsaw Technical University. The Workshop on Advanced Spectroscopy of Modern Materials, sponsored by the EC (contract GMA1-2002-72801) via ASPECT Centre of Excellence of Advanced Spectroscopy Applications in Physics, Modern Science, Biology and Environmental Protection, was organized as a part of the Symposium A of this Meeting, which was entitled Novel Wide Bandgap Materials for Optoelectronic and Electronic Applications.

The aim of the Workshop and of Symposium A was to review recent progress in the studies of such wide bandgap materials as nitrides and ZnO, as well as in relevant experimental techniques employed in these studies. Participants from several European countries were invited to participate in the Symposium. About 100 persons was registered as participants of this meeting.

Materials of the Symposium A are included, together with materials of the Symposium C, in this issue of *physica status solidi (c)*. From the 20 invited, 20 oral and 58 poster presentations accepted for the presentation, 50 papers (7 invited, 13 oral and 30 poster presentations) are included in these Proceedings, after a peer-reviewing procedure.

I am grateful to all participants who helped us in the reviewing procedure and who contributed to the success of the Meeting. We are very grateful to the EC and to the Polish Ministry of Science and Information Technology for financial support of the Workshop.

Marek Godlewski
Editor

Committees

Organisers

The European Materials Research Society
Institute of Physics, Polish Academy of Sciences, Warsaw, Poland
Centre of Excellence ASPECT at the Institute of Physics of the Polish Academy of Sciences, Warsaw, Poland
High Pressure Research Center, Polish Academy of Sciences, Warsaw, Poland

Prof. Marek Godlewski	Institute of Physics, Polish Academy of Sciences, Warsaw, Poland
Prof. Stanisław Krukowski	High Pressure Research Center, Polish Academy of Sciences, Warsaw, Poland
Prof. Andrii Matkovskii	Lviv Polytechnic, Lviv, Ukraine
Prof. Andrzej Suchocki	Institute of Physics, Polish Academy of Sciences, Warsaw, Poland
Dr. Adam J. Zakrzewski	Institute of Physics, Polish Academy of Sciences, Warsaw, Poland

Programme Committee

Prof. Jacek Baranowski	Institute of Experimental Physics, Warsaw University, Warsaw, Poland
Prof. Marek Godlewski	Institute of Physics, Polish Academy of Sciences, Warsaw, Poland
Prof. Adrian Kozanecki	Institute of Physics, Polish Academy of Sciences, Warsaw, Poland
Prof. Stanisław Krukowski	High Pressure Research Center, Polish Academy of Sciences, Warsaw, Poland
Prof. Andrii Matkovskii	Lviv Polytechnic, Lviv, Ukraine
Prof. Bo Monemar	Linköping University, Linköping, Sweden
Prof. Sylwester Porowski	High Pressure Research Center, Polish Academy of Sciences, Warsaw, Poland
Prof. Andrzej Suchocki	Institute of Physics, Polish Academy of Sciences, Warsaw, Poland
Prof. Tadeusz Suski	High Pressure Research Center, Polish Academy of Sciences, Warsaw, Poland

Sponsors

Sponsored by EU (Contract GMA 1-2002-72801) via "ASPECT" Centre of Excellence of Advanced Spectroscopy Application in Physics, Materials Science, Biology and Enviromental Protection (International Workshop on Advanced Spectroscopy of Modern Materials)

Preface

E-MRS 2003 Fall Meeting Symposium C
5th International Workshop on Molecular Beam Epitaxy and Vapour Phase Epitaxy Growth Physics and Technology

Workshops on Molecular Beam Epitaxy, Growth Physics and Technology (MBE GPT, for short) have been organized in Warsaw since 1994 every two years. In 2001 a decision was made to broaden the scope of these Workshops to encompass also other growth techniques of importance for low dimensional semiconductor structure fabrication, most notably the metalorganic vapor phase epitaxy (the acronym has been extended then to a lengthy MBE VPE GPT). One of the Workshops' aims was to bring together specialists in various aspects of the low-dimensional systems: crystal growers, engineers designing the equipment as well as physicists and electrical engineers who either characterize the physical properties or employ those structures in real life applications. The first Workshops were originally organized by the Institute of Vacuum Technology in Warsaw, then jointly by the Institute of Physics of the Polish Academy of Sciences, Institute of Electron Technology and the Institute of Experimental Physics of the Warsaw University. These institutions were joined later by the High Pressure Research Center of the Polish Academy of Sciences. All these institutions were leading in Poland in the techniques and methods that constituted the main subject of the Workshops. Warsaw as the Workshops venue was thought to be a suitable place for meetings of the specialists from all parts of the world. This was proven to be a right choice from that point of view. When the Center of Excellence CELDIS was established within one of the initiatives of the Fifth Framework Program of the European Community, the organization of the Workshops became one of its tasks. The most recent issue of the Workshop was held in Warsaw between 5–9 September 2003 and technically was a part of a larger meeting – the Fall Meeting of the European Materials Society (E-MRS) – moved to Warsaw from Strasbourg for the first time. Therefore it is not easy to precisely state how many participants did come to the Workshop – some of them were certainly session-hopping, taking advantage of the fact that other extremely interesting talks were presented in parallel. However, the number of scientists who officially registered for symposium C was 64. This number does not include a certain number of students. The present edition of the Workshop was co-sponsored by CELDIS and CEPHONA, a Center of Excellence active in the Institute of Electron Technology, and through them by the European Union and Polish State Committee for Scientific Research. Formally the auspices were extended also by the Polish Society for the Crystal Growth. Former Workshops had their proceedings published in Thin Solid Films. This is the first time that they are appearing in *physica status solidi* where we are happy to present papers accepted in a usual peer-reviewing procedure. Let me here express our gratitude to all the members of the program committee as well as to all referees without whose work the meeting itself and these proceedings would not be possible.

Warsaw, 21 December 2003
Jacek Kossut

Committees and Sponsors

Organisers

Centre of Excellence CELDIS at the Institute of Physics of the Polish Academy of Sciences
Centre of Excellence CEPHONA at the Institute of Electron Technology
Polish Society for Crystal Growth

Chairman:
Jacek Kossut   Institute of Physics, Polish Academy of Sciences, Warsaw, Poland

Co-chairmen:
Detlef Hommel   Institute of Physics, University of Bremen, Bremen, Germany
Zbigniew R. Żytkiewicz   Institute of Physics, Polish Academy of Sciences, Warsaw, Poland

Programme Committee

Jacek Baranowski   Institute of Experimental Physics, Warsaw University, Warsaw, Poland
Maciej Bugajski   Institute of Electron Technology, Warsaw, Poland
Detlef Hommel   Institute of Physics, University of Bremen, Bremen, Germany
Jacek Kossut   Institute of Physics, Polish Academy of Sciences, Warsaw, Poland
Maciej Oszwałdowski   Institute of Physics, Poznan University of Technology, Poznan, Poland
Sylwester Porowski   High Pressure Research Centre Unipress, Polish Academy of Sciences, Warsaw, Poland
Zbigniew R. Żytkiewicz   Institute of Physics, Polish Academy of Sciences, Warsaw, Poland

Sponsors

Centre of Excellence CELDIS (European Commission Contract ICA1-CT-2000-70018)
Centre of Excellence CEPHONA (European Commission Contract G5MA-CT-2002-04061)

physica pss status solidi a

www.pss-a.com

applied research

The following pages have been reprinted from
phys. stat. sol. (a) **201**, No. 2, 181–372 (2004) as part of the
Proceedings of the E-MRS 2003 Fall Meeting,
Symposia A and C,
held in Warsaw, Poland, 15–19 September 2003.

WILEY-VCH

Electronic structure of shallow impurities in GaN studied via bound exciton magnetooptics

R. Stępniewski[*,1,2], **A. Wysmołek**[1], and **M. Potemski**[2]

[1] Institute of Experimental Physics, Warsaw University, Hoża 69, 00-681 Warsaw, Poland
[2] Grenoble High Magnetic Field Laboratory, MPI/FKF-CNRS, BP166X, F-38042 Grenoble Cedex 9, France

Received 15 September 2003, revised 17 October 2003, accepted 17 October 2003
Published online 12 January 2004

PACS 71.20.Nr, 71.35.Ji, 71.55.Eq, 71.70.Ej, 78.55.Cr

We report the photoluminescence experiments on high quality GaN samples in high magnetic fields. A detailed analysis of the recombination due to excitons bound to neutral donors and acceptors is presented. Special attention is focussed on transitions for which the impurity is left in the excited state (so called two electron satellites). These results show a rich energetic structure of excited states of the impurity involved in such a recombination process. The magnetic field dependence of the energy structure of the shallow neutral donor and acceptor in GaN is then discussed. The validity of the effective mass approximation for shallow impurities in wide gap semiconductors is examined.

© 2004 WILEY-VCH Verlag GmbH & Co. KGaA, Weinheim

1 Introduction The possibility of the effective doping, via introducing the shallow electronic states into initially insulating materials, is an essential ingredient of the semiconductor physics. Impurities, in addition to the band structure determine the transport and optical properties of any semiconductor and make it potentially useful. A vigorous progress in the technology helps overcoming many barriers and, nowadays, any initially insulating material can be plausible transformed into a semiconductor. The effective p-doping of Gallium Nitride (GaN) is an example of such a problem which has been lately successfully resolved. The precise control of doping, so essential for appropriate design of semiconductor devices, continues to stimulate the basic studies of impurity centres in semiconductors. Understanding the shallow impurities, and in particular their electronic structure, persists to be one of the key subjects in semiconductor physics. A detailed knowledge of this structure (characteristic energies and symmetries of the ground and excited states) provides information about impurity location in the lattice and allows to identify its chemical origin.

A large variety of experimental techniques have been applied to study the energetic structure of impurities in semiconductors. Routine measurements of thermally activated resistivity is likely the simplest method which allow for rough estimation of the binding energy of the dominant impurity in lightly doped semiconductors. To obtain more precise information, the advanced optical studies have to be employed. In optical experiments the internal structure of the impurity states can be probed. Application of different external perturbations (strain, electric or magnetic field ...) which modify this structure is advantageous to obtained further information. The methods of magneto-optics are especially valuable. The magnetic field is a non-destructive probe and has a unique property, that influences both charge and spin excitations. Theory of the magnetic field effects is well elaborated, and B-field can be, for example, easily included in the models involving the effective mass approximation. The most typical method of the impurity investigations is the far infrared (FIR) magnetospectroscopy. In these experiments the intra-impurity transitions are directly observed in the transmission [1–3] or photoconductivity [4–7]. The

---

[*] Corresponding author: e-mail: Roman.Stepniewski@fuw.edu.pl, Phone: +48 22 55 32 166, Fax: +48 22 621 97 12

obtained results typically permit the observation of the dominant 1s-2p transition. Its energy can be used for the estimation of the ground state binding energy, while the splitting induced by the applied magnetic field gives precisely the effective mass value [1]. The absorption related to higher excited states is possible [5] but strongly limited due to the well-defined selection rules. Optionally, the electronic structure of the impurity can be studied with interband magneto-optics via probing more complex objects in which the impurity is involved. We would like to mention two examples of such experimental methods, both related to studies of donors in GaAs. The first one, selective magneto-luminescence spectroscopy of donor–acceptor pairs, allows to resolve a rich spectrum of the donor excitations [8], in consequence of the relaxed selection rules, due the presence of the axial electric field. Another method consist of measuring transitions related to the neutral-donor bound-exciton recombination. Here, the observation of the so called two electron satellites (TES), which occur when an exciton bound to a neutral donor – D^0X recombines and leaves the donor in an excited state, is advantageous. The two-electron energy is lower than that of the D^0X peak by the difference in the ground and excited state energies of the neutral donor. The observation of TES transitions gives the opportunity to reproduce a rich energetic structure of neutral impurity [9]. In this case, the selection rules are not as rigid as in the case of FIR spectroscopy. Similar method can be applied to the neutral acceptor when studying photoluminescence (PL) related to the exciton bound to a neutral acceptor (A^0X).

In this paper we review the recent results of TES magnetooptical studies of neutral donors and acceptors performed on high quality GaN samples.

2 Magnetospectroscopy of two-electron satellites of silicon and oxygen donors in GaN Freestanding GaN grown by hydride vapour-phase epitaxy (HVPE) on sapphire, and then separated from the substrate using laser lift-off technique, shows excellent structural [10] and transport properties [11]. Several groups investigate such samples, studying, in particular, shallow donor electronic structure. In FIR spectroscopy, two dominant donors has been identified as Si_{Ga} and O_N, with the binding energies equal to 30.16 meV and 33.2 meV, respectively [2]. The high optical quality of freestanding GaN samples allowed for the observation of different transitions due to recombination processes related to different shallow donors. A freestanding GaN sample used in our studies [12] was grown, separated, and polished at the Samsung Advanced Institute of Technology. All measurements were performed on the Ga face, which is the face farthest from the original Al_2O_3 substrate, and thus the face with the highest quality material. In the exciton range of the low temperature PL spectrum, shown in Fig. 1, three well-resolved lines $D_1^0X_A$, $D_2^0X_A$, $D_3^0X_A$, can be distinguished. When applying magnetic field all these lines split in a similar way showing dependencies typical for the neutral donor bound excitons in GaN [13]. In Fig. 2, the Zeeman splitting of these lines measured for the magnetic field direction perpendicular to the c-axis of the crystal (GaN crystallizes in wurzite symmetry) is presented. Thus, in the exciton range of the luminescence spectrum we can clearly resolve lines related to 3 different neutral donors D1, D2, and D3.

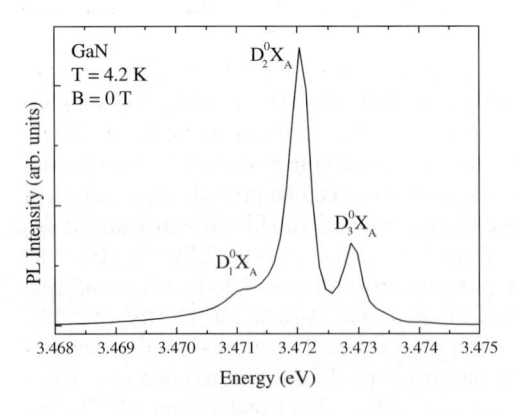

Fig. 1 Luminescence spectrum in the bound exciton recombination range measured at 4.2 K.

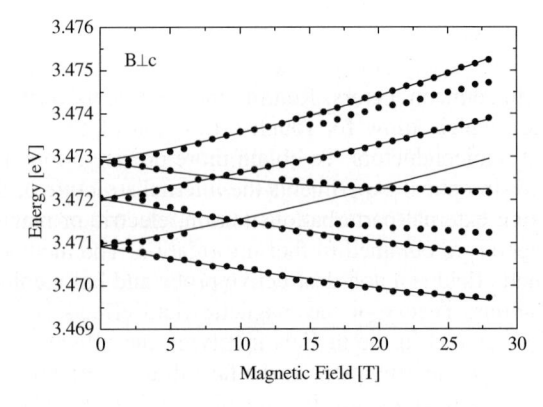

Fig. 2 Zeeman compoents of the $D_1^0X_A$, $D_2^0X_A$, $D_3^0X_A$ neutral donor recombiations, measured for B ⊥ c.

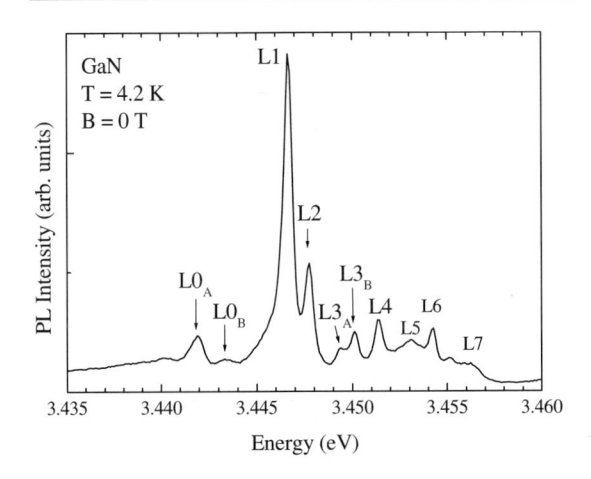

Fig. 3 Emission connected with two-electron satellite transitions measured at 4.2 K.

Figure 3 shows the PL spectrum of the freestandig GaN in the range of TES emission. A correct assignment between a given TES spectrum and its parent principal line is a central point for further interpretation of the magnetooptical results. We use time-resolved spectroscopy measurements [12] to assign groups of lines associated to a given neutral donor. The L3- and L4-lines correlate with $D_2^0X_A$ (transitions assigned to Silicon donor), whereas L5 is associated with $D_3^0X_A$. The chemical origin of a D3 donor is not identified yet. The case of the dominant L1 and L2 doublet is more controversial. We assume this doublet is associated with the low energy, $D_1^0X_A$ line (seen Fig. 1), although this later line is barely visible in time-resolved experiment [12]. L1 and L2 lines have not been observed in the homoepitaxial layers investigated previously by our group [13]. We assign these lines to the oxygen donor, which is

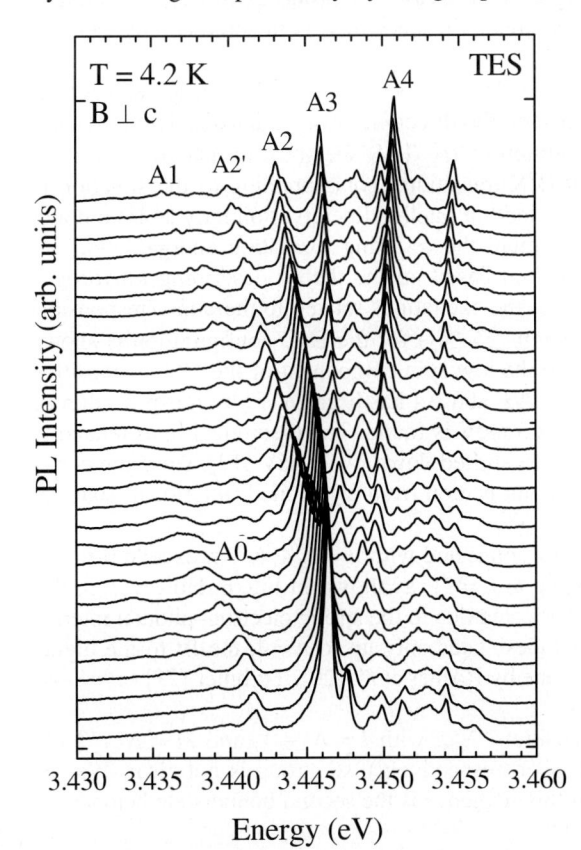

Fig. 4 Two-electron emission at 4.2 K for $B \perp c$, with B varying from 0 to 28 T.

Fig. 5 Energies of excited states of the oxygen donor with respect to the 1s ground state obtained from measurements (points) and compared with theoretical calculations (lines) based on a hydrogenlike model for the B⊥c configuration.

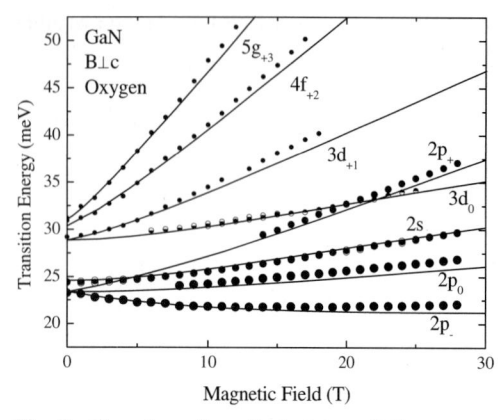

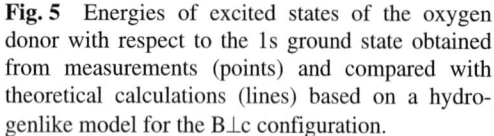

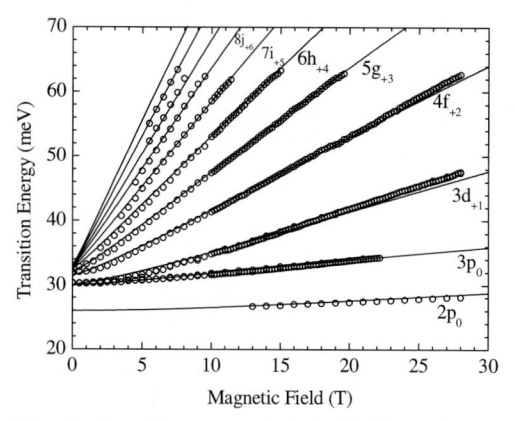

Fig. 6 Transition energy to the highly excited states of the oxygen, as a function of the magnetic field, as derived from TES spectra measured for B∥c.

unlikely present in the MOCVD homoepitaxial layers. This interpretation is different from the one proposed in Ref. [14]. Our point is that the splitting pattern shown in Fig. 2 clearly eliminates the possibility of the interpretation of the $D_1^0X_A$ line as due to the ionised donor bound exciton, which is, in contrary, pursued in Ref. 14. If related to ionised donor bound exciton, this line should show the zero-field splitting due to the electron–hole exchange interaction [15], but this splitting is clearly absent as can be seen from our data in the B ⊥ c configuration shown in Fig. 2. In any case, the proposed in Ref. 30 "shifted" assignment of $D_2^0X_A$ and $D_3^0X_A$ to oxygen and silicon donors, respectively, is in some contradiction to the results of time-resolved experiments [12].

When applying magnetic fields, the transitions observed in TES spectrum split. The observed splitting pattern depend, as usual for wurtzite symmetry crystals, on the direction of the magnetic field in respect to the crystal axis. In Fig. 4, the magnetic field evolution of the TES spectrum measured for B⊥c is shown. The energy difference between the principal D^0X transition, (corresponding to the situation in which the donor is left in the ground state), and its two-electron satellite line gives directly an intra-donor transition energy. Using the effective mass approximation, the resulting excitation energies obtained from experimental data can be compared with calculation based on the theory describing a hydrogen atom in a magnetic field [16]. In Fig. 5, the fit made for the excitations related to the donor responsible for the bound exciton with the largest localization energy ($D_1^0X_A$) for the B⊥c configuration is shown [12]. The experimental points are taken with respect to the low spin component of the $D_1^0X_A$ transition (Fig. 2). A set of lines marked in Fig. 4 as A0, A1, A2, A3, and A4 has been fit. From this procedure, a binding energy of the oxygen donor in GaN equal to 32.3 meV is found. The theoretical lines related to the 2p components, split by the magnetic field have been shifted down by 1 meV [12], since the corresponding experimental points are related to the TES recombination from the excited (rotational) states of the neutral donor bound exciton [17, 18].

The fit is very good also for the higher excited states connected with $n = 3, 4, 5$. An excellent fit has been found for $3d_{+1}$, $4f_{+2}$, and $5g_{+3}$, and higher states, up to $n = 11$, as shown in Fig. 6. However, in this energy range the simple hydrogenic model has to be corrected due to the strong electron–phonon interaction [19]. Since the calculations given in Ref. 16 includes states only up to $n = 4$, the fit to the higher states has been performed according to a calculation done by Rosner et al. [20], Turbiner [21] ($n = 6-8$), and using some crude approximation for higher states.

The selection rule that makes these particular transitions Nl_m, with $l = N - 1$, and $m = N - 2$ (N-principle, l – orbital, and m – magnetic quantum number, respectively), so strong is not clear. We can notice that in the high field limit, a given state within this sequence is the second bound state below each Landau level (the lowest one with the odd parity).

© 2004 WILEY-VCH Verlag GmbH & Co. KGaA, Weinheim

Similar analysis can be performed for the transitions related to the silicon donor ($D_2^0 X_A$, L3 and L4 lines) that gives the binding energy for this neutral donor to be equal 29.5 meV.

3 Fine structure of the shallow acceptor center in GaN

The shallow impurity ground state, within the effective mass approximation, should reflect the symmetry of the top of the corresponding conduction or valence band [22]. It is generally accepted that in GaN the conduction band is almost spherical and has Γ_7 symmetry. The valence band structure has been deduced from reflectance studies [23], where three excitons A, B and C, which correspond to the split valence bands of Γ_9, Γ_7 and Γ_7 symmetries, respectively. This valence band splitting is described by the crystal field and spin orbit interaction parameters, which are of the same order [24]: $\Delta_{cf} = 10.2$ meV, and $\Delta_{so} = 18.1$ meV. Therefore, one can expect that the six-fold degenerated ground state of the shallow, neutral acceptor in GaN, also splits into three doublets of Γ_9, Γ_7 and Γ_7 symmetries, respectively. The Γ_9 symmetry of the acceptor ground state should lead to a strong anisotropic variation in the external magnetic field. This is a consequence of the crystal field interaction, which for Γ_9 states fixes the angular momentum direction parallel to the c-axis. Such a doubly degenerate state should not split in a magnetic field perpendicular to the c-axis. On the contrary, the valence band holes localized on acceptor states show very weak anisotropy when, for example, studied in EPR experiments [25].

Recent experimental studies of the angle-resolved magneto-luminescence of excitons bound to neutral acceptors in samples with low doping concentrations clarify the controversy concerning the symmetry of acceptor states in GaN [26]. The samples used for these investigations were intentionally undoped, homoepitaxial GaN layers grown by MOCVD on GaN bulk crystals along the wurtzite c-axis. In the experiments, the low temperature (helium bath), near band-edge luminescence spectra in magnetic fields up to 25 T, for different orientations of the magnetic field with respect to the crystal c-axis have been measured. As illustrated in Fig. 7, the typical near band-edge emission spectrum of homoepitaxial GaN [13, 27, 28] shows distinct transitions related to both the exciton bound to a neutral donor (D^0X) and the exciton bound to a neutral acceptor (A^0X), as well as a weak feature related to the free exciton (X_A). The observation of narrow luminescence lines underlines the high sample quality. Here we concentrate on the A^0X emission, which shows, at zero magnetic field, a characteristic doublet structure with a dominant transition on the high energy side and less intense low energy satellite (0.8 meV below the dominant line). The sample-independent intensity ratio between both components of the A^0X doublet [27, 28] is an

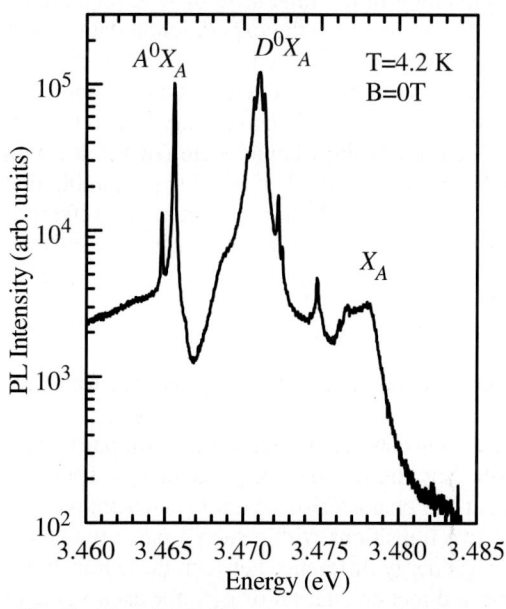

Fig. 7 Photoluminescence spectrum of homoepitaxial GaN in the range of excitonic transitions.

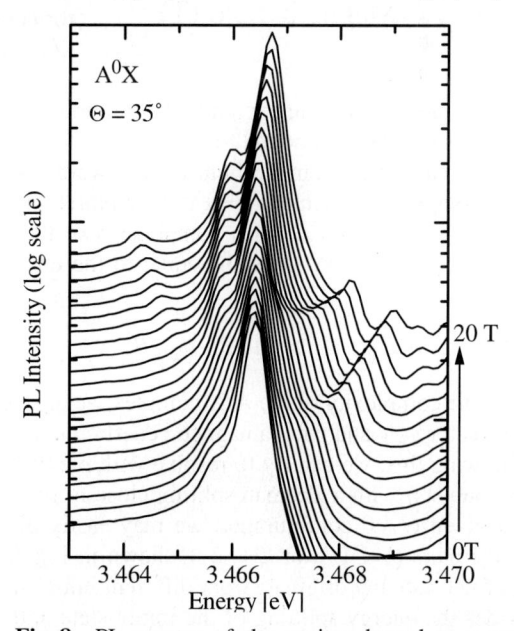

Fig. 8 PL spectra of the exciton bound to a neutral acceptor (A0X) in GaN recorded at different values of magnetic field oriented at 35° to the c-axis.

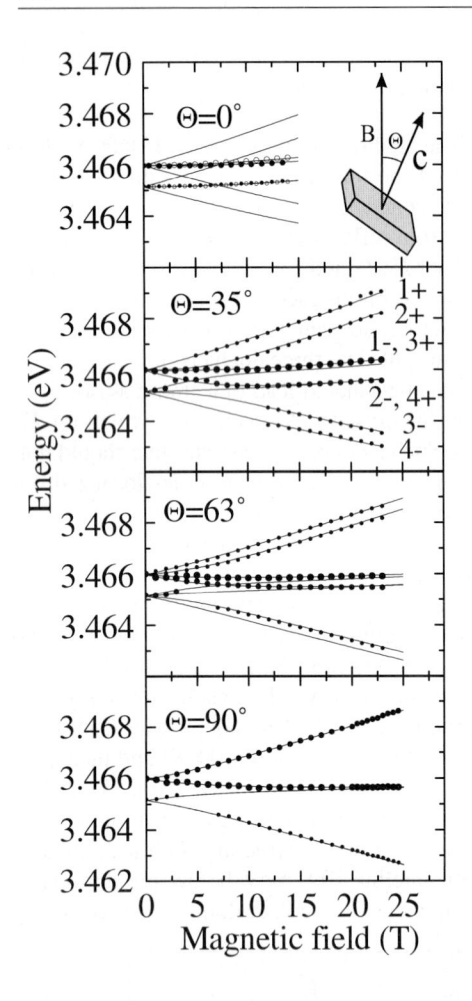

Fig. 9 Points-energies of the Zeeman components of the A^0X recombination, as a function of the magnetic field, measured for different orientations of B with respect to the c-axis of the sample, solid lines – calculated results according to the proposed model.

indication that this doublet structure is the property of the recombination related to a single impurity center. Representative traces of the A^0X-luminescence, measured as a function of the magnetic field applied along a direction tilted from the c-axis by $\Theta = 35°$ are shown in Fig. 8. The field evolution, and in particular the observed Zeeman pattern of the A^0X doublet is remarkably anisotropic. This can be clearly seen in Fig. 9, where the energies of the observed magneto-luminescence peaks are plotted for four different angles Θ. For the $\Theta = 0°$-configuration, only four central, circularly-polarized lines are observed [28]. At $\Theta \neq 0$, additional doublets (marked by (1+, 2+) and (3–, 4–) in Fig. 9), forbidden at $\Theta = 0°$, appear at energies higher and lower with respect to the central components. These "forbidden" doublets gain in intensity upon increasing Θ, but at the same time the splitting between the doublets' components becomes smaller and smaller. In order to obtain information about energy structure of the acceptor center, remaining in the final state of A^0X recombination, one has to extract, from the experimental data shown in Fig. 9, the information related to the initial state involved in the transition under study. Within the A^0X complex three particles (one conduction band electron and two holes) are bound to an acceptor potential. In spite of its possible complexity, it can be supposed to react to a magnetic field as if it were a single electron (weakly) bound state (two holes with opposite angular momentum $j_z = \pm 3/2$ are paired off). We follow such a simple model [15] assuming that the energies $E^{\pm}$ of the Zeeman components of the initial state (measured with respect to the reference energy $E_{A^0X} = 3.466$ eV set to be equal to the energy of the main A^0X line at zero magnetic field) are given by:

$$E^{\pm} - E_{A^0X} = \pm g_{A^0X} \mu_B B + d_{A^0X} B^2 . \tag{1}$$

Due to the complexity of the A^0X complex we have to assume the possible angular dependence for effective g-factor g_{A^0X} and the diamagnetic coefficient d_{A^0X}.

Knowing this, we can try to resolve within experimental points shown in Fig. 9 transition pairs (Zeeman doublets) with a Zeeman spliting close to that described by the free elecon g-factor $g_e = 1.95$ [25] For a given $\Theta \neq 0$ configuration we may easily distinguish the characteristic pairs of such transitions, e.g., the pairs (1+, 1–), or (2+, 2–), shown in Fig. 9. The two transitions within such a pair involve the same final state but originate from different initial states. The energy difference between these transitions provides the energy splitting of the initial state and allows a direct evaluation of g_{A^0X} for each value of $\Theta \neq 0$ (and extrapolation of g_{A^0X} for the $\Theta = 0$ case). Similarly, the d_{A^0X} coefficient can be evaluated, assuming that the effect of the diamagnetic shift of the observed transitions is entirely due to the initial

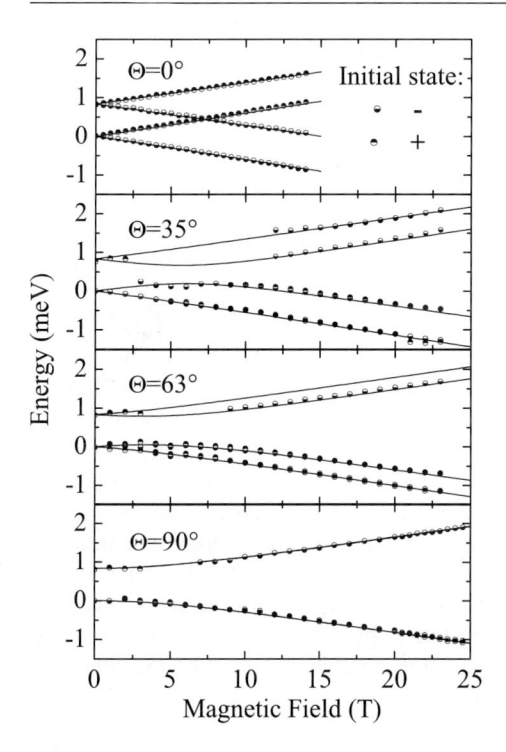

Fig. 10 Energies of the Zeeman components of the ground and first excited state doublets of the neutral acceptor in GaN as a function of the magnetic field, for different orientations of B with respect to the *c*-axis of the sample. The upper-(lower-) half-filled points were obtained through recalculation from the transition energies related to + (−) Zeeman components of the initial A⁰X ground state. A full point is seen when such two results coincide. Solid lines – results of calculations (Eq. 2).

state. The sensitivity of the initial state parameters to the field configuration, obtained from such analysis, is discussed in Ref. [26].

Finally, when the initial state of the A⁰X recombination is described with Eq. 1, using experimental data shown in Fig. 9, reconstruction of the energetic structure of the final state (neutral acceptor) involved in this transition can be done [26]. The resulting energies of the neutral acceptor Zeeman components obtained for different values of Θ are displayed in Fig. 10. The observed dependencies can not be described within simple approximation that account for the effective *g*-factor for a given Zeeman doublet. The model that includes the complexity of the valence band has to be taken into account. The effective Hamiltonian describing the six-fold degenerated ground state of the neutral acceptor can be written as following [26]:

$$H = E_c - \Delta'_{cf} l_z^2 - \tfrac{2}{3} \Delta'_{so} \boldsymbol{l} \cdot \boldsymbol{s} + g_0 \mu_B \boldsymbol{B} \cdot \boldsymbol{s} + g_1 \mu_B \boldsymbol{B} \cdot \boldsymbol{l} \tag{2}$$

where: E_c-Coulomb energy, spin-orbit Δ'_{so} and crystal-field Δ'_{cf} parameter, which describe the zero-field energies of three Zeeman doublets of Γ_9, Γ_7 and Γ_7 symmetries, respectively. Effective *g*-factors g_0 and g_1 express the coupling of the spin and angular momentum to the magnetic field, respectively.

Using the Hamiltonian given by Eq. 2 the experimental results displayed in Fig. 10 can be reproduced [26], assuming, however, a strong reduction of the spin–orbit interaction for the hole bound at the neutral acceptor. Solid lines shown in Fig. 10 have been calculated with the value $\Delta'_{so} = 1.30(2)$ meV as compared to $\Delta_{so} = 18.1$ meV deduced from studies of free excitons [24]. It was found that Δ'_{cf} has to be significantly larger than Δ'_{so}, but our experiment is not sensitive to its particular value. In calculations, the value appropriate for the free exciton $\Delta_{cf} = 10.2$ meV [24] was used. The value of $g_0 = 2.003$ as for the free electron was taken. From the fitting $g_1 = 0.041(5)$ was found. The calculations show, as expected, that the neutral acceptor ground state is of Γ_9 symmetry with $|+1, +1/2\rangle$ and $|-1, -1/2\rangle$ states, while the first excited state doublet is of Γ_7 symmetry with almost pure $|+1, -1/2\rangle$ and $|-1, +1/2\rangle$ states (the first quantum number corresponds to the orbital and second to the spin angular momentum, respectively). The strong reduction of the spin–orbit parameter resulting from the calculations cannot be explained by the conventional effective mass theory applied to acceptor states [29].

The small separation between the ground and the first excited state of the neutral acceptor makes its energy structure extremely sensitive to local axial fields in directions not aligned with the wurtzite *c*-axis.

Such a perturbation can be simulated by adding a term $\Delta_{ax} (l_{x^2} - l_{y^2})$ to the Hamiltonian given by Eq. (2). It has been verified that for Δ_{ax} of the order of 1 meV, the symmetry of the acceptor states is significantly modified, implying an effective quenching of orbital momentum effects [30]. As a consequence, for the samples with the higher concentration of impurities, the Zeeman splitting of such a perturbed centre is almost spherical, as observed in EPR and ODMR experiments [25].

4 Conclusions We have shown that the extensive magneto-luminescence studies of the bound exciton recombination permit to investigate intrinsic, energetic structure of both neutral donor and acceptor states. In the case of the O_N neutral donor in GaN two electron satellites spectrum resolved in high magnetic field allowed us to reconstruct rich structure of the excited donor states. Their energy dependence as a function of the magnetic field is very well described with calculations based on the theory of the hydrogen atom in a magnetic field. However, when the transition energy becomes comparable with optical phonon energy, both resonant and nonresonant, corrections due to electron–phonon interaction have to be included. In the case of the neutral acceptor in GaN, the samples chosen and the applied experimental technique allowed us to probe well-isolated acceptor states, which indeed are found to reflect the valence band symmetry of GaN. However, the experimental data are only well reproduced in calculations if a large reduction of the spin–orbit interaction for the hole localised on an acceptor as compared to the free hole is assumed. The axial symmetry of the neutral acceptor states, are very likely not visible in experiments on samples with high (intentional) doping concentrations in which acceptor centres are influenced by additional local symmetry effects caused, for instance, by the presence of neighbouring impurities. The reported magnitude of the reduction of the spin orbit interaction cannot be explained by the conventional effective mass theory applied to acceptor states.

Acknowledgements This work was partially supported by supported by Polish Committee of Science Grant No 2P03B 011 22, European Committee Project GMA1-2002-72503 "CEMOS" and NATO PST CL grant N° 978522.

References

[1] A. M. Witowski, K. Pakuła, J. M. Baranowski, M. L. Sadowski, and P. Wyder, Appl. Phys. Lett. **75**, 4154 (1999).
[2] W. J. Moore, J. A. Freitas, Jr., S. K. Lee, S. S. Park, and J. Y. Han, Phys. Rev. B **65**, 081201 (2002).
[3] S. Huant, A. Mandray, and B. Etienne, Solid State Commun. **93**, 435 (1995).
[4] J.-P. Cheng, B. D. McCombe, J. M. Shi, F. M. Peeters, and J. T. Devreese, Phys. Rev. B **48**, 7910 (1993).
[5] A. v. Klarenbosch, T. O. Klassen, W. Th. Wenckebach, and C. T. Foxon; J. Appl. Phys. **67**, 6323 (1990).
[6] M. Grynberg, S. Huant, G. Martinez, J. Kossut, T. Wojtowicz, G. Karczewski, J. M. Shi, F. M. Peeters, and J. T. Devreese, Phys. Rev. B **54**, 1467 (1996).
[7] R. A. Lewis, Y.-J. Wang, and M. Henini, Phys. Rev. B **67**, 235204 (2003).
[8] A. Wysmolek, phys. stat. sol. (b) **235**, 48 (2003).
[9] V. A. Karasyuk, D. G. S. Beckett, M. K. Nissen, A. Villemaire, T. W. Steiner, and M. L. W. Thewalt, Phys. Rev. B **49**, 16381 (1994).
[10] S. S. Park, I.-W. Park, and S. H Chou, Jpn. J. Appl. Phys. **39**, L1141 (2000).
[11] D. C. Look and J. R. Sizelove, Appl. Phys. Lett. **79**, 1133 (2001).
[12] A. Wysmołek, K. P. Korona, R. Stępniewski, J. M. Baranowski, J. Błoniarz, M. Potemski, R. L. Jones, D. C. Look, J. Kuhl, S. S. Park, and S. K. Lee, Phys. Rev. B **66**, 245317 (2002).
[13] R. Stępniewski, A. Wysmołek, M. Potemski, J. Łusakowski, K. Korona, K. Pakuła, J. M. Baranowski, G. Martinez, P. Wyder, I. Grzegory, S. Porowski, phys. stat. sol. (b) **210**, 373 (1998).
[14] J. A. Freitas, Jr., W. J. Moore, B. V. Shanabrook, G. C. B. Braga, S. K. Lee, S. S. Park, and J. Y. Han, Phys. Rev. B **66**, 233311 (2002).
[15] D. G. Thomas and J. J. Hopfield, Phys. Rev. **128**, 2135 (1962).
[16] P. C. Macado and N. C. McGill, J. Phys. C **19**, 873 (1986).
[17] B. J. Skromme, MRS Internet J. Nitride Semicond. Res. **4**, 15 (1999).
[18] G. Neu, M. Teisseire, E. Frayssinet, W. Knap, M. L. Sadowski, A. M. Witowski, K. Pakuła, M. Leszczynski, and P. Prystawko, Appl. Phys. Lett. **77**, 1348 (2000).
[19] A. Wysmołek et al., to be published.
[20] W. Rosner, G. Wunner, H. Herold, and H. Ruder, J. Phys. B **17**, 29 (1984).
[21] A. V. Turbiner, J. Phys. A, Math. Gen. **17**, 859, (1984).

[22] J. Bernholc, S. T. Pantelides, Phys. Rev. B **15**, 4935 (1977).
[23] R. Dingle, D. D. Sell, S. E. Stokowski, and M. Ilegems, Phys. Rev. B **4**, 1211 (1971).
[24] R. Stępniewski, M. Potemski, A. Wysmołek, J. Łusakowski, K. Pakuła, J. M. Baranowski, G. Martinez, P. Wyder, I. Grzegory, and S. Porowski, Phys. Rev. B **60**, 4438 (1999).
[25] E. R. Glaser, W. E. Carlos, G. C. B. Braga, J. A. Freitas, Jr., W. J. Moore, B. V. Shanabrook, R. L. Henry, A. E. Wickenden, D. D. Koleske, H. Obloh, P. Kozodoy, S. P. DenBaars, and U. K. Mishra, Phys. Rev. B **65**, 085312 (2002), and references threin.
[26] R. Stępniewski, A. Wysmołek, M. Potemski, K. Pakuła, J. M. Baranowski, I. Grzegory, S. Porowski, G. Martinez, and P. Wyder, Phys. Rev. Lett. **91**, 226404/1–4 (2003).
[27] K. Kornitzer, T. Ebner, K. Thonke, R. Sauer, C. Kirchner, V. Schwegler, M. Kamp, M. Leszczynski, I. Grzegory, and S. Porowski, Phys. Rev. B **60**, 1471 (1999).
[28] A. Wysmołek, M. Potemski, R. Stępniewski, J. Łusakowski, K. Pakuła, J. M. Baranowski, G. Martinez, P. Wyder, I. Grzegory, S. Porowski, phys. stat. sol. (b) **216**, 11 (1999).
[29] A. V. Malyshev, I. A. Merkulov, A. V. Rodina, Solid State Physics **40**, 917 (1998); Russ. Fiz. Tver. Tela **40**, 1002 (1998).
[30] C. Kittel, in "Introduction to Solid State Physics", (John Wiley & Sons Inc., New York, 1996) p. 427.

Surprisingly low built-in electric fields in quaternary AlInGaN heterostructures

S. Anceau[*, 1, 2], **P. Lefebvre**[2], **T. Suski**[1], **S. P. Łepkowski**[1], **H. Teisseyre**[1], **L. H. Dmowski**[1], **L. Konczewicz**[2], **A. Kaminska**[3], **A. Suchocki**[3], **H. Hirayama**[4], and **Y. Aoyagi**[4]

[1] UNIPRESS, High Pressure Research Center, Sokolowska 29/37, 01-142 Warszawa, Poland
[2] GES, CNRS- Université Montpellier II, 2 Place Eugène Bataillon, CC 074. F-34095 Montpellier, France
[3] Institute of Physics, Polish Academy of Sciences, al. Lotnikow 32/46, 02-668 Warsaw, Poland
[4] Institute of Physical and Chemical Research, RIKEN, 2-1 Hirosawa, Wako-shi, Saitama, Japan

Received 15 September 2003, revised 1 November 2003, accepted 1 November 2003
Published online 19 January 2004

PACS 78.47.+p, 78.55.Cr, 78.67.De

Measurements of (i) the time-resolved photoluminescence and (ii) pressure dependence of the emitted light energy in two series of quaternary (AlInGa)N multiquantum wells have been used to point out the small magnitude of internal electric field as promoting highly efficient luminescence in these specially designed heterostructures. Findings of the experimental examination give the magnitude of built-in electric field few times smaller than that predicted theoretically. Electrical transport measurements support the explanation of the obtained results consisting of the screening of the built-in electric field by means of unintentional doping.

© 2004 WILEY-VCH Verlag GmbH & Co. KGaA, Weinheim

1 Introduction

In the past few years, large amount of research and development efforts have been devoted to light emitters operating in UV-region of the spectrum and based on (AlGa)N and (InGa)N ternary nitrides. As in case of other semiconductor optoelectronic devices, all efficient nitride based light emitters consist of single or multiquantum wells (MQWs) and barriers [1]. However in wurtzite nitride heterostructures we always deal with a built-in electric field which is detrimental with respect to the performance of light emitting devices. This internal electric field is induced by the large spontaneous and piezoelectric polarizations present in this semiconductor family. The resulting tilting of the conduction and valence band profiles results in a red shift of the transition energy and causes the separation of the electron and hole wavefunctions in the QWs. Thus, the efficiency of radiative recombination is drastically reduced through so called Quantum Confined Stark Effect (QCSE) [2].

Application of quaternary AlInGaN heterostructures in UV optoelectronics represents very interesting concept since lattice constant and energy gap in both QWs and barriers can be tuned independently. A proper choice of the composition enables a strain engineering. In particular, the piezoelectric polarization induced by the strain at the interfaces between the well and the barrier may compensate the spontaneous polarization which is mainly dependent on Al-content in the alloy. In that case the built in electric field can be reduced significantly. However, it is difficult to achieve the total cancellation of the internal electric field because the In- and Al-content are interrelated within the epitaxial growth procedures [3]. Consequently, effective electric fields present in the quantum structures are different from those predicted theoretically.

This work aims at clarifying the magnitude of built-in electric field, F, in two series of AlInGaN quaternary MQWs showing intense photoluminescence, PL, between 3.1 eV and 4.13 eV [3, 4]. The only

---

[*] Corresponding author: e-mail: anceau@unipress.waw.pl, Phone: +48 22 837 42 05, Fax: +48 22

difference between the used structures is the amount of Al in the barrier material. To examine magnitude of F hydrostatic pressure photoluminescence (HPPL) and the time resolved photoluminescence (TRPL) measurements were performed. It is well established now that higher fields lead to strong reduction of the pressure coefficient of the emitted light energy dE_E/dP [5], and (ii) to drastic increase of the PL decay time, τ [6, 7]. These effects are better visible in structures with wider QWs (≥ 3 nm). We also compare the magnitude of F found experimentally with its amplitude obtained by means of theoretical calculations.

2 Samples The samples (Series A and B) were grown by low-pressure (76 Torr) MOVPE on 400 nm thick AlGaN buffer layers, which were deposited on the Si face of an on-axis 6H–SiC (0001) substrates. The samples consist of $3 \times$ QWs of AlInGaN of variable width embedded between barriers (widths of 5 nm) of the quaternary alloy with higher (lower) content of Al (In). Details of the growth have been given in previous papers [3, 4].

In the series A the compositions of the wells/barriers is $Al_{0.2}In_{0.05}Ga_{0.75}N/Al_{0.3}In_{0.02}Ga_{0.68}N$ respectively. The QWs thickness in 5 prepared samples varies between 1.3 and 4 nm. In 7 samples of the series B the compositions of the QWs is the same as in the series A whereas the composition of the barriers is $Al_{0.43}In_{0.02}Ga_{0.55}N$. The QWs thicknesses change from 0.9 to 4.7 nm. Additionally, 2 reference samples containing thick epitaxial layer of the composition of the wells and barriers (80 nm) were also grown on the same structure. The alloy compositions were measured by Rutherford backscattering spectrometry for the series A and then for the series B, the composition was estimated from technological procedure and lattice constant measurements in the reference samples.

The purpose of such a design of samples was twofold. First, the significant increase of Al content (in Series B in comparison with Series A) leads to the increase of the band offsets, i.e., quantum confinement effect in MQWs with sufficiently narrow QWs. This should lead to much better luminescence intensity at room temperature in case of Series B. Second, higher Al-content means an increase of spontaneous and piezoelectric polarization contributions to the internal electric field. The latter effect results from the increase in the lattice mismatch (strain) at the interfaces wells-barriers and can reduce the emitted light intensity through QCSE. We should mention also that we have performed already studies of internal electric field in the Series A using hydrostatic pressure and TRPL techniques [7, 8]. Lack of F or its very low magnitude (< 0.1 MV/cm) has been found.

3 Continuous wavelength photoluminescence (cw-PL) measurements Figure 1 shows the low temperature spectra of the PL in both series of QWs. Compared to the series A, for the series B the bright emission of quaternary compounds covers much wider range of UV spectrum. One can notice quite significant blue shift of emitted light energy, E_E, for narrow QWs belonging to Series B. We associate this

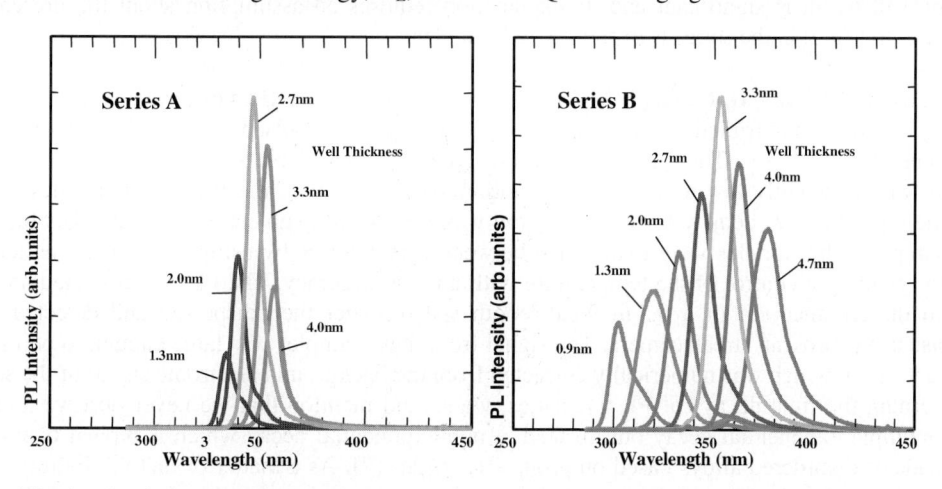

Fig. 1 Continuous wavelength photoluminescence measurements for samples of series A and series B.

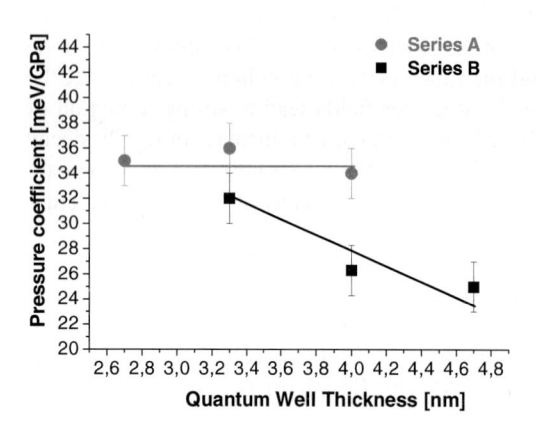

Fig. 2 Pressure coefficients of light emission as a function of MQW width for the widest samples of series A and B.

observation with an effect of quantum confinement enhanced in this set of QWs. In the three wider wells of Series B, with the thickness of 3.3, 4 and 4.7 nm, the red shift of the wavelength is visible which may suggest presence of the internal electric field in these samples. Below we will verify this hypothesis.

4 Hydrostatic pressure studies of the photoluminescence High pressure experiments were performed at $T = 80$ K in a low-temperature diamond anvil cell filled with liquid argon as a pressure transmitting medium. In the case of high pressure measurements the PL was excited by a He–Cd laser ($h\nu = 3.8$ eV) with power of 2 mW. The emission from the sample was collected in a backscattering geometry, dispersed by a SPEX500M spectrometer and detected by a GaAs photomultiplier. It is worthwhile to point out that in quantum structures without electric field, dE_E/dP corresponds usually to the blue-shift of the band gap induced by hydrostatic pressure, i.e., dE_G/dP. dE_G/dP in GaN and AlGaN alloys is about 40 meV/GPa and changes between 40 and 6–9 meV/GPa from GaN and InN [5]. Electric field present in QWs with a thickness above about 3 nm causes considerable lowering of dE_E/dP with respect to dE_G/dP. Built-in biaxial strain and thus piezoelectric component of F increase with application of hydrostatic pressure [9].

In Fig. 2 we compare the measured pressure coefficients (dE_E/dP) of the emitted light energy in the two studied series of samples. For Series A of MQWs, dE_E/dP changes between 34 meV/GPa to 36 meV/GPa for the samples with wells of 2.7–4 nm width. Taking into account the experimental error of 2 meV/GPa there is practically no change of the pressure coefficient with the quantum well width. On the contrary, for the series B the pressure coefficient changes from 32 meV/GPa for the 3.3 nm width MQW to 25 meV/GPa for the 4.7 nm width MQW. This decrease of the pressure coefficient with the quantum well width is significant and its explanation requires an assumption about the presence of a built-in electric field in the series B of the studied samples.

5 Time-resolved PL of (Al, In, Ga)N MQWs Time-resolved PL (TRPL) experiments were carried out by using the frequency-tripled radiation from a titanium-sapphire pulsed laser, thus with an excitation photon energy of 4.77 eV. The typical duration of the pulses was 2 ps, and the repetition frequency of 82 MHz was sufficiently low to avoid quasi-continuous excitation of the PL for all samples presented here. The typical power density for the excitation was 100 W/cm^2, subject to some changes for the different samples. The samples were maintained between 8 and 300 K by using a closed-cycle helium refrigerator yielding a control of the temperature with a 0.1 K accuracy. The PL was collected and focussed onto the entrance slit of a 0.5 m focal length spectrometer then dispersed and detected using a Hamamatsu synchroscan streak camera. The signal from these samples was large enough to permit direct analog detection, which was numerically corrected from the background electronic signal of the set-up.

Concerning the time decays of PL intensities, we should mention that we never observe a single or even a multiple exponential decay but, instead, a non-exponential decay which is typical of layers and QWs made of disordered alloys based on group III nitrides [7]. As a matter of fact the Indium fluctuations of composition in these quaternary compounds prevent carriers to recombine nonradiatively into

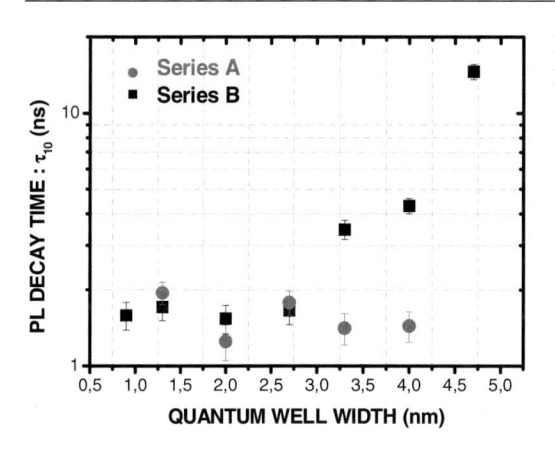

Fig. 3 PL decay time as a function of QW width for samples of series A and B.

the transversal dislocations for example. This is the reason why we have chosen to characterize each decay by the time, τ_{10}, necessary to divide the maximum intensity by a factor of 10.

Figure 3 shows the measured values of τ_{10} for the samples from series A and B. Generally, in case of QW with a large internal electric field [6], the PL decay time showed a nearly exponential dependence upon the well width. It was explained by the QCSE, i.e., the separation of the electron and hole wavefunctions towards the opposite sides of the QW. In the case of the present QWs, we observe a somewhat similar tendency (Fig. 3), for the three widest wells of the series B only. Compared to the series A where there is no variation of the characteristic time there is an increase of τ_{10} with the well width for the series B over one order of magnitude. From this result, using envelope-function calculations we can estimate the electric field to be less than 0.1MV/cm for the series A and between 0.6 and 0.7 MV/cm for the series B which are the values compatible with the decay time variation. In fact for the series B the field does not affect the wave functions for ultra-narrow QWs where the slight variations of the radiative lifetime with the QWs width are only a matter of different penetrations of electron and hole wave-functions into the barriers. These electric field values can only provide an indication for the case of the present quaternary QWs because the numerical parameters are not so well known and it is hardly possible to determine the electric field value with a higher accuracy than 0.1 MV/cm experimentally. This result supports the conclusion given by high hydrostatic pressure photoluminescence that there is a very small built-in electric field in our quaternary samples.

In the final part of experimental studies we analyzed the temperature evolution of the normalized intensity for samples from Series A and B with the same QW width. As a general tendency we observe much weaker changes of PL intensity with temperatures for the Series B. There is a real improvement of the PL intensity at room for series B of the studied MQWs. This is probably due to the quantum confinement effect because a large band-offsets has been reached between the wells and the barriers for the series B compared to the series A. We believe that because of a small magnitude of the built-in electric field in the series B this effect is not influencing the temperature characteristics of the PL emission intensity.

6 Theoretical calculations Using recent results of ab-initio calculations of the spontaneous and piezoelectric polarizations in ternaries compound [10] we performed calculations described in [7]. We obtain an internal electric field between 1.3 and 1.7 MV/cm for the series A and between 3.2 and 3.7 MV/cm for the series B. The calculation includes the nonlinear dependence of the piezoelectric polarization on strain as well as the nonlinear dependence of the spontaneous polarization on the alloy composition [10]. One can notice that there is an apparent contradiction between theoretical estimations and experimental value of the internal electric field in our QWs.

7 Conclusions Measurements of (i) time-resolved photoluminescence and (ii) of the hydrostatic pressure dependence of emitted light energy in two series of quaternary (AlInGa)N MQWs have allowed us to study an effect of highly efficient luminescence in these specially designed heterostructures. It has been proved that the internal electric field in all studied samples is very small compared to the theoretical predictions. For the first series of samples, both the pressure coefficient dE_F/dP and the PL decay time do not change with the thickness of the QWs. For the series B there is a small decrease of dE_F/dP and an increase of τ_{10} for the three widest QWs. These observations can be interpreted as corresponding to the internal electric field of less than 0.1 MV/cm for the series A and about 0.6 MV/cm for the series B. This finding gives the magnitude of built-in electric field few times smaller than predicted theoretically. Our electrical transport measurements support the most natural explanation of the obtained results, i.e., screening of built-in electric field (originating from piezoelectric and spontaneous polarizations) by means of unintentional doping.

Acknowledgements The work was supported by the European Commission, Project: "Support for Centers of Excellence" No. ICA1–CT–2000–70005.

References

[1] S. Nagahama, T. Yanamoto, M. Sano, and T. Mukai, Jpn. J. Appl. Phys. Part 2, **40**, L 788 (2001).
[2] S. F. Chichibu, Y. Kawakami, and T. Sota, in: Introduction to Nitride Semiconductor Blue Lasers and Light Emitting Diodes, edited by S. Nakamura and S. F. Chichibu, (Taylor & Francis, London, 2000), pp. 153–270.
[3] H. Hirayama, Y. Enomoto, A. Kinoshita, A. Hirata, and Y. Aoyagi, Appl. Phys. Lett. **80**, 1589 (2002).
[4] H. Hirayama, A. Kinoshita, T. Yamabi, Y. Enomoto, A. Hirata, T. Araki, Y. Nanishi, and Y. Aoyagi, Appl. Phys. Lett. **80**, 207 (2002).
[5] P. Perlin, I. Gorczyca, T. Suski, P. Wisniewski, S. P. Łepkowski, N. E. Christensen, A. Svane, M. Hansen, S. P. DenBaars, B. Damilano, N. Grandjean, and J. Massies, Phys. Rev. B **64**, 115319 (2001).
T. Suski, S. P. Łepkowski, P. Perlin, T. Kitamura, Y. Ishida, H. Okumura and S. F. Chichibu, Appl. Phys. Lett. **81**, 232 (2002).
[6] P. Lefebvre, A. Morel, M. Gallart, T. Taliercio, J. Allègre, B. Gil, H. Mathieu, B. Damilano, N. Grandjean, and J. Massies, Appl. Phys. Lett. **78**, 1252 (2001).
[7] P. Lefebvre, S. Anceau, P. Valvin, T. Taliercio, L. Konczewicz, T. Suski, S. P. Łepkowski, H. Teisseyre, H. Hirayama, and Y. Aoyagi Phys. Rev. B **66**, 195330 (2002).
[8] H. Teisseyre, T. Suski, S. P. Lepkowski, S. Anceau, P. Perlin, P. Lefebvre, L. Konczewicz, H. Hirayama, and Y. Aoyagi, Appl. Phys. Lett. **82**, 1541 (2003).
[9] G. Vaschenko, D. Patel, C. S. Menoni, H. M. Ng, and A. Y. Cho, Appl. Phys. Lett. **80**, 4211 (2002).
[10] V. Fiorentini, F. Bernardini, and O. Ambacher, Appl. Phys. Lett. **80**, 1204 (2002).

Site symmetry of erbium centers in GaN

V. Glukhanyuk*,1, H. Przybylińska1, A. Kozanecki1, and W. Jantsch2

1 Institute of Physics, Polish Academy of Sciences, Warsaw, Poland
2 Institute of Semiconductor and Solid State Physics, Johannes Kepler University, Linz, Austria

Received 15 September 2003, revised 24 October 2003, accepted 24 October 2003
Published online 19 January 2004

PACS 71.70.Ej, 78.55.Cr

In this work site selective excitation at an energy range corresponding to the $^4I_{15/2} \rightarrow {}^4I_{9/2}$ transition of Er^{3+} was applied in order to determine the site symmetry of luminescent erbium centers in cubic and hexagonal GaN layers. From the analysis of the Stark splitting it was found that in zinc blende GaN Er occupies a nitrogen interstitial site. A small deviation from cubic symmetry, best described by a tetragonal distortion, was observed, possibly related to the layer strain. In wurtzite GaN Er was found to occupy substitutional Ga sites, with C_{3v} symmetry.

© 2004 WILEY-VCH Verlag GmbH & Co. KGaA, Weinheim

1 Introduction Rare earth doped GaN is widely studied as a possible material for optoelectronic applications, such as full color displays [1, 2] and optical data storage devices [3]. Erbium doped GaN is of particular interest as a potential, temperature stable light source operating at 1.5 μm, i.e., in the minimum loss window of optical fiber communication [4–8].

Photoluminescence- (PL) and photoluminescence excitation (PLE) spectroscopy have been the main techniques used to characterize the Er implanted GaN system. Those techniques make it possible to distinguish between the numerous, optically active Er^{3+} centers observed in the GaN host and to gain insight into their excitation mechanisms [4–7]. Due to the large variety of Er centers, however, it was so far impossible to determine the site symmetry and lattice location of Er with use of optical methods. The latter information was provided only for hexagonal GaN from extended X-ray absorption fine structure (EXAFS) [9] and Rutherford backscattering (RBS) [10] studies, where it was found that Er occupies substitutional Ga sites. In zinc-blende GaN no microscopic structure studies have been reported.

In this paper we present the first identification of the site symmetry of Er implanted into cubic and hexagonal GaN based on the analysis of the Stark splitting of the $^4I_{9/2}$ excited multiplet of Er^{3+}. This was possible to achieve using high purity material and very low Er implant doses, which ensured that only one dominant Er luminescent center was produced. The analysis of the crystal field parameters allowed us also to conclude on the lattice location of Er in cubic and hexagonal GaN.

2 Experiment The GaN samples studied were cubic layers grown by MBE on GaAs substrates and high-purity wurtzite layers grown by MOCVD on sapphire. Er was implanted at 600 keV with doses varying from 10^{12} cm^{-2} to 10^{13} cm^{-2} into each material. The implantation damage was removed by annealing for 30 min in nitrogen flow at temperatures of 900 and 1000 °C.

Photoluminescence spectra were excited with a tunable Ti–Sa laser, dispersed through a monochromator and detected with a liquid-nitrogen cooled Ge detector.

3 Results In all the samples studied one dominant center is observed in photoluminescence, both under above band-gap as well as under intra-4f-shell excitation. With increasing implant dose and increasing

---

* Corresponding author: e-mail: vgluh@ifpan.edu.pl, Phone: +48 22 843 66 01 3326, Fax: +48 22 843 09 26

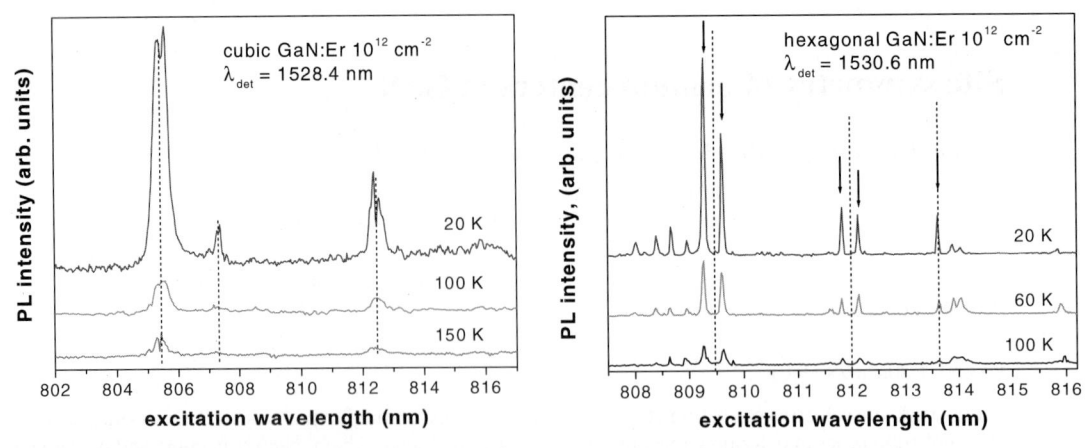

Fig. 1 Photoluminescence excitation spectra of Er^{3+} in cubic and hexagonal GaN. Dashed lines mark the positions of PLE lines in purely cubic symmetry, assuming that the deviation from cubic symmetry is small.

annealing temperature a slight broadening of the emission and excitation lines is observed. In hexagonal GaN, moreover, additional PLE lines are observed, the intensities of which increase relative to the dominant lines with increasing Er content. Apparently, other Er centers are present at a lower concentration, from which the excitation energy can be transferred to the center observed in photoluminescence. At higher temperatures the energy transfer mechanism gains in importance, indicating that it might be phonon assisted. No PL spectra related to the other Er centers were detected. The photoluminescence excitation (PLE) spectra to the $^4I_{9/2}$ manifold of Er^{3+} in cubic and hexagonal GaN, as measured in samples implanted with the smallest Er dose (10^{12} cm^{-2}) and annealed at 900 °C, are shown in Fig. 1. The measurement temperature is indicated in the figure.

In cubic symmetry the $J = 9/2$ manifold splits into three states, two quartets (Γ_8) and one doublet (Γ_7). In sites of lower symmetry the quartets will be further split and altogether 5 PLE lines can be expected. As seen in Fig. 1, in cubic GaN a small splitting of the Γ_8 states is observed, indicating a distortion from purely cubic symmetry. In hexagonal GaN:Er the splittings of the quartet states are slightly larger and the Er absorption lines extremely narrow (of the order of 0.05 nm). The PLE lines of the dominant center are indicated in the figure with arrows. Dashed lines in Fig. 1 mark the expected positions for transitions in purely cubic symmetry, assuming that the distortion is small, i.e., the center of gravity is taken for the position of the quartet states.

4 Calculations In the first approximation the cubic crystal field potential was taken into account. Assuming that distortions from cubic symmetry are small the energy positions of the Γ_8 states can be taken as centers of gravity of the split lines, as shown in Fig. 1. The cubic symmetry crystal field Hamiltonian, for the quantization axis along one of the fourfold cube axes, can be expressed as:

$$H_{cub} = B_4 O_4 + B_6 O_6 = B_4 (O_4^0 + 5 O_4^4) + B_6 (O_6^0 - 21 O_6^4), \quad (1)$$

where O_l^m are equivalent operators in the notation of Stevens [11], B_4 and B_6 are 4th and 6th order cubic crystal field parameters. If the quantization axis is taken along the threefold rotation axis the cubic Hamiltonian takes the form:

$$H_{cub}^{(2)} = B_4 \left(-\frac{2}{3} O_4^0 + \frac{40\sqrt{2}}{3} O_4^3 \right) + B_6 \left(\frac{16}{9} O_6^0 + \frac{140\sqrt{2}}{9} O_6^3 + \frac{154}{9} O_6^6 \right). \quad (2)$$

The parameters B_4 and B_6 are related to crystal field coefficients A_4 and A_6 by [12]:

$$B_4 = A_4 \langle r^4 \rangle \beta, \qquad B_6 = A_6 \langle r^6 \rangle \gamma \quad (3)$$

Table 1 Crystal field parameters (in the units of cm^{-1}) for the for the Stark splitting of the $J = 9/2$ state.

cubic GaN	hexagonal GaN
$T_d : B_4 = 0.0227; B_6 = 2.08 \cdot 10^{-4}$	$T_d : B_4 = -6.8 \cdot 10^{-4}; B_6 = 2.12 \cdot 10^{-4}$
$D_{2d} : B_2^0 = 0.11; B_4^0 = 0.0226; B_4^4 = 0.114$	$C_{3v} : B_2^0 = 0.27; B_4^0 = 4.3 \cdot 10^{-4}; B_4^3 = -0.012;$
$B_6^0 = 2.1 \cdot 10^{-4}; B_6^4 = 4.36 \cdot 10^{-3}$	$B_6^0 = 3.6 \cdot 10^{-4}; B_6^3 = 4.75 \cdot 10^{-3}; B_6^6 = 3.48 \cdot 10^{-3}$

where β and γ are multiplicative factors dependent on L, S, and J of the free ion state, and for the $J = 9/2$ state of Er^{3+} are both positive.

The splittings of the $J = 9/2$ level were then fitted with a crystal field Hamiltonian of tetragonal (D_{2d}):

$$H_{D_{2d}} = B_2^0 O_2^0 + B_4^0 O_4^0 + B_4^4 O_4^4 + B_6^0 O_6^0 + B_6^4 O_6^4 \tag{5}$$

and trigonal (C_{3v}) symmetry:

$$H_{C_{3v}} = B_2^0 O_2^0 + B_4^0 O_4^0 + B_4^3 O_4^3 + B_6^0 O_6^0 + B_6^3 O_6^3 + B_6^6 O_6^6 \tag{6}$$

in cubic and hexagonal GaN, respectively. The crystal field parameters obtained from the fit are summarized in Table 1. The calculated energies energies agree with the experimental ones within an accuracy of 0.1 cm^{-1}, as shown in Fig. 2.

5 Discussion and summary We have found that Er implanted into hexagonal GaN occupies a position with C_{3v} point symmetry, while in cubic GaN the site occupied has tetragonal point symmetry. Since the distortions from cubic symmetry are relatively small (especially in the case of cubic GaN) we can attempt to draw conclusions on the location of Er in the lattice, based on the analysis of cubic crystal field parameters.

The crystal field coefficients A_4 and A_6 can be calculated in a point charge model. For the crystal field potential at the center of a regular tetrahedron of negative point charges A_4 is negative, while for that at the center of an octahedron of negative point charges A_4 is positive. In both cases $A_6 > 0$. If Er would occupy a substitutional Ga site, with four nearest neighbor N ions the ratio of $B_4/B_6 = (A_4\beta)/(A_6\gamma)$ should be negative, since both $\beta > 0$ and $\gamma > 0$. This is the case observed for the Er^{3+} center in hexagonal GaN, which is in agreement with the substitutional position determined, e.g., in EXAFS [9] and RBS [10] experiments. However, in the cubic GaN B_4/B_6 was found to be positive. Such a situation would occur if Er was placed at an interstitial site in the latice, surrounded by 4 nearest neighbor Ga ions and 6 octahedrally coordinated, next nearest neighbor N ions (about 15% further away). In this case the fourth order potentials of both coordination shells would be positive and add up, whereas the sixth order potentials will partially cancel. This would also explain the much greater value of B_4 in cubic GaN as compared to

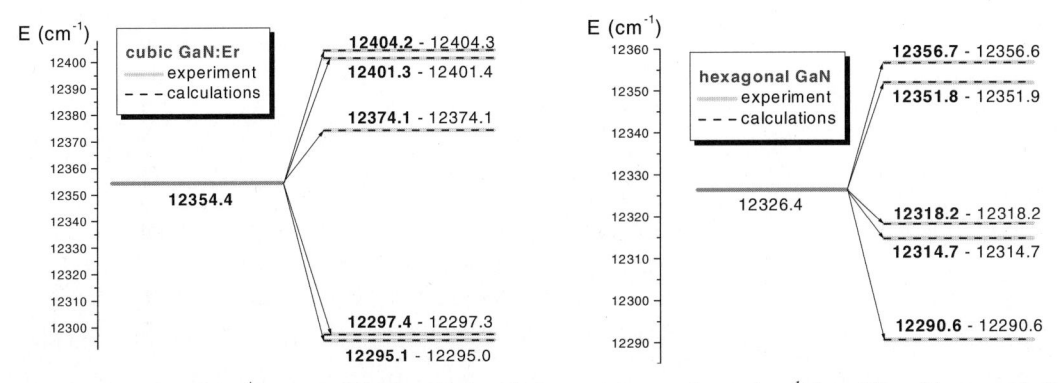

Fig. 2 Energies of the $^4I_{9/2}$ crystal-field multiplet relative to the ground state (cm^{-1}) in cubic and hexagonal GaN.

hexagonal GaN. Interstitial location of Er was also found in other zinc blend semiconductors, such as ZnSe [13] and Si [14]. The small tetragonal distortion might be related to strain in the layer ([100] is the growth direction), though charge compensation by an extra defect cannot be excluded at this stage.

Acknowledgements Work supported in part by Polish State Committee for Scientific Research and EU grant G6MA-CT-2002-04021 and Bundesministerium für Bildung, Wissenschaft und Kultur, Austria.

References

[1] Y. Q. Wang and A. J. Steckl, Appl. Phys. Lett. **82**, 502 (2003).
[2] D. S. Lee and A. J. Steckl, Appl. Phys. Lett. **80**, 1888 (2002).
[3] B. K. Lee, C. J. Chi, I. Chyr, D. S. Lee, F. R. Beyette, and A. J. Steckl, Opt. Eng. **41**, 742 (2002).
[4] S. Kim, S. J. Rhee, D. A. Turnbul, E. E. Reuter, X. Li, J. J. Coleman, and S. G. Bishop, Appl. Phys. Lett. **71**, 231 (1997).
[5] Myo Thaik, U. Hömmerich, R. N. Schwartz, R. G. Wilson, and J. M. Zavada, Appl. Phys. Lett. **71**, 2641 (1997).
[6] S. Kim, S. J. Rhee, X. Li, J. J. Coleman, and S. G. Bishop, Appl. Phys. Lett. **76**, 2403 (2000).
[7] H. Przybylinska, W. Jantsch, and A. Kozanecki, Mater. Sci. Eng. B **81**, 147 (2001).
[8] D. S. Lee, J. Heikenfeld, A. J. Steckl, U. Hömmerich, J. T. Seo, A. Braud, and J. M. Zavada, Appl. Phys. Lett. **79**, 719 (2001).
[9] P. H. Citrin, P. A. Northrup, R. Birkhahn, and A. J. Steckl, Appl. Phys. Lett. **76**, 2865 (2000).
[10] E. Alves, M. F. DaSilva, J. C. Soares, J. Bartels, R. Vianden, C. R. Abernathy, and S. J. Pearton, MRS Internet J. Nitride Semicond. Res. 4S1, G11.2 (1999).
[11] K. W. H. Stevens, Proc. Phys. Soc. Lond. A **65**, 209 (1952).
[12] M. T. Hutchings, Solid State Phys. **16**, 227 (1964).
[13] J. D. Kingsley and M. Aven, Phys. Rev. **155**, 235 (1967).
[14] H. Przybylinska, W. Jantsch, Yu. Suprun-Belevitch, M. Stepikhova, L. Palmetshofer, G. Hendorfer, A. Kozanecki, R. J. Wilson, and B. J. Sealy, Phys. Rev. B **54**, 2532 (1996).

Luminescence transients in highly excited GaN grown by hydride vapor-phase epitaxy

S. Juršėnas[*,1], **S. Miasojedovas**[1], **G. Kurilčik**[1], **A. Žukauskas**[1], and **P. R. Hageman**[2]

[1] Institute of Materials Science and Applied Research, Vilnius University, Saulėtekio al. 9-III, LT-2040 Vilnius, Lithuania
[2] University of Nijmegen, Fac. of Science, Dept. of Exp. Solid State Physics III, Toernooiveld 1, 6525 ED Nijmegen, The Netherlands

Received 15 September 2003, accepted 24 September 2003
Published online 29 December 2003

PACS 72.20.Jv, 78.47.+p, 78.55.Cr

Luminescence transients has been studied in GaN grown by hydride vapor-phase epitaxy (HVPE) under intense photoexcitation conditions. The HVPE grown GaN layer exhibited luminescence decay time of 205 ps, that implies the room-temperature free-carrier lifetime of 420 ps. The obtained carrier lifetime of HVPE-grown GaN is significantly higher that the typical values of carrier lifetime measured for GaN heterolayers grown by metalorganic chemical vapor deposition, what suggests high potential of HVPE growth technique for light-emitting diode and blue laser applications.

© 2004 WILEY-VCH Verlag GmbH & Co. KGaA, Weinheim

1 Introduction Group-III nitrides have recently been of great interest mainly because of the realization of short-wavelength light emitters, in particular laser diodes [1]. Growth of high-quality bulk-like GaN layers is of great importance since they are used as substrates for further epitaxy of light emitting structures and can facilitate efficient light extraction from light emitting diodes [1, 2]. Conventional heteroepitaxial growth of GaN on low-temperature GaN or AlN buffer layers deposited on Al_2O_3 and SiC substrates results in films containing high dislocation densities ($10^8 - 10^{10}$ cm^{-2}) due to lattice mismatch between the film and the substrate. These threading dislocations affect both optical and electrical properties and are believed to hinder the improving of device performance. The best quality GaN epilayers are obtained by using homoepitaxial growth over bulk GaN crystals [3, 4]. However at present, the size and growth rate of the high-pressure crystals are limited. Hydride vapor phase epitaxy (HVPE) offers large growth rates with relatively good quality of crystals [5, 6].

Highly-excited smiconductor under deep-trap saturation regime, can provide a common electron and hole lifetime, which is one of the crucial indicators of the materials quality [7, 8]. Here we present estimations of carrier lifetime in HVPE grown GaN under high excitation and its comparison to the carrier lifetime values obtained in the GaN epilayers grown by metalorganic chemical vapor deposition (MOCVD) under similar experimental conditions.

2 Experimental GaN film was deposited in a home built, horizontal HVPE reactor using a MOCVD-grown GaN layer (~1.9 μm) on a sapphire substrate as a template [9]. The HVPE reactor was equipped with a two-zone furnace and operated at atmospheric pressure. Nitrogen was used as a carrier gas (1 slm) and NH_3 as a nitrogen precursor (400 sccm). The reactor was designed to obtain laminar flow and allows mixing of the NH_3 and the GaCl just above the susceptor. The GaCl growth species were in-situ synthesized by passing pure HCl over liquid gallium (7N) at 800 °C. At a growth temperature of 955 °C, we

---

[*] Corresponding author: e-mail: saulius.jursenas@ff.vu.lt, Phone: + 3705 2 366027, Fax: +3705 2 366059

obtained a layer thickness of 105 μm for a growth time of 1 hour using 40 sccm HCl. The equilibrium electron density was estimated to be $n_0 = 5 \times 10^{16}$ cm^{-3}.

The sample was excited by the fourth harmonic (photon energy $h\nu_g = 4.66$ eV) of the actively-passively mode-locked YAG:Nd^{3+} (yttrium aluminum garnet) laser (pulse duration of $\tau_g = 20$ ps, repetition rate 2.7 Hz, maximum pump energy 20 μJ). The size of the excitation spot was approximately 1 mm. Luminescence was collected in backward geometry and dispersed by a 0.4-m grating monochromator. Toluene optical Kerr shutter was used for temporal resolution (20 ps) of the luminescence. The experiments were carried out at room temperature.

3 Results and discussion

Figure 1 shows some typical time-integrated luminescence spectra of the HVPE grown GaN for various excitation energy density. The spectra are seen to contain one broad emission band peaked in the vicinity of the bandgap energy (~3.4 eV). The band becomes broader and blueshifts for highest pump. The observed luminescence dynamics is typical of the radiative recombination of high-density electron–hole plasma (EHP) [7, 10] that occurs under intense photoexcitation with the excitonic states screened by the carrier system (at $I_g = 1.25$ mJ/cm^2, for a thickness of the excited region of $d_g \approx 0.1$ μm [10], the estimated carrier density is ~10^{19} cm^{-3}, i.e. essentially higher than the Mott density [10, 11]). Increase in the pump energy density ($I_g > 1.25$ mJ/cm^2) leds to a carrier heating effect, and enhaced recombination dynamics owing to stimulated transitions in degenerated electron–hole plasma [7].

For moderate excitation of $I_g = 1.2$ mJ/cm^2, carrier heating and stimulated emission effects can be significantly reduced, however, the created carrier density can be sufficient to maintain the deep-trap saturation regime [8]. In such experimental conditions, when the nonequilibrium electron and hole concentrations are large compared to the density of deep traps and equilibrium carrier density ($\Delta p \cong \Delta n \gg N_t, n_0$), for typical asymmetry of the electron and hole capture cross sections (say, $b_h > b_e$), the traps are saturated by holes. Here carrier recombination is controlled by the electron capture time $\tau_e = (b_e N_t)^{-1}$ provided that the bimolecular recombination rate is negligible ($b_r n \ll b_e N_t$). Since holes are captured only by empty traps, the hole lifetime is increased up to the electron lifetime. This means that EHP can be characterized by a common carrier density $n \approx p$ and a common lifetime $\tau_e \approx \tau_h$. Consequently, an exponential decay with a time constant of $\tau_{LU} \cong \tau_e/2$ has to be observed. With further increase of the plasma density ($b_r n \gg b_e N_t$), the luminescence decay becomes nonexponential and plasma-density dependent [7].

Figure 2 shows luminescence transients obtained for moderate excitation $I_g = 1.2$ mJ/cm^2 at the peak position (3.4 eV) of HVPE grown GaN epilayer. For comparison, results on GaN homo and heterolayers grown by MOCVD in UNIPRESS centre, Poland are shown [7]. It is evident, that luminescence decays almost

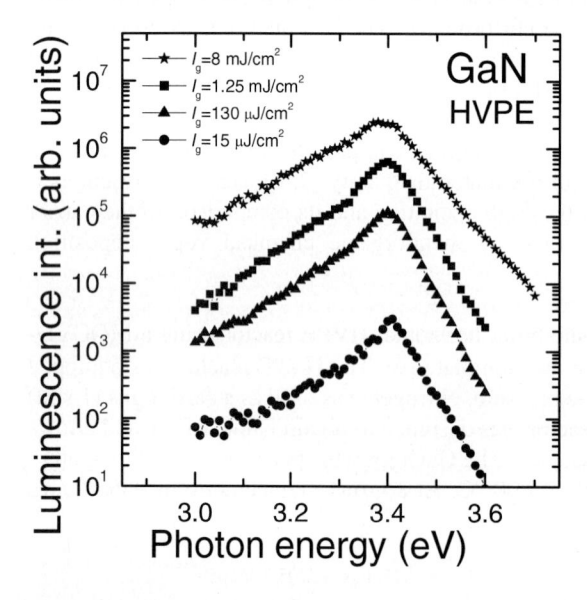

Fig. 1 Time-integrated luminescence spectra of the HVPE-grown GaN for various excitation energy density.

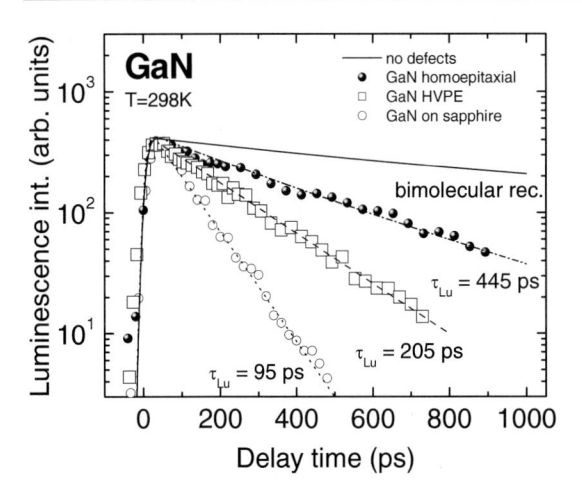

Fig. 2 Transient behavior of the normalized luminescence at the peak position of the emission band (3.4 eV) of HVPE GaN: squares. For comparison: solid and open circles correspond to homo- and heterolayers, grown by MOCVD, respectively. Points, experiment; lines, calculation for various electron capture times: dotted line, $\tau_e = 195$ ps; dashed line, $\tau_e = 420$ ps; dot-dashed line, $\tau_e = 970$ ps; solid line, intrinsic band-to-band recombination (no capture).

exponentially and the characteristic decay time varies in samples prepared by different growth procedures. The largest luminescence decay time $\tau_{LU} = 445$ ps is shown by homoepitaxial GaN epilayer grown by MOCVD (solid circles); GaN epilayer grown under similar conditions over sapphire substrates shows significantly lower value $\tau_{LU} = 95$ ps (open points). Low quality GaN epilayers might have luminescence decay time below 10 ps. Our study shows, that thick GaN epilayers (100 μm) grown by hydride vapor-phase epitaxy can have rather large luminescence decay time of 205 ps.

Lines in Fig. 2 show the results of calculation of the EHP spontaneous luminescence decay, obtained by solving a system of three coupled rate equations for nonequlibrium electrons, holes, and recombination centers, respectively, under standard assumptions [8, 12, 13] and for various electron capture times: dotted line, $\tau_e = 195$ ps; dashed line, $\tau_e = 420$ ps; dot-dashed line, $\tau_e = 970$ ps. If one fixes electron recombination coefficient at $b_e = 1 \times 10^{-8}$ cm^3/s [8], the density of the deep traps can be estimated: $N_t = 5.1 \times 10^{17}$ cm^{-3} for GaN epilayer grown on sapphire, $N_t = 2.1 \times 10^{17}$ cm^{-3} for HVPE grown GaN and $N_t = 1.0 \times 10^{17}$ cm^{-3} for homoepitaxial GaN. These deep traps most likely are related to a larger number of threading dislocations occurring at the epilayer substrate interface of heterolayers. Values of τ_e are slightly larger than $2\tau_{LU}$, because of the impact of bimolecular recombination.

Solid line in Fig. 2 shows luminescence decay of intrinsic band-to-band recombination for $b_r = 3 \times 10^{-11}$ cm^3/s (an average of the data reviewed in Ref. [14]). It is evident that although homoepitaxy significantly reduces the number of threading dislocations, the quality of crystals can still be improved, to reach luminescence decay time of about 1.6 ns expected for pure bimolecular recombination at $\sim 10^{19}$ cm^{-3} carrier density.

In conclusion, luminescence decay transient in a HVPE grown GaN was shown to be in good agreement with a model of saturated centers of nonradiative recombination with a trap density of $\sim 10^{17}$ cm^{-3} and carrier recombination coefficients of $\sim 10^{-8}$ cm^3/s. In such a regime, the luminescence decay time of 205 ps yields a common lifetime of 410 ps for both electrons and holes. This value is smaller than that in homoepitaxialy grown samples, but it is considerably higher than that in high-quality MOCVD grown epilayers grown on sapphire [7]. This suggests high potential of HVPE grown GaN for short-wavelength optoelectronics applications.

Acknowledgements The research at Vilnius University was partially supported by the Lithuanian State Science and Education Foundation and European Commission supported SELITEC centre Contract No.G5MA-CT-2002-04047. A. Ž. acknowledges the Lithuanian Ministry of Education and Science for his Fellowship.

References

[1] S. Nakamura and G. Fasol, The Blue Laser Diode: GaN Based Light Emitters and Lasers (Springer, Berlin, 1997).
[2] A. Žukauskas, M. S. Shur, and R. Gaska, Introduction to Solid-State Lighting (Wiley, New York, 2002).

[3] C. Kirchner, V. Schwegler, F. Eberhard, M. Kamp, K. J. Ebeling, K. Kornitzer, T. Ebner, K. Thonke, R. Sauer, P. Prystawko, M. Leszczynski, I. Grzegory, and S. Porowski, Appl. Phys. Lett. **75**, 1098 (1999).
[4] M. Leszczynski, B. Beaumont, E. Frayssinet, W. Knap, P. Prystawko, T. Suski, I. Grzegory, and S. Porowski, Appl. Phys. Lett. **75**, 1276 (1999).
[5] G. A. Bunea, W. D. Herzog, M. S. Ünlü, B. B. Goldberg, and R. J. Molnar, Appl. Phys. Lett. **75**, 838 (1999).
[6] P. W. Yu, C. S. Park, and S. T. Kim, J. Appl. Phys. **89**, 1692 (2001).
[7] S. Juršėnas, G. Kurilčik, N. Kurilčik, A. Žukauskas, P. Prystawko, M. Leszczynski, T. Suski, P. Perlin, I. Grzegory, and S. Porowski, Appl. Phys. Lett. **78**, 3776 (2001).
[8] S. Juršėnas, S. Miasojedovas, G. Kurilčik, A. Žukauskas, and P. R. Hageman, Appl. Phys. Lett. **83**, 66 (2003).
[9] V. Kirilyuk, P. R. Hageman, P. C. M. Christianen, M. Zielinski, and P. K. Larsen, Appl. Phys. Lett. **79**, 4109 (2001).
[10] F. Binet, J. Y. Duboz, J. Off, and F. Scholz, Phys. Rev. B **60**, 4715 (1999).
[11] S. Hess, R. A. Taylor, J. F. Ryan, B. Beaumont, and P. Gibart, Appl. Phys. Lett. **73**, 199 (1998).
[12] O. Brandt, J. Ringling, K. H. Ploog, H.-J. Wünsche, and F. Henneberger, Phys. Rev. B **58**, R15977 (1998).
[13] O. Brandt, H. Yang, and K. H. Ploog, Phys. Rev. B **54**, R5215 (1996).
[14] A. Dmitriev and A. Oruzheinikov, J. Appl. Phys. **86**, 3241 (1999).

Self-heating and microwave noise in AlGaN/GaN

L. Ardaravičius[*,1], **J. Liberis**[1], **A. Matulionis**[1], **L. F. Eastman**[2], **J. R. Shealy**[2], and **A. Vertiatchikh**[2]

[1] Semiconductor Physics Institute, A. Goštauto 11, 2600 Vilnius, Lithuania
[2] Cornell University, 425 Philips Hall, Ithaca, NY 14853, USA

Received 15 September 2003, revised 28 October 2003, accepted 28 October 2003
Publised online 19 January 2004

PACS 65.90+i, 72.20.Ht, 72.70+m, 73.40.Kp

Equivalent noise temperature and spectral intensity of current fluctuations at 10 GHz frequency are investigated for $Al_{0.15}Ga_{0.85}N$/GaN channels. Dependence on electric field strength and voltage pulse duration is studied, channel self-heating effects on noise are shown. The lattice temperature is estimated to reach 730 K at 14.1 kV/cm electric field, the heat drain time is found to be 400 ns.

© 2004 WILEY-VCH Verlag GmbH & Co. KGaA, Weinheim

1 Introduction Nitride field-effect transistors with two-dimensional electron gas (2DEG) channels are promising for microwave high-power applications. The channels can operate at elevated lattice temperatures, the electron density is high (sheet density exceeds 10^{13} cm^{-2}), the saturation velocity is close to 1.2×10^7 cm/s [1, 2]. However, the velocity and the resultant channel performance are limited by self-heating [3, 4]. The local increase in the lattice temperature can be estimated by time-resolved noise-thermometry method [5, 6] and in other ways [7–9]. The dependence of electronic noise on self-heating has been studied at low frequencies [10]. Our goal is to investigate the effect of self-heating on the noise temperature and spectral intensity of current fluctuations at 10 GHz frequency.

2 Experimental
2.1 Samples The investigated Si_3N_4/AlGaN/GaN/Al_2O_3 heterostructure contained a 1 μm thick GaN layer overgrown with a nominally undoped 25 nm layer of $Al_{0.15}Ga_{0.85}N$. The conductive channel was located in the GaN layer close to the AlGaN/GaN interface, the 2DEG was induced by spontaneous polarization and piezoelectric fields. The electron sheet density was $n_{2D} = 5 \times 10^{12}$ cm^{-2}; the low-field mobility was $\mu(293 \text{ K}) = 1100$ cm^2/(Vs), and $\mu(373 \text{ K}) = 850$ cm^2/(Vs).

2.2 Equivalent noise temperature and spectral intensity of current fluctuations Two-terminal samples supplied with coplanar ohmic Ti/Al/Ti/Au electrodes were investigated. The channel length and the electrode width were, respectively, $L = 7$ μm and $w = 120$ μm. The average applied electric field E in the channel was estimated according to $E = (V - IR_c)/L$ where V is the applied voltage, I is the current and R_c is the contact resistance. The microwave noise measurements were carried out in the frequency band near 10 GHz where contributions due to generation–recombination noise and 1/f fluctuations were negligible. The sample was placed into a waveguide and subjected to the pulsed electric field applied parallel to the interface for investigation of longitudinal current fluctuations in

---
[*] Corresponding author: e-mail: linas@pfi.lt, Phone: 370 5 261 81 01, Fax: 370 5 262 71 23

© 2004 WILEY-VCH Verlag GmbH & Co. KGaA, Weinheim

the bias direction. The gated modulation-type radiometric setup was used for time-dependent noise power measurements under pulsed bias [6].

For matched impedances of the sample and the input of the noise measuring circuit, the equivalent noise temperature T_n is determined by the noise power ΔP_n [6]:

$$\frac{\Delta P_n}{\Delta f} = k_B T_n, \qquad (1)$$

where Δf is the frequency bandwidth around the radiometer frequency f ($\Delta f \ll f$) and k_B is the Boltzmann constant.

The sample impedance depends on the applied electric field and time. Thus, the correction is needed when the available noise power is determined from the measured noise power. The effect of mismatch and the contribution of noise sources located outside the 2DEG channel (due to the contact resistance and non-ideal elements of the waveguide circuit) were taken into account in the known way [12].

The spectral intensity of current fluctuations S_I can be expressed in terms of the noise temperature and the small-signal admittance Z^{-1} [6]:

$$S_I = 4k_B T_n \operatorname{Re}\{Z^{-1}\}. \qquad (2)$$

3 Noise temperature and lattice temperature

3.1 Time-dependent noise temperature The effect of channel self-heating on hot-electron noise is evident from time-resolved measurements. Figure 1 illustrates the results on the excess noise temperature $\Delta T_n = T_n - T_0$ at room temperature ($T_0 = 293$ K) plotted as function of the delay time. The voltage pulse duration is 30 μs, the sampling pulse duration is 100 ns. A weak, if any, effect is observed at 2.5 kV/cm field (open squares). A higher electric field causes a considerable thermal walkout due to the self-heating (closed symbols). In particular, at 14.1 kV/cm (triangles), the noise temperature decreases from 4000 K to less than 2000 K as the delay time increases.

The lattice temperature T_L of the channel can be measured after the voltage switch-off in the following way [5, 11]. When the electric field is off, the electron temperature almost immediately relaxes to the lattice temperature − the electron energy relaxation time is in the picosecond−subpicosecond range [12]. Later, the electrons find themselves in equilibrium with the lattice. Under these conditions the noise temperature equals the lattice temperature. Open triangles in Fig. 1 illustrate the decay of the lattice temperature after the voltage pulse ($E = 14.1$ kV/cm). The exponential dependence of the excess lattice temperature $\Delta T_L = T_L - T_0$ on the delay time t is observed:

$$\Delta T_L(t) = \Delta T_L(0) \exp(-t/\tau). \qquad (3)$$

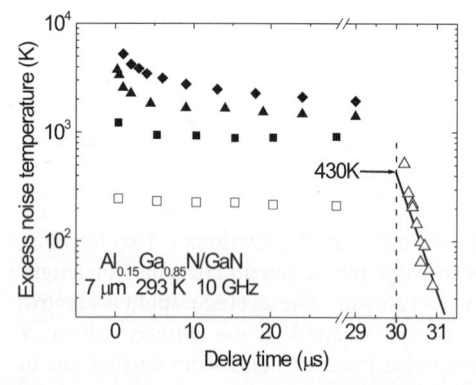

Fig. 1 Dependence of excess noise temperature at 10 GHz on time under bias: 2.5 kV/cm (open squares), 10.3 kV/cm (closed squares), 14.1 kV/cm (closed triangles) and 24.5 kV/cm (diamonds), and after 14.1 kV/cm pulse (open triangles). Pulse duration is 30 μs.

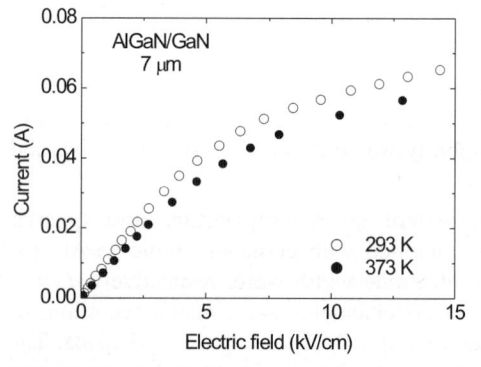

Fig. 2 Current−field characteristics for Al$_{0.15}$Ga$_{0.85}$N/GaN channel at 293 K and 373 K ambient temperatures. Pulse duration is 0.5 μs.

Solid line in Fig. 1 stands for Eq. (3) where the relaxation time $\tau \approx 400$ ns and the initial excess lattice temperature $\Delta T_n(0) \approx 430$ K. In conclusion, the lattice temperature reaches 730 K at $E = 14.1$ kV/cm.

3.2 Noise at elevated ambient temperatures The electron mobility decreases as the lattice temperature increases as illustrated by dependence of the current on the electric field (Fig. 2). As a result the electron heating becomes less efficient at a higher lattice temperature. The experiment confirms this. Fig. 3 presents dependence of the hot-electron noise temperature $T_n(E)$ on the applied electric field, measured for 0.5 μs voltage pulses at two ambient temperatures, 293 K and 373 K. As expected, the hot-electron noise temperature at 373 K ambient temperature is lower than that at 293 K in the field range 2.5 kV/cm $< E <$ 25 kV/cm. This effect is called thermal quenching of hot-electron noise. At a high electric field, the channel self-heating comes into play. As a result, the dependence on the ambient temperature tends to diminish at high electric fields.

3.3 Intersubband transfer noise In the investigated range of electric fields, the hot-electron noise is caused by three sources of fluctuations: the main source is due to the hot-electron temperature, the other two sources come from fluctuations in the hot-electron temperature and from intersubband occupancy fluctuations. The latter source manifests itself at $E > 8$ kV/cm at 293 K ambient temperature [13, 14].

The onset field for the intersubband fluctuations depends on the ambient temperature. This statement can be illustrated by presenting the dependence of the spectral intensity of current fluctuations on electric field (Fig. 4). Two maxima are resolved at $T_0 = 293$ K (open circles): the first maximum forms when LO-phonon emission by hot-electrons becomes important for the electron energy dissipation, the second maximum results from combined action of the hot-electron intersubband transfer and the thermal quenching of hot-electron noise. When the ambient temperature is increased, the onset of the second maximum shifts towards higher electric fields. The intersubband transfer fluctuations become the dominant source of hot-electron noise in the field range $E > 15$ kV/cm at 373 K.

3.4 Characteristic time of heat drain The characteristic time of the heat drain across the 300 μ thick saphire substrate is estimated $\tau_{\text{saph}} = 4.5$ ms [3]. Thus, the heat drain across the sapphire substrate is not important in our experiments with 30 μs voltage pulses. With this in mind, our microwave noise technique with the sampling time of 100 ns, is applicable for experimental investigation of heat dissipation inside the AlGaN/GaN heterostructure. The measured relaxation time constant of $\tau \approx 400$ ns characterizes the heat drain out of the 2DEG channel. This experimental value of the characteristic time is longer than the reported ones: 23 ns [3], 80–90 ns [15].

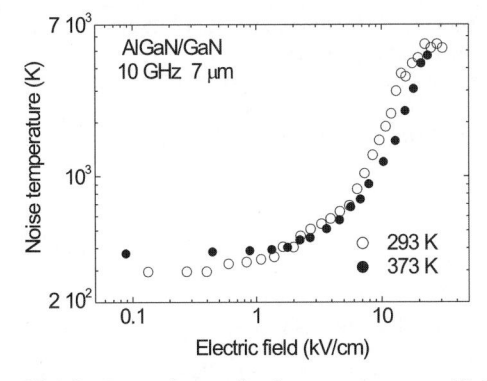

Fig. 3 Dependence of noise temperature at 10 GHz on electric field for the $Al_{0.15}Ga_{0.85}N$/GaN channel at two ambient temperatures: 293 K (open circles) and 373 K (closed circles). Electric field pulse duration is 0.5 μs.

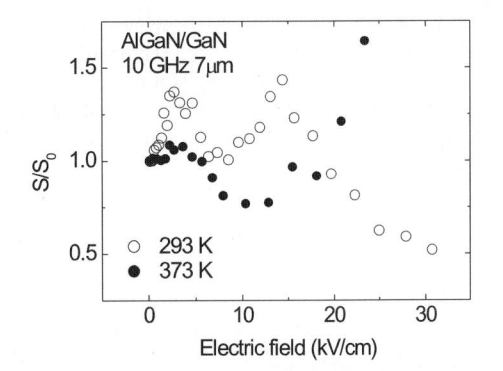

Fig. 4 Dependence of the spectral intensity of current fluctuations at 10 GHz on electric field for $Al_{0.15}Ga_{0.85}N$/GaN channel at 293 K (open circles) and 373 K (closed circles). Electric field pulse duration is 1 μs.

4 Conclusion Time-resolved microwave noise technique is applied for investigation of the self-heating effect in biased AlGaN/GaN 2DEG channels. The channel self-heating is responsible for the observed thermal walkout in the hot-electron noise temperature under bias. The lattice temperature of the 2DEG channel is estimated, the characteristic time of heat drain out of the 2DEG channel is measured. The onset of intersubband noise shifts towards higher electric fields at a higher ambient temperature.

Acknowledgements Vilnius group acknowledges support from ONR Award N00014-03-1-0558, Cornell group acknowledges support from ONR Contract N00014-01-1-0300.

References

[1] L. F. Eastman, V. Tilak, N. Kaper, J. Smart, R. Thomson, B. Green, J. R. Shealy, and T. Prunty, phys. stat. sol. (a) **194**, 433 (2002).
[2] Y. F. Wu, D. Kapolnek, J. P. Ibbetson, P. Parikh, B. P. Keller, and U. K. Mishra, IEEE Trans. Electron Devices **48**, 405 (2001).
[3] S. A. Vitusevich, S. V. Danylyuk, N. Klein, M. V. Petrychuk, A. Yu. Avksentyev, V. N. Sokolov, V. A. Kochelap, A. E. Belyaev, V. Tilak, J. Smart, A. Vertiatchikh, and L. F. Eastman, Appl. Phys. Lett. **82**, 748 (2003).
[4] L. Ardaravičius, A. Matulionis, J. Liberis, O. Kiprijanovic, M. Ramonas, L. F. Eastman, J. R. Shealy, and A. Vertiatchikh, Appl. Phys. Lett. **83**, 4038 (2003).
[5] M. de Murcia, E. Richard, J. M. Perraudin, A. Boyer, A. Benvenuti, and J. Zimmermann, Semicond. Sci. Technol. **10**, 515 (1995).
[6] H. Hartnagel, R. Katilius, and A. Matulionis, in: Microwave Noise in Semiconductor Devices (Wiley, New York, 2001), p. 86–90.
[7] H. M. Kim, J. S. Yoon, S. H. Lee, J. E. Oh, W. S. Lee, K. I. Chun, K. W. Chung, H. W. Chung, and T. W. Kang, Jpn. J. Appl. Phys. **39**, L1087 (2000).
[8] N. Shigekawa, K. Shiojima, and T. Suemitsu, J. Appl. Phys. **92**, 531 (2002).
[9] M. Kubal, S. Rajasingam, A. Sarua, M. J. Uren, T. Martin, B. T. Hughes, K. P. Hilton, and R. S. Balmer, Appl. Phys. Lett. **82**, 124 (2003).
[10] S. A. Vitusevich, M. V. Petrychuk, V. N. Sokolov, S. V. Danylyuk, A. M. Kurakin, N. Klein, A. Yu. Avksentyev, V. A. Kochelap, A. E. Belyaev, V. Tilak, J. Smart, A. Vertiatchikh, and L. F. Eastman, in: 27th Workshop on Compound Semiconductors and Integrated Circuits–WOCSDICE'2003 (Furigen, Switzerland, 2003), pp. 25.
[11] A. Matulionis, J. Liberis, I. Matulionienė, M. Ramonas, L. F. Eastman, J. R. Shealy, V. Tilak, and A. Vertiatchikh, Phys. Rev. B **68**, 035338 (2003).
[12] A. Matulionis, J. Liberis, L. Ardaravičius, M. Ramonas, I. Matulionienė, and J. Smart, Semicond. Sci. Technol. **17**, L9 (2002).
[13] A. Matulionis, J. Liberis, L. Ardaravičius, M. Ramonas, T. Zubkute, I. Matulionienė, L. F. Eastman, J. R. Shealy, J. Smart, D. Pavlidis, and S. Hubbard, phys. stat. sol. (b) **234**, 826 (2002).
[14] J. Liberis and A. Matulionis, in: Hot-electron noise in nitride 2DEG channels, Proc. of the 17th Int. Conf. on Noise and Fluctuations, edited by J. Sikula (CNRL s.r.o., Brno, 2003), pp. 241–246.
[15] R. Gaska, A. Osinsky, J. W. Yang, and M. S. Shur, IEEE Electron. Device Lett. **19**, 89 (1998).

In-depth and in-plane profiling of light emission properties of InGaN-based laser diode

M. Godlewski[*,1,2], **E. M. Goldys**[3], **M. R. Phillips**[4], **T. Böttcher**[5], **S. Figge**[5], **D. Hommel**[5], **R. Czernecki**[6], **P. Prystawko**[6], **M. Leszczynski**[6], **P. Perlin**[6], **I. Grzegory**[6], and **S. Porowski**[6]

[1] Institute of Physics, Polish Acadademy of Sciences, Al. Lotników 32/46, 02-668 Warsaw, Poland
[2] College of Science, Department of Mathematics and Natural Sciences, Cardinal S. Wyszyński University, Warsaw, Poland
[3] Division of Information and Communication Sciences, Macquarie University, Sydney, Australia
[4] Microstructural Analysis Unit, UTS, Sydney, Australia
[5] Institute of Solid State Physics, University of Bremen, 28334 Bremen, Germany
[6] High Pressure Res. Center (Unipress), Polish Acadademy of Sciences, Warsaw, Poland

Received 15 September 2003, accepted 17 September 2003
Published online 18 December 2003

PACS 78.60.Hk, 78.66.Fd, 85.30.De, 85.60.Bt, 85.60.Jb

We employ cathodoluminescence (CL) technique for evaluation of in-depth and in-plane instabilities of light emission in laser diode structure. We study light emission properties from laser structure and their relation to microstructure details. Large in-plane instabilities of light emission are also present for excitation densities, larger than the threshold densities for the stimulated emission, i.e., potential fluctuations are not fully screened in the active region of the laser even at large excitation densities.

© 2004 WILEY-VCH Verlag GmbH & Co. KGaA, Weinheim

1 Introduction

Basov and coworkers demonstrated the possibility of laser emission upon electron beam pumping [1]. This method was then successfully applied to ZnSe-based heterostructures [2–6] avoiding limitations of p-type doping of ZnSe [2]. Laser action at record low threshold currents was achieved [4–6]. The method turned out to be very useful in studying degradation mechanisms in laser diode (LD) structures [7, 8]. Recently we reported a stimulated emission in homoepitaxial GaN epilayers [9]. In this work we employed cathodoluminescence (CL) to the studies of in-plane and in-depth variations of CL intensity. We study light emission properties for a homoepitaxial GaN-based LD. Important role of potential fluctuations in active regions of the device is demonstrated.

2 Experimental

LD structure studied was Separate Confinement Heterostructure Device (SCHD) grown on bulk GaN substrate. Undoped structure was studied, used for observation of a stimulated emission under optical pumping [10]. GaN cap layer was made thinner on purpose, to reduce unwanted light absorption. The active region of the LD structure consisted of five $In_{0.09}Ga_{0.91}N/In_{0.01}Ga_{0.99}N$ quantum wells (QWs), embedded between two 0.1 μm thick GaN wave-guiding layers and two cladding layers. The lower cladding layer consisted of $GaN/Al_{0.15}Ga_{0.85}N$ superlattice (2.5 nm/2.5 nm) with 120 repetitions. The upper cladding layer was in the form of 0.36 μm thick $Al_{0.08}Ga_{0.92}N$, covered with a thin GaN cap layer. LD cavity length was $L = 300$ μm. The cavity was formed by cleaving, which, in the case of homoepitaxial structures, results in atomically flat surfaces. The so-obtained output mirrors have reflection coefficient of about $R = 16\%$.

---

[*] Correspnding author: e-mail: godlew@ifpan.edu.pl, Phone: +48 22 8436861, Fax: +48 22 8430926

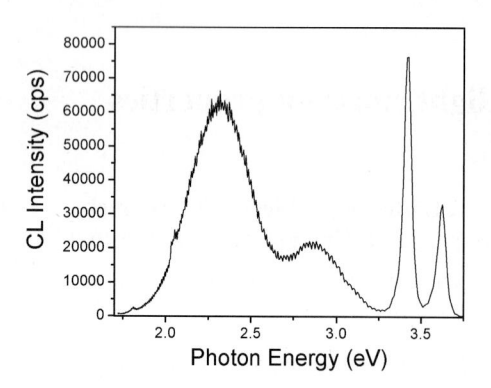

Fig. 1 CL spectrum of homoepitaxial LD structure measured at 10 kV accelerating voltage, 17.6 nA primary current density and at 2000 magnification.

The CL spectra and images were taken in a JEOL 35C scanning electron microscope which was equipped with a MonoCL2 CL system by Oxford Instruments. CL emission was detected using Hamamatsu R943-02 Peltier cooled photomultiplier. Emission was excited from top of the structure through a hole in a mirror, and was collected with a metal mirror mounted over the sample.

3 Results and discussion Relatively bright emission is observed under e-beam pumping, as shown in Fig. 1. Only the upper part of the device is excited at low accelerating voltage [11]. Several emission bands are observed, which we attribute to the CL from the upper GaN cap layer (likely with some contribution of underlying GaN wave-guiding layer), the edge emission from AlGaAs upper cladding layer,

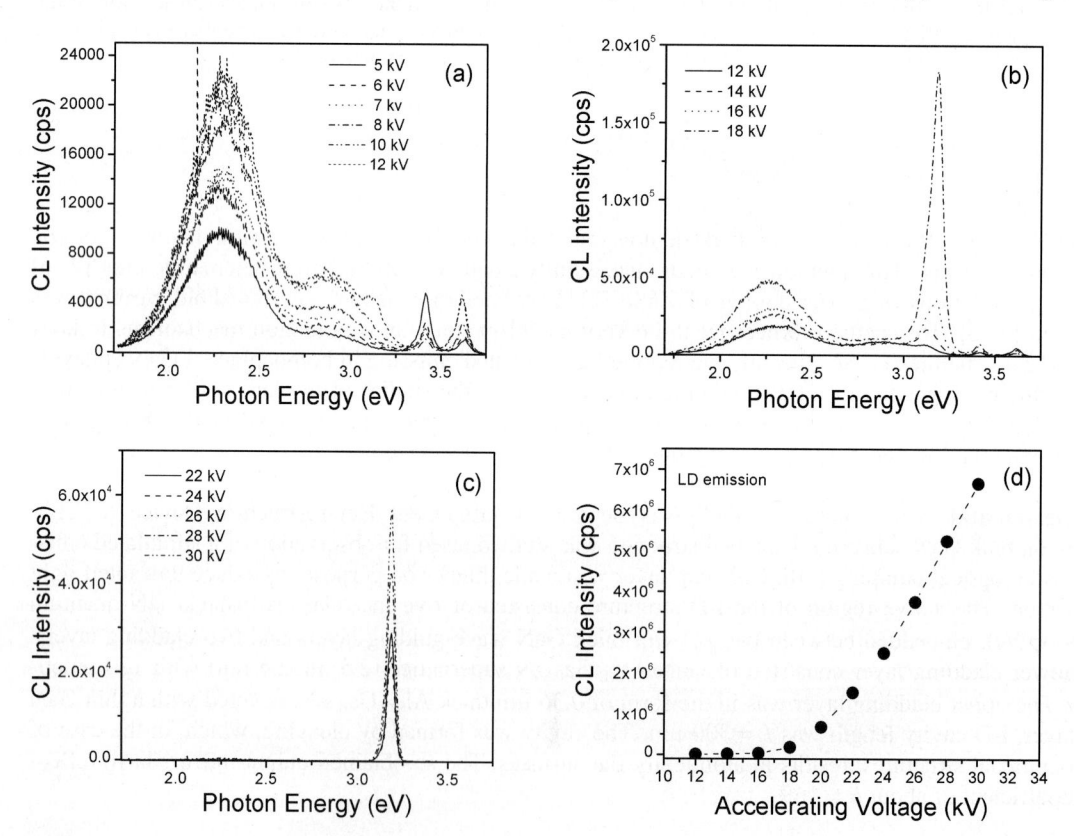

Fig. 2 (a–c) Results of depth-profiling CL investigations for the homoepitaxial LD structure measured at room temperature and accelerating voltage varied between 5 and 30 kV. (d) shows voltage dependence of the stimulated emission.

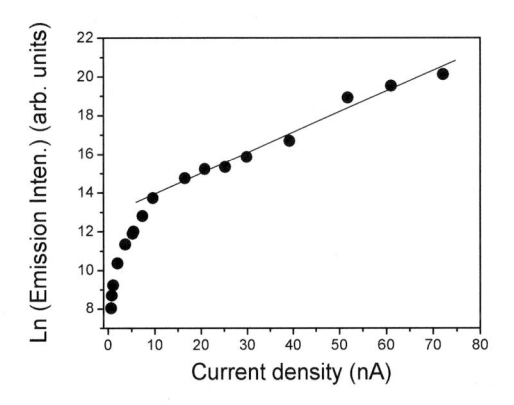

Fig. 3 Threshold dependence for the laser emission measured at 30 kV accelerating voltage at 2000 magnification.

and the two deep impurity-related (see Ref. 12 for explanation) yellow (YL) and blue (BL) emission bands. A better insight into origin of these CL bands was obtained from the depth-profiling CL measurements, as shown in Fig. 2 a–d. Basics of depth-profiling CL investigations are explained in [11].

Depth-profiling confirms that GaN excitonic emission mostly comes from the upper most cap layer. It decreases in intensity with increasing accelerating voltage, i.e., when CL is excited from deeper layers in the structure. The same emission is also excited from the GaN wave guiding layer, but is weaker.

GaN-related YL and BL emissions also originate from the wave-guiding region of the device. YL originates from GaN layers and is the brightest from the wave-guiding layer and from the GaN/AlGaN interface. Emission from the active region is not seen, when upper layers of the device are excited, which means that energy transfer, photon recycling or carrier diffusion are fairly inefficient.

A dramatic increase of the stimulated emission is observed for accelerating voltage exceeding 18 kV. Once conditions for excitation of this emission were optimized threshold current dependence could be

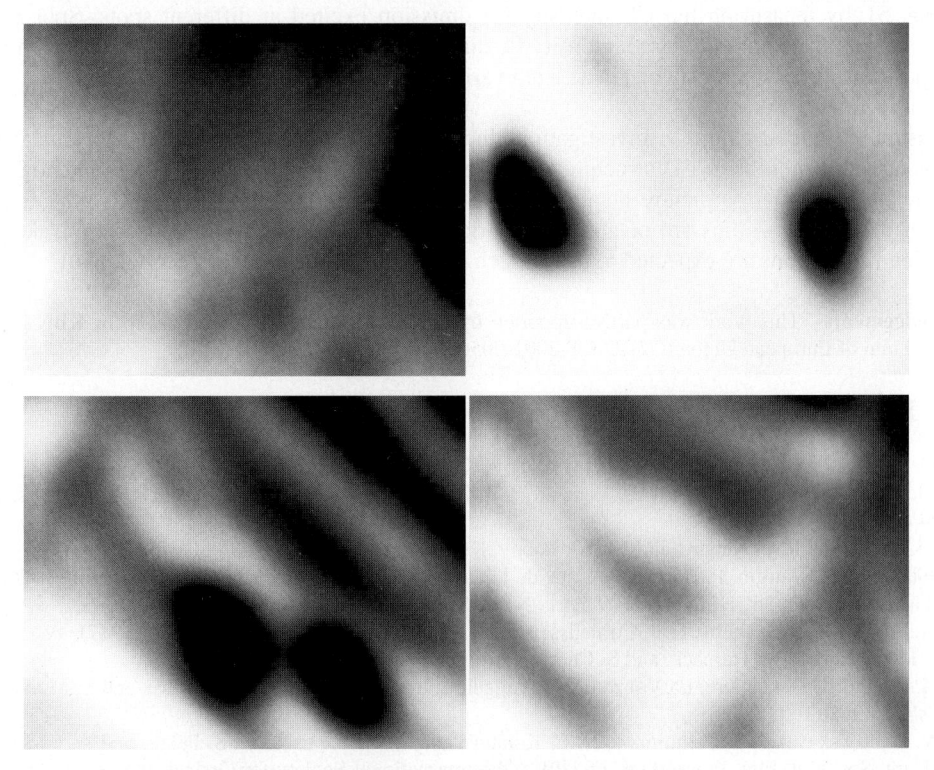

Fig. 4 In-plane fluctuations of the laser emission observed for excitation above the threshold value. Data from four different regions of the LD structure (at 6000 magnification) are shown.

© 2004 WILEY-VCH Verlag GmbH & Co. KGaA, Weinheim

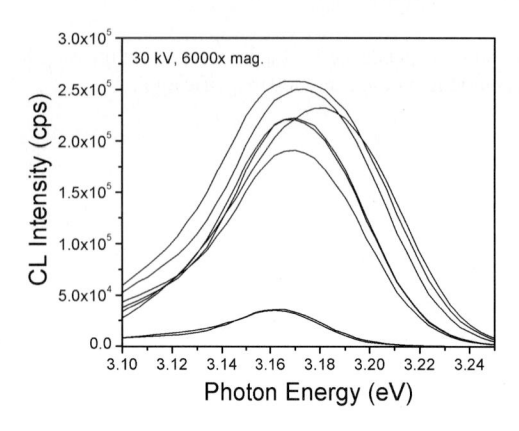

Fig. 5 Spot mode CL spectra measured for the detection set at the laser emission and excitation spot moved to regions of a different contrast in the CL images.

determined, as shown in Fig. 3. CL emission from QWs increases extremely rapid first nonlinearly, but then, for the current density above 10 nA, this increase is described by a power low dependence (Fig. 3). CL is dominated by a stimulated emission. CL line narrows at increased current density but we could not resolve laser modes, due to a limited resolution of our CL system. Some indication of these modes could however be seen. Other CL emissions were too weak (in comparison to the stimulated emission) to be detected.

In the further study we concentrated on evaluation of role of potential fluctuations on QW emission. We evaluated role of these potential fluctuations studying in-plane instabilities of CL intensity of the stimulated emission. Large contrast fluctuations are observed (Fig. 4). Growth steps are clearly resolved, together with totally dark area. Magnitude of these fluctuations we evaluated from the spot-mode CL study (Fig. 5), by measuring the CL intensity for emission excited at different spots. Spot mode CL measurements indicate not only large intensity fluctuations present in LD emission, but also a noticeable spectral position shift, which may originate from various strain conditions, etc…

4 Conclusion Summarizing, CL investigations indicate that potential fluctuations, so crucial in LED structures, are still present in LD structures, despite large excitation densities. These potential fluctuations correlate with micro-structure details, for example growth steps are seen in CL images showing contrast of the emission, thus are of a different origin than those observed/proposed by Nakamura. In composition fluctuations are expected to be of much smaller scale than those observed by us.

Acknowledgements This work was partly supported by the grant number 5 P03B 007 20 of KBN and by the DENIS program of European Union (G5RD-CT-2001-00566).

References

[1] N. G. Basov, V. A. Danilychev, and Yu. M. Popov, Kvantovaya Elektron. **1**, 29 (1971).
[2] D. Herve, R. Accomo, E. Molva, L. Vanzetti, J. J. Paggel, L. Sorba, and A. Franciosi, Appl. Phys. Lett. **67**, 2144 (1995).
[3] B. M. Lavrushin, R. F. Nabiev, and Yu. M. Popov, Kvantovaya Elektron. **15**, 78 (1988).
[4] B. Fitzpatric, J. Khurgin, P. Harnack, and D. de Leeuw, International Electron Devices Meeting 1986. Technical Digest (Cat. No. 86CH2381-2, New York, NY, USA) IEEE 1986, p. 630.
[5] D. A. Cammack, R. J. Dalby, H. J. Cornelissen, and J. Khurgin, J. Appl. Phys. **62**, 3071 (1987).
[6] B. J. Fitzpatrick, P. M. Harnack, and S. Cherin, Philips J. Res. **41**, 452 (1986).
[7] J.-M. Bonard, J.-D. Ganiere, La Vanzetti, J. J. Paggel, L. Sorba, A. Franciosi, D. Herve, and E. Molva, J. Appl. Phys. **84**, 1263 (1998).
[8] L. V. Borkovskaya, B. R. Dzhumaev, N. E. Korsunskaya, V. P. Papusha, G. S. Pekar, and A. F. Singaevsky, SPIE-Int. Soc. Opt. Eng. Proceedings of SPIE – the International Society for Optical Engineering, USA, **3724**, 244 (1999).

[9] M. Godlewski, V. Yu. Ivanov, E. M. Goldys, M. R. Phillips, T. Böttcher, S. Figge, D. Hommel, R. Czernecki, P. Prystawko, M. Leszczynski, P. Perlin, I. Grzegory, and S. Porowski, Acta Phys. Pol. A **103**, 689 (2003).
[10] V. Yu. Ivanov, M. Godlewski, H. Teisseyre, P. Perlin, R. Czernecki, P. Prystawko, M. Leszczynski, I. Grzegory, T. Suski, and S. Porowski, Appl. Phys. Lett. **81**, 3735 (2002).
[11] M. Godlewski, E. M. Goldys, M. R. Phillips, R. Langer, and A. Barski, J. Mater. Res. **15**, 495 (2000).
[12] M. Godlewski and E. M. Goldys, in III-Nitride Semiconductors: Optical Properties Vol. II (Optoelectronic Properties of Semiconductors and Superlattices), Hongxing Jiang and M. Omar Manasreh (Co-editors) (Taylor & Francis Books, New York, 2002), pp. 259.

Cathodoluminescence and atomic force microscopy study of n-type doped GaN epilayers

M. Godlewski[*,1,2], **E. Łusakowska**[1], **R. Bożek**[3], **E. M. Goldys**[4], **M. R. Phillips**[5], **T. Böttcher**[6], **S. Figge**[6], and **D. Hommel**[6]

[1] Institute of Physics, Polish Acadademy of Sciences, Al. Lotników 32/46, 02-668 Warsaw, Poland
[2] College of Science, Department of Mathematics and Natural Sciences, Cardinal S. Wyszyński University, Warsaw, Poland
[3] Institute of Experimental Physics, Warsaw University, Warsaw, Poland
[4] Division of Information and Communication Sciences, Macquarie University, Sydney, Australia
[5] Microstructural Analysis Unit, UTS, Sydney, Australia
[6] Institute of Solid State Physics, University of Bremen, 28334 Bremen, Germany

Received 15 September 2003, accepted 17 September 2003
Published online 18 December 2003

PACS 61.72.Ss, 61.72.Vv, 68.37.Ps, 78.60.Hk, 78.66.Fd

We evaluate influence of n-type doping and of the doping level on structural and optical quality of GaN epilayers. We confirm reported previously strong enhancement of light emission from Si doped samples and discuss possible mechanisms of this enhancement. Kelvin probe measurements indicate smoothing of potential fluctuations in the doped layers.

© 2004 WILEY-VCH Verlag GmbH & Co. KGaA, Weinheim

1 Introduction It was reported that n-type doping of GaN or InGaN epilayers enhances rate of radiative recombination [1]. It was observed that Stokes shift decreases in n-type doped structures, indicating weaker carrier/exciton localization [2]. Moreover, n-type doping apparently affects growth mode resulting in smoother interfaces in GaN/InGaN structures [4]. Three models were proposed to explain the observed effects. The first assumes that n-type doping results in screening of piezoelectric fields, the second is based on an improvement of morphology upon doping, the third supposes saturation of deep dislocation-related centers of a nonradiative recombination upon doping [2–4]. In this work we report results of the studies of two series of GaN samples grown with different dislocation density and different n-type doping levels. A dramatic enhancement of the emission intensity is documented and related to smoothing of potential fluctuations upon doping.

2 Samples Two sets of samples were studied grown by metalorganic vapour phase epitaxy (MOVPE). First template, denoted as g411, consisted of MOVPE-grown GaN/sapphire layer, with 1×10^{10} cm^{-2} dislocation density. Three 1500 nm thick samples were grown on this template by MOVPE doped with different concentrations of Si: g455 – undoped GaN layer, g456 – with silicon concentration of 10^{18} cm^{-3}, g457 – with silicon concentration 10^{19} cm^{-3}. The second template, denoted as g454, consisted of GaN/sapphire layer, with 2×10^{9} cm^{-2} dislocations. Three 1500 nm thick samples were grown on this template, which were analogous to those grown on the g411 template: g455 – undoped GaN, g456 – with silicon concentration of 10^{18} cm^{-3}, g457 – with silicon concentration 10^{19} cm^{-3}.

Structural and optical properties of the samples were studied with cathodoluminescence (CL) and scanning electron microscopy (SEM). The CL images were taken in a JEOL 35C scanning electron mi-

---

[*] Corresponding author: e-mail: godlew@ifpan.edu.pl, Phone: +48 22 8436761, Fax: + 48 22 8430926

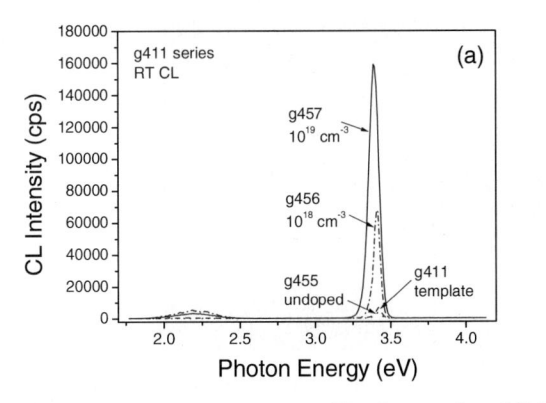

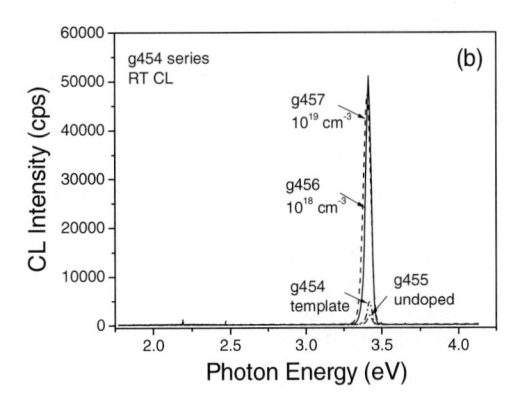

Fig. 1 a, b) Room temperature CL of two series of GaN epilayers measured at the same excitation conditions.

croscope equipped with a MonoCL2 CL system by Oxford Instruments. CL emission was detected using Hamamatsu R943-02 Peltier cooled photomultiplier. Quantitative information on microstructure of surfaces of the samples was drawn from the atomic force microscopy (AFM) investigations, using Multi Mode Scanning Probe Microscope, and from high-resolution SEM images taken in a JEOL JSM-6300F scanning microscope with a field emission gun. Potential fluctuations were evaluated with Kelvin probe of Nanoscan III AFM microscope of Digital Instruments.

3 Results

In Fig. 1a, b we show CL spectra observed for two series of the samples studied. For each of the series, the brightest CL emission is observed for heavily Si doped samples. We also observe that the edge CL (of excitonic origin) shifts down in the energy for these samples. Undoped layers (samples g455), which show a similar structural quality as substrate layers, show equally weak CL intensities. A parasitic yellow emission [5], observed for all samples studied, is slightly enhanced in heavily doped samples.

We performed depth-profiling CL investigations to analyse origin of the observed emissions. Basics of the CL depth-profiling are explained in our recent publication [6]. For low accelerating voltages CL is excited from the surface close region of the sample. For high voltages CL comes from the interface region [6]. CL depth-profiling indicates that for doped layers the brightest edge CL is observed from surface close layer of the film, as shown in Fig. 2. Then this CL is weaker in the volume of the sample and is slightly reduced in the intensity in the interface close layer of the sample. Depth-profiling indicates enhancement of the parasitic yellow emission from surface close layer of the film, which likely is due to an increased silicon concentration in the uppermost region of the sample. Yellow emission is weak in the volume of the sample and becomes enhanced in the interface region and in defected GaN substrate layer.

We evaluated dependence of the CL intensity on excitation beam density. These measurements were performed at 5 kV accelerating voltage and from 50×50 μm region of the film. A strongly nonlinear

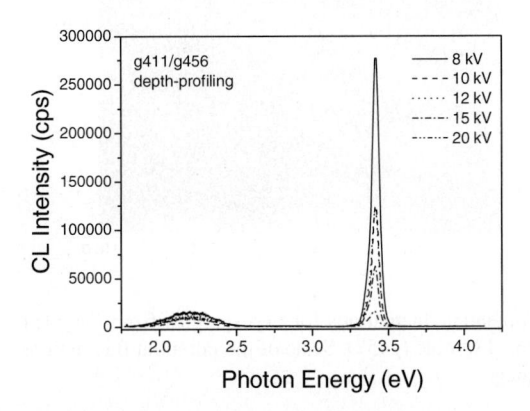

Fig. 2 Depth-profiling CL investigations at 2000 magnification for Si doped g411/g456 sample.

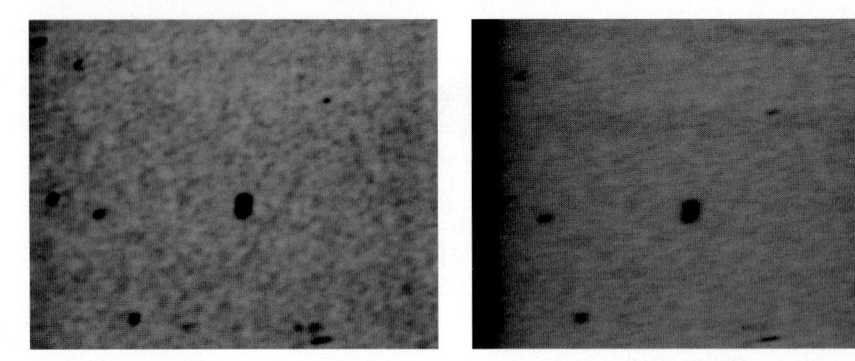

Fig. 3 Maps of the in-plane fluctuations of the edge CL for the two current densities (11.8 nA (left) and 4.3 nA (right)) for heavily Si doped g454/g457 sample. CL data were taken at room temperature, at 5 kV accelerating voltage and at 2000 magnification.

Table 1 Results of Kelvin probe measurements for two series of samples studied.

sample	RMS* (mV) light off	peak value** (mV) light off	RMS (mV) light on	peak value (mV) light on
G411 (template)	54	190	52	178 to 252
G455 (undoped)	94	350	64	231 to 334
G456 (10^{18} cm^{-3})	–	10	2	10
G457 (10^{19} cm^{-3})	4	15 to 23	2	5 to 20
G454 (template)	58	144 to 227	62	180 to 227
G455 (undoped)	55	107 to 128	17	45 to 132
G456 (10^{18} cm^{-3})	8	35 to 53	11	66 to 90
G457 (10^{19} cm^{-3})	2	19	5	17 to 20

* RMS – root mean square of the potential fluctuations from 10×10 μm^2 regions of the samples studied
** Peak values of potential fluctuations for different crossing through 10×10 μm^2 regions of the samples studied

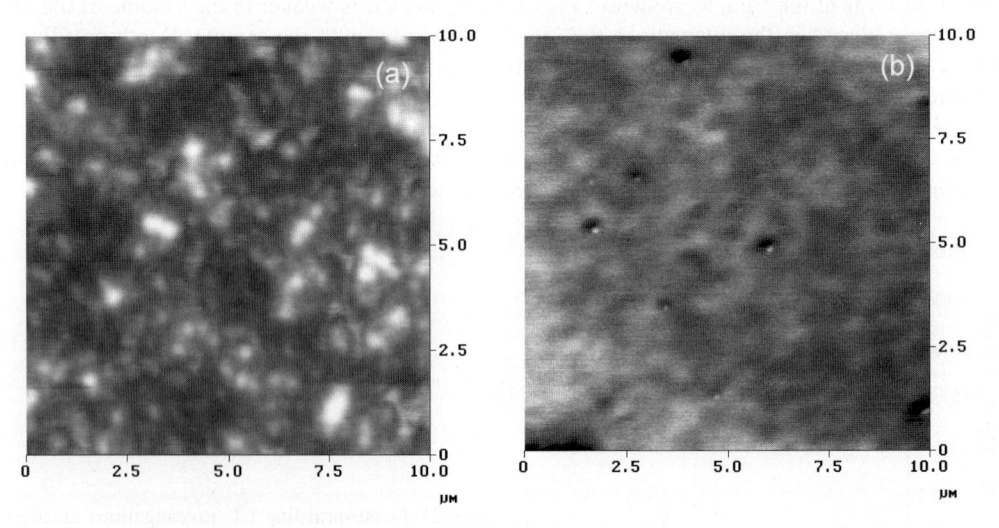

Fig. 4 Results of Kelvin probe measurements of potential fluctuations for two samples from the g411 series – for the template (a) and for the heavily Si doped sample (g457). Scale of the potential fluctuations is 300 mV for the (a) image and 50 mV for the (b) image.

current dependence was observed indicating possibility of a stimulated emission at highest excitation densities, as was recently reported by us for homoepitaxial GaN films [7]. Fig. 3a, b shows in-plane fluctuations of the edge CL, which was measured at 11.8 nA (a) and at 4.3 nA (b) density of a primary current. In-plane variations of the CL emission are also observed (even become enhanced) at larger current densities, which means that potential fluctuations, responsible for these fluctuations, are not screened at high excitation densities. Some of the data suggest an increase of carrier diffusion lengths at higher excitation densities, as was also concluded by us from another study [8].

As it was mentioned in the introduction, one of the models relates a strong enhancement of the GaN emission upon doping to a reduction of potential fluctuations present in layers. We verify such possibility evaluating potential fluctuations present in our samples with Kelvin probe of the AFM system. Results of the measurements are shown in Fig. 4a, b and are summarised in the Table 1. For the templates and for undoped layers potential fluctuations are fairly large and reflect details of microstructure of the samples. Doping dramatically reduces potential fluctuations (Table 1), which are reduced by more than one order in a magnitude to about 5 mV or even less. Amplitude of the observed fluctuations is smaller than 20 mV reported by Brandt and coworkers for n-type GaN [9]. No effect of illumination is observed for the templates, whereas for the undoped layers grown on the templates and also for doped ones photo-illumination smoothes potential fluctuations. Only for the heavily doped samples photo-illumination affects weakly the fluctuations. Surprisingly, they even increase in some cases, which we relate to charging of impurities decorating pin holes, charging of dislocations and of grain boundaries.

4 Conclusion Concluding, our results strongly favour model relating observed emission enhancement to a dramatic reduction of potential fluctuations in the layers. This reduction can also result in larger diffusion lengths of carriers, as concluded from the CL measurements, and also in increased rate of carrier binding and recombination, not discussed here.

Acknowledgements This work was partly supported by the grant numbers 5 P03B 007 20, 5 P03B 123 21 of KBN (Polish State Committee for Scientific Research) and by the DENIS program of European Union (G5RD-CT-2001-00566).

References

[1] S. Nakamura, T. Mukai, and M. Senoh, Jpn. J. Appl. Phys. (Part 2) **32**, 16 (1993).
[2] Y. Cho, J. Song, S. Keller, M. Minsky, E. Hu, U. Mishra, and S. DenBaara, Appl. Phys. Lett. **73**, 1128 (1998).
[3] M. Minsky, S. Chichibu, S. Fleischer, A. Abare, J. Bowers, E. Hu, S. Keller, U. Mishra, and S. DenBaars, Jpn. J. Appl. Phys. (Part 2) **37**, 1362 (1998).
[4] A. Munkholm, G. Stephenson, J. Eastman, O. Auciello, M. Ramana Murty, C. Thompson, P. Fini, J. Speck, and S. DenBaars, J. Cryst. Growth **221**, 98 (2000).
[5] T. L. Tansley, E. M. Goldys, M. Godlewski, B. Zhou, and H. Y. Zuo, in: Optoelectronic properties of semiconductors and superlattices, series editor M. O. Manasreh, Vol. 2, GaN and Related Materials, edited by S. Pearton (Gordon and Breach Publishers, 1997), p. 233–293.
[6] M. Godlewski, E. M. Goldys, M. R. Phillips, R. Langer, and A. Barski, J. Mater. Res. **15**, 495 (2000).
[7] M. Godlewski, V. Yu. Ivanov, E. M. Goldys, M. R. Phillips, T. Böttcher, S. Figge, D. Hommel, R. Czernecki, P. Prystawko, M. Leszczynski, P. Perlin, I. Grzegory, and S. Porowski, Acta Phys. Pol. A **103**, 689 (2003).
[8] M. Godlewski, E. M. Goldys, M. R. Phillips, T. Böttcher, S. Figge, D. Hommel, R. Czernecki, P. Prystawko, M. Leszczynski, P. Perlin, P. Wisniewski, T. Suski, M. Boskowski, I. Grzegory, and S. Porowski, Acta Phys. Pol. A **102**, 627 (2002).
[9] M. S. Brandt, P. Herbst, O. J. Homar, A. Stemmer, J. Appl. Phys. **84**, 1168 (1998).

Compensation mechanisms in magnesium doped GaN

M. Godlewski[*,1,2], **H. Przybylińska**[1], **R. Bożek**[3], **E. M. Goldys**[4], **J. P. Bergman**[5], **B. Monemar**[5], **I. Grzegory**[6], and **S. Porowski**[6]

[1] Institute of Physics, Polish Acadademy of Sciences, Al. Lotników 32/46, 02-668 Warsaw, Poland
[2] College of Science, Dep. of Mathematics and Natural Sciences, Cardinal S. Wyszyński University, Warsaw, Poland
[3] Institute of Experimental Physics, Warsaw University, Warsaw, Poland
[4] Division of Information and Communication Sciences, Macquarie University, Sydney, Australia
[5] Department of Physics and Meas. Techn., Linköping University, S-581 83 Linköping, Sweden
[6] High Pressure Research Center (Unipress), Polish Acadademy of Sciences, Warsaw, Poland

Received 15 September 2003, accepted 17 September 2003
Published online 18 December 2003

PACS 61.72.Ss, 61.72.Vv, 68.37.Ps, 78.55.Cr, 78.66.Fd

Compensation processes in magnesium doped GaN epilayers and bulk samples are studied. We demonstrate enhancement of potential fluctuations in Mg doped samples, from Kelvin probe atomic force microscopy measurements. Large- and small-scale light emission fluctuations are also demonstrated. Microphotoluminescence (PL) study indicates an unusual anti-correlation between the intensities of excitonic and defect-related emission processes in p-type doped structures and also the presence of the so-called hot-PL. Hot-PL observed in compensated p-type samples, we relate to the presence of strong potential fluctuations.

© 2004 WILEY-VCH Verlag GmbH & Co. KGaA, Weinheim

1 Introduction

In this work we discuss optical properties of Mg doped GaN samples and epilayers. Several conflicting observations on the mechanisms of radiative recombination in these samples are discussed in relation to compensation effects and to large in-plane variations of potential fluctuations in these samples. The fluctuations are enhanced by codoping, as we show in the present study.

2 Experimental

Bulk GaN samples used in the present study were grown in the High Pressure Research Center of the Polish Academy of Sciences, Warsaw, Poland. The as-grown samples were heavily oxygen contaminated and had a room temperature concentration of electrons reaching 10^{20} cm^{-3}. The samples doped in the melt with acceptors showed either weak n-type conductivity or (for heavily Mg doped samples, $6 \div 8 \times 10^{19}$ cm^{-3}) were highly resistive. We also studied several epilayers of GaN:Mg, GaN:Mg, Si and InGaN:Mg grown by metalorganic vapour phase epitaxy. These layers had different acceptor concentrations, controlled either by different times of annealing in nitrogen or by different intentional doping levels.

The photoluminescence (PL) experiments were performed at 2 K using a 351 nm UV line of an Ar$^+$ laser for excitation. Measurements of the PL or a photon counting photomultiplier system. Pulsed excitation was provided by a mode locked Ti: kinetics were taken with either a streak camera sapphire solid state laser with frequency doubling (λ = 340 nm) with a pulse length of 2 ps or a frequency tripled YAG:Nd laser (λ = 355 nm) with 2 ns pulses and a varying repetition rate. The reported measurements were performed at 2 K in a He bath cryostat.

---

[*] Corresponding author: e-mail: godlew@ifpan.edu.pl, Phone: +48 22 8436861, Fax: + 48 22 8430926

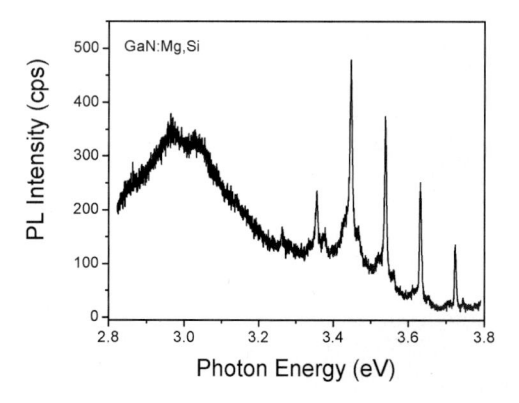

Fig. 1 Room temperature photoluminescence spectrum of GaN:Si,Mg epilayer observed under uv photo-excitation with He–Cd laser.

The micro-PL spectra were measured with a Renishaw micro-Raman system, with a resolution of 1.6 meV using a polarised 325 nm He–Cd laser beam at an excitation power density of 120 kW/cm^2. The micro-PL emission spectra were collected with a spatial resolution down to 2×2 µm^2 at room temperature.

Quantitative information on the microstructure of sample surfaces was drawn from atomic force microscopy (AFM) investigations, using a Multi Mode Scanning Probe Microscope. Potential fluctuations were evaluated with a Kelvin probe of a Nanoscan III AFM microscope of Digital Instruments.

3 Results In bulk samples as well as in doped epilayers two doping-related PL bands were observed – yellow luminescence (YL) and blue luminescence (BL) [1]. The PL decay characteristics of these two PL bands the fastest component of the decay) to about 1 µs (slow component of the decay). In turn, in were measured. In bulk metallic n-type samples, the observed PL decay times of YL (only YL is observed) were independent of the detection wavelength and relatively short – from single nanoseconds (for high resistivity Mg-compensated bulk samples detection wavelength-dependent decay times of YL and BL are observed. In these samples the decay time of YL is unusually long, up to about 500 µs, for the slow component of the decay. The PL decay times in GaN (InGaN) epilayers measured were in-between those observed for metallic and resistive bulk samples.

A characteristic feature of the samples compensated with magnesium is the appearance of sharp peaks in the high-energy wing of the PL spectra (Fig. 1), separated from the excitation line by multiples of LO phonon energy. Further on we will denote these peaks as hot-PL peaks. In Fig. 2 a, b we show spatially-resolved PL intensity fluctuations, for two selected hot-PL peaks. Large in-plane fluctuations in the intensities are observed, which are not correlated to each other. Another property of PL in Mg-compensated samples is the observation of large in-plane fluctuations of the edge PL and BL intensities, as we show in Fig. 3a, b. These fluctuations are not correlated with those observed for the hot-PL lines.

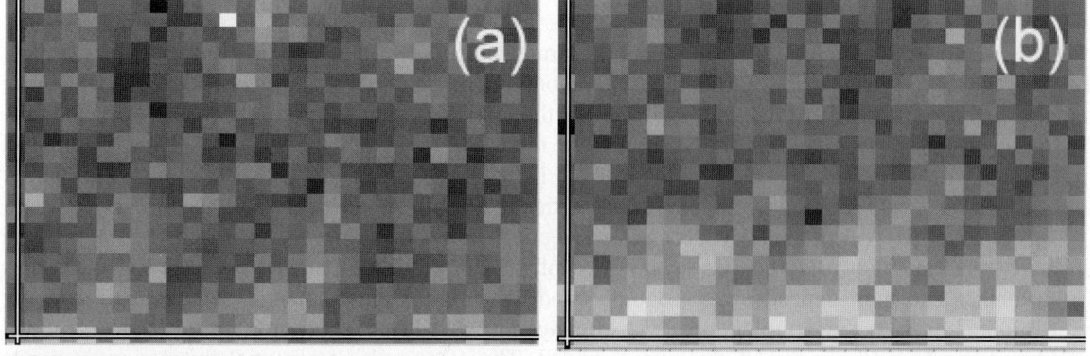

Fig. 2 Room temperature in-plane fluctuations of the intensity of two (a) the first line at the highest energy and b) the second line at the high energy wing) hot-PL lines. Micro-PL images were measured with 4 µm step. PL fluctuates between 1280 (dark spots) and 2760 (white spot) cps for the a) image and from 2550 to 4775 cps for the b) image.

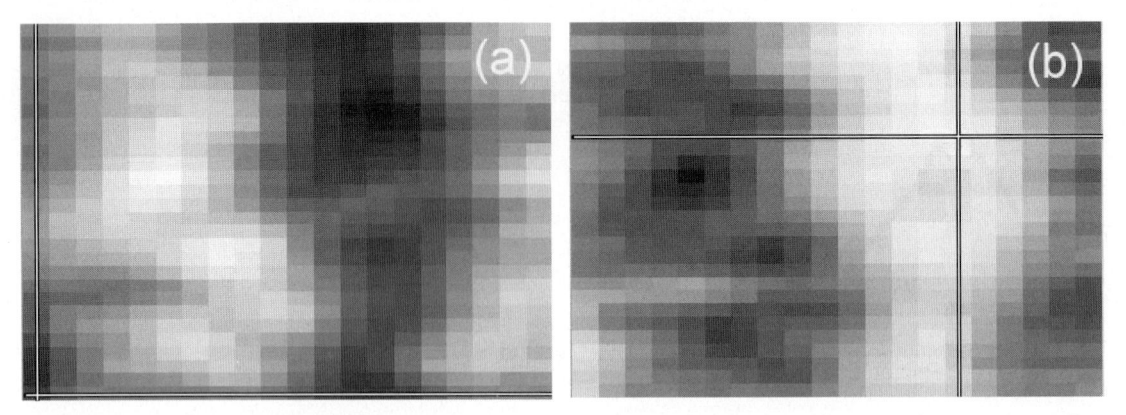

Fig. 3 Room temperature micro-PL (with 2 µm step) images of the in-plane fluctuations of the edge a) and BL b) intensity.

4 Discussion

Significantly different wavelength-dependent PL decay times suggest that two (or more) PL bands contribute to the YL. In fact, Colton et al. [2], using selective PL excitation, resolved the contributions of at least two PL bands to the yellow emission. They have related one of them to the recombination of distant donor-acceptor pairs (DAPs), with characteristic properties, as described by Thomas and Hopfield [3]. Experiments of Hofmann et al. [4] indicated participation of shallow donors in this transition. Colton et al. related the other contributing emission to the recombination of close DAPs [2].

Our time-resolved studies of YL in GaN crystals and epilayers, and also those of Seitz et al. [5], suggest another explanation of these observations. In metallic n-type bulk GaN, in which we observe wavelength-independent decay times of YL in the ns range, the concentration of shallow donors is above the Mott transition limit. We can, hence, expect the quasi free-to-bound transition to occur. We propose that quasi free-to-bound origin of this recombination process is related to recombination of delocalised electrons freely moving among donor sites with holes localized at acceptor sites. For Mg compensated samples the PL decay shows different properties. The fast and wavelength independent PL decay is followed by a slow decay component, with a time constant up to 500 s. The latter component of the PL decay is energy-dependent, with the longest PL decay times observed at the low energy tail of the PL emission, which is typical for DAP recombination processes.

The model presented above has a weak point. The change of the PL decay properties was observed for bulk samples with Mg concentration one order in magnitude smaller than that of oxygen-related shallow donors. Even if each magnesium compensates one of the donors, the concentration of the remaining shallow donors should still be well above the Mott limit, i.e., YL (BL) should still show properties of free-to-bound transitions, with electrons freely moving between shallow donor sites before recombining with holes localized at acceptors.

This discrepancy we relate to the presence of strong potential fluctuations in compensated samples. These fluctuations may affect electron diffusion among donor sites. Electrons can be localized by potential fluctuations resulting in DAP-like character of the recombination, even though shallow donor concentration is still above the Mott limit. Actually, the model of strong potential fluctuations was proposed by Oh et al. [6] to account for the BL properties in heavily Mg-doped samples. The model of potential fluctuations was also proposed to explain PL properties in heavily doped samples of GaAs [7].

We employed Kelvin probe AFM technique to verify this model. The present results and also those discussed separately for n-type doped layers [8] confirm the presence of strong potential fluctuations in GaN samples and their dependence on doping conditions. The largest potential fluctuations, reaching even 300 mV, were observed in undoped GaN epilayers. These layers show strong localization of carriers, as we concluded from the CL investigations [8]. The magnitude of potential fluctuations depends also on the compensation level, as we could observe for a range of Mg-doped samples with different compensation levels. The latter we controlled by varying the annealing time of Mg doped epilayers in nitrogen. Codoping with Mg increases potential fluctuations in GaN:Si epilayers (Fig. 4). The detected

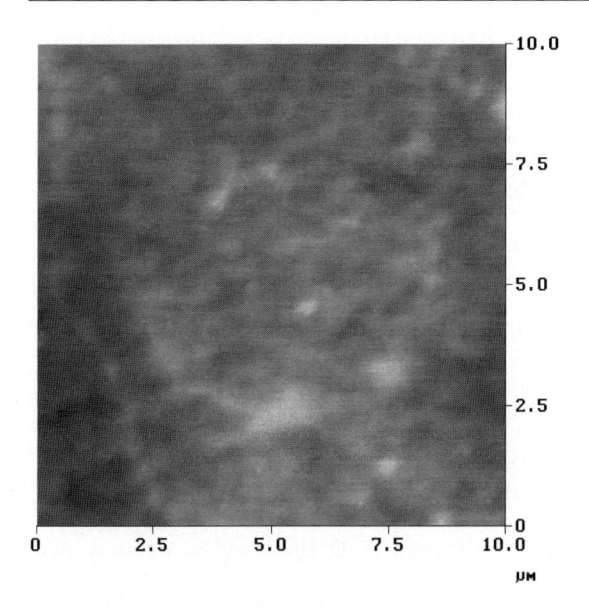

Fig. 4 Kelvin probe image of room temperature in-plane potential fluctuations observed for GaN:Mg,Si epilayer upon photo-excitation. Scale of fluctuations is 100 mV.

scale of fluctuations is larger than in those doped with Si only. Moreover, whereas for n-type samples photo-excitation in most of the cases decreases potential fluctuations (see [8]), these fluctuations surprisingly become enhanced upon illumination in Mg codoped samples. Their magnitude is of about 100 mV (Fig. 4), i.e., is few times larger than in n-type samples, and is similar to that given by Brandt and co-workers from the study of thermopower and the Seebeck coefficient [9].

In strongly compensated samples, were carriers are localized, the DAP recombination channel prevails in both YL and BL. Doping with silicon reduces potential fluctuation by more than one order of magnitude [8]. Then, YL and BL show increasingly free-to-bound character, since delocalized carriers can move among donor sites before recombining.

Another puzzling observation is the appearance of multiple and sharp lines at each LO-phonon energy below the laser line, as shown in Fig. 1. These can be resonant Raman or so-called hot-PL lines [10–13]. These two mechanisms can be distinguished from polarization studies. Information on excitation polarization is stored in the case of Raman lines, but is lost for the hot-PL lines. We observe that the higher order satellites progressively lose their degree of polarization, supporting the hot-PL hypothesis. At the same time, the satellites observed in the alternative polarization progressively increase in intensity. Such a behavior is not expected for a Raman scattering process.

In addition, we also observe in-plane intensity variations, uncorrelated with one another, for each of the sharp hot-PL peaks, as shown in Fig. 2a, b. This observation suggests an influence of disorder on the process of hot luminescence. Such effect was previously observed by Pelekanos et al. and by us for other systems [10–13]. Micro-PL studies show also that not only the intensity of hot-PL lines but also that of excitonic emissions fluctuate in-plane and that these fluctuations anti-correlate with fluctuations of the doping-related bands, as is shown for in Fig. 3a, b. These observations suggest that main reasons for the observed fluctuations of the PL intensity, and likely of the potential fluctuations, are not only micro-structures of the samples, but also inhomogeneities in distribution of impurities.

5 Conclusion Concluding, doping with silicon or magnesium measurably affects potential fluctuations present in GaN samples. This in turn results in change of recombination character, in the appearance of hot PL, and in strong in-plane variations of the PL intensities.

Acknowledgements This work was partly supported by the grant numbers 5 P03B 007 20, 5 P03B 123 21 of KBN and by the DENIS program of European Union (G5RD-CT-2001-00566).

References

[1] T. L. Tansley, E. M. Goldys, M. Godlewski, B. Zhou, and H. Y. Zuo, in: Optoelectronic properties of semiconductors and superlattices, series editor M. O. Manasreh, Vol. 2, GaN and Related Materials, edited by S. Pearton (Gordon and Breach Publishers, 1997), p. 233–293.
[2] J. S. Colton, P. Y. Yu, K. L. Teo, P. Perlin, E. R. Weber, I. Grzegory, and K. Uchida. In Proc. ICDS-20, Berkeley, 1999, Physica B **273–274**, 75 (1999).
[3] D. G. Thomas, J. J. Hopfield, and W. M. Augustyniak. Phys. Rev. **140**, A202, (1965).
[4] D. M. Hofmann, D. Kovalev, G. Stende, B. K. Meyer, A. Hofmann, L. Eckey, R. Heitz, T. Detchprom, H. Amano, and I. Akasaki. Phys. Rev. B **52**, 16702 (1996).
[5] R. Seitz, C. Gaspar, T. Monteiro, E. Pereira, M. Leroux, B. Beaumont, and P. Gibart, MRS Internet J. Nitride Semicond. Res. 2, art 36 (1997).
[6] E. Oh, H. Park, and Y. Park, Appl. Phys. Lett. **72**, 70 (1998).
[7] P. Hacke, H. Nakayama, T. Detchprohm, K. Hiramatsu, and N. Sawaki. Appl. Phys. Lett. **68**, 1362 (1996).
[8] M. Godlewski et al., this Proceedings.
[9] M. S. Brandt, P. Herbst, O. J. Homar, and A. Stemmer, J. Appl. Phys. **84**, 1168 (1998).
[10] N. Pelekanos, J. Ding, Q. Fu, A. V. Nurmikko, S. M. Durbin, M. Kobayashi, and R. L. Gunshor, Phys. Rev. B **43**, 9354 (1991).
[11] M. Godlewski, T. Wojtowicz, G. Karczewski, J. Kossut, J. P. Bergman, P. O. Holtz, and B. Monemar, Acta Phys. Pol. A **92**, 765 (1997).
[12] J. P. Bergman, B. Monemar, M. Godlewski, T. Wojtowicz, G. Karczewski, and J. Kossut, Proc. 7th Intern. Conf. on Shallow-Level Centers in Semiconductors, Amsterdam, The Netherlands, 1996, edited by C. A. J. Ammerlaan and B. Pajot, World Scientific Publ. Comp. (Singapore, 1997), pp. 221–226.
[13] E. M. Goldys, M. Godlewski, E. Kaminska, A. Piotrowska, G. Koley, M. G. Spencer, and L. F. Eastman, Conference on Optoelectronic and Microelectronic Materials and Devices, COMMAD 2000, Melbourne, Australia, 6–8 December 2000, edited by L. D. Broekman, B. F. Usher, J. D. Ridley (COMMAD 2000 Proceedings, 2000 IEEE), p. 539.

Impact of post-growth thermal annealing on emission of InGaN/GaN multiple quantum wells

Yung-Chen Cheng[1], **S. Juršėnas**[*,2], **Shih-Wei Feng**[1], **C. C. Yang**[1], **Cheng-Ta Kuo**[3], and **Jian-Shihn Tsang**[3]

[1] Graduate Institute of Electro-Optical Engineering and Department of Electrical Engineering, National Taiwan University 1, Roosevelt Road, Sec. 4, Taipei, Taiwan, R.O.C.
[2] Institute of Materials Science and Applied research, Vilnius University Saulėtekio al. 9-III, LT-2040 Vilnius, Lithuania
[3] Advanced Epitaxy Technology Inc., Hsinchu Industrial Park, Hsinchu, Taiwan, R.O.C.

Received 15 September 2003, accepted 17 September 2003
Published online 29 December 2003

PACS 78.47.+p, 78.55.Cr, 78.67.De

Photoluminescence, photoluminescence excitation and time-resolved luminescence study of three $In_{0.15}Ga_{0.85}N$/GaN multiple quantum wells (MQWs) with well-widths of 2, 3, 4-nm, upon thermal annealing at 800 °C for 30 min is presented. Blueshift of luminescence and pronounced changes in the absorbance indicate on remarkable interdiffusion of indium at the quantum well barrier interface for thin MQWs. While pronounced redshift in luminescence and increase in the excitation lifetime is attributed to prevalence of strain-induced field-effect for the thicker MQWs. Site-selective spectroscopy and microstructure analysis monitors improvement of the MQW interface quality of the thicker layers as well as reduction of disorder of in the quantum well layer upon post-growth thermal annealing.

© 2004 WILEY-VCH Verlag GmbH & Co. KGaA, Weinheim

1 Introduction InGaN-based multiple quantum wells (MQWs) are the key structures for production of high efficiency, long lifetime violet, blue, and green light emitting diodes and laser diodes [1, 2]. During the epitaxial growth of InGaN/GaN structures and during the fabrication of devices, active layers undergo several high-temperature treatments [1]. After these thermal treatments, the distribution of the indium composition and the strain in quantum wells may be changed due to both ordinary diffusion [3-5] and "up-hill", strain induced, diffusion [6–9] effects.

Usually thermal annealing, is related to indium and gallium interdiffusion across the interface of a quantum-well and a barrier [3–5]. This leads to corresponding blueshift in photoluminescence (PL) and absorbance spectra [5]. A complete disappearance of InGaN/GaN MQWs after annealing at high temperature has been observed [4]. On the other hand the interdiffusion of indium and gallium in InGaN alloys is complicated by the immiscibility of InN and GaN, which can result in phase separation in InGaN layers [6, 7]. Since InGaN alloys usually are grown at temperatures below 1473 K, the critical temperature above which InN-GaN system is completely miscible [10, 11], phase separation is expected in all InGaN alloys based on thermodynamics considerations. The phenomenon of InGaN phase separation [6–14] are subjects of considerable interest, since it has been reported that InN-rich regions play a significant role in the mechanism of emission of nitride-based light emitting diodes and laser structures [15–19]. Phase separation has been observed using several growth techniques, preferably in the films of larger thickness [6–9, 12]. The observed phase separation is evidently driven by strain due to the mixing of the two lattice mismatched components of the InGaN alloy system. Indium atoms are excluded from the InGaN lattice to form an alloy of different composition and reduce the strain energy of the system. It

---

[*] Corresponding author: e-mail: saulius.jursenas@ff.vu.lt, Phone: + 3705 2 366027, Fax: +3705 2 366059

© 2004 WILEY-VCH Verlag GmbH & Co. KGaA, Weinheim

is predicted that for standard growing temperature spinodal decomposition is expected for indium concentration larger than 20% [10, 11]. It has been reported, that upon post-growth thermal annealing, indium rich regions are formed in InGaN quantum wells [6–9, 13]. On the other hand, thermal annealing can lead to pronounced redistribution of the strain energy and reduce the phase separation, which is very sensitive to the strain effects. Both phase separation and strain effects are expected to enhance with increasing of the InGaN well width. Thus variation of indium compositional disorder and strain in as-grown samples of different well width can lead to changes in the dominating In separation mechanism upon thermal annealing.

Here we report on photoluminescence excitation (PLE) approaches combined with site-selective and time-resolved photoluminescence (PL) studies of the effect of post-growth thermal treatment on InGaN/GaN MQWs of different well thickness. We show that InGaN/GaN interface has a major impact on post-growth thermal-annealing related changes in the optical properties and material quality of InGaN/GaN MQWs of various thickness.

2 Experimental The samples were grown on c-plane sapphire by metal organic chemical vapor deposition. Following the deposition of 30 nm GaN buffer layer and a 2.3 µm GaN layer, QW structures consisting of five pairs of $In_{0.15}Ga_{0.85}N$ well and 10 nm-thick GaN barrier were grown. Three samples of different well widths at 2, 3, and 4 nm were prepared and referred to as samples w20, w30 and w40, respectively. The growth temperatures were 1010 and 720 °C for GaN and InGaN, respectively. As-grown samples were thermally annealed in a quartz tube furnace at 800 °C in nitrogen ambient for 30 min.

The samples were placed in a cryostat for the temperature-dependent measurements. PLE experiments were conducted using quasi-monochromatic excitation light source from a xenon lamp dispersed by a 0.15-m monochromator. A Verdi-pumped mode-locked Ti:sapphire laser with 100 fs pulse width and 76 MHz in pulse repetition rate was frequency doubled by a BBO crystal. The excitation pulses we used were at 3.177 eV. The luminescence from the sample was collected and focused into a spectrometer before imaging in a Hamamatsu streak camera. The temporal resolution of operation mode was 4.74 ps.

3 Results and discussion Figure 1 shows normalized PLE spectra, taken at the peak position, and normalized PL spectra of as-grown a) and annealed b) samples at 10 K. Main luminescence spectral properties are typical for low temperature emission of the localized exciton states in InGaN/GaN MQWs [16, 18, 20]. PLE spectra of as-grown w20 sample shows typical dependencies of a MQW structure, reflecting variation in a two dimensional density of states of QW (InGaN) and barrier (GaN) materials. Peaks at 3.103 eV and 3.503 eV can be related to the exciton positions in InGaN and GaN layers, correspondingly.

By the increase of the well-width, a pronounced changes in the region of GaN barrier layers can be distinguished. Instead of expected step-like dependencies, as it is seen for w20 structure, an abrupt dependence on the excitation photon energy for $h\nu > 3.50$ eV is observed. Changes of luminescence signal upon increase in the photon energy above the bandgap of the material can be attributed to the enhancement of nonradiative capture of the excited species with higher excess energy. In order to produce luminescence from QW states, excitons created by photoexcitation of the GaN barrier layer have to be captured by a well, thus to overcome the InGaN-GaN interface. Excitations with larger excess energy migrates for larger distances, thus have larger probability to be captured by nonradiative capture centers. Thus, narrowing of the PLE spectra are related to the density of the nonradiative capture centers, most of that are concentrated at the interfaces. As-grown samples show significant increase in the number of nonradiative centers, related to the GaN barrier with the increase in the well-width. This was also proved by site-selective PL experiments (not shown).

Differences in degree of disorder and strain of as grown samples of various well-width leads to different impact of thermal annealing on luminescence of localized states. Luminescence band of w20 MQWs blueshifts (about 20 meV) together with the absorbance-edge upon thermal annealing. This might be related to indium interdiffusion to the barrier region [3–5]. In contrast, luminescence band of w30 sam-

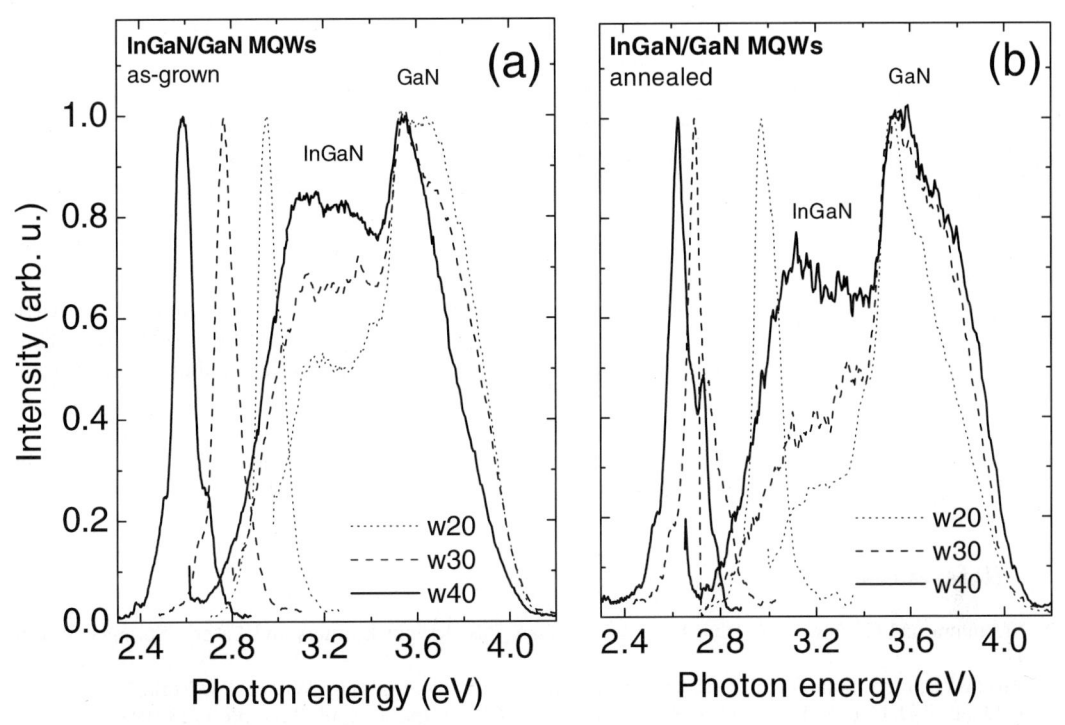

Fig. 1 Normalized at 3.50 eV PLE spectra of as-grown a) and annealed b) InGaN/GaN MQWs of various well width. Normalized PL spectra are shown for comparison on the left-hand side.

ple redshifts (71 meV) and that of w40 blueshifts (45 meV) while absorbance edge show little changes in both later cases. Such behavior might be affected by several impacts such as, indium diffusion, increase in disorder due to indium decomposition, strain redistribution and strain related spinodal decomposition.

The PLE spectral changes are most pronounced for the GaN barrier region. PLE spectra show rapid decrease in the PL signal for excitation photon energy larger than GaN bandgap. Thus InGaN-GaN interface related nonradiative recombination is significantly enhanced upon thermal annealing. Barrier layer related PLE spectral changes diminishes with increase of the well-width. Even improvement material quality is observed for w40 MQWs upon thermal annealing. As-grown and annealed samples show opposite trends in InGaN-GaN interface related defect concentration by increasing of well-width: as-grown structures shows the best quality for the thinnest QWs and InGaN-GaN interface quality decreases with the well-width, while annealed samples show the worse quality of the thinnest MQWs and improvement of the quality by increase of the well-width.

Variation of exciton lifetime at low temperatures with the well-width indicates on importance of the is built-in field effect [21–23]. The low temperature lifetime of as-grown samples increases from 1.8 ns to 9 ns and 137 ns for w20, w30 and w40 samples, respectively. Upon thermal annealing of the w20 sample low temperature lifetime is almost the same, what is in disagreement with the decrease in emission efficiency. However, capture process of delocalized excitons to nonradiative centers at the interface can be very fast (<1 ps) and probably cannot be detected because of low time-resolution of our system. After the excitons are localized they show the same lifetime. Thicker w30 sample shows increase in the lifetime from 9 ns to 41 ns upon thermal annealing. This is in line with the luminescence redshift and the same dynamic in the luminescence efficiency. Such trends might be accounted predominantly for the increase of the built-in field effect.

Summarizing, thinner samples, that initially are of the best optical quality, are almost destroyed by indium interdiffusion into barrier layer upon thermal annealing. Annealing leads to changes in indium content as it is seen by blueshift of the InGaN related absorption and formation of large amount of non-radiative defects at the interfaces. This is also monitored by microscopy images [24]. As consequence luminescence efficiency is significantly reduced. While the thicker samples are already very disordered due to indium segregation on growing surface. This leads to increase in the number of surface related defects in as grown samples. Annealing improves material quality, however for w30 sample annealing leads to formation more strain affected structure, what leads to decrease in efficiency due to built-in field effect. Phase separation in any alloy requires long range diffusion and thus correlation should exist between phase separation and period of the time required for the growth of the film. Therefore strain associated with thin InGaN layers could also stabilize the alloys against phase separation. For the thickest samples thermal annealing indeed improves material quality and leads to better luminescence efficiency. This imply that indium preferentially reorganized in the InGaN layer rather than diffused into GaN barriers. However as grown sample already possesses many defects on the growing surfaces of MQWs. This, together with increased built field effect, makes efficiency worse than that of the smaller thickness sample.

Acknowledgements This research was partially supported by the joint Lithuanian-Latvian-Taiwan grant and by National Science Council, The Republic of China, under the grants of NSC 90-2112-M-002-052, NSC 90-2215-E-002-027, and NSC 90-2215-E-002-041.

References

[1] S. Nakamura and G. Fasol, The Blue Laser Diode: GaN Based Light Emitters and Lasers (Springer, Berlin, 1997).
[2] A. Žukauskas, M. S. Shur, and R. Gaska, Introduction to Solid-State Lighting (Wiley, New York, 2002).
[3] J.-S. Tsang, J.-D. Guo, S.-H. Chan, M.-S. Feng, and C.-Y. Chang, Jpn. J. Appl. Phys. **36**, 1728 (1997).
[4] M. D. McCluskey, L. T. Romano, B. S. Krusor, N. M. Jonson, T. Suski, and J. Jun, Appl. Phys. Lett. **73**, 1281 (1998).
[5] C.-C. Chou, C.-M. Lee, and J.-I. Chyj, Appl. Phys. Lett. **78**, 314 (2001).
[6] R. Singh, D. Doppalapudi, T. D. Moustakas, and L. T. Romano, Appl. Phys. Lett. **70**, 1089 (1997).
[7] M. D. McCluskey, L. T. Romano, B. S. Krusor, D. P. Bour, N. M. Johnson, and S. Brennan, Appl. Phys. Lett. **72**, 1730 (1998).
[8] D. Doppalapudi, S. N. Basu, K. F. Ludvig, Jr., and T. D. Moustakas, J. Appl. Phys. **84**, 1389 (1998).
[9] Y.-T. Moon, D.-J. Kim, K.-M. Song, C.-J. Coi, S.-H. Han, T.-Y. Seong, and S.-J. Park, J. Appl. Phys. **89**, 6514 (2001).
[10] I. H. Ho and G. B. Stringfellow, Appl. Phys. Lett. **69**, 2701 (1996).
[11] S. Yu. Karpov, J. Nitride Semicond. Res. **3**, 16 (1998).
[12] K. Osamura, K. Nakajima, and Y. Murakami, Solid State Commun. **11**, 617 (1972).
[13] Y.-S. Lin, K.-J. Ma, C. Hsu, Y.-Y. Chung, C.-W. Liu, S.-W. Feng, Y.-C. Cheng, M.-H. Mao, C. C. Yang, H.-W. Chuang, C.-T. Kuo, J.-S. Tsang, and T. E. Weirich, Appl. Phys. Lett. **80**, 2571 (2002).
[14] S.-W. Feng, Y.-Y. Chung, C.-W. Liu, Y.-C. Cheng, C. C. Yang, M.-H. Mao, Y.-S. Lin, K.-J. Ma, and J.-I. Chyi, Appl. Phys. Lett. **80**, 4375 (2002).
[15] S. Chichibu, T. Azuhata, T. Sota, and S. Nakamura, Appl. Phys. Lett. **70**, 2822 (1997).
[16] S. Chichibu, T. Sota, K. Wada, and S. Nakamura, J. Vac. Sci. Technol. B **16**, 2204 (1998).
[17] K. P. O'Donnell, R. W. Martin, and P. G. Middleton, Phys. Rev. Lett. **82**, 237 (1999).
[18] Y.-H. Cho, T. J. Schmidt, S. Bidnyk, G. H. Gainer, J. J. Song, S. Keller, U. K. Mishra, and S. P. DenBaars, Phys. Rev. B **61**, 7571 (2000).
[19] P. R. Kent and A. Zunger, Appl. Phys. Lett. **79**, 1997 (2001).
[20] Y. Narukawa, Y. Kawakami, S. Fujita, and S. Nakamura, Phys. Rev. B **59**, 10283 (1999).
[21] F. Bernardini and V. Fiorentini, Phys. Rev. B **58**, 15292 (1998).
[22] P. Lefebvre, A. Morel, M. Gallart, T. Taliercio, J. Allègre, B. Gil, H. Mathieu, B. Damilano, N. Grandjean, and J. Massies, Appl. Phys. Lett. **78**, 1252 (2001).
[23] W. W. Chow, H. Amano, T. Takeuchi, and J. Han, Appl. Phys. Lett. **75**, 244 (1999).
[24] Y.-S. Lin, K.-J. Ma, Y.-Y. Chung, S.-W. Feng, Y.-C. Cheng, E.-C. Lin, C. C. Yang, C.-T. Kuo, and J.-S Tsang, J. Cryst. Growth **252**, 107 (2003).

Thermoelectric and thermal properties of AlInN thin films prepared by reactive radio-frequency sputtering

Shigeo Yamaguchi[*,1,2], **Ryohei Izaki**[1], **Yasuo Iwamura**[1], and **Atsushi Yamamoto**[2]

[1] Department of Electrical, Electronic and Information Engineering, Kanagawa Unversity, 3-27-1 Rokkakubashi, Kanagawa-ku, Yokohama, 221-8686, Japan
[2] Energy Electronics Institute, National Institute of Advanced Industrial Science and Technology, AIST Tsukuba Central 2 Umezono 1-1-1, Tsukuba 305-8568, Japan

Received 15 September 2003, revised 4 November 2003, accepted 4 November 2003
Published online 19 January 2004

PACS 72.20.Pa, 81.05.Ea, 81.15.Cd, 81.15.Kk

We have studied the thermoelectric and thermal properties of AlInN targeting a thermoelectric device using nitrides. In a $Al_{0.57}In_{0.43}N$ thin film, we obtained a minimum value of of 2.2×10^{-6} m^2/s for thermal diffusivity and a maximum value of 1.8×10^{-4} W/mK^2 for power factor.

© 2004 WILEY-VCH Verlag GmbH & Co. KGaA, Weinheim

1 Introduction Seebeck effect can cause solids to directly convert heat energy into electric energy without producing emissions such as carbon dioxide gas and radioactive substances. Moreover, using the reverse effect, Peltier effect, the simultaneous cooling and heating phenomena can be utilized to medical applications such as the local cooling or heating for the human body. There has recently been a great increase in the research and development of thermoelectric power generation systems that are designed to employ the vast resources of waste heat and environmentally sound cooling [1]. The realization of practical applications associated with such systems requires the achievement of a high efficiency characterized by electric resistivity, thermoelectric Seebeck coefficient and thermal conductivity. At this stage, no binary compounds better than Bi_2Te_3 and PbTe have been found. Accordingly, recent research has focused on analyzing the thermoelectric performance of many ternary systems, most of which are lanthanide and transition metal intermetallics. In the search for new thermoelectric materials, much effort has been directed toward the development of thermoelectric materials with improved characteristics. The thermoelectric materials have been evaluated using the figure of merit Z. The figure of merit Z, which indicates the thermal to electrical energy conversion efficiency of the material, is defined by $Z = P/\kappa$, $P = \alpha^2/\rho$, where P, α, ρ and κ are power factor, thermoelectric power, electrical resistivity and thermal conductivity, respectively. One of the criteria for the practical application of thermoelectric materials can be expressed as $ZT > 1$, where T is the absolute temperature. Generally, it is difficult to decrease ρ and κ, and increase α simultaneously, because α and ρ depend on many parameters such as the carrier concentration, carrier mobility and effective carrier mass. The power factor P indicates the electrical power generation capability, and is commonly used for materials of which the thermal conductivity is difficult to measure. Instead of using Z, in a given temperature gradient, the power factor P is often used as a good measure of the thermoelectric properties. In this study, we used P for the assessment of our samples as thermoelectric materials.

---

[*] Corresponding author: e-mail: yamags18@kanagawa-u.ac.jp

Table 1 Advantages of nitrides for a thermoelectric device.

Bi_2Te_3 system	nitride system	advantages
metals	nitrides	acid-resistant
heavy elements	lighter elements	low-cost
rare	abundant	low-cost
good performance at lower temperature	good performance at higher temperature	temperature-resistant
toxic	not toxic	not environmentally harm ful
low melting point	high melting point	widely used applications

We have recently studied the thermoelectric properties of $Al_{1-x}In_xN$ system in terms of substituting it for Be_2Te_3 and PbTe, which are considered to environmentally stress [2, 3]. When compared to the properties of Be_2Te_3, nitrides have the following advantages as listed in Table 1. For example, first, nitrides are not toxic and are mainly composed of lighter elements, which are important in terms of environmental problem and economic cost. In fact, the molecular weights of Be_2Te_3 and $Al_{0.5}In_{0.5}N$ are 800 and 85, respectively. Second, the conventional thermoelectric materials have better thermoelectric properties at intermediate temperatures, while nitrides do at higher temperatures.

2 Experiments The samples studied here were $Al_{1-x}In_xN$ prepared by the reactive radio-frequency sputtering method. They were grown on SiO_2 glass substrates at 100 °C using N_2 and Ar gases. The in-plane thermal diffusivity of the self-standing films (~10 μm) was measured using an ac calorimetric method in the temperature range of 300–673 K. The thermal diffusivity D is defined using the expression $\kappa = \delta c D$, where κ, δ and c are thermal conductivity, density and specific heat, respectively. Consequently, for a precise evaluation of thermal conductivity, the value of the thermal diffusivity is inevitable. As shown in Fig. 1, on the basis of the fact that in the attenuation of ac calorimetric method the ac temperature along the sample length can be related to the thermal diffusivity, we were able to estimate the values of the thermal diffusivity of our samples [4].

Figure 2 shows ac temperature of the sample surface as a function of the displacement from the origin of irradiated location of the sample as a parameter of the light chopper frequency f. We obtained $D = 2.2 \times 10^{-6}$ m^2/s for $Al_{0.57}In_{0.43}N$.

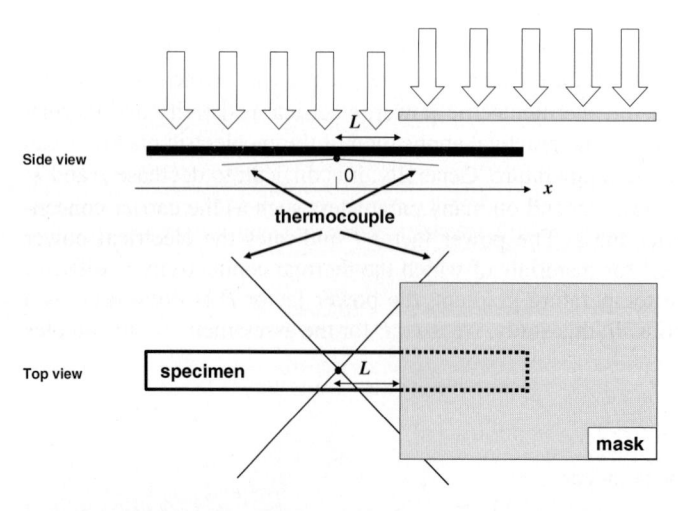

Fig. 1 Schematic structure of ac calorimetric method.

Fig. 2 L dependence of $\ln T_{ac}$ of $Al_{0.57}In_{0.43}N$ as a parameter of chopper frequency f.

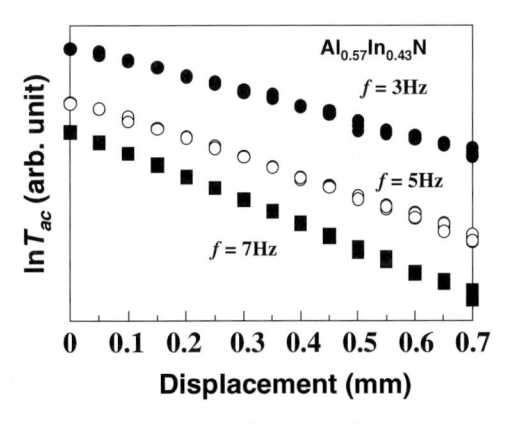

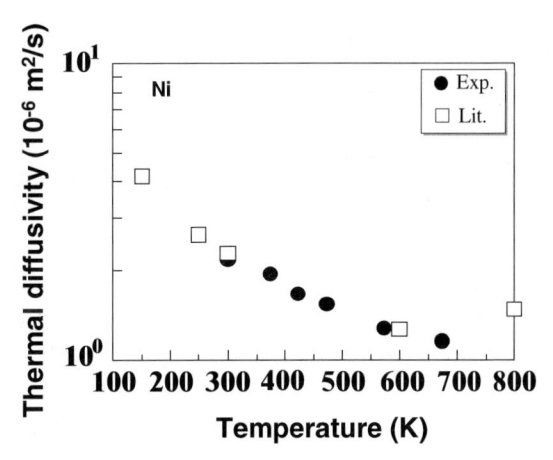

Fig. 3 Thermal diffusivity D of Ni metal as a function of temperature for reference.

Figure 3 shows thermal diffusivity D of the reference Ni metal as a function of temperature. The closed circles represent the experimental data, and the open squares represent the values derived from the previously reported values. As shown in Fig. 3, our experimental data have good precision and accuracy is within 5%, indicating that this ac calorimetric method has sufficient accuracy to enable discussion on the thermal diffusivity of our samples.

Figure 4 shows D of $Al_{0.57}In_{0.43}N$ films as a function of temperature. In a general trend, with increasing In content, the thermal diffusivity shows weaker temperature dependence. In general, D (T) of ordinary solids will have temperature dependence because the phonons are scattered by crystal imperfections such as point defects and dislocations. At 473 K, D has a minimum value of 7.53×10^{-7} m^2/s for $Al_{0.57}In_{0.43}N$.

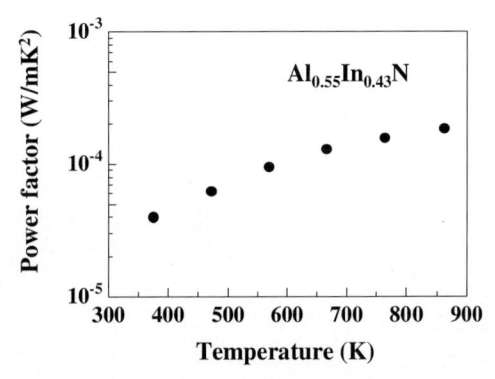

Figure 5 shows power factor, P, of $Al_{0.57}In_{0.43}N$ as a function of temperature. A maximum value of 1.8×10^{-4} W/mK^2 was obtained at 873 K.

Acknowledgements This work was partly supported by the Nippon Sheet Glass Foundation.

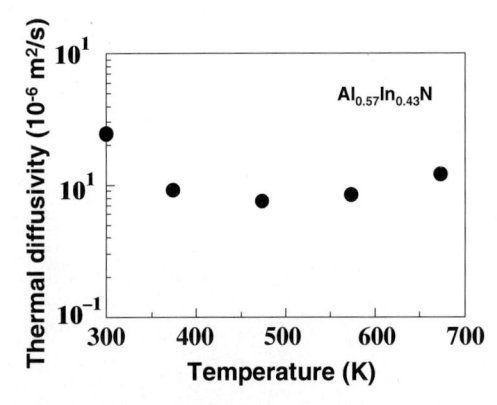

Fig. 4 Thermal diffusivity D of $Al_{0.57}In_{0.43}N$ as a function of temperature.

Fig. 5 Power factor of $Al_{0.57}In_{0.43}N$ as a function of temperature.

References

[1] S. Bhattacharya, A. L. Pope, R. T. Littleton, T. M. Tritt, V. Ponnambalam, Y. Xia, and S. J. Poon, Appl. Phys. Lett. **77**, 2476 (2000).
[2] S. Yamaguchi, Y. Iwamura, and A. Yamamoto, Appl. Phys. Lett. **82**, 2065 (2003).
[3] S. Yamaguchi, Y. Iwamura, and A. Yamamoto, Jpn. J. Appl. Phys. **41**, L1354 (2002).
[4] I. Hatta, Y. Sasuga, R. Kato, and A. Maesono, Rev. Sci. Instrum. **56**, 1643 (1985).

Cathodoluminescence properties of zinc oxide nanoparticles

M. R. Phillips[*,1], **O. Gelhausen**[1], and **E. M. Goldys**[2]

[1] Microstructural Analysis Unit, University of Technology, Sydney, Broadway, NSW 2007, Australia
[2] Division of Information and Communication Sciences, Macquarie University, North Ryde, NSW 2109, Australia

Received 3 November 2003, accepted 3 November 2003
Published online 12 January 2004

PACS 71.55.Gs, 78.60.Hk. 78.67.Bf

Zinc oxide nano-particles (25 nm) have been investigated by cathodoluminescence spectroscopy (300 nm–1700 nm) at 80 K and 300 K following thermal annealing in high purity H_2/N_2, N_2, O_2 and Ar gaseous atmospheres. The intensity of the ZnO near band edge peak was significantly increased after heat treatment in hydrogen. Conversely, thermal annealing in the other gas types decreased this emission. This effect is attributed to hydrogen passivation of competitive non-radiative defect centers, most likely bulk zinc vacancy centers. The appearance of a strong green emission centered at 2.4 eV following thermal annealing in all gas atmospheres is ascribed to the formation of bulk oxygen vacancy defects. A strong red shift of the near band edge emission with increasing beam current at 300 K is accredited to electron beam heating rather than to an increase in the carrier density. Electron beam heating is evidenced by the occurrence of a strong black body emission in the near infrared spectral region.

© 2004 WILEY-VCH Verlag GmbH & Co. KGaA, Weinheim

1 Introduction Fluorescent nano-particles have considerable potential in a wide range of technological applications because they exhibit a number of unique luminescent properties that result from quantum size effects. These include tuneable emission over a broad spectral range, high luminescence efficiency, tuneable emission lifetimes and stability to high power excitation as well as the ability to functionalize the surface for a specific purpose. This new luminescence behaviour occurs when the diameter of the nano-particle approaches the exciton Bohr radius. At these spatial dimensions, however, the surface area to volume ratio of the particle is extremely large compared with its bulk counterpart and luminescence quenching due to non-radiative surface recombination severely reduces the emission efficiency. Therefore, to produce a useful light output yield, passivation of surface charge traps is generally required to block these undesirable recombination channels.

ZnO is a particularly attractive material for luminescent nano-particle applications due to its wide band gap (3.37 eV), large exciton binding energy (60 meV), low refractive index (1.9), resistance to high energy irradiation, stability to intense ultra-violet illumination and low toxicity. It has been recently reported that the luminescence efficiency in both bulk and polycrystalline ZnO can be significantly enhanced via the incorporation of hydrogen using either hydrogen plasma treatment or heat treatment in hydrogen gas [1, 2]. The aim of the present study is to evaluate the utility of post growth thermal annealing in hydrogen gas to improve the cathodoluminescence (CL) efficiency of nano-granular ZnO. The role of hydrogen incorporation is of particular interest since it can, in principle, both passivate bulk and surface defects as well as increase the n-type conductivity by introducing a shallow donor state [3]. Heat treatment was also conducted in Ar, N_2 and O_2 gaseous atmospheres to determine their impact on the CL properties and establish which effects are directly related to the presence of hydrogen. The ZnO nano-particle results have been compared with CL measurement from high-quality ZnO thin films heat treated in the same manner.

---

[*] Corresponding author: e-mail: matthew.phillips@uts.edu.au, Phone: +612 9514 1620, Fax: +612 9514 1703

2 Experimental

ZnO particles 25 nm in diameter were chosen for this study to provide a large surface area to volume ratio without introducing quantum size effects: the Bohr radius of ZnO is 1.8 nm. The nanoparticles were obtained from Advanced Nano Technologies, Perth, Western Australia who fabricate the material using a mechanochemical-processing™ technique. The thin film ZnO was around 1 μm thick and was grown by molecular beam epitaxy on a sapphire substrate. Specimens were heat treated in a Stanton Redcroft STA-780 furnace in controlled gaseous atmospheres of high purity Ar, N_2, O_2 and $H_2(5\%)/N_2$. The thermal annealing cycle consisted of a 180 s ramp from room temperature to 780 °C, a 10 s dwell interval and exponential cooling for 20 minutes to 20 °C. The CL measurements were performed before and after heat treatment between 80 K and 300 K using an Oxford Instruments MonoCL2 system installed on a JEOL35C scanning electron microscope equipped with a liquid nitrogen cold stage. The CL spectra were measured from 300–900 nm using a Hamamatsu R943-02 Peltier cooled photomultiplier tube and a 1200 lines/mm grating blazed at 500 nm. Infrared CL spectra were collected using Hamamatsu R-5509-72 liquid nitrogen cooled photomultiplier and a 600 lines/mm grating blazed at 1600 nm. All CL spectra were corrected for system response. The SEM was operated at 10 kV and the electron beam current, I_b, was measured using a Faraday cup.

3 Results

3.1 Characterization of the "as received" specimens

The near band edge (NBE) CL of the thin film ZnO specimen at 80 K comprises a peak positioned at 3.37 eV attributed to an acceptor bound exciton (BE) emission overlapped with the A1 free exciton (FX) emission (Fig. 1). The FX longitudinal optical (LO) phonon replicas ($E_{LO} = 70$ meV) are present at 3.30 eV (1 LO-FX) and 3.23 eV (2 LO-FX). With increasing temperature to 300 K, the BE centers thermalize leaving the FX and 1 LO-FX peaks which red shift, widen and merge to produce a broad emission band centered at 3.24 eV [4]. Conversely, the untreated nano-powder exhibited a broad featureless emission at 80 K with no discernable phonon structure centred at 3.30 eV that red shifts to 3.17 eV at 300 K (Fig. 2). These data were collected with the same spectral resolution (1nm bandpass) as the thin film ZnO, however, the power density of the electron beam current had to be increased by a factor of three to produce an equivalent light emission intensity. Broad unstructured defect-related bands at 2.2 eV (Fig. 1) and 1.6 eV (Fig. 3) are observed in the thin film and nano-granular specimens, respectively. The 2.2 eV emission intensity was weak compared with thin film NBE whereas the 1.6 eV emission was significantly stronger than the powder NBE.

The green emission at 2.2 eV has been assigned a large number of defect centers including V_O^+ [5], $V_{Zn}^- - V_O^+$ [6], Zn_i^{2+} [7], $Zn_i^{2+} - V_{Zn}^{2-}$ [8], Zn_i^+ to Zn_i^{2+} transition [9], and a donor-acceptor pair transition $Cu^+ - A^-$ [10] for a highly structured emission. However, to date there is no consensus on the chemical origin of this emission peak. Two other emission bands have also been reported in the literature. One at 1.8 eV attributed to a recombination at neutral nitrogen acceptors N_O^0 [11] and another at 2.1 eV assigned to radiative recombination at a O_i^- defect [12].

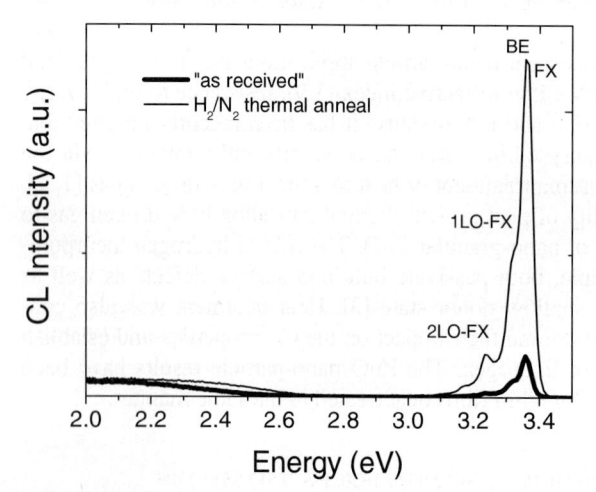

Fig. 1 CL spectra at 80 K of thin film ZnO (10 kV, 20 nA, 1 nm bandpass and scan area 170 μm × 130 μm) before and after heat treatment in H_2/N_2 gas.

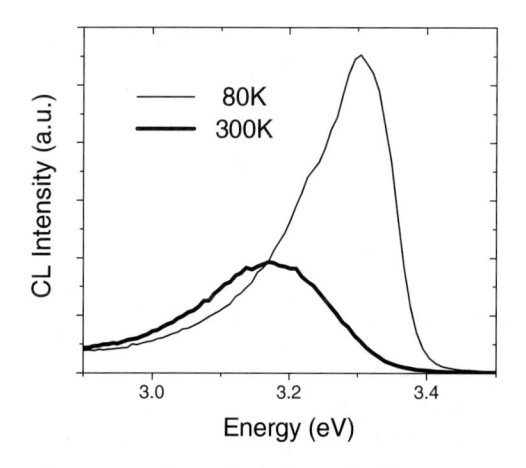

Fig. 2 CL spectra ZnO nano-particles at 80 K and 300 K (15 kV, 40 nA, 1 nm bandpass and scan area 170 μm × 130 μm) showing no discernable phonon structure.

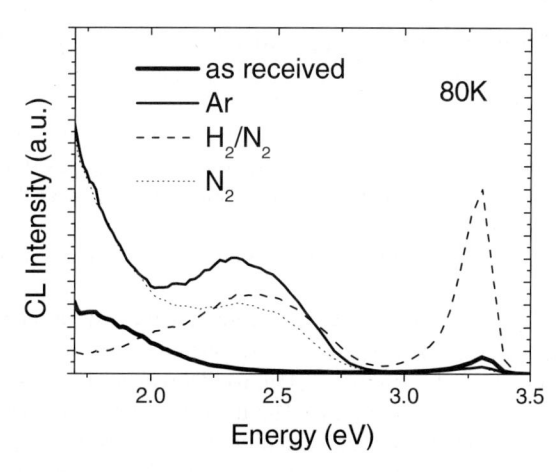

Fig. 3 CL spectra ZnO nano-particles at 80 K and 300 K (10 kV, 10 nA, 5 nm bandpass and scan area 170 μm × 130 μm). Thermal annealing in hydrogen significantly increases the NBE emission whereas heat treatment in Ar and N_2 reduces this emission.

3.2 Controlled gaseous atmosphere heat treatment

The NBE emission in both the nano-powder and thin film ZnO increased by a factor of 10 following thermal annealing in hydrogen (Fig. 1 and Fig. 3). This effect has also been reported following hydrogen plasma treatment and hydrogen heat treatment of thin film and polycrystalline ZnO. The heat treatment also has a significant effect on the defect emission. In the nano-powder specimen, a strong broad emission at 2.4 eV is produced and the emission intensity of the 1.6 eV peak is reduced. The existing weak defect emission at 2.2 eV in the thin film specimen is unaffected by the hydrogen treatment instead inducing a weak emission also peaked at 2.4 eV (Fig. 1).

Thermal annealing in Ar, O_2 and N_2 quenched the NBE emission in the ZnO nano-powder (Fig. 3) with the O_2 causing the greatest decrease (not shown). A strong defect-related emission in the near infrared is produced following heat treatment in all three gases, contrary to the H_2 annealing which reduces this emission (Fig. 3). However, an intense deep level emission at 2.4 eV was induced by heat treatment in Ar, N_2, O_2 and $H_2(5\%)/N_2$ gaseous atmospheres.

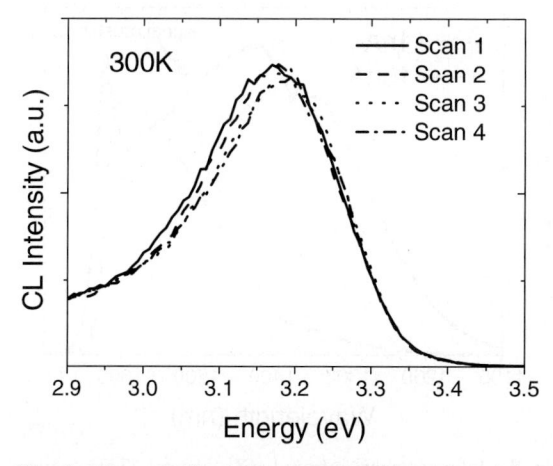

Fig. 4 Time resolved CL spectra ZnO nano-particles at 300 K (10 kV, 20 nA, 1 nm bandpass and scan area 170 μm × 130 μm).

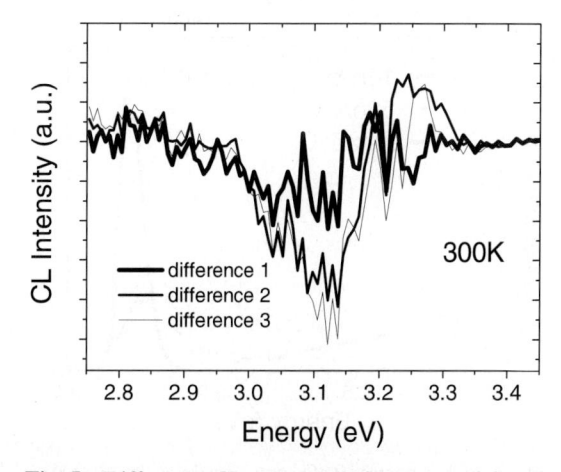

Fig. 5 Difference CL spectra ZnO nano-particles at 300 K (10 kV, 20 nA, 1 nm bandpass and scan area 170 μm × 130 μm) indicating the electron beam quenching of an emission peak at 3.13 eV.

3.3 Electron beam irradiation effects Time-resolved spectra at 80 K and 300 K from hydrogen annealed nano-powder specimens collected at 10 kV, 20 nA for 360 seconds/scan are shown in Fig. 4. Difference spectra reveal that the broad NBE peak at 300 K is comprised of a number of components that respond differently to electron beam irradiation (Fig. 5). At 300 K, a peak positioned at 3.13 eV decreases with irradiation whereas another peak centered at 3.27 eV slightly increases. Irradiation at 80 K, however, reduces the overall intensity of the emission peaked at 3.3 eV without changing the shape of the emission profile. The same effect is observed with the hydrogen heat treated thin film ZnO specimens but is absent in the "as received" nano-particle and thin film samples and those thermally annealed in O_2, N_2 and Ar.

3.4 Effect of electron beam power density With increasing beam current at 300 K, the NBE emission of the nano-powder specimen exhibited a strong red shift and an increase in the CL emission in the near infrared was also observed (Fig. 6). However, the peak position of the NBE in the thin film material was fixed for all beam currents. The nano-granular ZnO also displayed an intense featureless radiative emission in the infrared which steadily increased from the visible up to the infrared detection limit at 0.75 eV and strongly blue shifted with increasing beam current. The intensity of this emission exhibited a supralinear dependence on the excitation power of the electron beam with a power exponent above 9. This emission was completely quenched by a modest decrease in specimen temperature below room temperature, and was not observed in the thin film ZnO under all excitation conditions. A similar emission was also observed from nano-granular TiO_2, Al_2O_3 and ZrO_2. Based on these observations, the observed IR emission is attributed to black-body radiation due to electron beam heating (Fig. 7). The low thermal conductance of the ZnO nano-particles as well as the absence of heat loss by convection due to the high vacuum SEM environment leads to very efficient electron beam heating of the ZnO nano-particles and dominate radiative heat loss mechanism. Blackbody emission has also been reported for silicon nano-powders under laser irradiation [13, 14]. The observed red shift of the NBE is therefore attributed to its dependence on temperature [15] rather than band gap renormalization due to the increase in the carrier density.

A power-law model ($I_{CL} \alpha I_B^m$) was used to analyse the excitation power dependence of NBE emission in both forms of ZnO at 80 K and 300 K. At room temperature, the thin film ZnO exhibited a power law exponent greater than $m > 1$, characteristic of FX and BX recombination channels. However, sub-linear values, $m = 0.5$, were measured for the nano-powder ZnO material at 300 K which increased to $m = 1$ at 80 K, indicating that beam heating has a significant influence on these measurements at room temperature. The position of the defect emission peaks in both the thin film and nano-powder specimen were

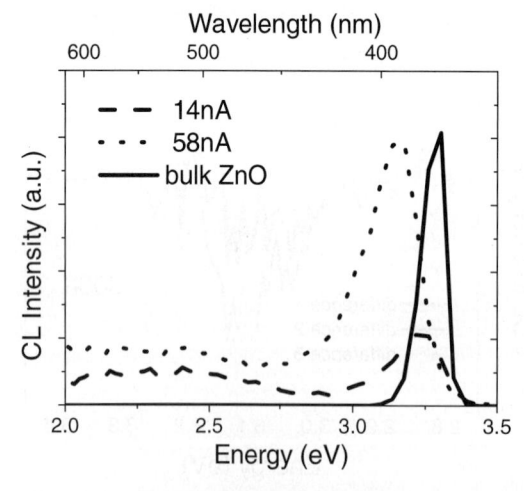

Fig. 6 CL spectra ZnO nano-particles at 300 K (10 kV, 40 nA, 5 nm bandpass and scan area 170 μm × 130 μm). The red peak shift is attributed to band gap broadening due to electron beam heating.

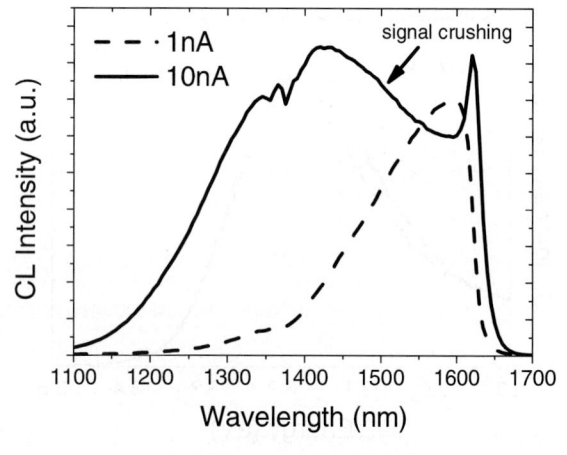

Fig. 7 Intense near infrared CL spectra ZnO nano-particles at 300 K (10 kV, 10 nm bandpass and scan area 170 μm × 130 μm) attributed to incandescence arising from electron beam heating.

fixed with increasing beam current. It is important to note that the absence of a blue shift does not automatically exclude a donor acceptor pair (DAP) recombination process as barely discernible shifts are characteristic of DAP centers with deep-donor – deep-acceptor pairs in ionic materials.

4 Discussion The significant rise in the NBE emission intensity in both the thin film and nano-powder ZnO following the hydrogen heat treatment is due the incorporation of hydrogen. Conversely, quenching of the NBE emission following heat treatment in O_2, N_2 and Ar can be attributed to loss of hydrogen from the ZnO. The increase in the NBE following thermal annealing in hydrogen can be attributed to one or more of the following mechanisms: First hydrogen passivation of competitive non-radiative recombination channels in the bulk and/or at the surface, second the introduction of an additional efficient NBE radiative transition via the incorporation of hydrogen as a donor itself or as a hydrogen-related complex and third the elimination of competitive radiative recombination channels by the formation of hydrogen complexes with acceptors and donors.

Reduction of the near infrared emission following hydrogen annealing demonstrates that hydrogen can indeed cluster with defects removing competitive radiative recombination pathways. However, another radiative deep level emission at 2.4 eV is produced by the same process, which suggests that the third mechanism is not the reason for the increase of the NBE emission. The second explanation can be also ruled out as the NBE emission is significantly enhanced at room temperature. Hydrogen has been theoretically predicted to form a shallow donor level in ZnO 35 meV below the conduction band [16] and an experimental donor level around this energy has also been reported [17]. Consequently, a large portion of the hydrogen donors will be thermally ionized at room temperature and not available for radiative recombination. Similarly, an emission at 3.22 eV assigned to a hydrogen donor and a deep acceptor pair [18] cannot explain the improved NBE emission at 300 K because the hydrogen donor must be in a neutral state for the DAP transition to occur.

Time-resolved electron beam irradiation experiments reported in this work have identified a weak radiative transition at 3.15 eV that is introduced in both the thin film and nano-powder ZnO specimens following thermal annealing in hydrogen. An optical absorption transition has been observed by other researchers in "as grown" materials and attributed to a valence band to intrinsic donor level at approximately 0.16 eV below the conduction band [19]. The present work suggests that this donor level is a hydrogen-related defect center since it is only observed following thermal heating in hydrogen. However, this hydrogen induced emission does not explain the increase in the NBE emission as its intensity is very weak compared with the 1 LO-FX emission at 300 K. After ruling out mechanism two and three above, we contend that the passivation of non-radiative surface states and/or bulk defects provides the best explanation for the significant increase in the NBE emission following hydrogen thermal annealing. The luminescence enhancement is much more pronounced in the thin film ZnO compared with the ZnO nano-particle specimens, suggesting that hydrogen passivation of bulk defects rather than surface states is the dominant mechanism. Thermal annealing in gases other than hydrogen leads to a decrease in the NBE emission when the non-radiative centers are re-activated following thermal dissociation of hydrogen and its subsequent out-diffusion from the nano-particle.

The chemical origin of the non-radiative center passivated by hydrogen in both the thin film and nano-particle ZnO is unknown at present. Oxygen or zinc vacancies seem to be most likely candidates as they are theoretically predicted to have the lowest formation energy in ZnO and should be the most abundant native defects [20, 21]. The zinc vacancy is the more likely candidate since hydrogen is predicted to exist in ZnO as a proton, H^+, state so should cluster with bulk V_{Zn}^{2-} defects in preference to V_O^+ and V_O^{++} centers.

The above explanation for the improvement of the NBE emission is supported by the thermal annealing behaviour of the 2.4 eV emission. The appearance of the 2.4 eV peak following heat treatment in all gas environments, O_2, Ar, N_2 and H_2/N_2, indicates that the induced emission arises from a bulk thermal effect rather than a chemical process at the surface of the nano-particle. Indeed, heat treatment in the inert gas argon produces a stronger 2.4 eV emission intensity than annealing in the acceptor-like and donor-like gases. Although oxygen and zinc vacancies will both form at high temperature, similar arguments to those present above suggest that hydrogen should cluster with zinc vacancies and quench the emission. Because the opposite effect is experimentally observed, at this stage the 2.4 eV emission is attributed to an oxygen vacancy or oxygen vacancy-related center. Recent studies have reported that the green emission in ZnO nano-particles is quenched when the nano-crystals are dispersed in a postassium iodide host crystal allowing the iodine ions to fill and passivate surface located oxygen vacancies [22]. A similar effect would be expected in this work after heat treating in O_2 and N_2 gaseous atmospheres if the oxygen vacancies are positioned at the surface of the nano-particle. This indicates that the oxygen vacancy defects produced by thermal annealing are most likely present in the bulk of the nano-particle.

5 Conclusion Thermal annealing in hydrogen was found to significantly increase the near band edge cathodoluminescence emission at 80 K and 300 K from ZnO nano-particles and ZnO thin films. The increase is tentatively attributed to hydrogen passivation of bulk zinc vacancy defects which act as competitive non-radiative recombination centers. Due to the poor thermal conductance of nano-particles, electron beam heating effects must be considered when interpreting CL data measured from these nano-structured materials.

References

[1] N. Ohashi, T. Ishigaki, N, Okada, N. Okada, T. Sekiguchi, I. Sakaguchi, and H. Haneda, Appl. Phys. Lett. **80**, 2869 (2002).
[2] A. Y. Polyakov, N. B. Smirnov, A. V. Govorkov, K. Ip, E. Overberg, Y. W. Heo, D. P. Norton, S. J. Pearton, B. Luo, F. Ren, and J. M. Zavada, J. Appl. Phys. **94**, 400 (2003).
[3] B. Theys, V. Sallet, F. Jomard, A. Lusson, J.-F. Rommeluère, and Z. Teukam, J. Appl. Phys. **91**, 3933 (2002).
[4] L. Wang and N. C. Giles, J. Appl. Phys. **94**, 973 (2003).
[5] K. Vanheusden, W. L. Warren, C. H. Seager, D. R. Tallant, J. A. Voigt, and B. E. Gnade, J. Appl. Phys. **79**, 7983 (1996).
[6] S. A. Studenikin and M. Cocivera, J. Appl. Phys. **91**, 5060 (2002).
[7] M. Liu, A. H. Kitai, and P. Mascher, J. Lumin. **54**, 35 (1992).
[8] E. G. Bylander, J. Appl. Phys. **49**, 1188 (1978).
[9] F. I. Vergunnas and G. A. Konivalow, J. Exp. Theoret. (USSR) **23**, 712 (1952).
[10] N. Y. Garces, L. Wang, L. Bai, N. C. Giles, L. E. Halliburton, and G. Cantwell, Appl. Phys. Lett. **81**, 622 (2002).
[11] X. T. Zhang, Y. C. Liu, Z. Z. Zhi, J. Y. Zhang, Y. M. Lu, D. Z. Shen, W. Xu, X. M. Fan, and X. G. Kong, J. Lumin. **99**, 149 (2002).
[12] X. L. Wu, G. G. Siu, C. L. Fu, H. C. Ong, Appl. Phys. Lett. **78**, 2285 (2001).
[13] J. Costa, P. Roura, J. R. Morante, and E. Bertran, J. Appl. Phys. **83**, 7879 (1998).
[14] P. Roura, J. Costa, M. López-de Miguel, B. Garrido, J. Fort, J. R. Morante, and E. Bertran, J. Lumin. **80**, 519 (1999).
[15] W. S. Shi, O. Agyeman, and C. N. Xu, J. Appl. Phys. **91**, 5640 (2002).
[16] C. G. Van de Walle, Phys. Rev. Lett. **85**, 1012 (2000).
[17] D. M. Hoffmann, A. Hofstaetter, F. Leiter, H. J. Zhou, F. Henecker, B. K. Meyer, S. B. Orlinskii, and P. G. Baranov, Phys. Rev. Lett. **88**, 045504 (2002).
[18] K. Thonke, Th. Gruber, N. Thefilov, R. Schönfelder, A. Waag, and R. Sauer, Physica B **308–310**, 945 (2001).
[19] V. Srikant and D. R. Clarke, J. Appl. Phys. **83**, 5447 (1998).
[20] A. F. Kohan, G. Ceder, D. Morgan, and C. G. Van de Walle, Phys. Rev. B **61**, 15019 (2000).
[21] F. Oba, S. Nishitani, S. Isotani, H. Adachi, I. Tanaka, J. Appl. Phys. **90**, 824 (2001).
[22] Y. Harada and S. Hashimoto, Phys. Rev. B **68**, 045421 (2003).

Influence of the substrate on the structural properties of sputter-deposited ZnO films

Hyoun Woo Kim[*] **and Nam Ho Kim**

School of Materials Science and Engineering, Inha University, Incheon 402-751, Republic of Korea

Received 15 September 2003, revised 6 October 2003, accepted 6 October 2003
Published online 18 December 2003

PACS 61.10.Nz, 68.37.Hk, 68.55.Jk, 81.05.Dz, 81.15.Cd

We have deposited ZnO films on various substrates using the r.f. magnetron sputtering method. X-ray diffraction and scanning electron microscopy coincidentally revealed that the larger grain size and the higher crystallinity were attained when the ZnO films were deposited on sapphire substrates, compared to the films on Si substrates. The XRD analysis revealed that the c-axis lattice constant decreased and increased, respectively, by thermal annealing for the ZnO films deposited on Si and sapphire substrates. Atomic force microscopy indicated that the surface roughness was higher for the films deposited on the sapphire substrates.

© 2004 WILEY-VCH Verlag GmbH & Co. KGaA, Weinheim

1 Introduction Zinc oxide (ZnO) thin films received much attention due to its unique piezoelectric and piezooptic properties made suitable for surface acoustic devices [1], low loss optical waveguides [2], optoelectronic devices [3, 4], and gas sensors [5]. Additionally, ZnO has been developed as a promising alternative to transparent conducting indium tin oxide and tin oxide coating due to its low cost and non-toxicity [6]. ZnO have been prepared by various physical vapor deposition techniques such as evaporation, r.f./d.c. sputtering, ion beam sputtering, atomic layer epitaxy etc., as well as a plethora of chemical methods. Among these methods, r.f. magnetron sputtering has gained importance due to its potential for low-cost and low temperature process. Although many researchers have grown ZnO films by various techniques, the effects of substrate material on the structural characteristics of ZnO thin films have not been sufficiently investigated [7, 8].

In this paper, we have deposited the c-axis oriented ZnO films by the r.f. magnetron sputtering method. We have studied the dependence of the structural properties on the substrate material, i.e. silicon (Si) and sapphire.

2 Experimental A schematic description of the r.f. sputtering system is provided elsewhere [9]. ZnO films were deposited by an r.f. magnetron sputtering system using a ZnO (99.99% purity) target with a diameter and a thickness of 75 mm and 6 mm, respectively. The sputtering was carried out in a 30 sccm Ar (99.99% purity) gas atmosphere by supplying 80–250 W r.f. power. The magnetron is capacitively powered with 13.56 MHz radio-frequency. In ordet to exclude the thickness effect, we have fixed the ZnO film thickness to about 500–600 nm.

The ZnO thin films were sputtered on Si and sapphire substrates. The Si substrate was p-type silicon with (001) orientation and the resistivity was 1–30 ohm–cm. The Si substrate was rinsed in acetone for 10 min, HF (20 : 1) for 10 seconds and then rinsed by deionized water for 1 minute before loading into the sputtering system. The sapphire substrate was of (001) orientation and were cleaned in acetone for 10 min.

---

[*] Corresponding author: e-mail: hwkim@inha.ac.kr, Phone: +82 32 860 7544, Fax: +82 32 862 5546

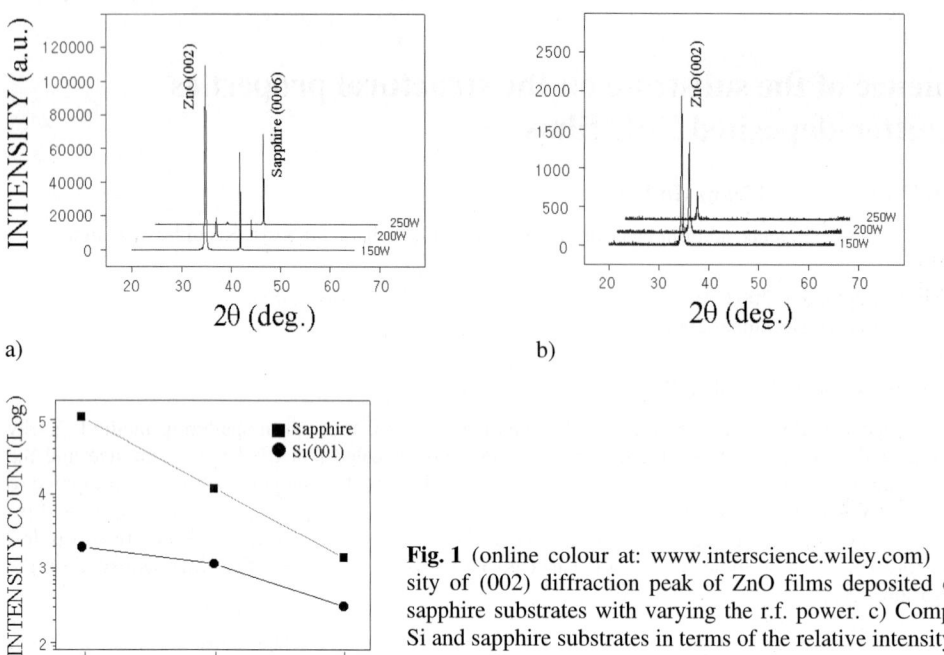

Fig. 1 (online colour at: www.interscience.wiley.com) Relative intensity of (002) diffraction peak of ZnO films deposited on a) Si and b) sapphire substrates with varying the r.f. power. c) Comparison between Si and sapphire substrates in terms of the relative intensity (log scale).

The chamber was down to 6×10^{-6} torr using a turbo-molecular pump before introducing the Ar sputtering gas into the chamber. The ZnO film was grown at room temperature at a pressure of 5.0×10^{-2} torr. The thermal annealing was performed in a furnace using quartz tube reactor in the temperature of 500–900 °C in oxygen gas flow of 3 standard liter per minute for 10 minutes. The structural properties of the films (crystalline structure or microstructure) were analyzed by X-ray diffraction (XRD) using CuKα1 radiation ($\lambda = 0.154056$ nm), by scanning electron microscopy (SEM) (Hitachi S–4200), and by atomic force microscopy (AFM) (Digital Instruments Nanoscope III) with a scan size of 10 μm × 10 μm.

3 Results and discussion

In order to compare the structural properties of ZnO films on Si and sapphire substrates, we have deposited ZnO films with varying the r.f. power for each substrate. The dependence of growth rate on the r.f. power indicates that the film growth rate is almost linearly proportional to the r.f. power, regardless of the substrate material. In XRD analysis, the θ–2θ scan data of ZnO films exhibit a strong 2θ peaks at 34.53°, corresponding to the (002) peaks of ZnO. Since the c-axis (002) diffraction

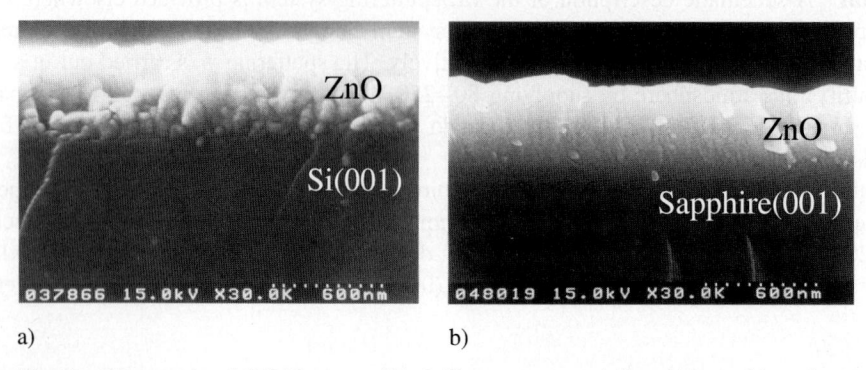

Fig. 2 Cross-sectional SEM images of ZnO films grown on a) Si and b) sapphire substrates.

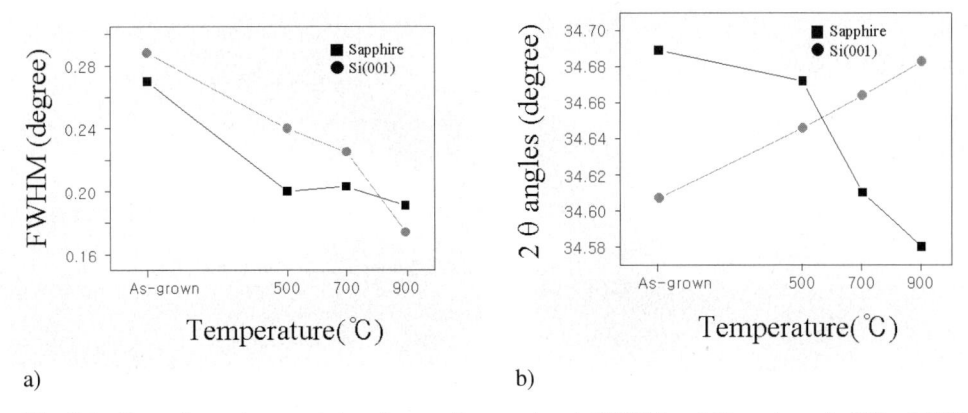

Fig. 3 (online colour at: www.interscience.wiley.com) a) FWHM and b) peak angle (2θ) of (002) diffraction peak from ZnO films deposited with an r.f. power of 150 W.

peaks were distinguishably observed in the grown ZnO films, we suppose that the c-axis oriented ZnO film is obtained, regardless of the substrate material. Figure 1 shows the relative intensity of ZnO film with the (002) orientation, deposited on Si and sapphire substrates, as a function of r.f. power ranging from 150 to 250 W, indicating that the relative intensity is significantly higher when ZnO films are deposited on sapphire substrates.

We have compared the cross-sectional SEM images of ZnO films grown on Si and sapphire substrates in Fig. 2, in which the RF power was set to 150 W for the same growth time. The images indicate that the crystallinity of the film on sapphire substrate is higher than that on Si substrate, agreeng with the XRD intensity data. We suppose that relatively little lattice mismatch between ZnO and sapphire substrates helps to grow the higher crystalline films.

In order to analyze the effect of substrate on the structural properties of ZnO films, we have annealed the samples at the temperature of 500–900 °C. Figure 3a and 3b, respectively, show the full width at half maximum (FWHM) and the peak angle (2θ) of (002) diffraction peak measured in ZnO thin films when deposited with the r.f. power of 150 W. Figure 3a indicates that the FWHM decreases with increasing the annealing temperature, regardless of substrate material. This observation also indicates that the grain size becomes larger and thus the crystallinity may be improved with increasing the annealing temperature.

Figure 3b indicates that the 2θ angle of the ZnO films with respect to the (002) diffraction peak grown on Si(001) and sapphire(001) substrates, respectively, are 34.61° and 34.69°. Since the XRD angle (2θ) of the bulk ZnO are 34.44° and by calculating with the Bragg law [10], we reveal that c-axis lattice constant becomes larger when deposited at room temperature, regardless of the substrate material. It is noteworthy that the 2(angle increases with increasing the annealing temperature for the ZnO films grown on Si substrates but decreases with increasing the annealing temperature for the films grown on sapphire substrates. The different substrates will lead to different crystal mismatch and thermal expansion between substrates and ZnO films, and the crystal lattice mismatch and difference of thermal expansion may cause different strains and stress in the films. In case of ZnO films on Si substrates, we suppose that increasing the annealing temperature increases the compressive stress along the c–axis direction and thus reduces lattice plane spacing and increases the 2(angle of XRD. On the other hand, in case of ZnO films on sapphire substrates, increasing the annealing temperature reduces the compressive stress along the c-axis direction and thus increases the lattice plane spacing and reduces the 2(angle of XRD. Further systematic study is necessary to reveal the detalied mechanism.

Figure 4 shows the AFM images of ZnO films deposited with the r.f. power of 150 W and subsequently annealed at 700 °C, indicating that the surface roughness (root-mean-square) of the films grown on sapphire substrates is higher than that on Si substrates. We have the similar results for the samples annealed at other temperatures and for the as-grown samples. Accordingly, the AFM data agree with the

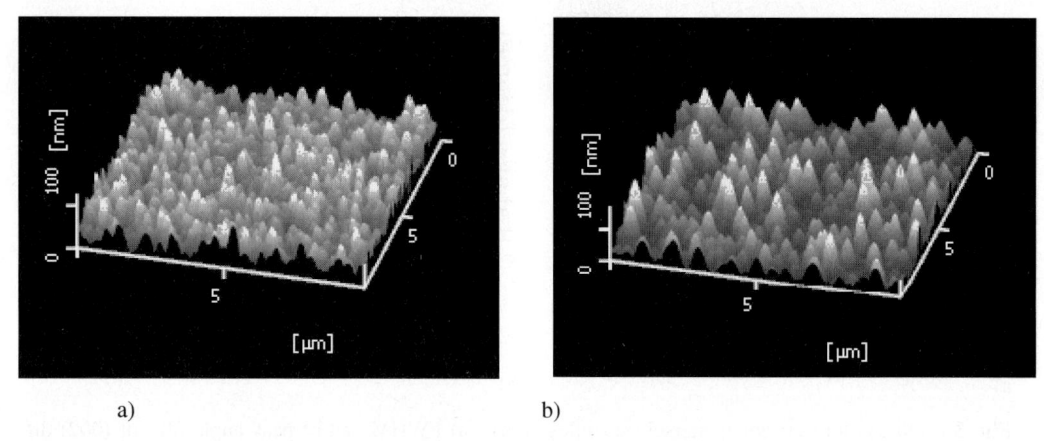

Fig. 4 (online colour at: www.interscience.wiley.com)   AFM images of ZnO films deposited a) on Si(001) and b) on sapphire(001) substrates with the r.f. power of 150 W and subsequently annealed at 700 °C.

XRD data (Fig. 3a), implying that larger grain size corresponds to the higher surface roughness. The different substrates will lead to different crystal mismatch between substrates and ZnO films. We conclude that for the ZnO films grown on sapphire substrates, due to their relatively smaller lattice mismatch, the atoms have more chances to occupy the correct site in the crystal lattice and grains with the lower surface energy will become larger. Then the growth orientation develops into one crystallographic direction of the lower surface energy.

4 Conclusions   ZnO thin films have been grown on Si and sapphire substrates at room temperature using the r.f. magnetron sputtering technique. The XRD of all ZnO films showed c-axis orientation, regardless of substrate material. The relative intensity of (002) diffraction peak is higher for ZnO films grown on sapphire substrates than for ZnO films on Si substrates. The FWHM of (002) diffraction peak decreases with increasing the annealing temperature, regardless of substrate material. The XRD (002) peak angle increases and decreases, respectively, with increasing the annealing temperature for ZnO films deposited on Si and sapphire substrates. The grain size is larger and the surface is more rough for ZnO films grown on sapphire substrate. We believe that the approach to investigate the very low temperature growth and annealing behavior of ZnO films depending on the substrate material are a step toward the efficient production of ZnO devices.

Acknowledgements   This work was supported by INHA UNIVERSITY Research Grant (INHA–22098).

References

[1] F. S. Hickemell, IEEE Trans. Sonics Ultrason. **SU–32**, 62 (1985).
[2] D. Persegol, E. Pic, and J. Plantier, J. Appl. Phys. **62**, 2563 (1987).
[3] D. F. L. Jenkins, M. J. Cunningham, G. Velu, and D. Remiers, Sens. Actuators A **63**, 135 (1997).
[4] A. S. Juarez, A. T. Silver, A. Oritz, E. P. Zironi, and J. Richards, Thin Solid Films **333**, 135 (1998).
[5] J. S. Zobinski, J. Corneille, S. V. Prasad, N. T. McDevitt, and J. B. Bultman, J. Mater. Sci. **32**, 5313 (1997).
[6] L. Stolt, J. Hedstrom, J. Kessler, M. Ruckl, K. U. Velthaus, and H. W. Schock, Appl. Phys. Lett. **62**, 597 (1993).
[7] S. H. Bae, S. Y. Lee, H. Y. Kim, and S. Im, Appl. Surf. Sci. **168**, 332 (2000).
[8] T. Oshima, R. K. Thareja, T. Ikegami, and K. Ebihara, Surf. Coat. Technol. **169–170**, 517 (2003).
[9] K. S. Kim, H. W. Kim, and N. H. Kim, Physica B **334**, 343 (2003).
[10] B. D. Cullity, Elements of X-ray Diffraction (Addison-Wesley, 1978), p. 100.

Ion beam fabricated silicon light emitting diodes

M. A. Lourenço[*,1], **M. S. A. Siddiqui**[1], **G. Shao**[2], **R. M. Gwilliam**[1], and **K. P. Homewood**[1]

[1] School of Electronics and Physical Sciences, University of Surrey, Guildford, Surrey, GU2 7XH, UK
[2] School of Engineering, University of Surrey, Guildford, Surrey, GU2 7XH, UK

Received 7 October 2003, accepted 7 October 2003
Published online 29 December 2003

PACS 61.12.Lk, 78.55.Ap, 78.60.Fi, 85.30.De, 85.60.Bt, 85.60.Jb

The addition of efficient optical emission to the functionality of crystalline silicon, because of its already dominant position for use in electronic devices and for integrated circuit production, is highly desired. Applications include optical interconnects on chip and between chips – needed to solve future data transfer limitations associated with current copper technology, cheap fibre-to-the-home transceivers, and many others. A solution using ion implantation is the optimum route given the ubiquitous role of this process in silicon technology. In this paper the use of a new method to fabricate silicon based light emitting devices, dislocation engineering, that allows to circumvent the fundamental problem caused by the indirect gap in silicon, is described. This approach is entirely compatible with current ULSI technology, a key requirement given the very high tool-up costs associated with silicon integrated circuit production.

© 2004 WILEY-VCH Verlag GmbH & Co. KGaA, Weinheim

1 Introduction

There has been a recognition in the last few years that new semiconductor systems will need to be developed to replace some of the more conventional materials used currently for semiconductor device production [1]. The current semiconductor market is dominated by the semiconductor silicon because of its widespread use in microelectronics. The electronics market is currently worth around 1500 billion dollars per annum with an underlying growth of around 7% per annum. The semiconductor contribution to this is around 20% or around 300 billion dollars per annum predicted to increase to around 40% by 2010. The silicon fraction of the semiconductor market is around 95% with the remainder dominated by III–V semiconductors. The III–V sector although much smaller is crucial because it provides devices of superior performance and additional functionality than silicon for several key device areas, for example laser diodes for optical communication systems.

Silicon is an indirect band gap semiconductor and as such is considered fundamentally unsuited for use as an optical emitter, one of the essential requirements for replacing III–V devices. Despite this intrinsic problem there has been massive effort world wide to develop routes to efficient light emission in silicon based devices [2–9]. A very significant driver for developing optical emitters in silicon is the interconnect problem in computer chips. The electronics and computing sectors have been driven by the exponential growth in processor power and speed. This growth has been achieved by scaling of the transistor dimensions enabling more and more transistors to be formed in the same area of silicon so increasing the speed and power of the chip. However, the transistors need to be connected together by the metallisation. The time delay associated with electron transport in the metallisation does not scale. Consequently a limit to classical scaling is reached within the next decade when computer chips will stop getting faster. The solution to this problem is the replacement of at least some of the metal interconnects with optical data transfer on chip [10]. The development of optical pin outs and interchip and interboard optical connections are also seen to be required. An integrated silicon light emitter is the preferred solution.

---

[*] Corresponding author: e-mail: m.lourenco@surrey.ac.uk, Phone: +44 1483 682298, Fax: +44 1483 689404

Given the huge tool up costs in the microelectronics industry, for example fabrication facilities are currently 2–3 billion dollars to set up, the industry is very conservative about new processes and materials. It is therefore a requirement that new approaches are closely compatible with silicon ultra large scale integration (ULSI) technology. The key technology for the fabrication of silicon chips is ion implantation and so solutions using this technology are essential.

We describe a new approach – Dislocation Engineering [11] – using, crucially, just conventional ULSI technology, to make efficient light emitting diodes (LEDs) in silicon operating at room temperature. Because of its indirect band gap silicon is fundamentally a poor emitter of light. The main reason is that fast non-radiative recombination routes dominate over slower radiative routes in this material. Indeed, in bulk silicon, at room temperature, radiative emission is normally entirely absent. However, if recombination through the non-radiative routes can be prevented, the radiative emission could in principle be enhanced. Non-radiative recombination is the result of diffusion of carriers to point defects in the silicon where efficient non-radiative recombination then occurs. Despite the already low defect concentrations in good quality silicon this non-radiative route always completely dominates. A way of enhancing the radiative efficiency would be to prevent the carrier diffusion. The method described here uses the controlled introduction of dislocation loops by conventional ion implantation and thermal processing to modify the silicon band gap, thus providing spatial confinement in three dimensions.

2 Theory The implantation of boron is made in to a device grade n-type silicon wafer to form the p-type top layer of the diode and to form dislocation loops. After the initial implant the silicon is disordered. A subsequent annealing step is undertaken to electrically activate the implanted boron dopants. The activation involves placing the boron atom on to a silicon lattice site. The consequence of this is that the silicon atom that was on that lattice site is now present as an interstitial atom. Within the implanted region there is now present an excess population of interstitial silicon atoms equal to the original implant dose of the boron. If a suitable anneal schedule is applied then the silicon atoms can diffuse, aggregate and form into small crystalline planes of silicon atoms, one atom thick, known as dislocation loops. The size and distribution of these loops is a function of the implant and anneal parameters and so can be engineered. If we wish to form dislocation loops independently of doping the material, we can use a silicon in to silicon implant. The basic steps of dislocation loop formation are shown in Fig. 1.

A dislocation loop is just an island, a single atom thick, of bulk crystalline silicon whose rim is formed of a simple single edge dislocation turned around on its self. Figure 2 is an schematic diagram showing how the dislocation loop leads to modification of the band gap energies. Inspection of Fig. 2b, showing a dislocation loop inserted into the silicon lattice, reveals that just outside the loop the silicon interatomic spacing is increased and just inside it is decreased. So, just outside the loop the silicon lattice is placed under negative hydrostatic pressure and just inside it is placed under positive hydrostatic pressure. The stress distribution across a diameter of the loop is shown schematically in Fig. 2c. The magnitude and

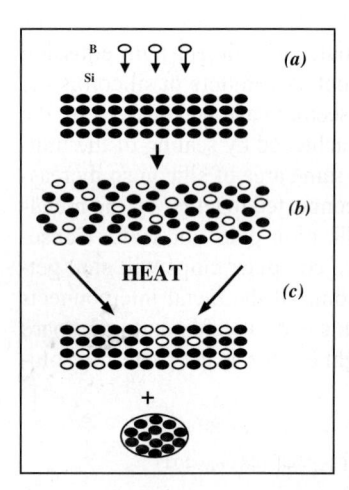

Fig. 1 A schematic showing the formation of dislocation loops by ion implantation: a) the implant process, b) the disordered lattice after implantation and c) the annealing step showing activation of the implant and formation of the loop from the excess interstitials.

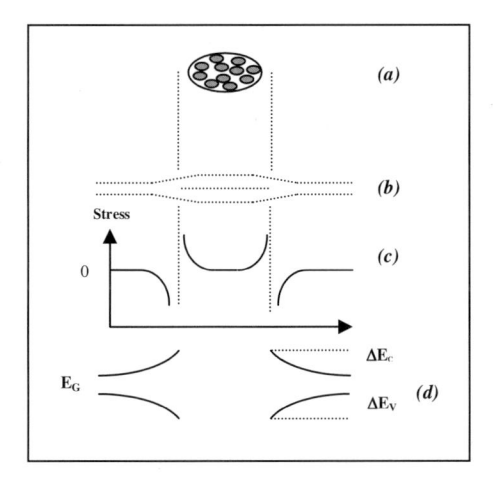

Fig. 2 A schematic diagram showing how the dislocation loop leads to modification of the band gap energies: a) looking down on a loop; b) a loop inserted into the silicon lattice – just outside the loop the lattice silicon atoms are forced further apart than the bulk value; c) the stress distribution across the loop; d) illustration of the band gap modification.

form of the stress field can be calculated using standard elastic theory of dislocations [12] which shows that the stress field decays inversely with distance and reaches a maximum stress of 25 to 50 GPa. This uses the known values of Poisson's ratio and Young's modulus for silicon, 0.42 and 113 GPa, respectively. The importance of this stress field for our applications is that the band gaps of semiconductors are pressure dependent and, in the case of silicon, the band gap decreases with increasing pressure and increases under negative pressure. A schematic of the dependence of the conduction and valence band energies is shown in Fig. 2d. The total change in the band gap, $\Delta E_G = \Delta E_C + \Delta E_V$, just outside the loop, can be easily calculated using the known pressure coefficient of the band gap for silicon. However, the distribution of this band gap change between the conduction and valence bands is still uncertain. The total band gap change is from 0.325 to 0.75 eV, a significant fraction of the band gap itself which is 1.1 eV.

The key to our device operation is the utilisation of this band gap change to modify the carrier flow after injection. In a conventional silicon diode, after injection the carriers diffuse and eventually recombine. The average linear distance before recombination occurs is just the diffusion length and in silicon this can be tens of microns although often the surface is reached first and recombination occurs here. The long diffusion lengths in silicon are the result of the low density of recombination centres in silicon and the fact that the material is indirect gap. This means that the band to band recombination is forbidden or in practice the radiative band to band transition is characterised by a long lifetime in the range of tens of microseconds, or even much longer in very pure material. The faster non-radiative recombination routes therefore also dominate. However, if the carrier diffusion can be prevented or limited then the carriers will eventually recombine radiatively across the band. Here we make use of an array of dislocation loops, or rather the blocking potential that they produce, placed in a plane just beyond the depletion region edge to prevent carrier diffusion in to the bulk of the device. The injected carriers then recombine radiatively across the band. It is noted that any injected carrier diffusing back across the junction is always dynamically replaced by a carrier moving in the other direction and is not a source of loss. This means we can have tight spatial localisation of injected carriers but with the total absence of any quantum confinement. A band diagram indicating the operation of the device under zero and forward bias is shown in Fig. 3.

3 Experimental details To demonstrate how the dislocation engineering method can be used to fabricate efficient silicon light emitting devices operating at room temperature, and also to show how this method can be employed to reduce the luminescence thermal quenching in other systems, two sets of devices were fabricated. For the first set, boron was implanted into [100] n-type silicon substrates (2–7 Ωcm, Phosphorus doped), followed by a high temperature anneal at 950 °C for 20 minutes in nitrogen ambient. The boron implant energy was in the range 20–70 keV and the dose varied with energy to maintain a peak dopant concentration of 10^{20} cm^{-3}, as shown in Table 1. The boron implantation served as a means of introducing the dislocation loops as well as the p-type dopant to form a p–n junction. Oh-

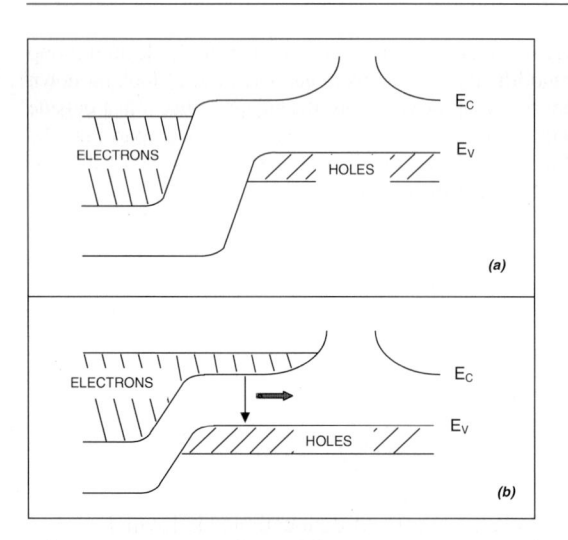

Fig. 3 A schematic of the band diagram of the dislocation engineered diode. The potential change is due to an array of dislocation loops formed parallel to the depletion region edge at that point. a) The device at zero bias and b) under forward bias.

mic contacts were formed by vacuum evaporation of Al and AuSb eutectic on the p-type region and n-type substrate, respectively and sintered at 360 °C for 2 minutes. Subsequently, the samples were mesa etched to isolate the p–n junction; the device area was 8×10^{-3} cm^{-2}.

The second set of samples consisted of Er/Si structures. Two different batches of erbium/silicon devices were fabricated. For the first one (referred to as ^{0}Er:Si) Er ions were implanted into a standard silicon substrate (2×10^{13} Er cm^{-2} at 0.4 MeV). The second batch (^{D}Er:Si) was fabricated by Er implantation (same dose and energy as the previous batch) into wafers previously implanted with boron (10^{15} B cm^{-2} at 30 keV and annealed at 950 °C for 20 minutes). The Er implants were followed by a rapid thermal anneal at 950 °C for 1 minute in nitrogen ambient to activate the Er, for both batches. Ohmic contacts, when formed, followed the same procedure described before.

The devices were characterised by electroluminescence (EL) and photoluminescence (PL) measurements. The EL and PL experiments were performed in the temperature range 80–300 K. In both cases the samples and devices were mounted in a continuous-flow liquid nitrogen cryostat placed in front of a conventional half-metre spectrometer. A liquid nitrogen cooled germanium p-i-n diode was used for detection of the EL and PL. The PL experiments were performed at 150 mW excitation laser power using a wavelength of 514 nm. The EL measurements were performed under forward bias through a window at the back of the samples.

4 Results and discussion The samples implanted with boron only (range 20–70 keV) showed similar PL and EL emissions, peaking, at room temperature, at ~1154 nm. Figures 4 and 5 shows the PL and EL integrated intensity plotted against measurement temperature for all samples. No temperature quenching is observed; for all devices the integrated intensity increases with increasing temperature. The PL integrated intensity increases with decreasing implant energy. At higher energies the dislocation loops are formed further from the surface so a greater volume of non-spatially confined silicon is available for absorption of photons before reaching the loops. The EL emission didn't show any dependence upon the energy range and doses studied here.

Table 1 Implant energy and dose for samples implanted with boron only. All samples were subsequently annealed at 950 °C for 20 minutes.

sample	boron energy (keV)	boron dose (cm^{-2})
GN#1	70	1.6×10^{15}
GN#2	50	1.37×10^{15}
GN#3	40	1.2×10^{15}
GN#4	30	1.0×10^{15}
GN#5	20	7.6×10^{14}

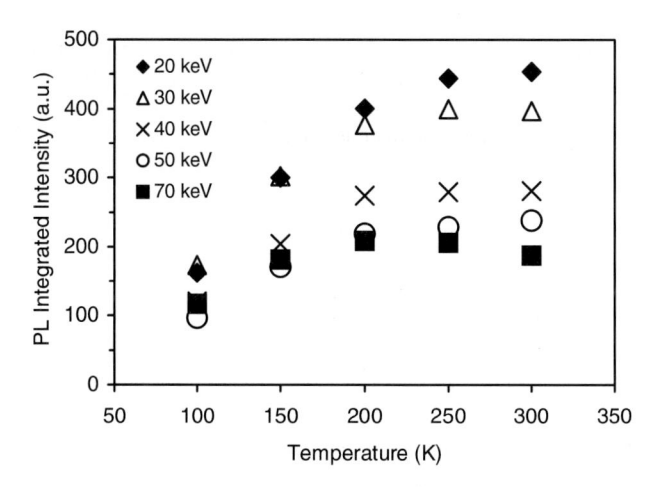

Fig. 4 Photoluminescence integrated intensity as a function of measurement temperature for devices implanted with boron at several energies.

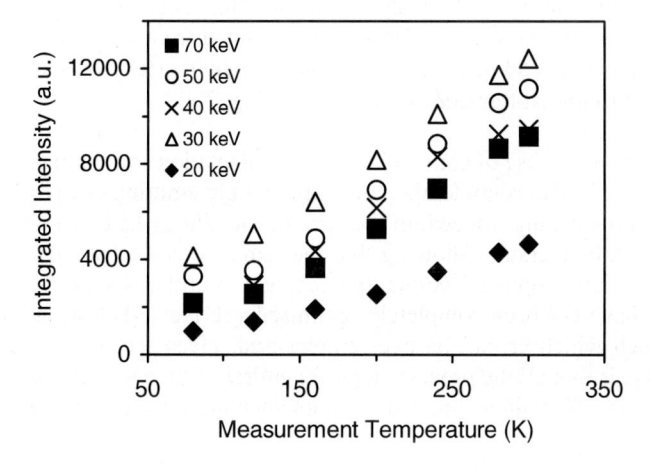

Fig. 5 Electroluminescence integrated intensity as a function of measurement temperature for devices implanted with boron at several energies.

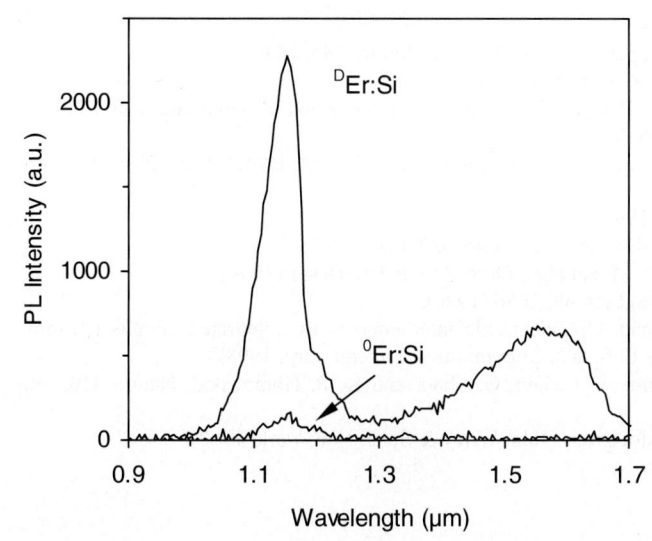

Fig. 6 Room temperature photoluminescence spectra of ^{0}Er:Si and ^{D}Er:Si structures.

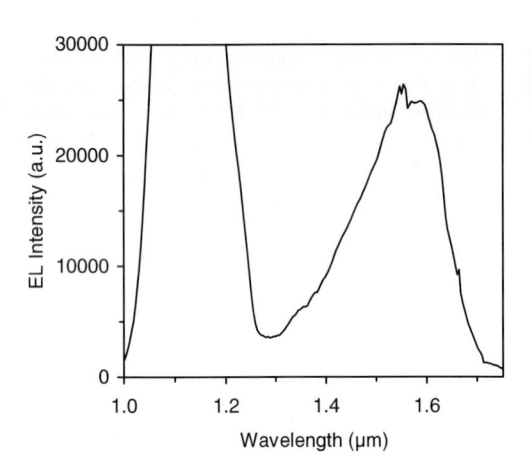

Fig. 7 Room temperature electroluminescence spectrum of a light emitting device fabricated from substrate ^{D}Er:Si. The peak at 1.1 μm is the silicon band edge emission and those at ~1.5 μm are the erbium emission peaks.

The room temperature PL spectra of the Er implanted structures is shown in Fig. 6. Spectra obtained from sample ^{0}Er:Si show only a weak Si band edge emission, while for samples ^{D}Er:Si a broad peak centred at ~1.5 μm can be observed, in addition to a strong Si band edge emission. This long wavelength peak is probably related to the Er centre. Indeed, room temperature EL spectra obtained from light emitting devices fabricated from substrate ^{D}Er:Si showed, in addition to the Si band edge, two clear peaks, at 1.555 μm and 1.588 μm, corresponding to two Er internal transitions, as shown in Fig. 7.

5 Conclusions Dislocation engineering, which makes use of conventional ion implantation and thermal processing, is a viable route to the fabrication of efficient room temperature silicon light emitting devices and to the control of the luminescence thermal quenching. Indeed, in the case of the silicon LED's, the luminescence integrated intensity increased with temperature showing that the radiative carrier population was completely decoupled from the non-radiative routes. Thermal quenching was still observed for the Si/Er structures – however, those devices have not been completely optimised yet. Nevertheless, an important technological aspect of this approach which cannot be over emphasised, given the massive "tool-up" costs in the microelectronics industry, is that all the process steps are entirely conventional and are completely compatible with ULSI technology. This allows immediate implementation on a standard fabrication line.

References

[1] Y. Makita, Proc. First NREL Conf., 3 (1977).
[2] K. D. Hirschman, L. Tysbekov, S. P. Duttagupta, and P. M. Fauchet, Nature **384**, 338 (1996).
[3] Z. H. Lu, D. J. Lockwood, and J. M. Baribeau, Nature **378**, 258 (1995).
[4] T. Komoda, J. Kelly, F. Cristiano, A. Nejim, P. L. F. Hemment, K. P. Homewood, R. Gwilliam, J. E. Mynard, and B. J. Sealy, Nucl. Instrum. Methods B **96**, 387 (1995).
[5] B. Zheng, J. Michel, F. Y. G. Ren, L. C. Kimerling, D. C. Jacobson, and J. M. Poate, Appl. Phys. Lett. **64**, 2842 (1994).
[6] L. Vescan and T. Stoica, J. Lumin. **80**, 485 (1999).
[7] D. Leong, M. Harry, K. J. Reeson, and K. P. Homewood, Nature **387**, 686 (1997).
[8] L. Tybeskov, K. L. Moore, D. G. Hall, and P. M. Fauchet, Phys. Rev. B **54**, R8361 (1996).
[9] E. Osveinbjörnsson and J. Weber, Appl. Phys. Lett. **69**, 2686 (1996).
[10] European Commission, Technology Roadmap, Optoelectronic interconnects for integrated circuits (Luxembourg: Office for Official Publications of the European Communities, Luxembourg, 1998).
[11] W. L. Ng, M. A. Lourenço, R. M. Gwilliam, S. Ledain, G. Shao, and K. P. Homewood, Nature **410**, 192 (2001).
[12] J. P. Hirth and J. Lothe (eds.), Theory of Dislocations (John Wiley & Sons, New York, 1982).

The Si(001)/C$_2$H$_2$ interaction to form a buffer layer for 3C-SiC growth

A. Goryachko[*,1], **Y. Yeromenko**[1], **K. Henkel**[1], **J. Wollweber**[2], and **D. Schmeißer**[1]

[1] Brandenburg Technical University Cottbus, Department of Applied Physics/Sensorics, Postfach 101344, 03013 Cottbus, Germany
[2] Institute for Crystal Growth, Department of Vapour Growth (IKZ), Max Born Str. 2, 12489 Berlin, Germany

Received 15 September 2003, accepted 22 October 2003
Published online 29 December 2003

PACS 68.35.Ct, 68.35.Fx, 68.55.Ac, 79.60.Jv, 81.15.Gh

The intial stages of Si(001) interaction with C$_2$H$_2$ were investigated with XPS and surface profilometry. The carbon enriched layer serves as a buffer between Si substrate and the 3C-SiC film aimed at growing on Si(001). Such parameters as substrate temperature and process duration come into a complicated interplay of C deposition, its reaction with Si and out-diffusion of the later from the substrate. Three distinct regimes, namely graphite dominated, carbide dominated and roughness dominated are found to result from this interplay.

© 2004 WILEY-VCH Verlag GmbH & Co. KGaA, Weinheim

1 Introduction The 3C polytype of SiC is especially desired for electron devices applications due to the highest electron mobility and saturated carriers velocity, but production of entire 3C-SiC wafers is not yet possible. However, 3C-SiC can be grown at relatively low temperatures on top of Si(001) substrates [1], thus making 3C-SiC/Si(001) combination interesting for commercial device production. Growing such films constitutes however a certain problem due to significant lattice constant mismatch. Therefore, a buffer layer is usually grown by C reacting with Si(001) surface, after which a main growth of 3C-SiC begins. To prevent Si outdiffusion from the substrate, this buffer layer must cover the Si surface as smooth and complete as possible.

There are many ways to grow a buffer layer on Si(001): Kitabatake [2] used conceptually the simplest option, namely the C flux in a temperature ramp window 400 °C – 1050 °C in the MBE system with UHV background. Kosugi et al. [3] and Crescenzi et al. [4] used C$_2$H$_4$ at 650 °C, Fuyuki et al. [5] C$_2$H$_2$ in a temperature ramp window of 400 °C – 970 °C, all in the UHV background. In this work we have investigated the buffer layer growth in the process of Nagasawa and Yagi [6]. It uses C$_2$H$_2$ diluted in H$_2$ (H$_2$/C$_2$H$_2$ = 10:1) at ~10^{-1} mbar as a C source, thus achieving high growth rates and having the highest relevance for commercially viable 3C-SiC/Si(001).

2 Experimental The growth was performed in turbo-pumped CVD system with base pressure of 10^{-6} mbar on direct current heated Si(001) substrates. The latter were treated with HF to remove the Si oxide before introduction into the chamber. Acetylene (C$_2$H$_2$) was diluted in a carrier gas (H$_2$) and supplied to the sample's surface at 10^{-1} mbar (H$_2$/C$_2$H$_2$ = 10:1). The X-ray photoelectron spectroscopy (XPS) can provide valuable information on the chemical state of C and Si species in buffer layer as well as detect the presence of impurities, e.g. oxygen. Furthermore, if performed with a synchrotron radiation of varying photon energy, the XPS can show depth distributions of the relevant species in the near-

---

[*] Corresponding author: e-mail: goryach@tu-cottbus.de, Phone: +49-(0)355-69-49-10, Fax: +49-(0)355-69-39-31

© 2004 WILEY-VCH Verlag GmbH & Co. KGaA, Weinheim

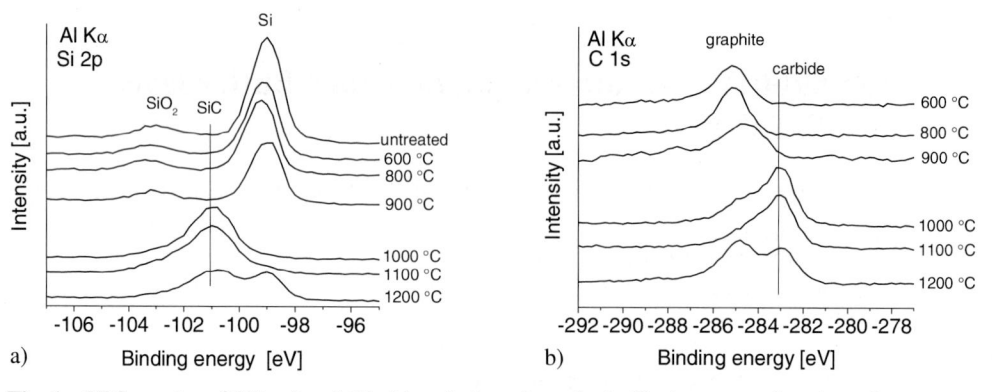

Fig. 1 XPS spectra of Si2p a) and C1s b) emissions from the buffer layer as a function of process temperature. Treatment duration was 5 min, H_2/C_2H_2 ratio was equal 10.

surface region (smaller photon energies produce surface specific spectra). The XPS measurements were performed with Al K_α or Mg K_α radiation as well as on the U49/2-PGM2 beam line at Bessy-II (Berlin, Germany) synchrotron radiation facility. The details of the latter experimental setup are given elsewhere [7]. The surface roughness was investigated with a Talystep profilometer supplied by Rank Taylor Hobson.

3 Results

Figure 1 demonstrates the influence of substrate temperature on buffer layer growth, which lasted for 5 minutes with H_2/C_2H_2 ratio of 10. The Si 2p and C 1s emissions obtained with Al K_α radiation are shown in Figs. 1a and 1b respectively. The onset of SiC at 1000 °C is clearly visible by appearing SiC-related components and reduction of Si substrate- and graphite-related components. The latter two show up again at 1200 °C due to oversupply of carbon and Si outdiffusion from the substrate (this leads to surface roughening and visibility of bulk Si in XPS). Optimum conditions should exist leading to maximum carbide/graphite ratio and minimum surface roughness. This is confirmed by a set of height profiles obtained on the surface by stylus profilometry technique (Fig. 2a), where a clear minimum in roughness is achieved for 1100 °C. Finally, Fig. 2b shows the partial contributions of carbide-related C and Si in C1s and Si2p signals respectively (calculated from the data of Fig. 1b) simultaneously with root mean square (RMS) roughness determined from the profiles in Fig. 2a. The maximum of 2.2 carbide/graphite ratio (as calculated from carbide and graphite contributions in C1s signal) is achieved at 1100 °C, which stands in perfect agreement with a minimum on the RMS roughness curve at the same temperature.

The depth distribution of phases was studied by XPS with synchrotron radiation of variable photon energy. Fig. 3 shows the Si 2p a) and C1s b) emissions obtained with different photon energies for

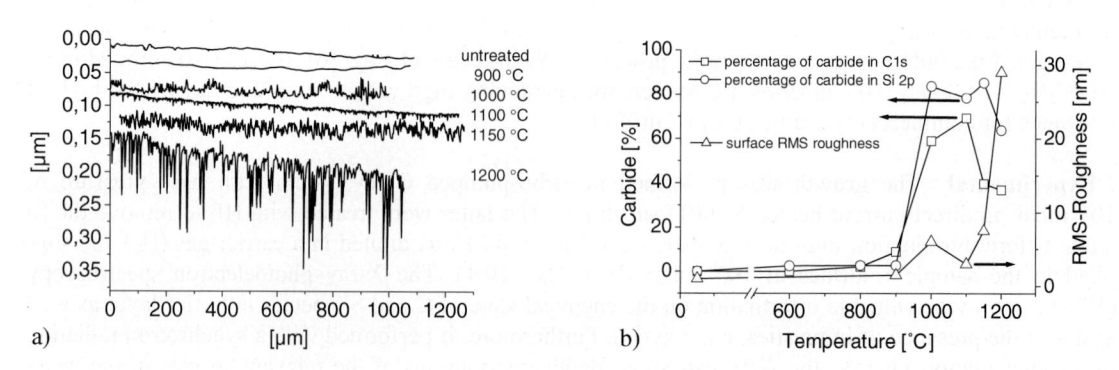

Fig. 2 a) Visualisation of height profiles along a straight line on the surface. b) Partial contribution of SiC-related carbon and silicon in the overall C1s and Si2p signals respectively; root-mean-square (RMS) roughness determined from lines profiles in a).

© 2004 WILEY-VCH Verlag GmbH & Co. KGaA, Weinheim

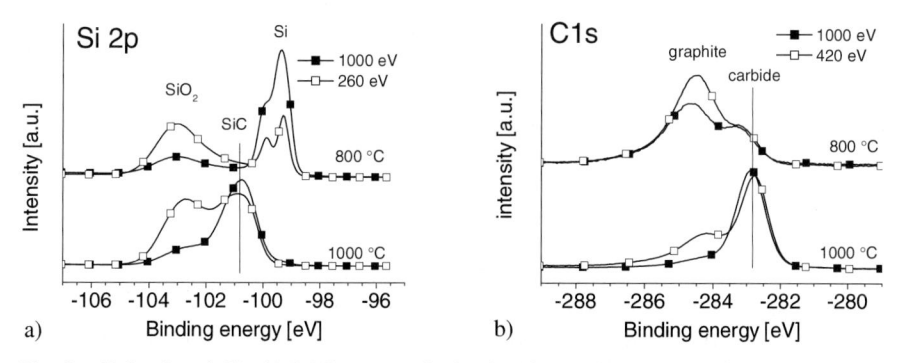

Fig. 3 Si 2p a) and C1s b) XPS spectra obtained with synchrotron radiation photon energy of 1000 eV, 420 eV, and 260 eV, after treatment at 800 °C and 1000 °C. Treatment duration was 5 min, H_2/C_2H_2 ratio was equal 10.

treatments at 800 °C and 1000 °C. Due to high sensitivity and resolution in our spectra we already observe the first traces of SiC in the C1s emission after 800 °C process. In this case carbon is already deposited on the surface, but the temperature is not high enough for effective reaction with Si. Therefore, a graphite is formed in a quantity larger than carbide. This situation reverses at 1000 °C, when there is a strong and clear SiC-related component both in spectra of Si2p a) and C1s b). Finally, in Fig. 4a we show the O1s/Si2p signals ratio plotted for various treatment temperatures performed on samples with preliminary HF treatment and without it. There is a clear tendency of reduction of overall O content in samples which were subjected to HF pre-treatment. The same holds for the part of oxidised Si component in the overall Si2p emission (Fig. 4a). Figure 4b shows the O1s/Si2p signals ratio plotted for various treatment durations at 900 °C and 1100 °C. We would also like to point out that we didn't find significant influence of H_2/C_2H_2 ratio on the oxygen content.

4 Discussion According to the data presented above we can distinguish three buffer layer growth regimes. Up to 900 °C the reaction between C and Si is very slow, therefore graphite is predominantly formed on the surface. Due to a very high sensitivity of our synchrotron-based XPS measurement we can detect the first traces of SiC already at 800 °C, but this SiC layer is still incomplete with a nominal thickness of less than 1 nm (Fig. 3b). Therefore, Si is still exposed to and oxidised by residual oxygen in the chamber and the ambient environment during transfer to analysis system (Fig. 3a). This leads to high O content in the buffer layer independent on process duration (Fig. 4b). We call this regime *graphite-dominated*. Starting at 1000 °C there is an intensive reaction leading to SiC layer formation which is

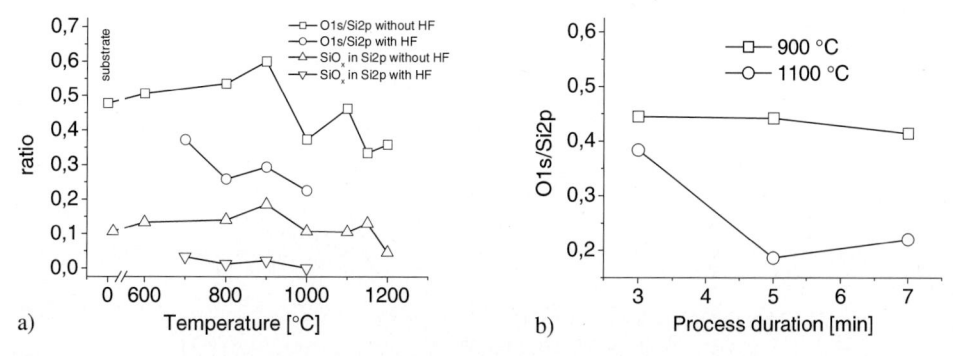

Fig. 4 a) Total O1s/Si2p signals ratio (with elemental sensitivity factors taken into account) and the fraction of oxidised component in the Si2p signal as functions of process temperature, with or without HF treatment; b) Total O1s/Si2p ratio (with elemental sensitivity factors taken into account) as a function of process duration without preliminary HF treatment.

more than 1 nm thick (Fig. 3). The graphite and SiO$_2$ still remain visible in surface-specific spectra of Fig. 3. Thus, we can conclude that graphite and SiO$_2$ are on top of the SiC and not embedded into it. At these temperatures for a small process duration there is still not enough carbide to continuously cover and protect the bulk Si from oxidation (Fig. 4b). However, if process duration is long enough, a complete SiC layer is formed. This regime is called *carbide-dominated*. Further increase of process temperature or treatment time produces an oversupply of carbon, leading to strong surface roughening and bulk Si oxidation (Figs. 2a, 4b). This regime is designated as *roughness-dominated*. The presence of initial SiO$_2$ layer on the surface hampers the reaction between Si and C$_2$H$_2$, leading to graphite accumulation and Si outdiffusion. In that case the surface becomes rough giving Si more chances to get oxidised. If, however, the initial SiO$_2$ was removed by HF etching, the reaction between Si and C$_2$H$_2$ starts immediately without any stopper. Therefore HF etching is important step before the main treatment, leading to reduction of overall O content on the Si(001)/SiC interface (Fig. 4a).

5 Conclusion In conclusion we have investigated an intricate interplay between C deposition on the Si(001) substrate in the H$_2$/C$_2$H$_2$ (10:1) flow at 10^{-1} mbar, outdiffusion of Si from the substrate, reaction of C with Si, and oxidation of Si by the rest oxygen in the CVD chamber. Three regimes of the buffer layer growth were found: graphite-, carbide-, and roughness-dominated. Further 3C-SiC growth experiments are needed to determine the buffer layer on top of which the best quality film can be grown. A very promising option seems to be a buffer layer grown for 5 minutes in the carbide-dominated regime at 1100 °C. This treatment leads to a SiC layer with minimum roughness, and graphite or oxygen impurities.

Acknowledgements The experimental help of G. Beuckert and Y. Burkov as well as technical assistance of the Bessy-II staff is widely acknowledged.

References

[1] Y. Narita, T. Inubushi, M. Harashima, K. Yasui, and T. Akahane, Appl. Surf. Sci. **216**, 575 (2003).
[2] M. Kitabatake, phys. stat. sol. (b) **202**, 405 (1997).
[3] R. Kosugi, Y. Takakuwa, K.-S. Kim, T. Abukawa, and S. Kono, Appl. Phys. Lett. **74**, 3939 (1999).
[4] M. De Crescenzi, R. Bernardini, M. Cardella, R. Grunnella, P. Castrucci, R. Pizzoferrato, and M. Casalboni, Surf. Sci. **521**, 57 (2002).
[5] T. Fuyuki, T. Hatayama, and H. Matsunami, phys. stat. sol. (b) **202**, 359 (1997).
[6] H. Nagasawa and K. Yagi, phys. stat. sol. (b) **202**, 335 (1997).
[7] P. Hoffmann, D. Schmeißer, G. Roters, and Z. Nenyei, Thin Solid Films **428**, 216 (2003).

Temporal response of CVD diamond detectors to modulated low energy X-ray beams

G. Conte[*,1], **M. C. Rossi**[1], **S. Salvatori**[1], **F. Spaziani**[1], **Y. Avigal**[2], **R. Kalish**[2], **P. Ascarelli**[3], and **D. Trucchi**[3]

[1] INFM and Electronic Engineering Department, University "Roma Tre", Via Vasca Navale, 84 – 00146 Rome, Italy
[2] Physics Department and Solid State Institute, Technion, Haifa 32000, Israel
[3] National Research Council (CNR-IMIP), Via Salaria km 29, 300 – Monterotondo Scalo, Rome, Italy

Received 15 September 2003, revised 3 November 2003, accepted 3 November 2003
Published online 12 January 2004

PACS 61.72.Mm, 68.55.Ln, 73.40.Ty, 73.50.Pz

Metal–diamond–metal vertical structures were realized and tested under a modulated 8.06 keV X-ray beam. AC photocurrent versus applied voltage curves were used to evaluate the mobility-lifetime product of collected charges whereas the density of states was achieved from the modulated photocurrent in the 7–900 Hz range. The influence of the modulation frequency and electric field strength on the device performance is discussed in the framework of multi-trapping carriers transport.

© 2004 WILEY-VCH Verlag GmbH & Co. KGaA, Weinheim

1 Introduction Polycrystalline CVD diamond has been used to realize UV photoconductive devices [1] and recently proposed for developing dosimeters for radiotherapy [2]. The interest for such a material stems on the ability to keep under control the "electronic quality" of deposited films. That is, it is possible to tailor the average microstructure of the system grain/grain boundary by using the control on the CVD growth parameters. On the other hand, it is well known that the grain boundary may produce a continuous distribution of gap states and thus to control the electronic properties. As a consequence, the development of detectors sensitive to over gap carriers generation might suffer the presence of shallow traps that reduce the speed of response, and recombination centres that depress the device signal. Among techniques proposed to study the density of states in semiconductors, modulated photocurrent (MPC) has been demonstrated useful both in amorphous and polycrystalline materials [3, 4]. To the aim of developing dosimeters for radiotherapy and infer the influence of defect states on the photoconductive processes under real bulk and over gap generation, we have undertaken the analysis of metal-diamond-metal (MDM) vertical structures by using a modulated 8.06 keV X-ray beam. In this paper we report on the AC photocurrent, mobility-lifetime product and distribution of electronic states within the diamond bandgap as evaluated on the basis of MPC theory.

2 Experimental 40–60 µm thick diamond films deposited by a hot-filament (HF) or a microwave (MW) assisted CVD technique have been used in this work. Dual side silver contacts have been made on the top and bottom faces of the diamond samples by thermal evaporation. The study under X-ray irradiation has been performed by using the K_α line (8.06 keV) of a Ni filtered Cu source. The modulation frequency has been changed by using a modified Stanford SR540 chopper unit. The AC component of the photocurrent has been amplified in the mV range with an EG & G 5182 I/V converter and measured by using an EG & G 5207 phase sensitive amplifier.

---

[*] Corresponding author: e.mail: gconte@ele.uniroma3.it, Phone: +39 06 5517 7268, Fax: +39 06 5579 078

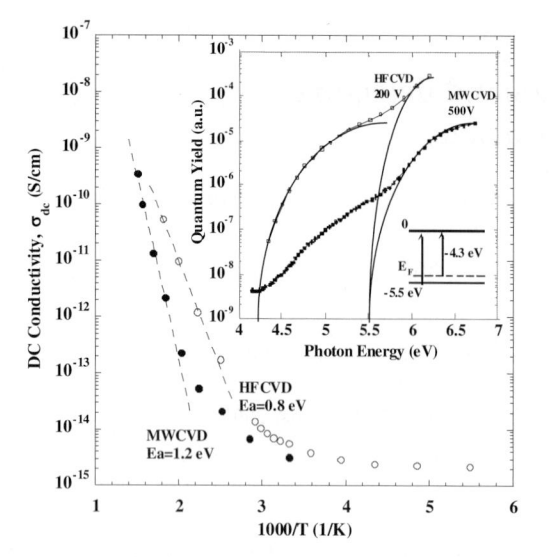

Fig. 1 Arrhenius plot of DC conductivity in the dark of both MDM structures. Dashed lines indicate the region where E_a was calculated. *Inset*: Photoemission Quantum Yield vs photon energy of the same samples. Lines are a visual aid to evidence the photo-ionisation thresholds.

Fig. 2 X-ray photocurrent of the HF specimen versus the applied voltage at two different modulation frequencies. Continuous lines are the best fit with Hecht relation. *Inset*: evaluated $\mu\tau$ vs frequency.

3 Results and discussion

Current versus voltage (I–V) measurements in the dark showed the ohmic and symmetric behavior of both devices up to 20 kV/cm. However, a different non-linear increase was observed at higher field strength. HF showed Frenkel–Poole and Poole's transport mechanisms [5] whereas space charge limited transport was evidenced in the MW one. The DC conductivity Arrhenius plot is shown in Fig. 1. It is noteworthy that both devices approach the same conductivity value around 2×10^{-15} S/cm at low temperature, whereas a different exponential trend is observed at higher T values. The calculated activation energy, E_a, is 0.8 ± 0.1 eV and 1.2 ± 0.1 eV for the HF and MW specimen, respectively. In the latter case, 1.2 eV is in good agreement with literature results and approaches values also observed in natural diamond, while the low E_a value calculated for HF doesn't corresponds to the Fermi level position, E_f, but is representative of a large density of defect states near the Fermi level. The lower ionization threshold evaluated from photoemission quantum yield measurements on this sample (see the inset of Fig. 1) leads to a value around 4.3 eV. It is clear that the threshold at 4.3 ± 0.1 eV has to be associated to E_f. A large density of occupied states is evident over the valence band in the case of HF device in respect to the MW one. The low E_a value calculated for HF, thus, may represent the result of a band-bending at the grain surface as a consequence of holes accumulation at the grain boundaries. The I_{ph}–V curves of the HF device at two different chopping frequencies are reported in Fig. 2. The signal saturation is achieved at higher voltages as the chopping frequency is increased. The signal saturation can be interpreted on the basis of a reduction in the transit time, with the bias increasing, leading to a more efficient charge collection. The collection efficiency, η_c, and thus the mobility-lifetime product, $\mu\tau$, has been evaluated by using the Hecht relation

$$I_{ph}(V) = I_s \eta_c(V) = I_s \frac{\mu\tau}{d^2} V \left[1 - \exp\left(-\frac{d^2}{\mu\tau V}\right)\right] \quad (1)$$

where I_S is the saturation photocurrent and d the distance between collecting contacts. Expression (1) was used to fit experimental data of I_{ph} measurements as a function of the modulation frequency of X-ray beam. The evaluated $\mu\tau$ product values are reported as inset of Fig. 2. Instead of the film thickness, an

© 2004 WILEY-VCH Verlag GmbH & Co. KGaA, Weinheim

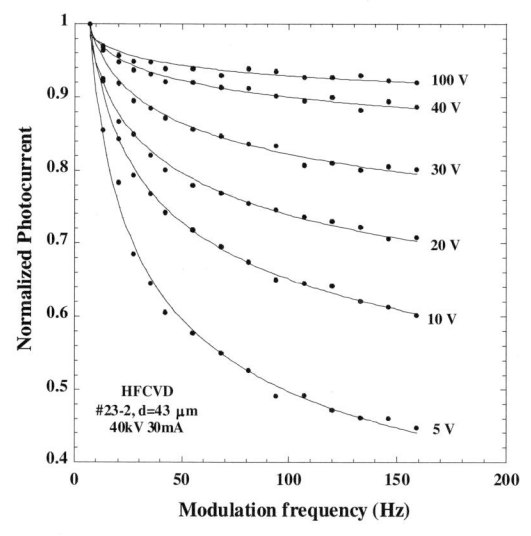

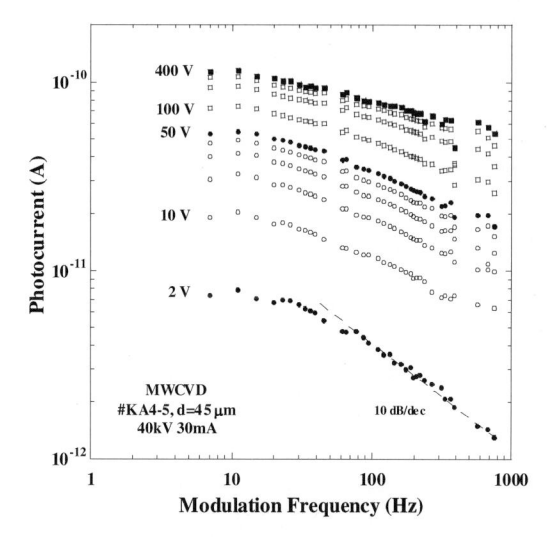

Fig. 3 Normalized photocurrent versus chopping frequency of the HF device at different voltages.

Fig. 4 Photocurrent vs frequency of the MW device as a function of the applied voltage.

effective distance $d = 10$ μm – the average grains dimension – estimated by SEM and AFM observations was assumed to perform the fit. The $\mu\tau$ product of HF device decreases with the frequency according to the $\mu\tau = 2.7 \times 10^{-7}/f^{0.3}$ relationship addressing to a dispersive carriers transport possibly associated to a continuous distribution of electronic states within the gap. The presence of shallow traps is confirmed by the analysis of I_{ph}–f curves in Fig. 3. The decay curve shows that with increases in frequency the photo-generated carriers gradually become trapped until a constant and non-zero value, corresponding to equal rates of generation and recombination, is reached.

At first, the decay is fast and possibly controlled by band-to-band recombination mechanisms, depending only on the lifetime of majority carriers (holes). The slow decay region is believed to be due to thermal emission of trapped holes into the valence band. The frequency dependent photocurrent of the MW device is reported in Fig. 4. It is worth noting the plateau at low frequency and the exponential decrease at higher values with a rolloff of 10 dB/dec, as expected for a photoconductor controlled by surface recombination. Indeed, carriers trapping appears to be less efficient in this device in respect to the HF one. I_{ph}–V characteristics (not shown) have been used to evaluate the $\mu\tau$ values considering the presence of ohmic contacts [6] and by using the real device thickness for d. As for HF specimen the mobility-lifetime product decreases with a frequency increase. The dependence is smaller then that observed on the HF device and the $\mu\tau = 7.6 \times 10^{-7}/f^{0.1}$ relationship has been evaluated.

The frequency dependent photoconductivity has been then used to evaluate the DOS on the basis of negligible re-trapping for both samples. In the high-frequency regime, where photocurrent is mainly determined by the interaction of free carriers with localized states at E_ω, for which the emission time equals the period of the signal, MPC theory gives the following expression [3]

$$\frac{v\sigma N_t(E_\omega)}{\mu} = \frac{2e}{\pi kT} \Phi_{ac} \frac{FA}{I_{ph}(\omega)} \sin(\varphi) \tag{2}$$

where e is the elementary charge, v the thermal velocity, σ denotes the capture cross-section, k the Boltzmann constant, Φ_{ac} the alternative photon flux, F the electric field, A the device area and φ the photocurrent phase shift. MPC and phase shift curves achieved at lower voltages have been used to evaluate the reduced DOS of both devices as shown in Fig. 5. The energy reported in abscissa has been estimated by using the relationship $E - E_v = kT \ln(v_0/\omega)$, being v_0 the attempt to escape frequency, assumed equal to 10^{12} Hz, and E_v the valence band edge.

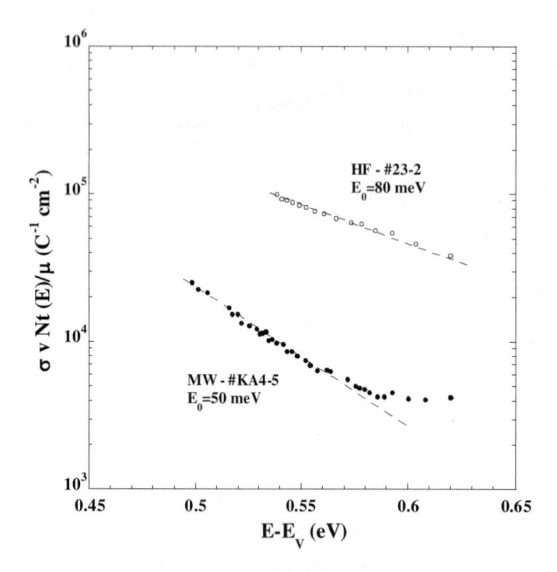

Fig. 5 Reduced density of states versus energy of tested devices. Dashed lines are the exponential fit to evaluate the Urbach tail slope E_0.

HF sample shows a high DOS value compared to the MW one. This result is in good agreement with the occupied DOS evaluated by photoemission measurements (see the inset of Fig. 1). Moreover, Urbach tail slopes equal to 50 and 80 meV, for MW and HF respectively, have been estimated. These values are in the range reported for similar polycrystalline specimens and in good agreement with the observed $\mu\tau$ trend. Indeed, smaller the DOS higher the $\mu\tau$ value and smaller the dependence on frequency. If we use $\mu = 150$ cm^2/Vs (estimated by independent steady state and transient UV photoconductivity measurements), $v = 10^7$ cm/s and $\sigma = 10^{-16}$ cm^{-2} we can calculate the trap density $N_t(E)$ assuming an energy independent σ, at least in this restricted energy range. Values in the $5 \times 10^{14} - 2 \times 10^{15}$ cm^{-3}eV^{-1} range have been estimated for MW sample. The plateau starting around 0.58 eV might indicate the grain interior DOS whereas the contribution of grain boundaries may contribute to the exponential tails.

Acknowledgement This work has been carried out in the framework of EU Project G5RD-CT-2001-00603.

References

[1] S. P. Lansley, O. Gaudin, H. Ye, N. Rizvi, M. D. Whitfield, R. D. McKeag, and R. B. Jackman, Diam. Relat. Mater. **11**, 433 (2002).
[2] S. Ramkumar, C. M. Buttar, J. Conway, A. J. Whitehead, R. S. Sussman, G. Hill, and S. Walker, Nucl. Instr. Meth. A **460**, 401 (2001).
[3] H. Oheda, J. Appl. Phys. **52**, 6693 (1981).
[4] R. Herberholz, T. Walter, and H. W. Schock, J. Appl. Phys. **76**, 2904 (1994).
[5] G. Conte, M. C. Rossi, S. Salvatori, P. Ascarelli, E. Cappelli, and D. Trucchi, J. Appl. Phys. **93**, 6078 (2003).
[6] S. Salvatori, M. C. Rossi, and F. Galluzzi, Carbon **37**, 811 (1999).

Photonic applications of lithium niobate crystals

L. Arizmendi[*]

Dept. Física de Materiales, Universidad Autónoma de Madrid, 28049 Madrid, Spain

Received 15 September 2003, accepted 19 September 2003
Published online 12 January 2004

PACS 42.70.Mp, 42.70.Nq, 78.20.–e, 81.40.Tv

In this paper the most important properties of lithium niobate crystals for photonic applications are reviewed. We will summarize how acting on the stoichiometry, sample structure, doping and domain structure of the crystals, these properties are governed, and can be taylored for each specific application. Finally we present a review of photonic applications in a wide spectrum of fields as lasers and non-linear optics, optical communications, optical memories, and diffractive optics.

© 2004 WILEY-VCH Verlag GmbH & Co. KGaA, Weinheim

1 Introduction

Lithium niobate is one of the most used crystalline dielectric materials at the present days. This is because this compound presents a very interesting combination of properties and characteristics that make possible to tailor its behavior for useful devices. The basic material properties of lithium niobate are by themselves very promising for applications. It is a material that, not being present in the nature, permits to be grown in quite large crystals by the use of a not much difficult technique. Then, this is a non expensive starting material for devices. Most common lithium niobate crystals present a large concentration of intrinsic defects. These crystals can be also doped with not few amounts of a good number of different impurities. The control of both intrinsic point defects and impurities offers a wide range of variation of responses. Another not less important factor to be considered from the practical aspect in this overview of the material is the possibility of producing different sample structures. Bulk, fibers and thin films have to be completed with surface waveguide structures, non-equivalent crystallographic orientations and ferroelectric domain structures. This exceptional situation provides the frame for many present and future uses in quite different physical branches. We shall see that the photonic applications are privileged in this context.

In this paper we will go from the relevant initial properties of lithium niobate to the more recent photonic applications. Some potentialities for near future uses will be also pointed out. In part 2 the basic material properties are analyzed attending principally to photonic aspects. This part is also devoted to review the role played by stoichiometry, point defects and impurities on these properties. Next, in part 3 we enter in the field of structural forms for optical devices. Part 4 is devoted to show some examples of currently demonstrated photonic applications. It is not intention of this paper to make an extensive review of all the reported applications of this singular material, but present some examples to manifest the vast and variedness spectrum of photonic areas that benefit of the results of this highly technological material. Finally a consideration about the near future possibilities of this material will be outlined in part 5.

2 Properties of lithium niobate relevant for photonic applications

The starting point is to consider that lithium niobate in its crystalline form is a solid material chemically very stable at room temperature, quite insensitive to humid surroundings. Moreover, it has enough

---

[*] e-mail: luis.arizmendi@uam.es, Phone: +34 91 379 5026, Fax: +34 91 397 8579

Table 1 Summary of basic properties of lithium niobate crystals.

general properties:	
melting point:	~1260 °C
crystal symmetry:	trigonal, point group 3 m
method for single crystal growth:	Czochralski, large crystals
hardness	~5 Mohs
stoichiometry:	nonstoichiometric, ~6% Li deficiency easy to dope in high concentration
ferroelectricity	Curie temperature T_C = 1150 °C spontaneous polarization P_s(RT) ~96 C/m^2 180° domains
piezoelectric	
piroelectric	
optical related properties:	
transparency region:	VIS/NearIR (~350 nm - ~5 µm)
optical anisotropy:	uniaxial, c-axis
refractive indices:	n_o = 2.286, n_e = 2.203. (at 632.8 nm)
optical homogeneity	$\Delta n \sim 5 \times 10^{-5}$
optical effects:	acoustooptic electrooptic second order nonlinearity bulk photovoltaic effect optical damage – photorefractive effect

hardness to be regularly manipulated [1]. The most extended method to grow single crystals of this compound, the Czochralski method and its variants, permits the fabrication of very homogeneous crystal even of several kilograms and more that three inches in diameter. A summary of the main crystal properties is presented in Table 1. The material is ferroelectric below a very high Curie temperature ($T_C \sim 1150$ °C). Ferroelectric domains with 180° orientations are present in the samples. In many applications a single domain or another specific domain structure will be required.

The crystalline structure of this material at room temperature corresponds to the trigonal 3 m point group. This makes the crystal to be optically uniaxial, with two refractive indices n_e and n_o. Both indices have values exceeding of 2 in all the transparency region, which makes in many cases necessary to fit antireflecting coating layers on the optical surfaces. The optical homogeneity of commercial crystals is as good as $\Delta n \sim 10^{-5}$ [2]. Crystal symmetry presents a lack of inversion center which is a necessary condition for presenting important properties such as linear electrooptic effect.

The pure (and not reduced) material is transparent from the band gap edge absorption at about 320 nm (~3.9 eV) up to the first infrared vibrational absorptions at a wavelength of about 5 µm (0.25 eV), covering all the visible and near infrared spectral regions. This provides a wide spectral window for photonic applications.

A relatively high acustooptic effect combined with the piezoelectric effect gives the possibility of produce surface acoustic ultrasonic waves (SAW) to control light beams. The electrooptic effect, with moderately high coefficients, is a very useful property for devices where an electric field serves to control light propagation along the medium. Beside this, a not negligible second order non-linear optical effect also constitutes a valuable property, giving the possibility of producing light waves of new frequencies [3], as will be commented below.

Another very interesting property of this material is the bulk photovoltaic effect [4]. A feature expressed by some impurities in this material consisting in the generation of photoinduced currents or voltages under certain light incidence. This rare property has some relevance in quite interesting applications presented in part 5.

For an important group of lithium niobate applications, the so called 'optical damage' is an undesired effect that encouraged many researchers during years. This consists in a semi-permanent refractive index change occasioned by the light beams passing through the samples. It is produced even at relatively low intensities and gives rise to beam distortion. This property is valuable for other applications, and then is called photorefractive effect.

Additionally to the above mentioned basic properties, we have to take into account the crystal composition or stoichiometry. This aspect has a relevant influence on the values of many material coefficients, and plays an important role on the incorporation of impurities in the crystal structure. This is a topic of very high interest in lithium niobate.

2.1. Stoichiometry of lithium niobate crystals

The phase diagram of the system $LiO_2-Nb_2O_5$ (Fig. 1) [5] presents a solid solution region close to the 50% of the component cations. The solid-liquid curve shows an eutectic at 48.45% mol. of Li [6]. Starting a crystal growth process with this melt composition, known as congruent, the crystal grows exactly with the same cationic ratio, and the liquid composition remains unaltered along the process. This has been along the material history the preferred composition for most research and applications. Starting to grow crystals with another liquid composition leads to compositional inhomogenity along the pulling direction [1].

This important defect of lithium ions in congruent lithium niobate derives directly in the presence of point defects, which have to accommodate this composition to the crystalline structure. Much effort has been done to investigate this point. After some speculations and with not very clear experimental facts, it seems very generally accepted that a main defect is the niobium ion located in a lithium position [7], niobium antisite, or $Nb^{5+}{}_{Li}$. The existence of this defect forces to the presence of others for charge neutrality of the material. Nowadays still there is not unanimity about the nature of the charge compensating defects [8]. In any case, the congruent crystal is quite internally disordered, leading to spectroscopic impurity broad bands and peaks. A large amount of impurities can accommodate in this spongelike structure. Impurity concentrations of several percents are accepted in some cases.

Several techniques have been used to obtain crystals of a composition as close as possible to the stoichiometry. This was stimulated by the intention of reduce defects and internal disorder, trying to avoid optical damage, and gaining in optical quality. The Li content of samples can be raised by prolonged

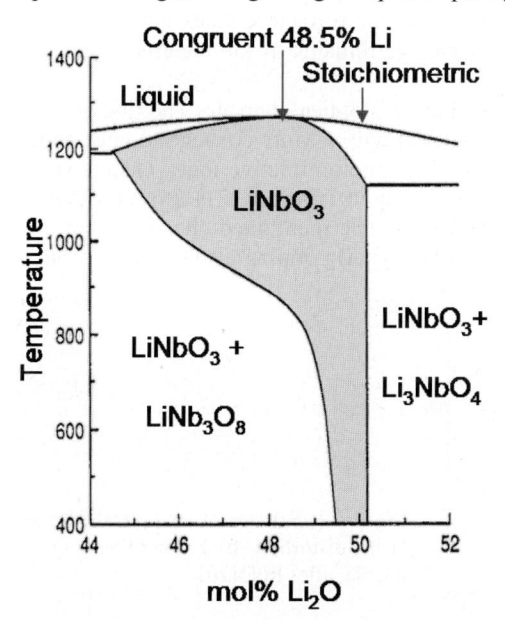

Fig. 1 Phase diagram of the $LiO_2-Nb_2O_5$ system (after Ref. [5]).

vapor transport equilibrium (VPE) annealing in Li-rich atmospheres at high temperature (~1100 °C). Samples with more than 49.9% of Li have been obtained by this technique [9]. With regard to single crystal growth, Kitamura et al. [10] have modified the Czochralski method by means of a double crucible to increase the Li content and avoid crystal inhomogeneity. The method is rather complicate and does not produce very stoichiometric crystals. Another bulk method using the addition of more than 6% of K_2O to the melt provides crystals very close to 50% Li and the incorporation of potassium to the crystal is negligible [11, 12]. It has been observed that using a flux of 8 mol% K_2O the crystals grow directly in the paraelectric phase, and starting with a single domain seed produced a single domain crystal [13].

Stoichiometric crystals obtained by the last mentioned method and doped with small amounts of impurities present very much narrow EPR lines than congruent samples [14], indicating more uniform impurity surroundings and less intrinsic defects and crystalline disorder in the stoichiometric ones. Moreover, the coercitive electric field for domain inversion is reduced drastically in these crystals. The coercitive field value for congruent samples is about 22 kV/mm, which frequently makes impossible to produce single domain samples at room temperature because of dielectric breakdown. In stoichiometric samples this field goes down to ~3 kV/mm [15]. This is also very interesting for nonlinear applications where periodic domain structures with spacing of few micrometers are needed. Another interesting fact is that stoichiometric crystals present reduced optical damage compared with congruent ones [16, 17]. It has been also observed in crystal doping that the impurity solubility limit and the segregation factor of many impurities become lower in stoichiometric samples [18].

2.2. Optical damage resistance

Optical damage is manifested when a light beam passes through a lithium niobate sample [19]. The exit beam presents a clear distortion more extended in parallel to the crystal c-axis (Fig. 2). A more detailed inspection of the beam results in that the extraordinary polarized light component is clearly more distorted than the ordinary part of the beam [20]. The distortion is due to an induced change in the refractive indices, more important in the extraordinary one. This change is not induced by a nonlinear optical effect because it is not instantaneously established and remains some time after the illumination is removed. At the present this photorefractive effect is considerably well understood [21]: the light beam produces the optical ionization of some uncontrolled donor impurities, the released charge moves by some transport process (diffusion, drift or photovoltaic) to the dark part of the sample where it is trapped by an acceptor center. The spatial distance between ionized donor and charged acceptor give rise to an internal electric field. The electric field by means of the electrooptic effect produces the refractive index change. As far as the charges are spatially separated the non uniform index change remains in the material. This is a very harmful effect for many optical applications.

By fortune some impurities have been found to markedly reduce this optical damage. The first discovered impurity having such effect is magnesium. Doping the material with a MgO concentration of about 5% or greater the optical damage is considerably reduced [22]. The photorefractive index change is reduced about 10 times and the optical erasure rate is increased also about 10 times [21]. It seems that in these crystals the Fe^{3+} does not act as electron acceptor center [23]. The mentioned changes occur in a stepwise form around a Mg^{2+} concentration from 4 to 6 mol% [21, 24]. Mg doping also influences many

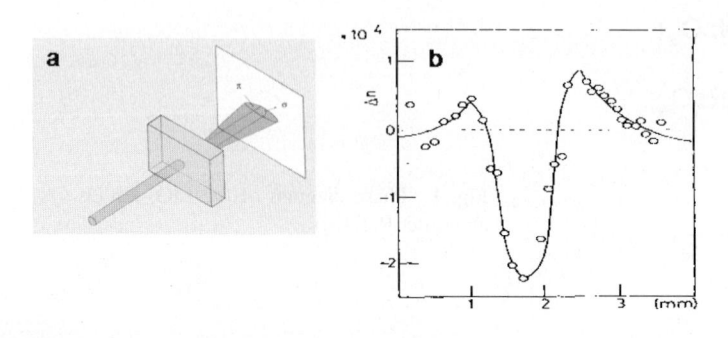

Fig. 2 a) Scheme of the optical damage beam distortion. b) Extraordinary index profile, after Ref. [20].

other properties of lithium niobate. The optical damage reduction obtained makes this material suitable for nonlinear optical applications [23]. In fact the optical damage threshold for 10 ns pulses of 1064 nm light passes from 250 kW/cm^2 to 5 GW/cm^2 [2].

Other impurities have been found to produce optical damage resistance similar to that happened with Mg^{2+}. A concentration of Zn greater than the 6 mol% threshold produces the same effect of Mg doping [25]. Zn enters in the lithium niobate lattice as Zn^{2+}. The trivalent impurities Sc^{3+} and In^{3+} also act as optical damage inhibitors [26, 27]. In these cases the threshold concentration is quite reduced, only about 1.5 mol% is required. The valence of the impurity seems to play an important role in this respect. Very recently a new optical damage impurity has been reported [28]. This is the tetravalent hafnium ion, Hf^{4+}, in concentration as low as 0.5 mol%.

Most nonlinear optical devices, and specially those which have to support very intense optical fields as in waveguide cases, are constructed on the basis of lithium niobate doped with one of these optical damage inhibitor impurities. Mg dopant has been the best studied of these impurities and the most accepted for practical uses.

2.3. Acoustooptic effect

The basic acoustooptic effect consists in that a sound wave travelling through the material induces a moving periodic refractive index change, i.e. a moving refractive index grating. A light beam propagating at the proper Bragg condition direction will be diffracted by such a grating. The acoustooptic figure of merit M is a measure of the refractive index change produced by the sound in the material. M is related to the material coefficients by [29]

$$M = \frac{n^6 p_{eff}^2}{\rho v_S^3}, \tag{1}$$

where n is the material average refractive index, p_{eff} is its effective photoelastic coefficient which for lithium niobate and common orientation is about 0.15. ρ is the mass density, and v_S is the sound velocity in the material. For lithium niobate ultrasonic acoustic waves the sound velocity has a value of about 4×10^3 m/s [30]. The resulting value of M is $\sim 9 \times 10^{-15}$ in MKS units, which is of the same order but a little larger than for other optical crystals [29].

The diffraction efficiency of a light wave at the bragg angle is given by [29]

$$\eta \cong \sin^2\left(2.2 \frac{L}{\lambda} \sqrt{M I_{acou}}\right), \tag{2}$$

where L is the interaction length and I_{acou} the acoustic wave intensity. Values close to unity can be obtained with the required interaction length and acoustic intensity. This dependence on the acoustic intensity can be useful in light intensity modulation devices [31]. The acoustic wave is generated from an

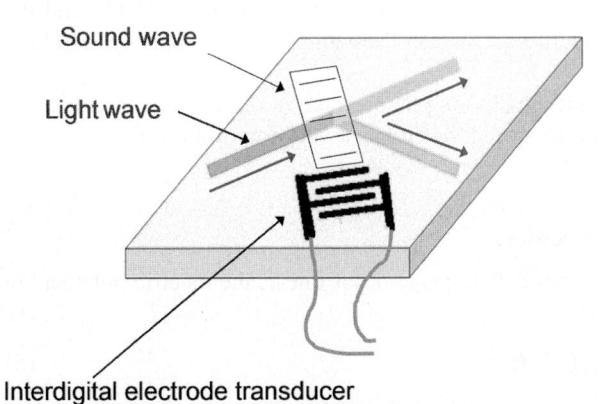

Fig. 3 Interdigital electrode transducer for surface acoustic wave (SAW) generation in lithium niobate.

electric signal by means of an interdigital electrode transducer, as shown in Fig. 3, using the converse piezoelectric effect of the lithium niobate.

As a consequence of the moving character of these gratings, the diffracted beam is frequency shifted in the amount of the sound wave frequency. This has been utilized to fabricate frequency modulators and shifters. The angle of diffraction also increases with the acoustic frequency. Using RF frequencies, this property can be used to produce high speed light beam deflectors, scanners and multiport switches [31]. Some example acoustooptic practical devices based in these features will be introduced in section 4.2.

2.4. Linear electrooptic effect

The linear electrooptic effect, also known as Pockels effect, is a linear change in the refractive index ellipsoid induced by an applied electric field. This change is produced by the first power of the field, in contrast with the quadratic or Kerr electrooptic effect. The change in the refractive index ellipsoid is expressed as

$$\Delta(1/n^2)_{ij} = \sum_k r_{ijk} E_k \qquad (3)$$

where E_k are the components of the electric field and r_{ijk} represents the electrooptic tensor which nonzero independent elements are determined by the crystal symmetry of the material. For example, the extraordinary index change derived from (3) when the electric field is applied along the crystal symmetry axis (z-axis), is $\Delta n_e = -\frac{1}{2} n_e^3 r_{33} E_3$. The non vanishing electrooptic tensor elements and their values for lithium niobate are listed in Table 2. These values are similar to those presented by other optical materials such as KH_2PO_4 (KDP), but substantially lower that those measured in others like $BaTiO_3$ [29]. In any case, they are high enough for many applications. The electrooptic applications will be described in section 4.3.

2.5. The bulk photovoltaic effect

This effect consists in the presence of photoinduced electric currents and voltages in the bulk dielectric material. It appears when the incident light ionizes impurities in the material. These ionized charges move in a particular direction inside the crystal without the presence of an external electric field. If the sample is uniformly illuminated and kept in open circuit, then a voltage between the faces is observed. In short-circuit a current is passing through the sample. This current density can be described by the following linear expression [4]:

$$j_i = \sum_{jk} \alpha G_{ijk} I e_j e_k^* . \qquad (4)$$

Here α is the absorption coefficient at the corresponding wavelength, I is the light intensity, e_j and e_k^* are the light polarization unit vectors, and finally G_{ijk} is the photovoltaic tensor, also known as Glass tensor. G_{ijk} nonvanishing components depend on the crystal symmetry of the material. The values depend on the specific crystal and the impurity involved. In Table 2 the nonvanishing elements for $LiNbO_3$, and the corresponding values for this crystal with Fe impurity are listed. The values for Cu impurity are about one forth of those obtained for Fe [32].

This particular property has interesting consequences in the photorefractive response of the material. No devices based directly in this effect have been proposed, but it has a very important influence in the photorefractive applications presented in section 4.5.

2.6. Second order nonlinear optical properties

Considering that the material response to electric fields is in general not linear, the electric polarisation can be expanded in powers of the field in the form:

$$p_i = \sum_j \varepsilon_0 \chi_{ij} E_j + \sum_{jk} 2 d_{ijk} E_j E_k + \sum_{jkl} 4 \chi_{ijkl}^{(3)} E_j E_k E_l + ... \qquad (5)$$

Table 2 Electrooptic, photovoltaic, and second order nonlinear optic tensors of lithium niobate.

Linear electrooptic effect: $\Delta(1/n^2)_{ij} = \sum_k r_{ijk} E_k$

Electrooptic tensor: $r_{ijk} = \begin{pmatrix} 0 & -r_{22} & r_{13} \\ 0 & r_{22} & r_{13} \\ 0 & 0 & r_{33} \\ 0 & r_{51} & 0 \\ r_{51} & 0 & 0 \\ -r_{22} & 0 & 0 \end{pmatrix}$

Values: r_{13} = 10 pm/V, r_{33} = 32.2 pm/V, r_{22} = 6.7 pm/V and r_{51} = 32.6 pm/V for static electric field and light of 633 nm at RT [57, 58].

Bulk photovoltaic effect: $j_i = \sum_{jk} \alpha\, G_{ijk}\, I\, e_j\, e_k^*$,

Photovoltaic tensor $G_{ijk} = \begin{pmatrix} 0 & 0 & 0 & 0 & G_{15} & -G_{22} \\ -G_{22} & G_{22} & 0 & G_{15} & 0 & 0 \\ G_{31} & G_{31} & G_{33} & 0 & 0 & 0 \end{pmatrix}$

Values for LiNbO3:Fe with light of 3 eV are [26]:
G_{22} = 0.3 × 10^{-9} A cm/W; G_{31} = 3.3 × 10^{-9} A cm/W; G_{33} = 2.7 × 10^{-9} A cm/W
For Cu doped samples the coefficients are about one forth of those for Fe.

Second order nonlinear optical effect:
Second order nonlinear polarization: $P_i^{\omega_3} = d_{ijk}^{\omega_3,\,\omega_1,\,\omega_2} E_j^{\omega_1} E_k^{\omega_2}$
Second order nonlinear susceptibility tensor:

$d_{ijk} = \begin{bmatrix} 0 & 0 & 0 & 0 & d_{15} & -2d_{22} \\ -d_{22} & d_{22} & 0 & d_{15} & 0 & 0 \\ d_{31} & d_{31} & d_{33} & 0 & 0 & 0 \end{bmatrix}$

Some values [2, 23]: For SHG d_{eff} = 5.7 pm/V or ~14.6 d_{36}(KDP) @1300 nm
For OPO d_{eff} = 5.5 pm/V or ~13.6 d_{36}(KDP) @1064 nm
For QPM-SHG d_{eff} = 17.6 pm/V or ~45.0 d_{36}(KDP)
Index reduction: 11–1, 22–2, 33–3, 23:32–4, 31:13–5, 12:21–6 [57]

The first right term is the usual linear dependence. The second term corresponds to the second order or quadratic nonlinearity, and next is the third order dependence. It is very straightforward to find that only noncentrosymmetric crystals, as it is the case of lithium niobate, can possess a nonvanishing d_{ijk} tensor [29]. Usually in these materials the second order processes are much more important in magnitude than those of higher orders for the moderate electric field commonly present in the materials.

The nonvanishing d_{ijk} second order nonlinear tensor elements for lithium niobate are presented in Table 2. The values depend on the frequencies involved. Some values of the effective coefficients for specific applications are also given and compared with those for KDP. As it can be seen, the second order coefficients of lithium niobate are quite competitive.

A number of nonlinear optical effects are observed as a result of the second order interaction. For accounting of this effects it is common to use the following nonlinear polarization expression: $P_i^{\omega_3} = d_{ijk}^{\omega_3;\,\omega_1,\,\omega_2} E_j^{\omega_1} E_k^{\omega_2}$. When the two field components are of the same wave of frequency ω, the nonlinear polarization will have a frequency 2ω, giving rise to the generation of a second harmonic new wave (SHG). Two different optical fields of frequencies ω_1 and ω_2 can be added to generate the sum frequency wave with $\omega_3 = \omega_1 + \omega_2$. In parametric amplification two waves of frequencies ω_1 and ω_2 with $\omega_1 > \omega_2$ and $I_1 \gg I_2$ give rise to a new wave of frequency $\omega_3 = \omega_1 - \omega_2$, and the amplification of the ω_2 wave. In optical parametric oscillation (OPO) a wave of frequency ω_1 is decomposed in frequencies ω_2 and ω_3 such that $\omega_1 = \omega_2 + \omega_3$, provided the resonance of the system for at least one of the created frequencies. All these effects and others have been observed in lithium niobate and gave rise to applications that will be presented in section 4.4.

In order to obtain a practical intensity of the resulting wave, a phase matching condition is required for the new wave to be growing efficiently along the material in a coherent way. Phase matching conditions can be obtained in anisotropic materials for those propagation directions and polarizations for which the speed of waves of different frequencies are equal. For SHG the exact birefringent phase matching condition is $n_e^{2\omega} = n_o^{\omega}$ [29]. This is a very restrictive condition that can be fulfilled only for certain frequencies, propagation directions and sample temperature. By fortune there is another possibility, more easy to fulfil, to obtain efficient wave generation by the so called quasi phase matching (QPM) disposition. This can be obtained in materials where a property has a spatial periodic change of the proper period. In lithium niobate this can be easily obtained with periodic domain structures, as we will described in section 3.1.

2.7. Photorefractive effect

This effect consists in a non-instantaneous and non-local refractive index change induced by light in the crystal [33]. In fact, this is due to a combination of some other effects. First, one has to account with the presence of certain impurities or color centers which could be optically ionisable. In lithium niobate the most active photorefractive impurities are Fe, Cu and Mn. When they are present in the crystal, appear simultaneously in two valence states: Fe^{2+}/Fe^{3+}, Cu^+/Cu^{2+}, and Mn^{2+}/Mn^{3+} [34]. The reduced valence ions act as charge donors whereas the oxidized valence ions act as acceptors. Second, inhomogeneous light of the proper wavelength produce ionisation of some donors. Third, the released charges move along the crystals by means of one of the charge transport processes: diffusion in the band, photovoltaic current, or drift in an external electric field. Forth, charge is trapped in acceptors. As the trapping occurs in a different place than ionization (darker places in average), a distribution of internal electric field is created. Fifth, finally this electric field distribution induces a refractive index change distribution via the electrooptic effect.

The charge distribution remains in the crystal for some time when it is not illuminated. In this case, only the crystal dark conductivity contributes to redistribute the charge. By contrast, upon uniform illumination the charge is redistributed homogeneously after some time depending on light intensity and spectrum, due to photoconductivity. Consequently the refractive index change is erased.

Photorefractive effect is very dangerous for most optical applications because the light itself modifies with time the refractive index of the material just in its own way path. The result is a perturbation of propagation, as for example a fanning effect. Then, usually it is known as optical damage of the material. It can be avoided by propagation at high temperature or by doping with damage resistant impurities (see sec. 2.2).

Considered this effect positively, one can make good use of it. This can be used to write with light and store for some time information in the material. This is what is done to produce optical memories. Alternatively one can produce a useful optical element, as a phase grating, by changing the distribution of refractive indices. For lithium niobate, a thermal fixing process is known [34] to make these changes quasi-permanent at room temperature. Then, the optical element produced by this effect can be included in a practical setup.

The uncontrolled residual photorefractive impurities and point defects provide of enough number of donors and acceptors to produce the unwanted damage effect. In contrast, crystals are intentionally doped with photorefractive impurities when one seeks photorefractive applications. Most common is iron doping, and the usual concentration varies depending on the application from 0.01 mol% up to about 0.2 mol%.

3 Sample structures relevant for photonic applications

In this section we will review the different sample structures used to improve or to obtain the necessary conditions for competitive applications. The structure of the material can be changed by the way used to produce the material: bulk, fibre or thin film, or by transforming spatially a particular property, as in the case of tailoring the ferroelectric domain structure, or in the formation of surface waveguides.

3.1. Quasi-phase matching and the fabrication of periodic domain structures

Quasi phase matching by a periodic material structure is a technique to maintain the relative phase between the interacting waves in the material. In fact this method corrects the relative phase mismatch at regular intervals by means of a structural periodicity of the nonlinear medium. In the case of $LiNbO_3$, periodically alternated domain structures have been produced by different methods.

In second harmonic generation a fundamental wave of frequency ω_1 and wavelength λ produces a polarization wave of frequency $2\omega_1$ through the second order nonlinear susceptibility. This polarization wave is forced by the fundamental wave, and then it travels with the same velocity $v_1 = c/n_1$. The polarization wave radiates at each point of the material a free second harmonic optical wave, which travels with a velocity $v_2 = c/n_2$. Because of the normal dispersion of the material, in general is $n_2 > n_1$, so that the fundamental and the second harmonic waves travel at different phase velocities. The sign of the energy transfer from one wave to the other is determined by the relative phase between them. The continuous increment of relative phase difference leads to an alternation in the direction of the flow of power between the waves. This produces an alternating growth and decay of the power of the second harmonic wave, which in general never gets a practical intensity. Along a propagation distance over which the relative phase changes in π the power is transferred to the second harmonic wave, and in the next equal distance the phase changes to 2π and the power is transferred to the fundamental wave. This distance, known al the coherence length L_c, is given by $L_c = \lambda/4(n_2 - n_1)$. Of course, when the refraction indices are equal for both waves, for example using the birefringence properties of the material, this coherence length becomes infinite. In this situation of phase matching the electric field of the second harmonic wave grows linearly with the interaction length, and thus the intensity grows quadratically (see Fig. 4).

Another possibility to obtain a final practical second harmonic wave is to produce a repeated inversion of the relative phase between the waves. One way to invert the phase of the generated wave is to change the sign of the nonlinear coefficient. This is done in lithium niobate by inversion of the direction of the ferroelectric domains. The highest efficiency of second harmonic generation in this situation is obtained when the inversion is done every coherence length, i.e. a change period of $2L_c$. This situation has been illustrated in Fig. 4b. Other longer periods can lead to a lower growth of the harmonic power.

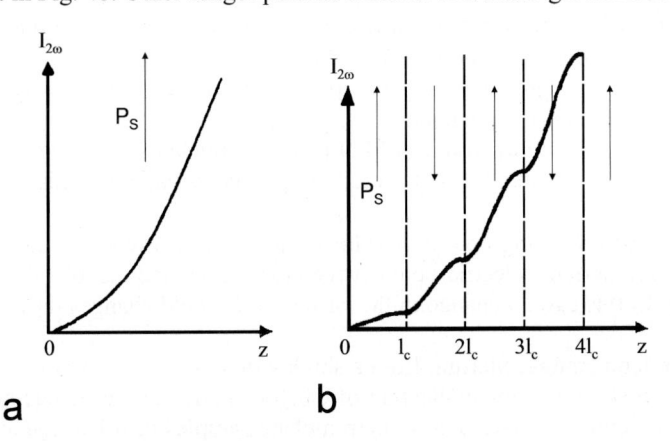

Fig. 4 Intensity of second harmonic generation as a function of the interaction length, a) for exact birefringent phase matching, and b) for quasi-phase matching in a PPLN of period 2Lc [35].

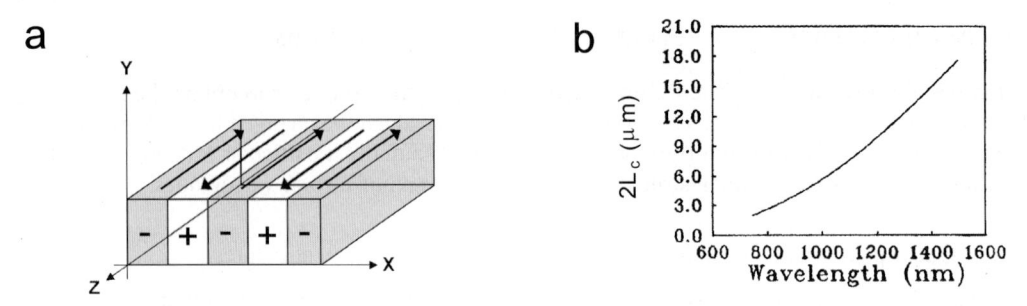

Fig. 5 Required values of the 2*Lc* period for quasi-phase matching SHG in lithium niobate periodic domain structures using the nonlinear coefficient d_{33}, as a function of the pump wavelength. Obtained from equation for *Lc* and Sellmeier formula for n_e (after Ref. [37]).

As it can be seen in Fig. 4, the second harmonic power grows more slowly using quasi-phase matching than it does with birefringent phase matching. The gratest advantage of QPM lies in its ability to produce phase matching in situations which would otherwise be impossible, for example the use of nonlinear coefficients coupling waves of the same polarization. Non-linear coefficients of highest value can be then addressed. As the domain period can be changed, another advantage is the possibility of design a device to operate at the desired wavelengths and temperature [35]. Additionally, the acceptance bandwidths can be significantly enhanced, leading to less critical phase matching. These structures can also be used to phase match other second order effects as parametric amplification or oscillation.

Attending to use for SHG the largest nonlinear coefficient, d_{33}, with propagation along the *a*-axis, both the fundamental and the generated waves have to be extraordinarily polarized. Using the corresponding Sellmeier equation for that refractive index [36] it is possible to calculate the required period as a function of the wavelength of the fundamental wave [37]. This is shown in Fig. 5b, where one can see that the practical range of periods for QPM-SHG is roughly from 3 to 18 µm.

Several methods have been developed to produce periodic domain structures in lithium niobate (PPLN). The most common of them are here summarized:

(1) Bulk PPLN structures by off-axis Czochralski method: A modification of the Czochralski growth method can be used to produce bulk PPLN structures with practical dimensions and period as small as 2 µm, and where the period is controllable by adjusting growing parameters. In this method the crystal is grown using as pulling direction the crystallographic *a*-axis, with an axial temperature gradient. In most cases a trivalent impurity helps to get a better quality domain structure [38]. After some years of experience using this method, growing with 0.5 mol% of yttrium as a dopant, practical structures for QPM-SHG of about 5 µm of period have been obtained. The samples had a period fluctuation limited to 5% with more than 300 periods [37]. Very recently regular structures as long as 3 mm with periods of about 6.5 µm have been reported [28]. Usually the impurity has a double role in these structures. First, the impurity helps to create the periodic structure during crystal growth, and second the impurity is a trivalent rare earth optical active ion for nonlinear applications [39].

An advantage of Czochralski bulk method to produce PPLN is that its thicker sample size permits the angular tuning of phase matching. In this method it is also easier to produce smaller periods, of the order of 2 µm [37].

(2) Periodic electric field application during crystal growth: It has been also reported the successful growth of PPLN by application of a periodic electric field between the crystal and the melt during the *a*-axis pulling crystal growth [28]. In this case, a change in the period of the field changes directly the period of the structure.

(3) Domain inversion at room temperature: Starting from a single-domain crystal it has been reported the production of periodic domain structures by application of electric fields at room temperature (RT) [40]. In order to produce these structures in congruent lithium niobate samples of 0.1 mm thickness, an pulsed electric field of about 24 kV/mm was successfully applied between a grid electrode of the desired domain spacing and a back plane electrode. More recently the technique was improved by the use of a dc

Fig. 6 Some aperiodic domain structures produced for nonlinear optical generation. a) a chirped structure [44], b) a fan-out structure [47], and c) a 2D hexagonal structure [48].

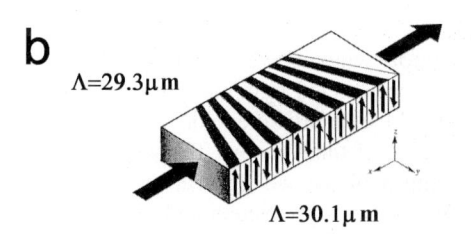

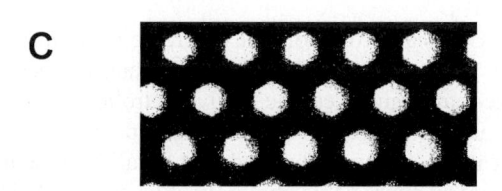

field component allowing to increase the thickness of the sample to 0.2 mm [41]. The results were domain structures of even as small as 2.8 µm periods and walls perpendicular to the z-cut faces. Most last years papers on PPLN structures obtained by an electrical poling at room temperature use commercial samples of 0.5 mm thickness, and interaction lengths of few cm are commonly reported. Thicker samples cannot be used because of dielectric breakdown appears before domain inversion. A method to obtain good quality domain structures consists of using lithographic techniques to produce a photoresist grating of the desired period, used as a mask for applying the electric field with a liquid electrolyte electrode [42]. More than 50 mm long periodic structure was produced with this method.

Recently it has been reported the decrease of the ferroelectric coercitive field by one order of magnitude between congruent and stoichiometric crystals. This make possible the electric field fabrication of periodic structures at RT in few millimetres thick samples [15].

(4) Electron beam writing: Domain inversion to create periodic structures has been also produced by an electron beam scanning [43]. This technique was successful producing domain structures of as small as ~3 µm period in 0.5 mm thick samples. The width of each domain inverted region on the c+ end was about twice the domain width on the c-face.

Aperiodic engineered domain structures have also been reported [44]. In an aperiodic chirped domain structure the distance of domain inversion is changing along the structure. A photograph of such structure is shown in Fig. 6a. The potential of aperiodically poled domain structures resides in the possibility of make efficient conversion processes for several frequencies simultaneously. The phase matching wavelength detuning curve for a nonlinear process is very narrow in a periodic domain structure. In an aperiodic structure the Fourier transform of the distribution of domain inversion determines the bandwidth of the detuning curve. Then, the QPM tolerance band can be slightly overdimensioned to absorb tuning, angle or thermal mismatches, at the expenses of some reduction of effective nonlinear coefficient. Either the tolerance band can be made to consist of several isolated bands corresponding to different nonlinear processes. These structures have been recently produced by changing conditions during the crystal growth of bulk PPLN [45], and successfully used to produce several laser emissions in self pumped frequency generation systems [44, 46].

Fan-out domain structures, as the one sketched in Fig. 6b, have been used to obtain QPM tuning in continuously tunable optical parametric oscillators [47].

Two dimensional domain structures have been also fabricated by electric poling by means of lithographic technique and electrolyte liquid electrode [48]. The hexagonal structure (Fig. 6c) had a period of 18 μm. This structure showed multiple quasi phase matching for high efficient frequency conversion by different mixing second order processes.

3.2. Surface optical waveguides

A waveguide is created on the surface of a dielectric sample when a layer of higher refractive index and the required thickness is produced on the surface. Then the injected light can propagate confined in the waveguide. Many methods have been developed to produce such waveguides on the lithium niobate surface, principally by doping a region close to the surface of the samples. The waveguide can be produced on all the plane surface of a sample, constituting a planar waveguide, or as stripe or channel waveguides. The channel waveguides confine the light in two dimensions and give the possibility to produce optical circuits. Here we will present only those which have been used extensively in practical devices. We will focus this on present briefly their distinct characteristics but not the details of fabrication processes. Table 3 summarizes their main features.

By titanium in-diffusion at high temperatures, a Gaussian profile of index increase is produced in both the ordinary and extraordinary indices. The index change is of the order of 10^{-2}, and the depth in the sample, of course depending on the diffusion time-temperature, can be easily controlled to be of a few microns [49]. This allows the formation of single mode waveguides for visible and near infrared beams, which is an important requirement for many applications. They can be produced for all X, Y, and Z sample cuts [50]. Unfortunately these waveguides present an appreciable photorefractive damage [51].

Table 3 Summary of properties of most common optical waveguides produced on lithium niobate.

type of guide	treatment / temperature	indices change	profile	cut	particular characteristics
Ti^{3+} in-diffusion (Ti)	900 – 1150 °C diffusion + 400 °C annealing in O_2	$\Delta n_e \sim 0.03$ $\Delta n_0 \sim 0.01$	Gaussian ~4 μm	X, Y, Z	optical damage
proton exchange (PE)	180 – 250 °C exchange + 400 °C annealing	$\Delta n_e \sim 0.1$ $\Delta n_0 \sim -0.04$	steplike and gaussian	X, Z	mainly benzoic acid+ Li benzoate, Several structural phases
TiPE	1100 °C (Ti) + annealing in O_2 at 400 °C + 220 °C exchange (H^+)	$\Delta n_e \sim 0.1$ $\Delta n_0 \sim 0.005$	steplike, up to 10μm	X, Y, Z	prevents Y-cut surface damage with exchange
Zn in-diffusion	550 °C in Zn vapor + 800 °C diffusion in air	$\Delta n_e, \Delta n_0$ ~ 0.004	exponential ~2 μm	X, Y, Z	good optical quality <0.5 dB/cm losses, high optical damage resistance
Ion implantation He^+	2 MeV, 2×10^{16} ions/cm²	$\Delta n_e \sim -0.08$ $\Delta n_0 \sim -0.04$	Opt. barrier ~5 μm	Z	same nonlinear coefficients as in bulk material
Ion implantation H^+	1.5 MeV, 4×10^{16} ions/cm²	$\Delta n_e \sim -0.04$ $\Delta n_0 \sim -0.04$	Opt. barrier ~15 μm	Z	same nonlinear coefficients as in bulk material
Ion implantation Si^+ and heavy ions	3 MeV, 4×10^{14} ions/cm²	$\Delta n_e \sim 0.008$ $\Delta n_0 \sim -0.006$	~5 μm	X, Z	optical barrier only for n_0

In the proton exchange method [52] the lithium niobate samples are immersed in benzoic acid at moderate temperatures (180 – 250 °C), producing a exchange of H$^+$ by Li$^+$. This process introduces a very high concentration of protons in the first layers of the sample and many times different structural phases appear [53]. An annealing treatment is necessary to relax the layer stress. With this method a very high increase (~0.1) of the extraordinary index value is produced. In contrast, the ordinary index is decreased (~(−0.04)). Because of this feature only the extraordinary waves can be guided in proton exchanges layers. The large extraordinary index change can be used to produce some optical elements as lenses inside waveguides produced by other methods. Only X and Z cuts can be used because of the poor quality of Y-cut surface obtained.

A combination of Ti in-diffusion and proton exchange techniques has some advantages [54]. First, the presence of Ti makes possible to perform proton exchange in Y cut samples without surface damage, giving the chance to use also this sample orientation. Second, the birefringence at the surface region can be precisely controlled. And finally, when the proton exchange is deep enough, also ordinary waves can be guided. The fabrication of these waveguides starts with the Ti in-diffusion, followed by the proton exchange. The presence of Ti slows down the exchange process and prevents the Y-cut surface from damage. Final index change is practically the superposition of those obtained by separated processes.

Another impurity that has been in-diffused to produce practical waveguides is Zn [55]. This dopant has the additional advantage that, as we mentioned, also induces optical damage resistance. Both indices are increased the moderate amount of $\Delta n \sim 5 \times 10^{-3}$, and so both TE and TM modes are supported. Very recently a new two step fabrication process starting from Zn vapour diffusion has been presented [56, 57]. The results were low loss (<0.5 dB/cm) reproducible waveguides, free of optical damage, and with a very good surface quality.

Ion implantation was used to produce optical waveguides in many materials, also in lithium niobate. In this case the waveguide is obtained basically by generating a low refractive index optical barrier. The nonlinear properties of the material are preserved because the barrier is the zone damaged by implantation, whereas light is guided by the surface unmodified region. Barrier position and shape are well controlled by the implantation parameters. In LiNbO$_3$ implantation of H$^+$ or He$^+$ produce a barrier reducing both refractive indices in about 0.01 [58]. The optical damage is avoided by using Mg doped substrates. The nonlinear coefficients of the material are well preserved in the waveguide [59].

Implantation of heavy ions at very low dose produces a waveguide by change of the refractive index of the surface layer [60]. The extraordinary index increases whereas the ordinary one experiences a small decrease.

3.3. Other structures

A few works have been published about the fabrication and application of crystalline lithium niobate fibers. These fibers had a diameter of tenths of a millimetre and a length of few tens of millimetres. Most of them were grown by the laser-heated pedestal method. In particular, fibers with periodic domain structures of few microns period were grown with Mg doped lithium niobate composition [61]. They presented very high green light SHG efficiency.

Good quality crystalline thin films of LiNbO$_3$ have been successfully grown on lithium niobate commercial z-cut wafers by the liquid phase epitaxy method (LPE), using Li$_2$O–V$_2$O$_5$ [62] and Li$_2$O–B$_2$O$_3$ [63] flux systems. The films, of a few micron thickness, had high crystallinity, flat surface and low light propagation loss. The ferroelectric polarization is inverted respect to the substrate. In the case of using Li$_2$O–V$_2$O$_5$ flux, the vanadium incorporated to the film. In spite of this, the film quality was better using this flux. Doping with Mg is done by adding MgO to the flux. Optical damage resistance is then obtained [64]. The growth of these thin films on PPLN substrate has been very recently reported [65]. The film presented also a PPLN structure, but additionally the surface presented a periodic striated structure due to different growth velocities between positive and negative domain ends.

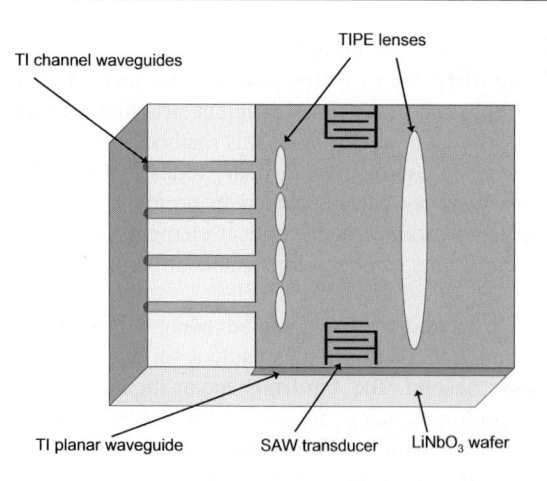

Fig. 7 Integrated lenses produced by PE in an acoustooptic integrated device [66].

4 Devices and applications

Thousands of papers on practical devices made with lithium niobate have been already published. Of course we cannot address all of them here. We have elected a few, we hope significant, examples of devices making use of the variety of material properties presented above. It is our purpose to briefly describe how they work in a simple way. Although in most cases they could work with bulk propagating light, many devices have been developed to work on a waveguided form because of the effectiveness obtained with confined light.

4.1. Simple guided optical elements

(a) Lenses can be formed on planar Ti waveguides by proton exchange using masks to obtain the desired form. Due to the relatively large index change given by proton exchange short focal length can be obtained. Both individual lenses and microlenses arrays have been formed [66]. In Fig. 7 these lenses are integrated in a acoustooptic device. Due to the index change properties of proton exchanged regions, these lenses are only operative for extraordinary polarized light.

(b) Linear polarizers can also be implemented using the anisotropic index changes produced by proton exchange. The device presented by Hemppelmann et al. [67], and reproduced in Fig. 8, works by coupling away the TE polarized light propagating in a Ti waveguide by means of coupling to PE lateral regions, in a X-cut sample with Y propagation. The TM polarization does not couple to those guides and remains localized in the channel Ti waveguide.

(c) Bragg gratings can also be constructed for the extraordinary waves propagating in a Ti waveguide by PE formation [68]. After deposition of a Si_2O layer on the sample surface, a Bragg grating pattern mask is produced by photolithographic techniques. The usual proton exchange method generate the

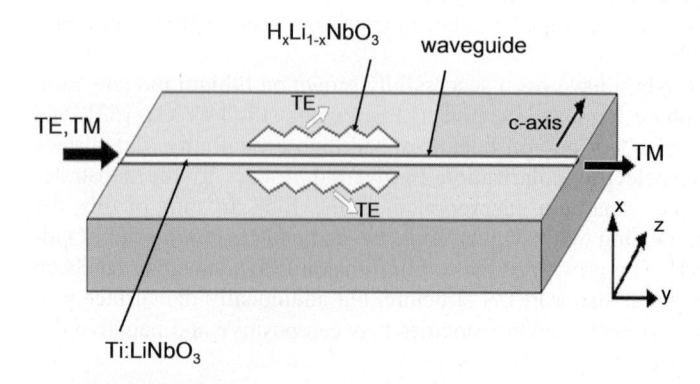

Fig. 8 Integrated waveguide optical polarizer (after Ref. [67]).

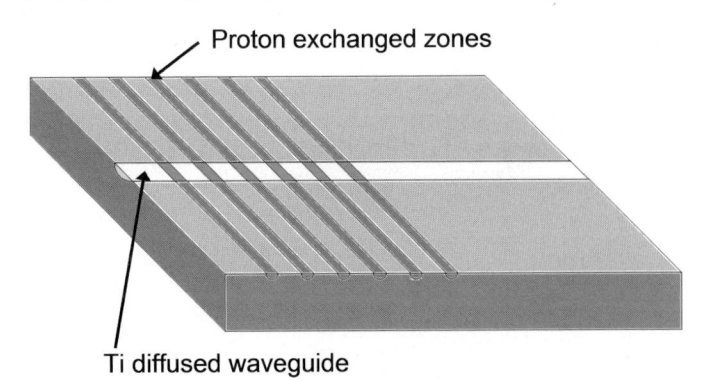

Fig. 9 Proton exchanged Bragg grating integrated in a channel Ti waveguide device [68].

Bragg grating in the Ti waveguide, Fig. 9. For fringe spacing of 8 μm and 4 mm long grating, a reflection of 94% with a bandwidth of 2 nm was obtained. These gratings can be used as distributed Bragg reflectors in Wavelength Division Multiplexing (WDM) communication technology.

4.2. Acoustooptic devices

Many acoustooptic (AO) integrated devices and circuits have been designed and fabricated on the basis of planar Ti waveguides, and incorporating lenses produced by PE. They work with surface acoustic waves (SAW) generated by interdigital electrode transducers. These circuits are usually end-coupled to optical fibers for input and output. We will comment some characteristics of the most common reported AO devices in lithium niobate [31]:

(a) Multiport deflectors and switches are devices that use the light deflexion produced by acoustooptic Bragg diffraction, combined with integrated microlens arrays to collimate the input light from a N channel waveguides array, and a large aperture lens to focus the light on M channel waveguides used for light output, Fig. 10. Switching times of 30 ns have been obtained. A 8×8 witch has been constructed and demonstrated [69].

(b) Frequency shifter modules use two consecutive Bragg diffractions to avoid misalignment produced by the change of propagation direction of the diffracted beam by the acoustic wave. A prototype device [31] had 76% final light power efficiency for a RF power of 200 mW, and a frequency shift of about 1 GHz. Used as a light frequency modulator it has a bandwidth of 1 GHz with a center frequency of the same order.

(c) Multichannel processor modules use the property of proportionality of the diffracted beam power to the power of the incident beam and also to the power of the acoustic wave, and the summing property of the collecting lens, to perform computing operations [31].

(d) Combined AO and EO modules are used to build very useful interferometric optical spectrum analyzers for RF frequency variations [31].

(e) Tunable waveguide optical filters are also produced [31]. The characteristics are about 300 nm tuning range with a 4 nm bandwidth for a wavelength of 1.3 μm.

In summary a variety of acoustooptic devices can be designed for competitive use in optical communications and optical computing.

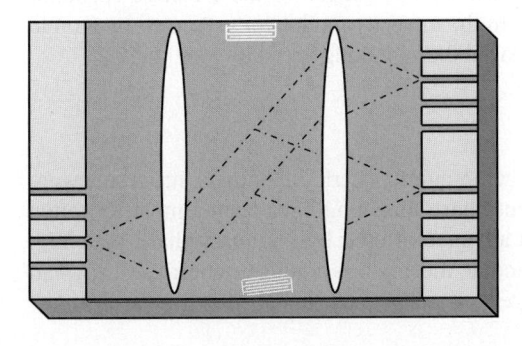

Fig. 10 Acoustooptic integrated switch with input and output multichannel waveguides [31].

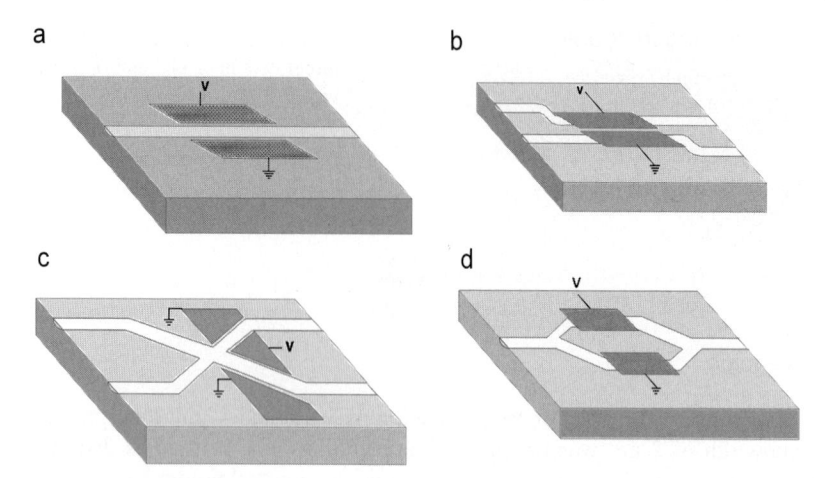

Fig. 11 Some different electrooptic waveguide devices. a) Phase modulator, b) Directional coupler amplitude modulator, c) X-switch, and d) Mach-Zehnder interferometer amplitude modulator [70, 71]

4.3. Electrooptic devices

Modulators, switches, couplers, interferometers and mode conversors are common electrooptic (EO) devices build on channel waveguides [70]. Some of them are represented in Fig. 11. The simpler one is the phase modulator (Fig. 11a) which is formed by two electrodes that upon an applied voltage change slightly the refractive index in the waveguide producing controlled wave retardation. The other EO devices include this one as a part. Directional couplers and switches work changing the refractive index in the region between two close channel waveguides (Fig. 11b and c), in such a form that the coupling among them can be controlled. In the Mach-Zehnder interferometer, Fig. 11d, a channel waveguide splits into two branches for some extent, and joint together after that. Electrodes on both branches permit to change the relative optical paths. Light divided to propagate by the branches will interfere when joint thereafter. Depending on the relative phase mismatch the interference will be constructive, and the wave passes throughout, or destructive and there is no exit wave, or intermediate transmission for intermediate phase difference.

All these EO devices need of single mode fibres to properly work. Usually the control voltage is in the 2–10 V range. The control frequency is limited by the capacitor response time. Using the so called travelling wave electrodes [70] the limit is about 20 GHz, or 10 Gbit/s for digital operation. These devices present very good behaviour in different environments, lifetimes longer than ten years, ease for changing designs, and possibility of integration in multi-device chips [71]. Hundreds of these elements can be integrated for parallel processing. Since more than ten years ago there are commercial electrooptic lithium niobate devices for communications systems. The device-manufacturing technology has progressed to enable extensive deployment within digital and analog communications systems.

New advances in this field tend to reduce the switching times, to be competitive for high speed communications systems. New designs reduce the switching times by a reduction of the effective capacitor time constant through a reduction of the material thickness. In the intensity modulator presented by Kondo et al. [72], Fig. 12, the micro-machined substrate back increased the switching speed up to 40 Gb/s.

4.4. Nonlinear optical devices

Nonlinear optical devices in $LiNbO_3$ have gained most developing after periodic domain structures were available. The initial experiments of SHG with birefringent phase matching have been almost forgotten from the practical point of view. This is due to the great advance offered by the quasi-phase matching technique for freely used different polarizations. In particular the highest nonlinear coefficient d_{33} can now be addressed. So, here we will refer almost exclusively to quasi phase matched nonlinear devices.

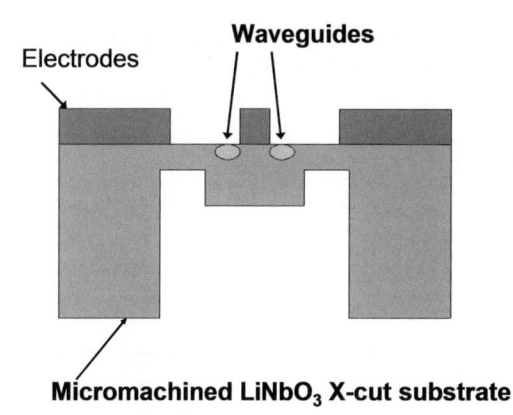

Fig. 12 Section of a back micromachined optical modulator for 40 Gb/s operation (after Ref. [72]).

In most of these devices the optical power to generate a practical second order beam must be moderately intense. To avoid the undesired effects of optical damage the material is usually doped with optical damage resistant impurities, mainly Mg and more recently Zn.

Among these devices one has to distinguish those that produce the nonlinear effect in PPLN bulk materials and those which make use of waveguides. The first group has the advantage of make possible fine tuning of phase matching by slight angular tilting, keeping the device at room temperature. In the second group the waveguide impose the light path, and then the phase matching tune have to be obtained by a temperature adjust. We will present separately these two groups.

The solid state lasers produced by doping the waveguide with laser active impurities, and many times with the laser cavity produced in the lithium niobate sample itself, although they do not make use of the polarization nonlinear terms, will be treated in this section too. A few of them involve a nonlinear intracavity device producing effects such as self doubling laser generation.

4.4.1. Nonlinear optical devices in bulk LiNbO$_3$

(a) Second harmonic generation devices: PPLN samples are used directly for QPM-SHG as single pass devices located in the pump beam path. Of course, the domain period has to be matched to the pump wavelength, according to the graph on Fig. 5b. When this is not exact, incidence angle and temperature can serve to fine tuning of matching condition. Usually the measurement of doubling efficiency is used to calculate the effective nonlinear coefficient, as a measure of the quality of the periodic structure. Of course, the conversion efficiency is a function of the interaction length and the pump power, as it is a non-linear effect [29]. Using electrically poled PPLN of 2.8 μm period, and an interaction legth of 3 mm, using a pump wave of 196 mW at 852 nm, a doubled emission power of 21 mW was obtained [28]. Using bulk PPLN with 1064 nm light an efficiency of 20% was obtained for 3 mm of interaction length [28]. Other authors [37], in bulk PPLN of 5.2 μm period and 1.6 mm of interaction length, reported an efficiency of 24% doubling a wave of 980 nm.

(b) Optical parametric oscillators: In the optical parametric oscillators an incident laser pump wave of frequency ω_3 in a nonlinear medium generates two waves, the signal of frequency ω_1 and the idler of frequency ω_2. By energy conservation it is clear that must happen that $\omega_3 = \omega_1 + \omega_2$. The quasi-phase matching condition can be expressed as momentum conservation:

$$\frac{n_p}{\lambda_p} - \frac{n_s}{\lambda_s} - \frac{n_i}{\lambda_i} - \frac{1}{\Lambda} = 0, \qquad (6)$$

where the subscripts p, s, and i stand for pump, signal and idler respectively, n are the refractive indices at the corresponding wavelengths λ, and Λ is the period of the domain structure.

The nonlinear material has to be inside an optical cavity with longitudinal modes resonant for the signal wave, the idler wave, or for both of them. Single resonant OPO is when the condition is provided

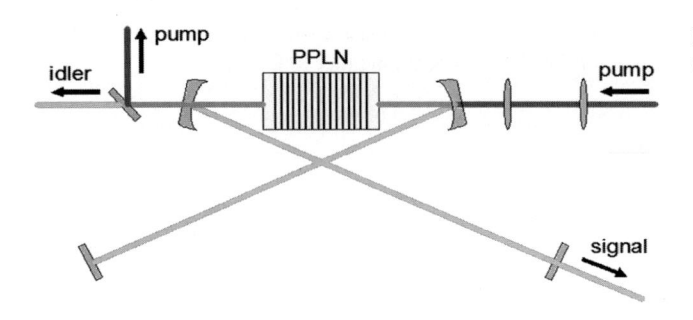

Fig. 13 An example of optical parametric oscillator (OPO) setup.

only for one wave or doubly resonant when it happens for both. When the pump wave overpass a threshold power, which is related with the losses of the signal and idler waves, the oscillation begins. The great interest of OPO's using PPLN comes from the possibility to produce efficiently laser radiation of different frequencies from the initial pump frequency, and with the possibility of tuning these frequencies by the phase matching conditions. An example of OPO setup is presented in Fig. 13. One of the biggest advantages of OPO's is the tuning ability. In general tuning can be achieved by temperature, by angle change, by an applied electric field, or by a change in the period of the PPLN structure.

Myers et al. [73, 74] have demonstrated efficient OPO emission in bulk PPLN with a single resonant cavity, in pulsed as well as CW operation. The three waves had extraordinary polarization, making use of the highest effective nonlinear coefficient $d_{eff} = (2/\pi) d_{33} = 17$ pm/V for 1064 nm pump. The PPNL crystal was electrically poled, 28 mm long and had a 30 μm period. Using a CW Nd:YAG laser of 13.5 W (three times the observed threshold) they obtained 1.6 W of a signal wave of 1.57 μm, and 3.6 W of an idler wave of 3.3 μm. These authors [75] build an OPO system with possibility of tuning from 1.3 μm to 5 μm, using a multigrating PPLN crystal for coarse tuning and temperature for fine adjustment.

Continuous tuning was obtained by Powers et al. [47] using a fan-out patterned PPLN sample. The periodicity of the domain structure changed from 29.3 mm at one side to 30.1 mm to the other side (see Fig. 6b). The tuning range was from 1.53 to 1.62 μm for the signal and from 3.1 to 3.5 μm for the idler wave. Electrical tuning in a wide wavelength range was obtained by Gross et al. [76] by means of the electrooptic effect in a special sample with two PPLN regions separated by a monodomain part. With an electric field up to 1230 V the signal wavelength could be tuned from 1560 to 1660 nm.

Visible OPO efficient operation has been demonstrated by Huang et al. [77]. These authors used a 532 nm pulsed pump to produce a signal wave of 589 nm, and an idler wave of about 5 μm.

Selfpumped OPO devices combine the laser emission and the parametric oscillation inside the same PPLN crystal. The lithium niobate sample is then doped with a laser active impurity which is pumped to one of its transition lines by the external wave. At the same time the sample has a periodic domain structure which induces the parametric oscillation. Capmany et al. [78] demonstrated this system operating with Yb^{3+} in continuous wave producing a signal wave of 1360 nm. A similar system working with Nd^{3+} and pulsed light has been also reported [79].

(c) Other frequency converters: Difference frequency generation has been used to obtain mid-infrared laser lines for high resolution molecular spectroscopy and sensitive detection of contaminant molecules. Laser emission in the range of 2 to 5 μm can be efficiently obtained by the difference of frequency of two lasers. For example, Petrov et al. [80] produced the frequency difference of a tunable diode laser around 808 nm and a Nd:YAG diode pumped laser at 1064 nm in a 1 cm long PPLN structure of 21 μm period, to obtain a laser emission at about 3.4 μm, in the range of some methane and water vapour resonances. CO_2 was also detected by a difference frequency system emitting at 4.25 μm [81].

Aperiodic domain structures where used to produce visible light sources by SHG. In the work by Birkin et al. [82], the aperiodic structure was designed to enlarge the spectral acceptance bandwidth of the structure to accommodate to the spectral bandwidth of a gain switched diode laser with 30 ps pulses. An average blue output power of 3.6 mW was obtained with a 83 mW at 980 nm pump. Compression of femtosecond pulses was observed in QPM-SHG in aperiodic structures [83].

When the lithium niobate domain structure is also doped with a laser active impurity, both laser emission and nonlinear frequency conversion are obtained in the same material and cavity system. Multicolor laser emission has been obtained [44, 46] with Nd^{3+} doped aperiodic domain structure (with a variation from 6.5 μm to 3 μm period), by simultaneous SHG and sum of frequencies of two infrared laser emissions of the Nd^{3+} ion which were obtained together. This result could be relevant for high brightness display applications.

4.4.2. Nonlinear optical guided PPLN devices

Nonlinear optical devices formed by waveguides in periodic structures have the advantages of combining the high optical power density inside the waveguides, which is greatly important for efficient nonlinear effects, with the periodic modulation of nonlinear coefficients, which gives the benefit of quasi phase matching techniques. Thus, in these situations we could expect high efficiency frequency conversion. In contrast, the impossibility of angular tuning makes sometimes necessary to appeal to temperature tuning. Except for the cases of self pumped devices, there is no need of an optical cavity to take advantage of these processes, as they work in single pass.

(a) Second harmonic generation: In particular visible laser light in the green-blue region from small size sources could be of great interest for applications such as optical data storage, displays, color printing and medicine. Such a compact device could be constructed with PPLN guided doubler attached to a laser diode.

Green and blue laser light was obtained by SHG-QPM of light of either 1064 nm Nd:YAG laser, or 864 nm of a Ti:Al$_2$O$_3$ laser, or 865 nm laser diode, in PE singlemode waveguides, in the experiments reported by Kintaka et al. [84]. The PPLN structure of this device had a total length of 3 mm. Preliminary efficiencies of about 5% in single pass were achieved. Webjörn et al. [85] using Ti indiffused waveguides and PPLN of ~6 μm period, 1 cm long, obtained an efficiency of about 30% doubling a wavelength of 1.1 μm. Good beam quality, stability and low noise are also the characteristics displayed by these devices.

Because of the much higher efficiency of visible light for generating optical damage distortions, some authors found less power deterioration and better long term results using Mg doped substrates [86], or even using waveguides produced by Zn indiffusion [87].

(b) Cascade difference frequency generation: In this nonlinear process the laser light of a new frequency is generated after two consecutive second order nonlinear steps: first, the pump laser is frequency doubled, and second there is a frequency subtraction of this generated wave and a signal wave of frequency close to that of the pump. The result is a new wavelength also close to those initially provided: $\omega_{out} = 2\omega_p - \omega_s$.

For example, in the work by Chou et al. [88], a pump with $\lambda_p = 1562$ nm and a signal of $\lambda_s = 1554$ nm, produced a new wave of $\lambda_{out} = 1569$ nm. The output power is approximately given by

$$P_{out} \approx (1/4)\, \eta^2 L^4 P_p^2 P_s, \tag{7}$$

where P_p and P_s are the powers of the pump and signal waves respectively, L is the interaction length, and η is the effective efficiency of the process. η has an estimated value of about 0.0015 mW^{-1} cm^{-1}. The strong dependence with the interaction length makes easy to obtain an output power equal to that of the signal wave. The constructed device had a 4 cm long PPLN structure of 15 μm period, and a PE waveguide of 12 μm wide and 0.7 μm depth. Simultaneous wavelength conversion was obtained by using several signal waves of different wavelengths, all of them in the 1.5 μm band. The result can be seen in Fig. 14 where four signal waves are used to generate four new lines. Each line can be modulated by modulating the corresponding signal wave. No crosstalk was observed with the other lines. This is very interesting to provide new channel wavelengths for DWDM systems.

4.4.3. Waveguide lasers

Several laser systems have been developed using surface channel waveguides in LiNbO$_3$. In particular are of special importance those produced in Er doped material. The particular Er^{3+} emission properties make possible to build integrated amplifiers and lasers in the 1.53 to 1.62 μm spectral range. It is evident

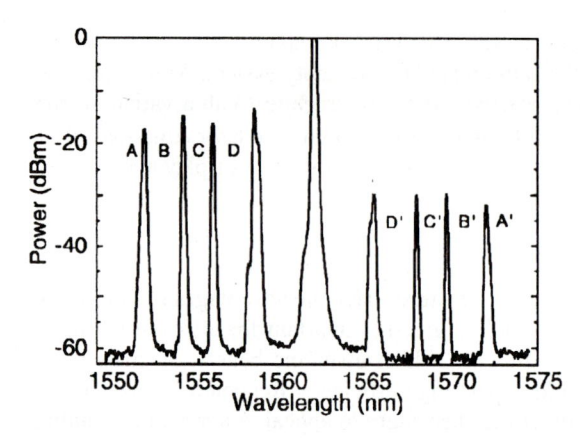

Fig. 14 Multiline generation by a cascade nonlinear process, from Ref. [88].

the importance of these devices for optical communication technology in the 3rd optical window. The waveguide technology permits a convenient connection to the optical fibers, the integration with other devices such as modulators to form integrated optical circuits, and to use intracavity elements to produce mode-locking, Q-switching or tunable laser systems [42]. Most of these systems were developed by the group of Professors Suche and Sohler at Paderborn University.

Fabrication of mode locked and Q-switched waveguide lasers is a long process that have almost the same steps for both systems [42]. First, deposition and diffusion of Er at 1130 °C on all the upper surface of the substrate is produced, intending to get high impurity concentration. Second, deposition and diffusion of 7 μm wide stripe of Ti at 1060 °C to form the single mode waveguide. Next a SiO_2 wafer layer is deposited on the upper face to avoid interference of electrodes in guiding properties. Following is the Ti/Au electrode deposition. Finally deposition of the high reflectivity rear end multilayer mirror,

The stabilizer mode-locked laser [42] obtained had a sidemode suppression ratio of 55 dB for 10 GHz pulse repetition rate suitable for high speed optical communications. Q-switched lasers, which scheme is showed in Fig. 15, presented a high extinction ratio and an electrooptic intracavity switch that allowed emission of pulses of 2.1 ns width and repetition rate of 1 kHz, with peak power up to 2.5 kW [42]. This laser has been presented commercially for applications such as distance measurement and collision avoidance sensing. Self-doubled Er doped lasers with PPLN structure present emission at the fundamental 1531 nm and the double 756 nm lines. This laser is investigated for intracavity cascade frequency generation for DWDM.

Very recently it has been presented a new Er doped laser which incorporates two thermally fixed holographic Bragg reflectors at both ends of the waveguide [89]. With these distributed Bragg reflectors (DBR) resulted in a single frequency laser. Using a pump light of 120 mW at 1490 nm, it delivers a CW single peak at 1561 nm with a bandwidth of around 0.006 nm and not optimized power of 1.1 mW. This is also a laser very interesting for DWDM applications.

4.5. Photorefractive holographic applications

Holographic recording by photorefractive effect in $LiNbO_3$ starts illuminating the crystal with the fringe pattern produced by interference of coherent light beams, in a conventional two wave mixing setup,

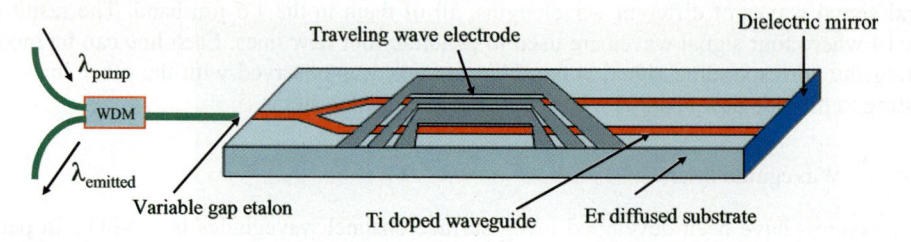

Fig. 15 Scheme of the Q-switched laser developed with a $LiNbO_3$ waveguide structure, fabricated by Becker et al. [42].

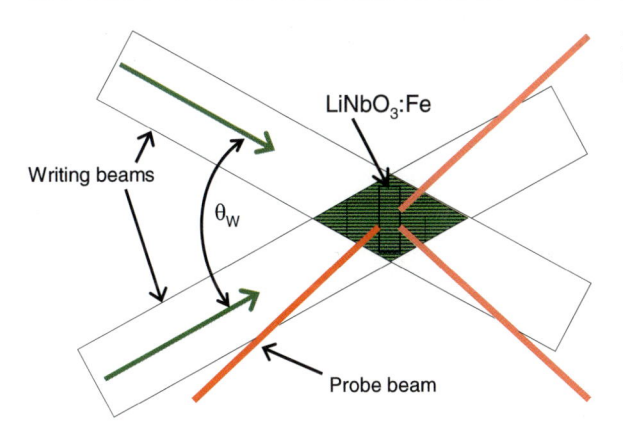

Fig. 16 Scheme of two wave mixing holographic recording.

Fig. 16. Usually one of the beams carries information (signal beam), whereas the other serves to interfere with the signal beam to produce the fringes (reference beam). The beam wavelength is required to be in the spectral range of sensitivity of the photorefractive impurity present in the crystal. Although this range depends on the particular impurity involved, usually in lithium niobate this is the green-blue spectral region. In many experiments the 514.5 nm green line of the Ar$^+$ laser is used. Few milliwatt per square centimeter are sufficient for recording. After some time a phase volume hologram will be stored. For inspection of the recording process, a second (probe) beam of different wavelength (for example the 632.8 nm line of He–Ne laser) could be used at its corresponding angle. This beam is diffracted and the diffraction efficiency can be monitored.

Some important properties of these holograms are derived from their volume and phase characteristics [33]. First, the diffraction efficiency can reach 100%. This is important for applications where the transformed wave needs to have high intensity. Second, there is Bragg selectivity. This means that to read the hologram with the same recording wavelength, the incidence angle must be the same as during recording. Third, there is the ability of multiplexing holograms. Each hologram is characterized by its fundamental grating vector. Two holograms stored with different recording angle, or with a different wavelength, will have different grating vector. Several holograms with different grating vectors can be stored in the same material volume and they will be not mixed. Each hologram can be read separately. All these properties are of capital importance for application in optical memories.

Other relevant properties for holographic applications of lithium niobate are the following: To be readable, no hologram processing is required after recording. To read a hologram containing information, just the same recording wavelength must be used. All holograms in a piece of material can be optically erased at once by a homogeneous light beam. Indeed, the readout beam can appreciably erase the hologram while reading. A single hologram can be erased selectively making use of the holographic system.

The holograms can be made quasi permanent at room temperature by means of a thermal fixing process [34]. Fixed holograms cannot be erased optically. This thermal process consists in heating the recorded sample at a temperature in the range 120 to 180 °C for some time. At those temperatures the internal electric field distribution is not destroyed but compensated by mobile ionic charges. Cooling back the sample to room temperature the compensating ions are frozen in their new positions. Using then homogeneous light, the original space charge distribution is partially removed. In this way the ionic space charge distribution, replica of the initial one, is now uncompensated and produces a refractive index distribution equal to that initially recorded. It is demonstrated [90] that the ions that form the fixed holograms in LiNbO$_3$:Fe are H$^+$, or protons. Protons are ever present in lithium niobate crystals in concentrations of the order of 10^{18}–10^{19} cm^{-3}. Quite more than the amount needed for hologram fixing.

4.5.1 Holographic memories

Holographic digital memories are very promising applications of lithium niobate crystals. Much effort has been done to obtain practical holographic memory systems in photorefractive materials with com-

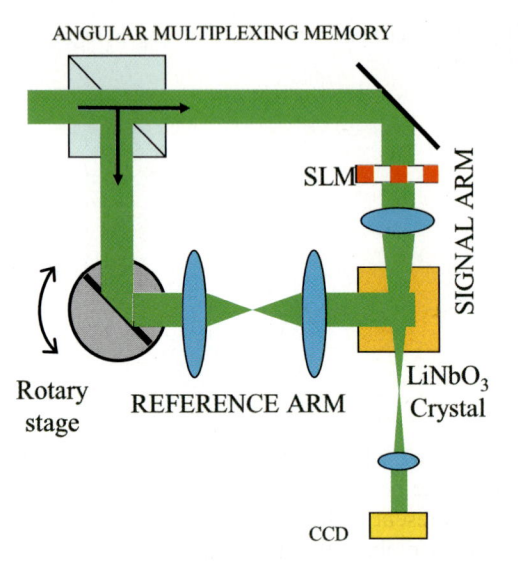

Fig. 17 Scheme of an angle multiplexed holographic memory.

petitive characteristics [91]. These memories are based on the recording of a great number of holographic pages of bits in the medium. Each signal beam that is spatially modulated with data is recorded by interference with a different reference beam. So, the reference beam is providing the key for encoding each hologram data page. Several possibilities of encoding such as angle of arrival of the reference beam, spatial phase distribution of the reference beam, or wavelength of both recording beams can be considered in the system. Bragg selectivity allows independent retrieval of each page provided the corresponding key reference beam. Readout of the desired hologram page is performed at once, when the required reference beam is incident on the medium. The part of this beam diffracted by the hologram is finally detected in a CCD device.

Figure 17 presents a scheme of a practical holographic memory system. In this case the multiplexing encoding system elected is the angular reference beam encoding. This is provided by a 4f telescopic system and a mirror mounted on an electronically driven rotary platform. A supplementary phase encoding possibility can be considered by including a phase spatial light modulator (PSLM) in the reference beam path. In the case of wavelength multiplexing the addressing mechanism must be provided by a tunable laser.

In this memory system, the lithium niobate crystal has a cubic shape and is oriented with its ferroelectric c-axis in the diagonal between faces. The signal and reference beams enter in the medium by orthogonal faces. This 90° sample configuration has been considered optimum for angular multiplexing and to take better profit of the material volume [92].

Something that in the past constituted the bottleneck of these memory systems were the required devices for information input and output, i.e. the amplitude spatial light modulator (ASLM) to introduce information in the signal beam and the CCD detector to convert the retrieved data page to electronic signal. Now, the advance of video display projection elements provides low cost good resolution spatial light modulators. The same history has happen with the CCD detectors which have been very much improved due to the advance of the digital photography market. Nowadays 1024 × 1024 data pixels are implemented in holographic memories [93].

In order to know the real capabilities of these memories, in comparison with solutions to mass digital information storage systems delivered by other technologies, one have to determine the real capacity and recording and readout velocities available. Lithium niobate compared with other recording photorefractive media has a handicap in its relatively low sensitivity. The sensitivity is defined as

$$S = \left(\frac{\partial \sqrt{\eta}}{\partial t}\right)_0 \frac{1}{I_R L}, \tag{8}$$

where η is the diffraction efficiency, I_R the recording intensity, and L the crystal length. The derivative represents the initial recording slope of the square root of the diffraction efficiency. For LiNbO$_3$:Fe this

Compact holographic memory

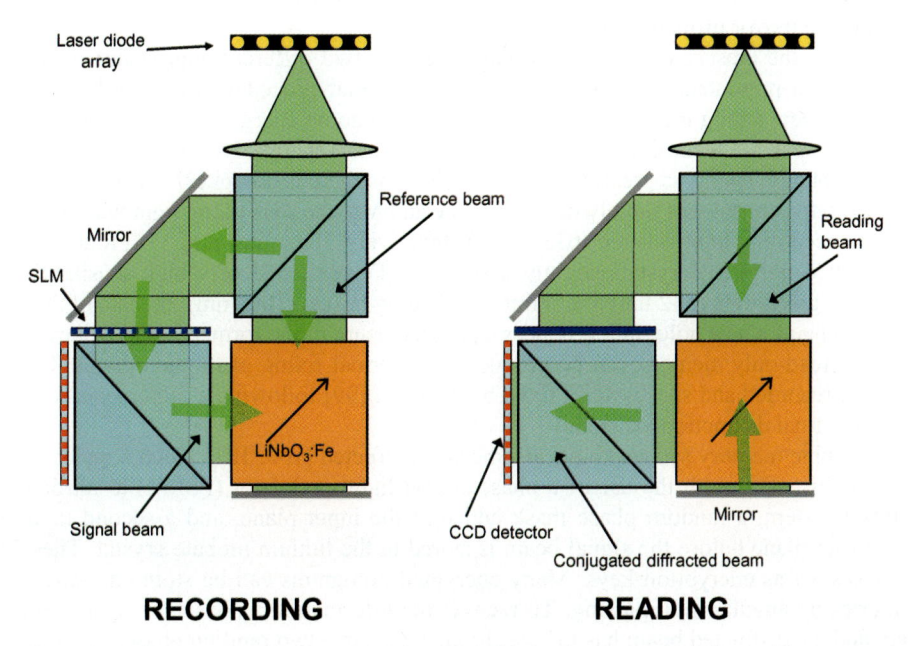

Fig. 18 Schemes of recording a) and readout b) in the compact holographic memory proposed by Liu et al. [95].

factor is about $S \sim 0.3$ cm/J, whereas for BaTiO$_3$ this is about ten times larger [33]. In contrast, the maximum achievable index change in lithium niobate is $\sim 10^{-3}$, about 50 times of that measured in barium titanate [33]. In a memory, each hologram has not to reach the maximum diffraction efficiency, but as higher it is the achievable maximum index change larger is the number of holograms that can be stored with the required efficiency.

In 1995 Hong et al. [94] stored 20.000 high quality holograms in a piece of lithium niobate of $10 \times 10 \times 22$ mm using a combination of angular and spatial multiplexing. With a doubled Nd:YAG laser and beam powers of 8 and 2 mW, the time required to write one of these holograms was 0.3 s.

In consideration of the capacity of a volume holographic memory, theoretically the fundamental limit is a density of bits of $1/\lambda^3$, which for green recording wavelength corresponds to 6×10^{12} bits/cm^3, an enormous quantity that in practice is very much reduced by other limitations as it is the aperture diffraction. A recent work [93] demonstrated experimentally data storage with an areal density of 38 Gpixel/cm^2 with a good bit error rate. The authors used a $15 \times 15 \times 8$ mm sample, recording holograms of 1 Mbit each. In the volume this represents only about 1% of the theoretical limit. Even so, this is already a competitive recording material. The time to record one hologram was 0.34 s with 6 mW of signal beam and 15 mW of reference beam ($\lambda = 532$ nm).

Compact lithium niobate based memory systems, with a reduced number of optical elements and suitable of integration in computer, have been also reported. The system proposed by Liu et al. [95] is reproduced in Fig. 18. It is composed only by a diode laser array as light source, a collimating lens, two beam splitter prisms, two plane mirrors, and of course the crystal, a transmission amplitude SLM, and the CCD array detector. To obtain such economy of elements it makes a very good use of the phase conjugation property of the volume holograms. The signal beam distortion made by the SLM when recording is completely restored during readout with a inverted reference beam. So, placing the SLM and the CCD in the same virtual position by means of a cube beam splitter, just the same spatial amplitude distribution produced by the SLM during recording is reconstructed on the CCD during readout, with magnification equal to one. In spite of its simplicity, this system can store and reproduce pixels of a size smaller than 1 μm.

Another even more compact is the system reported by Mitsuhashi et al. [96], using only one beam splitter cube and one mirror instead of two of each.

In order to avoid the erasure of holograms while readout, two different approaches of material improvement have been presented. Both of them propose a similar procedure for recording and readout. During recording a new gating incoherent light beam of high photon energy is homogeneously illuminating the crystal, whereas the recording beams are of lower photon energy. During readout only the reference beam of low photon energy is present, and in this situation there is no optical erasure. In the approach by Buse et al. [97], the crystal was doubly doped with Fe and Mn, the sensitizing light was near ultraviolet (365 nm) and the recording light was red (633 nm). Alternatively, Hesselink et al. [98] found that the near stoichiometric (49.9 mol% Li) crystals slightly doped with Fe presented very high sensitivity when recorded with gating green light (532 nm) and infrared recording beams (1064 nm). In spite of this, in any of these cases recording of a new hologram produces a partial erasure of the formerly stored ones.

A non-volatile read-only memory can be produced by thermal fixing of the holograms. Up to 10000 holograms where recorded and successfully fixed by An et al. [99] following a incremental fixing schedule to improve the final diffraction efficiency.

Secure holographic memory systems have also been constructed [100, 101]. Such a memory works by encrypting the information with the use of a phase spatial light modulator (PSLM) to introduce the information in the system, a random phase mask added at the input plane, and a second random phase mask at the Fourier plane before the signal beam is stored in the lithium niobate crystal. These two random phase masks act as encryption keys. Many encrypted holograms can be stored as in regular holographic memories by angular multiplexing. To recover the information a phase conjugate readout beam must be used, and the diffracted beam has to pass through the same two random phase masks (or an exact copy of them) at the same optical position, to recover the phase image initially produced by the PSLM. The recovered phase data image can be converted for detection to intensity image with the help of an interferometric setup.

4.5.2 Fixed grating applications

The particular characteristics of fixed holographic gratings in lithium niobate make them very attractive for a number of interesting applications. Among the most interesting characteristics are the following:

The space charge field of a fixed hologram can reach up to about 70% of the originally stored field [102]. Because for a given hologram space charge field the diffraction efficiency depends strongly on the interaction length, the diffraction efficiency of a fixed hologram can reach a value close to 100% provided the necessary sample length L. With the required conditions of doping, impurity oxidation state and H^+ concentration, this maximum efficiency can be reached in less than 1 mm of sample thickness [103].

Due to thermal diffusion of H^+ ions, the fixed holograms have a non-infinite lifetime at room temperature. In any case this lifetime is too much longer to make RT direct measurements, and extrapolation from higher temperature data is required. It is found that hologram lifetime is a function of the H^+ ion concentration, H_0, of the concentration of iron and relative valence ratio through the parameter $N_t = [Fe^{2+}] \, [Fe^{3+}]/[Fe]$, and of the diffusion coefficient of H^+ ions D_H, by the expression [90]

$$\tau_F = \left(\frac{\Lambda}{2\pi}\right)^2 \left[D_H\left(1+\frac{H_0}{N_t}\right)\right]^{-1}, \qquad (9)$$

where Λ stands for the hologram fringe spacing. So, low hydrogen concentration and large N_t value are required for long lasting holograms. Note that there is a limit when $1 \gg (H_0/N_t)$. The diffusion coefficient for hydrogen contains the temperature dependence. Figure 19 presents the extrapolated lifetime as a function of H^+ concentration for two different grating spacing of interest [90]. The values indicate that at room

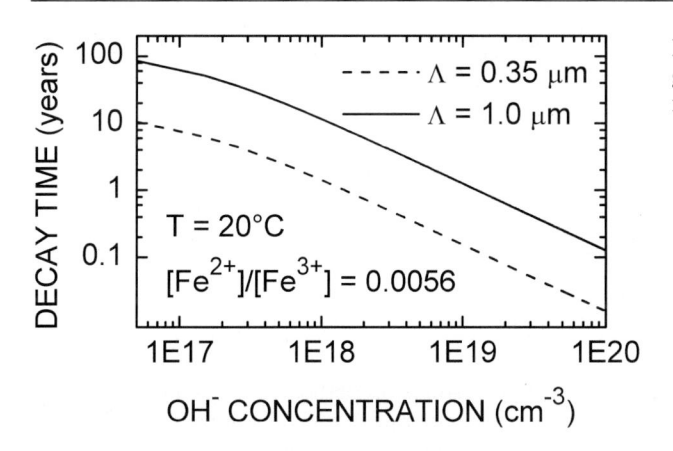

Fig. 19 Extrapolated lifetime of fixed holograms in LiNbO$_3$:Fe at room temperature, as a function of H$^+$ concentration.

temperature is possible to have hologram lifetimes longer than ten years for a wide range of hologram grating spacings. This is well compatible with the requirements for commercial applications.

When Δn large values are required for high diffraction efficiency with relatively thin samples, recording and fixing can be done at once at the fixing temperature (150 °C for example). Due to the unwanted random effects of air convection during recording, it is very convenient to stabilize actively the fringe pattern on the sample. An active procedure for this is obtained by acting on the phase of one of the writing beams with the use of a mirror mounted on a piezoelectric actuator and using as a reference signal the light interference of transmitted and diffracted recording beams [104].

4.5.2.1 Narrow band interference filters in bulk lithium niobate

Very narrow reflecting band filters have been produced by holographic recording and reading of holographic gratings in LiNbO$_3$:Fe in reflection geometry [105, 106]. In this geometry the crystal has its ferroelectric c–axis perpendicular to the entrance faces. During recording each writing beam impinges the sample by opposite face, and then the index change fringes are recorded parallel to the faces. A light reading beam incident perpendicular to the refractive index fringes will be reflected only when its wavelength is $\lambda_R = 1/2n\,\Lambda$. The wavelength selectivity is given by $\Delta\lambda = \lambda_R^2/2nL$, where L is the filter thickness. Of course, to constitute practical devices for long use time these filters must by fixed.

The reported prototype [106] filtering at 518.5 nm and showing a reflection peak higher than 30%, had a FWHM bandwidth of 0.05 nm for 2 mm thick crystal. In fact, the reflecting wavelength of the filter could be adjusted to the required value in the range 400 – 700 nm by changing the recording wavelength and/or the recording beam angle. The reflecting wavelength could also be finely tuned by changing the work temperature, with a variation slope of 5 pm/K. An applied electric field could also be used for tuning or reflection modulation, with a variation slope of 6 pm/kV. Multiwavelength filters can be produced by superposition of gratings.

Leyva et al. [107] presented a different recording procedure which permits extend the reflecting range to the infrared, and in particular to the 1.55 μm spectral region. This is based on the use of a thick crystal and recording in transmission geometry. They obtained 98% reflectivity at 1548 nm with a bandwidth of 0.18 nm.

These Bragg filters can be advantageously applied in high-resolution spectroscopy, solar and stellar astronomy, LIDAR, terrestrial remote sensing, or dense wavelength division multiplexing (DWDM) communication system, or standard filters for laser wavelength locking.

4.5.2.2 Holographic reflection filters in lithium niobate channel waveguides

Permanent refractive index gratings in LiNbO$_3$ waveguide devices are of considerable interest for DWDM communication systems and integrated waveguide lasers. Other applications include grating couplers and optical sensors. Such filters have been deeply studied and produced by Hukriede et al. [108 and references there in].

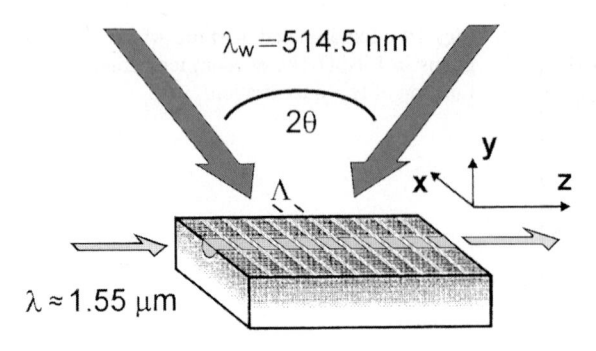

Fig. 20 Scheme of geometries for recording and reading holographic Bragg reflection filters in surface channel waveguides [108].

In order to take advantage of both waveguide and Bragg gratings properties, x- or y-cut samples for light propagation along z direction are used. Preparation start with rare earth active ions indiffusion if it is intended a laser device. Next, Ti indiffusion will produce the single-mode channel waveguide. A final diffusion of Fe in the zone reserved for the holographic Bragg gratings is performed. The order of these processes is determined by the differences in diffusion coefficients. Filter recording and fixing is produced at high temperature (about 180 °C) by two interfering green beams impinging upon the surface at the corresponding crossing angle θ_w. A scheme of the recording and readout geometries is presented in Fig. 20. Active phase stabilization must be used for recording. After going back to room temperature the grating has to be developed with homogeneous light for proper operation. These authors [108] produced filters of 15 mm grating length, showing more than 90% of power peak reflection, with 0.05 nm of bandwidth, and working as well for both polarizations. Several gratings of these characteristics, for reflection of some discrete wavelengths at the same time, were also superimposed in the same device portion. A distributed Bragg reflecting laser provided with two of these reflectors at both cavity ends has been successfully produced recently [90]. This laser presented excellent single-mode emission characteristics.

4.5.2.3 Other applications of fixed holograms in lithium niobate

A fixed high efficiency hologram can be advantageously used in confocal microscopes [109]. The holographic grating has in this case the role of selecting and discriminating the light coming from the objective focus respect to light coming from other aside points or from other distance. This function usually performed by a pinhole in commercial confocal microscopes, is demonstrated to be improved by the Bragg selectivity of volume holographic gratings. Objetive aberrations can be additionally corrected.

Correlation of problem intensity patterns with other stored images can be also successfully implemented with photorefractive holograms in an angular multiplexed memory. Holograms formed by interference of the Fourier transform of the images and the reference plane beam, are recorded to constitute the memory. To obtain the correlation of a new image with those stored in the memory, the crystal is read with the signal beam containing the Fourier transform of the problem image, instead of reading with the plane reference beam. Only a diffracted correlation beam will be produced at the angle of the reference beam used for recording the image matching with the problem one. Such a pattern correlator recorded and fixed in $LiNbO_3$:Fe [110] has been demonstrated to work without degradation under intense illumination.

Fixed gratings can be also used as wavelength filters for demultiplexing information channels of wavelengths separated by few nanometers. Breer et al. [111] and very recently Nee et al. [112] developed such kind of device based on the superimposed fixed holographic gratings in $LiNbO_3$. A demultiplexer for up to sixteen 0.4 nm spaced channels has been characterized.

5 Conclusions

Laser emission can be generated in lithium niobate at any wavelength in its transparency spectral range. This is achieved directly by laser action inside the material or by means of a variety of demonstrated nonlinear processes. Both pulsed and continuous wave emissions are produced. Intensities are increasing from day to day provided the improving in optical resistant dopants.

Devices to control the light: electrical and optical driven switches and modulators, narrow bandwidth filters, couplers, are also implemented. The fabrication techniques of masking, deposition, thermal diffusion, are quite similar to those well extended in semiconductor industry.

Solutions for the DWDM communications technology are offered. Almost all steps from the electrical RF signals to the injection in a fiber, and the inverse way, from the fiber to the separated electrical signals, can be performed in integrated devices on lithium niobate.

Therefore, many lithium niobate competitive waveguide devices seem to be enough mature for industrial implementation. Few of them, such as electrooptic filters and modulators have been produced since several years ago. For others the pass to industrial production is still to come.

In the field of optical memories the combat is very hard. Lithium niobate can offer its stability and material inertness, in comparison with other holographic recording materials such as polymeric and organic ones. Thematic search by holographic correlation can be implemented in angular multiplexed high capacity memory systems. Lithium niobate could have a good chance in this application.

In specific permanent holographic optical elements this material has a real opportunity since the thermal fixing in the adequate conditions provides a long enough lifetime, and maximum efficiencies are reached. The production processes of this kind of devices are well controlled at the present. Now it is the time to include advantageously this kind of optical parts in new optical systems.

The field of optical information encryption seems to be very promising for photorefractive materials. Phase conjugation furnishes the way for precise information decode. Lithium niobate could have its opportunity in this field in both storage medium for phase encoding keys and encoded information memory roles.

As a general concluding reflection one can say that Lithium Niobate can be considered as a Decathlon Winner, it is not the best material for a particular photonic application but could be among the best considered in general.

Acknowledgments The author is very grateful to Prof. A. Suchocki, Prof. M. Carrascosa and Dr. V. Bermudez for their helpful discussions and comments. This work has been partially supported by C.I.C. y T. under grant No. TIC2001–0605.

References

[1] A. Räuber, Chemistry and physics of lithium niobate, in: Current Topics in materials Science I, pp. 481–601, (North Holland, Amsterdam, 1978).
[2] For crystal quality and other parameters of commercial crystals see for example: Beijing Gospel OptoTech Co., http://www.bblaser.com/toriatukai/got/got_ln.htm.
[3] A. M. Prokhorov and Yu. S. Kuz'minov, Physics and chemistry of crystalline lithium niobate, (Adam Hilger, Bristol, 1990).
[4] B. Sturman and V. M. Fridkin, The photovoltaic and photorefractive effects in noncentrosymmetric materials, Ferroelectric and related phenomena Vol. 8, (Gordon and Breach Science Pub., Philadelphia, 1992).
[5] L. O. Svaasand, M. Eriksrud, G. Nakken, and A. P. Grande, Solid-solution range of $LiNbO_3$, J. Cryst. Growth **22**, 230 (1974).
[6] H. M. O'Bryan, P. K. Gallagher, and C. D. Brandle, Congruent composition and Li-rich phase boundary of $LiNbO_3$, J. Am. Ceram. Soc. **68** (9), 493 (1985).
[7] S. C. Abraham, Stoichiometry, defect structure and composition dependence of lithium niobate in: Properties of luthium niobate EMIS Datareview Series No. 5, Inspec (1989).
[8] F. Agulló-López and J. García-Solé, Point defects in $LiNbO^3$, p. 15 in: Properties of lithium niobate, edited by K. K. Wong, EMIS Datareviews Series No. 28, INSPEC, London (2002).
[9] P. F. Bordui, R. G. Norwood, D. H. Hundt, and M. M. Frejer, Preparation and characterization of off-congruent lithium niobate crystals, J. Appl. Phys. **71**, 875 (1992).
[10] K. Kitamura, J. K. Yamamoto, N. I. S. Kimura, and T. Hayashi, Stoichiometric $LiNbO_3$ sigle crystal growth by double crucible Czochralski method using automatic powder supply system, J. Cryst. Growth **116**, 327 (1992).
[11] G. I. Malovichko, V. G. Grachev, E. P. Kokanyan, O. F. Schirmer, K. Betzler, B. Gather, F. Jermann, S. Klauer, U. Schlarb, and M. Wöhlecke, Characterization of stoichiometric $LiNbO_3$ grown from melts containing K_2O, Appl. Phys. A **56**, 103 (1993).

[12] M. D. Serrano, V. Bermudez, L. Arizmendi, and E. Diéguez, Determination of the Li/Nb ratio in LiNbO$_3$ crystals grown by Czochralski method with K$_2$O added to the melt, J. Cryst. Growth **210**, 670 (2000).
[13] V. Bermúdez, P. S. Dutta, M. D. Serrano, and E. Diéguez, On the single domain nature of stoichiometric LiNbO$_3$ grown from melts containing K$_2$O, Appl. Phys. Lett. **70**, 729 (1997).
[14] G. Malovichko, V. Grachev, and O. Schirmer, Interrelation of intrinsic and extrinsic defects – congruent, stoichiometric, and regularly ordered lithium niobate, Appl. Phys. B **68**, 785 (1999).
[15] V. Bermúdez, L. Huang, D. Hui, S. Field, and E. Diéguez, Role of stoichiometric point defect in electric-field-poling lithium niobae, Appl. Phys. A **70**, 591 (2000).
[16] Y. Furukawa, M. Sato, K. Kitamura, Y. Yajima, and M. Minakata, Optical damage resistance and crystal quality of LiNbO$_3$ crystals with various [Li]/[Nb] ratios, J. Appl. Phys. **72**, 3250 (1992).
[17] M. Fontana, K. Chah, M. Aillerie, R. Mouras, and P. Bourson, Optical damage resistance in undoped LiNbO$_3$ crystals, Opt. Mat. **16**, 111 (2001).
[18] V. Bermudez and E. Dieguez, private communication.
[19] A. Askin, G. D. Boyd, J. M. Dziedzic, R. G. Smith, A. A. Ballman, J. J. Levinstein, and K. Nassau, Optically-induced refractive index inhomogeneities in LiNbO$_3$ and LiTaO$_3$, Appl. Phys. Lett. **9**, 72 (1966).
[20] F. S. Chen, Optically induced change of refractive indices in LiNbO$_3$ and LiTaO$_3$, J. Appl. Phys. **40**, 8 (1969).
[21] L. Arizmendi and F. Agulló-López, LiNbO$_3$: A paradigm for photorefractive materials, MRS Bulletin **19** (3), 32 (1984).
[22] G.-G. Zhong, J. Jian, and Z.-K. Wu, 11th Int. Quantum Electron. Conf., IEEE cat. no. 80CH1561-0, p. 631 (1980).
[23] T. R. Volk, M. A. Ivanov, N. M. Rubinina, A. I. Kholodnykh, and H. Metz, Non-photorefractive LiNbO$_3$: Mg as the effective material for the nonlinear optics. Ferroelectrics, **95**, 121 (1989).
[24] J. Koppitz, O. F. Schirmer, M. Wöhlecke, A. I. Kuznetsov, and B. C. Grabmaier, Threshold effects in LiNbO$_3$: Mg caused by change of electron-lattice coupling, Ferroelectrics **92**, 627 (1989).
[25] T. R. Volk, V. I. Pryalkin, and N. M. Rubinina, Optical damage resistant LiNbO$_3$: Zn crystal, Opt. Lett. **15**, 996 (1990).
[26] J. K. Yamamoto, K. Kitamura, N. Iyi, S. Kimura, Y. Furukawa, and M. Sato, Increased optical damage resistance in Sc$_2$O$_3$-doped LiNbO$_3$, Appl. Phys. Lett. **61**, 2156 (1992).
[27] T. Volk, N. Rubinina, and M. Wöhlecke, Optical damage resistant impurities in lithium niobate, J. Opt. Soc. Am. B **11**, 1681 (1994).
[28] E. P. Kokanyan, V. G. Babajanyan, G. G. Demirkhanyan, J. B. Gruber, and S. Erdel, Periodically poled structures in doped lithium niobate crystals, J. Appl. Phys. **92**, 1544 (2002).
[29] A. Yariv and P. Yeh, Optical waves in crystals, Wiley series in pure and applied optics, (John Wiley and Sons, New York, 1984).
[30] V. Hinkov, Velocity of SAW's on lithium niobate: Anisotropy, temperature and composition dependence, in: Properties of luthium niobate EMIS Datareview Series No. 5, Inspec (1989).
[31] C. S. Tsai, Integrated acoustooptic circuits and applications, IEEE Trans. Ultrason. Ferroelectr. Freq. Control, **39**, 529 (1992).
[32] H. G. Festl, P. Hertel, E. Krätzig, and R. von Baltz, Investigatin of the photovoltaic tensor in doped LiNbO$_3$, phys. stat. sol. (b) **113**, 157 (1982).
[33] P. Günter and J.-P. Huignard Ed., Photorefractive materials and their applications I, Topics, in: Applied Physics, Vol. 61, (Springer-Verlag, Berlin, 1988).
[34] W. Phillips, J. J. Amodei, and D. L. Staebler, Optical and holographic storage properties of transition metal doped lithium niobate, RCA Review **33**, 95 (1972).
[35] M. M. Frejer, G. A. Magel, D. H. Jundt, and R. L. Byer, Quasi-phase-matched second harmonic generation: tuning and tolerances, IEEE J. Quantum Electron. **28**, 2631 (1992).
[36] U. Schlarb and K. Betzler, Refractive indices of lithium niobate as a function of wavelength and composition, J. Appl. Phys. **73**, 3472 (1993).
[37] Y. Lu, Y. Lu, X. Cheng, C. Xue, and N. Ming, Growth of optical superlattice LiNbO$_3$ with different modulation periods and its application in second harmonic generation, Appl. Phys. Lett. **68**, 2781 (1996).
[38] N. B. Ming, J. F. Hong, and D. Feng, The growth striations and ferroelectric domain structures in Czochralski-grown LiNbO$_3$ single crystals J. Mater. Sci. **17**, 1663 (1982).
[39] J. Capmany, V. Bermudez, and E. Diéguez, Bulk periodically poled lithium niobate doped with Yb^{3+} ions: growth and characterization, Appl. Phys. Lett. **74**, 1534 (1999).
[40] M. Yamada, N. Nada, M. Sayito, and K. Watanabe, First order quasi-phase matched LiNbO$_3$ waveguide periodically poled by applying an external field for efficient blue second-harmonic generation, Appl. Phys. Lett. **62**, 435 (1993).

[41] M. Yamada and M. Saitoh, Fabrication of a periodically poled laminar domain structure with a pitch of a few micrometers by applying an external electric field, J. Appl. Phys. **84**, 2199 (1998).
[42] C. Becker, T. Oesselke, J. Pandavenes, R. Ricken, K. Rochhausen, G. Schreiber, W. Sohler, H. Suche, R. Wessel, S. Balsamo, I. Montrosset, and D. Sciancalepore, Advanced Ti:Er:LiNbO$_3$ waveguide lasers, IEEE J. Sel. Top. Quantum Electron. **6**, 101 (2000).
[43] A. C. G. Nutt, V. Gopalan, and M. C. Gupta, Domain inversion in LiNbO$_3$ using direct electron-beam writing, Appl. Phys. Lett. **60**, 2828 (1992).
[44] J. Capmany, V. Bermudez, D. Callejo, J. García-Solé, and E. Diéguez, Continuous wave simultaneous multi-self-frequency conversion in Nd^{3+}-doped aperiodically poled bulk lithium niobate, Appl. Phys. Lett. **76**, 1225 (2000).
[45] V. Bermudez, D. Callejo, R. Vilaplana, J. Capmany, and E. Diéguez, Engineering of lithium niobate domain structure through the off-centered Czochralski growth technique, J. Cryst. Growth **237–239**, 677 (2002).
[46] J. Capmany, Simultaneous generation of red, green and blue continuous wave laser radiation in Nd^{3+}-doped aperiodically poled lithium niobate, Appl. Phys. Lett. **78**, 144 (2001).
[47] P. E. Powers, T. J. Kulp, and S. E. Bisson, Continuous tuning of a continuous wave periodically poled lithium niobate optical parametric oscillator by use of a fan-out grating design, Opt. Lett. **23**, 159 (1998).
[48] N. G. R. Broderick, G. W. Ross, H. L. Offerhaus, D. J. Richardson, and D. C. Hanna, Hexagonally poled lithium niobate: A two dimensional nonlinear photonic crystal, Phys. Rev. Lett. **84**, 4345 (2000).
[49] W. K. Burns, P. H. Klein, E. J. West, and L. E. Plew, Ti diffusion in Ti:LiNbO$_3$ planar and channel waveguides, J. Appl. Phys. **50**, 6175 (1979).
[50] G. J. Griffiths and R. J. Esdaile, Analysis of titanium diffused planar optical waveguides in lithium niobate, IEEE J. Quantum Electron. **QE-20**, 149 (1984).
[51] R. A. Becker, Thermal fixing of Ti-diffused LiNbO$_3$ channel waveguides for reduced photorefractive susceptibility, Appl. Phys. Lett. **45**, 121 (1984).
[52] J. L. Jackel, C. R. Rice, and J. J. Veselka, Proton exchange for high index waveguides in LiNbO$_3$, Appl. Phys. Lett. **41**, 607 (1982).
[53] Y. N. Korkishko and V. A. Fedorov, Relationship between refractive indices and hydrogen concentration in proton exchanged LiNbO$_3$ waveguides, J. Appl. Phys. **82**, 1010 (1997).
[54] V. Hinkov and E. Ise, Control of birefringence in Ti:LiNbO$_3$ optical waveguides by proton exchange of lithium ions, J. Lightwave Technol. **4**, 444 (1986).
[55] W. M. Young, M. M. Frejer, M. J. F. Digonnet, A. F. Marshall, and R. S. Feigelson, Fabrication, characterization and index profile modeling of high damage resistance Zn-diffused waveguides in congruent and MgO:LiNbO$_3$, J. Lightwave Technol. **10**, 1238 (1992).
[56] R. Nevado and G. Lifante, Low loss, damage resistantoptical waveguides in Zn-diffused LiNbO$_3$ by a two step procedure, Appl. Phys. A **72**, 725 (2001).
[57] R. Nevado, C. Sada, F. Segato, F. Caccavale, A. Kling, J. C. Soares, E. Cantelar, F. Cussó, and G. Lifante, Compositional characterization of Zn-diffused lithium niobate waveguides, Appl. Phys. B **73**, 555 (2001).
[58] N. Hamelin and Y. T. Chow, Guided-type second harmonic generation in ion implanted MgO:LiNbO$_3$, J. Mod. Opt. **45**, 2125 (1998).
[59] A. Boudriona, J. C. Loulergue, F. Laurell, and P. Moretti, Nonlinear optical properties of (H$^+$, He$^+$)- implanted planar waveguides in z-cut lithium niobate: annealing effect, J. Opt. Soc. Am. B **18**, 1832 (2001).
[60] H. Hu, F. Lu, F. Chen, B-R. Shi, K-M. Wang, and D-Y. Shen, Monomode optical waveguide in lithium niobate formed by MeV Si$^+$ ion implantation, J. Appl. Phys. **89**, 5224 (2001).
[61] D. H. Jundt, G. A. Magel, M. M. Frejer, and R. L. Byer, Periodically poled LiNbO$_3$ for high-efficiency second-harmonic generation, Appl. Phys. Lett. **59**, 2657 (1991).
[62] H. Tamada, A. Yamada, and M. Saitoh, LiNbO$_3$ thin-film optical waveguide grown by liquid phase epitaxy and its application to second-harmonic generation, J. Appl. Phys. **70**, 2536 (1991).
[63] A. Yamada, H. Tamada, and M. Saitoh, Liquid phase epitaxial growth of LiNbO$_3$ thin film using Li$_2$O–B$_2$O$_3$ flux system. [Optical waveguide fabrication], J. Crys. Growth **132**, 48 (1993).
[64] A. Yamada, H. Tamada, and M. Saitoh, Photorefractive damage in LiNbO$_3$ thin-film optical waveguides grown by liquid phase epitaxy, J. Appl. Phys. **76**, 1776 (1994).
[65] D. Callejo, V. Bermúdez, M. D. Serrano, and E. Diéguez, Lithium niobate films on periodic poled lithium niobate substrates prepared by liquid phase epitaxy, J. Cryst. Growth **237–239**, 596 (2002).
[66] C. S. Tsai, D. Y. Zang, and P. Le, Acoustooptic Bragg diffraction in LiNbO$_3$ channel-planar composite waveguide with application to optical computing, Appl. Phys. Lett. **47**, 549 (1985).
[67] U. Hempplemann, H. Herrmann, G. Mrozynski, V. Reimann, and W. Sholer, Integrated optical proton exchanged TM-pass polarizers in LiNbO$_3$: modelling and experimental performance, J. Lightwave Technol. **13**, 1750 (1995).

[68] B. E. Benkelfat, R. Ferrière, B. Wacogne, and P. Mollier, Technological implementation of Bragg grating reflectors in Ti:LiNbO3 waveguides by proton exchange, IEEE Photonics Technol. Lett. **14**, 1430 (2002).

[69] A. Kar-Roy, and C. S. Tsai, 8 × 8 symmetric nonblocking integrated acoustooptic space switch module in $LiNbO_3$, IEEE Photonics Technol. Lett. **4**, 731 (1992).

[70] R. C. Alferness, Waveguide eletrooptic modulators, IEEE Trans. Microw. Theory Tech. **30**, 1121 (1982).

[71] E. L. Wooten, K. M. Kissa, A. Yi-Yan, E. J. Murphy, D. A. Lafaw, P. F. Hallemeier, D. Maak, D. V. Attanasio, D. J. Fritz, G. J. McBrien, and D. E. Bossi, A review of lithium niobate modulators for fiber-optic communications systems, IEEE J. Sel. Top. Quantum. Electron. **6**, 69 (2000).

[72] J. Kondo, A. Kondo, K. Aoki, M. Imaeda, T. Mori, Y. Mizuno, S. Takatsuji, Y. Kozuka, O. Mitomi, and M. Minakata, 40 Gb/s X-cut $LiNbO_3$ optical modulator with two step back-slot structure, J. Lightwave Technol. **20**, 2110 (2002).

[73] L. E. Myers, R. C. Eckardt, M. M. Frejer, R. L. Byer, W. R. Bosemberg, and J. W. Pierce, Quasi-phase-matched optical parametric oscillators in bulk periodically poled $LiNbO_3$, J. Opt. Soc. Am. B **12**, 2102 (1995).

[74] L. E. Myers and W. R. Bosemberg, Periodically poled lithium niobate and quasi-phase-matched optical parametric oscillation, IEEE J. Quantum Electron. **33**, 1663 (1997).

[75] L. E. Myers, R. C. Eckardt, M. M. Frejer, R. L. Byer, and W. R. Bosemberg, Multigrating quasi-phase matched optical parametrc oscillator in periodically poled $LiNbO_3$, Opt. Lett. **21**, 591 (1996).

[76] P. Gross, M. E. Klein, H. Ridderbusch, D. H. Lee, J. P. Meyn, R. Wallenstein, and K. J. Boller, Wide wavelength tuning of an optical parametric oscillator through electrooptic shaping of the gain spectrum, Opt. Lett. **27**, 1433 (2002).

[77] Y. C. Huang, A. C. Chiang, Y. Y. Lin, and Y. W. Fang, Optical parametric generation covering the sodium D1, D2 lines from a 532-nm pumped periodicallypoled lithium niobate (PPLN) crystal with ionic-nonlinearity enhanced parametric gain, IEEE J. Quantum Electron. **38**, 1614 (2002).

[78] J. Capmany, D. Callejo, V. Bermúdez, E. Diéguez, D. Artigas, and L. Torner, Continuous wave self-pumped optical parametric oscillator based on Yb^{3+}-doped bulk periodically poled $LiNbO_3$ (MgO), Appl. Phys. Lett. **79**, 293 (2001).

[79] L. Barraco, A. Grisard, E. Lallier, P. Bourdon, and J. P. Pocholle, Self-optical parametric oscillation in periodically poled neodymium-doped lithium niobate, Opt. Lett. **27**, 1540 (2002).

[80] K. P. Petrov, S. Waltman, E. J. Diugokencky, M. Arbore, M. M. Frejer, F. K. Tittel, and L. W. Hollberg, Precise measurement of methane in air using diode-pumped 3.4-μm difference-frequency generation in PPLN, Appl. Phys. B **64**, 567–72 (1997).

[81] D. Mazzotti, P. De Natale, G. Giusfredi, C. Fort, J. A. Mitcheli, and L. W. Hollberg, Difference-frequency generation in PPLN at 4.25 μm: an anlysis of sensitivity limits for DFG spectrometers, Appl. Phys. B **70**, 747–50 (2000).

[82] D. J. L. Birkin, E. U. Rafailov, G. S. Sokolovskii, W. Sibbett, G. W. Ross, P. G. R. Smith, and D. C. Hanna, 3.6 mW blue light by direct frequency doubling of a diode laser using an aperiodically poled lithium niobate crystal, Appl. Phys. Lett. **78**, 3172 (2001).

[83] P. Loza-Alvarez, M. Ebrahimzadeh, W. Sibbett, D. T. Reid, D. Artigas, and M. Missey, Femtosecond second harmonic pulse compression in aperiodically poled lithium niobate: a systematic comparison of experiment and theory, J. Opt. Soc. Am. B **18**, 1212 (2001).

[84] K. Kintaka, M. Fujimura, T. Suhara, and H. Nishihara, High efficiency $LiNbO_3$ waveguide second-harmonic generation devices with ferroelectric domain inverted gratings fabricated by applying voltage, J. Lightwave Technol. **14**, 462 (1996).

[85] J. Webjörn, S. Siala, D. K. Nam, R. G. Waarts, and J. Lang, Visible laser sources based on frequency doubling in nonlinear waveguides, IEEE J. Quantum Electron. **33**, 1673 (1997).

[86] N. Hamelin, and Y. T. Chow, Guided type second harmonic generation in ion implanted $MgO:LInbO_3$, J. Mod. Opt. **45**, 2125 (1998).

[87] M. Domenech, R. E. DiPaolo, G. Lifante, and F. Cussó, Blue light by SHG in diode pumped $LiNbO_3$ waveguides, phys. stat. sol. (a) **192**, 135 (2002).

[88] M. H. Chou, I. Brener, M. M. Frejer, E. E. Chaban, and S. B. Christman, 5 μm-band wavelength conversion based cascaded second order nonlinearity in lithium niobate waveguides, IEEE Photonics Technol. Lett. **11**, 653 (1999).

[89] B. K. Das, H. Suche, and W. Sohler, Single frequency Ti:Er:LiNbO3 distributed Bragg reflector waveguide laser with thermally fixed photorefractive cavity, Appl. Phys. B **73**, 439 (2001).

[90] E. M. de Miguel-Sanz, M. Carrascosa, and L. Arizmendi, Effect of the oxidation state and the hydrogen concentration on the lifetime of thermally fixed holograms in $LiNbO_3$:Fe, Phys. Rev. B **65**, 165101 (2002).

[91] L. Hesselink and M. C. Bashaw, Optical memories implemented with photorefractive media, Opt. Quantum Electron. **25**, S611 (1993).

[92] J. F. Heanue, M. C. Bashaw, and L. Hesselink, Volume holographic storage and retrieval of digital data, Science **265**, 749 (1994).
[93] G. W. Burr, C. M. Jefferson, H. Coufal, M. Jurich, J. A. Hoffnagle, R. M. Macfarlane, and R. M. Shelby, Volume holographic data storage at an areal density of 250 gigapixels/in.^{2}, Opt. Lett. **26**, 444 (2001).
[94] J. H. Hong, I. McMichael, T. Y. Chang, W. Christian, and E. G. Paek, Volume holographic memory systems: techniques and architectures, Opt. Eng. **34**, 2193 (1995).
[95] W. Liu and D. Psaltis, Pixel size limit in holographic memories, Opt. Lett. **24**, 1340 (1999).
[96] H. Mitsuhashi and M. Obara, Compact holographic memory system using a one-beam geometry in photorefractive crystal, Appl. Phys. Lett. **79**, 7 (2001).
[97] K. Buse, A. Adivi, and D. Psaltis, Non-volatile holographic storage in doubly doped lithium niobate crystals, Nature, **393**, 665 (1998).
[98] L. Hesselink, S. S. Orlov, A. Liu, A. Akella, D. Lande, and R. R. Neurgaonkar, Photorefractive materials for non-volatile volume holographic data storage, Science **282**, 1089 (1998).
[99] X. An, D. Psaltis, and G. W. Burr, Thermal fixing of 10000 holograms in $LiNbO_3$:Fe, Appl. Opt. **38**, 386 (1999).
[100] O. Matoba and B. Javidi, Encrypted optical memory system using three-dimensional keys in the Fresnel domain, Opt. Lett. **24**, 762 (1999).
[101] X. Tan, O. Matoba, T. Shimura, K. Kuroda, and B. Javidi, Secure optical storage that uses fully phase encryption, Appl. Opt. **39**, 6689 (2000).
[102] E. M. de Miguel, J. Limeres, M. Carrascosa, and L. Arizmendi, Study of developing of termal fixed holograms in lithium niobate, J. Opt. Soc. Am. B **17**, 1140 (2000).
[103] A. Méndez and L. Arizmendi, Maximum diffraction efficiency of fixed holograms in lithium niobate, Opt. Mat. **10**, 55 (1998).
[104] S. Breer K. Buse, K. Peithmann, H. Vogt, and E. Krätzig, Stabilized recording and thermal fixing of holograms in photorefractive lithium niobate crystals, Rev. Sci. Instum. **69**, 1591 (1998).
[105] G. A. Rakuljic and V. Leyva, Volume holographyc narrow-band optical filter, Opt. Lett. **18**, 459 (1993).
[106] R. Müller, M. T. Santos, L. Arizmendi, and J. M. Cabrera, A narrow-band interference filter with photorefractive $LiNbO_3$, J. Phys. D, Appl. Phys. **27**, 241 (1994).
[107] V. Leyva, G. A. Rakuljic, and B. O'Conner, Narrow bandwidth volume holographic optical filter operating at the Kr transition at 1547.82 nm, Appl. Phys. Lett. **65**, 1079 (1994).
[108] J. Hukriede, D. Runde, and D. Kip, Fabrication and applications of holographi Bragg gratings in lithium niobate channel waveguides, J. Phys. D, Appl. Phys. **36**, R1 (2003).
[109] G. Barbastathis, M. Balberg, and D. J. Brady, Confocal microscopy with a volume holographic filter, Opt. Lett. **24**, 811 (1999).
[110] E. M. de Miguel-Sanz, M. Tebaldi, S. Granieri, N. Bolognini, and L. Arizmendi, Photorefractive fixed optical correlator, Appl. Phys. B **70**, 279 (2000).
[111] S. Breer and K. Busse, Wavelength demultiplexing with volume phase holograms in photorefractive lithium niobate, Appl. Phys. B **66**, 339 (1998).
[112] I. Nee, O. Beyer, M. Muller, and K. Buse, Multichannel wavelength-division-multiplexing with thermally fixed Bragg gratings in photorefractive lithium niobate crystals, J. Opt. Soc. Am. B **20**, 1593 (2003).

Phase relations in the growth of stoichiometric lithium niobate

K. Polgár[*,1], **Á. Péter**[1], and **M. Ferriol**[2]

[1] Research Institute for Solid State Physics and Optics, Hungarian Academy of Sciences, Budapest, Hungary
[2] University of Metz, Laboratoire de Chimie et Applications – EA 3471, Saint-Avold, France

Received 15 September 2003, accepted 23 October 2003
Published online 29 December 2003

PACS 64.70.Dv, 81.10.Dn, 81.30.Dz

The conditions of the formation of stoichiometric and near-stoichiometric LiNbO$_3$ single crystals from high temperature solution have been investigated in the K$_2$O–Li$_2$O–Nb$_2$O$_5$ ternary system. The liquidus-solidus tie lines were established along different sections of the LiNbO$_3$ phase field. Solidification with constant and near 50 mol% Li$_2$O composition was found only at the LiNbO$_3$–K$_2$O isopleth in the temperature range 1114–1000 °C.

© 2004 WILEY-VCH Verlag GmbH & Co. KGaA, Weinheim

1 Introduction   LiNbO$_3$ is one of the most common perovskite-type crystals for optical applications. It is a typical non stoichiometric material, melts congruently at 48.6 mol% Li$_2$O content and can be grown easily by the Czochralski method from its melt. The Li deficiency is a serious drawback causing a defect structure with about 1% antisite Nb^{5+} ions in the lattice. Therefore stoichiometric crystals (50 mol% Li$_2$O, or [Li]/[Nb] = 1) are expected to show improved performance for a number of applications. In this structure the amount of antisite Nb^{5+} is below 0.01% which results in a blue-shift of the UV absorption edge [1, 2], and the narrowing of several spectral lines (EPR, NMR, [3], IR OH$^-$ vibration lines [4].) An important reduction of the electric field (200 V instead of 20 kV) required for ferroelectric domain inversion in the stoichiometric crystal [5] makes it easier to produce periodically polarized structures for quasi-phase matching.

Several methods exist for the preparation of stoichiometric LiNbO$_3$ single crystals. The one which yields the composition closest to 50% Li$_2$O is the high temperature top seeded solution growth method from the K$_2$O–Li$_2$O–Nb$_2$O$_5$ ternary mixture. K$_2$O containing flux is especially good for the growth of stoichiometric LiNbO$_3$ crystals, since the K$^+$ ion does not enter the lattice, and significantly lowers the temperature of crystallization [6]. In this work the role of the phase relations of the ternary system in the growth of stoichiometric lithium niobate is investigated. The impact of the [Li]/[Nb] ratio in the flux on the crystal's stoichiometry is discussed.

2 Experimental methods   A series of K$_2$O–Li$_2$O–Nb$_2$O$_5$ mixtures were prepared along the 10, 20, 25 and 30 mol% K$_2$O isopleths in the range 25–67.5 mol% Nb$_2$O$_5$. We measured the DSC curves of each sample and so determined the temperature of the melting point and other phase changes. The phases occurring in the samples were identified by X-ray phase analysis.

Crystals were grown from starting compositions located at the 10 mol% K$_2$O isopleth (Table 1) by the TSSG technique. The sample preparation and crystal growth method are described in [6, 7].

The Li/Nb ratio in the crystals was determined by the UV shift of the fundamental absorption edge [8].

---

[*] Corresponding author: e-mail: polgar@szfki.hu, Phone: +36 1392 2222, Fax: +36 1392 2223

Table 1 Starting flux compositions used for growth experiments and calculated yield of the LiNbO$_3$ phase.

growth label	K$_2$O [mol%]	Nb$_2$O$_5$ [mol%]	Li$_2$O [mol%]	yield [%]
g$_0$	10	44	46	32
g$_1$	10	45	45	41
g$_2$	10	48	42	64
g$_3$	10	50	40	61

3 Results and discussion Figure 1 shows the polythermal projection of the LiNbO$_3$ (LN) liquidus surface in the ternary system. The presented phase boundaries were established by our DSC analyses discussed elsewhere [9]. The liquidus surface beyond the LiNbO$_3$–KNbO$_3$ join is only estimation. The single-phase liquidus region of LN is limited by four monovariant lines (e.g. Liq. + Li$_3$NbO$_4$ + LiNbO$_3$, Liq. + KNbO$_3$ + LiNbO$_3$, Liq. + KLN + LiNbO$_3$ and Liq. + LiNb$_3$O$_8$ + LiNbO$_3$). During the crystallization of LN phase, the point representing the liquid in equilibrium with the solid moves on the liquidus surface towards one of the univariant lines limiting the liquidus surface. Since K$_2$O is insoluble in lithium niobate, the compositional changes, which occur during solidification, may be traced. The composition of the solid (LN) in first approximation is assumed to be the same at all temperatures, thus the liquid composition evolves along the straight line joining the starting composition, LN, and the one corresponding to the point on the univariant line (see "g" lines in the Fig. 1). When the liquid composition reaches a point on one of the univariant lines with further decrease of the temperature, two solids will separate: on the line e_1–E_t: LiNbO$_3$ + Li$_3$NbO$_4$ and on e_2–P: LiNbO$_3$ + KLN solid solution. Between points E_t and P the co-crystallization of LN + KN occurs. However, solidification cannot be completed until the liquid composition reaches the ternary eutectic point E_t at 997 °C.

To establish the tie lines linking the corresponding liquidus and solidus points, growth experiments were conducted from starting compositions located at the different "g" lines (Table 1). From the phase diagram [9] the yield of the LN phase is calculated starting from the 10 mol% K$_2$O isopleth. In the experiment labeled as g$_0$, g$_1$ and g$_2$–g$_3$, the Li/Nb ratio of the starting liquid composition was above, equal and below 1 respectively. For g$_1$, g$_2$ and g$_3$ the vertical sections of the phase diagram are shown in Fig. 2 a, b and c.

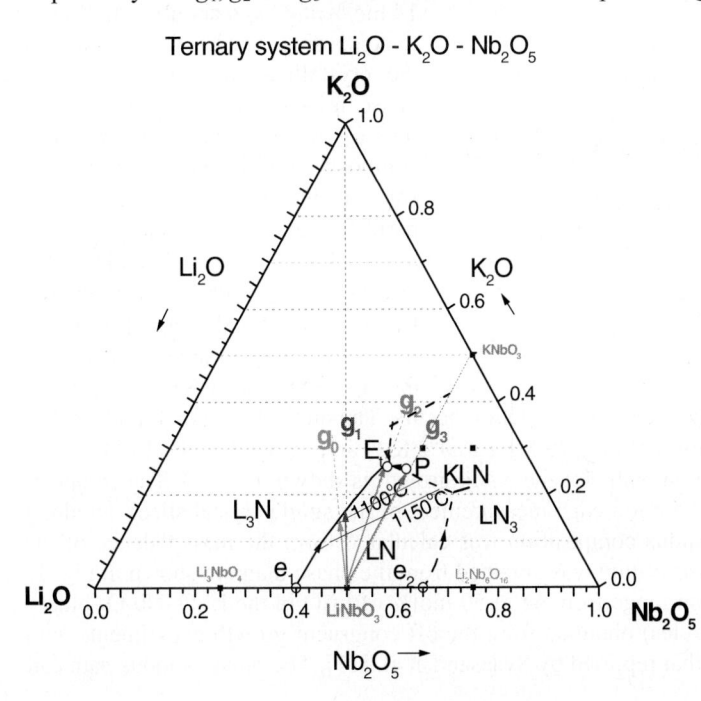

Fig. 1 Polythermal phase diagram of the K$_2$O–Nb$_2$O$_5$–Li$_2$O ternary system showing the LiNbO$_3$ liquidus region. E_t: ternary eutectic at 997 °C, e_1 and e_2 binary eutectic at 1160 °C and 1200 °C respectively. P: quasi-peritectic point at 1050 °C, the g$_0$, g$_1$, g$_2$, g$_3$ lines locate the starting compositions of the growth experiments. LN: LiNbO$_3$, L$_3$N: Li$_3$NbO$_4$, LN$_3$: LiNb$_3$O$_8$, KLN: tungsten bronze-type solid solution.

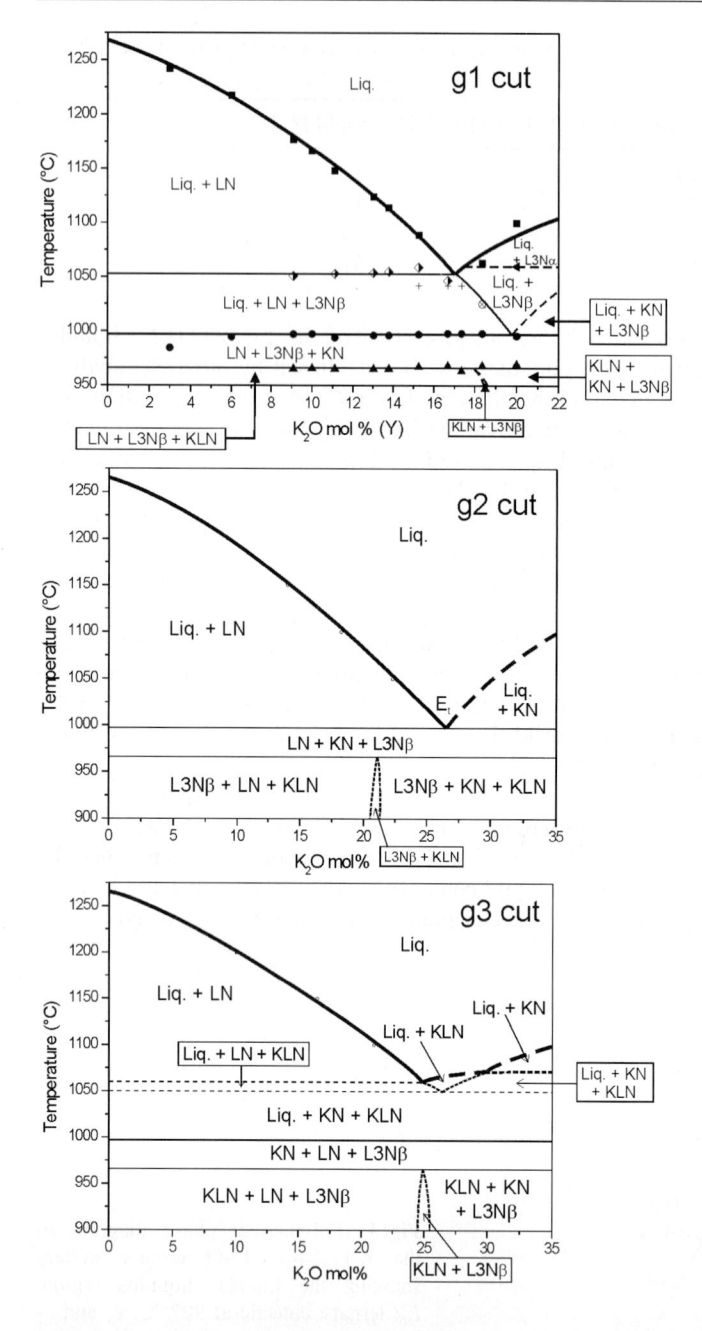

Fig. 2 Vertical sections of the LiNbO$_3$ phase field: LiNbO$_3$–K$_2$O: g_1 cut, LiNbO$_3$–E_t: g_2 cut and LiNbO$_3$–KNbO$_3$: g_3 cut. LN: LiNbO$_3$, L3N: Li$_3$NbO$_4$, KN: KNbO$_3$, KLN: tungsten bronze-type solid solution.

The vertical section along the LiNbO$_3$–K$_2$O join (shown in Fig. 2a) describes the evolution of the liquidus during the growth experiments conducted on the g_1 line. In previous works [6, 7, 10] the growth of sLN single-crystals was usually experienced from starting compositions located on this line and above 10 mol% K$_2$O. There the crystallization of the LN phase is limited by the e_1–E_t univariant line at 18 mol% K$_2$O and 1050 °C. The vertical cut along g_0 is not shown here since the constituent phases are the same as shown for g_1 and slope of the surface is similar too. Along g_0 the LN liquidus reaches the e_1–E_t eutectic valley at 14 mol% K$_2$O and about 1100 °C.

According to the phase diagram the crystallization yield of the LN phase is maximal along the LN–E_t line (g_2 vertical cut, Fig. 2b). The composition of the eutectic liquid (E_t) was deduced from growth experiments and was found: 44.5 ± 1.5 mol% Nb$_2$O$_5$, 28 ± 1.5 mol% K$_2$O and 27.5 ± 1.5 mol% Li$_2$O. LiNbO$_3$–KNbO$_3$ (g_3 vertical section) is shown in Fig. 2c. Close to the intersection of the g_3 and Liq. + KLN + LiNbO$_3$ monovariant lines at 1050–1055 °C a quasi-peritectic reaction occurs. The quasi-peritectic liquid (P) has a composition roughly equal to 49 ± 1 mol% Nb$_2$O$_5$, 25.5 ± 1 mol% K$_2$O and 25.5 ± 1 mol% Li$_2$O.

The evolution of the solidus composition paths for the g_0–g_3 growth is shown in Fig. 3. The composition of the solidus was deduced from absorption edge measurement of the single crystal slices cut along the growth axes. The corresponding liquidus composition was calculated from the mass balance of the pulled amount, and the crystallization temperature was obtained from the phase diagram shown in Fig. 2.

LiNbO$_3$ has a fairly large solid solution range from 44 to 50 mol% of Li$_2$O in the Li$_2$O–Nb$_2$O$_5$ binary system. The binary LN solidus (open circles) obtained from the off congruent growth experiments with excess lithium in the melt are close to that reported by Svaasand et al. [11]. The binary solidus path can

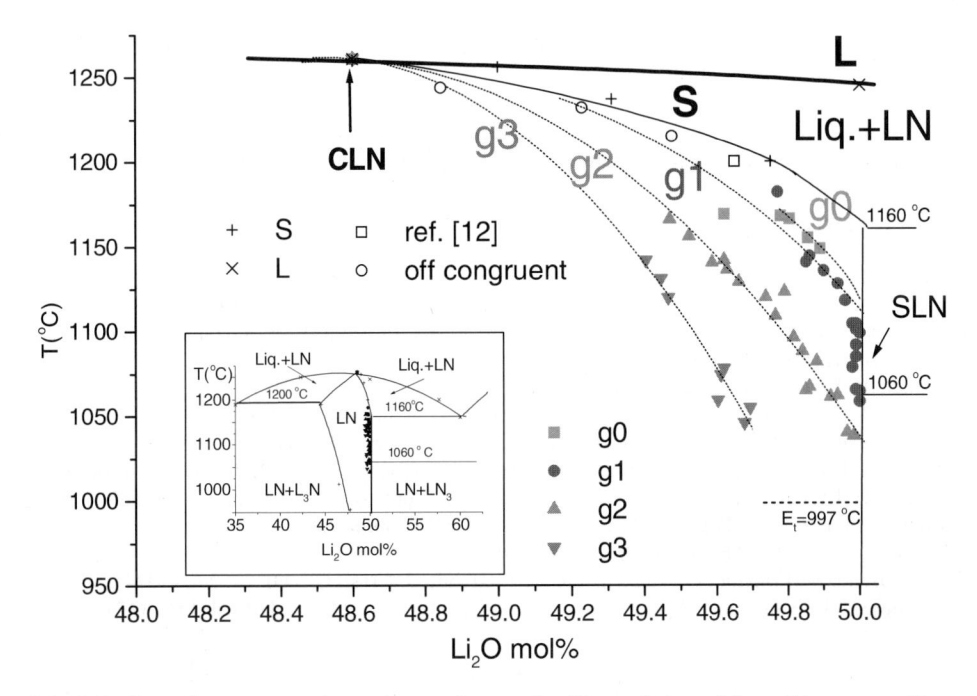

Fig. 3 (online colour at: www.interscience.wiley.com) The evolution of the solidus composition paths in the Li_2O–Nb_2O_5 binary system for liquidus paths along the g_0–g_3 lines of the K_2O–Nb_2O_5–Li_2O ternary system. The inset shows the binary Li_2O–Nb_2O_5 phase diagram from 35 to 60 mol percent Li_2O according to Ref. [11] with the indication of our points.

be deduced down to (e_1) 1160 °C. The lowest reported value (open square) growth experiments are at 1200 °C (obtained by TSSG growth [12] from melt with 58 mol% Li_2O using a continuous powder supply system. By growing crystals in the ternary system the solidus paths can be extended down to 1000 °C.

The solidus lines obtained from the crystals grown along the investigated "g" lines were different. Along the g_1 line [Li]/[Nb] = 0.999 – 1.000 can be reached and in this case the [Li]/[Nb] ratio stays constant (uniform) all along the whole crystal if the starting liquidus temperature is ≤1114 °C as already reported in [7]. Working on the g_2 and the g_3 lines, the [Li]/[Nb] ratio in the flux is <1 and on g_2 we can approach the stoichiometry to [Li]/[Nb] = 0.992, with monotonously changing composition along the growth between 0.979 and 0.992 (filled up triangles in Fig. 3). Growing crystals along the g_3 line the limit in composition is about [Li]/[Nb] = 0.984 with shifting composition as well (down triangles in Fig. 3). Growing with excess Li_2O ([Li]/[Nb] > 1) in the flux the solidus path converges to the binary line (see g_0, filled squares in Fig. 3). On the g_0 line crystals are grown also with shifting composition from [Li]/[Nb] = 0.984 down to 0.999. Stoichiometric composition can be obtained only at the end of the crystallization range, when the liquidus practically reaches the monovariant line (1100 °C). In the ternary system only the g_1 solidus composition path is likely to have a constant segment, very close to the [Li]/[Nb] = 1 composition. Thus from the point of view of the growth of stoichiometric $LiNbO_3$ crystals the choice of starting compositions along the g_1 line, above 13 mol% K_2O is advisable. For doped crystals, to assure both the right incorporation and the near stoichiometry of the crystals, starting from points with compositions [Li]/[Nb] < 1 have to be considered.

Conclusion The tie lines linking the corresponding liquidus and solidus points were determined along 4 vertical sections of the K_2O–Nb_2O_5–Li_2O ternary system. The compositional path of the solidus lines is traced down to 1000 °C. A constant (near 50 mol% composition) Li_2O solidus line segment was found only for [Li]/[Nb] = 1 liquidus composition (g_1 line) in the temperature range 1114–1050 °C.

The conditions of the formation of stoichiometric and near-stoichiometric LiNbO$_3$ single crystals from high temperature solution are optimal in this range.

Acknowledgment This work was supported by the Hungarian Scientific Research Fund (OTKA) Nos. T034176, T034262 and by the European Union under Grant CMRC No.ICA1-CT-2000-70029.

References

[1] I. Földvári, K. Polgár, and A. Mecseki, Acta Phys. Hung. **55**, 321 (1984).
[2] I. Földvári, K. Polgár, R. Voszka, and R. N. Balasanyan, Cryst. Res. Technol. **19**, 1659 (1984).
[3] G. I. Malovichko, A. V. Grachev, L. P. Yurchenko, V. Ya. Proshko, E. P. Kokanyan, and V. T. Gabrielyan, phys. stat. sol. (a) **133**, K29 (1992).
[4] L. Kovács, V. Szalay, and R. Capelletti, Solid State Commun. **52**, 1029 (1984).
[5] A. Grisard, E. Lallier, K. Polgár, and Á. Péter, Electron. Lett. **36**, 1043 (2000).
[6] K. Polgár, Á. Péter, L. Kovács, G. Corradi, and Zs. Szaller, J. Cryst. Growth **177**, 211 (1997).
[7] K. Polgár, Á. Péter, L. Pöppl, M. Ferriol, and I. Földvári, J. Crystal Growth **237–239**, 682 (2002).
[8] L. Kovács, G. Ruschhaupt, K. Polgár, G. Corradi, and M. Wöhlecke, Appl. Phys. Lett. **70**, 2801 (1997).
[9] M. Cochez, M. Ferriol, L. Pöppl, K. Polgár, and Á. Péter to be published.
[10] K. Polgár, Á. Péter, and I. Földvári, Opt. Mater. **19/1**, 7 (2002).
[11] L. O. Svaasand, M. Eriksrud, G. Nakken, and A. P. Grande, J. Cryst. Growth **22**, 230 (1974).
[12] K. Kitamura, J. K. Yamamoto, N. Iyi, S. Kimura, and T. Hayashi, J. Cryst. Growth **116**, 327 (1992).

Optical performance of Yb^{3+} in LiNbO$_3$ laser crystal

L. E. Bausá[*], **M.O Ramírez**, and **E. Montoya**[**]

Dpto. Física de Materiales, Universidad Autónoma de Madrid, 28049-Madrid, Spain

Received 10 October 2003, revised 17 November 2003, accepted 17 November 2003
Published online 12 January 2004

PACS 42.60.Jf, 42.62.Fi, 42.65.Ky, 42.70.Hj, 78.30.Hv, 78.45.+h

Yb^{3+} doped crystals are recently attracting much attention as potential solid state laser materials due to the particular characteristics provided by the energy level scheme of this ion. Laser action from Yb^{3+} doped systems shows several interesting properties such as the possibility of a certain tunability range. The most relevant results obtained in the Yb^{3+}:LiNbO$_3$ system will be shown. We show the main spectroscopic parameters of this system: polarized absorption and luminescence spectra at low and room temperature, and electron–phonon coupling. The laser performance under Ti:sapphire and diode pumping is presented. Stable laser action in the near infrared region and coherent green radiation by self-frequency doubling have been obtained simultaneously in this non-linear host. The results have been obtained for single domain and periodically poled lithium niobate crystals.

© 2004 WILEY-VCH Verlag GmbH & Co. KGaA, Weinheim

1 Introduction There is an increasing interest in the development of compact or miniaturized laser devices emitting in the green and blue region of the spectrum due to the requirements of a great variety of fields in science and technology, such as high density optical data storage, laser printing, undersea communications, medicine, laser show and display applications, which require the use of compact coherent visible light sources [1, 2].

Though semiconductor lasers are nowadays able to provide laser emission in the visible spectral region, our attention is focused on the so-called solid state lasers (those based on optically active centers in insulating materials) which offer some advantages over semiconductor lasers, as it is a better quality mode and a better frequency stability.

Most solid-state lasers in the green region have used Nd^{3+}-doped systems from which coherent green radiation is obtained by second harmonic generation of the fundamental laser emission by using additional non-linear crystals (intra or extra cavity) to that of the laser element [3, 4].

However, the use of nonlinear crystals as host matrices for optically active ions is an attractive field. The aim is the development of compact systems from multifunction optical materials in which the fundamental laser emission and, simultaneously, coherent visible radiation by self-frequency doubling (SFD) can be obtained without additional elements to that of active laser crystal.

Indeed, several Nd^{3+}-doped nonlinear crystals have proved to be efficient lasers in the near infrared region and simultaneously coherent visible radiation has been obtained by efficient self doubling [5, 6]. In our work we have chosen LiNbO$_3$ crystal, a non-linear material, as a host for Yb^{3+} active ions.

Laser action from Yb^{3+} activated crystals was demonstrated early in the seventies, in particularly, laser action from Yb^{3+} doped YAG was reported in 1971 [7]. However, until relatively recently, ytterbium could not compete with neodymium because it was less suited to pumping with flash lamps and, owing to its quasi-three level laser transition, it required high pump intensities. The advent of high brightness diode lasers in the last 10 years has removed these disadvantages. In particular, the development of

---

[*] Corresponding author: e-mail: luisa.bausa@uam.es, Phone: +34 91497 5028, Fax: +34 91497 8579
[**] Actual address: Saint-Gobain Recherche 39, quai Lucien Lefranc, Aubervilliers, 93303, France

strained InGaAs diode lasers emitting near 980 nm, where the Yb^{3+} ion has a strong absorption peak, makes these systems efficient diode-pumpable solid-state laser materials, which is one of the first requirement for compactness.

In fact, Yb^{3+} is now the most promising ion that can be used in a non-Nd^{3+} laser in the same range of emission wavelength. Yb^{3+} has some advantages over the Nd^{3+} as laser emitting ion, which makes it particularly attractive. These advantages concern its simple energy level scheme, its high quantum efficiency and the low quantum defect, among others. Additionally, the smaller ionic radius of Yb^{3+} compared to that of other rare earth ions often favors its incorporation into the host crystals allowing higher dopant concentration, better crystal quality and thus, shorter gain elements. Finally, its special electronic configuration ($4f^{13}$) makes the $4f$ electrons less shielded than in other ions of the lanthanide series, showing a higher tendency to interact with the lattice and neighbour ions [8, 9]. Coupling between electronic transitions of Yb^{3+} ions and phonons of the matrix in which they are embedded leads to vibronic broadening of the optical bands which, from the viewpoint of laser applications, allows the achievement of short and ultra-short pulses and the possibility of tunability of the laser wavelength, as demonstrated in several matrices [10, 11].

As a disadvantage, Yb^{3+}-doped lasers operate in a quasi-three level scheme, which leads to reabsorption losses and temperature related effects.

Novel laser designs, especially the thin-disc laser, have been developed to exploit the properties of Yb^{3+} activated systems. As an example, Yb:YAG has emerged as a practical laser medium since it can be more highly doped with ytterbium, so it is possible to use much thinner laser crystals and heat dissipation is less of a problem. Today, up to 60 W average output power from Yb:YAG thin-disk laser in the pico-second and femtosecond regime can be obtained [12]. Another interesting material is the Yb:YAB laser crystal, from which efficient and tunable cw laser operation at around 1040 nm has been demonstrated under diode pumping. Moreover, cw coherent green radiation inside the tunability range 513 to 546 nm, has been obtained with high quality beam and high output power of 1.1 W. To the best of our knowledge, this value corresponds to the highest green power achieved in a self-frequency doubling crystal. More recently, laser light generation in the 560–570 nm yellow spectral region has also been obtained in a coupled cavity Yb:YAB microchip laser [13, 14].

In this work, $LiNbO_3$:MgO has been chosen as a non-linear host for Yb^{3+} ions due to its excellent properties. The actual advantage offered by this system when compared to conventional laser systems such as YAG is its multifunctional character and the widespread use of this crystal in optoelectronics, which gives this material great commercial importance. Its large electro-optic and acousto-optic coefficients make it the most commonly used material for waveguide substrates, Pockel cells, Q-switches, and so on.

From the viewpoint of self-frequency doubling laser, $LiNbO_3$:MgO has a large nonlinear coefficient for type I second harmonic generation at 1.06 μm (d_{eff} = 5.6 pm/V). It can be easily doped with rare earth and transition metal ions and allows the fabrication of passive and active optical waveguides. Photorefractive damage can be minimized by the addition of about 6% MgO.

2 Results and discussion

2.1 Optical characterization Different Yb^{3+}:$LiNbO_3$ and Yb^{3+}:MgO:$LiNbO_3$ crystals were grown in our laboratory along the c-axis from the congruent melt, [Li]/[Nb] = 0.942, by the Czochralski method. More details on the experimental system can be found elsewhere [15–19].

The study of the absorption and emission spectra allowed us to determine not only the pumping region and potential laser wavelength but also the character of the involved transitions (electric/magnetic dipolar) between the different Stark sublevels of ytterbium in our system.

Figure 1a) shows the polarized absorption spectra (σ and π) of Yb:$LiNbO_3$ for a 1 mol% Yb^{3+} concentration, at two different temperatures: 15 K and 300 K. The polarized emission spectra for these temperatures are shown in Fig. 1b).

In $LiNbO_3$ crystals RE^{3+} ions usually enter Li^+ lattice sites and the local symmetry is C_3. For the case of Yb^{3+} ions in this crystal, it is expected that this C_3 symmetry splits the two, $^2F_{7/2}$ and $^2F_{5/2}$ free ion energy states into four and three doubly degenerate crystal field levels respectively, their character being E_1 or E_2, as follows from group theory. As a matter of fact, the spectra shown in these figure could be

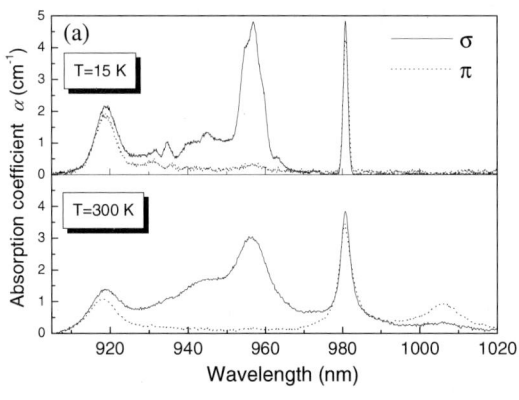

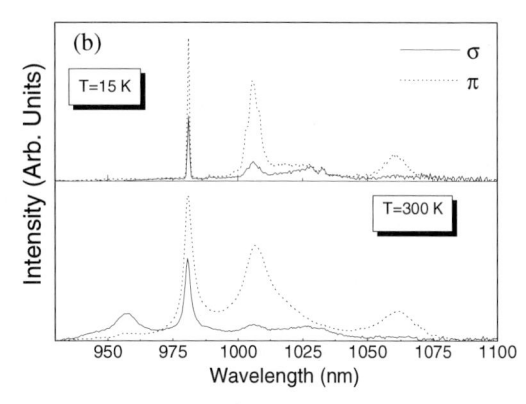

Fig. 1 a) Polarized absorption spectra and b) polarized emission spectra of Yb^{3+} doped LiNbO$_3$. Continuos line: σ configuration; dashed line: π configuration.

well analyzed on the basis that the local symmetry at the Yb^{3+} site is C$_3$ [18]. Fig. 2 shows the energy position and the character of the different Stark levels for Yb^{3+} activated LiNbO$_3$ crystals.

A more detailed inspection of the LT absorption and emission spectra revealed a certain substructure. This structure can be related to strong electron-phonon coupling of Yb^{3+} ions in this system. As it was pointed out, most of the Raman modes of LiNbO$_3$ (except for that at higher energy, 625 cm^{-1}, related to Nb–O vibrations) effectively couple with the $^2F_{7/2}(0) \leftrightarrow {}^2F_{5/2}(0)$ zero phonon absorption and emission transitions [17].

As an example, Fig. 3 shows a comparison between Raman and absorption spectra. As can be observed, the transition ending at the $^2F_{5/2}(1)$ level, shows a strong vibronic mixing and in fact, it is particularly enhanced by the resonance between vibrations and the involved Stark level. The same feature occurs for the transition terminating at the $^2F_{7/2}(1)$ level in the emission spectra.

2.2 Laser performance

2.2.1 Single domain Yb^{3+}:MgO:LiNbO$_3$ crystals

Two different orientations of the crystal have been investigated. One to study the laser performance along the crystallographic c-axis, and the other to investigate the laser properties and performance along the phase matching (PM) direction to generate green laser radiation by SFD [15, 16]. In this work we will show the most relevant experiments along the PM direction.

Lasing along the c-axis direction leads to oscillation of the σ polarization alone. However, when lasing along the PM direction, oscillation can be achieved in both the ordinary and the extraordinary polarizations. The ordinary polarization corresponds to σ polarization, while the extraordinary one is a combination of both σ and π polarizations.

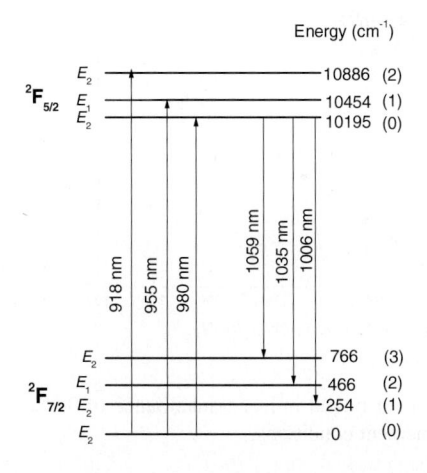

Fig. 2 Crystalline field energy level scheme of Yb^{3+} ion in LiNbO$_3$.

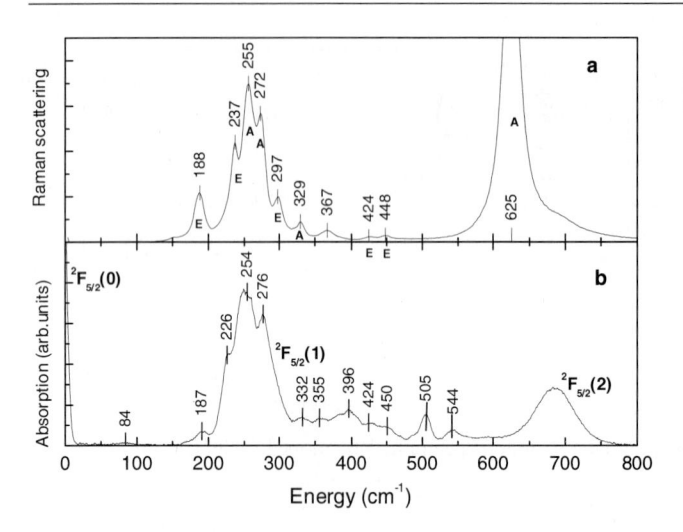

Fig. 3 RT polarized X(ZZ)X Raman spectrum and LT absorption spectrum shifted in energy.

Then, the extraordinary effective emission cross section, σ_e, depends on the propagation direction as

$$\sigma_e^2 = \sigma_\sigma^2 \cos^2(\theta) + \sigma_\pi^2 \sin^2(\theta)$$

where σ_σ and σ_π are the polarized emission cross sections and θ is the propagation angle measured from the c-axis. In our case, the crystal was cut for type I second harmonic generation at a PM angle of 83° and the extraordinary effective cross section value at 1064 nm was 5.43×10^{-21} cm^2.

With regard to SFD, LiNbO$_3$ is a negative uniaxial crystal ($n_e < n_0$) and thus, second harmonic generation in type I geometry ($n_e(2\omega) = n_0(\omega)$) can only be achieved for the fundamental ordinary beam which corresponds to σ polarization. Accordingly, end pumping in σ polarization can provide a better spatial overlap between pump and ordinary fundamental laser modes, lacking walk-off.

2.2.1.1 IR laser performance under Ti:Sapphire pumping To get maximum green radiation by SFD optimum values of the fundamental infrared laser emission are needed. Figure 4a) shows the laser performance by end pumping in σ configuration (optimum pumping configuration). A threshold value of 350 mW with a slope efficiency of 62% was obtained for an output coupler transmission of 0.6%. An almost Gaussian mode with an M^2 value of 1.2 was obtained indicating that photorrefractive damage was well stabilized by MgO codoping for the laser wavelength (1.06 μm) even at the highest laser powers. The cavity was adjusted to obtain high values of the slope efficiency, then high threshold values were obtained, though they were similar to those reported for other Yb^{3+} activated materials [16].

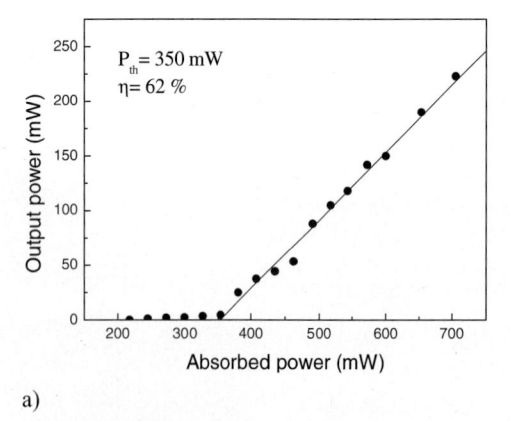

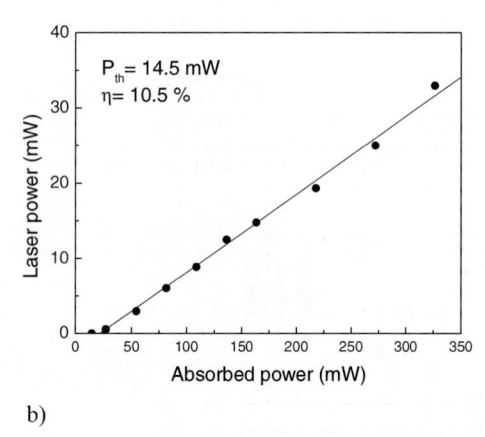

Fig. 4 1.06 μm laser power as a function of absorbed pump power for an output mirror transmitance of 0.6%. a) High slope efficiency alignement conditions b) minimum threshold alignement conditions.

© 2004 WILEY-VCH Verlag GmbH & Co. KGaA, Weinheim

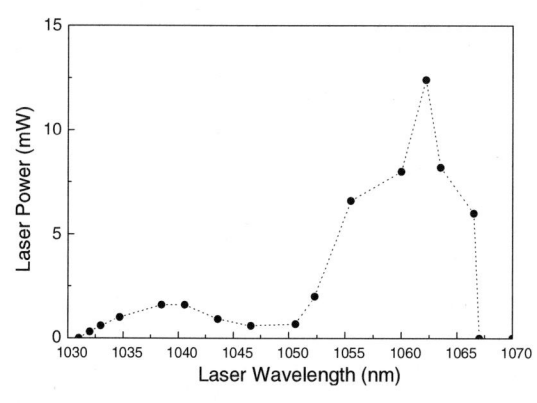

Fig. 5 Tunability of the infrared laser output.

Fig. 6 Pumping spectrum.

As suggested by modeling [20, 21], optimization of threshold and slope efficiency can not be obtained simultaneously. Optimizing one or the other can be achieved through suitable pump and laser mode overlap by independent control of the focusing system and of the cavity length. Minimum threshold values can be achieved by fine adjustment of the cavity length that produces a smaller laser beam waist and thus a lower threshold. Figure 4b) shows the fundamental output power versus the absorbed pump power under these conditions. A minimum threshold power of 14 mW with a slope efficiency of 10.4 % was obtained. It is close to that usually obtained for Nd^{3+}–based lasers under similar conditions and it evidences the potential of this material for low power devices and integrated optics.

The tuning performance of our system is shown in Fig. 5. The experiments were carried out with 900 mW of σ polarized absorbed pump power at 980 nm along the PM direction. Broad band tuning was obtained over 40 nm from 1030 to 1070. This range is comparable to those obtained for other Yb^{3+} doped laser crystals [16].

The pumping spectrum monitoring the fundamental laser emission wavelength ($\lambda = 1.06$ µm) is shown in Fig. 6. It corresponds to the sigma configuration and it displays a similar shape as the absorption spectrum. This confirms that efficient laser action can be obtained by pumping either in a broadband region (900–960 nm), useful when diode pumping, or around 980 nm.

2.2.1.2 Infrared laser performance under diode pumping An optimization of the laser action performance from a diode-pumped $Yb^{3+}:LiNbO_3:MgO$ crystal has been obtained very recently. A fiber coupled laser diode at 980 nm was launched longitudinally into a double-pass 10 cm long quasi-hemispherical cavity. Figure 7 shows the laser gain curve under diode pumping. Efficient laser action at 1060 nm achieving laser slope efficiencies as high as 74% has been obtained for a 10% transmission output coupling [19].

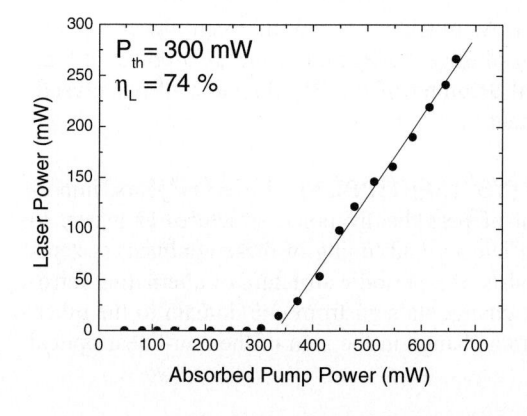

Fig. 7 1.06 µm laser power as a function of absorbed pump power. Output mirror transmittance was $T = 10\%$.

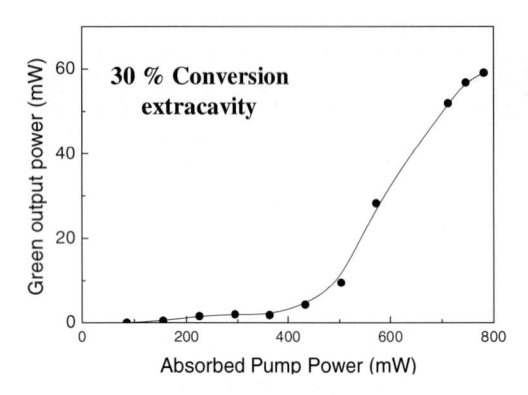

Fig. 8 Self frequency doubling (532 nm) output power as a function of absorbed pump power.

2.2.1.3 Second harmonic generation
The experiments of self-frequency doubling have been performed by pumping with a Ti–Sapphire laser.

Taking advantage of the high circulating laser intensity in the cavity it is possible to get a significant self-doubling power by second harmonic conversion of a small fraction of the intra-cavity laser power. The efficiency of second harmonic generation can be obtained from the expression [25]:

$$\frac{I_{2\omega}(L)}{I_\omega(0)} = \tanh^2\left\{\left[8d_{eff}^2\eta^3\omega^2 I_\omega(0)\right]^{\frac{1}{2}}\frac{L}{2}\right\} \times \frac{\sin^2(\Delta\kappa L/2\pi)}{(\Delta\kappa L/2\pi)^2}$$

for beams propagating collinearly. Here $I_w(0)$ corresponds to the intensity of the fundamental wave transmited into the crystal and $I_{2w}(L)$ corresponds to the intensity of the second harmonic generated wave at a length L into the crystal.

The first factor contains the contribution of the characteristics of the material (d_{eff} is the non-linear coefficient and η the optical impedance of the medium). The second factor takes into account the phase difference between the wave vectors of the fundamental and the second harmonic generated waves.

When the system meets the phase matching condition the electric field of the second harmonic generated beam $E(2\omega)$ linearly grows with the propagation through the non linear medium and its power increases in a quadratic way. If the system does not fulfill the phase matching condition the second harmonic power undergoes periodic oscillations as a function of thickness, with a coherence length given by [26]:

$$l_c = \pi/\left(\kappa^{2\omega} - 2\kappa^\omega\right)$$

The phase matching condition requires that the refractive index of the fundamental wavelength be the same as that of the second harmonic generated one. Special care was taken to determine the PM direction because in LiNbO$_3$ the expected lasing wavelength around 1060 nm is close to the second harmonic generation cut-off wavelength.

Figure 8 shows the input-output plot of the SFD power. As much as 58 mW of green power was obtained, which corresponds to a SFD conversion efficiency of 30%. Saturation of the SFD power at high pumping rates was observed. It can be related to thermal detuning of the PM direction of our crystal, whose orientation was maintained throughout the experiment.

2.2.2 Periodically polled Yb^{3+}:MgO:LiNbO$_3$ crystals (Yb^{3+}:MgO:PPLN)
In recent years, important efforts have been directed towards the development of periodically poled structures as a way to increase the potential for frequency conversion both in efficiency and tuning of these nonlinear systems by quasi-phase matching. As known, a PPLN crystal consists of a periodic structure of alternating ferroelectric domains in which the electric polarization vector change its sign from one domain to the other. This change of sign of the polarization vector gives rise to a change in the sign of the non-linear optical coefficients of the crystal.

In these structures it is possible to get SHG by the so called quasi-phase matching technique.

This technique consists in the control of the wave vector mismatch Δk between the interactive waves by a periodic modulation of the refractive index or the non linear coefficient via the periodic domain structure. If the half-period of the modulation is related to the coherence length of the second harmonic generated wave for a certain wavelength we can get efficient second harmonic generation. Under these conditions, the changes of sign of the non linear coefficient change the sign of the phase of the second harmonic wave and the interference between the fundamental and second harmonic wave is again constructive.

It is possible to demonstrate that the effective non linearity is lower in QPM than in BPM by a factor of $(2/m\pi)^2$ for a same value of the non linear coefficient. However, the use of quasi-phase matching in PPLN structures provides some advantages over Yb^{3+} doped single domain $LiNbO_3$ crystals [26]:

Firstly, these structures offer the possibility of using π polarized infrared fundamental laser oscillation for second harmonic generation. As previously mentioned, for single domain crystals the π polarized beam gain cross section is considerably higher than that corresponding to the σ polarized beam, which is the only polarization required for type I birefringent phase matching.

Secondly, lasing along a quasiphase matching direction with $E \| c$ allows access to the d_{33} component of the second order non linear tensor increasing the effective nonlinear coefficient from 5.6 pm/V to 21.6 pm/V and eliminates Pointing vector walk-off.

The crystals were obtained by the off-centered Czochralski method along the a axis by adding 1 mol% of Yb^{3+} and 5 mol % Mg^{2+} in the form of oxides. During the crystal growth the rotational axis is displaced away from the symmetry axis of the temperature field. This produces a periodic temperature fluctuation able to create alternating ferroelectric domains. The period length was controlled via rotation and pulling rates. In principle the rotation and pulling rates were adjusted to generate a 50% duty cycle of alternating domains with a period of 6.8 µm (single domain length of 3.4 µm) as required for first order quasi-phase matching second harmonic generation of a 1060 nm fundamental wavelength. The longest region containing a homogeneous periodic domain structure in our crystals was of the order of 1 mm [27].

The spectral dependence of quasi-phase matching efficiency (detuning curve) was checked by external second harmonic generation using a tunable optical parametric oscillator. Figure 9 shows the relative external second harmonic generation when tuning the fundamental wavelength between 980 nm and 1200 nm. In our sample conversion efficiency peaks at 524 nm, which does not match the optimum wavelength value for our system. The spectral bandwidth is 20 nm (766 cm^{-1}). This broadening is due to the inhomogeneous domain size in our 1 mm crystal and it takes place at the expense of the maximum effective non linear coefficient achievable. The inhomogeneity in domain size which follows from modeling and from this figure is about 7% leading to an effective non-linear coefficient of 6.5 pm/V. It can be further enhanced by improving the homogeneity in the domain structure.

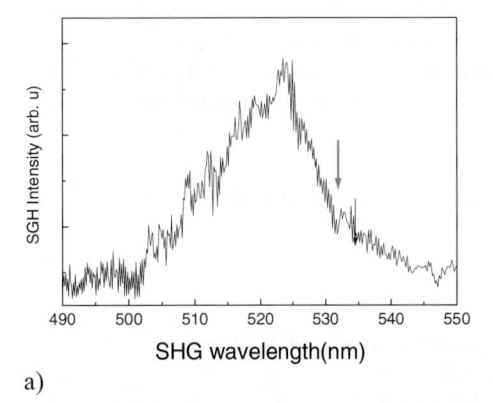

Fig. 9 a) Quasiphase matched external second harmonic generation detunning curve. b) SEM micrography revealing the domain structure.

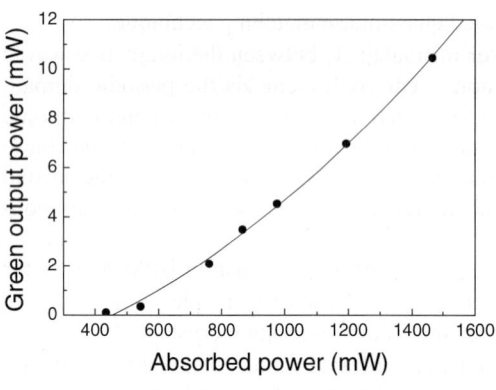

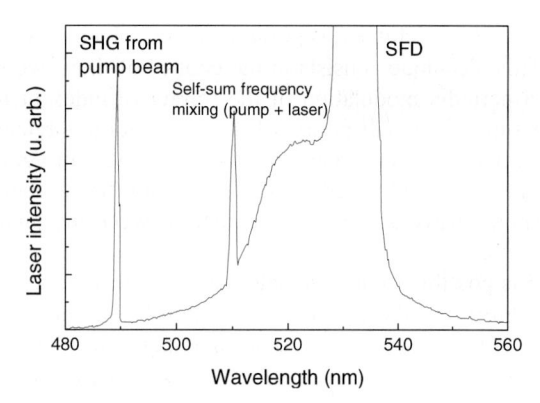

Fig. 10 Self frequency doubling output power as a function of absorbed pump power in a Yb doped PPLN crystal.

Fig. 11 Detail of the SFD emission spectrum.

Figure 10 shows the SFD performance in TEM$_{00}$ mode. This represents an overall efficiency of 15% which is relatively high and compares well to other SFD laser systems. Up to 10 mW of green output power was obtained for a total output power of 58 mW.

The green emission peaks at 532 nm and the width of the second harmonic generated signal is around 50 cm^{-1} which is substantially wider than that obtained in the case of SFD with single domain crystals (30 cm^{-1}). In a detail of the laser spectrum it is possible to see, in addition to the SFD radiation, other low intensity coherent emission at 490 nm and at 511 nm. They correspond, respectively, to second harmonic generation from the pump beam at 980 nm and self-sum frequency mixing from the pump and laser beam. (See Fig. 11). In fact, the presence of a certain distribution of the domain period length in our PPLN structures allows for a broader wavelength quasi-phase matching tolerance than in the case of type *I* birefringent phase matching in single domain LiNbO$_3$ crystals.

3 Summary We have demonstrated efficient and stable continuous wave laser and SFD at room temperature in Yb^{3+} doped single domain and PP LiNbO$_3$:MgO crystals with a good quality mode. High slope efficiencies (higher than 70%) and low threshold values (14.7 mW) have been achieved for single domain crystals. Tunability of the infrared laser action has also been obtained. In the case of PPLN samples, in accordance to the lower threshold associated to its higher gain, π polarization is observed to occur naturally in this crystal favoring the use of d_{33} effective non linear coefficient. In the case of Yb^{3+} single domain doped crystals σ and π polarization are obtained, the sigma configuration being the one used to generate second harmonic generation. High SFD efficiencies are obtained for both type of systems. As an advantage over the single domain SFD crystals a small dispersion in the domain length occurring naturally during the crystal growth process allows for improved spectral, thermal and angle tolerances at the expense of only a small reduction of the effective nonlinear coefficient. The efficiency reported here should be easily increased by matching the SFD peak efficiency to the fundamental wavelength.

Acknowledgements This work has been supported by the Comunidad Autonoma de Madrid (CAM) under project No. 07N/0020/2002 and by the Spanish CICyT under project No. MAT2001-0167.

References

[1] J. Han, L. He, R. L. Gunshor, and A. V. Nurmikko, IEEE Circuits Devices Mag. **10**, 18 (1994).
[2] T. F. Wiener and K. Sherman, IEEE Trans. Comm. **COM-28**, 1602 (1980).
[3] D. Jaque, J. J. Romero, and J. García Sole, J. Appl. Phys. **92**, 3436 (2002).
[4] D. Jaque, J. J. Romero, Y. Huang, and Z. D. Luo, Appl. Opt. **41**, 6394 (2002).
[5] D. Jaque, J. Capmany, and J. García Sole, Appl. Phys. Lett. **75**, 325 (1999).

[6] J. Capmany, D. Jaque, and J. Garcia Solé, **75**, 2722 (1999).
[7] A. R. Reinberg, L. A. Risenberg, R. M. Brown, R. W. Whacker, and W. C. Holton, Appl. Phys. Lett. **19**, 11 (1971).
[8] A. Ellens, H. Andre, M. L. H. Ter Heerdt, R. T. Wegh, A. Meijerink, and G. Blasse, J. Lumin. **66–67**, 240, (1996).
[9] A. Lupei, V. Lupei, V. N. Enaki, C. Presura, and A. Petraru, Spectrochim. Acta A **58**, 773 (1999).
[10] F. Druon, S. Chenais, P. Raybaut, F. Balembois, P. Georges, R. Gaume, G. Aka, B. Viana, D. Vivien, J. P. Chambaret, S. Mohr, and D. Kopf, Opt. Mater. **19**, 73 (2002).
[11] S. Chenais, F. Druon, F. Balembois, P. Georges, R. Gaume, P. H. Haumesser, B. Viana, G. P. Aka, and D. Vivien, J. Opt. Soc. Am. B **19**, 1083 (2002).
[12] E. Innerhofer, T. Sdmeyer, F. Brunner, R. Hring, A. Aschwanden, R. Paschotta, C. Hnninger, M. Kumkar, and U. Keller, Opt. Lett. **28**, 367 (2003).
[13] P. Dekker, P. A. Burns, J. M. Dawes, J. A. Piper, J. Li, X Hu, and J. Wang, J. Opt. Soc. Am. B **20**, 4 (2003).
[14] P. A. Burns, J. M. Dawes, P. Dekker, J. A. Piper, J. Li, and J. Wang, Opt. Commun. **207**, 315 (2002).
[15] E. Montoya, J. Capmany, L. E. Bausá, T. Kellner, A. Diening, and G. Huber, Appl. Phys. Lett. **74**, 3113 (1999).
[16] E. Montoya, J. A. Sanz-García, J. Capmany, L. E. Bausá, A. Diening, T. Kellner, and G. Huber, J. Appl. Phys. **87**, 4056 (2000).
[17] E. Montoya, F. Agulló-Rueda, S. Manotas, J. García Solé, and L. E. Bausá, J. Lumin. **94–95**, 701 (2001).
[18] E. Montoya, A. Lorenzo, and L. E. Bausá, J. Phys.: Condens. Matter **11**, 311 (1999).
[19] M. O Ramirez, D. Jaque, J. A. Sanz-García, L. E. Bausá, and J. E. Muñoz Santiuste, Appl. Phys. B **77**, 621–623 (2003).
[20] W. P. Risk, J. Opt. Soc. Am. B **5**, 1412 (1988).
[21] T. Taira, W. M. Tulloch, and R. L. Byer, Appl. Opt. **36**, 1868 (1997).
[22] F. Augé, F. Balembois, P. Georges, A. Brun, F. Mougel, G. Aka, A. Kahn-Harai, and D.Vivien, Appl. Opt. **38**, 976 (1999).
[23] M. Karszewski, U. Brauch, A. Giesen, I. Johansen, U. Schiegg, C. Stewen, and A. Voss, in: Advanced Solid State Lasers, 235 (1997).
[24] A. N. P. Bustamante, D. A. Hammons, R. E. Peale, B. H. T. Chai, M. Richardson, and A. Chin, Opt. Commun. **19**, 309 (2001).
[25] B. A. E. Saleh and M. C. Teich, (eds.) Fundamental of photonics, (John Wiley & Sons, New York, 1991).
[26] M. M. Fejer, G. A. Magel, D. H. Jundt, and R. L. Byer, IEEE J. Quantum Electron. **28**, 2631 (1992).
[27] J. Capmany, E. Montoya, V. Bermudez, D. Callejo, E. Dieguez, and L. E. Bausá, Appl. Phys. Lett. **76**, 1374 (2000).

Cr³⁺ ions in hydrogenated and proton exchanged lithium niobate crystals

A. Kamińska[*,1], **L. Arizmendi**[2], **A. Barcz**[1], **E. Łusakowska**[1], and **A. Suchocki**[1]

[1] Institute of Physics, Polish Academy of Sciences, Al. Lotników 32/46, 02-668 Warsaw, Poland
[2] Departamento de Fisica de Materiales, Universidad Autonoma de Madrid, Cantoblanco, 28049 Madrid, Spain

Received 15 September 2003, accepted 24 September 2003
Published online 19 January 2004

PACS 61.72.Ji, 62.50.+p, 78.55.Hx

We studied influence of annealing in water vapour and proton exchange on high-pressure low-temperature luminescence spectra of near stoichiometric LN:Cr, MgO crystals, containing both Cr_{Li} and Cr_{Nb} centers by using diamond-anvil cell technique. The observed changes of the spectra of Cr^{3+} ions are associated with the larger inhomogeneous broadening and increase of the splitting of the ^2E level of the so called δ Cr_{Nb} center. This shows that neither increase of hydrogen concentration in LN by almost two orders of magnitude (from 10^{18} to almost 10^{20} cm^{-3}) nor proton-exchange procedure do not create any new optically active Cr^{3+} centers in those crystals. This does not support the hypothesis about the close compensation of Cr^{3+}_{Nb} sites in LiNbO₃ crystals by hydrogen ions.

© 2004 WILEY-VCH Verlag GmbH & Co. KGaA, Weinheim

1 Introduction Lithium niobate (LN) crystals remain of the interest for applications as well as for fundamental studies. The defect structure of those crystals is one of the main subjects of research. Although recently stoichiometric or near-stoichiometric crystals are grown, typically congruent lithium niobate crystals are produced with a relatively large deficiency of lithium of about 6% [1]. Thus niobium anitisities are the main defect centers in LN crystals. Deficiency of lithium and existence of niobium antisities can be compensated either by lithium or niobium vacancies. It has been shown that deficiency of oxygen does not occur in LN [2]. Recently model of charge compensation by lithium vacancies gained firm experimental support [3]. However still some groups prefer the model of niobium vacancy for explanation of properties of LN crystals [4].

Doping of the LN crystals can also create additional defects. LN can accommodate various dopants with several valency states. Most of the dopants are typically located at Li crystallographic sites. Codoping with ions of different valency can force them to be transferred into Nb sites [5]. Accommodation of ions with different valency state than the ion to be substituted requires charge compensation, which can be provided by structural defects or by codopants. One would rather expect that necessary compensation is similar to that which occurs in pure crystals, though this is to be proven experimentally.

The Cr^{3+} ions turned out to be a very good probe of the defect structure of LN since they substitute for both Li and Nb ions in LN host. Several studies with use of EPR and ENDOR techniques have been performed that helped to identify several Cr^{3+} centers both in Li (Cr_{Li}) and Nb (Cr_{Nb}) sites [6, 7]. Also high-pressure low-temperature optical spectroscopy was very successful in the identification of optically active Cr^{3+} centers [8, 9]. Application of high-pressure transforms low-strength crystal field experienced by most of the Cr^{3+} centers in LN into high-strength crystal field. The Cr^{3+} ions in the latter ones exhibit sharp R-lines in the luminescence spectra, that replaces broad unresolvable emission from the Cr^{3+} cen-

---

[*] Corresponding author: e-mail: kaminska@ifpan.edu.pl, Phone: +48 228 436 601, Fax: +48 228 475 223

ters in the low-strength crystal field. This considerably increases spectral resolution of the optical measurements and allows for easier characterization of the luminescence centers.

In the case of doping of LN with Cr^{3+} ions one would expect that effect of self-compensation by simultaneous replacement of both Li^+ and Nb^{5+} ions by two Cr^{3+} ions may occur. However such an effect has not been found in LN. The Cr^{3+} usually substitute Li ions. However by codoping with MgO (or some other codopant as for example ZnO [10]) at proper concentration higher than certain threshold, dependent on stoichiometry, the Cr^{3+} ions are transferred into Nb sites. The problem of necessary compensation for doping with Cr^{3+} ions in Li or Nb sites was a subject of studies in several recent papers. Malovichko et al. on the basis of EPR and ENDOR results claim that the charge compensation for Cr^{3+}_{Li} site is provided by Nb vacancies [7]. This somehow contradicts Li vacancy model of defect structure of LN.

Also recently, on the basis of EPR and ENDOR, it has been suggested that interstitial hydrogen ions or additional Li^+ ions in structural vacancies in $LiNbO_3$ can provide required charge compensation for chromium ions in niobium sites (Cr_{Nb}) [6]. It is known that LN can accommodate large amount of hydrogen and its concentration can be controlled either by annealing crystals in water vapor or in a dry atmosphere (thermal diffusion process) [11]. The hydrogen ions are usually located in the interstitial positions at the oxygen crystallographic planes [12]. Another possibility is provided by so called proton-exchange, where by annealing crystals in various acids large amount of lithium ions can be replaced by H^+ in relatively thin surface layer of the crystal [13].

We studied the influence of both annealing in water vapour and proton exchange on high-pressure low-temperature luminescence spectra of near stoichiometric LN:Cr,MgO crystals, containing both Cr_{Li} and Cr_{Nb} centers by using diamond-anvil cell technique. The results of these studies are compared with the spectra obtained for as grown LN:Cr, MgO crystals (not processed for increasing of hydrogen content).

2 Samples and experimental procedures The samples used in these studies have been grown by high temperature top seeded solution growth (HTTSSG) method at the Universidad Autonoma de Madrid. The solution consisted of congruent $LiNbO_3$ + 6 w% of K_2O. This gives rise to a near-stoichiometric crystal with a composition of 49.7% mol. of Li [14]. Chromium in a concentration of about 0.2 mol.% has been added to the starting flux. Another sample doped with Cr and Mg was cut from a crystal grown from a solution with K_2O but also containing about 1.5% of MgO, although the amount of MgO introduced into the crystals was only equal to 0.2%. Those "as grown" crystals were previously studied by high-pressure low temperature spectroscopy [9]. The first sample with characteristic green color contained chromium ions in the Li sites, in the second one, pink in colour, the Cr^{3+} ions were located mainly in the Nb sites [9].

The concentrations of OH^- complexes in the samples were established from the absorption data around the wavelength of 2.82 μm, where the absorption bands associated with the OH^- band streching occur. The samples were crystallographically oriented and measured in polarized light in a proper orientation [15]. The concentrations of OH^- established this way were equal to 7.8×10^{18} cm^{-3} and 1.1×10^{18} cm^{-3} in the as grown samples doped only with Cr and doped both with Cr and MgO, respectively.

The crystals were annealed in water vapour flow at temperature of 800 °C for about 2 hours. This procedure increased the concentrations of OH^- up to about 2.7×10^{19} cm^{-3} in the sample doped only with chromium and up to 8.1×10^{19} cm^{-3} in the sample doped with Cr and MgO. This means an increase of OH^- concentration of about of two orders of magnitude in the sample doped with Cr and MgO, as compared with only threefold increase in the sample without MgO.

Several pink in colour near-stoichiometric LN:Cr, MgO samples underwent the procedure of proton exchange, among them was also a z-cut sample. The samples were thinned, polished and annealed in sulphuric acid at a temperature of about 190 °C for 16 hours. Then they were annealed at temperature of 360 °C in air for 4 hours for equilibration of hydrogen content. The thickness of the samples were about 60 μm after the treatment. The procedure caused some damages to the surfaces of the samples, with corrugation of about 1.5 μm, estimated with use of atomic force microscope (AFM). An example of

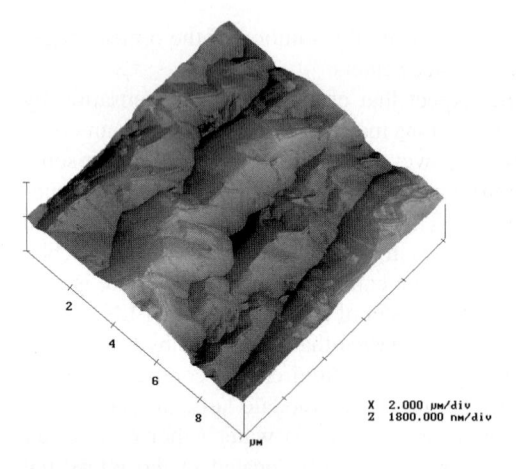

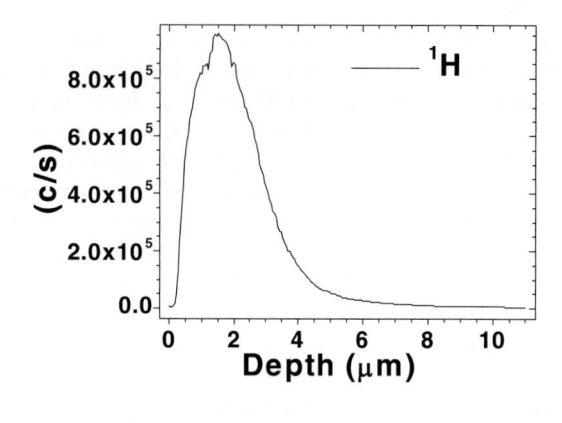

Fig. 1 AFM image of the surface of the z-cut near-stoichiometric LiNbO$_3$:Cr, MgO sample after proton exchange procedure in sulphuric acid (details of the procedure are given in the text).

Fig. 2 SIMS profile of hydrogen in proton exchanged near-stoichiometric LiNbO$_3$:Cr,MgO crystal.

surface damages of the proton-exchanged sample is shown in Fig. 1. Then the relative concentration of hydrogen in the sample was measured by SIMS technique with use of CAMECA IMS 6f equipment. The hydrogen profile in the proton-exchanged sample is shown in Fig. 2. The hydrogen concentration is strongly increased up to about 5 μm of depth, which means that about 15% of sample total thickness underwent a procedure of proton exchange.

Continuous wave emission spectra were obtained using 514.5 nm line of an argon-ion laser as the excitation source. The spectra were measured with use of GDM-1000 double grating monochromator equipped with a cooled photomultiplier (EMI 9684B) with S1-type cathode and an SR530 lock-in amplifier. The spectra were corrected for the quantum efficiency of the photomultiplier. The high-pressure measurements were performed with use of low temperature diamond anvil cell (Diacell Products MCDAC-1). The argon was used as a pressure transmitting medium. The diamond anvil cell was mounted into an Oxford 1204 cryostat equipped with temperature controller for low temperature measurements. The R$_1$-line ruby luminescence was used for pressure calibration [16, 17]. The thin samples were loaded into the cell along with a small ruby ball. To measure the luminescence the argon-ion laser line was focused either on the measured LiNbO$_3$ sample or on ruby. The changes of pressure were done at room temperature in order to minimize non-hydrostatic effects that are known to exist in diamond-anvil cells especially at higher pressure. The hydrostatic conditions could be partially monitored by recording the half-width of the ruby emission. In our measurements we observed increase of the half-width of ruby luminescence with increasing pressure. The half-width of the R$_1$ ruby luminescence did not exceed 6 cm^{-1} at high pressures (2.5 cm^{-1} at ambient pressure). This means that the non-hydrostatic effects were rather weak.

The decay kinetics of the luminescence have been measured with use of SR430 Multichannel Scaler. A large number of decays have been collected in order to obtain good signal-to-noise ratio. The exciting laser beam has been chopped by accoustooptic modulator with the transient time below 10 ns. The decay times of luminescence were calculated by fitting the decay kinetics with single- or double-exponential curves.

3 Experimental results Figure 3 presents the pressure dependence of luminescence, recorded at a temperature of 10 K, of LN:Cr^{3+}, MgO near-stoichiometric samples in the region of the R-lines, for samples as grown, hydrogenated by thermal diffusion and proton exchanged. The observed changes are very small, i.e. in the hydrogenated samples the ratio of the intensity of the R-line associated with so called center δ(Cr^{3+}Nb) to the intensity of the line associated with center γ(main Cr^{3+}Li center) is

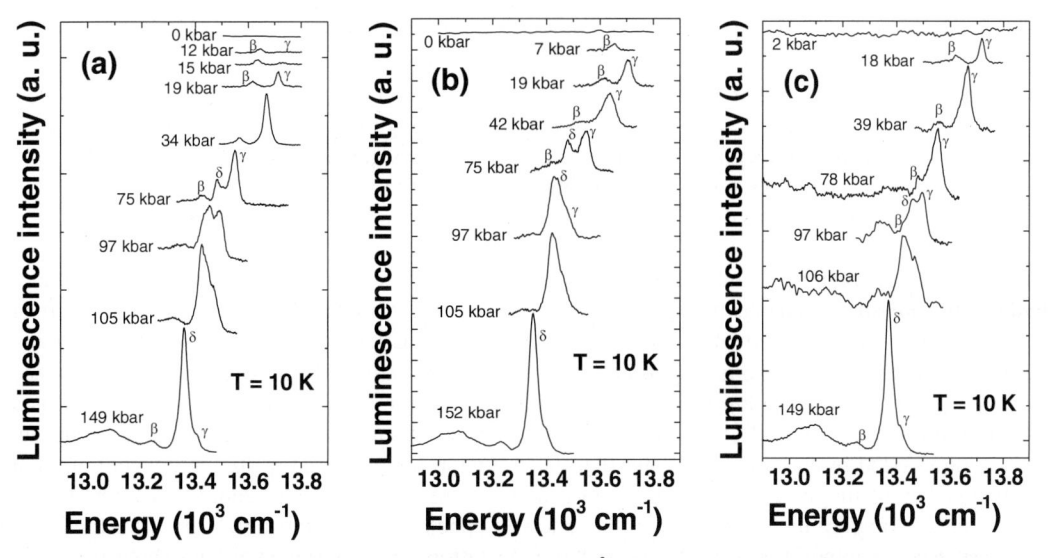

Fig. 3 Pressure dependence of luminescence of LiNbO$_3$:Cr^{3+}, MgO crystals in the region of the R-lines. a – as grown sample; b – sample hydrogenated by thermal diffusion; c – proton exchanged sample.

slightly different at a similar pressure. We did not observe any new R-lines associated with the increased hydrogen concentration by two orders of magnitude in the sample. No appreciable changes were neither observed in the proton-exchanged sample.

Larger changes are observed in the splitting of the R-lines associated with the center δ, however only for the sample hydrogenated in the proton-exchange process. They are presented in Fig. 4. The splitting of the R-lines is due to the splitting of the first excited state ^2E in the strong crystal field scheme. It is observed in the luminescence as the thermal population of the upper split ^2E level and increase of luminescence intensity of the R$_2$ line with increase of temperature. In the sample as grown and hydrogenated by thermal diffusion this splitting is equal to about 19 ± 3 cm^{-1} [9], in contrast to the splitting in the proton-exchanged sample where it is much larger and equal to about 26 cm^{-1}. No change in the splitting of the R-lines was observed for the main Cr$^{3+}_{Li}$ γ center.

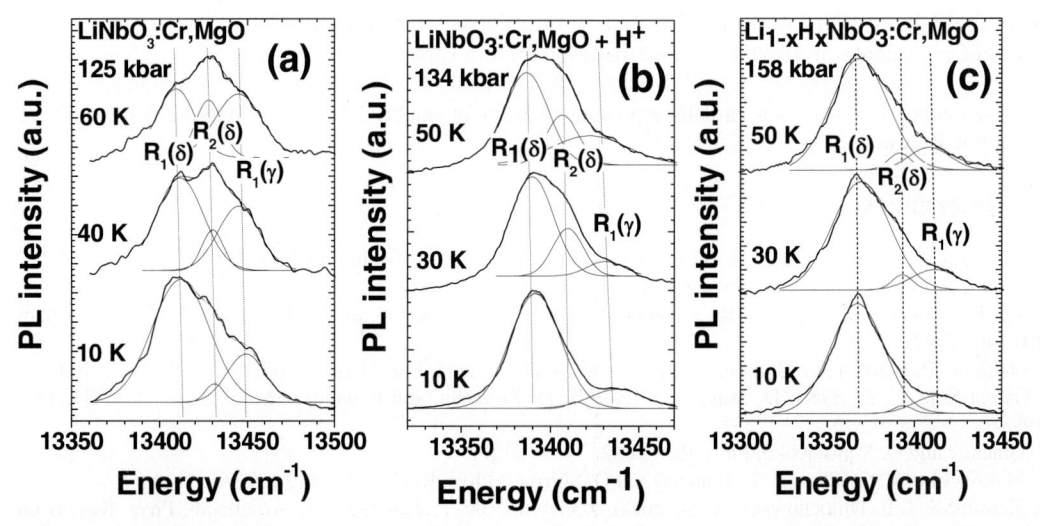

Fig. 4 Temperature dependence of the luminescence under pressure of the near-stoichiometric LiNbO$_3$:Cr, MgO crystals in the region of the R-lines associated with the δ–Cr^{3+} center (Cr$^{3+}_{Nb}$). a – as grown sample; b – sample hydrogenated by thermal diffusion; c – proton exchanged sample. Pressures of measurements are given in the graphs.

Relatively small changes were observed in the width of the R-lines of the Cr^{3+}. The line-widths in the hydrogenated samples are larger than in the as grown crystals. Changes, though also small, are seen in the relative ratio of the particular R-lines to the total intensity of luminescence emitted by the samples. They may suggest that during thermal annealing or proton-exchange procedure the lattice undergoes a small expansion. Therefore conversion from the weak strength crystal field to the strong crystal field scheme occurs at higher pressures in the hydrogenated samples. The same behaviour was also observed in the pressure dependencies of the decay times of the luminescence of particular centers.

4 Discussion and conclusions The observed changes of the luminescence spectra of Cr^{3+} ions under high hydrostatic pressures induced by hydrogenation processes are rather weak. The concentration of Cr^{3+} in our samples is in the order of 10^{19} cm^{-3}. It is even more difficult to know accurately how many ions are in the δ center. However one may assume that that number should be at least 5×10^{18} cm^{-3} since the absorption and high pressure luminescence data show that in the samples codoped with MgO it is a dominating luminescence center [9]. This means that in the as grown crystals only very few Cr^{3+} ions can be closely compensated by the protons. Water thermal diffusion increases the hydrogen concentration at of about two orders of magnitude, proton exchange is even more efficient. This should create new Cr^{3+} centers with close compensation and change the observed spectra, that should be observed in the high-pressure luminescence experiments. Contrary to the expectations the experimental data show that used hydrogenation procedures do not create new luminescence Cr^{3+} centers. It means that the hydrogen ions are not likely to be closely located to the Cr^{3+} luminescence centers. This is not in agreement with the results Malovichko et al. [6]. However they also stated that instead of compensation by hydrogen in interstitial position another type of compensation may be in agreement with their ENDOR results, i.e. compensation by Li^+ ions in structural vacancy close by to the Cr^{3+}_{Nb} ions. Our results show that this second possibility is rather more likely than the hypothesis about close compensation by hydrogen ions.

The hydrogenation by thermal diffusion and by proton exchange should lead to different results in crystallographic position of incorporated hydrogen ions in LN lattice. The hydrogen ions introduced by thermal diffusion are located in the oxygen planes in between the oxygen ions [12]. On the other hand the proton exchange is a replacement of Li to H [13]. The crystallographic structure of $HNbO_3$ is different from $LiNbO_3$ [8], however not much is known about the local structure around the Cr^{3+} ion in Nb site in $HNbO_3$. Most probably it does not change considerably under proton exchange since no new Cr^{3+} centers are observed in the luminescence after this procedure. The difference is seen mainly in the change of the splitting of the R-lines of the δ center. It increases considerably after proton exchange. The splitting of the R-lines is a probe of local distortion from ideal octahedral symmetry. This means that the symmetry of Cr^{3+}_{Nb} center in $HNbO_3$ is more distorted than in $LiNbO_3$.

Acknowledgements This work was partially supported by the grant number 2 P03B 054 24 of the Polish Committee for Scientific Research.

References

[1] P. F. Bordui, R. G. Norwood, C. D. Bird, and G. D. Calvert, J. Cryst. Growth **113**, 61 (1991).
[2] P. Lerner, C. Legras, and J. P. Dumas, J. Cryst. Growth **3/4**, 231 (1968).
[3] N. Iyi, K. Kitamura, F. Izumi, J. K. Yamamoto, T. Hayashi, H. Asano, and S. Kimura, J. Solid State Chem. **101**, 340 (1992).
[4] F. Lhomme, P. Bourson, G. Boulon, Y. Guyot, and M. D. Fontana, Eur. Phys. J. Appl. Phys. **20**, 29 (2002).
[5] J. Garcia Sole, L. E. Barsa, D. Jaque, E. Montoya, H. Murrieta, and F. Jaque, Spectrochimi. Acta **54**, 1571 (1998).
[6] V. Grachev and G. Malovichko, Phys. Rev. B **62**, 7779 (2001).
[7] G. Malovichko, V. Grachev, E. Kokonyan, and O. Shirmer, Phys. Rev. B **59**, 9113 (1999).
[8] A. Kaminska, J. E. Dmochowski, A. Suchocki, J. Garcia-Sole, F. Jaque, and L. Arizmendi, Phys. Rev. B **60**, 7707 (1999).
[9] A. Kaminska, A. Suchocki, M. Grinberg, L. Arizmendi, D. Callejo, and F. Jaque, Phys. Rev. B **62**, 10802 (2000).

[10] F. Abdi, M. Ailerie, M. Fontana, P. Bourson, T. Volk, B. Maximov, S. Sulyanov, N. Rubinina, and M. Wöhlecke, Appl. Phys. B **68**, 795 (1999).
[11] R. Müller, L. Arizmendi, M. Carrascosa, and J. M. Cabrera, Appl. Phys. Lett. **60**, 3212 (1992).
[12] M. Wöhlecke and L. Kovacs, Crit. Rev. Solid State Mater. Sci. **25**, 1 (2001).
[13] C. E. Rice and J. L. Jackel, J. Solid State Chem. **41**, 308 (1982).
[14] M. D. Serrano, V. Bermudez, L. Arizmendi, and E. Dieguez, J. Cryst. Growth, **210**, 670 (2000).
[15] J. M. Cabrera, J. Olivares, M. Carrascosa, J. Rams, R. Müller, and E. Dieguez, Adv. Phys. **45**, 349 (1996).
[16] G. J. Piermarini, S. Block, J. D. Barnett, and R. A. Forman, J. Appl. Phys. **46**, 2774 (1975).
[17] R. A. Noack, W. B. Holzapfel, in: High Pressure Science and Technology, edited by K. D. Timmerhaus and M. S. Barber (Plenum, New York), Vol. 1, p. 748.
[18] C. E. Rice, J. Solid State Chem. **64**, 188 (1986).

Elementary energy bands in band structure calculations of some wide-bandgap crystals

M. Sznajder[*,1], **D. M. Bercha**[1,2], and **K. Z. Rushchanskii**[2]

[1] Institute of Physics, University of Rzeszów, Rejtana 16a, 35-310 Rzeszów, Poland
[2] Institute of Physics and Chemistry of Solid State, Uzhgorod National University, 54 Voloshin Str. 88000 Uzhgorod, Ukraine

Received 15 September 2003, revised 10 October 2003, accepted 10 October 2003
Published online 29 December 2003

PACS 71.15.–m, 71.20.Nr

Semiconducting Tl_3AsS_4 crystal was chosen as an example to show that its valence band is composed of the elementary energy bands. Their topology and symmetry obtained in the empty-lattice approximation is confirmed by *ab initio* band structure calculations. It was shown that these elementary energy bands correspond to the Wyckoff position c in a unit cell. Earlier predictions that the largest electron density distribution is focused in the vicinity of this position, similarly to the case of $YAlO_3$ and SbSI crystals, is confirmed. A conclusion on the common topology and symmetry of the elementary energy bands in electronic and phonon spectra is presented.

© 2004 WILEY-VCH Verlag GmbH & Co. KGaA, Weinheim

1 Introduciton Contemporary solid state physics deals with the investigation of complicated crystals (with a large amount of atoms in the unit cell) which are the basis materials for a wide application in technics. Since the calculated (via various methods) energy spectrum of such crystals is usually very complicated there arises a necessity to find the simplest building parts of the spectrum. From the papers of Zak, Michel [1–4] (and references therein) and Bercha [5] (and references therein) follows that these simplest building parts are the elementary energy bands. Zak has associated the elementary energy bands from the k space with the Wyckoff positions in a unit cell, in the real space. However, the presence in a unit cell a few Wyckoff positons for every space group (except some groups where the only one general Wyckoff position exists) leads to an ambiguity in a choise of such position which corresponds to the real topology and symmetry of the elementary energy bands creating the band structure of a crystal. Our papers [5, 7] present a final conclusion which says that the empty lattice approximation and the general data concerning a semiconducting crystal (the existence of the forbidden energy gap, lattice constants, number of valence electrons in the unit cell, space symmetry group) are sufficient to distinguish the only possible elementary energy bands from the viewpoint of symmetry and topology. This conclusion has been proved for the orthorhombic crystals so far and it has been illustrated in [5] for the wide bandgap $YAlO_3$ (YAP) and SbSI crystals described by the space symmetry group D_{2h}^{16}. It was shown that the elementary energy bands obtained in the empty-lattice approximation and confirmed by the *ab initio* band structure calculations of the aforementioned crystals correspond to such Wyckoff position where an accumulation of the electron density distribution can be found. This paper presents topics concerning the elementary energy bands for Tl_3AsS_4 crystal (space group D_{2h}^{16}) whose band structure has been unknown so far. Experimental investigations of ternary thallium chalcogenides show that they are candidates for use in quantum electronics [6].

---

[*] Corresponding author: e-mail: sznajder@univ.rzeszow.pl, Phone: +48 17 8625 628 ext. 1017, Fax: +48 17 85 26 792

2 Results and discussion We have discussed the band structure of accousto-optic wide bandgap Tl_3AsS_4 crystal in the empty-lattice approximation in [5]. We have shown that its valence band consists of the elementary energy bands which are described in the Γ point of the Brillouin zone by the following irreducible representations (in Kovalev's notation [8]):

$$12(\Gamma_1, \Gamma_4, \Gamma_5, \Gamma_8) + 7(\Gamma_2, \Gamma_3, \Gamma_6, \Gamma_7). \qquad (1)$$

It should be noted that in [5] only s and p electrons of all the atoms where taken into account as the valence ones. From [8] follows that the elementary energy bands (1) correspond to the Wyckoff position $c(x, 1/4, z)$ in the unit cell where the structural unit as the coordination tetrahedron $[AsS_4]$ and Tl atom are placed. (In crystallography, a Wycoff position consists of all those points in the Wigner-Seitz cell whose local symmetry is determined by a subgroup of the space group of a crystal).

Below we present *ab initio* band structure calculations of the Tl_3AsS_4 crystal using the full-potential nonorthogonal local-orbital minimum basis (FPLO®-3) scheme [9] within the local spin density approximation. In these scalar relativistic calculations we used the exchange and correlation potential of Perdew and Wang [10]. We utilized the following values of the lattice constants: $a = 8.86$ Å, $b = 10.85$ Å, $c = 9.18$ Å and position of atoms presented in [11]. There are 32 atoms in the unit cell. We have chosen the following configuration as a minimum-basis set for the valence states: Tl $6s$, $6p$, $5d$; As $4s$, $4p$, $3d$; S $3s$, $3p$, $3d$. All states lying lower were treated as core states. The As and S $3d$ states were added to increase the quality of the basis set. A k mesh of 245 points in the Brillouin zone (48 in the irreducible part) was used. For a comparison, we have carried out independent calculations of the band structure of this crystal at s and p configuaration for all atoms since it is enough to analyse the upper part of the valence band. These calculations were done in LDA by means of the ABINIT code [12] which allows to establish the symmetry of the wave functions and to find the electron density distribution in the unit cell. We have chosen the value of 20 Ry for the kinetic energy cutoff which results in approximately 9000 plane waves, $4 \times 4 \times 4$ k mesh, and the Troullier-Martins pseudopotentials [13]. As a result, we obtained a good agreement between these calculations and, moreover, we have confirmed the predicted symmetry and topology of the valence band of this crystal which is described in the center of the BZ by Eq. (1). The sequence of 76 energy states, starting from the bottom of the valence band, is the following (without Tl d states): $\Gamma_1, \Gamma_4, \Gamma_5, \Gamma_8, \Gamma_7, \Gamma_3, \Gamma_2, \Gamma_6, \Gamma_5, \Gamma_4, \Gamma_1, \Gamma_8, \Gamma_5, \Gamma_1, \Gamma_8, \Gamma_4, \Gamma_1, \Gamma_5, \Gamma_4, \Gamma_8, \Gamma_5, \Gamma_6, \Gamma_8, \Gamma_3, \Gamma_4, \Gamma_7, \Gamma_8, \Gamma_1, \Gamma_5, \Gamma_4, \Gamma_7, \Gamma_3, \Gamma_8, \Gamma_5, \Gamma_1, \Gamma_2, \Gamma_4, \Gamma_2, \Gamma_1, \Gamma_6, \Gamma_1, \Gamma_5, \Gamma_4, \Gamma_8, \Gamma_2, \Gamma_3, \Gamma_7, \Gamma_1, \Gamma_5, \Gamma_8, \Gamma_2, \Gamma_1, \Gamma_6, \Gamma_4, \Gamma_5, \Gamma_6, \Gamma_1, \Gamma_7, \Gamma_8, \Gamma_3, \Gamma_4, \Gamma_3, \Gamma_4, \Gamma_5, \Gamma_8, \Gamma_2, \Gamma_6, \Gamma_7, \Gamma_4, \Gamma_1, \Gamma_2, \Gamma_6, \Gamma_5, \Gamma_3, \Gamma_8, \Gamma_7$. The first four conduction states, starting from the bottom of the conduction band, have the symmetry: $\Gamma_8, \Gamma_1, \Gamma_4, \Gamma_3$.

Figure 1 displays the band structure of the Tl_3AsS_4 crystal calculated by means of the FPLO®-3 code (including d states). As can be seen, the whole valence band in the energy range $(-17$ eV, 0 eV) is created by 34 elementary energy bands which, in turn, are composed of 4 branches. It can be seen that the band gap is a indirect one. The maximum of the valence band is located in the Γ point of the BZ and the minimum of the conduction band can be observed along the $\Gamma-X$ direction. The smallest width of the energy gap equals ~ 1.1 eV, i.e. approximately 2/3 of its experimental value ($Eg = 1.8$ eV [14]) what is typical for the band structure calculations in the LDA [15] of most ionic compounds. From the orbital-resolved density of states calculations we can state that the edge of the valence band in the range $(-3$ eV, 0 eV) is created by p electrons of all sulphur atoms. On the bottom of the valence band $(-17$ eV, -11 eV), there are 4 separated elementary energy bands (of a weak but visible dispersion) which originated from $4s$, $4p$ arsenic and $3s$, $3p$ sulphur electrons. The bottom of the valence band $(-17$ eV) is created by $4s$ arsenic electrons. In the energy range $(-10.7$ eV, -10.2 eV) there are 15 elementary energy bands (of a small dispersion) created from $5d$ thalium electrons with an account of $3s$ sulphur electrons S(2). $4s$, $4p$ electrons of As and $3p$ electrons of S(1), S(3) contribute to the valence bands also in the energy range from -9 till -5 eV. These results are in a good agreement with the experimental data of X-ray spectra presented in [14]. In the Wyckoff position $c(x, 1/4, z)$ in the unit cell there are atoms of a structural unit $[AsS_4]$ and Tl(1). The results presented above indicate that all of these atoms contribute to the creation of the whole valence band of this crystal. Left part of Fig. 2 presents an intersection of the valence electron density distribution by the

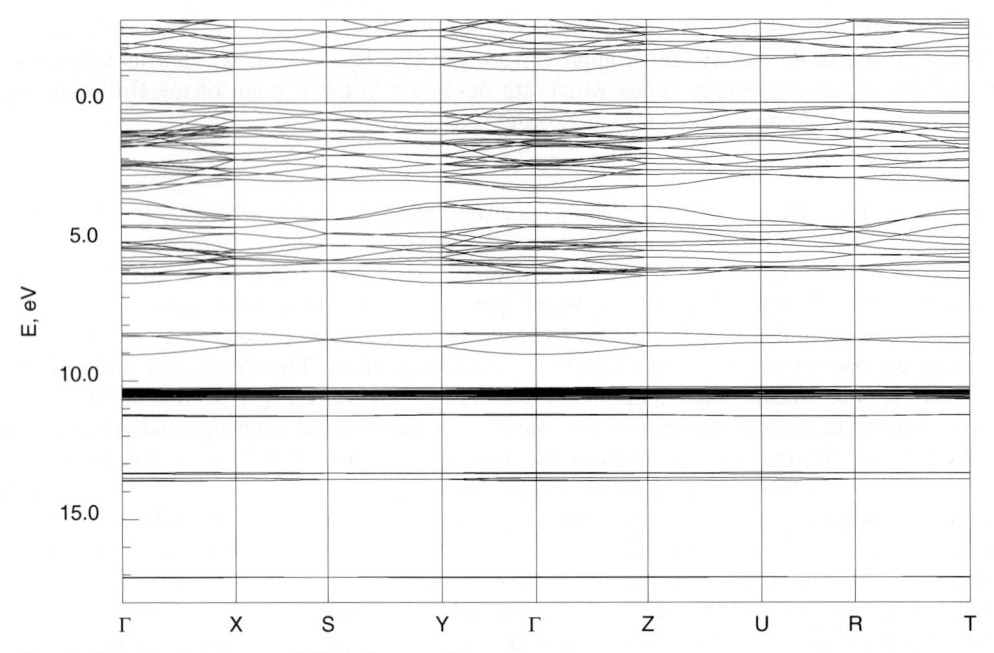

Fig. 1 The band structure of Tl₃AsS₄ crystal.

plane $y = 0.25$ which just corresponds to the Wyckoff position c, while the right part shows an intersection by the plane $y = 0.5$ where the Wyckoff positions a_i and b_i are situated. Again, our calculations show that in the plane $y = 0.25$ one can observe the largest electron density distribution (in comparison with other positions), moreover, it is focused in the vicinity of the structural unit AsS_4^{-3}. It should be emphasized that this result was predicted in [5] already from the empty-lattice approximation. From Fig. 2 follows that there exist a covalent bond between sulphur and arsenic atoms and a ionic one between Tl and S atoms. The last result is in agreement with the analysis of interatomic distances S–As as well as Tl–S atoms [14, 16]. It should be noted that the electron density distribu-

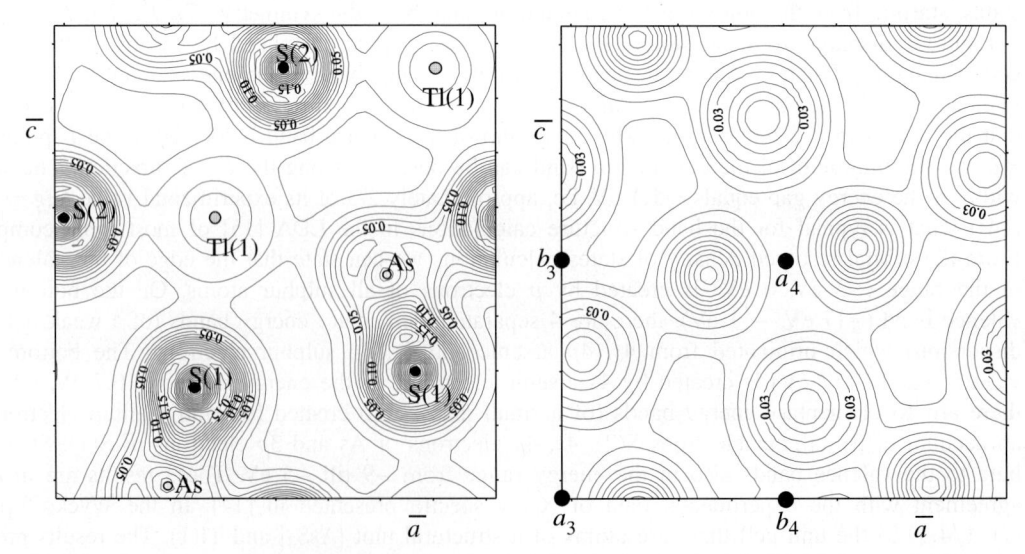

Fig. 2 Intersection of the valence electron density distribution with the plane left: $y = 0.25$, right: $y = 0.5$. d electrons of Tl were not taken into account.

tion from Fig. 2 corresponds only to the upper part of the valence band since none of d electrons were taken into account in these calculations.

One should be aware that the elementary energy bands build every single-particle energy spectra, including the phonon spectrum. One can establish that the vibrational spectrum of Tl_3AsS_4 crystal has the following symmetry in the Γ point:

$$14(\Gamma_1(A_g) + \Gamma_4(B_{1u}) + \Gamma_5(B_{2g}) + \Gamma_8(B_{3u})) + 10(\Gamma_2(A_u) + \Gamma_3(B_{1g}) + \Gamma_6(B_{2u}) + \Gamma_7(B_{3g})).\qquad(2)$$

As should be expected, the symmetry of the phonon spectrum is described by the representations which are present in the elementary energy bands Eq. (1). The results of paper [17] confirm that the normal vibration of free complexes AsS_4^{-3} which were placed in the Tl_3AsS_4 crystal field split into two sets of vibrations with the symmetry: $(A_g, B_{2g}, B_{1u}, B_{3u})$ and $(B_{1g}, B_{3g}, A_u, B_{2u})$. These two sets were observed in the Raman or infra-red spectra. It is evident that the symmetry of vibrational sets corresponds to the symmetry of elementary energy bands Eq. (1).

Acknowledgements We express our thanks to P. Zieba for his help in instalation of FPLO®-3 code.

References

[1] J. Zak, Phys. Rev. Lett. **45**, 1025 (1980).
[2] J. Zak, Phys. Rev. B **25**, 1344 (1982).
[3] L. Michel and J. Zak, Phys. Rev. B **59**, 5998 (1999).
[4] L. Michel, in: Proceeding of the 5th International School on Theoretical Physics, Symmetry and Structural Properties of Condensed Matter, edited by T. Lulek, B. Lulek, and A. Wal, (World Scientific, Singapore 1997), pp. 3–17.
[5] D. M. Bercha, K. Z. Ruschanskii, M. Sznajder, A. Matkowski, and P. Potera, Phys. Rev. B **66**, 195203 (2002).
[6] V. Yu. Slivka, Yu. V. Voroshilov, and D. V. Chepur, Kvantovaya Elektron. **20**, 74 (1981).
[7] D. M. Bercha, M. Sznajder, and K. Z. Rushchanskii, in: Proceedings of the 6th International School on Theoretical Physics, Symmetry and Structural Properties of Condensed Matter, Myczkowce, 2000, edited by T. Lulek, B. Lulek, A. Wal, (World Scientific, Singapore, 2001), pp. 376–382.
[8] O. V. Kovalev, in: Representations of the Crystallographic Space Groups. Irreducible Representations, Induced Representations and Corepresentations, edited by Harold T. Stokes and Dorian M. Hatch, (Brigham Young University, Provo, 1993).
[9] FPLO®-3 [scalar-relativistic, improved version of the original FPLO® code by K. Koepernik and H. Eschrig, Phys. Rev. B **59**, 1743 (1999)].
[10] J. P. Perdew and Y. Wang, Phys. Rev. B **45**, 13244 (1992).
[11] I. J. Fritz, T. J. Isaacs, M. Gottlieb, and B. Morosin, Solid State Commun. **27**, 535 (1978).
[12] The ABINIT code is a common project of the Universite Catholique de Louvain, Corning Incorporated, and other contributors (URL http://www.pcpm.ucl.ac.be/abinit).
[13] N. Troullier and J. L. Martins, Phys. Rev. B **43**, 1993 (1991).
[14] A. A. Lavrentyev, A. N. Gusatinskii, I. Ya. Nikiforov, A. P. Popov, and V. V. Ilyasov, J. Phys.: Condens. Matter **5**, 1447 (1993).
[15] M. S. Hybertsen and S. G. Louice, Phys. Rev. B **30**, 5777 (1984).
[16] Yu. V. Voroshilov and V. Yu. Slivka, Kvantovaya Elektron. **19**, 38 (1980).
[17] V. A. Stefanovich, V. S. Gerasimenko, Yu. V. Voroshilov, and Yu. Slivka, Fiz. Tverd. Tela **21**, 843 (1979) [Sov. Phys.-Solid State **21**, 493 (1979)].

Growth and characterization of AgIn$_5$S$_8$ and CuIn$_5$S$_8$ thin films

Liudmila V. Makhova[*], I. Konovalov, and R. Szargan

Wilhelm-Ostwald-Institut für Physikalische und Theoretische Chemie, Universität Leipzig (WOI),
Linnéstr. 2, 04103 Leipzig, Germany

Received 15 September 2003, revised 28 October 2003, accepted 28 October 2003
Published online 12 January 2004

PACS 68.35.Np, 68.37.Ps, 79.60.Dp, 81.15.Ef, 81.65.Cf

Spinel AgIn$_5$S$_8$ and CuIn$_5$S$_8$ materials are perspective for application in optoelectronic semiconductor devices. For the first time, thin films of these compounds were grown and investigated. The films were deposited by a sequential process on Mo-coated soda lime glass. This process consists of sulphur vapour sulphurization of a metallic alloy precursor. We found that a good adhesion to the substrate (Mo-coated glass) could be achieved by introduction of a Ti layer between the substrate and the spinel layer. The spinel layers have been investigated by XRD and XPS methods. Moderate roughness of the films and a crystallite size of about 200–400 nm were observed in the AFM images.

© 2004 WILEY-VCH Verlag GmbH & Co. KGaA, Weinheim

1 Introduction The interest in semiconductor thin films of ternary chalcogenides of copper and indium, for example, CuInS$_2$ and CuInSe$_2$ with a chalcopyrite structure is motivated by a successful application of these materials as light absorber of solar cells. During the preparation of CuInS$_2$ chalcopyrite layers, the formation of CuIn$_5$S$_8$ phase was observed, as for instance, after sulphurization of Cu–In alloy films in slightly In-rich CuInS$_2$ [1] and in the CuInS$_2$ (CIS) absorber layer prepared on Cu-tape by electrochemical deposition of In layer and sulphurization in sulphur vapour [2]. The efficiency of Cu–In–S solar cells, consisting of CuInS$_2$ and CuIn$_5$S$_8$ phases with p- and n-type conductivity, respectively, and a p-type CuI buffer layer reached 9.1% [3]. Despite the fact that CuIn$_5$S$_8$ phases can exist in chalcopyrite layers, the properties of pure spinel films were never studied in detail. A study of the physical properties of AgIn$_5$S$_8$ and CuIn$_5$S$_8$ single crystal spinel demonstrated the possibility to use these materials in semiconductor devices [4].

The aim of this investigation is the preparation of pure AgIn$_5$S$_8$ and CuIn$_5$S$_8$ spinel thin films and investigation of their physical and chemical properties.

2 Sample preparation and characterization Growth of AgIn$_5$S$_8$ and CuIn$_5$S$_8$ was realized by a sequential process: formation of M–In (M = Ag or Cu) alloy films and sulphurization of these films in sulphur vapour. Precursor alloy films were prepared by evaporation and deposition of metals on soda lime glass or Mo/Ti-coated glass in a vacuum chamber (10^{-6} mbar). The amount of the metals was monitored using a quartz microbalance. The process of sulphurization of M–In alloy films was carried out in sulphur vapour in a vacuum chamber (10^{-2} mbar) for 10 min at 400 °C. The heating was provided by a halogen lamp.

Characterization of the surface spinel layers was realized by X-ray photoelectron spectroscopy (XPS) using a VG ESCALAB 220iXL with Al K$_\alpha$ radiation. Core-level binding energies were referenced to C 1s peak at 285.0 eV of the adventitious carbon. The surface of Mo/Ti–coated glass was investigated before the deposition process and after deposition of spinel layer and its mechanical removal by an adhesive tape. The phase composition of the samples was studied by X-ray diffraction (XRD) analysis using

---

[*] Corresponding author: e-mail: makhova@rz.uni-leipzig.de, Phone: +49 341 –97 36 457, Fax: +49 341- 97 36 399

Table 1 Coefficients of linear thermal expansion (α) of spinel structures and substrates [4, 5].

materials	AgIn$_5$S$_8$	CuIn$_5$S$_8$	glass	Ti	Mo	W	Ge
$\alpha(K^{-1} \times 10^6)$	8.8	11.1	8.9	8.7	4.1	4.5	6.0

XRD 3000 (Seifert) with Cu K$_\alpha$ radiation. The sputter cleaning and smoothing of the spinel surface were performed using 1 keV Ar and N$_2$ ion beam at an impingement angle of 60°. Investigation of morphology of the spinel films and measurements of the root-mean-square (rms) roughness of the layers before and after sputtering were performed using the Atomic Force Microscope (AFM) Rasterscope 4000.

3 Results and discussion AgIn$_5$S$_8$ and CuIn$_5$S$_8$ spinel layers with a nominal thickness of 600 nm were prepared. For the first time, we studied the adhesion of spinel films to various substrates. According to our experiments, AgIn$_5$S$_8$ and CuIn$_5$S$_8$ layers generally show a poor adhesion to the usual conductive substrates. Ge, W, Mo, etc. were deposited on soda lime glass and were tested as substrates for growth of spinel films. After sulphurization at 400 °C, the films delaminated from the substrates. A good adhesion was observed in presence of a Ti interlayer (thickness ~50 nm) on Mo-coated glass. The quality of adhesion of the spinel layers may depend on differences between the thermal expansion coefficients of substrate materials and spinel materials (Table 1). A large difference can lead to a thermally induced stress and delamination of the layer during the growth process. For Ti (or glass) and the spinel materials, this difference is the smallest among all the substrate materials mentioned above.

We found good electrical contact between the Ti substrate and the spinel layers. The resistance of the junctions in the forward and the reverse directions is not greater than 10 Ω cm^2.

An investigation of the surface of the Mo/Ti-coated glass has been performed by the XPS method before preparation of spinel layers and after deposition and removal of spinel layers (Fig. 1). We also observed an additional component Ti-3 in Ti 2p$_{3/2}$ core level spectrum of the substrate surface after deposition and removal of spinel films (both of AgIn$_5$S$_8$ and CuIn$_5$S$_8$) compared to the core level spectrum of the initial Mo/Ti-coated glass surface (Fig. 1). The binding energy of this component is 455.2 eV and it does not match 456.5 eV reported for TiS$_2$ or 458.8 eV for TiO$_2$ [6]. The observed value is in between those for TiS (454.4 eV) and TiS$_2$ (456.5 eV), indicating that the formal oxidation state of Ti ions in these layers is apparently between +2 and +4.

XRD patterns of spinel films prepared on soda lime glass demonstrate the presence of only cubic phases AgIn$_5$S$_8$ and CuIn$_5$S$_8$ (Figs. 2a and b). XRD patterns of spinel layers prepared on Mo/Ti-coated glass show spinel phases and that of substrate materials Mo and Ti (Fig. 2c). Despite the formation of the thin titanium sulphide layer between the substrate and the spinel layer, which is concluded from data of the XPS analysis, the phase of TiS is absent in the bulk.

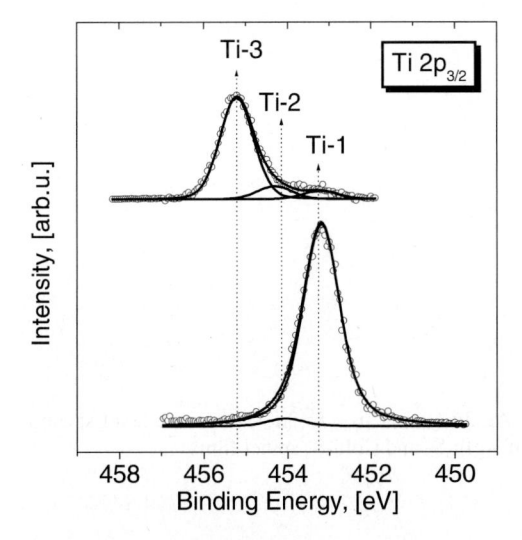

Fig. 1 Ti 2p$_{3/2}$ core level spectra of the Mo/Ti-coated glass: a) before deposition of the spinel layers, b) after deposition and removal of the spinel layers.

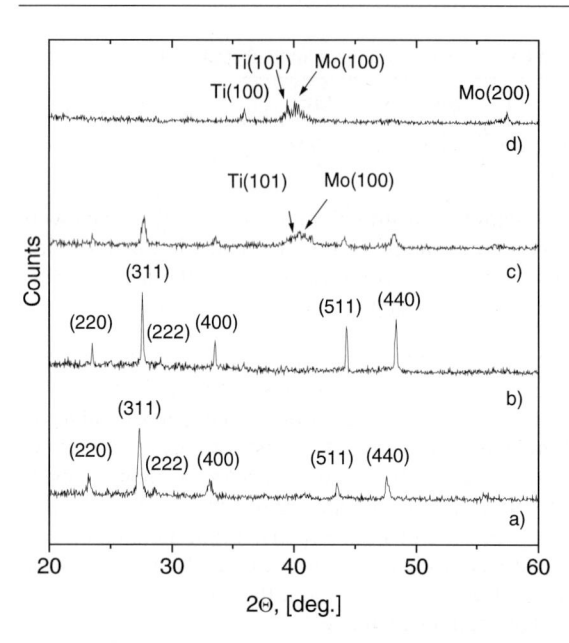

Fig. 2 XRD patterns of the spinel films. a) AgIn$_5$S$_8$ on soda lime glass; b) CuIn$_5$S$_8$ on soda lime glass; c) CuIn$_5$S$_8$ on Mo/Ti coated glass; d) Mo/Ti coated glass.

The investigation of AgIn$_5$S$_8$ and CuIn$_5$S$_8$ surfaces was performed by the XPS method. Ag 3d$_{5/2}$, Cu 2p$_{3/2}$, In 3d$_{5/2}$ and S 2p (doublet) core level spectra are shown in Fig. 3. The value of binding energy of peaks can not be referred to standards, because the spinel layers were investigated for the first time. However, the binding energies of these peaks are expected to be similar to those observed in binary sulphides or in CuInS$_2$ chalcopyrite layers with a similar atomic environment [7, 8]. We observed an additional component of the In 3d$_{5/2}$ core line (In-1) with energy corresponding to In in the metallic state ($E_B = 444.0$ eV for AgIn$_5$S$_8$ and $E_B = 443.7$ eV for CuIn$_5$S$_8$).

AFM images of the spinel films are demonstrated in Fig. 4. The morphology of AgIn$_5$S$_8$ and CuIn$_5$S$_8$ spinel films prepared on soda lime glass was studied. The root-mean-square (rms) roughness of the layers was determined typically as an average of three scans taken at sample different positions. A moderate roughness (~100 nm) as compared to thickness of layers was found to be characteristic of both types of

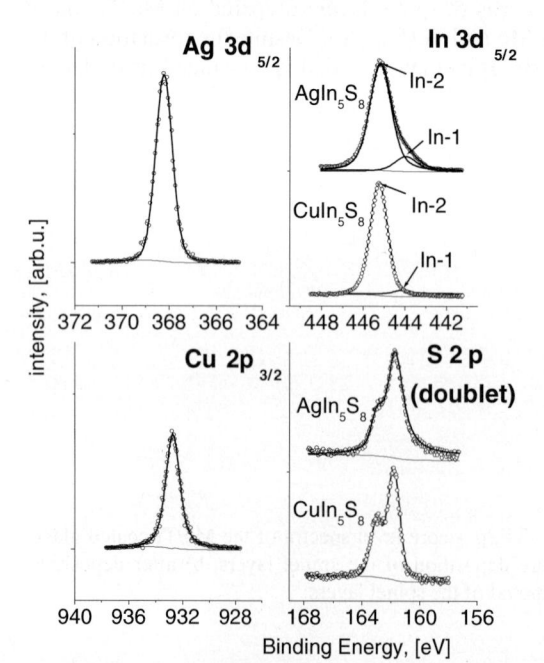

Fig. 3 Ag 3d$_{5/2}$, Cu 2p$_{3/2}$, In 3d$_{5/2}$, S2p core level spectra (XPS) of AgIn$_5$S$_8$ and CuIn$_5$S$_8$ spinel films.

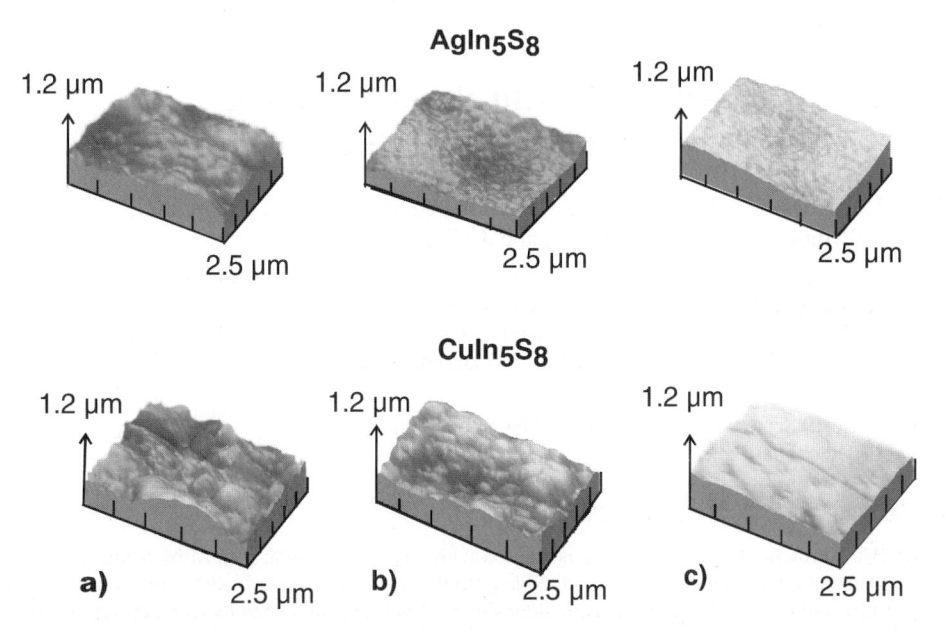

Fig. 4 (online colour at: www.interscience.wiley.com) AFM images of AgIn$_5$S$_8$ and CuIn$_5$S$_8$ spinel films: a) before sputtering; b) after sputtering by N$_2$; c) after sputtering by Ar.

spinel layers. We also found, that the size of crystallites of AgIn$_5$S$_8$ spinel is approximately twice as large as that of CuIn$_5$S$_8$. The surface cleaning and smoothing was done with 20 min long Ar and N$_2$ ion sputtering for both types of spinel layer and it was subsequently confirmed by the absence of the adventitious C 1s peak in XP survey spectra. The smoothing by Ar ion beam turns out to be more effective, as shown in the Fig. 4.

Finally, Hall effect measurements demonstrated the n-type of conductivity of the spinel layers.

4 Conclusions For the first time, growth of the pure AgIn$_5$S$_8$ and CuIn$_5$S$_8$ spinel layers by a sequential process is presented. Ti layers provide a good adhesion to the substrate due to its appropriate thermal expansion coefficient. XRD patterns demonstrate the presence of only cubic phases AgIn$_5$S$_8$ and CuIn$_5$S$_8$. An additional component of the In core line (In-I) is observed and corresponds to In in metallic state. The surface of Mo/Ti-coated glass was investigated after deposition and removal of spinel layers. The value of the binding energy of Ti-3 peak suggests the formal oxidation state of Ti ions in titanium sulphide to be between +2 and +4. The films are continuous and conductive. The roughness of the films is moderate (about 100 nm) as compared to the thickness of spinel layers.

Acknowledgements The authors thank R. Hesse for the program "UNIFIT" (fit procedure for XP spectra) and Dr. M. Heuer for technical support during XRD measurements. This work was supported by BMBF INNOCIS project.

References

[1] I. Konovalov, O. Tober, M. Winkler, and K. Otte, Sol. Energy Mater. **67**, 49 (2001).
[2] F. O. Adurodija, J. Song S. D. Kimj et al., Jpn. J. Appl. Phys. **37**, 4248 (1998).
[3] I. Konovalov and O. Tober, "PV in Europe" Conf., 7–11 October 2002, Rome, Italy (WIP-Munich, 2002) pp. 401–403.
[4] N. S. Orlova, I. V. Bondar, and E. A. Kudritskaya, Cryst. Res. Technol. **33**, 37 (1998).
[5] R. Chakrabarti et al., phys. stat. sol. (a) **160**, 67 (1997).
[6] J. Cheon, J. E. Gozum, and G. S. Girolami, Chem. Mater. **9**, 1847 (1997).
[7] K. V. Kaushik, J. Electron Spectrosc. Relat. Phenom. **56**, 273 (1991).
[8] R. Scheer and H. J. Lewerenz, J. Vac. Sci. Technol. A 12 (1), **56** (1994).

MOVPE growth and in situ characterization of GaN layers on sapphire substrates

H. Hardtdegen[*,1], **N. Kaluza**[1], **R. Schmidt**[1], **R. Steins**[1], **E. V. Yakovlev**[2], **R. A. Talalaev**[2], **Yu. N. Makarov**[2], and **J.-T. Zettler**[3]

[1] Institut für Schichten und Grenzflächen, Center of Nanoelectronic Systems for Information Technology, Forschungszentrum Juelich, 52425 Juelich, Germany
[2] Semiconductor Technology Research GmbH, P.O. Box 1207, 91002, Erlangen, Germany
[3] LayTec GmbH, Helmholtzstr. 13–14, 10587 Berlin, Germany

Received 19 September 2003, accepted 3 November 2003
Published online 12 January 2004

PACS 68.35.Ct, 68.55.Ac, 78.66.Fd, 81.15.Kk

MOVPE growth of GaN/sapphire was studied with respect to growth control by using in situ reflectometry. The main requirements for reproducible growth were identified to be the control of the substrate temperature and the avoidance of parasitic deposition. This is accomplished by using emissivity corrected pyrometry to monitor the growth temperature and modeling to find flow parameters which lead to strongly reduced parasitic deposition.

© 2004 WILEY-VCH Verlag GmbH & Co. KGaA, Weinheim

1 Introduction In the last decades, research in the field of GaN and related materials (AlN, InN and their ternary compounds) expanded exponentially. This is due to the fact that the group III-nitrides exhibit a wide range of direct band gaps ranging from 0.7 eV for InN [1] to 6.2 eV for AlN. At first progress was quite slow since there is a lack of native substrates on which homo-epitaxial MOVPE growth could be performed. Substrates were employed such as sapphire with a lattice mismatch of 16% on which the deposited nitrides exhibited not only high dislocation densities but also a high n-type background carrier concentrations.

In order to obtain device quality GaN on sapphire a two-step growth procedure was proposed by Amano et al. in 1986 [2]. This procedure included the deposition of a predominantly amorphous low-temperature (LT) buffer layer, in which nucleation crystallites form and the formation of islands during growth at high temperatures (HT) is observed which coalesce to develop the final desired two-dimensional growth mode. In situ monitoring of the surface morphology development by means of infrared pyrometry was reported first by Nakamura et al. [3, 4] and confirmed the proposed growth model. In the mean time in situ monitoring of growth development has become a basic method to understand the influences of growth parameters during the respective growth stages and the substrate pretreatment on the quality of the GaN bulk layer. A number of groups have reported their studies (for example [5–7]). Also systematic ex situ investigations were performed (for example [8–11]). Deposition strategies for device quality layers could be developed after a clear correlation between results obtained by in situ and those obtained by ex situ characterization methods was found. A low dislocation density is to be achieved. To this end all studies indicate that growth parameters and a substrate surface pretreatment are to be chosen with which GaN first grows three dimensionally to produce large crystallites. A transition from the three-dimensional to the two-dimensional growth mode arises when the crystallites coalesce. At the boundaries between the coalescing crystallites edge threading dislocations arise. Therefore the fewer the boundaries, the lower the

---

[*] Corresponding author: e-mail: h.hardtdegen@fz-juelich.de, Phone: +49 246 161 2360, Fax: +49 246 161 8143

dislocation density in the bulk GaN layer. The growth chemistry strongly influences the development of GaN surface morphology and needs to be understood to control growth properly.

Homogeneous as well as heterogeneous reactions take place in the growth of GaN. In the gas phase mainly the decomposition of TMGa (trimethylgallium) to MMGa (monomethylgallium) with an activation energy of 59.5 kcal/mol and its further decomposition to Ga with an activation energy of 77.6 kcal/mol [12] is important. Even though thermodynamics predicts that ammonia should be completely decomposed in H_2 and N_2 above 300 °C a high activation energy for its pyrolysis of 95 kcal/mol [13] was calculated. Therefore even at the growth temperature used for HT growth only a few percent of ammonia is decomposed. But this pyrolysis reaction is insignificant since the reaction products H_2 and N_2 do not play a role in GaN growth. Surface reactions however dominate GaN growth. GaN surfaces have a catalytic effect on the decomposition of ammonia adsorbed on the surface. The reaction products NH_x produce the active nitrogen necessary for GaN growth [14]. At low temperatures – 400–600 °C – used for nucleation layer deposition, growth is controlled by the catalytic decomposition of the absorbed NH_3 on the sample surface and by methyl radicals blocking Ga sites. At intermediate temperatures – 600–1050 °C – growth is limited by the transport of the group III species to the surface. At higher temperatures growth becomes thermodynamically controlled and is limited by Ga desorption from the sample surface. Since both the LT and the HT GaN growth are carried out in regimes where surface kinetics are important and where growth is extremely temperature dependent, substrate surface temperature is the most important growth parameter which needs to be controlled. Additionally since GaN deposits have a catalytic effect on especially NH_3 decomposition, the absence and the presence of deposits in the reactor chamber as well as on the substrate also will strongly influence growth. This paper deals with methods with which excellent growth control can be achieved.

2 Experimental Growth was performed in an AIX 200/4 RF-S horizontal reactor (AIXTRON) equipped with a separation plate in the gas inlet which allows the separate injection of the group III source TMGa through the upper channel and ammonia through the lower channel. Two different separation plates were employed: a straight plate and a plate tilted towards the ceiling of the reactor. Figure 1 presents the schematics of the two inlets. The gas streams meet and mix shortly before they reach the substrate. Growth was performed on inductively heated (0001) α-Al_2O_3 substrates. To improve uniformity in a horizontal reactor, the samples were rotated by gas foil rotation using nitrogen gas. The ambient in the reactor and the carrier gas were hydrogen. Reproducibility investigations were carried out with the following high temperature (HT) growth parameters: 200 mbar reactor pressure, a total flow of 8 slm, a TMGa flow rate of $1.10 \cdot 10^{-4}$ mol/min and an ammonia flow rate of $8.93 \cdot 10^{-2}$ mol/min. The HT growth was performed on a low temperature (LT) nucleation layer deposited at 540 °C with a thickness of about 25 nm. This LT layer was annealed, re-crystallized and etched by ramping up the temperature to 1050 °C for 5 min and holding this temperature for 3 min. The temperature was further ramped to HT epitaxy conditions for 1.5 min.

The reactor geometries as well as the high temperature growth parameters were employed as boundary conditions and starting points for the modeling investigations. The mathematical model is based on the solution of the 2D/3D Navier-Stokes conservation equations describing flow, heat transfer, and multicomponent diffusion of the gaseous mixture inside the reactor. Gas phase chemical reactions include the two-step pyrolysis of TMGa. The rate constants were taken from [12]. The surface chemistry was simulated by using a quasi-thermodynamical approach as suggested in [15]. The model is of a general charac-

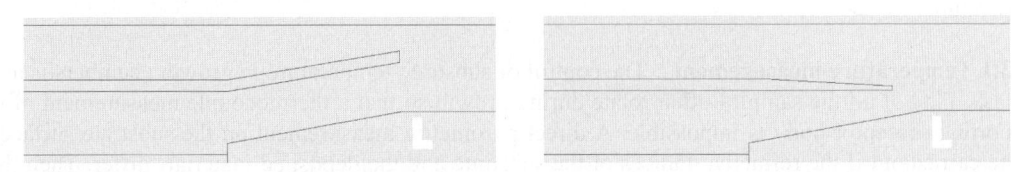

Fig. 1 Schematics of inlet configurations, with a tilted (left) and a straight separation plate, which allow the separate injection of TMGa and ammonia together with carrier gas.

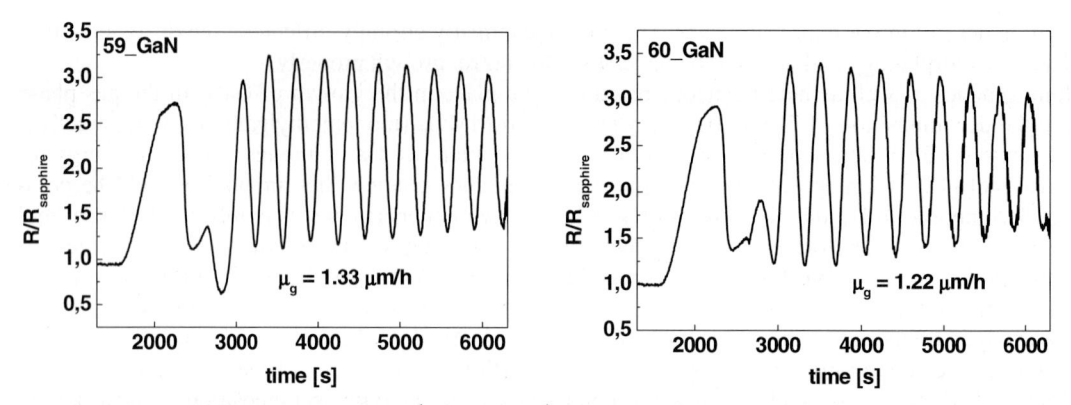

Fig. 2 In situ reflectance transients of 2nd (left) and 3rd (right) GaN/sapphire growth run using nominally identical growth conditions and an HT of 1110 °C. The growth rate in the HT range is higher for the 2nd run than for the 3rd indicating a drift in substrate temperature to higher values with increasing run number.

ter and applies to various operating conditions in the reactor since it combines thermodynamic and kinetic descriptions of surface processes,

Growth was observed by an EpiR-M-TT in situ optical sensor (LayTec). The detection wavelength was chosen to be 600 nm. At this wavelength Fabry-Perot growth oscillations are well observed even for thick layers and the signal is still very sensitive to morphological changes. The optical sensor includes an emissivity corrected pyrometer which detects the wafer temperature at 950 nm. At this wavelength, both the III-nitride layers and the sapphire substrates are optically transparent. Therefore the susceptor surface temperature below the wafer is measured through the sample and corrected for the formerly calibrated temperature difference to the wafer surface. Emissivity correction is performed in real-time by means of the 950 nm reflectance measured synchronously to the pyrometer reading.

3 Results and discussion The repetition of nominally the same growth run was observed by in situ reflectometry. Usually when the first growth run is carried out using a clean uncoated quartz reactor liner the transient differs quite extremely from that of the following runs. For the following investigations this run was not taken into account. It is quite obvious that the reactor needs a certain "conditioning" before series of experiments are performed. In a series of nominally identical growth runs a gradual systematic change of the transients from run to run is observed. At first the changes from run to run are small but the changes tend to increase strongly shortly before the parasitic deposition in the quartz reactor needs to be removed. This behavior is best resolved for two successive runs. Figures 2a and 2b present the transients of the 2nd and the 3rd run in the reactor. The spacing of the Fabry-Perot oscillations in the HT growth regime is narrower for the 2nd run than for the 3rd according to a higher growth rate. In the LT growth regime the ascent to the first oscillation is higher for the 3rd run than for the 2nd run according to a higher growth rate. Both observations can be explained by an increase in temperature in the LT and HT growth steps. The effect however is different. In the LT step, growth is kinetically controlled and the increase in temperature leads to a higher growth rate. In the HT step, growth is thermodynamically controlled so that an increase in temperature leads to a decrease in growth rate. These results indicate that the temperature controlled by the MOVPE equipment is not identical with the substrate temperature.

3.1 Temperature measurement The control of substrate temperature in growth chambers is a difficult task. First of all the samples often rotate during growth so that a thermocouple measurement of the supporting susceptor body is impossible. A direct pyrometric measurement on the substrate surface cannot be carried out if the refractive indices of the substrate and the deposited materials differ. Then the emissivity changes since Fabry-Perot oscillations are observed. Furthermore the measurement system should not interfere with growth and needs to be contact-less so as not to contaminate the growing surface.

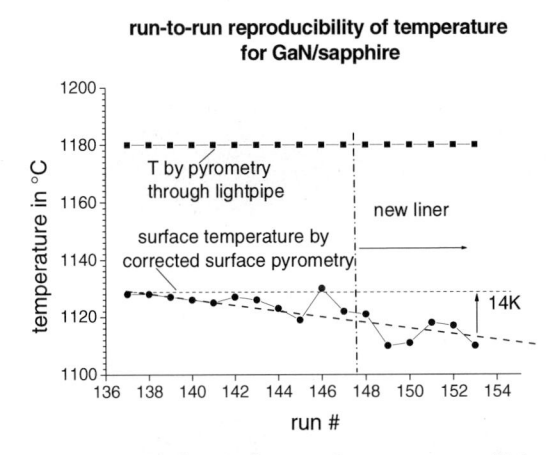

Fig. 3 Comparison of control temperature (light pipe) at 1180 °C with surface temperature for a series of successive growth runs. The surface temperature changes here in the range of 14 K and the absolute temperature difference is more than 50 K.

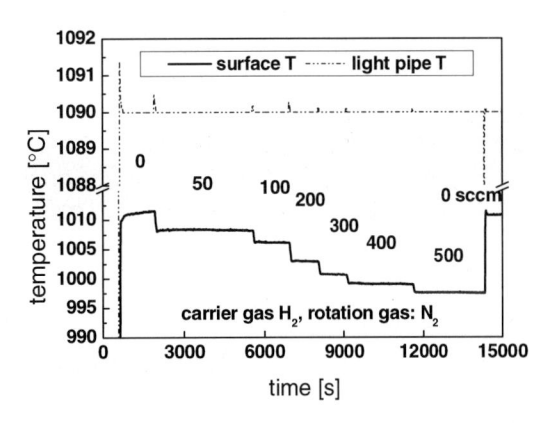

Fig. 4 Comparison of control temperature (light pipe) at 1090 °C with surface temperature for increasing rotation flows. The surface temperature drops by 15 K when the N_2 rotation gas flow is increased. Here an absolute deviation in temperature of up to to 95 K is observed.

Therefore emissivity corrected pyrometry was employed to check on the run to run reproducibility of the substrate temperature. The control and temperature measurement set-up is the following for the experiments: the light pipe sensor senses the radiation through a quartz bowl on the back side of a graphite disk. This disk is separated by rotation gas from the graphite disk which supports the wafer. This "light pipe" temperature is standardly used to control the inductive heating. The emissivity corrected pyrometer directly senses the substrate surface and the graphite disk underneath the substrate. The reproducibility of substrate surface temperature versus light pipe temperature (1180 °C) is presented in Fig. 3. The surface temperature varied in the high temperature regime by up to 14 K and an absolute temperature difference of more than 50 K was observed. Also the influence of hardware changes – here the exchange of a coated for a clean uncoated reactor liner is demonstrated. Apparently the substrate temperature changes by its surroundings. The surroundings that have an influence on growth are for example the parasitic deposition on the reactor walls and the rotation and ambient gases. The parasitic deposition changes the radiation away from the heated sample and also catalytically decomposes the ammonia on its surface so that the gas phase composition changes with the amount (area) of GaN coverage. The ambient itself and the rotation gas have an influence on the substrate surface temperature. The heat transfer from the inductively heated lower graphite disk to the upper substrate supporting disk occurs through this rotation gas

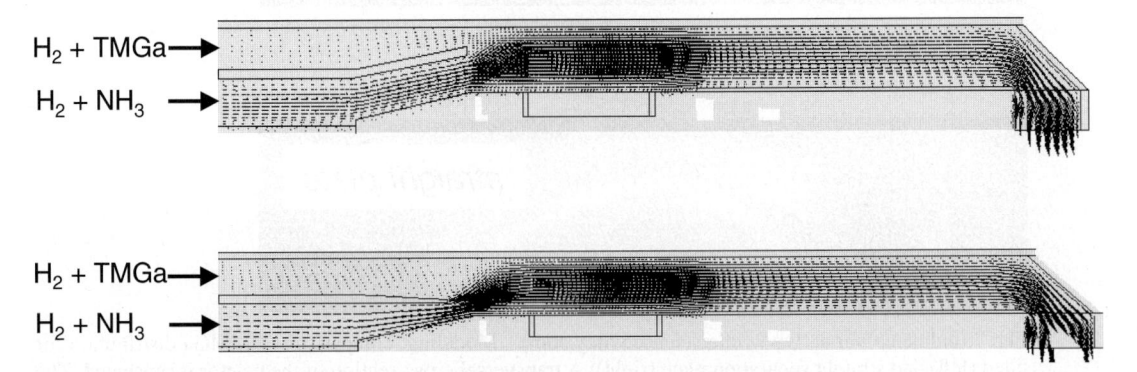

Fig. 5 Modeling: comparison of the flow patterns for a tilted (upper) and a straight (lower) separation plate: the gases mix earlier for the straight plate and the flow velocity is higher close to the ceiling for the tilted one.

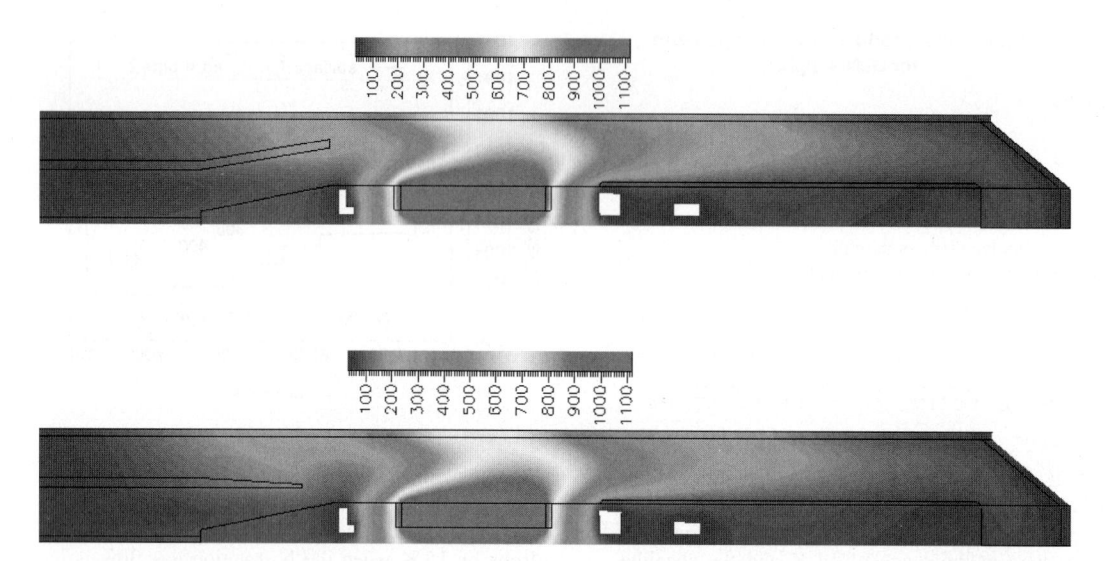

Fig. 6 (online colour at. www.interscience.wiley.com) Modeling: comparison of temperature distribution in the reactor for a tilted (upper) and a straight (lower) separation plate. The temperature distribution is independent on the inlet configuration.

flow and is influenced by its thermal conductivity. The thermal conductivity of the ambient gas will determine the temperature distribution and therefore the distribution of Ga-containing compounds in the gas phase [16]. Figure 4 demonstrates how the rotation gas (preferably nitrogen is used because of its inert nature even at high temperatures) changes the surface temperature. In the investigated flow range, substrate surface temperatures varied by 15 K. Since a rotation speed of about 60 rpm is to be achieved and flows vary to achieve this speed it is quite obvious how important true surface temperature monitoring and control is.

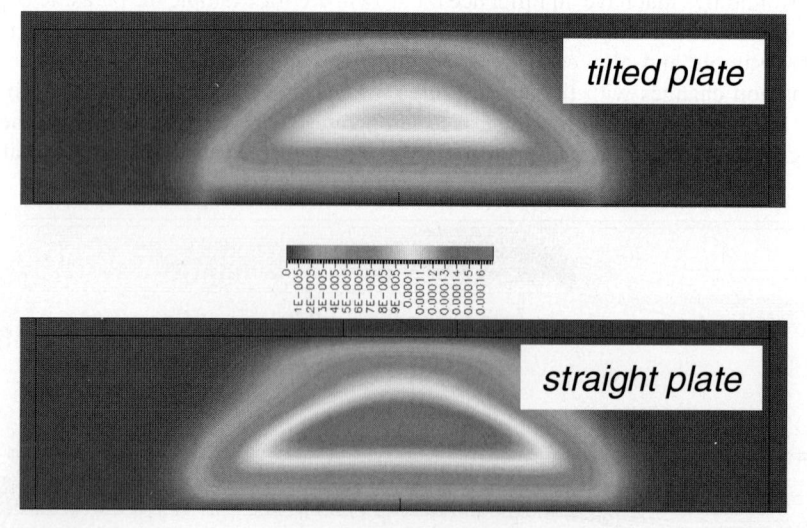

Fig. 7 (online colour at. www.interscience.wiley.com) Modeling: comparison of MMGa distribution for tilted (left) and straight separation plate (right). A transversal cross-section of the reactor is presented. The concentrations are translated into colors: blue for low and red for high concentrations. The scale is identical for both plates. The MMGa concentration is higher over the substrate for the straight plate.

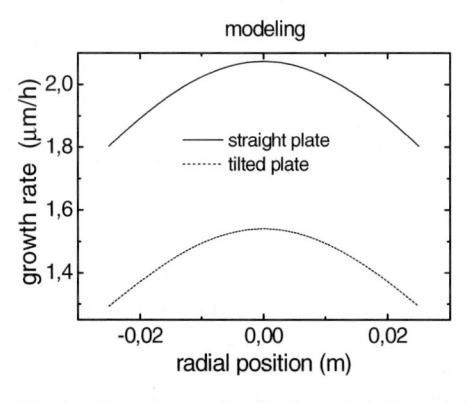

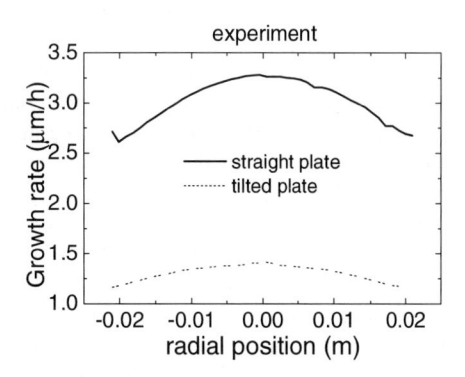

Fig. 8 Growth rate distribution of GaN over the cross section of a two inch wafer: modeling, left hand and experiment, right hand. The growth rate is higher for the straight plate.

3.2 Modeling of GaN growth The intention of the modeling experiments was to find growth conditions at which parasitic deposition would be suppressed. Parasitic deposition will mainly arise during HT growth since then the reactor walls especially the ceiling above the heated substrate and the side walls in the vicinity of the substrate will become warm enough to decompose the source materials. The parameters which influence the distribution of the growth governing species in the reactor chamber at a given set of HT growth parameters are on the one hand the choice of the carrier gas (here the influence is its thermal conductivity) [16] and on the other hand the flow distribution between the separate flow channels and the geometry of the gas inlet [17]. A systematic investigation was carried out for two different gas inlets (see 2. Experimental) at a given set of HT growth parameters and hydrogen as the carrier gas. With 1 slm of hydrogen TMGa was injected into the reactor through the upper channel. In the lower inlet channel 5 slm of hydrogen carried the ammonia into the reactor. Figure 5 presents the cross section of the reactor in which the susceptor with the wafer can be seen. A big difference in the flow patterns for the inlet configurations is observed. The flows mix earlier for the straight separation plate inlet than for the tilted one. A higher flow velocity is observed near the ceiling of the reactor for the tilted separation plate.

The heat transfer and therefore the temperature distribution in the reactor is not influenced by the inlet configuration as long as the total flows and the nature of the carrier gas is not changed. Figure 6 shows the temperature distribution comparison. Again the cross section of the horizontal reactor is presented. The

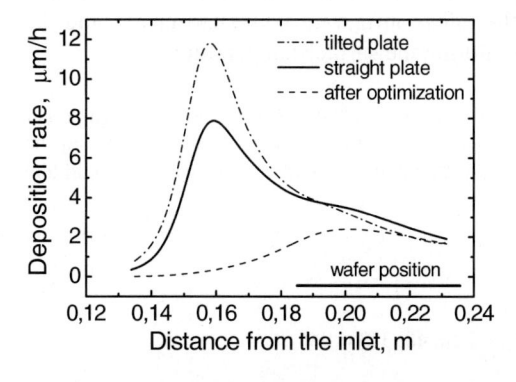

Fig. 9 Modeling: growth rate of deposition on the ceiling of the reactor chamber with respect to the distance from the inlet. The wafer position is marked. The gas inlet configuration, flows and carrier gas strongly influences the amount of parasitic deposition formed.

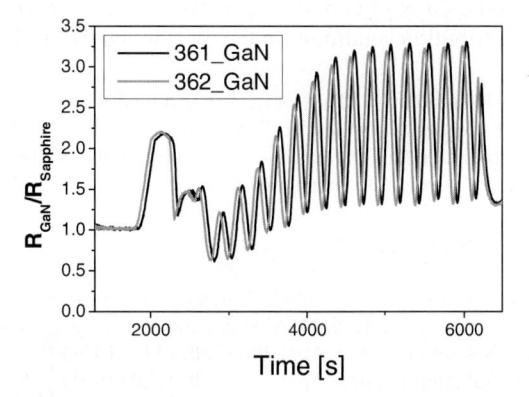

Fig. 10 Comparison of reflectance transients for two successive runs after process optimization. The transients do not vary discernibly any longer. Growth control is achieved.

temperatures are translated into colors ranging from blue for low and to red for high temperatures. For a given inlet configuration also the carrier gas flow distribution through the channels has a big influence on the ratio of the convective to diffusive flows of the Ga containing compounds and therefore influences the distribution of the sources in the reactor chamber [16]. Figure 7 presents a comparison of the MMGa in the reactor for both inlets with otherwise identical inlet flows as well as growth parameters. Here the transversal cross-section of the rectangular reactor tube is presented across the middle of the two inch wafer. The width of the tube is 13 cm. The concentration of the growth species MMGa is translated into colors ranging from blue for low to red for high concentration. Here again big differences are obvious. As can be expected the concentration of MMGa near the substrate is low in both cases since the further reaction to Ga takes place and MMGa is consumed for the growth there. For the straight plate the MMGa concentration is higher by a factor of about 1.5 than for the tilted plate. This leads to a higher growth rate on the substrate for the straight plate as can be seen in cross section over the wafer in Fig. 8 on the left side. On the right side in Fig. 8 the experimentally obtained growth rate is shown, which qualitatively reproduces the modeling data quite well. The reasons for the increased growth rate on the substrate for the straight plate are more favorable conditions for the transport of group-III species to the substrate and also that parasitic deposition is suppressed compared to the tilted plate. Figure 9 shows the calculated deposition rates on the ceiling of the reactor. The position of the wafer is marked beneath the curves. The growth experiments show, that growth is more reproducible for the straight plate compared to the tilted one. With further optimization of the inlet flow ratios and the carrier gas together with the inlet configuration parasitic deposition is nearly completely suppressed before the wafer where it would take influence on the growth chemistry. Its influence on growth should therefore be minimal. A detailed description of this optimization is presented elsewhere [18]. The effect on reproducibility is huge. Figure 10 shows the results of this optimization in which the reflectance transients do not differ discernibly any longer.

4 Summary and conclusions In this paper the control of growth was the central topic of discussion. By using in situ reflectometry the reasons for poor growth control were found and were associated to the surface chemistry and the growth development of GaN growth. It was observed that the surface temperature is strongly influenced by the surroundings of the wafer and that therefore surface temperature monitoring is a prerequisite for growth control. A suitable method is the use of emissivity corrected pyrometry. Also there is a strong indication that varying parasitic deposition is a reason for poor growth control. GaN growth modeling was performed with the aim to find growth parameters with which parasitic deposition is strongly suppressed. To this end especially the gas inlet configuration as well as the flow distribution were investigated. The growth efficiency on the substrate increases with decreasing parasitic deposition. Growth control is achieved after optimizing the inlet configuration, the carrier gas and the flow distribution conditions in the reactor and carefully monitoring the surface temperature.

Acknowledgements This work was performed within the trilateral cooperation "Group III nitrides for high power and high frequency applications", a cooperation between AIXTRON, the Technical University of Aachen and the Research Center Juelich financially supported by the North-Rhine Westfalian ministry of economics. We would like to thank K. Wirtz for technical assistance.

References

[1] V. Yu. Davydov et al., phys. stat. sol. (b) **229** (3), R1 (2002).
[2] H. Amano, N. Sawaki, I. Akasaki, and Y. Toyoda, Appl. Phys. Lett. **48**, 353 (1986).
[3] S. Nakamura, Jpn. J. Appl. Phys. **30**, 1348 (1991).
[4] S. Nakamura, Jpn. J. Appl. Phys. **30**, 1620 (1991).
[5] J. Han, T.-B. Ng, R. M. Biefeld, M. H. Crawford, and D. M. Follstaedt, Appl. Phys. Lett. **71**, 3114 (1997).
[6] P. Vennéguès, B. Beaumont, S. Haffouz, M. Vaille, and P. Gibart, J. Cryst. Growth **187**, 167 (1998).
[7] S. Figge, T. Böttcher, S. Einfeldt, D. Hommel, J. Cryst. Growth **221**, 262 (2000).
[8] L. Sugiura, K. Itaya, J. Nishio, H. Fujimoto, and Y. Kokubun, J. Appl. Phys. **82**, 4877 (1997).
[9] X. H. Wu, P. Fini, E. J. Tarsa, B. Heying, S. Keller, U. K. Mishra, S. P. DenBaars, J. S. Speck, J. Cryst. Growth **189/190**, 231 (1998).

[10] T. Böttcher, S. Einfeldt, S. Figge, R. Chierchia, H. Heinke, and D. Hommel, Appl. Phys. Lett. **78**, 1976 (2001).
[11] S. Einfeldt, T. Böttcher, S. Figge, and D. Hommel, J. Cryst. Growth **230**, 357 (2001).
[12] M. G. Jacko and S. J. M. Price, Can. J. Chem. **41**, 1560 (1963).
[13] D. Davidson, K. Kohse-Höinghaus, A. Chang, and R. Hansen, Int. J. Chem. Kin. **22**, 515 (1995).
[14] S. Liu and D. Stevenson, J. Electrochem. Soc. **125**, 1161 (1978).
[15] S. Yu. Karpov, V. G. Prokofyev, E. V. Yakovlev, R. A. Talalaev, and Yu. N. Makarov, MRS Internet J. Nitride Semicond. Res. **4**, 4 (1999).
[16] M. Dauelsberg, H. Hardtdegen, L. Kadinski, A. Kaluza, and P. Kaufmann, J. Cryst. Growth **223**, 21 (2001).
[17] H. Hardtdegen, A. Kaluza, D. Gauer, M. V. V. Ahe, M. Grimm, P. Kaufmann, and L. Kadinski, J. Cryst. Growth **223**, 15 (2001).
[18] H. Hardtdegen, N. Kaluza, R. Schmidt, R. Steins, E. V. Yakovlev, R. A. Talalaev, Yu. N. Makarov, to be published.

Growth optimisation of the GaN layers and GaN/AlGaN heterojunctions on bulk GaN substrates using plasma-assisted molecular beam epitaxy

C. Skierbiszewski[1,*], **Z. Wasilewski**[2], **M. Siekacz**[1], **A. Feduniewicz**[1], **B. Pastuszka**[1], **I. Grzegory**[1], **M. Leszczynski**[1], and **S. Porowski**[1]

[1] High Pressure Research Center, Polish Academy of Sciences, 01-142 Warsaw, Poland
[2] Institute for Microstructural Sciences, National Research Council, Ottawa, Canada

Received 19 September 2003, accepted 3 November 2003
Published online 12 January 2004

PACS 81.15.Hi, 81.05.Ea, 72.20.Fr, 72.80.Ey

Influence of growth conditions in plasma assisted molecular beam epitaxy on quality of GaN layers and GaN/AlGaN heterojunctions is studied. The growth diagram for step-flow growth mode and different nitrogen flux is presented. The low defect density of bulk GaN substrates together with very low impurity background concentrations resulted in high electron mobility for GaN/AlGaN heterojunctions: 109,000 cm^2/Vs at 1.5 K, and 2500 cm^2/Vs at 295 K.

© 2004 WILEY-VCH Verlag GmbH & Co. KGaA, Weinheim

Due to the large band gap GaN and its alloys are very promising materials for device applications such as light emitters in visible and ultraviolet range, field effect transistors, and heterojunction bipolar transistors. The main barrier limiting the performance of the optoelectronic devices is presence of defects, mainly due to the growth of GaN on substrates with large lattice mismatch such as sapphire or SiC. Recently considerable effort was put into reducing the dislocation density in so called GaN template epi-ready substrates, notably through epitaxial lateral overgrowth technique where reduction of dislocation density down to about 10^5 cm^{-2} has been demonstrated. Here we present study of GaN growth on pressure grown GaN bulk substrates as well as on epi-ready sapphire/GaN templates using plasma assisted MBE (PAMBE). Bulk GaN crystals demonstrate record low dislocation density at the levels down to 10^2 cm^{-2} [1]. The homoepitaxial growth of GaN on these substrates has not been studied extensively. However, preliminary works by Grandjean (ammonia MBE) [2] and Foxon [3] (PAMBE) showed great potential of this technique. On the other hand much work has been done on PAMBE growth of GaN on sapphire/GaN templates (Heying [4] and Adelman [5]). It was found that excess of Ga is necessary for successful two-dimensional step flow growth mode [4]. Northrup [6] and Zywiec [7] showed that bilayer Ga coverage of grown layer increases dramatically the lateral mobility of the atoms in this layer and helps low mobile N atoms to find a proper place in GaN crystal. The amount of excess Ga required was found to depend strongly on the growth temperature with activation energy E_A around 2.8 eV, close to the activation energy of desorption of Ga atoms from liquid gallium [4]: $\Phi_{Ga}^{EXCESS}(T) = \Phi_0 \cdot \exp(-E_A/kT)$.

This equation with $E_A = 2.8$ eV determines the temperature dependence of the maximum excess Ga flux above which Ga droplets are formed on the surface. However, evidences of formation of stable Ga bilayer were reported across a range of fluxes below this maximum value [5].

All GaN bulk substrates used in the present study were mechanically polished, etched by reactive ion etching and overgorwn with 1 µm undoped GaN layer in MOCVD reactor. The PAMBE growth was

---

[*] Corresponding author: e-mail: czeslaw@unipress.waw.pl

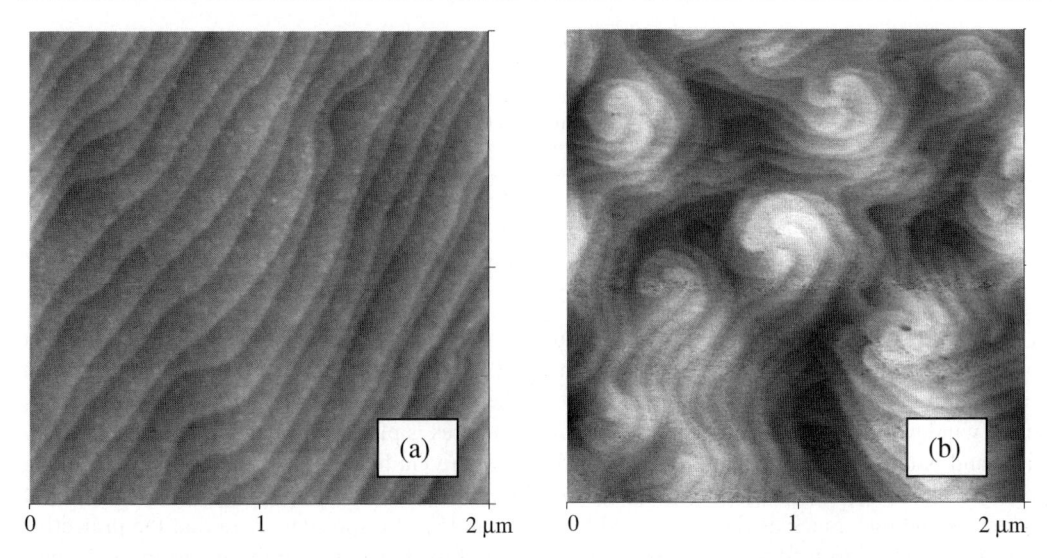

Fig. 1 (online colour at: www.interscience.wiley.com) AFM images of the GaN layers on bulk crystal a) and GaN/sapphire template b).

performed in Thermo VG VG90 MBE reactor. The Vecco "Unibulb" radio-frequency (RF) plasma source was used as a source of active Nitrogen. Laser reflectometry was used to monitor the growth rate while the temperature of the substrate was monitored with pyrometer. For precise tuning of the growth conditions we relied on the great sensitivity of RHEED specular spot intensity to Ga surface coverage. Similar to work by Heying [4] we can distinguish three different regime of GaN growth as a function of Ga flux. For low Ga flux 3D growth mode was observed. By adjusting the Ga flux we were able to move from 3D to 2D step-flow growth mode. Further increase of Ga flux results in gallium droplets formation on the surface. Comparison of the surface morphology on bulk crystal and GaN/sapphire templates in the step-flow mode is given in Fig. 1. For GaN/sapphire templates, shape of terraces is influenced strongly by presence of defects (Fig. 1b), leading to characteristic spirals around each threading dislocation. On the bulk GaN substrates situation is very different. There, formation of parallel atomic steps leads to atomically smooth surface on a macroscopic scale (Fig. 1a). In spite of much work in the area the growth kinetics of GaN in metal rich regime are not well understood. Previous literature reports concentrated on the influence of the substrate temperature on the surface morphology [4–6]. Here we study the sensitivity of growth kinetics to different N flux. In gallium rich regime the growth rate is controlled by the active N flux; by adjusting N flux we are able to vary GaN growth rate continuously up to the maximum of 1 µm/h. Fig. 2 illustrates influence of N flux on layer morphology.

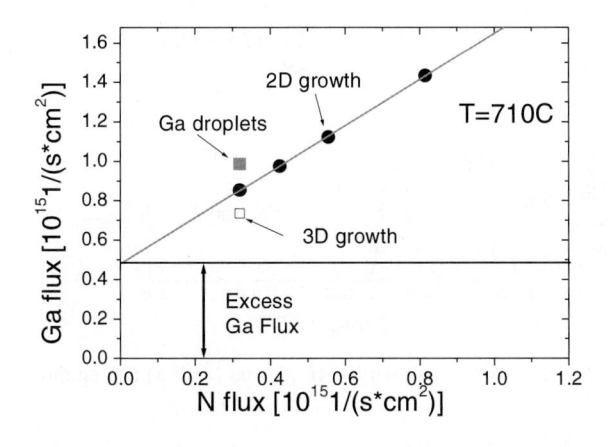

Fig. 2 (online colour at: www.interscience.wiley.com) Growth diagram of GaN for $T = 710$ °C. Dots presents data for 2D step flow mode. Open square and square are for 3D and Ga droplets regime respectively.

Table 1 Mobilities on 2D densities (n_{2D}) and conductivity at 295 K on different GaN/AlGaN heterojunctions.

sample	AlGaN spacer (nm)	Al content (%)	n_{2D} (cm^{-2})	mobility μ (cm^2/Vs)	$\mu \cdot n_{2D}$ (10^{16}/Vs)
P061B	30	10	2.3×10^{12}	1380	0.32
P051	23	9	7.7×10^{12}	2530	1.95
P069a	26	10	2.9×10^{13}	2600	7.54

Open and filled squares indicate values of Ga flux for 3D growth and 2D growth with Ga droplet formation respectively for N flux equal 3×10^{14} 1/cm^{2}s at $T = 710$ °C. Dots in this figure indicate growth conditions resulting in step-flow growth mode and excellent final layer morphology. We should stress here that the window of Ga fluxes for such high quality structures is placed just below the droplet formation region, and is much narrower than the zone of fluxes where Ga bilayer has been observed. Moving away from this tight window, but staying in the Ga bilayer mode, leads to morphology which is a mix of smooth plateaus and pits, much as those reported by Aldeman [5]. We speculate here that the protective Ga bilayer can be indeed formed in a fairly broad range of Ga flux, however mainly in a form of patches, while continuous coverage of the surface takes place only at fluxes just below the onset of droplets formation. The linear behavior of Ga flux vs. N flux for this optimum growth regime, visible in Fig. 2, indicates that discussed by Northrup Φ_{Ga}^{EXCESS} depends only on the growth temperature and can be obtained from equation $\Phi_{Ga} = \Phi_N + \Phi_{Ga}^{EXCESS}$ by extrapolation to zero nitrogen flux. This observation gives simple and very powerful tool to arrive at optimum Ga flux for arbitrary active N flux in wide range of growth temperatures after fairly straightforward calibration procedure. The photoluminescence data on MBE grown GaN layers is used as an insight into the layer quality. In MBE grown 1 μm GaN layer on bulk substrate we observed very narrow donor bound exciton lines (D^{0}X) with the half-width of 0.3–0.5 meV. Our results are comparable to the PL data from GaN layers grown on bulk crystals by Grandjean et al. [2] using ammonia MBE. However, in contrast to their work we do not observe acceptor bound exciton (A^{0}X) in our spectra. This result indicates considerably lower acceptor concentrations in our layers.

The GaN/AlGaN heterojunctions were grown at very high temperatures 720 °C. 1 μm GaN layer followed by 22–30 nm AlGaN spacer and 3 nm GaN cap layer was grown. In Table 1 growth parameters together with results of room conductivity and hall effect is presented. The growth on the semi-insulating bulk substrates results in high mobilities of 2D electron gas at low temperatures ($T = 1.5$ K): in the range from 15,000–109,000 cm^2/Vs for $n_{2D} = 2 \times 10^{12}$–$10^{13}$ cm^{-2}. The transport experiments were performed on etched hall-bar structures. The low acceptor and donor concentrations and low density of defects in our material resulted in mobility of 109,000 cm^2/Vs. At the moment is the highest mobility demonstrated in GaN/AlGaN heterojunction. In addition, mobility (2500 cm^2/Vs) and conductivity

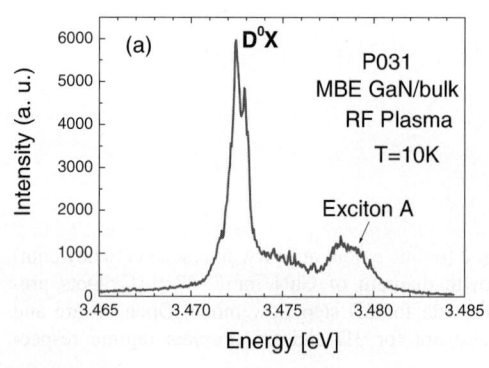

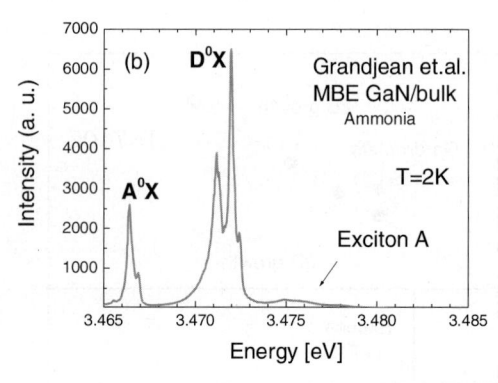

Fig. 3 (online colour at: www.interscience.wiley.com) PL on bulk GaN grown by RF plasma MBE a) and ammonia MBE b).

($>7 \times 10^{16}$/Vs) observed at room temperature are very promising for high current AlGaN/GaN HFETs based on the dislocation free bulk substrates. Experimentally observed mobility at room temperatures are higher than calculated mobility limits for LO optical phonon scattering, which presents new challenge for theory.

In conclusion, the growth diagram for step-mode under different N fluxes in Ga rich regime was studied in PAMBE. Growth of very high quality GaN and AlGaN/GaN layers was demonstrated and attributed to the dislocation free bulk GaN substrates, low background doping donor and acceptor concentrations introduced in our PAMBE process as well as precise control of growth conditions.

Acknowledgements Work was partially supported by the Project of the European Commission DENIS G5RD-CT-2001-00566 and Project SPUB-M of the Polish Ministry of Scientific Research and Information Technology.

References

[1] I. Grzegory and S. Porowski, Thin Solid Films **367**, 281 (2000).
[2] N. Grandjean, B. Damilano, and J. Massies in: Low-dimensional nitride semiconductors, edited by B. Gil, (Clarendon Press, Oxford, 2002) p. 121, (2002), Clarendon Press, Oxford.
[3] C. T. Foxon, T. S. Cheng, S. V. Novikov, D. Korakakis, N. J. Jeffs, I. Grzegory, and S. Porowski, J. Cryst. Growth, **207**, 1 (1999).
[4] B. Heying, I. Smorchkova, C. Poblenz, C. Elsass, P. Fini, S. Den Baars, U. Mishra, and J. S. Speck, Appl. Phys. Lett. **77**, 2885 (2000).
[5] C. Adelmann, J. Brault, D. Jalabert, P. Gentile, H. Mariette, Guido Mula, and B. Daudin, J. Appl. Phys. **91**, 9638 (2002).
[6] J. E. Northrup, J. Neugebauer, R. M. Feenstra, and A. R. Smith, Phys. Rev. B **61**, 9932 (2000).
[7] T. Zywietz, J. Neugebauer, and M. Scheffler, Appl. Phys. Lett. **73**, 487 (1998).

Kinetic and strain-driven growth phenomena on Si(001)

C. Schelling[1,2], **J. Myslivecek**[1,2], **M. Mühlberger**[1], **H. Lichtenberger**[1], **Z. Zhong**[1], **B. Voigtländer**[2], **G. Bauer**[1], and **F. Schäffler**[*,1]

[1] Institut für Halbleiterphysik, Johannes Kepler Universität, Linz/Austria
[2] Forschungszentrum Jülich/Germany

Received 19 September 2003, accepted 3 November 2003
Published online 14 January 2004

PACS 68.55.AC, 68.35.–p, 68.65.Hb, 81.15.Hi

Self-organization phenomena in semiconductors are usually based on strain-driven island growth during hetero epitaxial layer deposition. However, kinetic phenomena can become important and even dominating at the low growth temperatures usually employed during molecular beam epitaxy. We report on kinetic step bunching on Si(001), and identify the driving mechanism on the atomic scale via kinetic Monte Carlo simulations. Another phenomena discussed is facet formation during annealing of SiO_2-covered Si(001) nanostructures at the relatively low temperatures usually employed for oxide desorption. Both phenomena are combined to facilitate perfect ordering of self-assembled Ge dots on facetted Si(001) nanostructure templates.

© 2004 WILEY-VCH Verlag GmbH & Co. KGaA, Weinheim

1 Introduction While commercial silicon ultra-large-scale integrated circuits have reached physical gate lengths of 65 nm with optical lithography and other standard technologies, alternative (bottom-up) approaches are investigated to exploit self-organization or self-assembly schemes based on growth phenomena in strained heteromaterial systems. The most widely exploited mechanism is driven by a minimization of the total energy. This leads in many heterosystems to the so-called Stranski–Krastanov growth mode, where the stress in a strained epilayer is partly relieved by 3D island growth. However, in recent years it became more and more obvious that homoepitaxial kinetic growth phenomena at the relatively low temperatures typically employed for epitaxy are important, and can even become dominant under certain conditions. In the following we report three such phenomena on the single most important technical surface, namely Si(001).

2 Step bunching during Si homoepitaxy Recently, we could show that under homoepitaxial growth conditions vicinal Si(001) surfaces develop a purely kinetic step bunching instability in a relatively narrow temperature range [1, 2]. Figure 1 shows a typical AFM image of a step-bunched Si(001) surface with a miscut of 0.66° along the [110] direction. We also showed that this phenomenon is not strain-related, [3, 4] in contrast to equilibrium models discussed in the literature, which attributed the observed step-bunching to strain in the Si/SiGe heterosystem [5, 6]. Atomically resolved in-situ STM experiments revealed that the onset of the instability is correlated with the transition from island nucleation to step flow growth [7, 8]. The instability disappears when the surface reconstruction approaches the thermal equilibrium configuration with equal coverages of (2×1) and (1×2) reconstructed A and B terraces.

To identify the dominating mechanisms, 2D kinetic Monte Carlo simulations (KMC) were performed, with model potentials that allowed us to account for anisotropic diffusion, anisotropic sticking, and for additional step edge barriers [7]. Rather surprisingly, we found that diffusion anisotropy alone causes step bunching of the type, and with the pronounced temperature-dependence that was observed in the

---

[*] Corresponding author: e-mail: f.schaffler@hlphys.uni-linz.ac.at

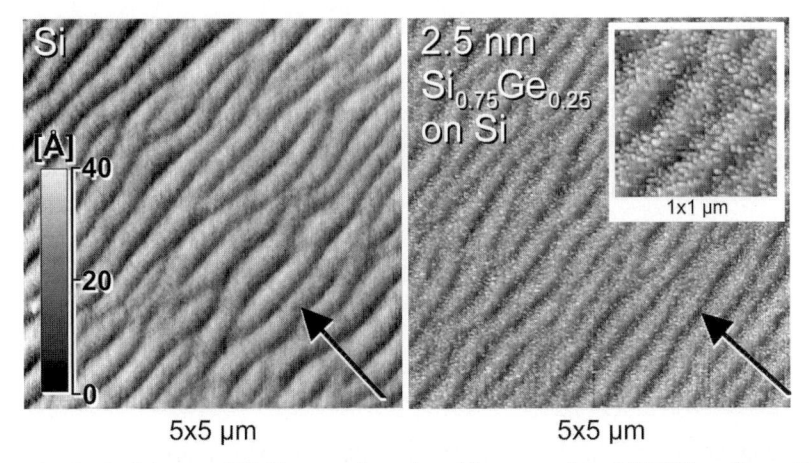

Fig. 1 Left image: AFM image of step bunching on a vicinal Si(001) surface with 100 nm of Si deposited at 490 °C and 0.02 nm/s. Right image: Same growth conditions except that additionally 2.5 nm of $Si_{0.75}Ge_{0.25}$ were deposited and annealed for 1 h, also at 490 °C. The step bunching morphology of the Si buffer is clearly preserved, whereas the $Si_{0.75}Ge_{0.25}$ layer disintegrates into 3D islands. The substrate miscut angle is 0.66° along a $\langle 110 \rangle$ azimuth, as marked by arrows.

STM experiments. Obviously, the dimer-reconstructed Si(001) is intrinsically unstable against step bunching under kinetic growth conditions. This effect can be enhanced by anisotropic sticking and/or additional step edge barriers, but anisotropic sticking alone always leads to stable step-flow growth.

To gain further insight, we extracted the surface mass current vector and its derivative with respect to the miscut [9]. In the temperature range of the instability we found an overall down-step mass current with positive derivative with respect to the miscut. This is exactly the combination of conditions that is associated with step bunching [9]. To identify the dominant contribution to the downward mass current we evaluated the ad-particle movements during the KMC simulations, and found that the overall mass current is mainly fed by thermal detachment of weakly bound ad-particles from straight segments of the S_A steps. Because of the diffusion anisotropy, this detachment current is preferentially released onto the adjacent downhill B terrace, where the fast diffusion direction is along the miscut direction [7]. This mechanism ceases, when the area coverages with A and B terraces approach the thermal equilibrium value of 50%.

Figure 2 shows the model potential for the KMC simulations, which accounts for the anisotropy of diffusion along and perpendicular to the dimers, and the difference in the binding energies at S_A and S_B steps. The latter is primarily caused by the much lower kink density at the straight S_A steps (dimers parallel to the step edge) in comparison with the rugged S_B steps (dimers perpendicular to the step edges). The

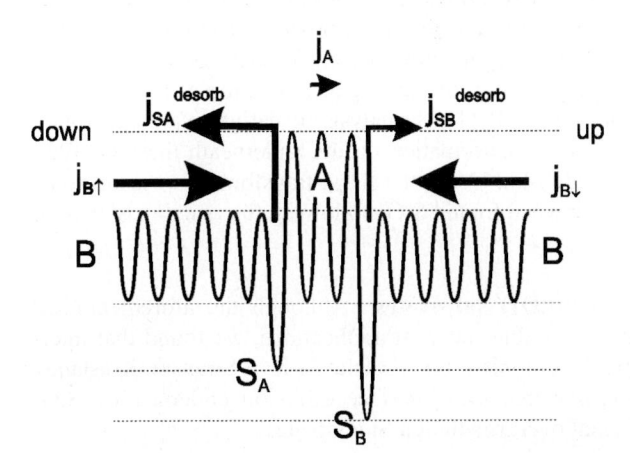

Fig. 2 Model potential used for kinetic Monte Carlo simulations. The arrows indicate the components and the relative contributions of ad-particle movement to the surface mass current.

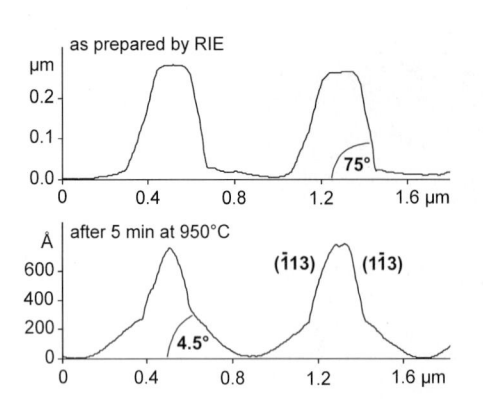

Fig. 3 AFM line scans from wire templates generated by reactive ion etching and covered with a native oxide layer before and after annealing at 950 °C for 4 min. The trapezoidal wire cross section was converted into a triangular cross section terminated by {311} facets, concomitant with a loss of 60% of the original peak-to-valley ratio. Note the different height scales.

dimer rows in Fig. 2 are in, and perpendicular to the paper plane on B, and A terraces, respectively. The contributions to the overall surface mass current as derived from the KMC simulations are indicated by arrows, the widths of which represent the respective amount of the current.

The newly discovered step-bunching mechanism caused by the surface diffusion anisotropy is a new generic mechanism which should be of general relevance for all surfaces of similar symmetry.

3 Transient-enhanced surface diffusion on Native-Oxide-Covered Si(001) With critical semiconductor devices in the nanometer range, the preparation and conservation of small morphological features becomes increasingly relevant. The surface free energy of a structured heterosystem is determined by composition, crystal orientation and strain. The interplay between these parameters leads to the aforementioned self-organization phenomena.

Here we report on the morphological integrity of such templates under chemical and thermal treatments typically employed during device processing. The samples consist of periodic wire arrays of rectangular cross section fabricated by holographic lithography and reactive ion etching on Si(001) substrates. Periods were varied between 400 and 800 nm at etch depths of typically 250 nm. After photoresist stripping the samples underwent an RCA clean with or without a final HF dip immediately before transfer into the UHV environment of an MBE system. Here the samples were radiatively annealed for 2–5 min at 900–950 °C, which corresponds to the standard oxide desorption step prior to epitaxial MBE growth. Samples that had the natural oxide removed by HF showed no significant morphological changes. However, on all samples covered by native oxide the originally rectangular profiles were transformed into {311} facetted trapezoids concomitant with a loss of up to 80% of the peak-to-valley modulation [10]. Figure 3 shows AFM line scans from samples before and after annealing. The shape transformation requires drastically enhanced mass transport, which occurs only in the presence of SiO_2 on the surface, and consequently ceases after complete oxide desorption. Figure 4 shows XTEM images of an originally rectangular wire profile before and during the shape transition process: The onset of {311} facets at the upper edges, and of a {111} facet at the lower right corner become clearly visible.

We followed by high-resolution XTEM imaging the kinetics of faceting as a function of the annealing conditions, and developed a decoration technique for the HRTEM analysis of native oxide desorption [10]. From these experiments we found that the shape transformation occurs underneath the SiO_2 film, and not in voids that form at an early stage in the oxide film. Hence the shape transformation seems to be related to the accumulation of SiO at the interface between Si and SiO_2 following the reaction $Si + SiO_2 \rightarrow 2\ SiO$.

4 Organized self-assembly on Nano-Structured Si(001) substrates Although the aforementioned structural transformation of Si nanostructures is undesirable for most applications, we found that intersects between low energy facets can work as nucleation sites for 3D islands in a Stranski–Krastanov growth mode [11, 12]. Besides faceting by transient enhanced interface diffusion underneath a SiO_2 layer, faceting can also be achieved by homoepitaxial overgrowth of a Si template.

© 2004 WILEY-VCH Verlag GmbH & Co. KGaA, Weinheim

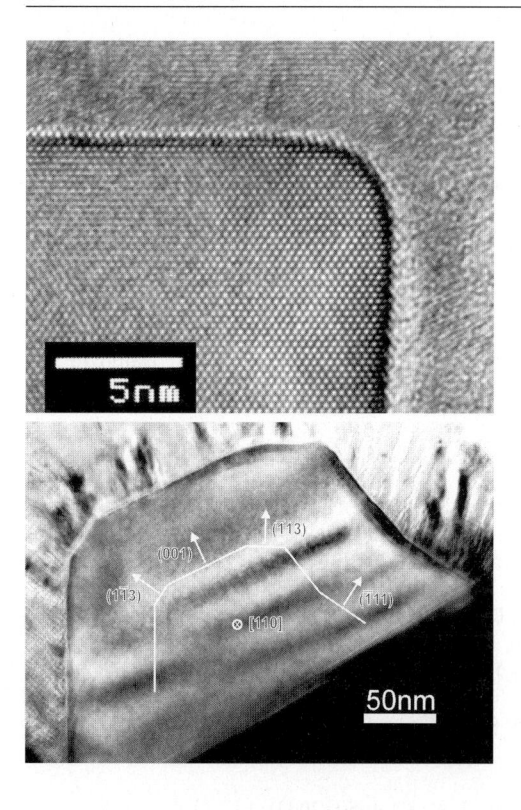

Fig. 4 XTEM image of wire cross sections before and after annealing. Upper image: HRXTEM image of the upper right corner of a wire after fabrication by reactive ion etching; corner rounding is <5 nm. Lower image: Complete wire cross section after annealed at 900 °C for 3 min. The onset of {311} and {111} faceting is clearly visible.

Si templates with facetted pits can solve a general problem of self-organization schemes, namely the improvement of lateral ordering [13–15]. We found that if the mean-free-path of Ge atoms or dimers on such a template with a periodic 2D array of facetted pits is kept larger than the period of the array, then only one Ge dot nucleates per unit cell of the template [16]. The nucleation site is at the intersect of the facets at the bottom of the pit, as shown in the XTEM images of Fig. 5, where the facets are indicated by dashed lines [12]. In addition, Fig. 5 shows an AFM image of an ordered array of Ge dots on a facetted Si template.

A further advantage of this kind of template-induced organized self-assembly is the fact that the area from which deposited Ge is funneled into such a single dot is constant in a periodic 2D array, which leads to a very homogeneous distribution of the dot sizes. Typical variations of the dot size can then be kept well below 4% [17]. Hence, a very high degree of ordering and a very narrow size distribution can be achieved simultaneously.

Ordering on facetted templates is of pure geometrical origin and is not caused by e.g. dot-dot interaction or by strain effects, mechanisms that are widely considered for increasing the degree of order. Therefore, the symmetry of the dot arrangement is, within the aforementioned mean-free-path-limitations, exclusively defined by the template, and does not depend on the crystallographic orientation. We had therefore no problems to implement square, hexagonal and triangular 2D dot lattices [17].

A further benefit of utilizing facetted Si templates rather than SiO_2 masks is the feasibility of epitaxial overgrowth after Ge self-assembly. This allows for Ge dot superlattices with perfect lateral ordering via the template, and vertical ordering via the well-known strain fields above each dot [18].

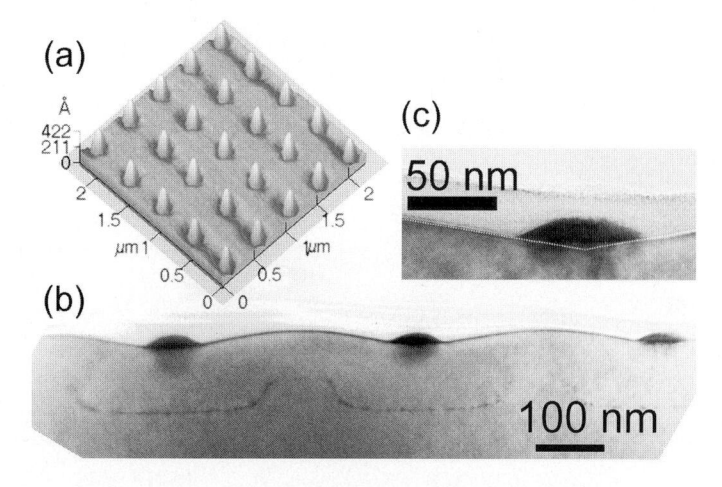

Fig. 5 a) AFM image of an ordered Ge dot array on a Si template with faceted pits. b) XTEM cross section through a dot row. c) Detail of b) with facets marked by dotted lines. The Ge dots nucleate at the facet intersection.

5 Conclusion Kinetic growth and reaction kinetics on the Si(001) surface become increasingly relevant as device dimensions shrink into the nano-scale regime. We identified the dominant mechanism responsible for kinetic step-bunching during Si homoepitaxy, and found a transient-enhanced surface diffusion mechanism on Si(001) surfaces covered by native oxide. Both mechanisms can be exploited in self-organization schemes. This has impressively been demonstrated by combining Stranski–Krastanov growth of Ge dots with nanostructured Si(001) substrates, which allow perfect lateral and vertical long range ordering of the dots [16]. However, for conventional applications it is as important to know that the kinetic instabilities can be suppressed by an adequate choice of growth and cleaning conditions.

Acknowledgements Work reported here was financially supported by FWF (Vienna) via projects P12143PHY, P16223-N08, and by a scholarship of the Macke Foundation. Expert technical assistance by A. Halilovic and G. Hesser are gratefully acknowledged.

References

[1] C. Schelling, G. Springholz, and F. Schäffler, Phys. Rev. Lett. **83**, 995 (1999).
[2] C. Schelling, G. Springholz, and F. Schäffler, Thin Solid Films **369**, 1 (2000).
[3] C. Schelling, G. Springholz, and F. Schäffler, Thin Solid Films **380**, 20 (2000).
[4] C. Schelling, M. Mühlberger, G. Springholz, and F. Schäffler, Phys. Rev. B **64**, 041301(R) (2001).
[5] J. Tersoff, Y. H. Phang, Z. Zhang, and M. G. Lagally, Phys. Rev. Lett. **75**, 2730 (1995).
[6] Feng Liu, J. Tersoff, and M. G. Lagally, Phys. Rev. Lett. **80**, 1268 (1998).
[7] J. Mysliveçek, C. Schelling, F. Schäffler, G. Springholz, P. Smilauer, J. Krug, and B. Voigtlander, cond-mat/0212331.
[8] J. Myslivecek, C. Schelling, F. Schäffler, G. Springholz, P. S. Smilauer, J. Krug, and B. Voigtländer, Surface Sci. **520**, 193 (2002).
[9] J. Krug, Adv. Phys. **46**, 139 (1997).
[10] H. Lichtenberger, M. Mühlberger, and F. Schäffler, Appl. Phys. Lett. **82**, 3650 (2003).
[11] Z. Zhong, A. Halilovic, M. Mühlberger, F. Schäffler, and G. Bauer, Appl. Phys. Lett. **82**, 445 (2003).
[12] Z. Zhong, A. Halilovic, M. Mühlberger, F. Schäffler, and G. Bauer, J. Appl. Phys. **93**, 6258 (2003).
[13] T. I. Kamins and R. S. Williams, Appl. Phys. Lett. **71**, 1201 (1997).
[14] O. G. Schmidt, N. Y. J. Phillipp, C. Lange, U. Denker, K. Eberl, R. Schreiner, H. Gräbeldinger, and H. Schweizer, Appl. Phys. Lett. **77**, 4139 (2000).
[15] L. Vescan and T. Stoica, J. Appl. Phys. **91**, 10119 (2002).
[16] Z. Zhong, A. Halilovic, M. Mühlberger, T. Fromherz, F. Schäffler, and G. Bauer, Appl. Phys. Lett. **82**, 4779 (2003).
[17] Z. Zhong et al., to be published.
[18] J. Tersoff, C. Teichert, and M. G. Lagally, Phys. Rev. Lett. **76**, 1675 (1996).

Kinetics of the heteroepitaxial growth of Ge layer at low temperature on Si(001) in UHV-CVD

M. Halbwax[*,1], **V. Yam**[1], **C. Clerc**[2], **Y. Zheng**[3], **D. Debarre**[1], **Lam. H. Nguyen**[1], and **D. Bouchier**[1]

[1] Institut d'Electronique Fondamentale, Bât 220, Universite Paris-Sud, CNRS UMR 8622, 91405 Orsay Cedex, France
[2] Centre de Spectrometrie nucleaire et Spectrometrie de masse, Bât 108, Universite Paris-Sud, 91405 Orsay Cedex, France
[3] Laboratoire de Mineralogie-Cristallographie, Universités Paris VI et Paris VII, CNRS UMR 7590, 4 place Jussieu, 75252 Paris Cedex 05, France

Received 19 September 2003, accepted 3 November 2003
Published online 12 January 2004

PACS 68.55.Ac, 81.15.Gh, 68.35.Fx, 61.14.Hg, 82.80.Yc

The Ge growth at 330 °C by ultrahigh vacuum chemical vapour deposition is investigated in real time by reflection high energy electron diffraction (RHEED) in combination with atomic force microscopy and Rutherford back scattering spectrometry (RBS). The Stranski-Krastanov-related 2D to 3D transition is avoided at low temperature and the major part of the relaxation process occurs during the deposition of the first two monolayers. The very low growth rate observed during this first step is related to the deposition of Ge on Si. Beyond 2 deposited MLs, the growth rate increases drastically due to a complete coverage of Si by Ge. Finally, the deposition of Ge at 330 °C results in an in-plane lattice parameter approaching 90% of that of Ge bulk and a flat surface with rms roughness of 0.6 nm for a film thickness lower than 30 nm.

© 2004 WILEY-VCH Verlag GmbH & Co. KGaA, Weinheim

1 Introduction Optical interconnects are considered as an attractive solution to overcome the limitations of metallic interconnects in future microelectronic circuits. For that, low-loss SOI microwaveguides at $\lambda = 1.3$ µm were previously demonstrated [1]. Fast and high responsivity integrated photodetectors are also needed to transform optical information into electrical signal. For this application, pure germanium appears as an excellent material, because of its compatibility with the silicon technology and its high absorption coefficient at 1.3 µm (~8000 cm^{-1}). Unstrained Ge films must then be deposited on Si, with the same thickness as optical microwaveguides, i.e. 300–400 nm. However, due to a lattice mismatch between Si and Ge of 4.17%, the growth of a thin unstrained Ge film with a low density of threading dislocations remains a challenge. In order to locate the misfit dislocations at the interface between Si and Ge, it is crucial to control the Ge relaxation at the very beginning of the film growth. At usual temperatures (400–700 °C), the Stranski–Krastanov growth mode is observed during ultrahigh vacuum chemical vapour deposition (UHV-CVD) of Ge on Si(001). The undesirable related 2D to 3D transition can be avoided by depositing a Ge buffer layer at a very low temperature [2] which minimizes the Ge diffusion and favours the formation of vacancies. This procedure allowed previously the realisation of thick Ge photodetectors [3–5].

This paper deals with the relaxation kinetics of the Ge/Si(001) growth at low temperature, investigated *in situ* by reflection high energy electron diffraction (RHEED). This study is combined with measurements by Rutherford back scattering spectrometry (RBS), atomic force microscopy (AFM), transmission electron microscopy (TEM) and X-ray diffraction (XRD).

---

[*] Corresponding author: e-mail: Mathieu.halbwax@ief.u-psud.fr, Phone: +33 1 69 15 40 47, Fax: +33 1 69 15 40 10

2 Experimental setup

Experiments are carried out in a UHV-CVD system with a base pressure of 1×10^{-10} Torr. Pure SiH_4 and GeH_4 diluted at 10% in H_2 are used as gas source. The sample temperature is measured with an infrared pyrometer and the total pressure with a capacitance gauge. The growth process is monitored by RHEED. The corresponding diffraction patterns are recorded by using a camera connected to a computer. The substrates are p-type Si(001) wafers. After a modified Shiraki chemical cleaning [6], the substrates are slowly annealed up to 650 °C, the pressure being maintained below 2×10^{-8} Torr. Afterwards, the chemical surface oxide is removed by flashing at 1050 °C. After deposition of a Si buffer layer at 700 °C, the Ge heteroepitaxy is initiated at a total pressure of 1.8×10^{-3} Torr and a temperature fixed at 330 °C.

Post deposition imaging of the film morphology is done by using AFM and TEM. The deposited Ge amount in each sample is measured in atoms number per surface unit by RBS. Structural characteristics are complementarily deduced from *in situ* RHEED and *ex situ* XRD measurements. The in-plane lattice parameter evolution at the growth temperature $a_\parallel$(330 °C) is evaluated by analysing RHEED patterns recorded successively on the Si substrate and on the Ge layer. The strain in the film resulting from cooling the sample at room temperature is given by X-ray experiments.

3 Results and discussion

Figure 1 represents typical RHEED patterns recorded during a one hour deposition experiment at 330 °C. The first one (Fig. 1a) is related to the Si substrate. A well developed 2×1 reconstruction (presence of 1/2 streaks) and Kikuchi lines can be observed. After 15 min of deposition (Fig. 1b), the RHEED pattern shows modulated streaks and weaker 1/2 lines which indicates a wavy surface. For longer deposition times, the Ge surface evolves towards a smoother surface and presents finally a 2×1 reconstruction but no Kikuchi lines. It is noteworthy that the Stranki–Krastanov related 2D to 3D transition, which should result in a spotty pattern, never occurred during that experiment. As an important feature, let us note also that the in plane lattice parameter deduced from the streak spacings in initial and final patterns increases by 3.4%, which indicates a nearly complete relaxation of the Ge film. The corresponding rms roughness, as deduced from AFM imaging, is found equal to 0.6 nm.

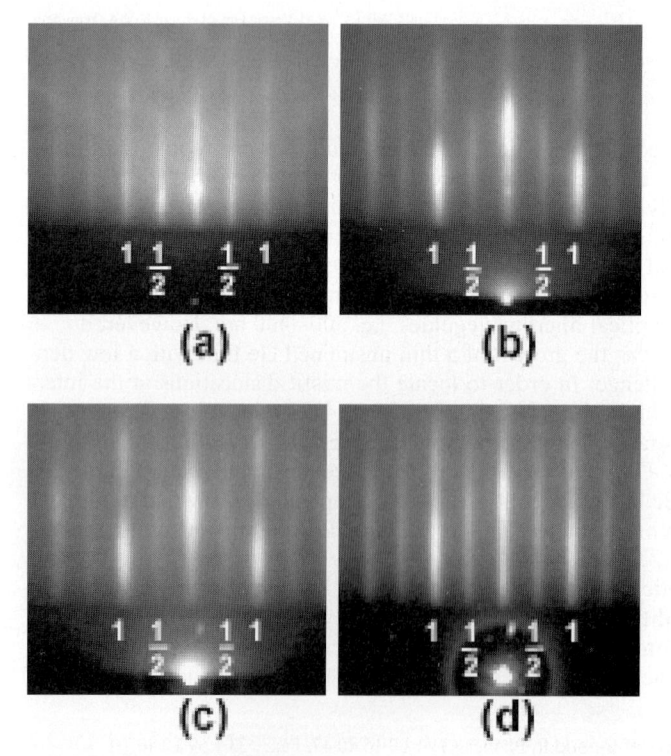

Fig. 1 RHEED patterns evolution at 330 °C a) Si(001) substrate surface, b) wavy Ge surface after 15 min of deposition c) Ge surface after 30 min of deposition and d) flat surface after 1 h of deposition.

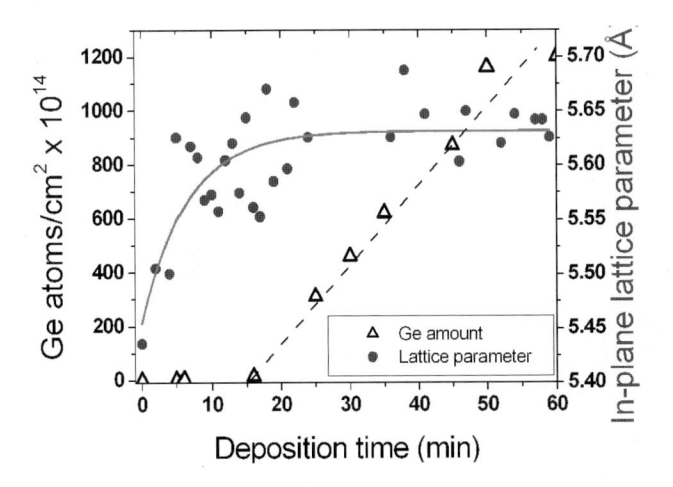

Fig. 2 (online colour at: www.interscience.wiley.com) Deposited amount of Ge in atoms/cm² from RBS and RHEED in-plane lattice parameter as a function of deposition time.

The surface diffusion of Ge atoms is the key parameter to explain these experimental results. The low growth temperature reduces of course the surface diffusion of Ge, but the hydrogen coverage is more probably the main restrictive factor to prevent the Stranski–Krastanov-related surface relaxation, as shown previously from molecular beam epitaxy (MBE) experiments performed under atomic hydrogen flow [7, 8].

In order to investigate more thoroughly the corresponding kinetics, a series of Ge films was made for deposition times varied between 5 and 60 min. These films were then systematically characterized by RBS and AFM. Figure 2 shows RBS measurements of the deposited Ge amount (in atoms number per cm²) as a function of the deposition time. During the first 15 min, i.e. for Ge amounts less than 1.4×10^{15} atoms cm^{-2}, or 0.3 nm in equivalent thickness, or 2 monolayers (MLs), the growth rate evolves very slowly with a main value of 0.02 nm/min. During this step, it is likely that Ge atoms do not cover completely the substrate surface, even for coverages larger than 1 monolayer ($>6.3 \times 10^{14}$ atoms cm^{-2}), and the observed growth rate is mainly related to the deposition of Ge atoms on a Si surface. From AFM measurements the surface is found to exhibit islands of 3 nm in height and about 40 nm in width, but with a very low density, near 2×10^7 cm^{-2} for a 6 min deposition. Such a low density lies below the RHEED sensitivity which gives streaky patterns in this range of deposition times. For a 6 min deposition time, the observed islands are constituted of about 1.7×10^{12} atoms cm^{-2}, i.e less than 0.5% of the deposited amount. So, we can consider that the growth proceeds mostly layer-by-layer. After 15 min of deposition, the surface has turned to wavy, as seen by RHEED (Fig. 1b) and confirmed by AFM. From spectroscopic ellipsometry experiments, Palange et al. [9] found a very high fraction of voids in films of similar thicknesses, which is in agreement with our observations. For deposition times above 15 min, the thickness increases linearly as a function of time with a constant growth rate of 0.61 nm/min. This can be interpreted as the deposition of Ge atoms on a Ge surface. This would indicate also that the substrate surface was completely covered by Ge as soon as about 2 MLs have been deposited.

The in plane lattice parameter at the growth temperature $a_{//}$(330 °C), as evaluated from RHEED patterns, is also plotted in Fig. 2 as a function of the deposition time. The value of $a_{//}$(330 °C) is found to evolve very sharply from the Si lattice parameter (5.436 Å at 330 °C) to reach the range of 5.62–5.64 Å at the end of the deposition. The more surprising is that 90% of the relaxation process seems to occur during the deposition of the two first monolayers.

After the growth, the sample temperature turns down to 20 °C and thermal strain develops as a result of a difference in thermal expansion coefficients between film and substrate. Since the thermal expansion coefficient of germanium is greater than that of silicon in the whole range of considered temperatures, the resulting contribution of thermal stress is tensile. Starting from the experimental value of $a_{//}$(330 °C), the calculation of $a_{//}$(20 °C) gives 5.626 ± 0.01 Å. From XRD measurements, the lattice parameter in the vertical direction at 20 °C is found equal to 5.678 Å, which corresponds to $a_{//}$(20 °C) = 5.631 Å by assuming a quadratic strain, which confirms the preceding evaluation from RHEED analysis. These two

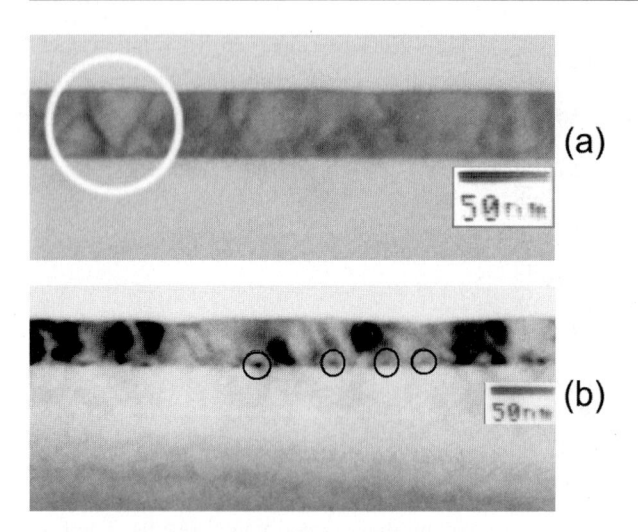

Fig. 3 Cross-sectional TEM images of Ge layer grown at 330 °C of 30 nm thickness, showing a) stacking faults in (111) planes and b) misfit dislocations loops.

evaluations of $a_{//}$ indicate a compressive strain which is not compensated by the cooling of the sample down to room temperature and relates to a global relaxation rate of 88%.

From cross-sectional TEM images, two relaxation modes can be observed. Emerging extended defects are identified as stacking faults in the (111) plane (Fig. 3a) and misfit dislocations (Fig. 3b) are located at the Ge/Si(001) interface. This result is confirmed by RBS in channelling (RBS-C) from which a χ_{min} value of 58% is derived for the $\langle 001 \rangle$ alignment. However this high value must be considered with caution because the RBS resolution is not good enough to discriminate the oxidized surface from the bulk in a 30 nm thick film. As a matter of fact a χ_{min} value of 4% can be found for thicker films made by further depositing germanium at a temperature of 600 °C up to a thickness of 800 nm. Infrared detectors realized from these films exhibit sensibilities similar to that obtain previously by Colace et al.

4 Conclusion By investigating the Ge growth kinetics at 330 °C, we found that the Stranski–Krastanov growth mode is avoided at low temperature and that the relaxation process occurs mostly during the deposition of the first two monolayers. This results in an in-plane lattice parameter approaching 90% of that of Ge bulk and a flat surface with rms roughness of 0.6 nm for a film thickness lower than 30 nm. In spite of stacking faults, that Ge films can act as good buffer layers for the growth of relaxed Ge layers of good quality, much thinner than graded SiGe pseudo-substrates (>1000 nm) [10, 11].

References

[1] S. Lardenois, D. Pascal, L. Vivien, E. Cassan, S. Laval, R. Orobtcouk, M. Heitzman, N. Bouzaid, and L. Mollard, Opt. Lett. **28**, 1150 (2003).
[2] L. Colace, G. Masinni, F. Galluzzi, G. Assento, G. Capellini, L. Di Gaspare, E. Palage, and F. Evangelisti, Appl. Phys. Lett. **72**, 3175 (1998).
[3] H.-C. Luan, D. R. Lim, K. L. Lee, K. M. Chen, J. G. Sandland, K. Wada, and L. C. Kimerling, Appl. Phys. Lett. **75**, 2909 (1999).
[4] L. Colace, G. Masinni, G. Assento, H.-C. Luan, K. Wada, and L. C. Kimerling, Appl. Phys. Lett. **76**, 1231 (1999).
[5] H.-C. Luan, K. Wada, L. C. Kimerling, G. Masinni, L. Colace, and G. Assento, Opt. Mater. **17**, 71 (2001).
[6] L. H. Nguyen, V. Le Thanh, D. Debarre, V. Yam, and D. Bouchier, Mater. Sci. Eng. B **101**, 199 (2003).
[7] S.-J. Kahang, J. Y. Park, K. H. Booth, J. Lee, Y. Khang, and Y. Kuk, J. Vac. Sci. Technol. A **15**, 927 (1997).
[8] D. Dentel, J. L. Bischoff, T. Angot, and L. Kubler, Surf. Sci, **402/404**, 211(1998).
[9] E. Palange, L. Di Gaspare, and F. Evangeliste, Thin Solid Films, **428**, 160 (2003).
[10] P. M. Mooney, F. K. LeGoues, and J. L. Jordan-Sweet, Appl. Phys. Lett. **65**, 2845 (1994).
[11] M. T. Currie, S. B. Samavedam, T. A. Langdo, C. W. Leitz, and E. A. Fitzgerald, Appl. Phys. Lett. **72**, 1718 (1998).

Kinetic Monte Carlo simulation of SiC nucleation on Si(111)

A. A. Schmidt[1], K. L. Safonov[1], Yu. V. Trushin[1], V. Cimalla[2], O. Ambacher[2], and J. Pezoldt*,[2]

[1] A. F. Ioffe Physico-Technical Institute, Politekhnicheskaya 26, 194021 St. Petersburg, Russian Federation
[2] FG Nanotechnologie, Zentrum für Mikro- und Nanotechnologien, TU Ilmenau, Postfach 100565, 98684 Ilmenau, Germany

Received 19 September 2003, accepted 3 November 2003
Published online 12 January 2004

PACS 02.70.Uu, 68.55.Ac, 68.43.Hn, 68.43.Jk, 81.15.Hi, 61.46.+w

Solid source molecular beam epitaxy was applied to create silicon carbide nanoclusters on silicon. The island size distribution can be controlled by appropriate substrate temperature, carbon fluxes and process times. Nevertheless, up to now the earliest stages of the carbon interaction with silicon are not well understood. To have a deeper insight into the early nucleation stages and to investigate the growth laws in such systems the kinetic Monte Carlo computer simulation method was applied. The simulation enabled us to estimate the values of the growth parameters and to obtain the cluster size distribution function and nanocluster concentration in agreement with the experimental obtained data.

© 2004 WILEY-VCH Verlag GmbH & Co. KGaA, Weinheim

1 Introduction Nanosized SiC-Si heterostructures offer applications such as tunneling barriers and antidot structures. With their specific properties they could be used in nano- and optoelectronics. The common technique to grow SiC on Si is a two-step process consisting of a carbonisation process, i.e. the conversion of Si into SiC by supplying carbon containing species to the silicon surface, followed by epitaxial growth. The grown layers suffer from the high lattice mismatch and thermal expansion coefficients discrepancy between these two materials, which lead to high residual stresses and lattice defect densities in the grown silicon carbide layer. During the early stages of the carbonization SiC clusters are formed on the surface. The density and the size distribution of these clusters mainly affect the quality of the grown SiC structures or layers. The simplest method to achieve better structural and morphological quality is to grow at low substrate temperatures under controlled flux conditions using high reactive carbon containing species. In this case the lowest suitable SiC formation temperature was determined to be approximately 500 °C due to the use of high reactive elemental carbon [1]. For a better control of the nanosized SiC nuclei shape and distribution an ultra high vacuum environment, i.e. solid source molecular beam epitaxy (SSMBE), has to be used. This technique was applied to study SiC cluster growth in a wide range of experimental parameters: temperatures, fluxes and deposition times [2].

Because of the complexity of the interaction process of carbon with the silicon substrate computer simulations are a useful tool to achieve an in-depth understanding of the early stage of SiC clusters nucleation. Moreover, it is possible to predict the kinetic parameters of silicon and carbon atoms on silicon surfaces with changing structure and morphology. The kinetic Monte Carlo (KMC) simulation method is the most suitable one for such a task, because in contrast to the molecular dynamics method it is able to simulate a quite large area of the substrate (around 100×100 nm^2) during a relatively long time (up to the tens of seconds) which still providing atomic scale data in contrast to the rate equations method.

---

* Corresponding author: e-mail: joerg.pezoldt@tu-ilmenau.de, Phone: +49 3677 693408, Fax: +49 3677 693355

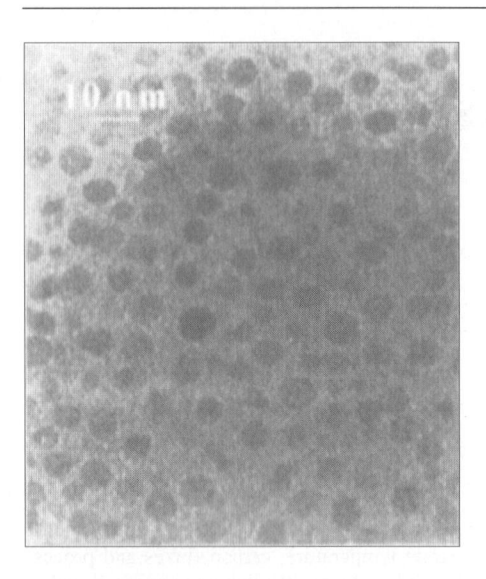

Fig. 1 TEM image of Si sample exposed to C flux of 5.5×10^{13} at/cm^{2}s for 55 s at the temperature of 660 °C [4].

2 The simulation model

For the simulation of the SiC nanoclusters growth on the Si(111) surface the so-called lattice kinetic Monte Carlo method was applied. In this approach silicon and carbon atoms are allowed to hop between the fixed sites of the simulation lattice. At each simulation step there is a number of processes that are currently possible, e.g. diffusion of a silicon or a carbon atom, deposition of a new carbon atom etc. The process to perform is chosen from the list of all possible processes for all atoms in the simulation cell according to their rate. Then, when the process is performed, lists of the possible processes are updated according to the lattice changes, and the simulation cycle is repeated.

3 Two-dimensional growth at low temperatures

The TEM micrograph (Fig. 1) shows flat circular silicon carbide clusters at low substrate temperature (660 °C) [3]. Thus, for the simulation of the growth at low temperatures the two-dimensional hexagonal lattice, which corresponds to the dangling bond symmetry of the (1 × 1)-Si reconstructed Si(111) surface, was used. The following processes were included into the simulation algorithm.

3.1 Deposition of carbon atoms onto the surface

The deposition rate is defined by the carbon flux F set according to the experimental conditions. A carbon atom is reflected if it hits a carbon adatom existing on the surface [4]. This is also considered by the model, thus, the flux is reduced with the deposition time.

3.2 Diffusion hops of free carbon adatoms on the silicon surface

The rate of the diffusion hops of the carbon adatoms is given by the Arrhenius law:

$$P_d = \nu \cdot \exp\left(-\frac{E_a}{kT}\right), \tag{1}$$

where ν is characteristic Debye frequency (assumed to be equal to 10^{13} Hz), k is the Boltzmann constant, T is the substrate temperature and E_a is the activation energy for carbon diffusion on the silicon surface.

3.3 Incorporation of adatoms into existing clusters

When the diffusing carbon adatom reaches an existing cluster it is incorporated. It is assumed, that at the early growth stages an excess of silicon atoms exists on the surface (some atoms come from the substrate and some are the weakly bonded atoms of Si(111)–(7 × 7) reconstruction cell) and so the impinging carbon atoms limit the cluster nucleation. This is apparently true until the coalescence of the clusters starts. Afterwards, the growth is limited by silicon atoms coming from the substrate via enhanced diffusion along the cluster boundaries. This is in agreement with the void formation in the silicon substrate usually observed beneath and near the SiC/Si interface [3].

3.4 Detachment of carbon atoms The carbon atom incorporated into the cluster is energetically more favoured than the free one positioned on the surface. So the detachment energy for the former atom increases with the number of neighbours (only the nearest are taken into account). The rate of the diffusion hops of the carbon atoms that are incorporated into the cluster is given by:

$$P_d^n = v \cdot \exp\left(-\frac{E_a + nE_b}{kT}\right), \tag{2}$$

where n is the number of nearest neighbours and E_b is the atom binding energy per neighbour.

To determine, which process has to be executed at each simulation step a modified Bortz–Kalos–Lebowitz algorithm was used [5].

First the probability of all possible processes normalized to unity was obtained. For this reason the total rate of all processes was calculated:

$$P = P_d N_0 + \sum_n P_d^n N_n + F N_{\text{free}}, \tag{3}$$

where N_0 is the number of free carbon adatoms not incorporated into clusters (having no neighbours), N_n is the number of carbon atoms with n neighbours, and N_{free} is the number of free surface sites (suitable for adatom deposition).

Then every process rate was divided by the total rate P, and the processes were represented as a sequence of numerical intervals of the lengths equal to the corresponding ratios, so having the whole sequence to be a [0, 1] interval. Then, a random number ranging from 0 to 1 was chosen to select the process to be performed. Finally, the atom (or the empty site in case of carbon atom deposition) was randomly chosen from the list of the atoms (sites) to perform the selected process.

In such a scheme there are no attempts that are rejected. The lists of the possible process rates after each step are updated only locally and so the speed is significantly increased in comparison to the conventional methods.

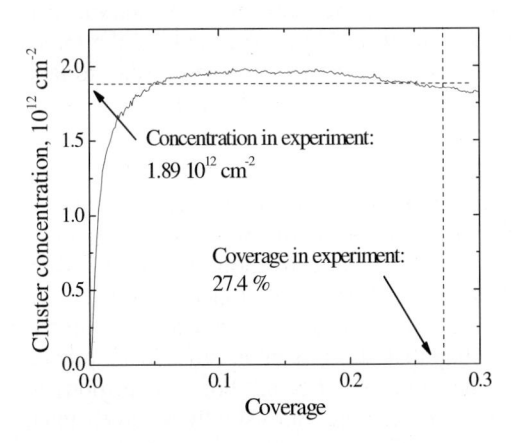

Fig. 2 Simulated cluster concentration versus coverage. $E_a = 0.6$ eV, $E_b = 0.4$ eV, $T = 660$ °C, $F = 5.5 \times 10^{13}$ cm^{-2}s^{-1}. The experimental coverage and cluster concentration is given at the intersection of the dashed lines.

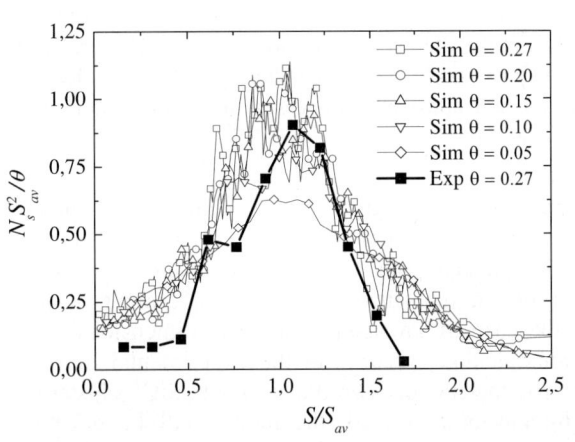

Fig. 3 The scaled size distribution functions obtained in the simulation for different coverages and the experimental data. N_S is the number (per site) of the clusters of size S, S_{av} is the average cluster size, θ is the coverage.

The surface diffusion activation barrier E_a is the main simulation parameter. It defines the cluster growth mode and strongly affects the concentration and the size distribution of the SiC nuclei. There is a number of studies dedicated to the carbon diffusion in silicon, but the surface diffusion activation barrier was determined only in [6], and the value was most likely overestimated. Thus, in the present work E_a was used as a fitting parameter. The binding energy E_b affects the shape of the islands (at $E_b > 0.5$ eV "snowflake"-shaped islands appeared in the simulations). Therefore, the value of E_b was determined by comparing the experimental and simulated nuclei shape and was in the range between 0.3 and 0.5 eV.

The surface concentration of the two-dimensional clusters from Fig. 1 is 1.89×10^{12} cm^{-2}. Good coincidence of the cluster concentration with the one calculated from Fig. 1 was obtained for a diffusion activation energy of 0.6 eV. In Fig. 2 the dependence of the simulated cluster concentration on the coverage is presented together with the experimental values. The slight decrease of the cluster concentration is due to a few casual cluster coalescences.

4 Simulation results A scaling analysis of the cluster size distribution function was performed in order to check the simulation procedure. The experimental data shown in Fig. 1 belongs to the case of sub-monolayer cluster nucleation described in [7]. Consequently, the cluster size distribution function should satisfy the scaling theory predictions, i.e. the scaled cluster size distribution function should be time (and coverage) independent. Fig. 3 shows the scaled size distribution function for different coverages ranging from 10 up to 27 percent as obtained by the simulation together with the experimental data. Except for the case of the 5% coverage (below the validity range of the scaling theory) the simulated size distribution functions seems to fall approximately onto a single curve. Besides, the scaled experimental size distribution function agrees qualitatively with the simulation results. The deviation of the simulated and experimental data at the region of small S/S_{av} ratios could be explained by the lack of precision of the TEM picture. The lower number of the big clusters (region of a high S/S_{av} ratios) obtained in the experiment compared to the simulation results is caused by the limitations in the TEM picture size which can not be sufficient large enough to give a statistically representative distribution. For this reason an in depth TEM study is in progress. Moreover, in the simulation two connected clusters were counted as a single bigger one (because otherwise the cluster counting algorithm consumes too much computational time), while in the experiment it is possible to separate clusters. Thus, the fitted E_a value allows to achieve not only the cluster concentration that agrees with the experimental data, but also the cluster size distribution function. At the same time, additional validation of E_a is necessary. The experimental data obtained at the higher substrate temperatures can help to make the estimation more precise. However, at the temperatures higher than 660 °C the three-dimensional nucleation takes place [8]. This is why now the model has to be developed further to take into account additional processes like adatom hops to the upper lattice levels and the diffusion of the silicon atoms.

5 Summary The lattice kinetic Monte Carlo method was used to investigate the early stages of SiC cluster nucleation on Si(111) substrate during SSMBE carbonization process. Comparison of the simulated data with the experimental results allow to estimate the activation energy of carbon diffusion on the silicon surface. The scaling of the cluster size distribution function was shown for the case of low-temperature two-dimensional cluster nucleation. Also the simulated size distribution function agrees with the experimental one, confirming that the model is capable to predict the parameters of two-dimensional SiC clusters on the silicon surface. For application to high-temperature growth the model will be developed to consider the intrinsic three-dimensional growth.

Acknowledgements This work is supported by the DAAD project "Internationale Qualitätsnetzwerke (IQN): Ilmenauer Koordinationszentrum", and the RFBR project N 02-02-17610, GR 01 890073799 and the scientific school project of the Russian Federation N 00-15-96812.

References

[1] V. Cimalla, Th. Stauden, G. Ecke, F. Scharmann, G. Eichhorn, S. Sloboshanin, J. A. Schäfen, and J. Pezoldt, Appl. Phys. Lett. **73**, 3542 (1999).
[2] F. Scharmann, P. Maslarski, Th. Stauden, W. Attenberger, J. K. N. Lindner, B. Stritzker, and J. Pezoldt, Thin Solid Films **380**, 92 (2000).
[3] W. Attenberger, J. Lindner, V. Cimalla, and J. Pezoldt, Mater. Sci. Eng. B **61/62**, 544 (1999).
[4] K. L. Safonov, D. V. Kulikov, Yu. V. Trushin, and J. Pezoldt, Proc. SPIE **4627**, 165 (2002).
[5] A. B. Bortz, M. H. Kalos, and J. L. Lebowitz, J. Comput. Phys. **17**, 10 (1975).
[6] F. Scharmann, D. Lehmkuhl, P. Maslarski, Th. Stauden, and J. Pezoldt, Proc. SPIE **4348**, 173 (2001).
[7] M. C. Bartlet and J. W. Evans, Phys. Rev. B **46**, 12675 (1992).
[8] V. Cimalla, J. Pezoldt, Th. Stauden, Ch. Förster, O. Ambacher, A. A. Schmidt, and K. Zekentes, submitted to J. Vac. Sci. Technol.

Modeling carrier dynamics in quantum-dot lasers

A. Markus[*,1] **and A. Fiore**[1,2]

[1] Institute of Quantum Electronics and Photonics, Ecole Polytechnique Fédérale de Lausanne EPFL, CH-1015 Lausanne, Switzerland
[2] Permanent address: Institute of Photonics and Nanotechnology, CNR, Via del Cineto Romano 42, 00156 Roma (Italy)

Received 19 September 2003, accepted 3 November 2003
Published online 14 January 2004

PACS 85.35Be, 42.55.Px

Theoretical analysis of carrier dynamics in quantum dot lasers is presented, particularly with regard to the effect of carrier relaxation. Two modeling approaches are applied, one is based on standard rate equations and the other on master equations of microstates, in order to consider random carrier capture. Both models show similar results and can be exploited to describe realistic devices quantitatively. In particular, a reliable estimation of the relaxation time can be obtained from comparison with experimental results. Finally, the modulation dynamics of real devices is modeled, which shows a suppression of relaxation oscillations on the ground state transition in dependence of the cavity length.

© 2004 WILEY-VCH Verlag GmbH & Co. KGaA, Weinheim

1 Introduction Due to the discrete nature of their density of states, quantum-dot (QD) lasers have certain advantages compared to quantum-well (QW) devices: lower threshold current densities [1] combined with decreased temperature sensitivity [2, 3], low chirp [4] and emission wavelengths that are suited for telecom applications (1.3 and 1.55 μm) [5].

Recently we demonstrated the importance of intersublevel relaxation on the performance of a QD laser [6, 7]. In order to capture the physical phenomena in such a device comprehensive modeling is essential. Yet it is desirable that such a model is compact enough to account for realistic (measured) device parameters. In general there are two semiclassical approaches to modeling of QD devices. Rate equation models (REMs) are widely-used in literature, inspired by the description of average carrier densities in bulk material, QWs or quantum wires. Master equation models (MEMs), as proposed by Grundmann et al. [8], are more complex but also better suited for QD systems, since they account for random carrier capture. So far a quantitative comparison of the two models has been given for QD systems involving spontaneous emission processes only [8]. Here, we apply both approaches on a QD laser system based on the same model assumptions and using identical parameters, corresponding to those measured in a three stack QD laser emitting at 1285 nm [6]. The objective is to explore the parameter space, in which the simpler REM can be applied safely.

2 Models Our model system is based on excitonic energy states. The QDs are assumed to confine eh-pairs in two discrete energy levels, a twofold degenerate ground state (GS) and a fourfold degenerate excited state (ES). Different QDs are interconnected by a reservoir (wetting layer or barrier), which gets filled by external carrier injection. The injection rate G is normalized by the total number of QDs N_D, i.e. $G = I/(eN_D)$. The reservoir is treated as a discrete energy state with a degeneracy

---
[*] Corresponding author: e-mail: alexander.markus@epfl.ch

© 2004 WILEY-VCH Verlag GmbH & Co. KGaA, Weinheim

much larger than N_D. From the reservoir eh-pairs are captured into the ES at a time constant τ_c and emitted back from the ES by thermal excitation at a time constant τ_e^{ES}. Furthermore intersublevel relaxation from the ES to the GS with a time constant τ_0 and thermal escape from the GS to the ES with a time constant τ_e^{GS} are considered. In both levels and in the reservoir the time constant for spontaneous recombination τ_r is set to be identical, corresponding to the experimentally found value [9] ($\tau_r \approx 1$ ns). Time resolved photoluminescence spectra on our QDs show a GS rise time of about 10 ± 2 ps [10], which accounts for the capture and the relaxation process, corresponding to a time constant ($\tau_c + \tau_0$).

2.1 Rate Equation Model

Figure 1 shows a schematic representation of the REM. The average level occupations of GS and ES are expressed in terms of the distribution functions $f_{GS} = N_{GS}/2N_D$ and $f_{ES} = N_{ES}/4N_D$. Here, N_{GS} (N_{ES}) is the total number of eh-pairs in the GS (ES). We describe the population of the reservoir with the term $n_{res} = N_{res}/N_D$, where N_{res} is the total number of eh-pairs in the reservoir. The rate equations are then given as:

$$\frac{\partial n_{res}}{\partial t} = G - \frac{n_{res}(1 - f_{ES})}{\tau_c} + \frac{4f_{ES}}{\tau_e^{ES}} - \frac{n_{res}}{\tau_r} \quad (1)$$

$$\frac{\partial f_{ES}}{\partial t} = \frac{n_{res}(1 - f_{ES})}{4\tau_c} - \frac{f_{ES}}{\tau_e^{ES}} + \frac{f_{GS}(1 - f_{ES})}{2\tau_e^{GS}} - \frac{f_{ES}(1 - f_{GS})}{\tau_0} - \frac{f_{ES}}{\tau_r} - 2BN_\varphi^{ES} N_D \left(f_{ES} - \frac{1}{2}\right) \quad (2)$$

$$\frac{\partial f_{GS}}{\partial t} = -\frac{f_{GS}(1 - f_{ES})}{\tau_e^{GS}} + \frac{2f_{ES}(1 - f_{GS})}{\tau_0} - \frac{f_{GS}}{\tau_r} - 2BN_\varphi^{GS} N_D \left(f_{GS} - \frac{1}{2}\right). \quad (3)$$

For simplicity we have neglected the effect of reservoir filling in the escape terms, which is justified by the large density of states in the reservoir. The time constants for the escape of eh-pairs are derived from Eqs. (1) and (R3), by assuming the system reaches a quasi-Fermi equilibrium in the absence of external excitation [11]:

$$\tau_e^{GS} = \frac{\tau_0}{2} e^{\frac{E_{ES} - E_{GS}}{k_B T}}, \quad \tau_e^{ES} = 4\tau_c e^{\frac{E_{WL} - E_{ES}}{k_B T}},$$

where $E_{WL} = 1.14$ eV, $E_{ES} = 1.04$ eV and $E_{GS} = 0.96$ eV are the respective transition energies. B is the radiative transition probability, assumed to be equal for GS and ES transitions.

N_φ^{GS} and N_φ^{ES} are the total number of photons in the cavity at the GS and ES transition energies respectively, divided by N_D. They are described by the two rate equations:

$$\frac{\partial N_\varphi^{GS}}{\partial t} = 4BN_\varphi^{GS} N_D \left(f_{GS} - \frac{1}{2}\right) - \frac{N_\varphi^{GS}}{\tau_\varphi} + \frac{2\beta f_{GS}}{\tau_r} \quad (4)$$

$$\frac{\partial N_\varphi^{ES}}{\partial t} = 8BN_\varphi^{ES} N_D \left(f_{ES} - \frac{1}{2}\right) - \frac{N_\varphi^{ES}}{\tau_\varphi} + \frac{4\beta f_{ES}}{\tau_r}. \quad (5)$$

Here, $\beta = 10^{-5}$ is the spontaneous emission coefficient, characterizing the fraction of spontaneously emitted photons that go into the laser mode. Note that the different numerical prefactors in the relaxa-

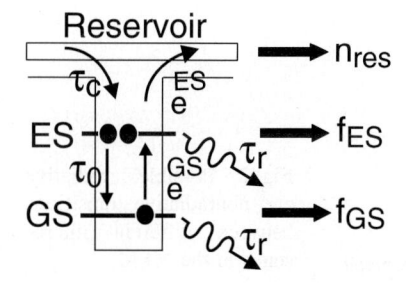

Fig. 1 Schematic representation of the rate equation model.

tion, escape and stimulated emission terms in Eqs. (1–5) are due to the different degeneracy in the GS, ES and reservoir. The photon lifetime τ_φ is derived from the measured laser characteristics [6]:

$$\tau_\varphi = \frac{n}{c(\alpha_i + \alpha_m)},$$

with the internal losses $\alpha_i = 2 \text{ cm}^{-1}$ and the mirror losses $\alpha_m = \ln(1/R)/L$ with $R = 0.32$ for as-cleaved facets, L being the cavity length. The Einstein B factor can be theoretically determined as

$$B = \frac{\lambda_0^4}{8\pi n^3 \Delta\lambda \, V \tau_r},$$

with the laser mode λ_0, the gain spectral linewidth $\Delta\lambda$ and the cavity volume V. This value was multiplied by a factor of 0.335, to give the experimentally determined saturation of GS gain. This reduced gain value may be related to non-radiative exciton states due to thermal spreading of holes in closely-space valence-band states [2].

2.2 Master Equation Model In MEMs [8, 12] the QD population is expressed in terms of microstates (i, j): one microstate describes all QDs with a certain number of electrons i in the GS ($0 \leq i \leq 2$) and j in the ES ($0 \leq j \leq 4$). The probability of a particular microstate is denoted n_{ij}. The average population of GS and ES is then expressed by the distribution functions $f_{GS} = \sum_{i,j} i \cdot n_{ij}$ and $f_{ES} = \sum_{i,j} j \cdot n_{ij}$. The microstates have to fulfill the normalization condition:

$$\sum_{i,j} n_{ij} = 1 \tag{6}$$

As possible transitions between the microstates we consider only processes, which involve a single eh-pair. The diagram in Fig. 2 summarizes all possible radiative and nonradiative transitions from one microstate to the other. In order to calculate the respective transition probabilities the occupations of the initial and final level for each process have to be considered, which are described by the distribution functions f_{in} and f_{fi} respectively $(0 \leq (f_{in}, f_{fi}) \leq 1)$. For example the emission of a photon from microstate $(1, 0)$ via the GS transition yields $f_{in} = 1/2$ and $f_{fi} = 1/2$. Here the transition probability is given by $2(1/\tau_r + BN_\varphi^{GS})f_{in}(1 - f_{fi})$, where the factor of 2 arises from the twofold GS degeneracy and the two addends account for spontaneous and stimulated emission processes. Table 1 summarizes all transition probabilities between two microstates. The master equations for microstates can be derived by applying the probability coefficients from Table 1 to the diagram in Fig. 2:

$$\frac{\partial n_{00}}{\partial t} = -n_{00}\left(\frac{n_{res}}{\tau_c} + 2BN_\varphi^{GS} + 4BN_\varphi^{ES}\right) + n_{10}\left(\frac{1}{2\tau_r} + \frac{B}{2}N_\varphi^{GS}\right) + n_{01}\left(\frac{1}{4\tau_r} + \frac{1}{\tau_e^{ES}} + \frac{B}{4}N_\varphi^{ES}\right)$$

$$\vdots \qquad \vdots \tag{7}$$

$$\frac{\partial n_{24}}{\partial t} = -n_{24}\left(\frac{6}{\tau_r} + \frac{4}{\tau_e^{ES}} + 2BN_\varphi^{GS} + 4BN_\varphi^{ES}\right) + n_{23}\left(\frac{n_{res}}{4\tau_c} + \frac{B}{4}N_\varphi^{ES}\right) + n_{14}\frac{B}{2}N_\varphi^{GS}.$$

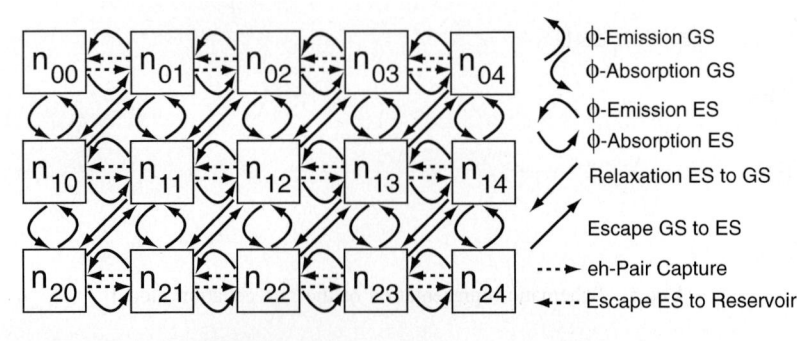

Fig. 2 Possible radiative and nonradiative transitions between different microstates in the MEM.

Table 1 Probabilities of transitions between two microstates.

radiative transitions	probability	nonradiative transitions	probability
ϕ-emission GS	$2(1/\tau_r + BN_\varphi^{GS})f_{in}(1-f_{fi})$	relaxation ES to GS	$4/\tau_0 f_{in}(1-f_{fi})$
ϕ-absorption GS	$2BN_\varphi^{GS} f_{in}(1-f_{fi})$	escape GS to ES	$2/\tau_e^{GS} f_{in}(1-f_{fi})$
ϕ-emission ES	$4(1/\tau_r + BN_\varphi^{GS})f_{in}(1-f_{fi})$	eh-pair capture	$1/\tau_c n_{res}(1-f_{fi})$
ϕ-absorption GS	$4BN_\varphi^{GS} f_{in}(1-f_{fi})$	escape ES to reservoir	$4/\tau_e^{ES} f_{in}$

The population of the reservoir is described by n_{res}, which is the total number of eh-pairs divided by the number of QDs N_D:

$$\frac{\partial n_{res}}{\partial t} = G - n_{res}\left(\frac{1}{\tau_r} + \frac{1}{4\tau_c}\sum_{i,j}(1-j)n_{ij}\right) + \frac{1}{\tau_e^{ES}}\sum_{i,j} j n_{ij}. \tag{8}$$

The number of photons per QD at both transition energies are given as:

$$\frac{\partial N_\varphi^{GS}}{\partial t} = -N_\varphi^{GS}\left(\frac{1}{\tau_p} + 2B\sum_{i,j}(1-i)n_{ij}\right) + \frac{\beta}{\tau_r}\sum_{i,j} i n_{ij}, \tag{9}$$

$$\frac{\partial N_\varphi^{ES}}{\partial t} = -N_\varphi^{ES}\left(\frac{1}{\tau_p} + 4B\sum_{i,j}(1-j)n_{ij}\right) + \frac{\beta}{\tau_r}\sum_{i,j} j n_{ij}. \tag{10}$$

In Ref. [12] and [6] the relaxation process has been treated in a more sophisticated way by assuming infinitely fast hole relaxation and by applying the relaxation time τ_0 for electron relaxation only. This approach leads to a similar MEM, where certain transitions in Fig. 2 are forbidden due to selection rules. However, the results obtained from the two MEMs are nearly identical in the regime explored in this letter.

3 Model comparison The two models have been applied to describe the carrier dynamics in lasers operated at high bias above threshold, which show dual-wavelength lasing. As shown experimentally in Ref. [6, 13] this behavior depends strongly on the intraband relaxation process, so it is a sensitive test of the accuracy of the models.

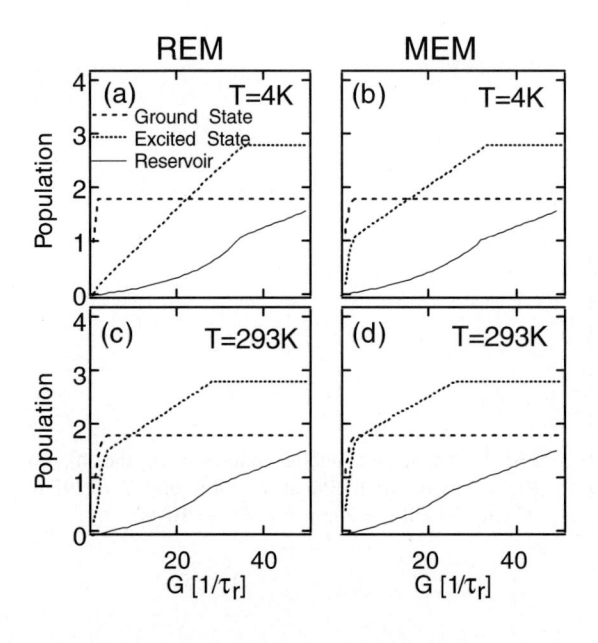

Fig. 3 Calculated level populations in the a), c) REM and b), d) MEM at $T = 4$ K and $T = 293$ K (Cavity length $l = 2$ mm, $\tau_0 = \tau_c = 10$ ps).

Stationary solutions are found numerically by equating Eqs. (1–5) and (2–10) to zero. In Fig. 3 calculated level populations using REM and MEM at two different temperatures are shown as a function of G (cavity length $L = 2$ mm, $\tau_0 = \tau_c = 10$ ps). The calculated results of the two models vary a lot at low temperature ($T = 4$ K) and at low carrier injection ($G < 2/\tau_r$): In the REM (Fig. 3a) ES population is negligible until GS population would reach its saturation value, whereas in the MEM (Fig. 3b) the ES population shows a strong increase already at low injection G, roughly at values for which GS reaches transparency. The difference between the two models in this regime is due to consideration of random carrier capture in the MEM, which allows some QDs to be filled with ES carriers while other QDs are empty, as discussed in Ref. [8] in detail. As a consequence, GS lasing threshold in the MEM is reached at slightly higher carrier injections ($G \approx 2.29/\tau_r$) as compared to the REM ($G \approx 2.02/\tau_r$). At GS threshold the GS population clamps, whereas ES population continues increasing, although at a lower rate. The stimulated emission process provides a much faster recombination channel for eh-pairs on the GS than spontaneous emission and therefore an increasing number of eh-pairs on the ES is required, due to the finite relaxation time τ_0 [14]. This incomplete clamping of ES population ultimately leads to the appearance of a second lasing line on the ES, as shown experimentally [6, 13]. The slope of the ES-population in this regime is lower in the MEM results. This leads to a similar ES threshold in both models, despite the much higher ES population at GS threshold ($G \approx 35.47/\tau_r$ in REM and $G \approx 33.11/\tau_r$ in MEM). At $T = 293$ K (Fig. 3a, b) thermalisation of eh-pairs leads to a similar behavior in both models. However, ES-population is still slightly favored in MEM. Figure 4 displays the calculated photon numbers, corresponding to the respective populations in Fig. 3. It is evident that for both temperatures both models have nearly identical results. At both temperatures the ES threshold is slightly lower in the MEM.

We stress that the dual-state lasing behavior modeled here corresponds very closely to the experimental light-current characteristics [6]. The only fitting parameters are the relaxation time τ_0 and the capture time τ_c. Figure 6a, b shows the impact of the relaxation time τ_0 on the ratio of ES and GS threshold currents for $T = 4$ K and $T = 293$ K ($\tau_c = 10$ ps). In this scale a difference between both models is not visible. In Fig. 6c, d threshold current ratios over different capture times are plotted for both models, $T = 4$ K and $T = 293$ K ($\tau_0 = 10$ ps and $\tau_0 = 100$ ps). Here, a small difference between REM and MEM results can be seen at faster relaxation times. Obviously the capture time shows a very small dependence on the threshold current ratios of ES and GS, in contrast to the big impact on the relaxation time. Thus a measurement of the lasing threshold currents can provide a reliable estimation of τ_0, which is otherwise difficult to measure [6].

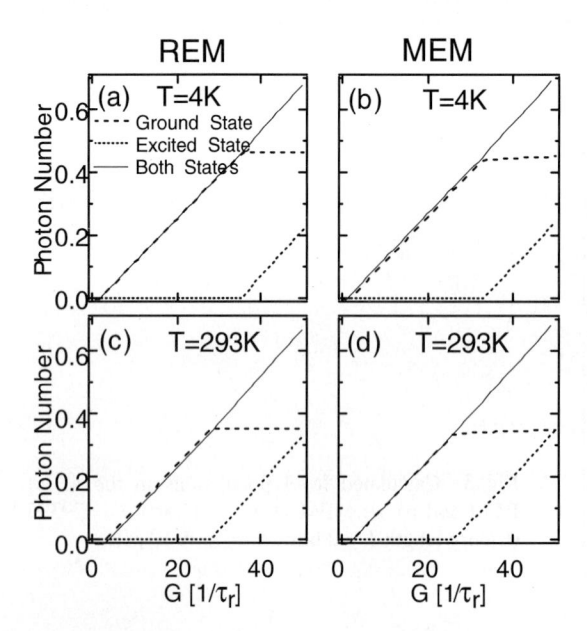

Fig. 4 Calculated photon numbers in the a), c) REM and b), d) MEM at $T = 4$ K and $T = 293$ K (Cavity length $l = 2$ mm, $\tau_0 = \tau_c = 10$ ps).

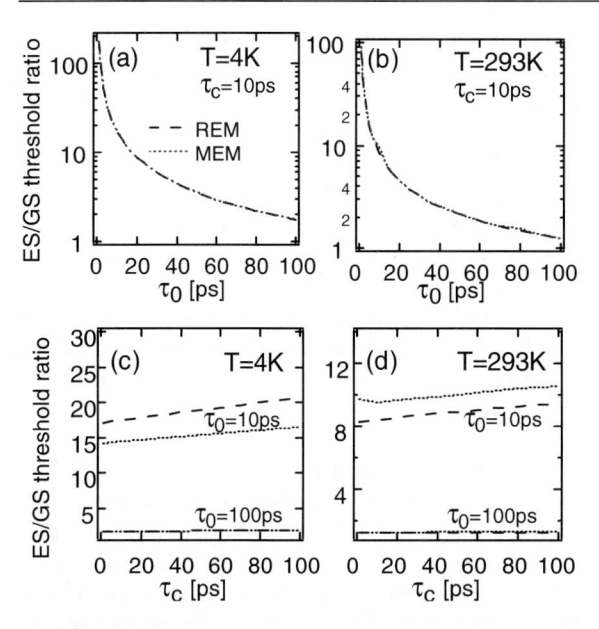

Fig. 5 Impact of the relaxation time τ_0 on the ratio for a) $T = 4$ K and b) $T = 293$ K ($\tau_c = 10$ ps) and of the capture time tau_c for (c) $T = 4$ K and (d) $T = 293$ K ($\tau_0 = 10$ ps and $\tau_0 = 10$ ps) on the ratio of ES and GS threshold current.

4 Lasing dynamics

As a further illustration of the interplay between GS and ES, we investigate GS/ES carrier dynamics under current modulation, by solving the differential rate Eqs. (1–5). Here we model two realistic devices with a cavity length of $L = 1.65$ mm and $L = 2$ mm, which we presented in Ref. [6]. Figure 5 depicts relaxation oscillations for both device length in response to a step-like increase in current. The step is between an injection current slightly above GS threshold for both device lengths ($G = 6$) to above the ES lasing threshold of the 2 mm long device ($G = 44$). The longer cavity shows a small relaxation oscillation (RO) at the GS transition with a rise time of about 0.3 ns, whereas in the shorter cavity the RO is completely damped and the rise time increases to about 0.8 ns (Fig. 4c and d). The reason becomes evident looking at the corresponding behavior of the GS population (Fig. 4a and b). In the longer cavity the sudden increase of injected carriers creates a

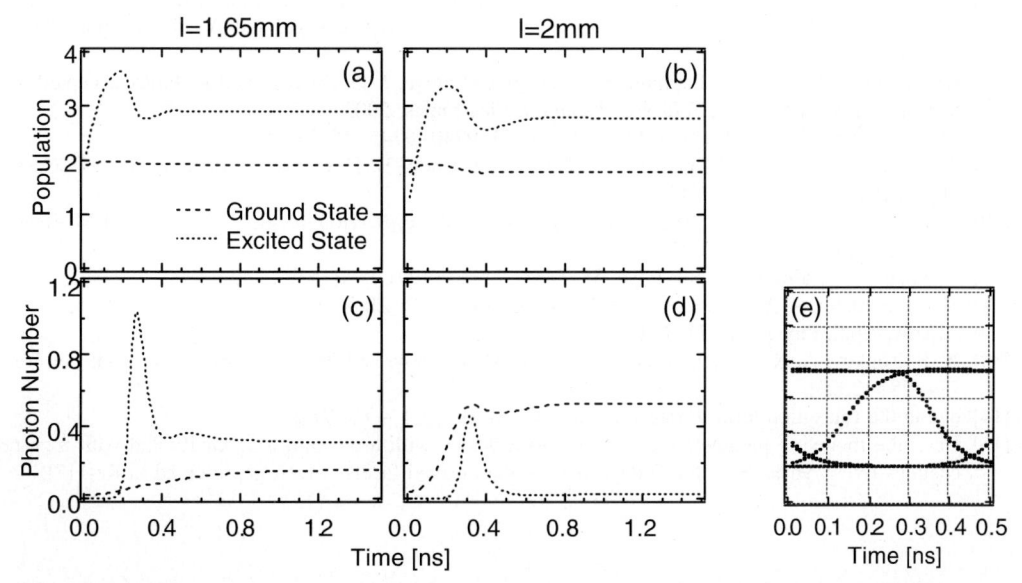

Fig. 6 Modulation dynamics of a a), c) 1.65 mm and b), d) 2 mm long cavity for a step-like pulse from $G = 6$ to $G = 44$ and e) calculated eye diagram at 2.5Gbit/s for a bit-sequence 101011100.

temporary population overshoot, which quickly declines to the equilibrium situation by the stimulated emission process and by increasing the photon number in the cavity. In the shorter cavity the GS threshold gain is clamped at a value closer to saturation and the lack of available states on the GS level constrains the relaxation process. In this situation gain saturation strongly damps the GS carrier dynamics and slows down the laser response. The injected carriers populate almost exclusively the ES, which undergoes strong ROs. The dynamics of the ES transition is similar for both lengths. The ROs show a rise time similar to the RO of the GS in the 2 mm long device. However, in contrary to the situation for the GS, ES lasing has to be turned on from an initial value of G below threshold. This includes a certain delay, which is not present for the GS. We note that the damping of ROs due to gain saturation can have a positive impact on transmission performance, since they result in a more symmetric eye diagram, as we observed experimentally [7]. As an example, Fig. 6e shows a calculated eye diagram for a bit sequence 101011100 at 2.5 Gb/s [15], which is in good agreement to the experiment.

5 Conclusion In summary we presented two approaches to model the dynamics of carriers in a QD laser, REM and MEM. The much simpler REM proved to give nearly identical results to the REM, even at low temperatures. Further on, interlevel dynamics was found to depend mostly on the relaxation time and the influence of the capture process can be neglected. This provides experimental access to a direct measurement of the relaxation time. Modeling of the modulation dynamics shows a slowed down response and a suppression of relaxation oscillations on the GS transition, for cavity lengths such that the GS gain is close to saturation. The excellent agreement of the two models with each other and with experimental results can be applied to provide quantitative predictions of static and dynamic characteristics of QD lasers.

References

[1] G. Park, O. B. Shchekin, D. L. Huffaker, and D. G. Deppe, IEEE Photonics Technol. Lett. **13**, 230–232 (2000).
[2] O. B. Shchekin and D. G. Deppe, Appl. Phys. Lett. **80**, 3277–3279 (2002).
[3] P. Bhattacharya and S. Ghosh, Appl. Phys. Lett. **80**, 3482–3484 (2002).
[4] H. Saito, K. Nishi, A. Kamei, and S. Sugou, IEEE Photonics Technol. Lett. **12**, 1298–1300 (2000).
[5] D. L. Huffaker, G. Park, Z. Zou, O. B. Shchekin, and D. G. Deppe, Appl. Phys. Lett. **73**, 2564–2566 (1998).
[6] A. Markus, J. X. Chen, C. Paranthoën, C. Platz, O. Gauthier-Lafaye, and A. Fiore, Appl. Phys. Lett. **82**, 1818–1820 (2003).
[7] A. Markus, J. X. Chen, C. Paranthoën, O. Gauthier-Lafaye, J. G. Provost, and A. Fiore, accepted for publication in Journal of Selected Topics on Quantum Electronics, 2003.
[8] M. Grundmann and D. Bimberg, Phys. Rev. B **55**, 9740–9745 (1997).
[9] A. Markus, A. Fiore, J. D. Ganière, U. Oesterle, J. Chen, B. Deveaud, M. Ilegems, and H. Riechert, Appl. Phys. Lett. **80**, 911–913 (2002).
[10] A. Fiore, P. Borri, W. Langbein, J. M. Hvam, U. Oesterle, R. Houdre, R. P. Stanley, and M. Ilegems, Appl. Phys. Lett. **80**, 911–913 (2002).
[11] H. Jiang and J. Singh, J. Appl. Phys. **85**, 7438–7442 (1999).
[12] M. Grundmann, A. Weber, K. Goede, V. M. Ustinov, A. E. Zhukov, N. N. Ledentsov, P. S. Kopev, and Zh. I. Alferov, Appl. Phys. Lett. **77**, 4–6 (2000).
[13] A. E. Zhukov, A. R. Kovsh, D. A. Livshits, V. M. Ustinov, and Zh. I. Alferov, Semicond. Sci. Technol. **18**, 774–781 (2003).
[14] due to the relaxation term in rate Eqs. (2) and (3) ($\propto f_{ES}(1 - f_{GS})/\tau_0$)
[15] Here, the modeling parameters correspond to a device with a 0.6 mm long cavity and with high-reflectivity coatings ($R_1 = 96\%$, $R_2 = 85\%$) deposited on the cleaved facets, which is presented in Ref. [7].

Quantum dot semiconductor lasers with optical feedback

G. Huyet[*,1], **D. O'Brien**[1], **S. P. Hegarty**[1], **J. G. McInerney**[1], **A. V. Uskov**[2], **D. Bimberg**[3], **C. Ribbat**[3], **V. M. Ustinov**[4], **A. E. Zhukov**[4], **S. S. Mikhrin**[4], **A. R. Kovsh**[4], **J. K. White**[5], **K. Hinzer**[5], and **A. J. SpringThorpe**[6]

[1] Physics Department, National University of Ireland, University College Cork, Cork, Ireland
[2] NMRC, National University of Ireland, University College Cork, Lee Malting, Cork, Ireland
[3] Institut für Festkörperphysik, Technische Universität Berlin, Germany
[4] A. F. Ioffe Physico-Technical Institute, Russian Academy of Sciences, St. Petersburg, Russia
[5] Bookham Technology, 3500 Carling Ave., Ottawa, On, Canada
[6] National Research Council Canada, Institute for Microstructural Sciences, Ottawa, On, Canada

Received 19 September 2003, accepted 17 November 2003
Published online 14 January 2004

PACS 42.55.Px, 78.67.Hc

We analyse the sensitivity of quantum dot semiconductor lasers to optical feedback. While bulk and quantum well semiconductor lasers are usually extremely unstable when submitted to back reflection, quantum dot semiconductor lasers exhibit a reduced sensitivity. Using a rate equation approach, we show that this behaviour is the result of a relatively low but nonzero line-width enhancement factor and strongly damped relaxation oscillations.

© 2004 WILEY-VCH Verlag GmbH & Co. KGaA, Weinheim

1 Introduction The performance of quantum dot devices is presently the subject of great interest [1]. In particular, several studies have concentrated on the sensitivity of quantum dot semiconductor lasers to the influence of optical feedback [2, 3]. Unlike bulk or quantum well semiconductor lasers, these devices have symmetric gain curves [1] and experiments [4] have demonstrated very small phase-amplitude coupling (α-parameter). As the extra sensitivity of semiconductor lasers to feedback is governed to a large extent by the non-zero value of α, it is thought they could have feedback characteristics different to bulk semiconductor and quantum well lasers. This has brought the possibility of designing directly modulated semiconductor lasers operating without optical isolators and therefore reducing the cost of these devices.

In this paper, we investigate quantum dot lasers with external optical feedback. We first review the properties of semiconductor lasers under the influence of optical feedback. Then we present measurements of the line-width enhancement factor and the relative intensity noise spectrum of these devices before showing the reduced sensitivity to optical feedback of quantum dot semiconductor lasers compared to quantum well semiconductor lasers. In the last part of the paper, we provide a Lang-Kobayashi type model and provide an explanation for the reduced sensitivity to optical feedback.

2 Dynamics of semiconductor lasers with optical feedback External cavity semiconductor lasers were first considered in order to narrow the natural laser linewidth. However it was also observed that, for some parameter regimes, the re-injection of light into the main resonator generates instabilities. The laser may, for example, operate with several external cavity modes and the relative intensity noise

---

[*] Corresponding author: e-mail: huyet@physics.ucc.ie, Phone: +353 21 490 23 81, Fax: +353 21 427 69 49

exhibits a series of peaks at multiples of the external round trip frequency. Feedback can also induce self sustaining relaxation oscillations and even Low Frequency Fluctuations (LFF) [5]. Since these early works the origin of the LFF regime has been the subject of great interest both theoretically and experimentally. Partly this interest stems from the original motivation of improving the laser characteristics, but is also due in part to a desire to understand the high-dimensional non-linear dynamics of the time-delayed systems. In the early 80's, experimental investigations have concentrated on linewidth properties [8–11] and radio frequency fluctuations with and without modulation of the injection current [12, 13]. It was also experimentally shown that LFF are closely related with fluctuations at higher frequency such as the external cavity frequency and the solitary laser relaxation rate [14]. In this last article, it was shown that the power drop-outs occur randomly and are followed by a slow recovery of about 10 steps, each of them corresponding to an external cavity round trip. The coherence properties were analysed by Lenstra et al. [15] who found a line broadening up to 25 GHz. For this reason, this instability is also refered to as the "Coherence Collapse" (CC) regime. A classification of laser behaviour vs the control parameters was first performed by Tkach and Chraplyvy [16] who identified five different regimes. Regime I corresponds to low feedback level were broadening or narrowing of the optical linewidth is observed depending on the feedback phase, regime II to mode hopping between several external cavity modes and regime III to stable emission for higher feedback level. Regime IV, which is observed for higher feedback levels, is associated with the Coherence Collpase and regime V to stable emission with a narrow linewidth at high feedback levels. However regime IV should be divided in three different sub-regimes: first the Low Frequency Fluctuation regime where power drop-outs can be distinctively observed; second the coherence collapse regime and third bistability between stable emission and the LFF and the CC regime. This last sub-regime, which was, for instance, reported in ref. [17–19] occurs at the boundary of regime V. The model commonly used to describe the dynamics of external cavity lasers is the model developed by Lang and Kobayashi [7]. This model describes the laser with rate equations for the complex amplitude of the electric field and for the carrier density (population inversion). The optical feedback was accounted for by adding a term in the field equation corresponding to the re-injection of a portion of the field, delayed by an external cavity round-trip. In this paper we will use the following dimensionless form of these equations:

$$\dot{E} = \kappa(1+i\alpha)(N-1)E + \gamma e^{-i\varphi_0}E(t-\tau),$$
$$\dot{N} = -\frac{1}{T}(N - J + |E|^2 N), \qquad (1)$$

where E is the complex amplitude of the electric field, N is the carrier density, κ is the field decay rate, T is the carrier decay time, α is the linewidth enhancement factor, γ represents the feedback level, J is the pumping parameter, φ_0 is the feedback phase when the laser emits at the solitary laser frequency ω_0, and τ is the external cavity round trip time, implying that $\varphi_0 = \omega_0 \tau$. It is worthwhile to note that these equations assume single laser mode operation and neglect re-injection of light after multiple reflections in the external cavity.

The steady-state solutions of these equations correspond to external cavity modes and other unstable solutions sometimes referred to as antimodes. They are associated with a destructive interference between the emitted field and the re-injected field (in the limit where $\alpha = 0$) and are therefore unstable. The number of external cavity modes depends on the parameter $C = \gamma\tau\sqrt{1+\alpha^2}$ [15]. If $C > 1$ there is more than a single solution and if $C \gg 1$ there are approximately $\frac{C}{\pi} + 1$ external cavity modes. The stability of these solutions has been addressed by several authors [20–24] and it was found that external cavity modes can become unstable as relaxation oscillations become undamped and the laser displays self-sustained pulsations. For example, Helms and Petermann defined the onset of self pulsations as when the damping rate of the relaxation oscillations is $\Gamma = \gamma\sqrt{1+\alpha^2}$, showing that the α parameter is not the only critical parameter defining the sensitivity to optical feedback. The behaviour of semiconductor lasers in the unstable regime has been the subject of many experimental,

© 2004 WILEY-VCH Verlag GmbH & Co. KGaA, Weinheim

numerical and theoretical investigations. Numerical simulations [25] of the LK equations have shown that the LFF regime is the result of a chaotic itinerancy between destabilised external cavity modes. It was later shown for this model that the power exhibits strong irregular picosecond light pulses at a rate associated with the relaxation oscillation frequency [26]. These results were explained by reducing the dynamics to a three dimensional dynamical system [27]. Fast pulses in the laser intensity have also been observed [28, 29] for both multimode [30] and single mode [31] lasers.

3 The experiment The sensitivity of quantum dot semiconductor lasers was investigated using semiconductor lasers emitting at around 1.3 μm. Both the QD and QW lasers were as-cleaved Fabry–Pérot ridge waveguide devices. The cavity length was 1.5 mm, and both devices operated predominantly in a single lateral mode. The QDL ridge width was 7 μm and the QWL ridge width was 2.4 μm. The active region of the QDL consisted of a five-fold stack of InGaAs QDs in a GaAs waveguide, while the quantum well laser had six 6 nm AlGaInAs QWs as its active region. The detailed QWL and QDL growth and fabrication conditions are described elsewhere [32, 33]. The line-width enhancement factor for both devices was measured using the Hakki–Paoli method. The lasers were run in the pulsed regime to avoid thermally induced wavelength shifts. From these measurements, we obtain an alpha factor of 1.6. The same measurement carried out with the quantum well devices gives alpha 3–5. The Relative Intensity Noise (RIN) was measured in the CW regime. The devices were pumped using an ILX LDX3235 current driver with stability of 50 ppm. All measurements were conducted at 20 °C, the laser mount temperature being maintained using a Newport 350 TEC. The light was coupled to a Newport (D30) detector after passing through an optical isolator providing isolation of 40 dB. The signal was amplified using a Miteq high gain amplifier and analysed with an Advantest R3172 electronic spectrum analyzer (ESA). The D30 had a bandwidth of 14 GHz, the ESA a bandwidth of 26 GHz and the Miteq amplifier had a bandwidth of 20 GHz and a noise figure of 3.5 dB. The QWL demonstrated an archetypal RIN spectrum with a well defined peak near the relaxation oscillation frequency. RIN spectra for different operating points of the QDL are presented in Fig. 1. The QDL never shows a clear relaxation oscillation at any operating point. The suppressed peak is clear evidence of overdamping of the relaxation oscillation, a conclusion supported by transient measurements on similar devices [34]. To analyse the effect of optical feedback on these devices, a broadband coated Al mirror was mounted 45 cm from the laser, the lens being adjusted so as to position the beam waist at the mirror for maximal coupling. In order to vary the level of feedback into the device, a variable neutral density filter (ND) was placed in the external cavity. A beam splitter was used to couple a fraction of the beam to the diagnostic instruments. Again, in order to avoid extraneous feedback from the diag-

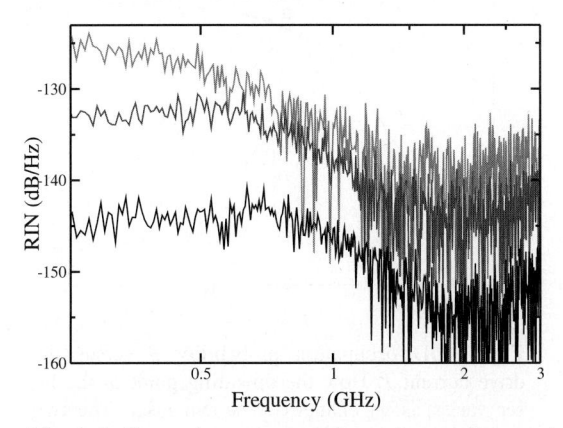

Fig. 1 (online colour at: www.interscience.wiley.com) RIN spectra of the QD lasers for different drive currents. From uppermost to lowermost: 130 mA, 140 mA, 190 mA.

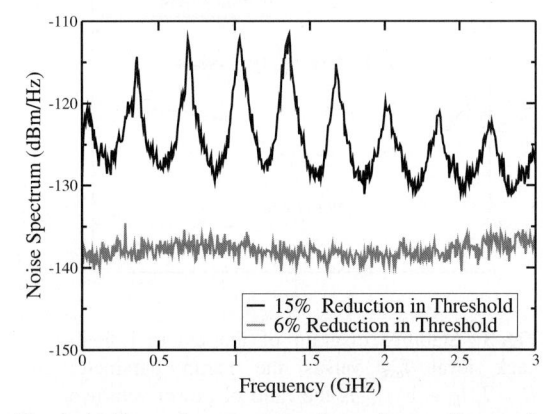

Fig. 2 (online colour at: www.interscience.wiley.com) Noise spectra of QD laser under different feedback conditions. Lower curve is for 6% reduction in threshold and upper curve is 1% reduction in threshold.

nostics, an isolator was positioned after the beamsplitter. A second lens coupled the diagnostics beam to a short length of fiber which guided the light to a Newport D30 fast detector. The electronic spectrum was measured at different feedback levels and pump currents. Stability of the output of the laser was assessed from the presence or absence of RF and microwave components in the photodiode signal. When present these peaks tended to be particularly pronounced at multiples of the round-trip frequency (~330 MHz). For strong feedback levels, the feedback strength was deduced from the amount of threshold reduction using a linear gain approximation, in a method similar to Olsson [35]. For weaker feedback levels, the transmittance of the external cavity filters was used to calculate the feedback level. In an effort to clarify the experimental data, a stability criterion was introduced. If any component of the microwave spectrum exceeded −130 dB m/Hz, the laser operating point was recorded as unstable. If no part of the spectrum exceeded this level, the operating condition was recorded as stable. This criterion corresponded to a maximum RIN < −136 dB/Hz when the laser emits 1 mW. Two sample microwave spectra are shown in Fig. 2. These are recorded for the same QDL, operating at 1.25 times threshold, with different threshold reductions. The cavity was kept the same for all data. In the lower curve (6% threshold reduction), the microwave spectrum is simple broadband noise (in fact amplifier noise), and the max level is < −135 dB m/Hz. Therefore this operating point is stable. For the upper curve (15% threshold reduction), the microwave spectrum consists of broadened peaks at every round trip frequency multiple within the trace bandwidth. As virtually every frequency exceeds the −130 dB m/Hz criterion this operating point is classed as unstable. This procedure was repeated as a function of the driving current and the feedback level for both the QDL and the QWL. In previous work that has carried out similar parameter space mappings, multiple regime classes have been used, including Low Frequency Fluctuations, Coherence Collapse and at very high feedback levels, linewidth narrowing [37]. In this experiment, where the laser facets had no anti-reflection coatings, the laser never recovered stability with the injection fixed and the feedback level increased. Also, all types of intensity instability were grouped together as an unstable operating condition. Thus, our stability diagram for each type of laser simplified to a boundary separating stable operation from unstable operation. This diagram is shown in Fig. 3. The result for the QWL is broadly in line with the literature (see for example [36, 37] and refs therein). The QDL however maintains its stability at much greater levels of feedback, up to 30 dB greater at some operating points.

4 The model In quantum dot semiconductor lasers, the carriers are first injected into a quantum well before being captured into the quantum dot as previously used in other models describing QD laser

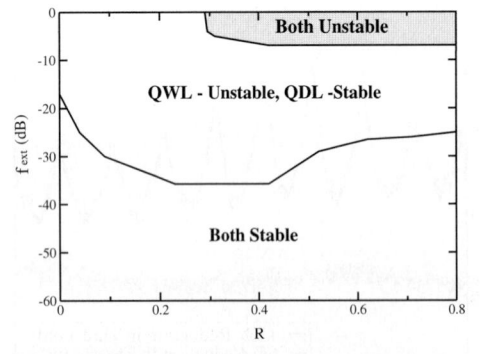

Fig. 3 Stability diagram of the external feedback level f_{ext} versus the pump parameter $R = I/I_{th} - 1$. f_{ext} is the ratio of power returned to the facet divided by the power emitted, including all coupling losses.

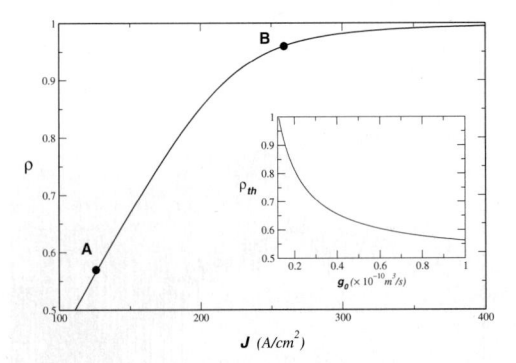

Fig. 4 QD occupation probability ρ versus the drive current J. How the operating point of the laser varies as g_0 changes is shown inset. The two different operating points investigated are shown, (A and B). $C = 10^{-20}$ m^4/s, $R_{esc} = 0$, $B = 0$, $N_d = 2 \times 10^{15}$ m^{-2}, $\vartheta = 2.4 \times 10^{22}$ m^{-3}, $\tau_n = \tau_d = 1$ ns and $\tau_s = 3$ ps.

dynamics [38, 40], we write one equation for the population in the dots and one equation for the population in the well. We neglect coupling between dots and the influence of inhomogeneous broadening as these are not essential features for the physics described in this paper. With these approximations, the system can be described using the following rate equations:

$$\dot{E} = -\frac{E}{2\tau_s} + \frac{g_0 \vartheta}{2}(2\rho - 1)E + i\frac{\delta\omega}{2}E + \frac{\gamma}{2}E(t-\tau),$$

$$\dot{\rho} = -\frac{\rho}{\tau_d} - g_0(2\rho - 1)|E|^2 + F(N, \rho),$$

$$\dot{N} = J - \frac{N}{\tau_n} - 2N_d F(N, \rho), \qquad (2)$$

where E is the complex amplitude of the electric field, N is the carrier density in the well, ρ is the occupation probability in a dot, τ_s is the photon lifetime, τ_n and τ_d are the carrier lifetime in the well and the dot respectively, N_d is the 2D density of dots and J is the pump. $g_0 = \sigma_{res} v_g$, where σ_{res} is the cross section of interaction of carriers in the dots with photons, v_g is the group velocity, $\vartheta = \frac{2N_d \Gamma}{d}$, where Γ is the confinement factor and d is the thickness of the dot layer [40]. The parameters γ and τ describe the feedback level and delay time. The function $F(N, \rho)$ describes the rate of exchange of carriers between the well and the dots. Carrier capture and escape from the dots can be mediated through carrier-phonon or through carrier-carrier (Auger) interaction. In its most general form, $F(N, \rho) = R_{cap}(1-\rho) - R_{esc}\rho$, where $R_{cap} = CN^2 + BN$. The constants B and C describe the carrier-phonon and Auger carrier capture respectively. R_{esc} is a temperature dependent function controlling carrier escape from the dots. From a phenomenological viewpoint, the refractive index of the device depends on the carrier densities in the well and/or dots. If we assume linear dependencies, the variation of the laser frequency with these densities reads $\delta\omega = \beta_1 N + \beta_2 \rho$. The coefficient β_1 describes the plasma effect from the carriers in the well, while β_2 describes the variations due to the population in the dots. If the dots are described by two-level atoms with symmetric line shape, then $\beta_2 = 0$ from the Kramers–Kronig relations. However more complex effects [39], such as Coulomb interactions can lead to nonzero values for β_2.

We first consider the properties of the laser without optical feedback. In this case the evolution of the intensity and carrier densities do not depend on the phase of the electric field and, as a result, Eq. (2) can be rewritten with the photon density S and the carriers densities. An important feature of these equations is that below threshold the dot population ρ, and thus the gain, saturates as the injection current is increased as shown on Fig. 4. This effect is important in many current quantum dot devices as gain per unit length can be very low requiring device operation close to saturation. Transparency is reached when $\rho = 0.5$, and the threshold condition is obtained when $\rho = \rho_{th}$, where

$$\rho_{th} = \frac{1}{2\tau_s g_0 \vartheta} + \frac{1}{2}. \qquad (3)$$

A linear stability analysis is performed about the lasing fixed solution to understand how the system reacts to perturbations under different operating conditions. The eigenvalue solutions of the system, λ_n, determine the nature of the fixed point. In this case the solutions consist of two complex conjugate eigenvalues $\lambda_{1,2} = -\Gamma_a \pm i\Omega_a$, associated with the relaxation oscillations with damping Γ_a and oscillation frequency Ω_a. The third eigenvalue $\lambda_3 = -\Gamma_b$ is real and negative and associated with the capture mechanism into the dots. The damping rate of the relaxation oscillations as a function of the threshold occupation probability ρ_{th}, is shown in Fig. 2 for different capture rates. In each curve, there are three main features. If the laser operates close to transparency, (i.e if $\rho_{th} > \simeq 0.5$) the damping rate decreases as ρ_{th} increases. When $\rho_{th} \simeq 1$, the damping rate decreases rapidly. In the intermediate range, the damping as a function of ρ_{th} reaches a local minimum and then begins to increase until it reaches a local maximum close to $\rho_{th} = 1$. We can approximate the damping rate for two limiting cases. The first limit, shown in bold solid line in Fig. 5, corresponds to the limit of very high capture

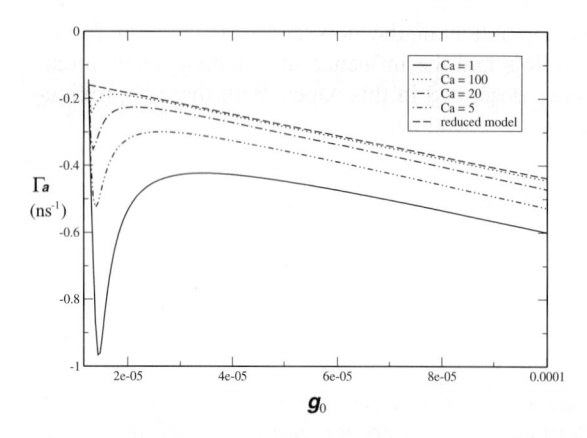

Fig. 5 Damping of relaxation oscillation as a function of g_0, for different values of the auger carrier capture rate C ($\times 10^{-20}$ m^4/s), and for the reduced equations in S and N, and S and ρ. For values of $g_0 \vartheta < \dfrac{1}{\tau_s}$ the laser can never meet the lasing condition. (Other values as for Fig. 4).

rate of carriers from the well into the dots. Here, the dynamics of the well can be adiabatically eliminated and the laser can be described by the coupled rate equations for S and ρ and the relaxation oscillations can be easily calculated. The second limit is in the case where ρ_{th} is very close to one. Here Pauli blocking causes the carrier lifetime to be dominated by the lifetime in the well. In this case the damping rate can be calculated via the adiabatic elimination of ρ. Between these two limit cases, we note that the damping rate of the relaxation oscillations has a maxima.

In order to understand the behavior of the laser when subjected to optical feedback, it is worthwhile to first consider the behavior in both limits described above. In these cases, Eq. (2) can be reduced to a pair of equations, one describing the evolution of the field and the other describing the carriers, either N or ρ. In the first limit when N is adiabatically eliminated, the equations can be written as the Lang–Kobayashi equations. In the second limit, a similar though more complicated set of equations written in terms of E and N will emerge. In the analysis below, this will create non-linear effects leading to a renormalization of the relaxation oscillation. As shown above, the damping rate of the relaxation oscillations is an extremely important parameter for the sensitivity of semiconductor lasers with optical feedback. For quantum dot lasers, the sensitivity to optical feedback will therefore depend on ρ_{th}.

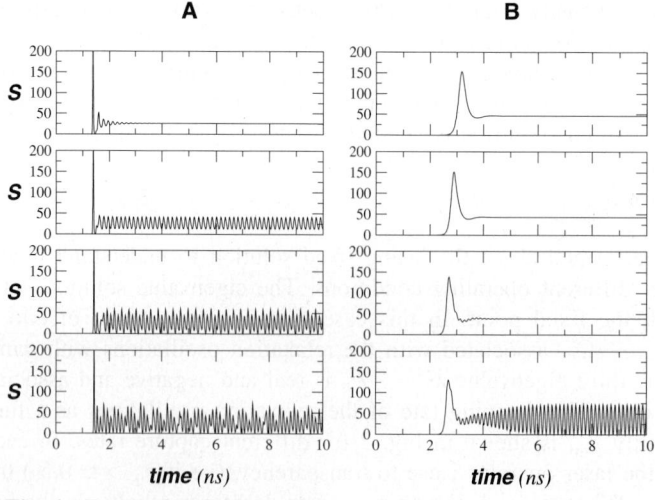

Fig. 6 Temporal behavior of lasers at different operating points subjected to different amounts of optical feedback, Photon density S ($\times 10^{18}$ m^{-3}) Vs time. Free running turnon transient top, with the feedback level increasing going down ($\gamma = 0, 0.01, 0.02, 0.025$ ps^{-1}). External cavity round trip time $\tau = 100$ ps. Operating point taken as 1.5 threshold. (Other values as for Fig. 4).

To illustrate this we choose two lasers operating with different ρ_{th}, labeled A and B on Fig. 4 and the rate Eq. (2) were numerically integrated. The lasing condition is reached at a slightly lower value of ρ_{th} with the introduction of optical feedback. The relaxation oscillations are more strongly damped for point B, which is operating close to the local maximum of the damping rate. The response to different amounts of optical feedback is shown in Fig. 6. In both cases the pump level was chosen to be 1.5 times the threshold current. The results presented in Fig. 6 were obtained for $\beta_1 = 0$ but similar results were obtained for $\beta_1 \neq 0$. In this case we took $\beta_1 \sim \rho_{th}/N_{th}$ to compare lasers with the same α factor at threshold. The turn on transients of the free running lasers as well as the response when subjected to optical feedback are shown in both cases. Laser A shows instabilities associated with optical feedback, displaying a self pulsating regime with a frequency of the external cavity round trip time. With increased feedback, period doubling of the undamped oscillation is observed and continues into a chaotic regime. Laser B is more stable as instabilities only appear for the highest feedback level shown here.

In summary, we have reviewed the main features of semiconductor lasers with optical feedback and presented new experimental results about the sensitivity of quantum dot semiconductor lasers with optical feedback. It was found that the damping rate of relaxation relaxation oscillations is the main source for the reduced sensitivity and this was illustrated with a rate equation model.

Acknowledgements This research was supported by Science Foundation Ireland under the grant SFI/01/F.1/CO13 the European Union in the framework of the IST-Dotcom, by the Irish Higher Education Authority under the Programme for Research in Third Level Institutions.

References

[1] D. Bimberg, M. Grundmann, and N. N. Ledentsov, Quantum Dot Heterostructures. Chichester, UK: Wiley, 1998.
[2] L. Zhang, R. Wang, Z. Zou, A. L. Gray, L. Olona, T. C. Newell, D. Webb, P. Varangis, and L. F. Lester, InAs Quantum dot DFB lasers on GaAs for uncooled 1310 nm fiber communication, OFC, (2003).
[3] D. O'Brien, S. Hegarty, G. Huyet, J. McInerney, D. Bimberg, C. Ribbat, V. Ustinov, A. Zhukov, S. Mikhrin, and A. R. Kovsh, Sensitivity to External Optical Feedback in Quantum Dot Lasers, submitted Elect. Lett., (2003).
[4] T. Newell, D. Bossert, A. Stintz, B. Fuchs, K. Malloy, and L. Lester, Gain and linewidth enhancement factor in InAs quantum-dot laser diodes, IEEE Phot. Tech. Lett. **11**, 1527 (1999).
[5] T. Morikawa, Y. Mitsuhashi, and J. Shimada, Elect. Lett. **112**, 435 (1976).
   C. Risch and C. Voumard, J. Appl. Phys. **48**, 2083 (1977).
[6] D. Lenstra, M. V. Vaalen, and B. Jaskorzynska, Physica C **125**, 255 (1984).
[7] R. Lang and K. Kobayashi, IEEE J. Quantum Electron. **QE-16**, 347 (1980).
[8] R. Miles, A. Dandridge, A. Tveten, H. Taylor, and T. Giallorenzi, App. Phys. Lett. **37**, 990 (1980).
[9] L. Goldberg, H. F. Taylor, A. Dandridge, J. F. Weller, and R. Miles, IEEE J. Quantum Electron. **QE-18**, 555 (1982).
[10] F. Favre, D. L. Guen, and J. C. Simon, IEEE J. Quantum Electron. **QE-18**, 1712 (1982).
[11] G. P. Agrawal, IEEE J. Quantum Electron. **QE-20**, 468 (1984).
[12] K. E. Stubkajaer and M. B. Small, IEEE J. Quantum Electron. **QE-20**, 472 (1984).
[13] T. Fujita, S. Ishizuka, K. Fujito, H. Serizawa, and H. Sato, IEEE J. Quantum Electron. **QE-20**, 492 (1984).
[14] H. Temkin, N. A. Olsson, J. H. Abeles, R. A. Logan, and M. B. Panish, IEEE J. Quantum Electron. **QE-22**, 286 (1986).
[15] D. Lenstra, M. V. Vaalen, and B. Jaskorzynska, Physica C **125**, 255 (1984).
[16] R. Tkach and A. R. Chraplyvy, J. Lightwave Technol. **4**, 1655 (1986).
[17] P. Besnard, B. Meziane, and G. Stephan, IEEE J. Quantum Electron. **QE-29**, 1271 (1993).
[18] M. Giudici, C. K. Green, G. Giacomelli, U. Nespolo, and J. R. Tredicce, Phys. Rev. E **55**, 6414 (1997).
[19] T. Heil, I. Fischer, and W. Elsaesser, Phys. Rev. A **58**, R2672 (1998).
[20] B. Tromborg, J. H. Osmundsen, and H. Olesen, IEEE J. Quantum Electron. **QE-20**, 1023 (1984).
[21] J. Helms and K. Petermann, IEEE J. Quantum Electron. **QE-26**, 833 (1990).
[22] A. Ritter and H. Haug, J. Opt. Soc. Am. B **10**, 130 (1993).
[23] A. Levine, G. van Tartwijk, D. Lenstra, and T. Erneux, Phys. Rev. A **52**, R3436 (1995).

[24] G. Lythe, T. Erneux, A. Gavrielides, and V. Kovanis, Phys. Rev. A **55**, 4443 (1997).
[25] T. Sano, Phys. Rev. A **50**, 2719 (1994).
[26] G. van Tartwijk, A. Levine, and D. Lenstra, IEEE Sel. Top. Quantum Electron. **1**, 466 (1995).
[27] G. Huyet, P. Porta, S. P. Hegarty, J. McInerney, and F. Holland, A low dimension dynamical system to describe low-frequency fluctuations in a semiconductor laser with optical feedback, Opt. Commun. **180**, 339 (2000).
[28] I. Fischer, G. van Tartwijk, A. Levine, W. Elsaesser, E. O. Gobel, and D. Lenstra, Phys. Rev. Lett. **76**, 220 (1996).
[29] G. Huyet, S. Hegarty, M. Giudici, B. de Bruyn, and J. G. McInerney, Europhys. Lett. **40**, 619 (1997).
[30] G. Vaschenko, M. Giudici, J. Roca, C. Menoni, J. Tredicce, and S. Balle, Phys. Rev. Lett. **81**, 5536 (1998).
[31] G. Huyet, J. K. White, A. J. Kent, S. P. Hegarty, J. V. Moloney, and J. G. McInerney, Phys. Rev. A **60**, 1534 (1999).
[32] A. SpringThorpe, T. Garanzotis, P. Paddon, G. Pakulski, and K. I. White, Strained 1.3 µm mqw algainas lasers grown by digital alloy mbe, Electron. Lett. **36**, 1031 (2000).
[33] V. Ustinov, A. Zhukov, N. Maleev, A. Kovsh, S. Mikhrin, B. Volovik, Y. G. Musikhin, Y. Shernyakov, M. Maximov, A. Tsatsul'nikov, N. Ledenstov, Z. I. Alferov, J. Lott, and D. Bimberg, 1.3 µm inas/gaas quantum dot lasers and vcsels grown by molecular beam epitaxy, J. Cryst. Growth **227–228**, 1155 (2001).
[34] M. Kuntz, N. N. Ledentsov, D. Bimberg, A. R. Kovsh, V. M. Ustinov, A. E. Zhukov, and Y. M. Shernyakov, Spectrotemporal response of 1.3 µm quantum-dot lasers, Appl. Phys. Lett. **81**, 3846 (2002).
[35] A. Olsson and C. L. Tang, Coherent optical interference effects in external-cavity semiconductor lasers, IEEE J. Quantum Electron. **17**, 1320 (1981).
[36] L. A. Coldren and S. W. Corzine, Diode lasers and photonic integrated circuits. New York: Wiley Interscience, 1995.
[37] B. Krauskopf and D. Lenstra, Fundamental issues of nonlinear laser dynamics. Melville, New York: AIP, 2000.
[38] M. Sugawara, K. Mukai, and H. Shoji, Effect of phonon bottleneck on quantum-dot laser performance, Appl. Phys. Lett. **71**, 2791 (1997).
[39] A. Uskov, I. Magnusdottir, B. Tromborg, J. Mork, and R. Land, Line broadening caused by Coulomb carrier-carrier correlations and dynamics of carrier capture and emission in quantum dots, Appl. Phys. Lett. **79**, 1679 (2001).
[40] A. Uskov, Y. Boucher, J. L. Bihan, and J. McInerney, Theory of a self-assembled quantum-dot semiconductor laser with Auger carrier capture: Quantum efficiency and nonlinear gain, Appl. Phys. Lett. **11**, 1499 (1998).

Formation and optical properties of Ge quantum dots selectively grown on patterned Si(001) substrates

Lam H. Nguyen[*,1], **V. Le Thanh**[2], **V. Yam**[1], **D. Débarre**[1], **M. Halbwax**[1], and **D. Bouchier**[1]

[1] Institut d'Electronique Fondamentale (IEF), Bat 220, Université Paris-Sud, 91405 Orsay Cedex, France
[2] CRMC2- CNRS, Campus de Luminy, Case 913, 13288 Marseille Cedex 9, France

Received 19 September 2003, accepted 3 November 2003
Published online 12 January 2004

PACS 73.21.La, 73.21.Ac, 81.10.Bk, 81.65.Cf

Single and stacked layers of Ge/Si quantum dots (QDs) were grown in windows patterned by electron-beam lithography on oxidized Si(001) substrates. A Si buffer layer was selectively deposited in the windows prior to Ge growth, that plays an important role for nucleation of Ge dots. The Si buffer layer evolves towards [113] faceted pyramids, which reduces the area of the topmost (001) surface available for Ge nucleation. By controlling the top facet area of the Si buffer layers, only one dot per circular window and a high cooperative arrangement of dots on a stripe can be observed. In stacked layers, the dot homogeneity can be improved through the adjustment of the Ge deposited amount in the upper layers. The optical properties of these structures measured by photoluminescence spectroscopy are also reported. In comparison with self-assembled QDs, we observed, both in single and stacked layers, the absence of the wetting-layer contribution and an energy blue shift, confirming therefore the dot formation by selective growth.

© 2004 WILEY-VCH Verlag GmbH & Co. KGaA, Weinheim

1 Introduction The growth of Ge QDs has attracted much interest in recent years due to the possibility for the realisation of novel quantum dot-based electronic and optoelectronic devices. The most widely used technique to produce Ge/Si QDs is based on Stranski–Krastanov growth mode. However, in view of applications one of the major drawbacks of self-assembled QD's is their random distribution over the substrate surface. Several techniques have been used for controlling the nucleation of Ge QDs. These techniques can be divided, according to the method of surface preparation, into two categories. The first, referred to as *ex-situ* methods, which are assisted by electron-beam lithography, consists of growth on patterned substrates [1, 2], or on Si mesas created by etching [3–6]. The second, named *in-situ* methods, includes growth on chemically prepared H-passivated surfaces [7], on thermally desorbed oxide surfaces [8–10], or growth on surfaces with controlled elastic or plastic strain distributions [11, 12]. In this work, we report on selective epitaxial growth (SEG) of single and aligned Ge dots in patterned windows, as well as multiple SEG of Ge dots layers.

2 Experimental The *p*-Si(001) substrates used in this study were first thermally oxidized to form about 300 nm-thick SiO_2. Then, circular windows with a diameter of about 500 nm and stripe windows with a width of 400 nm were patterned by electron-beam lithography and processed by reactive ion etching at the CEA/LETI. In this study, Si and Ge depositions were performed by ultra-high vacuum chemical-vapour deposition (UHV-CVD) at a pressure of about 10^{-4} Torr and a temperature of 650 °C. Pure SiH_4 and hydrogen-diluted (10%) GeH_4 flows of 5 sccm were used for the deposition of the Si buffer layer and for Ge growth. For multilayered samples, five bi-layers were grown, in which Ge layers were separated by a Si spacer layer of ~9 nm thick (6 min deposition). In order to form Ge dots of equal size along

---

[*] Corresponding author: e-mail: huu-lam.nguyen@ief.u-psud.fr, Phone: +33 1 69 15 40 47, Fax: +33 1 69 15 40 10

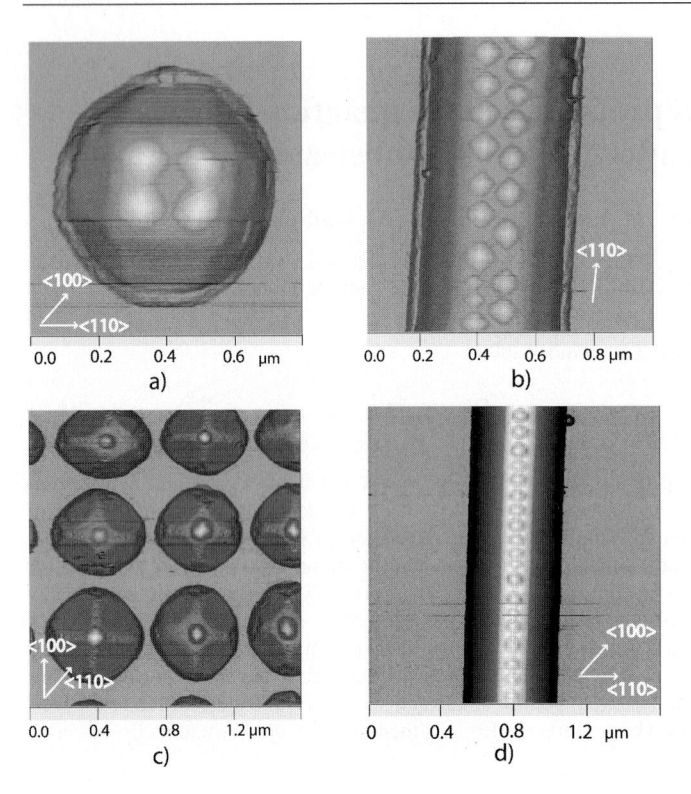

Fig. 1 (online colour at: www.interscience.wiley.com) AFM images illustrating the formation of four Ge dots in a circular window a), of two arrays of dots situated near borders of the stripe window b). For a longer Si deposition time, a single Ge dot is formed in a circular window c), and ordered Ge dots with period constant of 80 nm are formed in a stripe window d).

the vertical axis, the amount of Ge in upper layers was monitored by *in-situ* reflection high-energy electron diffraction (RHEED). After growth, the SiO$_2$ layer was removed for atomic force microscopy (AFM) studies. The photoluminescence (PL) measurement was carried-out in a liquid helium cryostat and the signal was detected by a liquid N$_2$ cooled Ge detector.

3 Results and discussions AFM images shown in Fig. 1 illustrate several surface morphologies of SEG of Ge dots obtained by varying the height of the Si buffer layer prior to Ge deposition. Images of Figs. 1 a) and c) concern growth on circular windows while those shown in Figs. 1 b) and d) relate to growth on stripe windows. As we have recently shown in Ref. [8], the Si buffer layer in the windows evolves towards (113) facetted truncated pyramids whose (001) top facet width can be controlled via the Si growth time. This implies that, by controlling the height of the Si buffer layer, one can adjust the area of the Si(001) top facet on which the nucleation of Ge dots takes place. An example illustrating this approach is depicted in Figs. 1a) and 1b). By adjusting the Si deposition so that the width of the Si(001) topmost facet approaches the range of 300 nm, four Ge dots are found to nucleate near corners of the facet in a circular window and two straight arrays of Ge dots are formed near borders of stripe windows. With further increase of the Si deposition time, the Si top facet area will become progressively smaller and when the area width reaches a value of about 200 nm, a single dot or a single array of Ge dots were formed as shown in Figs. 1c) and 1d) while the Ge deposited amount was kept constant. The aligned Ge dots in the case of stripe windows have a mono-modal size distribution and a cooperative arrangement with a constant period of about 80 nm. The base width and the height of dots, estimated from AFM line-profiles, are 80 nm and 6 nm, respectively. Our results support the previous works [1, 4, 5], in which Ge dots were regularly formed on the Si mesas.

The formation of centred single Ge dots when the top facet width of the buffer layer is reduced below 200 nm, may be attributed to an energetically preferential nucleation of Ge dots at the centre of the top plan. Kim et al. [1] found a corresponding dimension (for which only one Ge dot was formed) of about 300 nm. However, Jin et al. [2] pointed out that this strongly depends the total Ge deposited amount. The perfect alignment of Ge dots on Si stripe windows may arise from a combination of the elastic interac-

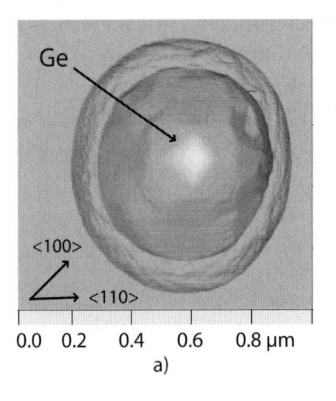

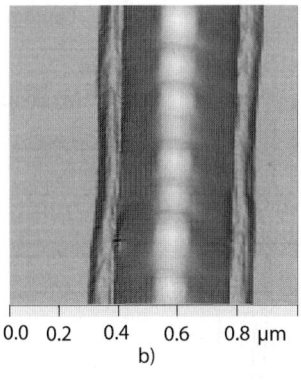

Fig. 2 (online colour at: www.interscience.wiley.com) AFM images of the stack of 5 Ge/Si layers. One Ge dot was formed on the highest layer of a circular window a) and an aligned Ge dots were grown in a stripe window b) showed a high vertical ordering between layers.

tion between the neighbouring dots and the Ge surface diffusion length, along which the total energy of the system is minimized [5].

The second part of the present work is devoted to selective epitaxial growth of stacked Ge/Si QDs in patterned windows. It is known that when the thickness of the Si spacer layers is low enough, the dots in the upper layers grow on the top of the buried ones, giving rise to a vertical correlation between dots along the growth direction [13]. The interest of multiple layers of Ge QDs in patterned windows is, on the one hand, positioning the dots for the realisation of devices based on vertical transport such as resonant tunnelling diodes, and, on the other hand, modifying the confinement effects of carriers in the structure. The mechanism leading to the vertical ordering of stacked QD's grown on a bare Si(001) substrate arises from the elastic strain field created by buried dots and mediated by Si spacer layers [13]. As previously shown [14], this strain field results also in a drastic reduction of the 2D–3D critical thickness when increasing the number of stacked layers. Consequently, the Ge amount in the upper layers must be reduced in relevant proportions in order to equalize the sizes of the stacked dots [15]. We tried to apply a similar approach in the case of SEG of QD's.

We studied the formation of multiple layers consisting of 5 Ge/Si periods with initial conditions resulting in the formation of single Ge dots in circular windows or single arrays of Ge dots in stripe windows. As observed in a single layer, AFM images show only one dot grown on the top facet in a circular window (Fig. 2a) and aligned Ge dots regularly formed along the top layer in a stripe window (Fig. 2b). RHEED measurements performed during growth enabled us to estimate the 2D-3D transition time versus the number of deposited layers. It must be noted that the investigated zone of the substrate was not patterned. A critical thickness value of 4 MLs was found for the first layer, which corresponds to Ge growth on a bare Si(001) substrate, while it was found to decrease down to 2 MLs in the upper ones, as previously reported [15]. As a result, the Ge deposited amount in each layer was not kept constant in order to form Ge dots with homogeneous sizes. Here, by growing 4 MLs of Ge in the upper layers as compared with 6 MLs in the first one, we found, from AFM measurements, average dimensions of 100 nm in width and 8 nm in height for the dots formed on the uppermost layer. Therefore, it can be observed that, in spite of a strong reduction of the deposited amount of Ge in the upper layers, the size of dots in the top layer is notably higher than that observed in the first one. This seems to indicate that the reduction of the top area of the layer due to its evolution in a pyramid shape has a strong influence on the critical thickness for the formation of SEG dots.

Photoluminescence (PL) measurements were performed at liquid-helium temperature using an Ar^+ laser to investigate the optical properties of these structures. Figure 3 presents the spectra of a 5 periods stacked structure of self-assembled (a) and SEG Ge QDs (b). We note that in both cases the growth conditions were the same: the same growth temperature of 650 °C, both structures containing 6 MLs in the first layer and 4 MLs in the upper layers. The narrow peaks located at energies higher than 1030 meV are attributed to the phonon-assisted recombination of the free-exciton in the Si substrate. We also note that compared to the self-assembled Ge dots, the absence of the PL component related to the wetting layers (WL) is correlated with selectively grown dots since in this case the area of the Ge wetting layers is very limited. The related quantum-dot peak of SEG dots is located at 850 meV while that of self-assembled

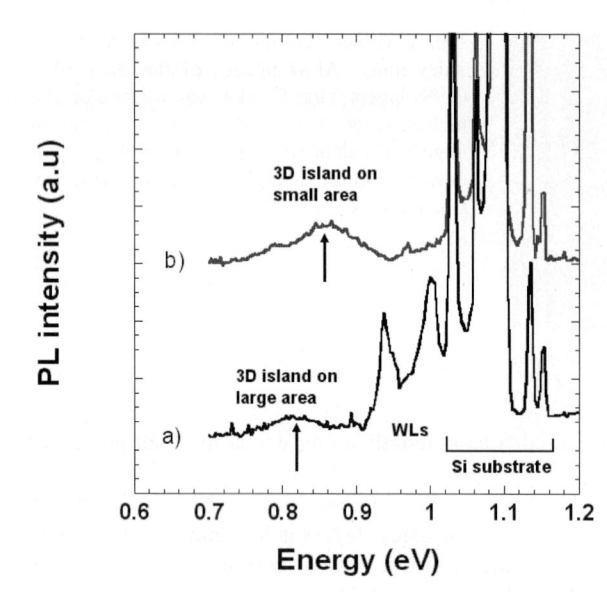

Fig. 3 Photoluminescence (PL) spectra recorded from 5 stacked Ge/Si layers obtained a) by self-assembled growth and b) by selective epitaxial growth. Measurements were carried out at 4.2 K.

dots is observed at 820 meV. We explain this blue shift as resulting from the reduction of the dot size in the case of the SEG QDs compared with the self-assembled Ge dots (dimension of 150 nm in width and 15 nm in height).

4 Conclusion In conclusion, we have investigated the selective growth and optical properties of Ge QDs in patterned windows. We have shown that the number, the shape and also the position of dots depend on the top area of the Si buffer layers. We have demonstrated that the thickness of the Si buffer layers is a relevant additional parameter for controlling the nucleation process of Ge dots. The formation of stacked QDs in these windows has been done by adjusting the Ge deposited amount. It showed a cooperative arrangement of Ge QDs on the top Si layer, the same result as compared with the single layer. The PL of the stacked Ge QDs showed the absence of WLs peaks caused by the selective growth of Ge dots in the windows. An energy blue shift of SEG Ge dots was observed due to the reduction of their size as compared with self-assembled Ge dots.

Acknowledgements The authors gratefully acknowledge Dr P. Boucaud, M. El Kurdi at IEF for their help in PL measurements and J-M. Hartmann at CEA/LETI for substrate preparation.

References

[1] E. S. Kim, N. Usami, and Y. Shiraki, Appl. Phys. Lett. **72**, 1617 (1998).
[2] G. Jin, J. L. Liu, and K. L. Wang, Appl. Phys. Lett. **76**, 3591 (2000).
[3] L. Vescan, J. Crystal Growth **194**, 173 (1998).
[4] T. Kitajima, B. Liu, and S. R. Leone, Appl. Phys. Lett. **80**, 497 (2002).
[5] T. I. Kamins and R. S. Williams, Appl. Phys. Lett. **71**, 1201 (1997).
[6] Z. Zhong, A. Halilovic, M. Muhlberger, F. Schaffler, and G. Bauer, Appl. Phys. Lett. **82**, 445 (2003).
[7] V. LeThanh, Thin Solid Films **321**, 98 (1998).
[8] Lam H. Nguyen, V. Le Thank, D. Débarre, V. Yam, and D. Bouchier, Matter. Sci. Eng. B **101**, 199 (2003).
[9] Y. Nitta, M. Shibata, K. Fujita, and M. Ichikawa, Surf. Sci. Lett. **462**, L587 (2000).
[10] K. Fujita, H. Watanabe, and M. Ichikawa, Appl. Phys. Lett. **70**, 2807 (1997).
[11] H. Omi, D. J. Bottomley, and T. Ogino, Appl. Phys. Lett. **80**, 1073 (2002).
[12] Y. H. Xie, S. B. Samavedam, M. Bulsara, T. A. Langdo, and E. A. Fitzgerald, Appl. Phys. Lett. **71**, 3567 (1997).
[13] J. Tersoff, C. Teichert, and M. G. Lagally, Phys. Rev. Lett. **76**, 1675 (1996).
[14] V. LeThanh, V. Yam, Y. Zheng, and D. Bouchier, Thin Solid Films **369**, 43 (2000).
[15] V. Le Thanh, V. Yam, Lam H. Nguyen, Y. Zheng, P. Boucaud, D. Débarre, and D. Bouchier, J. Vac. Sci. Technol. B **20**, 1259 (2002).

TEM evaluation of strain and stress in III–V semiconductor epitaxial structures

A. Rocher[*,1], **M. Cabié**[1], **A. Ponchet**[1], **A. Arnoult**[2], and **E. Bedel-Pereira**[2]

[1] CEMES-CNRS, BP 4347, F-31055, Toulouse, France
[2] LAAS-CNRS, 7 av. Colonel Roche, F-31055 Toulouse, Cedex 4, France

Received 19 September 2003, accepted 3 November 2003
Published online 14 January 2004

PACS 68.37.Lp, 81.15Hi, 81.05.Ea, 62.20.Dc, 68.55.Jk, 68.35.Dv

A new method of determination of the stress induced by a misfit between an epilayer and its substrate is presented. It is based on the experimental measurement of the specimen curvature induced on the bilayer by this misfit stress. The curvature radius, the misfit stress and the thicknesses of the epilayer and the substrate are related by the Stoney formula. Then from the determination of the radius of curvature and the thickness of the substrate we are able to determine the misfit stress. The originality of the method is to measure locally these parameters on specimen thinned for Transmission Electron Microscopy (TEM). For a thickness of TEM specimen of about 0.3 μm, the measured radius of curvature is about 100 μm for a lattice mismatch of 1% and an epilayer thickness of 10 nm. This method has been applied to the GaInAs/GaAs system with an In concentration of 20% and a lattice mismatch of 1.4%. The uniformity of the misfit strain is evaluated at the sub-micrometer scale. Experimental values of the misfit strain obtained by this method are 10 to 40% lower than the theoretical values. This difference is discussed in term of In segregation and chemical gradient at the level of the interface.

© 2004 WILEY-VCH Verlag GmbH & Co. KGaA, Weinheim

1 Introduction In semiconductor hetero-structures grown by epitaxy, the elastic strain constitutes an important parameter in the engineering of the opto- and micro-electronic components. Since it modifies the energy band gap, it allows the monitoring of the emission wave length of quantum wells. This elastic strain applied to an epitaxial layer is due to the accommodation of the lattice mismatch, f, existing between this epilayer and its substrate as shown Fig. 1. If the deposited layer is thinner than the critical thickness above which plastic relaxation occurs, all the models of epitaxy admit that the lattice mismatch is fully accommodated by a purely elastic deformation of the layer [1, 2]. In that case, the in-plane crystalline parameters of both the epilayer and the substrate are equal. The in-plane strain $\varepsilon_{//}$ is therefore directly related to f by:

$$\varepsilon_{//} = -f \tag{1}$$

Following the elasticity theory, the strain and stress are related by a tensorial equation through the elastic coefficients. The stress in the epilayer has a biaxial component in the plane normal to the growth direction, while the component parallel to the growth direction equals zero. It is worth to notice that the stress tensor is completely described by its in-plane component σ_{misfit}. The theoretical value of σ_{misfit} is related to the lattice mismatch, f, by the well-known formula:

$$\sigma_{misfit} = \frac{E_l}{1-\upsilon_l} \cdot \varepsilon_{//} = -M_l \cdot f \tag{2}$$

---

[*] Corresponding author: e-mail: andre.rocher@cemes.fr

© 2004 WILEY-VCH Verlag GmbH & Co. KGaA, Weinheim

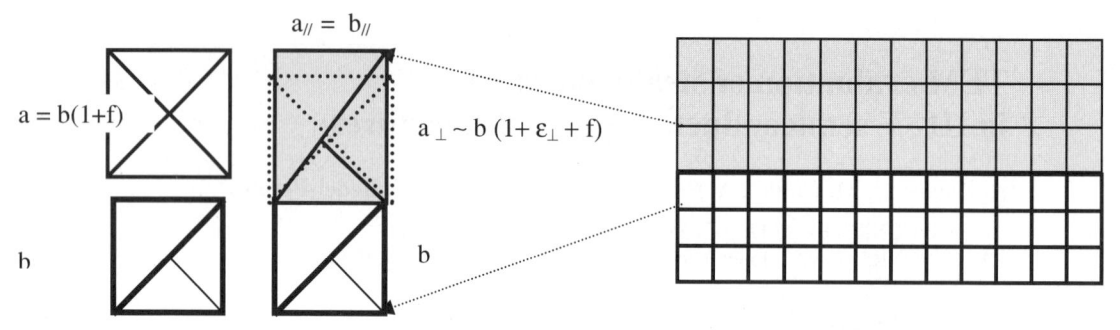

Fig. 1 Schema of a strained epitaxial structure: a) free cells of substrate and epilayer materials; b) strained unit cell; c) representation of a strained structure.

in which E_l, v_l and M_l are the Young modulus, the Poisson ratio and the biaxial elastic modulus of the epilayer, respectively. The use of the isotropic elasticity formalism is justified in the case of a cubic material grown on a {001} surface.

The object of this contribution is to describe how to determine experimentally by TEM (Transmission Electron Microscopy) the elastic strain and the stress due to a lattice mismatch. Experimental results are compared to the theoretical ones. From this comparison, the representation of the pseudomorphic system is discussed.

2 Position of the problem A pseudomorphic layer, i.e. mono-crystalline and homogeneously strained, is described by a strain field directly defined by the continuity of the atomic planes through the interface as shown in Fig. 1. The stress and strain due to a lattice mismatch are always considered as well described by the linear elasticity theory and a perfect epilayer/substrate interface. The absence of extended defects in the epitaxial structure is a condition sufficient to characterize it as a fully strained system. The problem is then to determine experimentally the elastic strain and the stress fields induced by the lattice mismatch in the epitaxial system.

TEM is a tool well adapted to study epitaxial structures at the nano-metric scale [3]. Three useful techniques of observation are classically performed to characterize epitaxial systems: (i) the one-beam image (bright or dark field) where defects such as dislocations, phases, grains, ... are observed; (ii) the multi-beam image showing the arrangement of atomic columns parallel to the electron beam, is very convenient to study interfacial structures and dislocation cores; (iii) electron diffraction which gives equivalent information to X-ray analysis but for nano-metric volume. The electrons, with an energy of 200 keV, have their elastic scattering factor about 10^{+4} larger than those for X-rays. This strong scattering factor plays an important role in the characteristic of the specimen investigated by TEM. Specimens of III–V compounds must be thin enough to be transparent for the electron beam, i.e. the signal/noise ratio needs to be favorable: 0.4 µm to image a dislocation by its strain field, about 10 to 20 nm to image the arrangement of $\langle 110 \rangle$ or $\langle 001 \rangle$ atomic columns of epitaxial systems. X-ray analysis is the best technique for lattice parameter determination of an homogeneous layer; TEM is very well adapted to study the breaks of the crystalline periodicity in materials at the atomic scale nearly.

3 Experiments The investigated specimens are constituted by a layer of GaInAs grown by MBE on a (001) GaAs substrate [4]. Two layer thicknesses with an In concentration of 20% have been investigated: 10 and 25 nm in order to study the behavior of the system by reference to the critical thickness calculated equal to 14 nm for 1.4% lattice mismatch [1]. Specimens for plane view observations have been prepared by chemical etching of the substrate from the back-side. The thinner parts of the sample prepared for TEM observations appear to be bent as shown in Fig. 2.

The TEM observations allow us to verify the homogeneity of the layer by following the evolution of the bend contour during the rotation of the specimen. The presence of threading and misfit dislocations

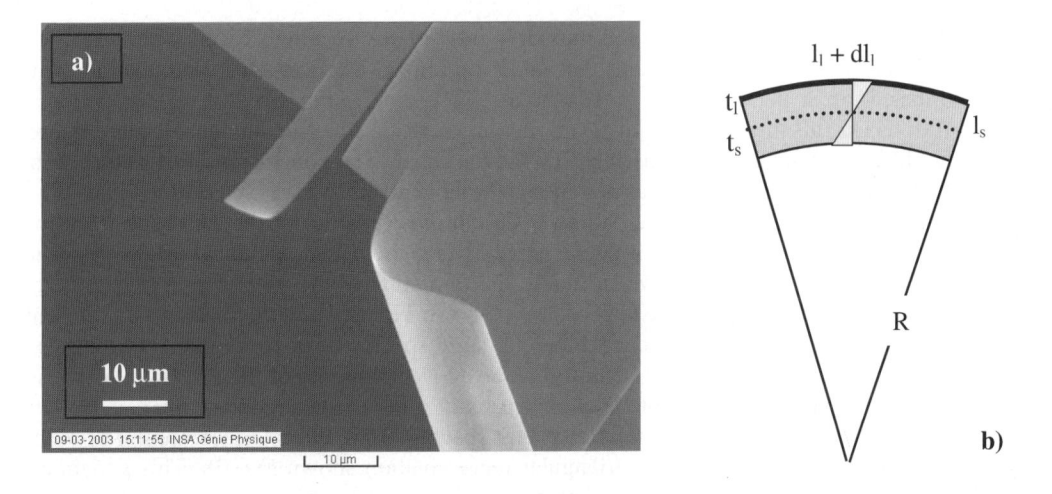

Fig. 2 Scanning Electron Microscopy images of the specimen ready for TEM observations. Note the curvature of the thin areas; a) general view; b) geometric representation of the curved specimen: continuous variation of the elastic strain along the growth direction; doted line is the neutral line corresponding to the same length of the specimen before bending.

are also checked by TEM: when no extended defects are seen, thin layers are considered as pseudomorphic, i.e. homogeneously strained and mono-crystalline.

4 Experimental results

4.1 Strain analysis The TEM and X-rays methods, based on diffraction patterns, determine the interplanar spacings of both the substrate and the epitaxial layer using Bragg reflections. From the experimental data, the elastic strain is determined by comparing the experimental values to an ideal crystal free of stress taken as a reference. The misfit stress responsible for the elastic strain is calculated assuming perfect mechanical properties of both the epilayer and the substrate.

A measurement technique of elastic strain has been proposed by Rocher et al. [5], using both image and diffraction pattern on plan-view specimens in order to image the interface normal to the electron beam. The idea was to compare by moiré technique, the 220 and 202 inter-planar spacings of both the strained epilayer and the substrate: for relaxed system, accurate measurements have been obtained for the

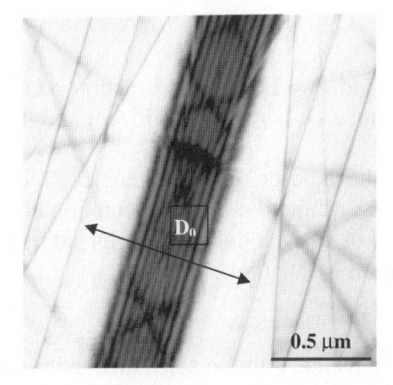

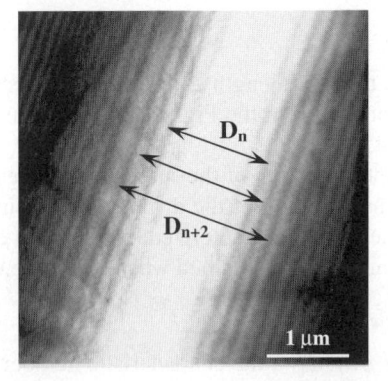

Fig. 3 TEM observations: 220 bend contour observed in brigth and dark field images: no dislocations or moiré patterns are observed on this image; the distance D_0, between the 440 bend contours allows us to determine the curvature radius of the TEM sample; the local thickness of the specimen is deduced from the serie of I_{220} intensity minima.

GaSb/(001)GaAs system: in this case, all measured inter-planar spacing fit well with the lattice parameters of bulk GaSb. This result indicates that under optimized MBE growth conditions the GaSb layer is fully relaxed by a perfect array of Lomer dislocations [5]. For the ternary compounds such as GaInAs/(001)GaAs, all the tentatives of strain measurement do not give any characteristic observations of specific diffraction spots or moiré fringes as shown Fig. 3. The conclusion of this work was that the crystalline structure of the epilayer was not enough homogeneous to give well defined inter-planar spacings [5]. In fact, the absence of moiré pattern is due to the curvature of the TEM specimen inducing a stress relaxation when its thickness becomes smaller than few μm. An image of a typical specimen thinned for TEM observation is shown in Fig. 2: the thin parts of the specimen appear to be bent with local curvature radius as low as some tens of μm as shown Fig. 2. The Fig. 2b shows a representation of the variation of inter-planar spacing due to the curvature specimen: the inter-planar spacing is equal to nominal lattice parameter only on the neutral line, taken in the middle of the substrate thickness. Above this line, the upper part of the substrate is in expansion in order to accommodate the positive lattice mismatch. The continuous variation of the inter-planar spacing along the interface normal of both the epilayer and the substrate is imaged by a triangular representation shown Fig. 2b, This variation explains well the absence of moiré pattern: the inter-planar spacings of the substrate and the epilayer are not well defined and do not correspond to the representation of Fig. 1b.

4.2 Stress determination by curvature analysis

A direct method of stress determination by TEM analysis has been proposed by Rocher et al [6] and described in details by Ponchet et al. [7]. It is based on the measurement of the curvature observed on thin samples prepared for TEM observations and the Stoney formula. This formula is a relationship between the radius of curvature, the biaxial stress induced by a lattice mismatch in a bilayer and the thicknesses of both the epilayer and the substrate [8]. When the substrate is rigid, with a thickness two order of magnitude larger than the thickness of the layer, the Stoney formula can be written as:

$$\sigma = \frac{M_s}{6} \cdot \frac{t_s^2}{t_l \cdot R} \qquad (3)$$

in which t_l and t_s are the epilayer and substrate thicknesses, R the radius of curvature induced by the misfit stress, M_s the biaxial elastic modulus of the substrate. This formula is applied to one single strained layer located at the top surface of a rigid substrate ($t_s \gg t_l$). The measured stress corresponds to the stress existing in the epitaxial structure at room temperature. This method consists to measure on the same zone: (i) the curvature induced by the misfit stress; (ii) the thickness of the substrate. The curvature radius, R, is determined by the measurement of the distance, D_0, between 440 bend contours such as seen Fig. 2a. A relationship between R and D_0 has been established [6, 7]:

$$R = D_0/(\lambda \cdot g_0) \qquad (4)$$

where $\lambda = 0.025$ Å, the wave length of the 200 keV electrons and $g_0 = 1$ Å^{-1} the diffraction vector related to the 440 bend contour. From the Fig. 2a, the measured distance D_0 is equal to 1.15 μm and the related curvature radius is 45 ± 2 μm.

The method of determination of the substrate thickness has been described elsewhere [6, 7]. It is based on the analysis of the variation of the diffracted intensity I_{220} as a function of the orientation of the electron beam versus the crystal characterized by its Bragg error. For a given thickness traversed by the electron beam, the I_{220} minima are obtained when the phase equation is satisfied:

$$t(n) = n \cdot \xi_g/(1 + w_0^2 \, (D_n/D_0)^2)^{0.5} \qquad (5)$$

where $w_0 = (\lambda \cdot g_0 \cdot g/2) \cdot \xi_g$, n the characteristic number of the fringe, g the selected diffraction vector, ξ_g the extinction distance, equal to 630Å for the 220 reflection in GaAs, D_0 defined in Fig. 2a and D_n the distance between the symmetrical I_{220} minima. The specimen thickness t is determined when this equation is simultaneously satisfied for n and at least three D_n, D_{n+1}, D_{n+2}. The specimen thickness t determined by this method in the zone shown Fig. 3 is equal to 0.20 ± 0.01 μm.

© 2004 WILEY-VCH Verlag GmbH & Co. KGaA, Weinheim

Table 1 Value of R, dl_s/l_s, dl_l/l_l as a function of the substrate thickness for $m = M_s/M_l = 1$, $t_l = 10$ nm and $f = 0.01$.

substrate thickness	0.1 μm	0.3 μm	300 μm
R	17 μm	150 μm	150 m
dl_s/l_s	3×10^{-3}	10^{-3}	10^{-6}
dl_l/l_l	3×10^{-4}	3×10^{-5}	3×10^{-10}

From the determination of the curvature radius R and the thickness of the TEM specimen, the stress is determined using the Stoney Formula (3). Nevertheless, the small value of R indicates that the curvature induced by the misfit modifies strongly the strain state of the system. According to Fig. 2b, the elongation dl_s of the top surface of the substrate due to the curvature is given by:

$$dl_s/l_s \sim t_s/2R \sim 3 \cdot m \cdot t_l/t_s \cdot f \tag{6}$$

in the same time the variation of elastic strain related to the epilayer, dl_l is given by:

$$dl_l/l_l \sim t_l/2R \sim 3 \cdot m \cdot (t_l/t_s)^2 \cdot f \tag{7}$$

in these equations, $m = M_l/M_s$ is the ratio of the biaxial elastic moduli, f the lattice mismatch and t_l et t_s the thicknesses of the epilayer and the substrate. Typical values are given on Table 1.

From this table we can see that the elastic strain at the Layer/Substrate interface is modified by the sample bending. For a substrate thickness of 300 μm, this elongation is only 10^{-6}. For substrate thicknesses of 0.1 and 0.3 μm, the elongation at the level of the interface is about 0.3% and 0.1% of the lattice mismatch. This elongation of dl_s indicates in fact a relaxation of the elastic misfit stress. The effect of the relaxation due to the bending of the specimen has been studied analytically by Freund et al. [9]. A correcting function $g(h)$ needs to be taken into account to determined the real misfit stress applied to a rigid substrate, i.e. with a thickness larger than 300 μm:

$$g(h) = (1 + h \cdot m \cdot (4 + 6h + 4h^2))/(1 + h) \tag{8}$$

where $h = t_l/t_s$ the ratio of the layer and substrate thicknesses and m, the ratio of the biaxial elastic moduli. Taking into account this correction, the Eq. (3) can be written as:

$$R = \frac{M_s}{6} \cdot \frac{t_s^2}{t_l \cdot \sigma} \cdot g\left(\frac{t_l}{t_s}\right) \tag{9}$$

where R and t_s are the experimental values determined by TEM. In this equation, the applied stress is the fit parameter in order to obtain that all experimental points are on the same analytical curve given by the

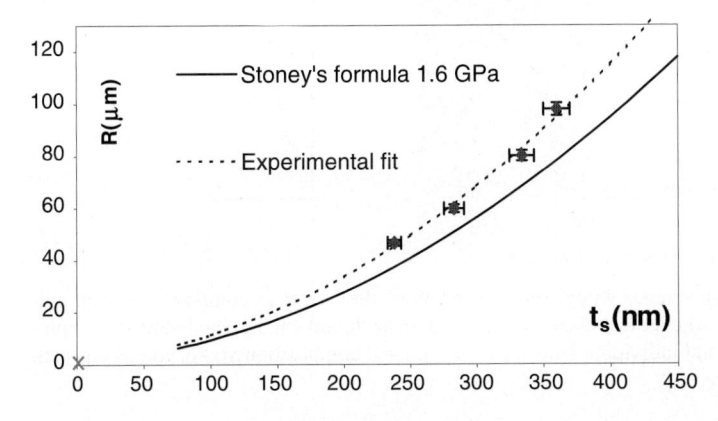

Fig. 4 (online colour at: www.intescience.wiley.com) Variation of the radius of curvature versus TEM specimen thickness, for $t_l = 25$ nm: a) doted line is the experimental fit to determine the epitaxial stress $\sigma_{exp} = 1.3$ GPa; the continuous line is the variation of R calculated for an epitaxial stress of 1.6 GPa.

Table 2 Comparison between theoretical and experimental values of the epitaxial stress.

epilayer thickness	measured stress	theoretical stress: $M_l \cdot f$
10 nm	0.9 GPa	1.6 GPa
25 nm	1.3 GPa	1.6 GPa

Eq. (9). The results are shown in Fig. 4 for the sample characterized by $f = 1.4\%$ and an epilayer thickness of 25 nm. Here the best fit is obtained for an epitaxial stress of 1.3 ± 0.2 GPa. The theoretical value of the misfit stress is 1.6 GPa.

The same experiment was also performed for a 10 nm thick epilayer with the same In concentration [7]. The measured stress was in this case $\sigma_{10\,nm} = 0.90 \pm 0.15$ GPa.

5 Discussion The elastic stress and the related strain created by a lattice mismatch in epitaxial structures have been studied by TEM. Their determination is usually performed by diffraction techniques which gives the inter-planar spacings. This strain needs to be defined by comparison to the reference taken as the cubic structure with a lattice parameter defined by the Vegard law for ternary compounds. The determination of the strained unit cell is most often based on the ideality of the system.

At the difference of the diffraction techniques, the Stoney method is based on the direct observation of the curvature created by the stress induced in the epilayer by a lattice mismatch. In this case, any hypothesis concerning the elastic properties of the epilayer is necessary to obtain significant result.

Our work has shown that at least for plan-view observation, the morphology of the TEM specimen is strongly modified when the thickness ratio t_s/t_l becomes smaller than 20. This curvature modifies completely the morphology of the strain field of the epitaxial system. This modification induced by the curvature of the TEM specimen is at the origin of the inefficiency of the moiré technique to characterize the misfit strain: in these plan-view specimens there are no well defined lattice parameters defined for both the epilayer and the substrate. The specimens studied by cross-section are also probably perturbed by the TEM preparation.

The curvature induced by the relaxation of the misfit stress when the substrate is thinned has been taken as a tool to determine the misfit stress. The specificity of the TEM technique described in this paper is the large sensitivity for determining the stress induced by a lattice mismatch for layer thinner than the critical thickness, i.e. 10 nm for 1% lattice mismatch. As shown in Table 1, the radius of curvature is about 100 μm for a substrate 0.3 μm thick compared to the 100 m for the bulk 300 μm thick substrate.

The experimental determination gives two different values of the misfit stress for the same type of layer (chemical composition and substrate) with different thicknesses (10 and 25 nm) as reported in

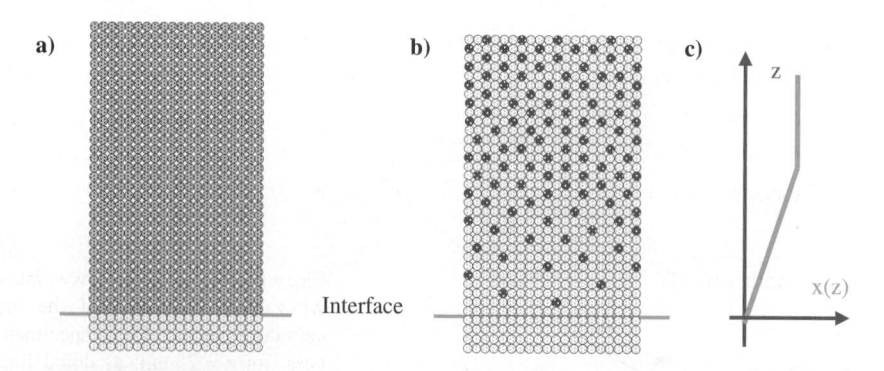

Fig. 5 (online colour at: www.interscience.wiley.com) Schema of the chemical composition of an epitaxial structure near the interface: a) classical elastic representation with uniform strained atoms; b) representation with segregation effect and individual atoms; c) the chemical composition $x(z)$ becomes constant after deposition of some atomic layers.

Table 2. The two experimental values, 0.9 and 1.3 GPa, are smaller than the theoretical one (1.6 GPa) calculated for GaInAs with an In concentration of 20%. The accuracy of this determination depends on the measurement of both the radius of curvature and the local thickness of the specimen. It has been evaluated to be about 15%. In this condition, the difference of the measured misfit stress between the two thicknesses need to be explained by an other physical argument than the incertitude of the technique.

The classical representation of a strained layer is shown Fig. 5a: the epilayer is uniformly strained with an uniform composition of III-type atoms. A well-known effect arises in the epitaxial growth of the ternary compounds such as GaInAs: the segregation of the In atoms when starts the growth of the GaInAs film [10]. It induces a compositional gradient of the III-type atoms as represented in Fig. 5. In this representation the variation of the epilayer composition corresponds to a lack of In atoms close to the interface. The result of this gradient is that the product $\sigma_l \cdot t_l$ could be written as:

$$\sigma_l \cdot t_l = M_l \cdot \delta a/a \cdot x \cdot t_l = M_l \cdot \delta a/a \cdot \int_0^{t_l} x(z) \cdot dz \qquad (10)$$

where $\delta a/a$ is the relative variation of the parameter between InAs and GaAs, and $x(z)$ is the In compositional profile in the epilayer. Under the segregation effect, this value is smaller than the value obtained for uniform In concentration shown Fig. 5a. This effect is more important for a layer thickness of 10 nm than 25 nm where a larger volume reaches its nominal chemical composition.

6 Conclusion This article has presented a new method of determination of stress in epitaxial system using plan-view TEM observations. This method is based on the curvature measurement of a specimen thinned to be transparent to the electron beam. The useful thickness of the substrate for TEM observation (t_s < 0.5 µm) give a radius of curvature of the order a few hundred µm which can be easily determined by TEM technique. In the same time, an accurate determination of the thickness of the specimen has been developed. These two values, introduced in the corrected Stoney Eq. (9), allow us to determine the misfit stress for nanometric layer thickness with an accuracy better than 15% and at the sub-micrometric scale.

Work is progress in some directions: (i) the experimental conditions of this determination need to be validated from the point of view of the elasticity theory; (ii) due to the good accuracy for evaluating misfit strain, it will be interesting to study more precisely the mechanism of elastic stress relaxation with a variation of the epilayer thickness. Incidentally, on the 25 nm layer no dislocations indicating a plastic relaxation are seen. This thickness is about two times larger than the critical thickness of GaInAs for $x = 20\%$. From this result, it is probably possible to study the mechanism of plastic relaxation and determine locally the strain released by misfit dislocations.

Acknowledgements The authors acknowledge Jean Luc Gauffier (INSA, Toulouse) for the nice SEM images and Jacques Crestou for his help for specimen preparation for TEM studies.

References

[1] J. H. van der Merwe, J. Appl. Phys. **34**, 117, (1963).
[2] J. W. Matthews and A. E. Blakeslee, J. Cryst. Growth **27**, 118, (1974).
[3] Transmission Electron Microscopy, D. B. Williams and C. B. Carter, Plenum, Publishing Corp. (New York, 1996).
[4] S. Blanc, A. Arnoult, H. Carrère, E. Bedel, G. Lacoste, C. Fontaine, M. Cabié, A. Ponchet, and A. Rocher, IEEE Proceedings Optoelectronics **150**, 64 (2003).
[5] A. Rocher, A. Ponchet, and C. Fontaine, Appl. Surf. Sci. **188**, 55 (2002).
[6] A. Rocher, M. Cabié, and A. Ponchet, to be published in the Proceedings of the MSM conference Cambridge, March (2003).
[7] A. Ponchet, M. Cabié, and A. Rocher, accepted for publication in EPJAP, (2004).
[8] R. C. Cammarata, J. C. Bilello, A. L. Greer, K. Sierdzki, and S. M. Yalisove, Feb. 1999, MRS Bulletin, p. 34–38.
[9] L. B. Freund, J. A. Floro, and E. Chason, Appl. Phys. Lett. **74** (14), 1987 (1999).
[10] J. M. Moisan, F. Houzay, F. Barthe, J. M. Gérard, B. Jusserand, J. Massies, and F. S. Turco-Sandroff, J. Cryst. Growth **111**, 141 (1991).

Optical properties of GaInNAs/GaAs quantum wells: character of optical transitions and carrier localisation effect

R. Kudrawiec[*,1], **J. Misiewicz**[1], **M. Fisher**[2], and **A. Forchel**[2]

[1] Institute of Physics, Wrocław University of Technology, Wybrzeże Wyspiańskiego 27, 50-370 Wrocław, Poland
[2] Institute of Physics, Würzburg University, Am Hubland, 97074 Würzburg, Germany

Received 19 September 2003, accepted 3 November 2003
Published online 12 January 2004

PACS 78.67.De, 78.55.Cr

The set of $Ga_{0.59}In_{0.41}N_xAs_{1-x}$/GaAs single quantum wells (SQWs) with high nitrogen content (up to 5.2%) has been investigated in photoreflectance (PR) and photoluminescence (PL) spectroscopies at low and room temperatures. Strong manifestation of the carrier localisation effect due to the increase in nitrogen content has been observed. The carrier localization at low temperatures leads to a large Stokes shift and a decrease of the efficiency of band bending modulation. The second phenomena causes a significant decrease of PR signal at low temperatures. In addition, the temperature induced shift of the energy band gap significantly depends on nitrogen content. We have observed that the shift between 10 and 300 K is 82, 73, 67, and 62 meV for the SQW with the nitrogen content of 0, 2.0, 3.2, and 5.2%, respectively.

© 2004 WILEY-VCH Verlag GmbH & Co. KGaA, Weinheim

The GaInNAs/GaAs quantum wells (QWs) are very promising for low-cost and efficient GaAs-based optoelectronic devices operating in the near-infrared [1]. However, the optical quality of GaInNAs deteriorates dramatically with increasing nitrogen concentration. The nature of the deterioration of the optical quality is still debated. Theoretical studies have shown that atomic configurations in (Ga, In)(N, As) compounds can significantly affect its optical properties [2, 3]. The different nitrogen nearest-neighbors (n-n) environments and nitrogen pairs or trimers probably are responsible for the strong carrier localisation and large Stokes shift. In addition, a little N content fluctuation causes a significant variation of band gap due to the changes of conduction band while the change of valence band can be neglected [4]. All these phenomena were investigated mostly for samples with relatively low N content (less than 2%). It is expected that the mentioned phenomena should be strongly manifested for structures with high N content. In this paper, we show PL and PR investigations of optical properties of $Ga_{0.59}In_{0.41}N_xAs_{1-x}$/GaAs SQWs with nitrogen content up to 5.2%.

The $Ga_{0.59}In_{0.41}N_xAs_{1-x}$/GaAs SQW structures have been grown by solid source molecular beam epitaxy (MBE) on unintentionally doped (100) GaAs substrates. Details of MBE machine and the growth process can be found in Ref. [5]. For this investigations we have selected four samples with In content of 41%, QW width of 8.2 nm and N concentration of: 0% (reference sample), 2.0, 3.7 and 5.2%. Standard PR and PL set-ups have been used to measure the optical spectra. Details of PR measurement can be found in Ref. [6].

Figure 1 shows a comparison of PR spectra recorded at room and low temperatures for the four samples in the vicinity of the fundamental QW transition. In addition, low temperature photoluminescence spectra recorded under a low excitation power (10 mW/cm^2) are shown in Fig. 1. The excited state transitions have been analysed in Ref. [7] and are not discussed in this paper. In this work we consider only the fundamental QW transition. It is seen that with the increase in nitrogen content, the ground state

---

[*] Corresponding author: e-mail: robert.kudrawiec@pwr.wroc.pl, Phone: +48 071 320 23 58, Fax: +48 071 328 36 96

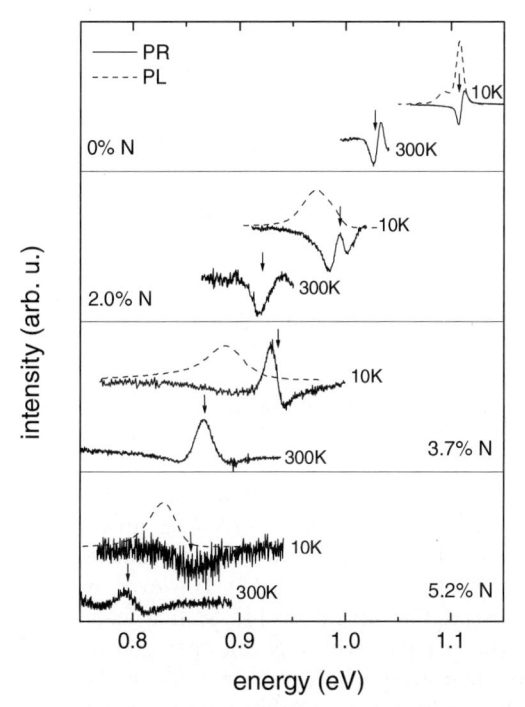

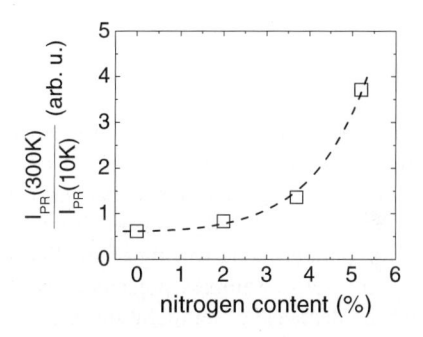

Fig. 1 Room and low temperature PR spectra (solid line). Low temperature PL spectra (dashed line) recorded under the low excitation power (10 mW/cm^2). In order to enhance details of every spectrum, all spectra are shown in different arbitrary units.

Fig. 2 Ratio of room and low temperature PR amplitudes versus the nitrogen content.

transition shifts towards lower energies and PR line becomes more complex and complicated for interpretation.

The red shift depends on nitrogen content linearly in the presented concentration range. The magnitude of the shift at room temperature is 44 meV/%. It is significantly less than in GaAsN layers (170 meV/% [8]) and also less than in GaInNAs/GaAs QWs with low indium content. Such result is consistent with Duboz et al. [9] analysis and it is understood in the framework of the band anticrossing (BAC) model [10] with an assumption that the matrix element V_{NM} in the model decreases with the increase in indium content. In this case, the assumption is justified and has been recently confirmed experimentally [11]. With the decrease of temperature we have observed strong decrease of PR signal intensity. The magnitude of the decrease increases with the increase in the nitrogen content. Figure 2 shows the ratio of room and low temperature PR amplitudes for the four samples. This phenomenon is attributed to a decrease of the modulation mechanism due to carrier localisation effect [12]. At low temperatures, while free carriers are strongly localised, the photomodulation of band bending is less efficient that at temperatures for which the localisation effect is negligible. The dependence in Fig. 2 shows that the increase in nitrogen content in GaInNAs/GaAs SQW leads to an increase in the carrier localisation effect and a decrease of the photomodulation efficiency at low temperatures. The phenomenon is responsible for difficulties in low temperature PR investigations of GaInNAs-based structures.

The other factor which complicates the analysis of PR spectra is associated with the complex character of PR transitions [3, 4, 13, 14]. Our experience with fitting of PR spectra by the Gaussian-like [15] or Lorentzian-like [16] PR line-shape support the thesis that in many cases the line is composed of a few resonances separated by energy which is comparable with the broadening of individual resonance [13]. It is the cause that individual resonances are weakly separated, especially at room temperature. The complex character of PR resonance is associated with a different value of band gap energy at the same compound content. It has been shown that multiple character of band gap is associated with different

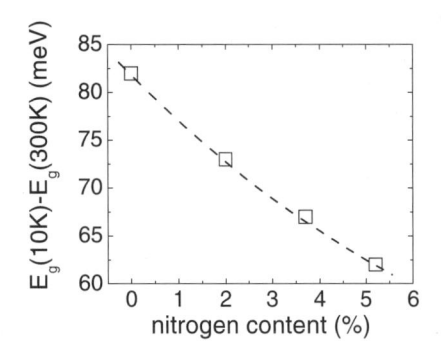

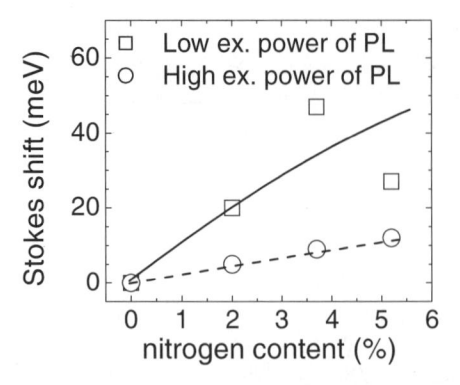

Fig. 3 The temperature shift of the fundamental QW energy versus the nitrogen content.

Fig. 4 Stokes shift obtained for the excitation power densities of 10 mW/cm^2 (squares) and 100 W/cm^2 (circles).

nitrogen n–n environments [3, 13]. The presence of different nitrogen n–n environments depends on the post growth annealing [3, 13] and also can depend on the growth conditions of QW (e.g. growth temperature of GaInNAs layer and GaAs barriers). In the case of QWs with the low nitrogen content, room temperature PR lines are well described by a single Gausian-like resonance [4, 17]. However, with the decrease of temperature a narrowing of PR line was not observed in some cases [4]. Similar features are observed for the set of samples presented in this paper (i.e. samples with high nitrogen content). We observe that the decrease of temperature does not lead to a narrowing of PR resonance like in the nitrogen free SQW (see reference sample in Fig. 1). In addition, the line-shape exhibits a complex character, which is associated with the presence of more than one nitrogen n–n environments. For not annealed samples, such as reported in this paper, we expected that the dominant atom configuration is the surrounding of nitrogen atom by fours Ga atoms. Such atom configuration results from non-equilibrium MBE growth which favours Ga–N bonds instead of In–N bonds [18]. The surface state is frozen during the growth and (Ga–N) + (In–As) configuration is preferred in terms of bond energy after growth. However, such formation leads to crystal structure with "small atom-small atom" plus "large atom-large atom" arrangement which is not preferable in terms of strain. Therefore, we suppose that during growth process also N–In bonds can appear for structures with the high content of indium. PR results presented in this paper (i.e. the complex character of PR resonance) seem to confirm our presumptions. In addition, these resonances are not associated with a free exciton absorption because they are too broad [4]. They should be rather attributed to a band-to-band absorption.

Figure 3 shows the temperature induced shift of the ground state transition versus the increase in nitrogen content taken from the Kramers–Krönig analysis [19] of PR data. We have found that the incorporation of nitrogen atoms into the QW leads to a decrease of the temperature shift of the fundamental QW transition with respect to the reference sample.

The decrease equals about 9, 15, and 20 meV for SQW with nitrogen content of 2.0, 3.2, and 5.2%, respectively. It confirms directly by PR (absorption-type experiment) that the nitrogen induces significantly decrease of the temperature shift of the energy band gap.

Figure 4 shows the Stokes shift obtained under low (squares) and high (circles) excitation power. It is seen that the Stokes shift increases with the increase in N content. However, it is noted that the PL spectrum strongly depends on the power density of PL excitation. In Fig. 1 we show PL spectra obtained for low excitation power (P_{ex} = 10 mW/cm^2). In this case the Stokes shift is the largest. We have observed that with the increase in excitation power the PL line-shape changes and shifts to blue. It leads to smaller Stokes shift (see circles in Fig. 4). In addition, changes in PL line-shape are different for all four SQWs. For example in Fig. 1 it is visible that PL emission from reference sample is composed of two lines at energies of 1.093 and 1.107 eV which are attributed to a bound and free exciton recombinations. With the increase in the excitation power the free exciton recombination becomes dominant in PL and the bound exciton line is not visible as the individual line. For SQW with 2% of nitrogen the PL emission is broad but also possesses a contribution of two lines. However, the high energy line is rather not associated with a free exciton recombination. In this case the recombination between localised electrons and free holes [20] is rather observed. Such recombination is justified in the framework of band gap diagram proposed in Ref. [4]. On the other hand, a recombination of localised excitons trapped by the potential fluctuations of band edge is often suggested [21, 22]. We suppose that the second type of recombination also can take place for our GaInNAs/GaAs SQWs. For SQW with the nitrogen content of 3.7 and 5.2% only one PL line has been observed in the whole range of excitation power. In the case of high excitation power regime the laser beam was focused and the excitation power was increased. Finally, the density of the excitation power was about four orders bigger than for the low excitation regime. It has been observed that the Stokes shift decreases significantly. This phenomenon is caused by a filling of localised state. However, the decrease of Stokes shift can be also caused by a heating of carriers in-plane of the SQW.

In summary, it has been observed that the increase in incorporation of nitrogen atoms in $Ga_{0.59}In_{0.41}N_xAs_{1-x}$/GaAs SQWs leads to red shift of the fundamental QW transition about 44 meV/% and simultaneously to degradation of optical properties. The carrier localisation effect is strongly manifested at low temperatures. This effect leads to the decrease of modulation efficiency and the large Stokes shift. The strong influence of nitrogen content on the temperature shift of fundamental QW energy has been confirmed by PR (absorption-type experiment). It has been found that the temperature shift for the $Ga_{0.59}In_{0.41}N_xAs_{1-x}$/GaAs SQWs decreases from 82 to 62 meV for temperatures between 10 and 300 K with the increase in nitrogen from 0 to 5.2%. PL from SQW with nitrogen at temperature of 10 K has been attributed to bound exciton recombination and/or localised electron and free holes recombination. PR lines at 10 and 300 K have been attributed to the band-to-band absorption.

Acknowledgements This research was supported in part by the Committee for Scientific Research in Poland under Grant No. T11B 00823 and Grant No. 2P03B 10825.

References

[1] M. Kondow, K. Uomi, A. Niwa, T. Kikatani, S. Watahiki, and Y. Yazawa, Jpn. J. Appl. Phys. **35** (2B), 1273 (1996).
[2] K. Kim and A. Zunger, Phys. Rev. Lett. **86**, 2609 (2001).
[3] P. J. Klar, H. Grüning, J. Koch, S. Schäfer, K. Volz, W. Stolz, W. Heimbrodt, A. M. Kamal Saadi, A. Lindsay, and E. P. O'Reilly, Phys. Rev. B **64**, 121203(R) (2001).
[4] R. Kudrawiec, G. Sek, K. Ryczko, J. Misiewicz, P. Sundgren, C. Asplund, and M. Hammar, Solid State Commun. **127**, 613 (2003).
[5] M. Fisher, D. Gollub, M. Reinhardt, M. Kamp, and A. Forchel, J. Cryst. Growth **251**, 353 (2003).
[6] J. Misiewicz, P. Sitarek, G. Sek, and R. Kudrawiec, Materials Science **21**, 263 (2003).
[7] J. Misiewicz, P. Sitarek, K. Ryczko, R. Kudrawiec, M. Fischer, M. Reinhardt, and A. Forchel, Microelectron. J. **34**, 737 (2003).
[8] M. Weyers, M. Sato, and H. Ando, Jpn. J. Appl. Phys. **31**, Part 2, L853 (1992).
[9] J.-Y. Duboz, J. A. Gupta, Z. R. Wasilewski, J. Ramsey, R. L. Williams, G. C. Aers, B. J. Riel, and G. I. Sproule, Phys. Rev. B **66**, 085313 (2002).
[10] W. Shan, W. Walukiewicz, J. W. Ager III, E. E. Haller, J. F. Geisz, D. J. Friedman, J. M. Olson, and S. R. Krutz, Phys. Rev. Lett. **82**, 1221 (1999).
[11] M. Hetterich, A. Grau, A. Yu. Egorov, and H. Riechert, J. Appl. Phys. **94**, 1810 (2003).
[12] R. Kudrawiec, G. Sek, J. Misiewicz, L. H. Li, and J. C. Harmand, Appl. Phys. Lett. **83**, 1379 (2003).
[13] R. Kudrawiec, G. Sek, J. Misiewicz, D. Gollub, and A. Forchel, Appl. Phys. Lett. **83**, 2772 (2003).
[14] M. Geddo, G. Guizzetti, M. Capizzi, A. Polimeni, D. Gollub, and A. Forchel, Appl. Phys. Lett. **83**, 470 (2003).
[15] O. J. Glembocki, Proc. SPIE **1286**, 2 (1990).
[16] D. E. Aspnes, Surface Sci. **37**, 418 (1973).
[17] G. Sęk, K. Ryczko, J. Misiewicz, M. Fischer, and A. Forchel, Thin Solid Films **380**, 240 (2000).
[18] T. Matsuoka, T. Sasaki, and A. Katsui, Optoelectron. Devices Technol. **5**, 53 (1990).
[19] K. Jezierski P. Markiewicz, J. Misiewicz, M. Panek, B. Ściana, R. Korbutowicz, and M. Tłaczała, J. Appl. Phys. **77**, 4139 (1995).
[20] B. Q. Sun, M. Gal, Q. Gao, H. H. Tan, and C. Jagadish, Appl. Phys. Lett. **81**, 4368 (2002).
[21] I. A. Buyanova, W. M. Chen, G. Pozina, J. P. Bergman, B. Monemar, H. P. Xin, and C. W. Tu, Appl. Phys. Lett. **75**, 501 (1999).
[22] I. A. Buyanova, W. M. Chen, B. Monemar, H. P. Xin, and C. W. Tu, Appl. Phys. Lett. **75**, 3781 (1999).

Raman scattering study of InGaAs/AlAsSb and InGaAs/AlAs/AlAsSb heterostructures

T. Mozume[*], **N. Georgiev**[**], and **J. Kasai**

Femtosecond Technology Research Association (FESTA), 5-5 Tokodai, Tsukuba 300-2635, Japan

Received 19 September 2003, accepted 3 November 2003
Published online 29 December 2003

PACS 78.30.Fs, 78.55.Cr, 78.67.De, 78.67.Pt, 61.10.Nz

We report here on a detailed Raman scattering study of InGaAs/AlAsSb superlattices (SLs) and quantum-wells (QWs) grown by molecular beam epitaxy (MBE). The interface termination procedure and AlAs interface control layer are shown to influence the Raman spectra significantry, as follows: (1) the Sb-terminated SL shows a clear InSb-related interface phonon at around 195 cm^{-1}, and (2) when 3-ML of AlAs is introduced, no features were observed below 200 cm^{-1}, indicating well-controlled interfaces.

© 2004 WILEY-VCH Verlag GmbH & Co. KGaA, Weinheim

1 Introduction

InGaAs/AlAsSb quantum wells (QWs) grown on InP substrates are recently attracting a great deal of attention for application to optical and electrical devices. We have reported on the near-infrared intersubband transitions resulting from the use of this material system [1]. We have also reported an ultra-fast absorption recovery of 0.69 ps and shown that this material system is suited for the ultra-high speed optical devices used in optical communication networks [2]. However, growth of this system has proven quite difficult. We have already reported that the As-interface termination procedure improves QW properties [3]. The PL spectra of Sb-terminated QWs are broadened and red-shifted relative to those of the equivalent As-terminated version. In order to improve interface quality further, we have introduced several mono-layers (MLs) of AlAs between the InGaAs well and the AlAsSb barrier [4]. Using the intersubband transitions in the InGaAs/AlAs/AlAsSb multiple quantum wells (MQWs), we have achieved an extremely low absorption saturation intensity of 3 $fJ/\mu m^2$, demonstrating that this system is suited for ultrafast all optical switching devices [5].

The assessment of the InGaAs/AlAsSb QWs is still in the earliest stages. Previously we presented a brief report [6] on the Raman scattering analysis of InGaAs/AlAsSb short-period superlattices (SLs) grown with different interface terminations, and characterized the effects of interface termination on interface quality.

In this article, we present a detailed Raman scattering study of the InGaAs/AlAsSb SLs and QWs to study the effects of interface termination and interface control layer on the structural quality.

2 Experimental

All samples were grown on semi-insulating InP (001) substrates by MBE. Elemental Ga, In, and Al were used for group-III growth species. For group-V growth species, either uncracked tetramer (Sb$_4$ and As$_4$) or cracked dimer sources (Sb$_2$ and As$_2$) were used. The growth temperatures were 490 °C and 460 °C. X-ray diffraction spectra were recorded using a four-crystal high-resolution diffractometer (Philips MRD system). Raman spectra were measured at room temperature and 60 K. The 514.5, 488, and 457.9 nm lines of an Ar$^+$-ion laser (Spectra Physics 165-09) were used. The Raman scattering measurements were performed with backscattering geometry from the (001) surface. The laser spot di-

---

[*] Corresponding author: e-mail: mozume@festa.or.jp, Phone: +81-29-847-5181, Fax: +81-29-847-4417
[**] present address: Institute of Ion Beam Physics and Materials research, Forschungszentrum Rossendorf

Fig. 1 (online colour at: www.interscience.wiley.com) Low-temperature (60 K) unpolarized Raman spectra of 30-period (InGaAs)$_7$(AlAsSb)$_7$ superlattices. The spectra were recorded in the backscattering configuration. Sample #570 was grown with As-termination, while sample #575 was grown with Sb-termination.

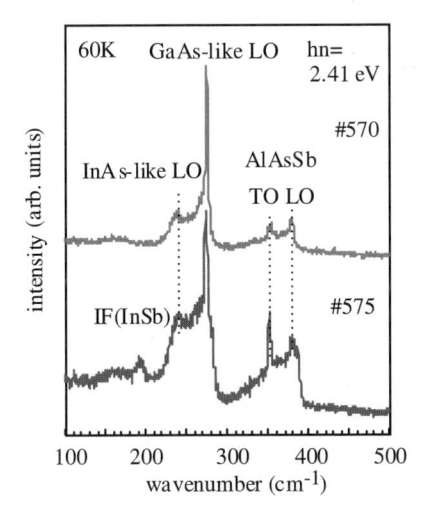

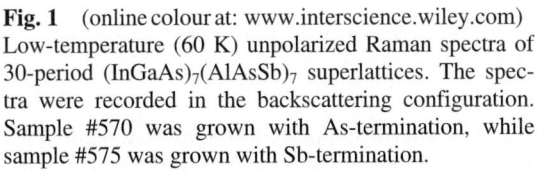

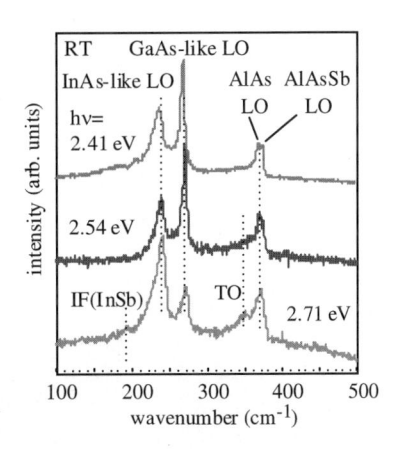

Fig. 2 (online colour at: www.interscience.wiley.com) Room temperature Raman spectra of an InGaAs/AlAsSb MQW structure recorded for different incident photon energies, hν, as indicated. The spectra were recorded with the polarization of the scattered light perpendicular to that of the incident light [$z(x, y)z$].

ameter on the sample is less than 1 µm. The scattered light was collected by the same objective lens, then dispersed into the triple monochromator (Jobin-Yvon T64000) and detected by a liquid-nitrogen-cooled charge coupled devices (CCD).

3 Results and discussions Shown in Fig. 1 are the low-temperature (60 K) unpolarized Raman scattering spectra for the short period superlattices (SLs) with 30 periods of (InGaAs)$_7$(AlAsSb)$_7$, where n is 7 MLs. The SLs were grown using uncracked tetramer group-V species. Optical excitation was performed using the 2.41 eV line of an Ar$^+$-ion laser. The substrate temperature was 490 °C. The heterointerfaces were terminated either by arsenic (sample 570) or antimony (sample 575) for 30 seconds. The Raman spectra consist of two main bands. The lower band, between 200 to 300 cm^{-1}, corresponds to the InGaAs optical vibrations, while the higher band, between 330 to 400 cm^{-1}, corresponds to the AlAsSb optical phonons. Besides these two main bands, the Sb-terminated sample (#575) shows a clear peak at around 195 cm^{-1}, which is attributed to scattering by the longitudinal InSb interface mode [IF(lnSb)] [7]. The lower InGaAs band shows two-mode behavior, where the Raman spectrum is defined by the characteristic features of constituent binaries, e.g. GaAs and InAs. This lower band is characterized by a sharp peak at 275 cm^{-1} attributed to a GaAs-like LO phonon and a smaller InAs-like LO mode at 236 cm^{-1}. Using line shape analysis, Estera et al. [8] identified additional peaks as the GaAs-like TO (255 cm^{-1}) mode and a crystalline disorder related optical phonon (244 cm^{-1}), both at room temperature. The convolution of these peaks gives the observed sharp profile at the high-energy side of the GaAs-like LO peak and a relatively broad valley-like spectrum at the lower energy side.

Contrary to the lower InGaAs band, which shows two-mode behavior, the higher AlAsSb band shows single-mode behavior, where the peak positions in the Raman spectra are defined by the appropriate alloy compositions. The AlAs$_x$Sb$_{1-x}$ ternary alloy is reported to exhibit a single-mode behavior [9, 10]. The presence of the TO mode could be attributed to the disorder-activated scattering from imperfections. The intensity of the TO mode becomes comparable to that of LO mode for x between 0.4 and 0.7 due to severely reduced symmetry [10].

These results are consistent with the X-ray diffraction measurements reported earlier [6], which showed that extended Sb-As intermixing at the interface takes place when the access of Sb atoms to the InGaAs interface is increased. The X-ray rocking curve simulation also suggested that high As composi-

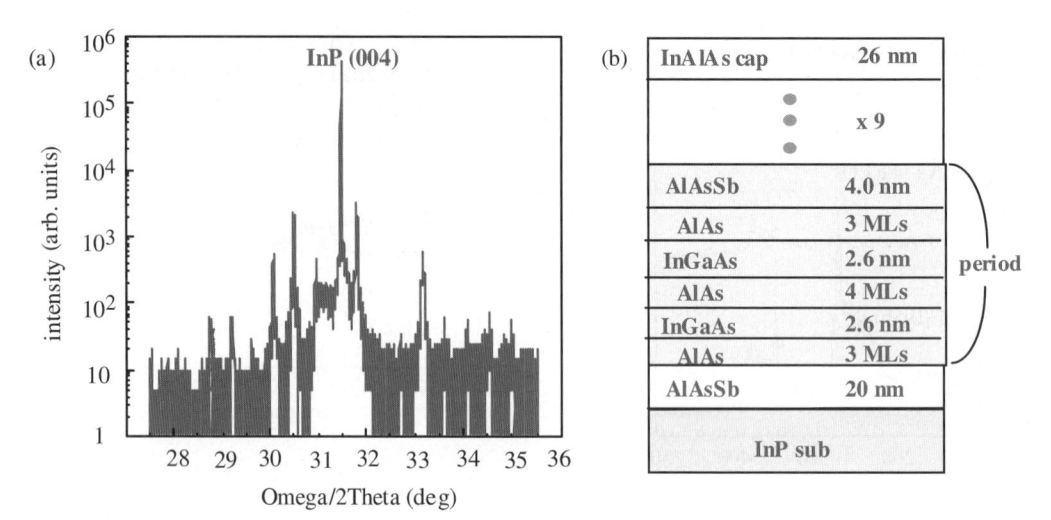

Fig. 3 (online colour at: www.interscience.wiley.com) a) shows an X-ray diffraction rocking curve for an InGaAs/AlAs/AlAsSb C-DQWs, and b) is the sample structure et al.

tion AlAs$_y$Sb$_{1-y}$ layers are formed on the As-rich interface. The interface quality of InGaAs/AlAsSb QWs can be improved when the cracked dimer group-V species were used instead of tetramer group-V species [11]. The photoluminescence line shape and peak energy as well as the photoreflectance critical point energies confirm the enhanced quality of QWs [11, 12]. Figure 2 shows room temperature Raman spectra of undoped 10 periods InGaAs/AlAsSb multiple quantum wells (MQWs) with 10 nm AlAsSb barriers and 10 nm InGaAs wells grown using dimer group-V species. The substrate temperature was decreased to 460 °C to reduce the As to Sb exchange reaction. The sample was grown using As interface termination of 10 s. This sample shows good quality with PL full width at half maximum of 10 meV at low temperature [12]. In the lower energy band, clear GaAs-like LO phonon and InAs-like LO phonon peaks are observed. Increasing the incident photon energy from 2.41 eV to 2.71 eV, which matches the InAs E_1 band gap resonance [13], the InAs one-LO phonon line is drastically enhanced in its intensity, as compared to other excitations. The InAs-LO peak width is greatly reduced from that of the sample shown in Fig. 1. The spectrum excited by 2.71 eV shows a small InSb-like interface LO peak. In the higher energy band, the AlAsSb TO phonon peak intensity is reduced, indicating improved crystalline quality. The AlAsSb LO peak splits into two peaks. As the InAlAs shows two-mode behavior, where InAs-like LO phonon and AlAs-like LO phonon peaks are observed at around 237 and 367 cm^{-1}, respectively. Therefore, the lower and higher energy peaks (368.3 cm^{-1} and 373.8 cm^{-1}) can be attributed to AlAs-like LO phonon of InAlAs and AlAsSb LO phonon, rspectively.

For application to all-optical switches using intersubband transitions, QWs must be doped heavily. The heavy doping to InGaAs/AlAsSb MQWs leads to extraordinary degradation of structures that is caused not only by the interchange reaction between As and Sb but also by the interdiffusion of group-III species [14]. We have introduced several monolayers of AlAs between the InGaAs well and the AlAsSb barrier to prevent interdiffusion [4]. All interfaces were terminated by arsenic. In order to accommodate misfit strain caused by the AlAs layers, the In composition of InGaAs is raised to 0.80 from lattice match composition to InP substrate. Figure 3a shows an X-ray diffraction rocking curve for the undoped InGaAs/AlAs/AlAsSb coupled double quantum well structure shown in Fig. 3b. The sample was grown at 460 °C using cracked dimer group-V species. The extended and sharp satellite peaks confirms the high quality. As shown in Fig. 3b, an AlAs diffusion-barrier layer thickness of 3 MLs were chosen for both sides of the heterointerfaces. The Raman scattering spectra for this structure are shown in Fig. 4. No InSb-like interface LO peak is observed even for 2.71 eV excitation. The Raman data reported by Wagner [15] shows that an InSb bond is formed at the heterointerfaces of MBE-grown InAs/AlAs/AlSb heterostructures with up to 3 MLs AlAs deposited at the interface. They also reported the formation of a

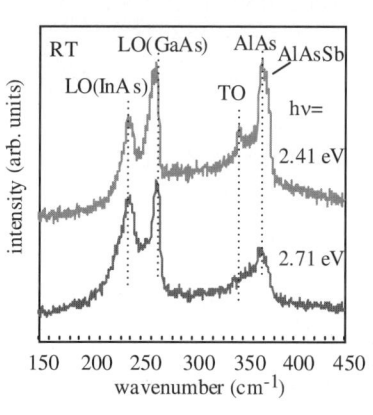

Fig. 4 (online colour at: www.interscience.wiley.com) Room temperature Raman spectra of an InGaAs/AlAs/AlAsSb MQWs sample recorded for different incident photon energies, hν, as indicated. The spectra were recorded with the polarization of the scattered light perpendicular to that of the incident light [$z(x, y) z$].

pseudoternary AlAsSb barrier region. Splitting of the AlAsSb LO peak (Fig. 4) can be attributed to the AlAs-like LO phonon peak of InAlAs cap as shown above. Present results indicate that the formation of InSb interface bonding will be affected by the MBE growth conditions, e.g. substrate temperature, beam flux, etc.

4 Summary In summary, we have shown by Raman measurements that the chemical bonds formed at the heterointerfaces of MBE-grown InGaAs/AlAsSb heterostructures are affected by the MBE growth conditions and the AlAs interface control layer. Present results demonstrate that well-controlled heterointerfaces are formed using dimer group-V sources with As-interface termination and by inserting 3 MLs of AlAs between the InGaAs and the AlAsSb.

Acknowledgements This work was supported by the New Energy and Industrial Technology Development Organization (NEDO), within the framework of Femtosecond Technology Project.

References

[1] T. Mozume, H. Yoshida, A. Neogi, and M. Kudo, Jpn. J. Appl. Phys. **38**, 1286 (1999).
[2] T. Akiyama, N. Georgiev, T. Mozume, H. Yoshida, A. V. Gopal, and O. Wada, Electron. Lett. **37**, 129 (2001).
[3] T. Mozume and N. Georgiev, Thin Solid Films **380**, 249 (2000).
[4] T. Mozume, J. Kasai, N. Georgiev, T. Simoyama, A. V. Gopal, and H. Yoshida, Proceedings of the Photonic West (2003).
[5] A. V. Gopal, H. Yoshida, T. Simoyama, N. Georgiev, T. Mozume, and H. Ishikawa, IEEE J. Quantum Electron. **39**, 299–305 (2003).
[6] N. Georgiev and T. Mozume, Appl. Surf. Sci. **159–160**, 520 (2000).
[7] I. Sela, C. R. Bolognesi, L. A. Samoska, and H. Kroemer, Appl. Phys. Lett. **60**, 3283 (1992).
[8] J. P. Estera, P. D. Stevens, R. Glosser, W. M. Duncan, Y. C. Kao, H. Y. Liu, and E. A. Beam III, Appl. Phys. Lett. **61**, 1927 (1992).
[9] I. Sela, C. R. Bolognesi, and H. Kroemer, Phys. Rev. B **46**, 16142 (1992).
[10] H. C. Lin, J. Ou, C. H. Hsu, W. K. Chen, and M. C. Lee, Solid State Commun. **107**, 547 (1998).
[11] T. Mozume and N. Georgiev, Physica E **13**, 361 (2002).
[12] T. Mozume, Proceedings of the 26th International Conference on the Physics of Semiconductors (ICPS26), Edinburgh, July 29 to Aug. 2, (2002).
[13] R. Carlels, N. Saint-Cricq, J. B. Renuccu, A. Zwick, and M. A. Renucci, Phys. Rev. B **22**, 6129 (1980).
[14] T. Mozume, N. Georgiev, and H. Yoshida, J. Cryst. Growth **227/228**, 577–581 (2001).
[15] J. Wagner, J. Shmitz, D. Behr, J. D. Ralston, and P. Koidl, Appl. Phys. Lett. **65**, 1293 (1994).

The Vast World of Nanomaterials. From Wiley-VCH

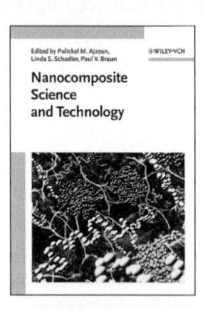

Nanocomposite Science and Technology

P. M. AJAYAN, and LINDA S. SCHADLER, both Rensselaer Polytechnic Inst., Troy, USA, and PAUL V. BRAUN, Univ. of Illinois, USA

FROM THE CONTENTS:
Introduction: Why Nanocomposites?; Bio-Nanocomposites: Nanocomposites from and inspired by Nature; Polymer based and Polymer filled Nanocomposites; Bulk Metal and Ceramics Nanocomposites; Active Structures; Nodeling of Nanoscale Composited Systems.

3527-30359-6   2003   238pp with 194 figs   Hbk   € 149.- / £ 105.- / US$ 135.-

Nanotechnology

MICHAEL KÖHLER, Ilmenau Technical Univ., Inst. of Physics, Ilmenau, Germany, and WOLFGANG FRITZSCHE, IPHT Jena, Germany

Written for both scientists and engineers, this book offers a fascinating insight into the "other side" of nanotechnology. It begins with the fundamentals of microtechnology and chemistry necessary for understanding the shaping of nanoscale structures, before going on to illustrate the fabrication of nanostructures from different materials through a variety of examples and finally presenting methods for characterization of the generated structures.

3527-30750-8   2004   approx 350pp with approx 150 figs and approx 30 tabs   Hbk
approx € 139.- / £ 90.- / US$ 149.-

Nanomaterials by Severe Plastic Deformation

Fundamentals – Processing – Applications

Edited by MICHAEL ZEHETBAUER, Inst. f. Materials Physics, Vienna Univ., Austria

These proceedings of the "Second International Conference on Nanomaterials by Severe Plastic Deformation" review the enormous scientific avalanche that has been developing in the field over recent years. A valuable reference for any scientist and engineer working in this emerging field.

3527-30659-5   2003   approx 450pp with approx 300 figs and approx 20 tabs   Hbk   approx € 199.- / £ 115.- / US$ 215.-

Multilayer Thin Films

Sequential Assembly of Nanocomposite Materials

Edited by GERO DECHER, Inst. Charles Sadron, Strasbourg, France, and JOSEPH B. SCHLENOFF, Florida State Univ., Tallahassee, USA

This detailed survey of the new technology of layer-by-layer adsorption from solution of molecularly ordered polymer or poly-electrolyte films is essential reading for specialists concerned with the assembly of several functional components in specific structures and composites.

3527-30440-1   2002   543pp with 298 figs, 5 in color, and 19 tabs   Hbk
€ 179.- / £ 105.- / US$ 145.-

Online Book

Molecular Devices and Machines

A Journey into the Nanoworld

VINCENZO BALZANI; MARGHERITA VENTURI and ALBERTO CREDI, all of Univ. of Bologna, Bologna, Italy

3527-30506-8   2003   511pp with 294 figs, 4 in color   Hbk
€ 99.- / £ 60.- / US$ 75.-

Colloids and Colloid Assemblies

Edited by FRANK CARUSO, Max Planck Inst. of Colloids and Interfaces, Potsdam, Germany

This book deals with the recent developments in the synthesis, modification, utilization and application of colloids, ranging from metal nanoparticles through to inorganic particles and polymer latexes. It outlines strategies for their modification to impart new properties and illustrates ordered assemblies derived from colloid particles as well as various applications for colloids.

3527-30660-9   2003   approx 500pp with approx 150 figs   Hbk
approx € 199.- / £ 115.- / US$ 199.95

Wiley-VCH, Customer Service Department, P.O. Box 101161, D-69451 Weinheim, Germany
Tel.: (49) 6201 606-400, Fax: (49) 6201 606-184, E-Mail: service@wiley-vch.de, www.wiley-vch.de

John Wiley & Sons, Ltd., Customer Services Department, 1 Oldlands Way, Bognor Regis, West Sussex, PO22 9SA England
Tel.:+44 (0) 1243-843-294, Fax:+44 (0) 1243-843-296, www.wileyeurope.com

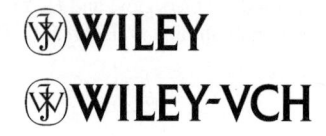

WILEY

WILEY-VCH

physica status solidi c

www.pss-c.com

conferences and critical reviews

WILEY-VCH

Optical detection of 2DEG in GaN/AlGaN structures – High magnetic field studies

B. Chwalisz[*,1], **A. Wysmołek**[1], **R. Stępniewski**[1], **M. Potemski**[2], **W. Knap**[3], **J. M. Baranowski**[1], **N. Grandjean**[4], **J. Massies**[4], **P. Prystawko**[5], and **I. Grzegory**[5]

[1] Institute of Experimental Physics, Warsaw University, ul. Hoża 69, 00-681 Warsaw, Poland
[2] High Magnetic Field Laboratory, CNRS-MPI/FKF, 25, Avenue des Martyrs, 38-042 Grenoble, France
[3] Groupe d'Etude des Semiconducteurs, CNRS-UMR 5650, Université Montpellier 2, cc074, 12 Place Eugene Bataillon, 24095 Montpellier, France
[4] CNRS-CRHEA, rue B. Gregory, Sophia Antiopolis, 06-560 Valbonne, France
[5] High Pressure Research Center, PAS, ul. Sokołowska 29/37, 01-142 Warsaw, Poland

Received 15 September 2003, revised 24 October 2003, accepted 24 October 2003
Published online 22 December 2003

PACS 73.21.–b, 78.20 Ls, 78.55.Cr, 78.67.De

Magneto-photoluminescence (PL) studies up to 28T of the GaN/AlGaN structure are presented. PL spectra in the energy range below the excitonic bulk transitions show periodic energy and line width oscillations what is interpreted as a fingerprint of two dimensional electron gas confined at GaN/AlGaN interface. Observed oscillations are a direct consequence of Landau quantization and known as optical Shubnikov de Haas Effect if related to the Fermi level oscillations. This observation allows us to determine the electron concentration to $3 \cdot 10^{12} \text{cm}^{-2}$. Appearance of the oscillations is discussed in terms of magnetic field induced potential redistribution in the structure and theory including many-body effects. Surprisingly, it was observed that magneto-PL intensity oscillates also in other energy regions of the spectra. It was found that this emission originates from the second electric energy level and the oscillations are a direct consequence of the interaction with the first electronic subband Landau levels.

© 2004 WILEY-VCH Verlag GmbH & Co. KGaA, Weinheim

1 Introduction

The possibility of fabrication of high-electron mobility transistor (HEMT) based on the GaN/AlGaN structures has stimulated electrical and transport measurements of this system, but there are only few reports showing the emission from two-dimensional electron gas (2DEG) on the GaN/AlGaN interface [1–3] especially in high magnetic field regime [4]. Optical experiments provide the information about states involved in recombination processes therefore are indispensable for understanding band structure at heterointerfaces. Application of magnetic field perpendicular to two-dimensional (2D) plane imposes the quantization of energy bands into Landau levels (LL) and changes in the density of states (DOS). The direct consequence of the carrier redistribution over LL is the observation of magnetic-field modulated oscillatory phenomena. They have been seen mostly in transport measurements, i.e. Shubnikov de Haas (SdH) effect, where they are a consequence of DOS singularities of 2DEG. In photoluminescence (PL) experiments one probes radiative recombination of photo created holes with 2D electrons. With monotonic increase of the magnetic field one can expect oscillatory behaviour of the emission lines intensity, energy position and width [5–9].

In this paper we report on the emission due to recombination of 2DEG with photo-excited holes at GaN/AlGaN interface in the high magnetic field, in the Integer Quantum Hall Effect regime. The

---

[*] Corresponding author: e-mail: Barbara.Chwalisz@fuw.edu.pl, Phone: +48 22 5532166, Fax: +48 22 6219712

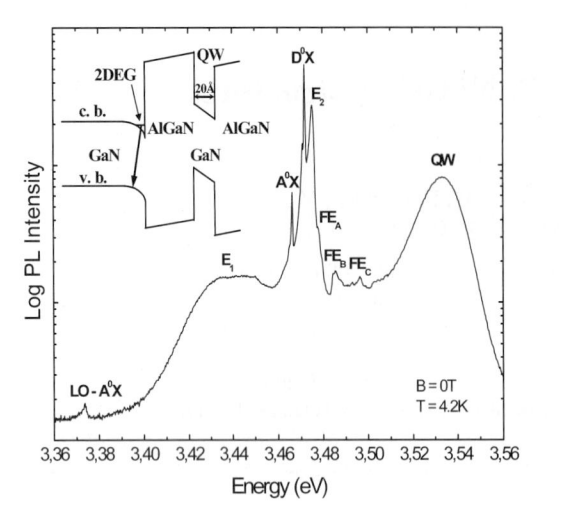

Fig. 1 PL spectrum of the sample. Inset: scheme of the sample structure.

emission energy as well as the line width shows well-resolved oscillations in magnetic field. This behaviour is discussed in terms of many-body interactions and/or electrostatic effects.

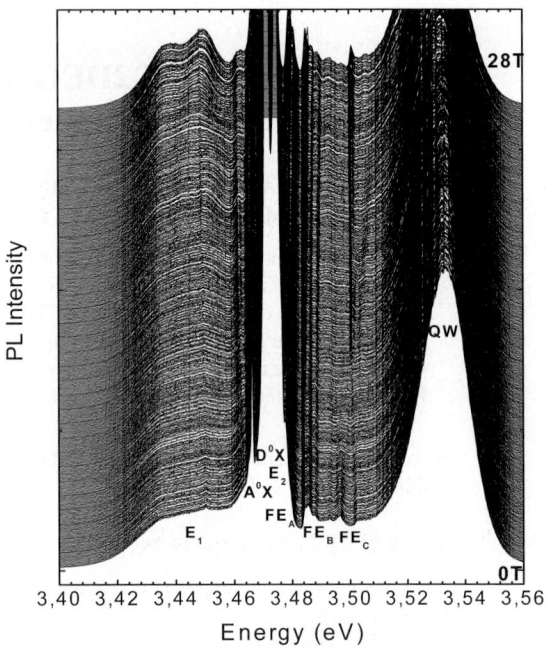

Fig. 2 Magnetic field evolution of the PL spectra.

2 Experimental details

The investigated structure was grown by molecular beam epitaxy (MBE) using n-type bulk GaN as a substrate. The complete structure consisted of 2nm GaN quantum well (QW) embedded in the $Al_{0.08}Ga_{0.92}N$ barriers grown on top of a thick ($5 \cdot 10^{17} cm^{-3}$ n-type doped) GaN buffer layer (see inset in Fig. 1). In this paper the GaN buffer layer – $Al_{0.08}Ga_{0.92}N$ barrier interface is under special investigation. The luminescence signal was excited well above the energy gap of $Al_{0.08}Ga_{0.92}N$ barriers using 325nm line from He–Cd laser. For detection standard CCD camera system equipped with single grating monochromator was used. Exciting and emitted light were transmitted using optical fibres. The experiment was performed at liquid helium temperature in magnetic field up to 28T in the Faraday configuration.

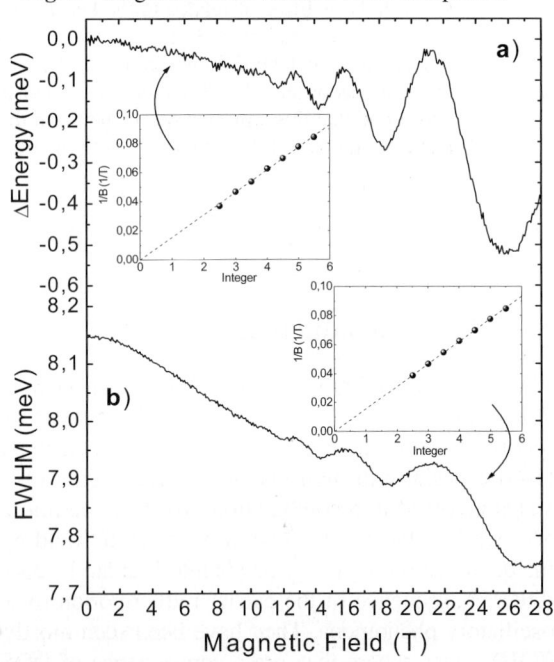

Fig. 3 Oscillations of a) mean energy and b) FWHM of E_1 emission. Insets: position of the extremes in 1/B vs. integer number.

3 Results and discussion

Zero-field PL spectrum of the investigated structure is shown in Fig. 1. The spectrum is dominated by sharp luminescence lines due to recombination of free excitons A, B and C (lines FE_A, FE_B and FE_C, respectively), donor bound exciton (line D^0X) and acceptor bound exciton (line A^0X) originating from the GaN buffer layer. The emission from the QW is observed at 3.53 eV. In the energy range 3.40–3.45 eV pronounced but relatively broad emission structure (E_1 line) is observed. It shows temperature and excitation power dependences expected for the recombination of 2DEG with photoexcited free holes at GaN/AlGaN interface [1–3]. Such recombination is known as *H*-band recombination (see inset of Fig. 1).

The evolution of the PL spectra in the magnetic field is illustrated in Fig. 2. At first sight no significant changes in the E1 emission are observed. However, interesting information can be obtained when calculating higher moments of the spectral line. First moment represents variation of the mean energy position (Fig. 3a), and the second is related to the line width at half maximum (FWHM) (Fig. 3b). It appeared that they oscillate as a function of magnetic field B. The oscillations are relatively regular and are periodic versus 1/B (see insets of Figs. 3a and b, respectively). Assuming that E1 luminescence originates mostly from the lowest LL, the observed oscillations can be explained as due to magnetically modulated changes of the DOS at Fermi energy (EF). In analogy to SdH effect, from the characteristic frequency of the optical SdH oscillations, 2DEG concentration of ne = 3.1 · 1012cm–2 can be determined. This value is in a good agreement with the concentration expected for GaN/ Al0.08Ga0.92N interface [10]. Taking this into account the position of the Fermi energy (EF = neh2/4πm*) in the structure of EF ≈ 31meV was determined. The oscillation extremes are observed at magnetic field values corresponding to integer filling factors. In both cases, i.e. for the mean energy and FWHM of the E1 emission, maxima occur in the vicinity of even filling factors, whereas minima are associated with odd filling factors. This is schematically marked at Fig. 4, in which electron cyclotron energy $\hbar\omega_c$ = 0.478meV/T was used [11]. Oscillatory behaviour of the observed emission can be explained in two different ways [5]. First, one can take into account many-body effects (exchange and correlation) in 2DEG [8, 12]. Such model explains successfully the appearance of maxima of the mean transition energy and the FWHM of the 2DEG emission line in the vicinity of even filling factors and minima in odd filling factors. Assuming that in the case of GaN the electron spin splitting (~ 0.1 meV/T) can be neglected as compared to cyclotron splitting, odd filling factors correspond to the situation when the Fermi level is located close to the maximum of DOS in the extended states region, whereas for even filling factors, the Fermi level is located in between the LLs at localized states. In the localized states region screening efficiency is limited. As a consequence, the scattering effects are increased and broadening of the emission line is observed, what was found in the experiment.

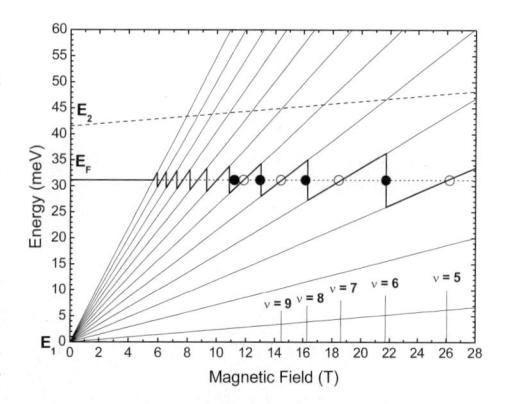

The oscillations of the mean transition energy can be also understood by considering electrostatic effects. In the investigated structure the Fermi level is located at the donor level in the bulk GaN layer. Therefore, assuming that the Fermi energy is fixed at the donor level at any magnetic field, the conduction and the valence band profiles change each time the depopulation of the subsequent LL takes place. This results in the change of the transition energy between 2DEG and photo-created holes. Such behaviour is oscillatory versus 1/B with the same frequency as changes of the DOS at the Fermi energy, what is observed in the experiment. Surprisingly, we do not observe any oscillations of the integrated intensity of the E1 emission, since the model also predicts changes in the transition probability. This suggests that the emission efficiency is mostly determined by kinetic properties of the photo-created holes.

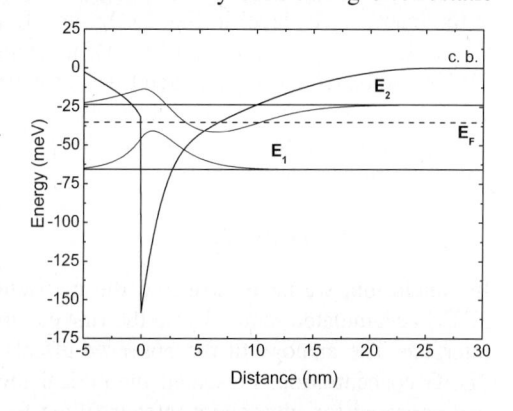

Fig. 4. Scheme of 2DEG LL fan chart. The evolution of E_F in magnetic field (in the limit of no impurity scattering) and second electronic level E_2 (dashed line) are marked. Solid circles: maxima, open circles: minima of observed E_1 line spectral moments oscillations.

Fig. 5 Calculated structure of the conduction band and electron wave functions at the interface. E_F, first and second electronic levels are marked.

© 2004 WILEY-VCH Verlag GmbH & Co. KGaA, Weinheim

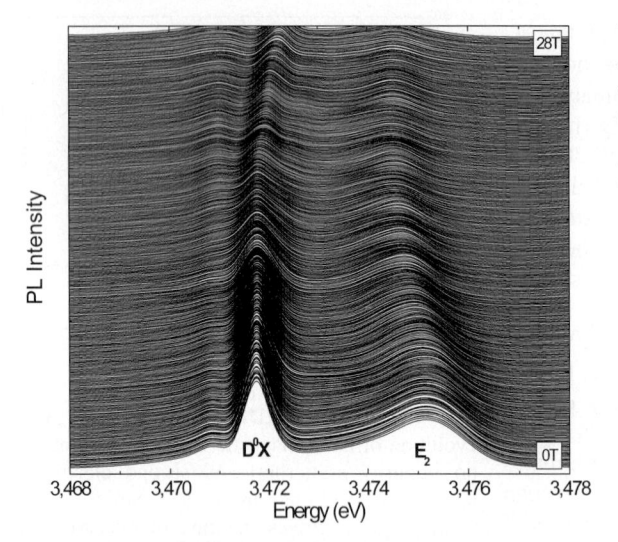

Fig. 6 Magnetic field evolution of the PL spectra in the excitonic bulk GaN region.

Figure 5 illustrates the result of a self-consistent calculation of the band energy potential and energy levels at the investigated GaN/AlGaN heterointerface. In the calculations the difference in GaN and $Al_{0.08}Ga_{0.92}N$ band gap energies was taken from the experiment as the energy difference between barrier and bulk emission lines. Electron mass $m^* = 0.242\ m_0$ [11] was used. Other structural parameters (as band offset value, spontaneous and polarization fields, charge density at the interface) were taken from elsewhere [10].

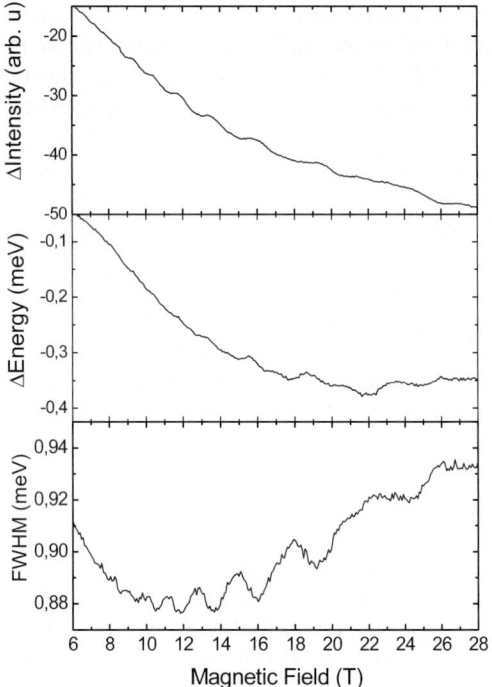

Fig. 7 Oscillations of the integrated intensity, energy and FWHM of E_2 emission line.

In the PL spectra additional interesting line appears. It is marked as E_2 in Fig. 1, approx. 40 meV above E_1 line. Field dependent evolution of the E_2 PL spectra is illustrated in Fig. 6. It can be seen that the intensity of E_2 emission slightly oscillates when increasing magnetic field. Spectral moments calculated for this line are presented in Fig. 7. Again, the oscillations are periodic in 1/B, but their frequency is higher than observed for E_1 line. Per analogy to the previous oscillations one can calculate the characteristic energy responsible for the observed period: $E_2 = 41.5$ meV (dashed line in Fig. 4), what well agrees with the expected energy distance between the first and the second electronic level in the potential notch at the interface (E_2 level in Fig. 5). In the PL spectra this corresponds to the energy difference between E_1 and E_2 lines (compare with Fig. 1). It seems that the oscillations in the intensity, the energy and the FWHM of the E_2 line appear each time when the successive LL from the first occupied subband cross the second electronic level. However, in such a complicated structure where the presence of the QW can not be neglected it is necessary to consider possible tunnelling effects and interactions between different layers existing in the structure.

4 Conclusions

In conclusion, we have measured the magnetic field dependence of emission due to recombination of 2DEG accumulated at first LL in the first electronic subband with photo-created holes at the GaN/AlGaN interface. The analogy of the observed effects to SdH oscillations allowed us for the estimation of the 2DEG concentration. Presented theoretical models permit for the explanation of oscillations extremes that appeared for subsequent integer filling factors. Moreover, we have observed the emission from the second, not occupied, electronic subband present at GaN/AlGaN interface. The PL intensity, energy position and line width of this emission were influenced by the LL in the first electronic subband.

© 2004 WILEY-VCH Verlag GmbH & Co. KGaA, Weinheim

Acknowledgements This work was partially supported by European Committee Project GMA1-2002-72503 "CEMOS" and NATO PST CL grant N° 978522.

References

[1] J. P. Bergman, T. Lundström, B. Monemar, H. Amano, and I. Akasaki, Appl. Phys. Lett. **69** (23), 3456 (1996).
[2] A. Raymond, S. Juillaguet, B. Couzinet, I. EL. Mezouar, A. Kamal-Saadi, M. Asif Khan, Q. Chen, and J. W. Yang, Mat. & Science Eng. B **43**, 211 (1997).
[3] G. Martinez-Criado, A. Cros, A. Cantarero, C. R. Miskys, O. Ambacher, R. Dimitrov, and M. Stutzmanjn, Mat. & Science Eng. B **82**, 200 (2001).
[4] P. A. Shields, R. J. Nicholas, K. Takashina, N. Grandjean, and J. Massies, Phys. Rev. B **65**, 195320 (2002).
[5] M. C. Smith, A. Petrou, C. H. Perry, J. M. Worlock, and R. L. Aggarwal, Proc. 17th Int. Conf. Phys. Semicond., San Fransisco (1984), edited by D. J. Chadi, W. A. Harrison (Springer-Verlag, New York, 1985), p. 547.
[6] I. V. Kukushkin, K. v. Klitzing, and K. Ploog, Phys. Rev. B **37** (14), 8509 (1988) and I. V. Kukushkin, V. B. Timofeev, Adv. Phys. **45** (3), 147 (1996).
[7] R. Stępniewski, W. Knap, A. Raymond, G. Martinez, T. Rötger, J. C. Maan, and J. P. André, in High Magnetic Fields Semiconductor Physics II, edited by G. Landwehr (Springer-Verlag, Berlin, 1989), p. 62.
[8] K. Meimberg, M. Potemski, P. Hawrylak, Y. H. Zhang, and K. Ploog, Phys. Rev. B **55** (12), 7685 (1997).
[9] W. Chen, M. Fritze, W. Walecki, A. V. Nurmikko, D. Ackley, C. Colvard, and H. Lee, Phys. Rev. B **45** (15), 8464 (1992).
[10] O. Ambacher et al., J. Phys.: Cond. Matter **14** (13), 3399 (2002).
[11] W. Knap, E. Frayssinet, M. L. Sadowski, C. Skierbyszewski, D. Maude, V. Falko, M. Asif Khan, and M. S. Shur, Appl. Phys. Lett. **75** (20), 3156 (1999).
[12] S. Katayama and T. Ando, Solid State Commun. **70** (1), 97 (1989).

Magnetotransport studies of Ga(Mn,Fe)N bulk crystals

C. Jastrzebski[*,1], **W. Gebicki**[1], **M. Zdrojek**[1], **M. Bockowski**[2], **B. Strojek**[3], **T. Szyszko**[3], **M. Kaminski**[3], and **S. Podsiadlo**[3]

[1] Faculty of Physics, Warsaw University of Technology, Koszykowa 75, 00-662 Warsaw, Poland
[2] High Pressure Research Center, Polish Academy of Science, Sokolowska 29/37, 01-142 Warsaw, Poland
[3] Faculty of Chemistry, Warsaw University of Technology, Noakowskiego 3, 00-664 Warsaw, Poland

Received 15 September 2003, revised 30 September 2003, accepted 30 September 2003
Published online 22 December 2003

PACS 72.80.Ey, 75.50.Pp, 81.05.Ea

Magnetoresistance and Hall effect in (Ga,Mn)N, (Ga,Mn,Si)N, and (Ga,Fe)N bulk crystals has been studied in the temperature range from 4.2 K to 300 K. No anomalous Hall effect contribution to the Hall voltage vs. magnetic curve has been found. The negative magnetoresistance has been found only in samples with metallic type of conductivity and is interpreted as resulting from weak localization of current carriers.

© 2004 WILEY-VCH Verlag GmbH & Co. KGaA, Weinheim

1 Introduction

The ferromagnetism in III-V semiconductors heavily doped with transition elements like Mn, Fe, Ni, Co, Cr has been studied since the preliminary reports on ferromagnetism in GaAs:Mn appeared in 1999. Dietl *at al.* [1] predicted on the basis of Zener model of ferromagnetism that some wide band semiconductors like GaN and ZnO heavily doped with manganese are ferromagnets with Curie temperature T_c above room temperature. This paper drove attention of many research teams and stimulated them to verify experimentally the results of Dietl's calculations.

In the paper [1] sine qua non conditions of growth of ferromagnetic material with $T_c > 300$ K have been formulated. The necessary concentration of manganese in GaN has been estimated about 5% with the concentration of holes above 3×10^{20} cm^{-3}. Both conditions are difficult to fulfill experimentally. It has been shown that high concentration of manganese could be achieved either by nitridization of pure metallic Ga in supercritical ammonia (bulk crystals) or low temperature MBE growth (thin films). Little is known about the solubility of the other transition metal ions in GaN. Probably the solubility of Fe as well as the other transition elements is much lower than the solubility of Mn and it doesn't exceed 0.5% [2, 3]. That high concentration of Mn ions in the crystals rises questions concerning formation of paramagnetic ions clusters and ferromagnetic inclusions. Dietl model assumed that ferromagnetism in doped III–V semiconductors comes from isolated paramagnetic ions with the exchange interactions transmitted through polarized spins of free current carriers. For the concentration of manganese ions exceeding 5% it is a challenging task to avoid clustering of the manganese ions in GaMnN crystals. In consequence inclusions as a factor responsible for ferromagnetic contribution to the magnetization have been intensively studied but the results are contradictory [3, 4].

The concentration of holes in GaN:Mn system needed to achieve ferromagnetic material with Tc exceeding room temperature is difficult to realize. Bulk crystals are typically n-type with high concentration of electrons (10^{19}/cm^{-3}). Uncontrolled defects like oxygen ions forming shallow donors are difficult

---

[*] Corresponding author: e-mail: cez_j@if.pw.edu.pl, Phone: +48 22 660 5181, Fax: +48 22 660 5447

to eliminate in bulk crystals technology. Many efforts have been done to compensate and convert the material to p-type through doping it intentionally with shallow acceptors ions. Unfortunately the ionization energy of typical acceptors in GaN is relatively high (ionization energy of Mg acceptor is about 200 meV). This factor makes difficult to achieve the desired concentration of holes. It has been shown in some recent publications that the existence of ferromagnetism in transition elements doped bulk crystals is possible also in n-type materials and compensated materials with more than 5% of manganese the low carrier concentration. The mechanism of ferromagnetism in such materials is not clear [2, 5].

The reports on ferromagnetism in GaN heavily doped with transition elements are contradictory. The possibility of growth GaN:Mn system with more than 5% of manganese has been verified experimentally both in bulk crystals and in MBE grown thin films. Early magnetization measurements showed a paramagnetic behavior of GaN:Mn bulk crystals but in the later papers a ferromagnetic type magnetization dependence on magnetic field has been reported. Several different T_c temperatures have been proposed [6, 7].

Transport measurements and particularly magnetoconductivity and Hall effect are useful tools of magnetic materials characterization and may be helpful in to conclude the controversy. Negative magnetoresistance is a sine qua non condition of ferromagnetism but it is also observed in strongly compensated, paramagnetic semiconductors due to the weak localization and s–d interactions [7, 8]. The coincidence of negative magnetoresistance with anomalous Hall effect is considered a hard test for bulk ferromagnetism.

2 Experimental

(Ga,Mn)N and (Ga,Fe)N bulk crystals have been grown by nitridization of pure metallic Ga in supercritical ammonia method described elsewhere [9]. To increase the concentration of free carriers samples (Ga,Mn,Si)N crystals have been used, because it is possible to (Ga,Mn,Si)N samples show higher concentration of Mn than (Ga,Mn)N crystals. The crystals were rather small and formed irregular plates or needles. Indium contacts were soldered to the samples to make the classic Hall geometry where possible or the four probe geometry to study only the magetoconductivity. The ohmicity of all the contacts has been carefully verified. Hall effect and longitudinal magnetoresistance have been measured at temperatures between 4.5 K and 300 K in magnetic field varying in the range ±7 T. Oxford Instruments Spectromag 4000 cryostat and standard Kithley DC Hall effect setup has been used. Because the crystals used for the measurements were quite irregular the Hall voltage and longitudinal voltage drop due to the sample magnetoresistance have been separated numerically. All the measured samples exhibit n-type conductivity.

3 Results and discussion

The results of the magnetoresistance measurements are presented at Fig. 1(a,b), (Ga,Mn,Si)N and (Ga,Fe)N respectively, and are correlated with Hall effect measurements (Fig. 2 and Fig. 3). The clear difference between the two systems (positive magetoresistance of (Ga,Mn,Si)N crystals and negative magneoresistance of (Ga,Fe)N system) is seen. At Fig. 2a and Fig. 2b Hall voltage vs. magnetic field and temperature dependence of free electron concentration vs. temperature in (Ga,Fe)N system are presented. The dependence of Hall effect on magnetic field and the electron concentration vs. temperature in (Ga,Mn,Si)N system are presented at Fig. 3a and 3b respectively. For both compounds no correction of the Hall voltage vs. magnetic field curve for anomalous Hall effect has been seen (Fig. 2a and Fig. 3a). Considerable difference between the temperature dependence of electron concentration in (Ga,Fe)N and (Ga,Mn,Si)N is observed. The relatively high concentration of electrons (~6×10^{19}/cm^3) weakly depends on temperature exhibiting metallic conduction above Mott transition. The lower concentration of carriers (~10^{19}/cm^3) in (Ga,Mn,Si)N depends exponentially on temperature. An important coincidence between the results of magnetoresistance and Hall effect measurements has been observed. The negative magnetoresistance coincides with the metallic conductivity of the material. The positive type of magnetoresistance coincides with temperature dependent type of conductivity characteristic for nondegenarate semiconductors. This coincidence has been observed in all the examined (Ga,Mn)N, (Ga,Mn,Si)N, and (Ga,Fe)N bulk crystals.

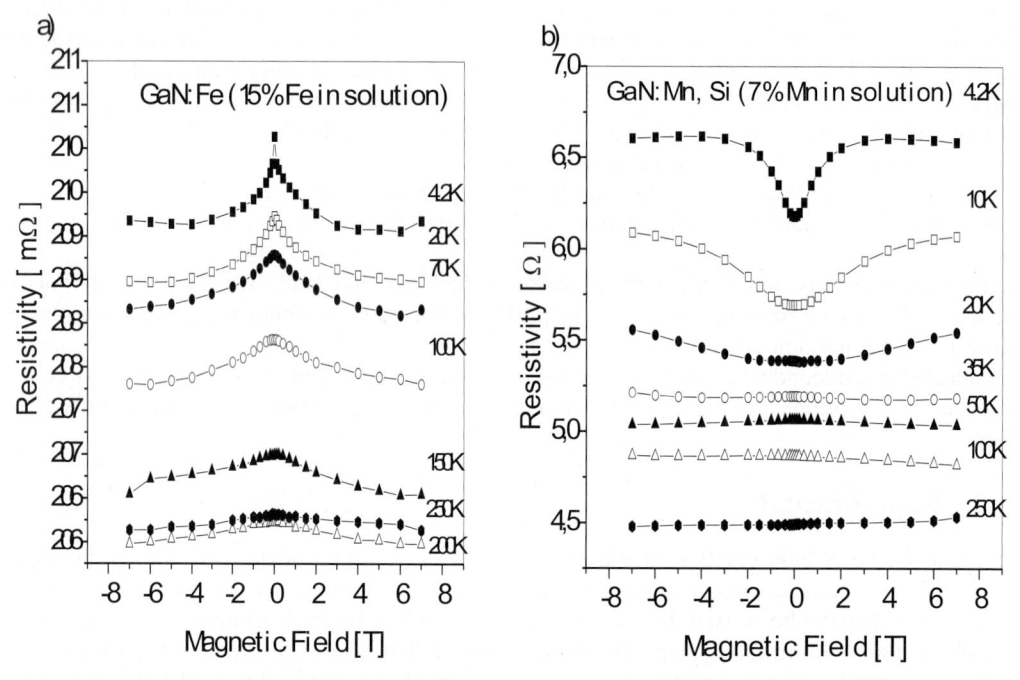

Fig. 1 Temperature dependence of magetoresistance in (Ga,Fe)N (a) and in GaMnSiN (b) system.

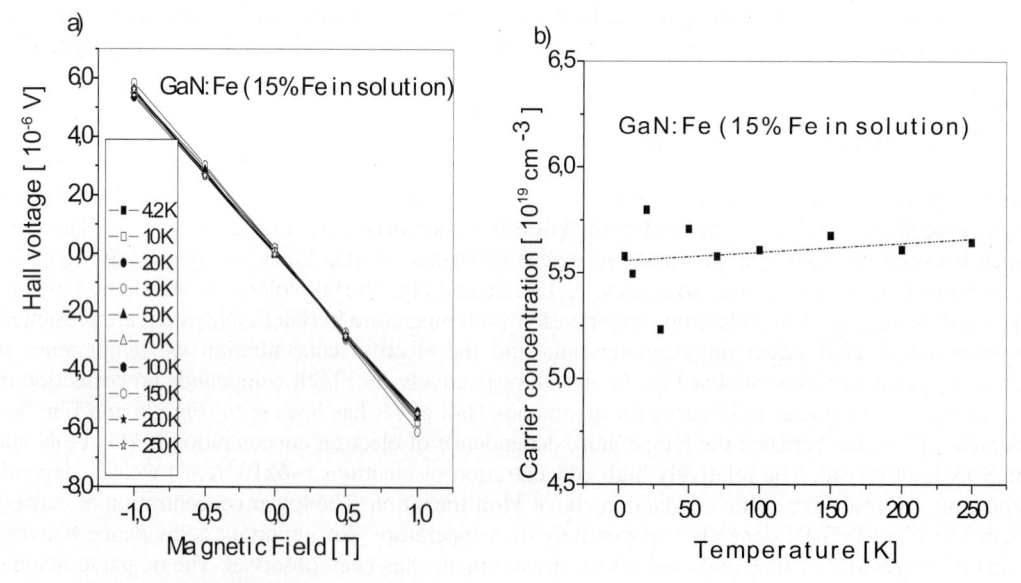

Fig. 2 Hall effect vs. magnetic field at various temperatures (a) and temperature dependence of electron concentration (b) in (Ga,Fe)N bulk crystals.

© 2004 WILEY-VCH Verlag GmbH & Co. KGaA, Weinheim

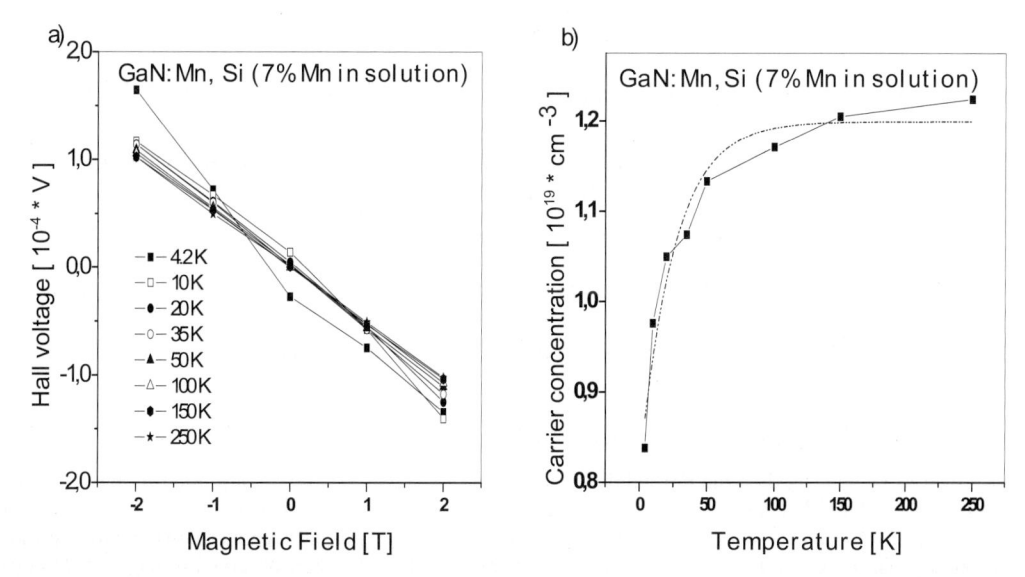

Fig. 3 Hall effect vs. magnetic field at various temperatures (a) and temperature dependence of electron concentration (b) in (Ga,Mn,Si)N bulk crystals.

This result strongly suggests that the negative magnetoresistance of (Ga,Fe)N crystal is related to the weak localization of carriers and to the quantum corrections to the magnetoresistance [10, 11], not to the ferromagnetic ordering of the magnetic moments of paramagnetic ions.

We conclude the paper as follows:

In all the (Ga,Mn,Si)N, (Ga,Mn)N and (Ga,Fe)N samples no anomalous Hall effect has been found in spite of relatively high concentration paramagnetic ions (Mn or Fe) and high concentration of free electrons. The negative magnetoresistance is related to the metallic type of conductivity and is not related to the ferromagnetic ordering of magnetic moments. Probably the results can be explained within the model taking into account weak localization and s–d interaction of free electrons.

This work was partially supported by KBN grant PBZ-KBN-044/P03/2001.

References

[1] T. Dietl, H Ohno, F. Matsukara, J. Cibert, and D. Ferrand, Science **287**, 1019 (2000).
[2] S. J. Pearton, C. R. Abernathy, M. E. Overberg, G. T. Thaler, D. P. Norton, N. Theodoropoulou, A. F. Hebard, Y. D. Park, F. Ren, J. Kim, and L. A. Boatner, J. Appl. Phys. **93**, 1 (2003) and citations therein.
[3] J.Gosk, M. Zając, M. Byszewski, M. Kamińska, T. Szczytko, A. Twardowski, B. Strojek, and S. Podsiadło, Journal of Superconductivity, Incorporating Novel Magnetism, **16**, 79 (2001).
[4] M. Zając, J. Gosk, M. Kamińska, A. Twardowski, T. Szczytko, and S. Podsiadło, Appl. Phys. Lett. **79**, 2432 (2001).
[5] S. Cho, S. Choi, G. Cha, S. H, Y. Kim, Y. Zhao, A. Freeman, and J. Ketterson, Phys. Rev. Lett. **88**, 257203 (2002).
[6] N. Theodoropoulou, A. F. Hebard, S. N. G. Chu, M. E. Overberg, C. R. Abernathy, S. J. Peraton, R. G. Wilson, and J. M. Zawada, Appl. Phys. Lett. **79**, 3452 (2001).
[7] H. Hori, S. Sonada, T. Sasaki, Y. Yamamado, S. Shimizu, K. Suga, and K. Kindo, Physica B **324**, 142 (2002).
[8] M. E. Overberg, C. R. Abernathy, S. J. Peraton, N. Theodoropoulou, K. T. McCarthy, and A. F. Hebard, Appl. Phys. Lett. **79**, 1312 (2001).
[9] G. Kamler, J. Zachara, S. Podsiadło, L. Adamowicz, and W. Gebicki, J. Cryst. Growth **212**, 39 (2000).
[10] S. Kawahara, K. Ishii, S. Hameda, T. Kondo, and H. Munekata, Physica E **10**, 223 (2001).
[11] M. Sawicki, T. Dietl, J. Kossut, J. Igalson, T. Wojtowicz, and W. Plesiewicz, Phys. Rev. Lett. **56**, 508 (1986).

Sub-micron InGaN ring structures for high-efficiency LEDs

H. W. Choi[*,1], **P. R. Edwards**[2], **C. Liu**[1], **C. W. Jeon**[1], **R. W. Martin**[2], **I. M. Watson**[1], **M. D. Dawson**[1], **S. Tripathy**[3], and **S. J. Chua**[3]

[1] Institute of Photonics, University of Strathclyde, Glasgow G4 0NW, UK
[2] Department of Physics, University of Strathclyde, Glasgow G4 0NG, UK
[3] Institute of Materials Research and Engineering, 3 Research Link, Singapore 117602

Received 15 September 2003, revised 24 October 2003, accepted 24 October 2003
Published online 15 January 2004

PACS 68.37.Ps, 78.30.Fs, 78.60.Fi, 78.60.Hk, 85.60.Jb

A novel technique based on the Fresnel diffraction effect for the fabrication of sub-micron ring structures in InGaN/GaN multi-quantum well structures has been demonstrated. The ring structures have an external diameter of 1.5 μm and a wall width of 500 nm. A 1 cm^{-1} Raman shift has been measured, signifying substantial strain relaxation from the fabricated structure. The 5 nm blue shift observed in the cathodoluminescence spectra can be attributed to band-filling and/or screening of the piezoelectric field. A light emitting diode based on this geometry has been demonstrated.

© 2004 WILEY-VCH Verlag GmbH & Co. KGaA, Weinheim

1 Introduction Various forms of the InGaN-based micro-scale light emitting diodes (LEDs) have been reported in recent years [1, 2], all of which indicate that an enhanced efficiency can be achieved with such microstructures. The overall external quantum efficiency of an InGaN/GaN LED is heavily dependent on the internal quantum efficiency and the extraction efficiency. However, the presence of internal piezoelectric fields and spontaneous polarization limits the quantum efficiency to approximately 30% [3], while the high refractive index of nitride materials means that most of the light emitted from the active region remains trapped in the mesa structure. A significant increase in extraction efficiency has already been demonstrated in the current generation of micro-LEDs, where the dimension of an individual element is typically larger than 4 μm. This is due to a higher surface area to light generation area ratio and reduced absorption. However, in order to achieve a reduction of the piezoelectric field in the InGaN quantum wells, the dimension of the micro-LED should be further reduced. In fact, Demangeot et al. observed significant strain relaxation in InGaN microstructures of less than 1 μm through micro-Raman scattering [4], suggesting that micro-LEDs should be scaled down to sub-micron dimensions in order to benefit from an increased in internal quantum efficiency.

In this paper, we report on the fabrication of sub-micron InGaN ring microstructures using standard micro-fabrication techniques. The objective of fabricating this structure is to produce a LED with high extraction efficiency and internal quantum efficiency. This may be achievable if partial or complete strain relaxation takes place in the sub-micron structures, leading to a reduction in the piezoelectric field. The fabricated ring structures, alongside with the as-grown wafer, micro-disks (with diameter of 12 μm) and micro-rings (with internal/external diameters of 12/20 μm) are investigated by Raman Spectroscopy and cathodoluminescence to detect any degree of strain relaxation. An LED based on this geometry has been demonstrated.

---

[*] Corresponding author: e-mail: anthony.choi@strath.ac.uk, Phone: +44 141 5793006, Fax: +44 141 5793048

2 Experimental details
Disk structures are patterned using AZ1805 photoresist (with a nominal thickness of 0.5 μm) as a masking material and standard photolithography techniques (using a 365 nm excitation source). The ring structures are spontaneously formed after photoresist development. They are subsequently transferred to the nitride material by inductively-coupled plasma (ICP) etching using a combination of Cl_2/Ar as etchant gases. The sample used is an 3-period InGaN (3 nm)/GaN (7 nm) multi-quantum well (MQW) structure with emission wavelength targeted for 470 nm. The extent of strain relaxation in the structures is evaluated by Raman Spectroscopy using the 514 nm line of an Ar^+ laser as an exciton source focused to a spot of less than 1 μm. The scattered light was detected in the backscattering geometry with a Jobin Yvon T64000 triple-grating spectrometer with a LN_2-cooled charge coupled device (CCD) detector. The ring structures were also examined by cathodoluminescence (CL) in an electron probe microanalyzer (Cameca SX100).

3 Experimental results

3.1 Assembly of ring structure
The ring structures are designed to have an external diameter of 1.5 μm and a wall width of about 500 nm. Such a pattern cannot be imaged directly using 365 nm lithography techniques. Instead, 1.5 μm micro-disk structures are patterned onto photoresist. Upon development, ring structures are spontaneously formed. The evolution of the ring structure during the development process is illustrated in the AFM images in Fig. 1. This phenomenon may be explained by the Fresnel diffraction effect [5]. The micro-disk array pattern on the mask, apart from playing the role of image transfer, also acts as a 3-dimensional diffraction grating. Consider the cross-section of the micro-disk array pattern on the photo-mask as shown in Fig. 2a. This is essentially a 2-dimensional N-slit diffraction grating with a slit width of 2 μm (separation of the micro-disks) and a slit spacing of 1.5 μm (diameter of the microdisks). During light exposure, the photoresist-coated sample is held in proximity to the photo-mask, as depicted in Fig. 2b. When monochromatic light (of 365 nm) passes through it, a diffraction pattern is formed. The simulated intensity plot of the diffraction pattern is shown in Fig. 2c. While the maxima occur at the centre of the slits, secondary peaks are observed at positions corresponding to the centre of the microdisks. These secondary peaks are a result of 1st and higher order diffraction patterns. As a result, the centres of the micro-disk regions are exposed and a hole is spontaneously formed upon development. The intensity of the secondary peaks is naturally lower then those of the primary peaks; this explains the lower rate of development of the hole in the micro-disk compared to the external exposed regions.

3.2 Strain relaxation in ring structure
The fabricated ring structures are subjected to examination by Raman spectroscopy. The E_2 mode frequency from the Raman spectra is used to monitor the strain component in the structure, since it is much more sensitive to strain compared to other Raman modes due to its independence from the free carrier concentration. The E_2 line will shift towards the higher-frequency region with increasing compressive strain by 2.9 cm^{-1}/GPa [6]. Raman spectra taken from the

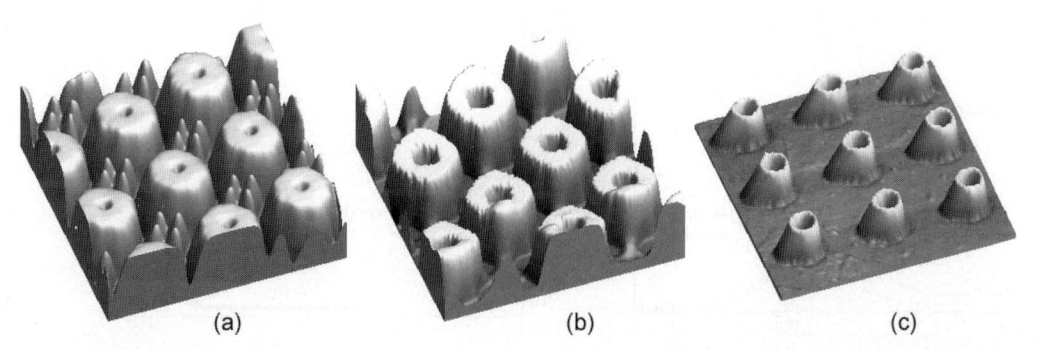

Fig. 1 (online color at: www.interscience.wiley.com) 10×10 μm AFM images illustrating the evolution of the ring structure during photoresist development at a) t = 15 s and b) t = 30 s. The ring pattern array transferred onto the nitride material by ICP etching is shown in c).

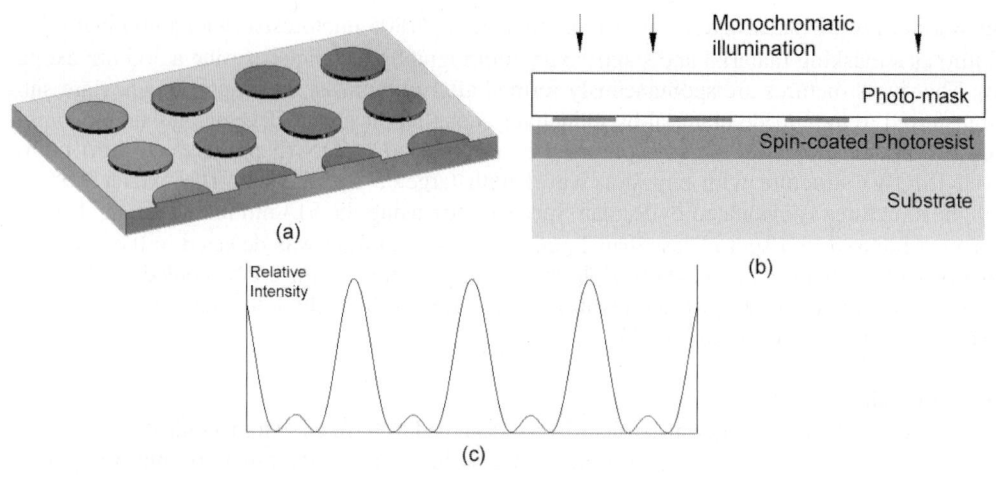

Fig. 2 a) Schematic image of the micro-ring mask pattern used. b) The photo-mask acting as a diffraction grating during exposure. c) Simulated intensity plot of the diffraction pattern.

as-grown sample, sub-micron ring, micro-ring and micro-disk structures are illustrated in Fig. 3. Apart from the sub-micron ring structure, the spectra have peaks at 568.4 cm^{-1}, typical of compressively-strained as-grown GaN. This is consistent with the findings of Demangeot et al., where no significant shift was observed for reactive-ion etched (RIE) pillars with dimensions greater than 1 μm. However, a significant shift was recorded in the spectra of the sub-micron ring structure. In fact, the centre frequency of 567.3 cm^{-1} was close to that of a spectrum measured from a piece of free-standing GaN (which can be assumed to be strain-free). We thus conclude that significant strain relaxation has taken place in the formation of the sub-micron ring structures.

3.3 Cathodoluminescence The fabricated structures were excited with a 20 keV – 1 nA electron beam, and the collected CL spectra are shown in Fig. 4. A consistent blue shift was observed as the dimension of the structures is scaled down, reaching a maximum of 5 nm for the sub-micron ring structure compared to the as-grown sample. The blue shift can be attributed to band-filling effects and/or screening of the piezoelectric field by the carriers, as the current densities in the various structures increased as the dimension reduces.

3.4 Electroluminescence An LED adopting the sub-micron ring structure has been fabricated. The light emission region consists of a high-density array of micro-ring mesa structures. The fabrication sequence is identical to that of the micro-ring LEDs [2, 7]. A micro-photograph of light emission from

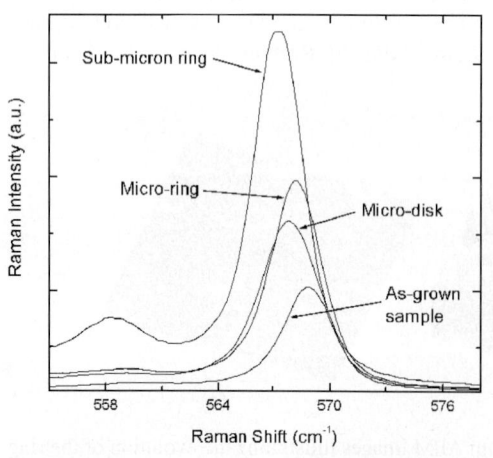

Fig. 3 Raman spectra for the as-grown, micro-disk, micro-ring and sub-micron ring structures.

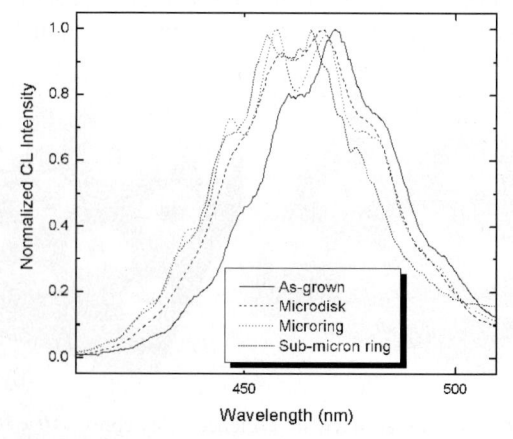

Fig. 4 CL spectra of the as-grown, micro-disk, micro-ring and sub-micron ring structures.

© 2004 WILEY-VCH Verlag GmbH & Co. KGaA, Weinheim

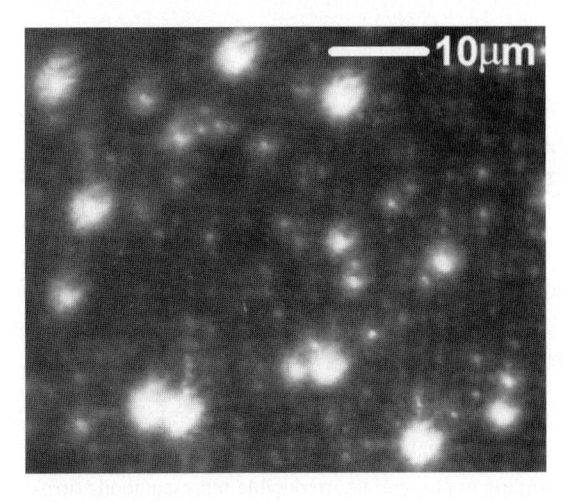

Fig. 5 (online color at: www.interscience.wiley.com) Microphotograph of light emission from a submicron ring LED. Each pixel of light represents light emission from a single micro-ring element.

the device is shown in Fig. 5. Each pixel of light represents light emission from a single micro-ring element. As can be seen from the figure, not all the micro-ring elements are operational. This is attributed to the complexity of the fabrication process, especially the difficulty in forming an ohmic contact to the individual sub-micron elements. Nevertheless, this represents a first demonstration of electroluminescence from a sub-micron scale LED. Full details of the operation of the device will be reported in the future.

4 Conclusions In summary, the fabrication of sub-micron ring structures with an external diameter of 1.5 µm and a wall width of 500 nm have been reported. The spontaneous assembly of the ring structures is based on Fresnel diffraction effects. Substantial strain relaxation in the structures has been observed through Raman spectroscopy. Blue-shifts in the CL spectra has also been observed, attributed to band-filling effects and/or screening of the piezoelectric fields. An LED based on the sub-micron ring structure has also been demonstrated.

References

[1] S. X. Jin, J. Li, J. Y. Lin, and H. X. Jiang, Appl. Phys. Lett. **77**, 3236 (2000).
[2] H. W. Choi, C. W. Jeon, M. D. Dawson, P. R. Edwards, and R. W. Martin, IEEE Photon. Tech. Lett. **15**, 510 (2003).
[3] M. R. Krames, J. Bhat, D. Collins, N. F. Gardner, W. Gotz, C. H. Lowery, M. Ludowise, P. S. Martin, G. Mueller, R. Mueller-Mach, S. Rudaz, D. A. Steigerwald, S. A. Stockman, and J. J. Wierer, phys. stat. sol. (a) **192**, 237 (2002).
[4] F. Demangeot, J. Gleize, J. Frandon, M. A. Renucci, M. Kuball, D. Peyrade, L. Manin-Ferlazzo, Y. Chen, and N. Grandjean, J. Appl. Phys. **91**, 6520 (2002).
[5] S. Wolf and R. N. Tauber, Silicon Processing for the VLSI Era, Vol.1 (Lattice Press, California, USA, 2000), pp. 560.
[6] F. Demangeot, J. Frandon, M. A. Renucci, O. Briot, B. Gil, and R. L. Aulumbard, Solid State Commun. **100**, 207 (1996).
[7] H. W. Choi, M. D. Dawson, P. R. Edwards, and R. W. Martin, Appl. Phys. Lett. **83**, 4483 (2003).

Multi-phonon processes in C_{6v}^4 (P6₃mc) hexagonal semiconductors: GaN, ZnO, ZnS, BeO, CdS, and Al₂O₃ – sapphire: D_{3d}^6 (R3c)

H. W. Kunert[*]

Department of Physics, University of Pretoria, 0001 Pretoria., South Africa

Received 15 September 2003, revised 4 November 2003, accepted 4 November
Published online 22 December 2003

PACS 02.20.–a, 63.20.–e, 78.30.Fs, 78.30.Hv

The vibrational states of crystals are classified according to appropriate irreducible representations (irrps) of wave vector space groups. Based on the irrps of the C_{6v}^4 space group the selection rules for one – two – and three – phonon processes are derived. Experimentally, we have performed inelastic light scattering experiments on several different GaN samples as well as ZnO. Some multi – phonon transitions present on our spectra have been verified by the derived multi – phonon selection rules.

© 2004 WILEY-VCH Verlag GmbH & Co. KGaA, Weinheim

1 Introduction

GaN and the nitride semiconductors have shown great promise as materials for light – emitting diodes (LEDs), lasers, high-speed, and high – power electronics. Ion implantation and post – implantation annealing in nitrides are processes of considerable technological importance as well as methods of producing controlled damage for studying the nature of order and disorder in these materials. The performance of current – injected laser diodes based on GaN require suitable thermodynamic properties. Phonons are the primary excitation which influence the thermodynamic properties. Raman spectroscopy (RS) provides useful information about vibrational states of as grown and treated semiconductors. The first – order phonon processes in GaN have been studied by many groups [1 and references 2, 3, 5–7, therein]. However, the second – third – and higher processes are rather poorly understood. In this paper we provide a useful method for analysis of higher phonon processes in semiconductors with the C_{6v}^4 space group: GaN/SA, ZnO, and others.

2 Phonons and their symmetries in GaN/SA

The number and symmetries of vibrational modes in any compound with a well established space group is determined by so – called reducible representation of lattice modes (LMRR) [2, 3], (also called as displacement representation DR [4]). The explicit form of the LMRR for hexagonal GaN has been recently derived [5]. Decomposing the LMRR onto irreducible representations (irrps) of high symmetry points Γ, A, K, L, M, H and lines ΔP, U, R, Σ, Q, S, Λ, and T of the first Brillouin zone for hexagonal GaN, we obtain all possible modes and their symmetries listed in Table 1. We used the CDML tables for irrps and Kronecker Products [6]. The frequencies of the listed modes can be determined by neutron and X-rays scattering nmethods. In these techniques the X-rays and neutrons are scattered by phonons with the wave vectors from the entire Briloouin zone.

---

[*] E-mail: hkunert@nsnper1.up.ac.za, Fax: + 27 12 362 52 88

Recently, Nipko et al [7] reported the measured phonon density of states of a bulk GaN powder by time – of – flight neutron spectroscopy. They have also calculated the phonon dispersion curves from fitting data with a rigid – ion model. Similarly, the phonon dispersion curves in wurtzite GaN structure have been determined by inelastic X-ray scattering and *ab initio* lattice dynamical calculation [1]. In both cases the group – theoretical assignment has not been taken into account. With the help of our Table 1 and compatibility relations the group theo9retical assignment of phonons in GaN follows. The assignment helps with identification of phonon symmetries and their degeneracy. From example, from Nipko's approach [7, Fig. 2] follows that at high symmetry point K there are twelve phonons, whereas from Ruf's calculations [1, Fig. 2] only eight K's phonons are possible. Our Table provides two K_1, two K_2 (non – degenerate) and four K_3 (two-fold degenerate) modes which is in accordance with Ruf's results. Moreover, Table 1 is valid for all semiconductors with the C_{6v}^4 space group and the same number of atoms in the unit cell as GaN; such as ZnO, ZnS, CdS, BeO and others. For the long wavelength modes ($\mathbf{k} \cong \mathbf{0}$) the neutron scattering method yields rather pure accuracy of phonon frequencies, whereas Raman spectroscopy (RS) provides exact energy values of these vibrational modes. However, it requires strict group-theoretical selection rules. These are discussed in this paper.

Table 1 Normal modes spanned by Lattice Modes reducible representation at critical high symmetry points and lines in C_{6v}^4 (P6$_3$mc) wurtzite crystals.

Γ: $2\Gamma_1 \oplus 2\Gamma_4 \oplus 2\Gamma_5 \oplus 2\Gamma_6$
A: $2A_1 \oplus 2A_4 \oplus 2A_5 \oplus 2A_6$
Δ: $2\Delta_1 \oplus 2\Delta_4 \oplus 2\Delta_5 \oplus 2\Delta_6$
H: $2H_1 \oplus 2H_2 \oplus 4H_3$
P: $2P_1 \oplus 2P_2 \oplus 4P_3$
K: $2K_1 \oplus 2K_2 \oplus 4K_3$
L: $4L_1 \oplus 2L_2 \oplus 2L_3 \oplus 4L_4$
M: $4M_1 \oplus 2M_2 \oplus 2M_3 \oplus 4M_4$
U: $4U_1 \oplus 2U_2 \oplus 2U_3 \oplus 4U_4$
R: $8R_1 \oplus 4R_2$
Σ: $8\Sigma_1 \oplus 4\Sigma_2$
Q: $6Q_1 \oplus 6Q_2$
S: $6S_1 \oplus 6S_2$
Λ: $6\Lambda_1 \oplus 6\Lambda_2$
T: $6T_1 \oplus 6T_2$

3 One-phonon processes: Raman active modes

In order to determine the Raman active modes we decompose the Symmetrized Kronecker Product (SKP) of the Vector Representation (VR) onto single value irrps of the C_{6v}^4 space group at the high symmetry point Γ: ($\Gamma_1, \Gamma_2, \Gamma_3, \Gamma_4, \Gamma_5, \Gamma_6$ CDML [6] labelling), which correspond to $A_1, A_2, B_1, B_2, E_1, E_2$ irrps, respectively, in the Bethe's notation. Only irrps contained in the SKP are Raman active. In most cases (230 space groups) the VR (D^v) is reducible and therefore its SKP can not be found in the CDML tables. For the GaN, VR is $D^v = A_1 \oplus E_2$, ($\Gamma_1 \oplus \Gamma_6$) and its SKP contains the following Raman active species $[(A_1 \oplus E_2) \otimes (A_1 \oplus E_2)]_{(2)} = 2A_1 \oplus E_1 \oplus E_2$, while the Antisymmetric KP (AnSKP) yields $\{(A_1 \oplus E_2) \otimes (A_1 \oplus E_2)\}_{(2)} = A_2 \oplus E_2$. Adding the SKP to the AnSKP we obtain: $2A_1 \oplus A_2 \oplus E_1 \oplus 2E_2$ equivalently to $2\Gamma_1 \oplus \Gamma_2 \oplus \Gamma_5 \oplus 2\Gamma_6$. This result can be extracted from the CDML tables ($\Gamma_1 \oplus \Gamma_6$) $\otimes$ ($\Gamma_1 \oplus \Gamma_6$) $= \Gamma_1 \otimes \Gamma_1 \oplus \Gamma_1 \otimes \Gamma_6 \oplus \Gamma_6 \otimes \Gamma_1 \oplus \Gamma_6 \otimes \Gamma_6$. The KPs on the right hand side of the last equation are given in the tables and agree with ours. Therefore, the Raman active modes which may appear on Raman spectra are A_1, E_1, E_2. It means that we deal with six first – order Raman active vibrational modes: A_1(TO), A_1(LO), E_1(TO), E_1(LO), E_2(low), E_2(high) [8].

4 Two-phonon processes: overtones and combinations

The case of a single infrared photon being absorbed and simultaneously producing two phonons has been examined in the past [9 and reference 17 therein]. The symmetry allowed overtones result from complete reduction of Symmetrized Squares (SymSq) of the normal modes spanned by LMRR at critical points and high symmetry lines. For crystals with the C_{6v}^4 space group the normal symmetry allowed modes are listed in the Table 1.

Table 2 Two – phonon processes in wurtzite C_{6v}^4 (P6$_3$mc) hexagonal crystals. The D indicates infrared absorption dipole allowed transition.

SPECIES	ACTIVITY	OVERTONS
$[\Gamma_1]_{(2)}$	D, R	2O (Γ_1): 2A$_1$(LO), 2A$_1$(TO)
$[\Gamma_5[_{(2)}$	D, R	2O (Γ_1): 2A$_1$(LO), 2A$_1$(TO)
		2O (Γ_5): 2E$_1$(LO), 2E$_1$(TO)
$[\Gamma_6]_{(2)}$	D, R	2O (Γ_1): 2A$_1$(LO), 2A$_1$(TO)
		2O (Γ_5): 2E$_1$(LO), 2E$_1$(TO)
$[^*K_1]_{(2)}$	D, R	2($^*K_1$): 2LO(K), 2TO(K)
$[^*K_2]_{(2)}$	D, R	2($^*K_1$): 2LO(K), 2TO(K)
$[^*K_3]_{(2)}$	D, R	2($^*K_1$): 2LO(K), 2TO(K)
		2($^*K_3$): 2LO(K$_3$), 2TO(K$_3$)
$[^*M_1]_{(2)}$	D, R	2($^*M_1$): 2LO(M), 2TO(M)
$[^*M_2]_{(2)}$	D, R	2($^*M_1$): 2LO(M), 2TO(M)
$[^*M_3]_{(2)}$	D, R	2($^*M_1$): 2LO(M), 2TO(M)
$[^*M_4]_{(2)}$	D, R	2($^*M_1$): 2LO(M), 2TO(M)
		COMBINATIONS
$^*K_1 \otimes ^*K_2$	R	($^*K_2$): LA(K) + TA (K)
$^*K_1 \otimes ^*K_3$	R, D	($^*K_3$): LO(K) + TO(K)
		K$_3$(low) + K$_3$(high)
$^*K_2 \otimes ^*K_3$	R, D	($^*K_3$): LO(K) + TO(K)
		K$_3$(low) + K$_3$(high)
$^*M_1 \otimes ^*M_2$	R, D	LO(M)+?(M), TO(M)+?(M)
$^*M_1 \otimes ^*M_3$	R	LA(M)+?(M), TA(M)+?(M)
$^*M_1 \otimes ^*M_4$	R	LA(M)+?(M), TA(M)+?(M)
$^*M_2 \otimes ^*M_3$	R	LA(M)+?(M), TA(M)+?(M)
$^*M_2 \otimes ^*M_4$	R	LA(M)+?(M), TA(M)+?(M)
$^*M_3 \otimes ^*M_4$	R, D	LO(M)+?(M), TO(M)+?(M)

If there is a symmetry allowed mode in the decomposition of the SymSq (Table 1) the two phonon – process is permitted. For example: from the Table 1 it follows that we deal with two K$_1$ modes; and the SymSq $[K_1 \otimes K_1]_{(2)} = \Gamma_1 \oplus \Gamma_3 \oplus K_1$, contains the allowed mode K$_1$. Therefore, the overtone of the K$_1$ phonon is permitted. The overtones of the phonons with large pseudomomenta $(h/2\pi)\mathbf{k}$ can possibly be measured by infrared absorption (IR), reflectance (indirectly), neutron scattering, and X-ray spectroscopy, but not by RS. The overtones of phonons originating from $\mathbf{k} \cong 0$ (long wavelength modes – point Γ) can only originate from Raman active modes determined by SymSq of the VR, $D^v = \Gamma_1 \oplus \Gamma_6$ and these are Γ_1, Γ_5, and Γ_6. If the SymSqs of these species contain an Raman active mode the appropriate overtone is permitted. Combinations are obtained from the decomposition of the KP of two different species listed in the Table 1. If the KP of two different modes contains an allowed mode in Table 1, the combination is permitted. For example, the combination of Γ_5 and M_3 modes results in $\Gamma_5 \otimes M_3 = M_3 \oplus$

M_4. It means, that we may measure the resulting frequency; $w(M_3) + w(M_4)$. Some possible two – phonon processes, overtones and combinations in hexagonal GaN are listed in Table 2.

5 Three-phonon processes

Similarly, the three – phonon overtones result from symmetrized cubes and the combinations from the complete reduction of the KP of different species. Some three phonon overtones for GaN are presented in the Table 3. There are many possible three phonon combinations such as: $K_m \otimes L_p \otimes M_s$ (m = 1, 2, 3, p,s = 1, 2, 3, 4). These combinations must firstly be determined by wave vector selection rules (pseudo-momenta conservation principles). A necessary, but not sufficient, condition for Raman active three phonon process is that the KP must contain first order phonons originating from Γ. For example the phonon states $^*M_1$ and $^*M_3$ may combine with Raman active states Γ_1, Γ_5, or Γ_6. Some possible three phonon Raman transitions from GaN/SA are presented in Figs. 1–3.

Table 3 Three – phonon processes in wurtzite C_{6v}^4 (P6$_3$mc) hexagonal crystals. The D indicates infrared absorption dipole allowed transition

SPECIES	ACTIVITY Overtons	TYPE
$[\Gamma_1]_{(3)}$	R, D	3O (Γ)
$[\Gamma_5]_{(3)}$	R, D	3O (Γ)
$[\Gamma_6]_{(3)}$	R	3O (Γ)
$[^*K_1]_{(3)}$	R, D	3TO(K) and 3LO(K)
$[^*K_2]_{(3)}$	R	3TO(K) and 3LO(K)
$[^*K_3]_{(3)}$	R, D	6TO(K) and 3LA(K)
$[^*M_1]_{(3)}$	R, D	3LO(M) and 3TO(M)
$[^*M_2]_{(3)}$	R	3LO(M) and 3TO(M)
$[^*M_3]_{(3)}$	R	3LA(M) and 3TA(M)
$[^*M_4]_{(3)}$	R	3LA(M) and 3TA(M)

6 Experimental

Two samples Mg:GaN and ZnO grown by plasma-assisted molecular beam epitaxy (MBE) on (0001) sapphire were investigated optically by means of the inelastic light scattering experiment. The Raman spectra were recorded in a back scattering arrangement using the 514 nm Ar+ laser at 77 and 300 K. In Table 4 we list the measured frequencies of ZnO, and GaN/Al$_2$O$_3$ first order active Raman modes observed from GaN and sapphire side. The frequencies of the modes are in a good agreement with previously reported [8]. The Al$_2$O$_3$ – sapphire, belongs to the D_{3v}^6 space group. The LMRR for this crystal yields: $2A_{1g} \oplus 2A_{1u} \oplus 3A_{2g} \oplus 2A_{2u} \oplus 5E_g \oplus 4E_u$ symmetry allowed vibrational modes at $k \cong 0$. The vector representation is $D^v = A_{2u} \oplus E_u$ [10]. Following the described group – theoretical procedure we have decomposed the SymSq of the D^v that results in $2A_{1g} \oplus 2E_g$ which means that only A_{1g} (two) and E_g(five) modes are Raman active [10]. It is interesting that we also observe the E_2(high- 569 cm^{-1}) mode belonging to GaN from sapphire side. However, at T= 300K this mode is no more present on spectrum. The second order (two – phonon processes) transitions from the GaN side are depicted on Fig. 1. Applying two– phonon Raman selection rules (see Table 2) we have attributed the mode at 1478 cm^{-1} to overtone of the A$_1$(LO- 739.2cm^{-1} see Table 4). The modes at 861, 917, and 1155cm^{-1} we tentatively attribute to overtones and combinations of 1st order phonons of sapphire, which belongs to the D_{3d}^6 space group. The Raman selection rules for this space group are consistent with our assignments.

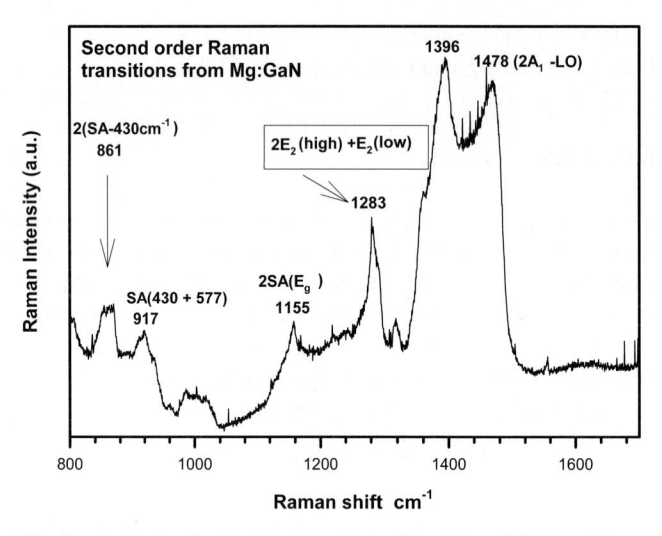

Fig. 1 Second order Raman transitions from Mg: GaN/Sapphire.

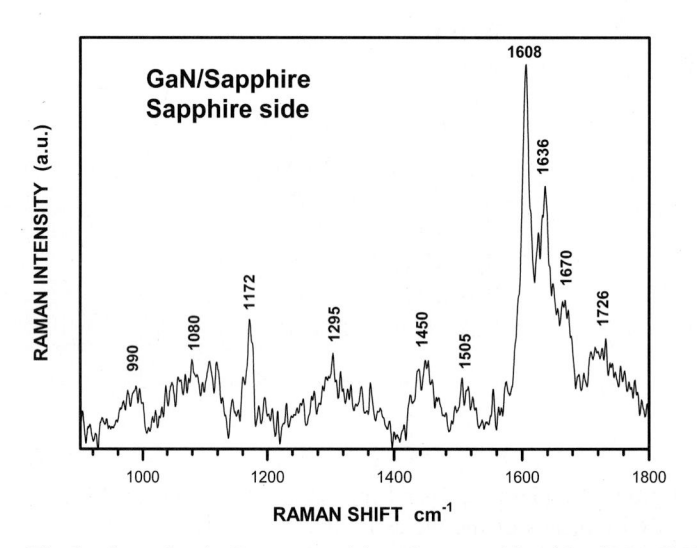

Fig. 2 Second order Raman transitions from sapphire side of Mg: GaN/Sapphire.

The mode at 1283cm^{-1} may result from symmetry allowed combination of three GaN phonons 2E$_2$(high 568.7 cm^{-1}) and E$_2$(144 cm^{-2}). All other modes presented on the Fig. 1 have been assigned to overtones or combinations of primary modes belonging to sapphire. The symmetry arguments and the energy conservation law have been taken into account. Despite of that our assignment may still have an ambiguity. Fig. 2 displays second-order phonon processes in the sample observed from sapphire side. There are several options for their assignments satisfying the energy conservation and symmetry selection rules. For example the band at 1608 cm^{-1} may involve three phonon process resulting in 3A$_1$(TO-536cm^{-1}) = 1608cm^{-1}. The symmetrized cube of the A$_1$(Γ$_1$) contains the Raman active first order mode A$_1$(Γ$_1$) and therefore the selection rules are satisfied (see Table 3). Similarly, the bands at 1636, 1670, 1720 cm^{-1} can be associated with three phonon processes such as SA-1636: SA(Eg 750 + 450 + 431),

SA;1670: SA(Eg 578 + 645 + 449 see Table 4) and SA-1720: SA(3E$_g$ 576 internal). The bands at 990, 1080, 1172, 1295, 1450, and 1505 cm^{-1} are the feature of two – phonon processes. The detailed analysis of these processes will be given elsewhere.

The Fig.3 exhibits the first and second order Raman transition from the ZnO crystal annealed at 400 °C [11–14], which belongs to the same space group C_{6v}^4 of GaN. Therefore, the selection rules for these two compounds are the same. We tentatively associate the band centred at 1156 cm^{-1} with an overtone of the A$_1$(longitudinal)-577 cm^{-1}. The band at 665 cm^{-1} seems to be an overtone of the 332 cm^{-1} which does not belong to the symmetry allowed primary modes in ZnO (see Table 4). This mode and modes at 545, 698, and 722cm^{-1} most presumably are the annealing feature. Consequently, with the help of our Tables it is possible to distinguish the treatment induced modes from the primary Raman symmetry allowed phonons.

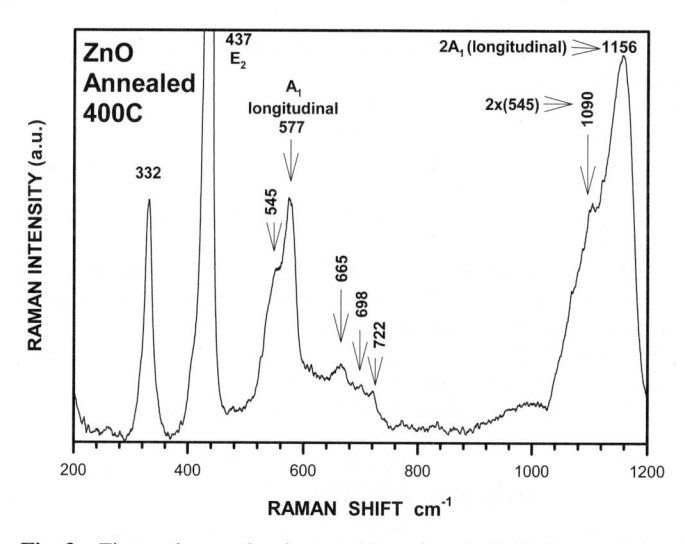

Fig. 3 First and second order transitions from bulk ZnO annealed at 400 °C

Table 4 Frequency (cm^{-1}) and symmetry of fundamental optical modes in GaN/SA, ZnO, and Al$_2$O$_3$ – Sapphire (SA)

GaN/SA C_{6v}^4 (P6$_3$mc)	ZnO C_{6v}^4 (P6$_3$mc)	Al$_2$O$_3$ D_{3d}^6 (R3c)
E$_2$ (low), Γ$_6$ 144 ± 4	E$_2$, Γ$_6$ 102 ± 4	378 ± 4 E$_g$(extern)
E$_2$ (high), Γ$_6$ 569 ± 4	E$_2$, Γ$_6$ 438± 4	418 ± 4 A$_{1g}$
A$_1$ (TO), Γ$_1$ 536 ± 4	A$_1$(transverse), Γ$_1$ 380± 4	432 ± 4 E$_g$(extern.)
E$_1$ (TO), Γ$_5$ 560 ± 4	E$_1$(transverse), Γ$_5$ 407± 4	451 ± 4 E$_g$(intern.)
A$_1$ (LO), Γ$_1$ 739 ± 4	A$_1$(longitudinal), Γ$_1$ 577±4	578 ± 4 E$_g$(intern.)
E$_1$ (LO), Γ$_5$ 744 ± 4	E$_1$(longitudinal), Γ$_5$ 580±4	645 ± 4 A$_{1g}$
		751 ± 4 E$_g$(intern.)
2A$_1$(LO)	2A$_1$(Long) and 2E$_1$ (Long)	
		SA modes observed in GaN/SA from GaN side at T = 77K 420, 433,450,578,750

7 Conclusions

Group theoretical Raman selection rules for one – two – and partly – three phonon processes have been explicitly derived for all crystals with the space groups C_{6v}^4 (listed in Tables 1–3). Particularly the Table 1 is very useful because it helps with assignments of phonons, and interpretations of multi – phonon processes spectra. It also can be useful in the determination of the dimension of dynamical matrices. Experimentally, using the standard RS we have found evidences for a number of multi – phonon processes in GaN, Sapphire and ZnO. Our approach can be very useful in the interpretation of Raman spectra of treated samples (irradiated, implanted, and annealed) where we deal with treatment induced new vibrational modes. These can only be recognized by knowing the number of primary (first – second – third-…) order phonons and their symmetries.

References

[1] T. Ruf, J. Serrano, M. Cardona, P. Pavone, M. Pabst, M. Krisch, M. D'Astuto, T. Suski, I. Grzegory, and M. Leszczynski, Phys. Rev. Lett. **86**, 906 (2001).
[2] S. Bhagarantan and T. Venkatarayudu, Proc. Indian Acd. Sci. **9A**, 224 (1939).
[3] P. Mathieu, Spectre de Vibration et Symmetrie (Hermann et Cie, Paris) 1945.
[4] H. W. Kunert, Cryst. Res. Technol. **38**, No. 3–5, 366–373 (2003).
[5] H. W. Kunert, Appl. Surf. Sci. **212–213**, 890–896 (2003).
[6] A. P. Cracknel, B. L. Davies, S. C. Miller, and W. F. Love, Kronecker Product Tables, Vol. **1–4**, IFI/Plenum Press, New York, Washington, London, 1979.
[7] J. C. Nipko, C.-K. Loong, C. M. Balkas, and R. F. Davis, Appl. Phys. Lett. **73**, 34–36 (1998).
[8] Jacques I. Pankove and Theodore D. Moustakas, in "Gallium Nitride (GaN) I Semiconductors and Semimetals, Vol. **50**, Academic Press, San Diego, London, New York, Sydney, Tokyo, Toronto.
[9] J. L. Birman, Phys. Rev. **131**, 1489–1496 (1963).
[10] S. P. S. Porto and R. S. Krishnan, J. Chem. Phys. **47**, 1009–1012 (1967).
[11] T. C. Damen, S. P. S. Porto, and B. Tell, Phys. Rev. **142**, 570–574 (1966).
[12] Masamichi Tsuboi and Akiyoshi Wada, J. Chem. Phys. **181**, 2615–2618 (1968).
[13] C. A. Arguello, D. L. Rousseau, S. P. S. Porto, Phys. Rev. **181**, 1351–1363 (1969).
[14] D. J. Brink and H. W. Kunert, phys. sat. sol. (b) **229**, 859–862 (2002).

Luminescent properties of wide bandgap materials at room temperature

M. Godlewski[*,1,2], **J. Szmidt**[3], **A. Olszyna**[4], **A. Werbowy**[3], **E. Łusakowska**[1], **M. R. Phillips**[5], **E. M. Goldys**[6], and **A. Sokołowska**[7]

[1] Institute of Physics, Polish Acad. of Sciences, Al. Lotników 32/46, 02-668 Warsaw, Poland
[2] College of Science, Dep. of Mathematics and Natural Sciences, Cardinal S. Wyszyński University, Warsaw, Poland
[3] Institute of Microelectronics and Optoelectronics, Warsaw Technical Univ., Warsaw, Poland
[4] Faculty of Mat. Sci. Eng., Warsaw Technical Univ., Warsaw, Poland
[5] Microstructural Analysis Unit, UTS, Sydney, Australia
[6] Division of Information and Communication Sciences, Macquarie Univ., Sydney, Australia
[7] Institute of Mat. Sci. Eng., Technical Univ. of Lodz, Lodz, Poland

Received 15 September 2003, accepted 19 September 2003
Published online 22 December 2003

PACS 68.37.Ps, 78.60.Hk, 81.05.Ea, 81.15.Gh

Properties of nanocrystalline thin films of selected nitrides are discussed as possible buffer materials for obtaining freestanding GaN wafers. These films are grown by impulse plasma deposition on silicon substrates. We demonstrate high smoothness of these films.

© 2004 WILEY-VCH Verlag GmbH & Co. KGaA, Weinheim

1 Introduction

Most of GaN epilayers studied by different groups and also commercial light emitting diodes (LEDs) are grown on lattice mismatched sapphire [1]. Sapphire is used despite of about 14 % lattice mismatch. This is due to its resistivity to ammonia, introduction of so-called buffer layers, but mostly due to lack of lattice matched substrates of required size and parameters. By buffer layer we mean here thin disordered layer initialising growth of ordered GaN layers. A two-step growth mode is commonly used to introduce buffer layers, in which a low-temperature AlN [2–4] or GaN [5] layer is deposited prior to GaN epitaxy. For sapphire substrates buffer can also be formed by initial nitridation of sapphire in an ammonia stream [6–8]. It was claimed that such nitridation results in formation of a thin AlN [9], BN [10], or AlNO [11] amorphous and elastic buffer layer, which initializes GaN growth.

The so-obtained GaN-based structures are good enough for applications in LEDs. However, development of high power and high stability laser diodes (LDs) requires use of structures with a reduced dislocation density, i.e., structures grown on lattice matched substrates. Homoepitaxial GaN layers and LDs show in fact superior properties [12, 13, 14]. Unfortunately, technology of bulk GaN substrates is very difficult and only relatively small and thin GaN wafers are available at present.

This fact led to concentrated efforts to produce so-called freestanding GaN wafers by etching (lifting off) them from lattice mismatched substrates [15–21]. In most of the approaches hydride vapour phase epitaxy (HVPE) was used to grow thick GaN films, with rather limited success. Several materials were used for initial deposition of GaN layer - sapphire, (111) GaAs, Si, NdGaO₃ were used together with different methods of GaN deposition (with or without buffer layer) and removing GaN layer – laser induced lifting off [22] or inductively coupled plasma reactive ion etching [23]. Up to 2 inch size free-

---

[*] Corresponding author: e-mail: aolszyna@meil.pw.edu.pl

standing GaN films of a few hundred micrometers thickness were achieved but not good enough for commercialisation.

2 Aim of the study

Recently we proposed another approach to obtain freestanding GaN wafers. Instead of attempting crystalline growth using high cost methods, we proposed to use extremely simple substrates and cheaper growth method. Lime glass and atomic layer epitaxy were used as substrates coated with thin films of ZnO, using modification of chemical vapour deposition method so called remote plasma enhanced laser induced CVD (LCVD) [24]. High flatness of ZnO/glass films allowed using them as substrates for GaN growth. The so-obtained GaN films were polycrystalline. Lime glass and ZnO buffer could easily be etched away. The obtained a polycrystalline freestanding layer of GaN could be recrystallized by a low temperature (below 570 °C) annealing.

Unfortunately, ZnO is not resistant to ammonia treatment when growth temperature is higher than 650 °C. This limited the range of methods which allow deposition of GaN on ZnO. Success of this approach, but also its limitations, motivated us to grow thin films of selected wide band gap nitrides, deposited on silicon, as potential substrate materials for GaN epitaxy. For a growth method we selected a modification of deposition method called either impulse plasma deposition (IPD) or impulse plasma assisted chemical vapour deposition (IPCVD). The method was developed by us [25,26]. This method shows several superior properties. First, it is not substrate limited, so films can be deposited on variety of substrates including lime glass, as in our initial attempt to get freestanding GaN, substrate can be at room temperature and, which is important for a present study, allows to obtain relatively oxygen free nitride films. IPCVD method uses a crystallization process occurring in nonisothermic conditions and pulses of entirely ionized plasma. The nucleation on ions leads to nanocrystallinity of deposited layers.

3 Experimental methods

We have grown thin films of GaN, AlN, sapphire and diamond on silicon, as potentially perfect substrates for GaN-based devices. All the materials used in the experiment were grown in the form of layers by impulse plasma assisted chemical vapour deposition method. A schematic of the apparatus and deposition parameters are shown in Fig. 1 and are given in Table 1.

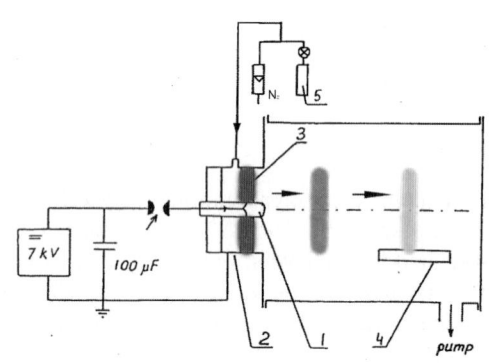

Fig. 1 Experimental set up used for the growth of nitride films. In the scheme equipment elements are denoted with: 1-internal electrode, 2-external electrode, 3-plasma disc, 4-substrate, 5-reactant vapours.

The photoluminescence (PL) experiments were performed at 2 K using a 351 nm UV line of an Ar+ laser for excitation or a frequency tripled YAG:Nd laser ($\lambda = 355$ nm) with 2 ns pulses and a varying repetition rate. The reported measurements were performed at 2 K in a He bath cryostat. The micro-photoluminescence (micro-PL) spectra were measured with either a Renishaw micro-Raman system, with a resolution of 1.6 meV using a polarised 325 nm excitation at an excitation power density of 120 kW/cm^2. The PL emission spectra at a spatial resolution down to 2x2 µm^2 and were measured at room temperature.

Table 1 Deposition parameters.

Energy released in single impulse (J)	900
Impulse duration (s)	$100 \cdot 10^{-6}$
Substrate temperature (K)	300
Carrier gas	N2
Reactant vapour	Ga(CH3)3, Al(CH3)3

Characterization techniques included electron diffraction, cathodoluminescence (CL) and atomic force microscopy (AFM). The CL spectra images were taken in a JEOL 35C scanning electron microscope equipped with a MonoCL2 CL system by Oxford Instruments. CL emission was detected using Hamamatsu R943-02 Peltier cooled photomultiplier. Quantitative information on microstructure of surfaces of the samples was drawn from the atomic force microscopy (AFM) investigations, using Multi Mode Scanning Probe Microscope, and from high-resolution SEM images taken in a JEOL JSM-6300F scanning microscope with a field emission gun.

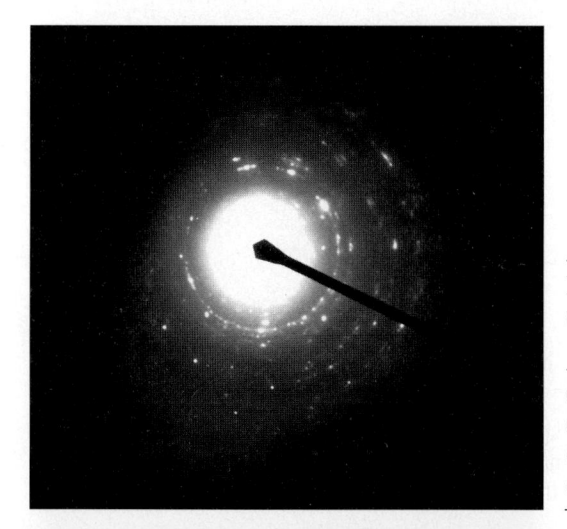

Experimental d (nm)	GaN PDF 4-1078 (ICDD) d (nm)
0,270	0,276
0,239	0,243
0,166	0,159
0,133	0,133

Experimental d (nm)	AlN PDF 8-262 (ICDD) d (nm)
0,268	0,269
0,177	0,182
0,153	0,155
0,132	0,132

Fig. 2 Electron diffraction patterns for (top) w-GaN and (bottom) w-AlN films.

4 Results and discussion

Electron diffraction patterns (shown in Fig. 2) confirm growth of thin films (100 nm thick) of AlN and GaN. In Fig. 3 (a, b) we show AFM images of GaN (a) and AlN (b) thin films deposited on silicon. Both these materials show quasimonomodal nanocrystalline microstructure with grains of about 140 nm size for GaN films and 100 nm for AlN films. Larger sizes of the grains, observed in the case of GaN, likely result in the increased roughness of the surfaces, which are indicated by the so-called RMS (root mean square) parameters. RMS is about 19.8 nm for GaN film and decreases from 1.57 nm (89 nm thick

AlN) to 0.44 nm for 120 nm thick film. AlN films are thus exceptionally smooth, even though they are very thin.

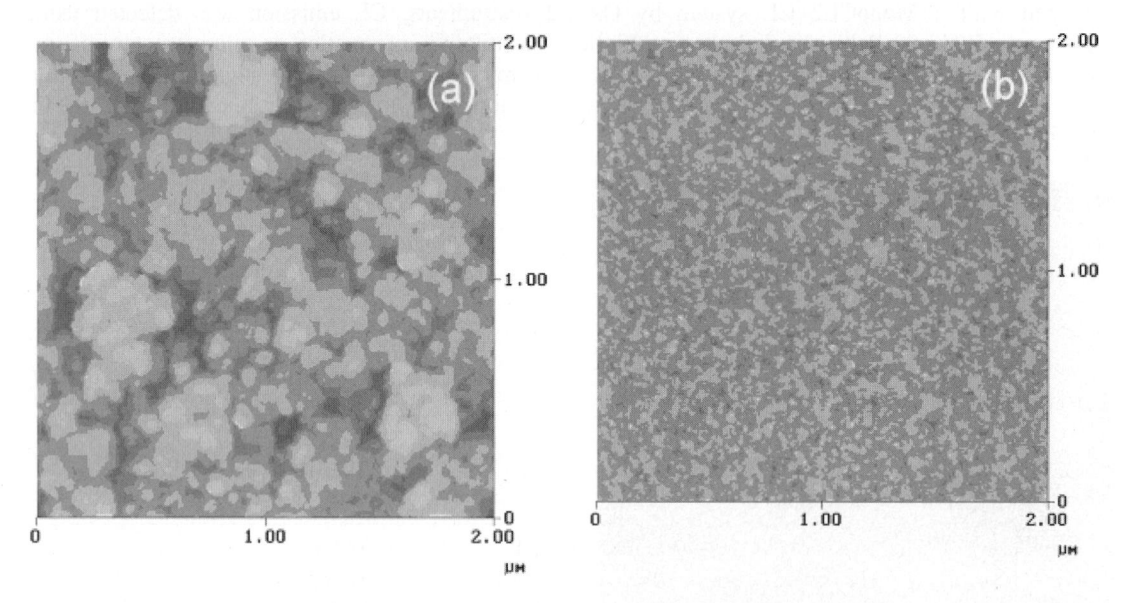

Fig. 3 AFM images for 100 nm thick GaN (a) and 120 nm thick AlN (b) layers deposited on silicon with IPD method. Maximal scale of height fluctuations is 200 nm for the (a) image and 10 nm for the (b) image.

Exceptionally good smoothness of the AlN film resulted also in the observation of sharp PL emission lines in UV in band gap region, not discussed here, and rather weak and broad PL/CL emission peaked at about 2.0 eV. This emission (a) and its in-plane homogeneity (b) are shown in Fig. 4.

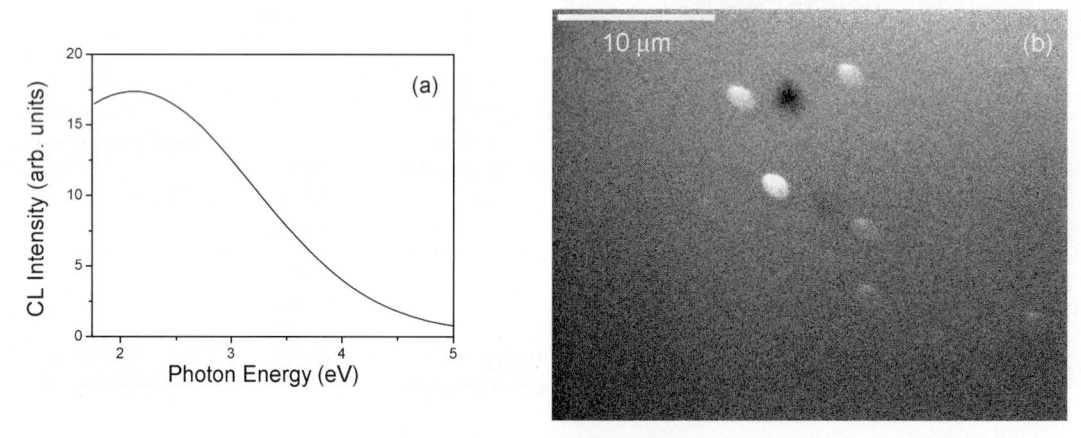

Fig. 4 Room temperature CL spectrum of 90 nm thick AlN layer (a). In (b) we show in-plane homogeneity of this emission.

For GaN broad PL/CL emission was observed, which was peaked in yellow color spectral range. This is a well-known parasitic emission of GaN, as discussed in the Ref. [1]. The relevant CL spectrum (a) and in-plane homogeneity of the emission (b) are shown in Fig. 5. The latter spectrum was measured with CL setting detection at the maximum of the emission and scanning exciting electron beam in a selected region of the sample. Micro-scale fluctuations are observed, reflecting granular structure of the film. GaN films shows also weak and rather broad edge emission peaked close to 3.5 eV and 3.2 eV. These emissions could also be observed under high density of excitation, realized in micro-PL study.

© 2004 WILEY-VCH Verlag GmbH & Co. KGaA, Weinheim

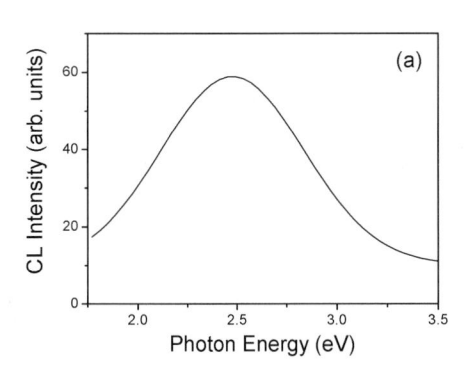

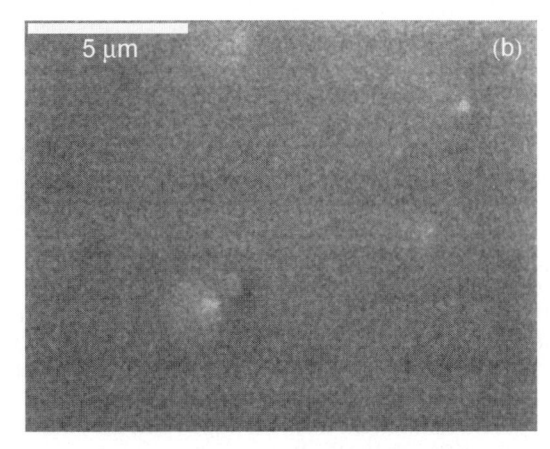

Fig. 5 Room temperature CL spectrum of 100 nm thick GaN layer (a). In (b) we show in-plane homogeneity of this emission.

Important property of our samples is their large-scale homogeneity. In addition to the one shown in Figs. 5 and 6 we studied PL homogeneity using micro-PL and scanning emission in larger regions with several micrometers steps. These measurements indicate that the growth method used by us enables homogeneous covering of large area substrates (silicon in our case). Use of silicon is fortunately not the process-related. Thin films of nitrides can be deposited on other substrates, including lime glass. Thus flexibility of the method and its low costs are highly encouraging.

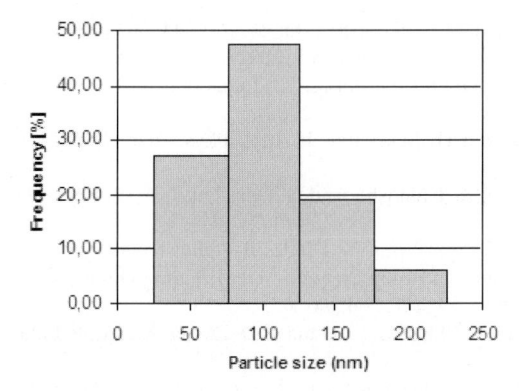

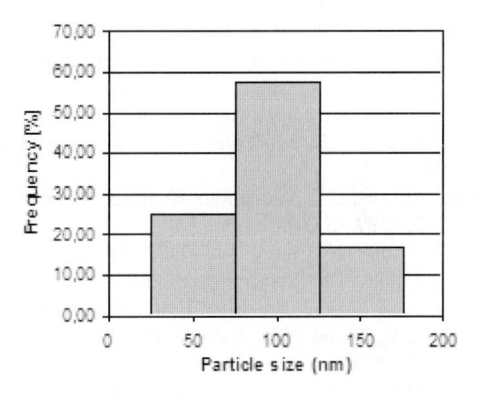

Fig. 6 Grain size distribution in 90 nm thick AlN layer (left) and in 100 nm thick GaN layer (right).

As already mentioned, micro-scale fluctuations of the PL and CL intensity are observed, reflecting granular structure of the films. From these intensity instabilities we can estimate grain size distributions in AlN (left) and GaN (right) layers. The relevant results are shown in Fig. 6, in which we show size distributions for the two nitrides studied. The observed size distributions are in the range of those concluded from AFM experiments. This indicates that from CL studies we can independently estimate grain sizes.

4 Conclusion

Thin films of selected nitrides are deposited using IPCVD method. We conclude that these films can be attractive substrates for GaN technology to obtain freestanding GaN layers.

Acknowledgements This work was partly supported by the grant number 8T11B 07319 of KBN.

References

[1] T.L. Tansley, E.M. Goldys, M. Godlewski, B. Zhou, and H.Y. Zuo, in Optoelectronic Properties of Semiconductors and Superlattices, series editor M.O. Manasreh, Vol. 2, "GaN and Related Materials", ed. S.J. Pearton, Gordon and Breach Publishers, 1997, p. 233–293, and references therein.
[2] H. Amano, N. Sawaki, I. Akasaki, and Y. Toyoda, Appl. Phys. Lett. **48**, 353 (1986).
[3] H. Amano, T. Asahi, and I. Akasaki, Jpn. J. Appl. Phys. **29**, L205 (1990).
[4] T. Sasaki and T. Matsuoka, J. Appl. Phys. **77**, 192 (1995).
[5] S. Nakamura, Jpn. J. Appl. Phys. **30**, L1705 (1991).
[6] H. Kawakami, K. Sakurai, K. Tsubougchi, and N. Mikoshiba, Jpn. J. Appl. Phys. **27**, L161 (1988).
[7] A. Yamamoto, M. Tsujino, M. Ohkubo, and A. Hashimoto, J. Cryst. Growth **137**, 415 (1994).
[8] K. Uchida, A. Watanabe, F. Yano, M. Kouguchi, T. Tanaka, and S. Minagawa, J. Appl. Phys. **79**, 3487 (1996).
[9] B. Pecz, M.A. di Forte-Poisson, F. Huet, G. Radnoczi, L. Toth, V. Papaioannou, and J. Stoemenos, J. Appl. Phys. **86**, 6059 (1999), and references therein.
[10] A.J. Ptak, K.S. Ziemer, M.R. Millecchia, C.D. Stinespring, and T.H. Myers, MRS Internet J. Nitride Semicond. Res. 4S1, G3.10 (1999).
[11] N. Ohshima, H. Yonezu, S. Yamahira, and K. Pak, J. Cryst. Growth **189/190**, 275 (1998).
[12] M. Godlewski and E. M.Goldys, chapter in III-Nitride Semiconductors: Optical Properties Vol. II (Optoelectronic Properties of Semiconductors and Superlattices), Hongxing Jiang and M. Omar Manasreh (Co-editors), Taylor & Francis Books, New York, 2002, pp. 259.
[13] V.Yu. Ivanov, M. Godlewski, H. Teisseyre, P. Perlin, R. Czernecki, P. Prystawko, M. Leszczynski, I. Grzegory, T. Suski, and S. Porowski, Appl. Phys. Lett. **81**, 3735 (2002).
[14] R. Stępniewski, A. Wysmołek, K.P. Korona, and J.M. Baranowski, chapter in III-Nitride Semiconductors: Optical Properties Vol. I (Optoelectronic Properties of Semiconductors and Superlattices), Hongxing Jiang and M. Omar Manasreh (Co-editors), Taylor & Francis Books, New York, 2002, p. 197.
[15] Y. Oshima, T. Eri, M. Shibata, H. Sunakawa, K. Kobayashi, T. Ichihashi, and A. Usui, Jpn. J. Appl. Phys. (Part 2) **42**, L1 (2003).
[16] K. Tomita, T. Kachi, S. Nagai, A. Kojima, S. Yamasaki, and M. Koike, phys. stat. sol. (a) **194**, 563 (2002).
[17] Y. Kumagai, H. Murakami, H. Seki, and A. Koukitu, J. Cryst. Growth **246**, 215 (2002).
[18] K. Motoki, T. Okahisa, S. Nakahata, N. Matsumoto, H. Kimura, H. Kasai, K. Takemoto, K. Uematsu, M. Ueno, Y. Kumagai, A. Koukitu, and H. Seki, J. Cryst. Growth **237–239**, 912 (2002).
[19] Y. Kumagai, H. Murakami, A. Koukitu, K. Takemoto, and H. Seki, Jpn. J. Appl. Phys. (Part 2) **39**, L703 (2000).
[20] A. Wakahara, T. Yamamoto, K. Ishio, A. Yoshida, Y. Seki, K. Kainosho, and O. Oda, Jpn. J. Appl. Phys. (Part 1) **39**, 2399 (2000).
[21] S.T. Kim, S.H. Chung, and D.C. Moon, in Blue Laser and Light Emitting Diodes II. (Ohmsha, Tokyo, Japan, 1998) pp.654-5. Proceedings of 2nd International Symposium on Blue Laser and Light Emitting Diodes. Chiba, Japan. 29 Sept.-2 Oct. 1998, Editors: Onabe K; Hiramatsu K; Itaya K; Nakano Y.
[22] J. Xu, R. Zhang, Y.P. Wang, X.Q. Xiu, B. Shen, S.L. Gu, Y. Shi, Z.G. Liu, and Y.D. Zheng, Materials Letters **56**, 43 (2002).
[23] In Jae Song, Seong Kuk Lee, Kyoyeol Lee, Sung Soo Park, and Jae Yong Han, Jpn. J. Appl. Phys. **41** (Part 2), L317 (2002).
[24] K.S.A. Butcher, Afifuddin, P.-T. Chen Patrick, M. Godlewski, A. Szczerbakow, E.M. Goldys, and T.L. Tansley, J. Cryst. Growth **246**, 237 (2002).
[25] A. Sokołowska, A. Olszyna, A. Michalski, K. Zdunek, Surface and Coat. Tech. **47**, 141 (1991).
[26] A. Olszyna, Thin Solid Films **79**, 277 (1996).

Study of long-term stability of ohmic contacts to GaN

E. Kaminska[1], K. Golaszewska[*,1], A. Piotrowska[1], A. Kuchuk[1], R. Kruszka[1], E. Papis[1], R. Szeloch[2], P. Janus[1,2], T. Gotszalk[2], and A. Barcz[3]

[1] Institute of Electron Technology, Al. Lotnikow 32/46, 02-668 Warsaw, Poland
[2] Wroclaw University of Technology, ul. Janiszewskiego 11/17, 50-372 Wroclaw, Poland
[3] Institute of Physics, Polish Academy of Sciences, Al. Lotnikow 32/46, 02-668 Warsaw, Poland

Received 15 September 2003, accepted 27 October 2003
Published online 22 December 2003

PACS 68.60.Dv, 73.40.–c, 81.05.Ea, 81.70.Pg, 85.30.–z

We report on low-resistivity thermally stable ohmic contacts to p-GaN using ZrN/ZrB$_2$ metallisation. Transport properties, thermal conductivity and long-term stability of contacts were examined. p-GaN/ZrN/ZrB$_2$ contacts show excellent stability upon aging in air, indicating their suitability for long-term operation at temperatures up to 150 °C.

1 Introduction GaN's intrinsic properties such as wide bandgap, high thermal conductivity, high melting temperature, high breakdown voltage and high saturation velocity make it a material of choice for high temperature and high power electronic devices [1]. During the last decade, significant progress in both material and processing technologies as well as in the design of new device structures has been made. Now the most important and challenging problem is to master the reliability of GaN-based devices.

In this paper we address the issue of long-term stability of ohmic contacts to GaN. Electrical contacts that are structurally stable at high temperature are the prerequisite for reliable operation of GaN electronic devices. Zr-based metallisation was chosen as a contact material because of high electrical conductivity, superior thermal stability and chemical resistance against corrosion, making it compatible with GaN. Zr/ZrN metallisation was reported effective in providing low-resistivity contacts to n-type GaN by depleting its superficial film from nitrogen, creating thus a highly doped subcontact region and a stable ZrN compound at the contact interface [2]. In our previous studies, ZrN/ZrB$_2$ metallisation was used for p-type GaN contacting purposes. The presence of Zr in the metallisation was a key factor for the formation of the ohmic contact by removal hydrogen from the subcontact layer and creation thereby a highly doped p$^+$ region [3]. Moreover, structural stability of ZrN/ZrB$_2$ metallisation deposited on GaN upon annealing up to 1100 °C in N$_2$ has been proven [4]. In the present study we have investigated the thermal conductivity of p-GaN/ZrN/ZrB$_2$ ohmic contacts and performed accelerated lifetime testing of these contacts. It is well established that efficient heat removal is critical to the performance of semiconductor devices. Thus for thermal management of the device, it is important to know accurately the value of the thermal conductivity of metallisation and semiconductor subcontact region. Thermal conductivity investigations were performed using Scanning Thermal Microscopy (SThM), enabling to couple the topographic image with the thermal conductivity information [5]. To age the contacts they were annealed in air at temperatures up to 150 °C over a period of 300 hours. The aging process was monitored by specific contact resistance measurements and compositional depth profiling of the contacts.

---

[*] Corresponding author: e-mail: krystyg@ite.waw.pl, Phone: (48-22) 5487 949, Fax: (48-22) 847 06 31

2 Experimental details

The substrate materials for this study were 2 μm thick Mg doped to ~3×10^{19} at.cm^{-3} GaN epilayers grown by MOCVD on a sapphire substrate. The concentration of holes from Hall effect measurements was $p = 5 \times 10^{17}$ cm^{-3}. Zr-based metallisation, consisting of ZrN(50nm)/ZrB$_2$(50nm) bilayer was deposited by sequential DC magnetron sputtering in Ar discharge from ZrN and ZrB$_2$ targets, respectively. The microstructure of both ZrN and ZrB$_2$ films was amorphous. Their resistivity, as determined from four point probe measurements, was 200 μΩcm and 150 μΩcm, respectively.

The patterns for contact resistivity evaluation were formed using photolithography and lift-off techniques. Contacts were annealed in a RTP system at 800 °C for 60 s. in flowing N$_2$. Subsequent to annealing, 100 nm thick Ag or Au overlayer was deposited for bonding.

Contact resistivity (r_c) was evaluated by circular transmission line method (cTLM) using test pattern with contact pads separation varying from 10 to 60 μm. Aging experiments were done using a Keithley 2400 source-meter equipped with a microprobe and hot-plate enabling thermal measurements in air. Depth profiling by secondary ion mass spectrometry (SIMS) was performed with a Cameca 6F instrument, using cesium primary ion beam and detection of CsX$^+$ secondary cluster ions.

Thermal conductivity (κ) measurements were done using a combined SThM/AFM system [6]. The calibration procedure made it possible to evaluate absolute values of local thermal conductivity coefficient with submicron spatial/depth resolution [7]. The system allowed measuring local temperature with thermal resolution of 5 mK and thermal conductivity with a resolution of 10^{-2} Wm^{-1}K^{-1}.

3 Results and discussion

The electrical characteristics of p-GaN/ZrN/ZrB$_2$/Ag ohmic contacts are shown in Fig. 1. Contact resistivities of 7×10^{-5} Ωcm^2 were obtained. The temperature dependence of r_c was also evaluated. As shown in Fig. 2, contact resistivity remains unaltered by temperature changes, which is indicative that tunnelling via metal/semiconductor interface is responsible for ohmic behaviour.

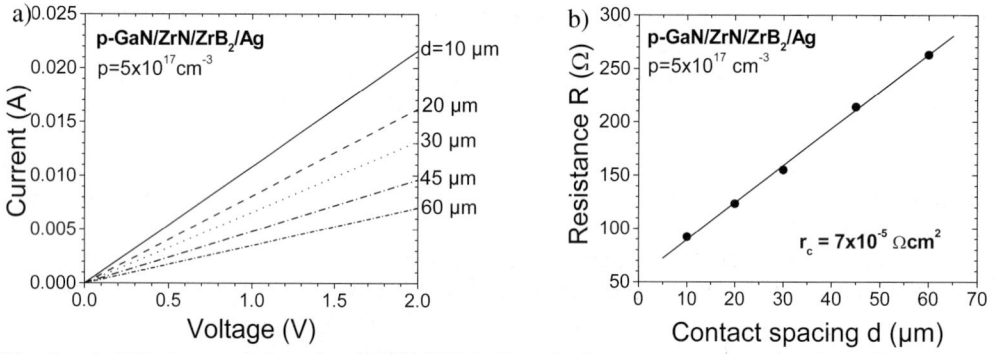

Fig. 1 a) I-V characteristics of p-GaN/ZrN/ZrB$_2$/Ag ohmic contacts as measured on five c-TLM pads, b) c-TLM results of contact resistivity measurements.

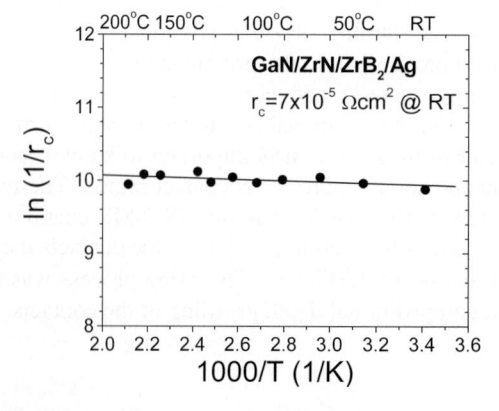

Fig. 2 The temperature dependence of contact resistivity for p-GaN/ZrN/ZrB$_2$/Ag ohmic contacts.

The topographic and the thermal conductivity images of a fragment of ZrN/ZrB$_2$ contact pad on p-GaN acquired simultaneously are juxtaposed in Figures 3a and 3b. In both images the area corresponding to p-GaN appears much darker than that of the Zr-based metallisation indicating that the thermal conductivity of 100 nm thick ZrN/ZrB$_2$ metallisation is significantly higher than that of p-GaN. The thermal conductivity of p-GaN was measured to be 118 Wm^{-1}K^{-1}. For comparison purposes, reported values of κ for GaN vary from 130 Wm^{-1}K^{-1} (400 mm thick HVPE grown GaN) [8] to 220 Wm^{-1}K^{-1} (bulk n-GaN crystals) [9]. The relatively low thermal conductivity of p-GaN can be attributed to extremely high concentration of Mg and high concentration of hydrogen [3]. As for ZrN/ZrB$_2$ metallisation, it is not thick enough to determine the absolute thermal conductivity. Nevertheless very good uniformity of the thermal conductivity should be noted. Thermal aging did not cause changes in the thermal conductivity image of p-GaN/ZrN/ZrB$_2$ contacts.

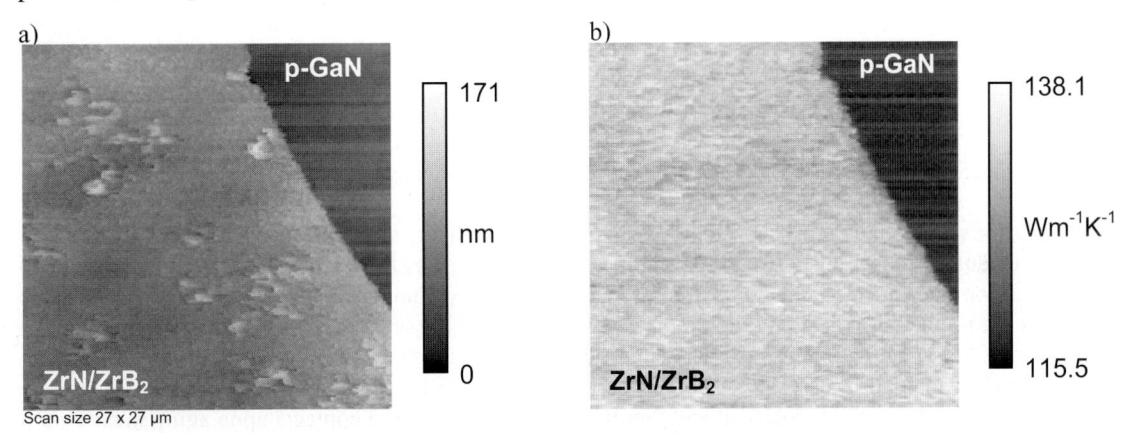

Fig. 3 Topographic (a) and thermal conductivity (b) images of a fragment of ZrN/ZrB$_2$ contact pad on p-GaN substrate.

To evaluate their long-term stability, p-GaN/ZrN/ZrB$_2$/Ag ohmic contacts were aged during two consecutive cycles: at 125 °C and next at 150 °C for 300 hours each. The contact resistivity was measured after each aging step. The results from aging experiments and compositional SIMS profiling of p-GaN/ZrN/ZrB$_2$/Ag contact performed subsequent to aging, are shown in Figures 4 and 5, respectively. The contact resistivity remains unchanged upon these two aging cycles. SIMS analysis confirms the stability of the contact system. Here, particularly significant is the sharpness of the Ag/ZrB$_2$ interface (Ag signal at this interface decreases by 4 orders of magnitude) testifying for the excellent behaviour of Zr-based metallisation as a diffusion barrier.

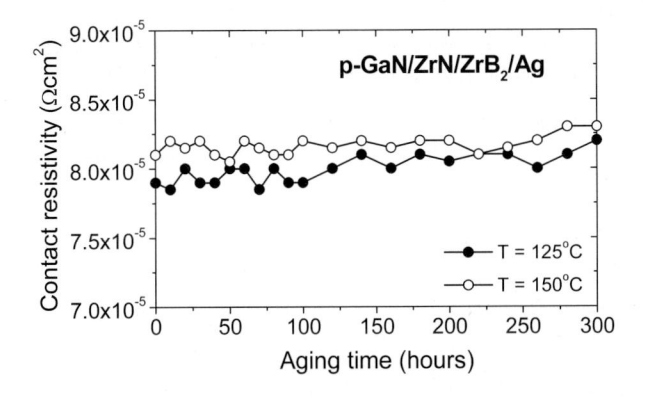

Fig. 4 The resistivity of p-GaN/ZrN/ZrB$_2$ ohmic contacts as a function of aging time.

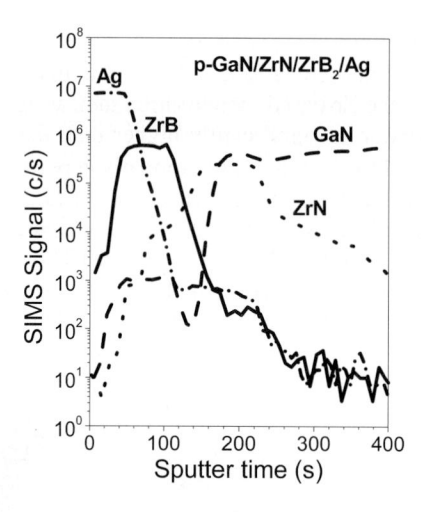

Fig. 5 SIMS profiles for p-GaN/ZrN/ZrB$_2$ ohmic contacts after aging.

3 Conclusions

We have fabricated low-resistivity ohmic contacts to p-GaN using ZrN/ZrB$_2$ metallisation and evaluated their transport and thermal properties, as well as their long-term stability. It has been found that the current transport across the contact interface is dominated by tunnelling of holes. This confirms our previous suggestions regarding the mechanism of the ohmic contact formation by enhanced activation of Mg and creation of a p$^+$-GaN superficial film [3]. Furthermore, we have proven the applicability of SThM method to assess the thermal conductivity of metal/semiconductor contacts. Finally we have demonstrated excellent electrical and thermal stability of Zr-based contacts upon aging in air over a period of 300 hours, indicating their suitability for long-term operation at temperatures up to 150 °C.

Acknowledgements Research is partially supported by grants from European Commission G5RD-CT-2001-00566-DENIS and the State Committee for Scientific Research 7T11B 009 20.

References

[1] S. P. Pearton, F. Ren, A. P. Zhang, and K. P. Lee, Mat. Sci. Eng. R **250**, 1 (2000).
[2] D. Wolter, B. P. Luther, S. E. Mohney, R. F. Karlicek, Jr., and R. S. Kern, Electrochem. and Solid-State Lett. **2**, 151 (1999).
[3] E. Kaminska, A. Piotrowska, A. Barcz, D. Bour, M. Zielinski, and J. Jasinski, Mat. Sci. Eng. B **82**, 265 (2001).
[4] J. Jasinski, E. Kaminska, A. Piotrowska, A. Barcz, and M. Zielinski, Mat. Res. Soc. Symp. Proc. **622**, T6.34.1, (2000).
[5] F. Ruiz, W. D. Sun, and F. H. Pollak, C. Venkatraman, Appl. Phys. Lett. **73**, 1802 (1998).
[6] R. F. Szeloch, T. P. Gotszalk, and P. Janus, Microelectron. Reliability **42**, 1719 (2002).
[7] P. Janus, Thermography in near- and far-field in microsystem diagnostics, PhD thesis, Wroclaw University of Technology, 2003.
[8] E. K. Sichel and J. I. Pankove, J. Phys. Chem. Solids **38**, 330 (1977).
[9] A. Jezowski, B. A. Danilchenko, M. Bockowski, I. Grzegory, S. Krukowski, T. Suski, and T. Paszkiewicz, Solid State Commun. **128**, 69 (2003).

Optical characterization of GaN doping superlattices: as grown, hydrogen implanted, and annealed

H.W. Kunert[*,1], **D.J. Brink**[1], **M. Hayes**[1], **J. Malherbe**[1], **L. Prinsloo**[2], **J. Barnas**[3], **A.G.I. Machatine**[1], and **M.W. Diale**[1]

[1] Department of Physics, University of Pretoria, 0001 Pretoria, South Africa
[2] Department of Chemistry, University of Pretoria, 0001 Pretoria, South Africa
[3] Department of Physics, Adam Mickiewicz University, ul. Umultowska 85, 61-614 Poznan, Poland

Received 15 September 2003, revised 3 October 2003, accepted 3 October 2003
Published online 22 December 2003

PACS 63.20.–e, 78.30.Fs

Hydrogen–ion implantation was studied for GaN doping superlattices. The samples were grown epitaxially with eight periods of n-GaN/p-GaN (50 nm/50 nm) doped with 10^{18} cm^{-3} (n-Si) and 10^{18} cm^{-3} (p-Mg), and with a thin Aluminium Nitride layer (AlN) nucleation layer (20 nm) on a (0001) sapphire substrate. Low temperature photoluminescence and Raman spectroscopy were used to characterize the as grown and hydrogen implanted (dose 10^{12}–10^{17} cm^{-2}, energy 1 MeV) samples. In the studied (relatively narrow) region of the laser power, the as grown samples show neither tunable effective band gap nor the subband structure. Implantation induces the yellow band (YL) which is not present in the spectra of as grown samples. The peak position of the YL band shifts towards higher energy with increasing implantation dose. Raman spectroscopy reveals implantation induced broad band centred at 1618 cm^{-1}, with many shoulders and several sharp peaks in the region of 2127–1411 cm^{-1}.

© 2004 WILEY-VCH Verlag GmbH & Co. KGaA, Weinheim

1 Introduction

Recently much attention has been paid to GaN quantum wells and superlattices. Good quality quantum wells with AlGaN barriers have been manufactured and characterized by Khan et al [1] and by Itoch et al [2] (see also the review article by Jain et al [3]). However, much less is known about GaN doping superlattices (DSL's). With our design parameters: $d_n = d_p = 50$ nm, and doping $N_d = 10^{18}$ cm^{-3} (Si), $N_a = 10^{18}$ cm^{-3} (Mg) one could expect similar tunable effects as in GaAs DSL's [4]. It is well known that in heavily doped semiconductors the band tails are formed due to concentration fluctuation. The effective band gap changes due to the band tails and also due to carrier interactions. The luminescence in GaAs doping superlattices is strongly affected by the spatial separation between carriers and by extremely long recombination lifetimes of excess electrons and holes. Consequently, the shape and position of the peaks in photoluminescence (PL) spectra are tunable quantities. That may not be the case for GaN DSL's. In this paper we demonstrate that optical properties of GaN DSL's are different from those of GaAs DSL's.

2 Photoluminescence spectroscopy of GaN DSL's

Figures 1 and 2 show the PL spectra of as grown GaN DSL's for different values of the laser power and different temperatures, respectively. The spectra consist of Donor–Acceptor-Pair (DAP) emissions in the region of 3.275–2.945 eV and several well resolved bands centred at 2.82, 2.7, 2.6, 2.467, 2.348, 2.23, and 2.09eV. Positions of the bands do not shift with the change of the laser power and temperature. Such

---

[*] Corresponding author: e-mail: hwkunert@postino.ac.za

a behaviour is not a usual feature of typical superlattices. For example, np GaAs (n-Te, p-Zn doping) DLS's exhibit a significant shift of the PL peak positions with temperature and laser power [4]. However, the only effect of such a treatment in the case of GaN DSL's is a drop of intensities for the corresponding transitions.

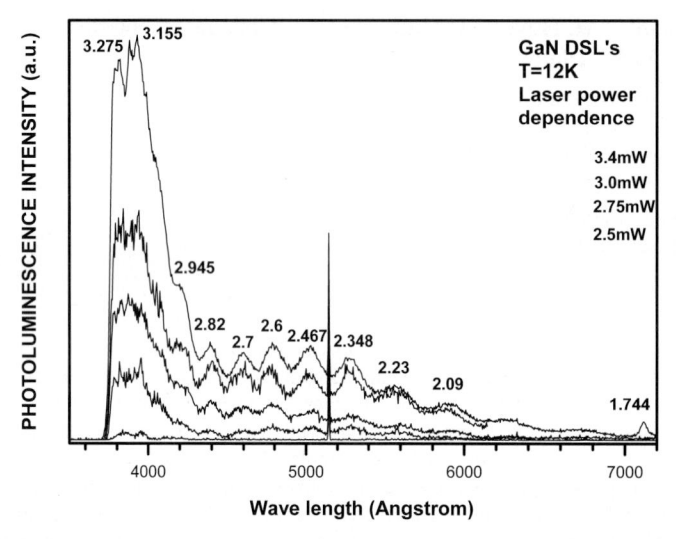

Fig. 1 Photoluminescence spectra from GaN doping superlattices: laser power dependence.

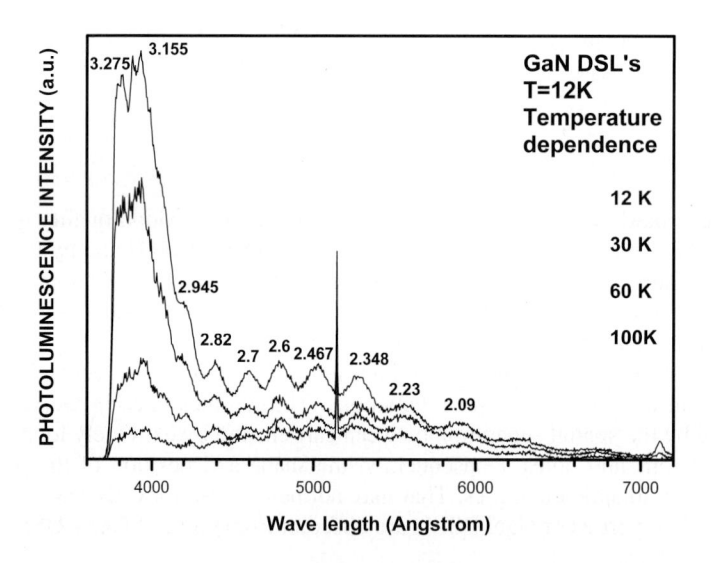

Fig. 2 Photoluminescence spectra from GaN doping superlattices: temperature dependence.

The PL spectra from hydrogen implanted samples are shown in Fig. 3. It is evident that the entire DAP band shifts towards lower energy. At the highest dose of hydrogen implantation, 10^{17} cm^{-2}, the DAP band shifts about 230 meV. The implantation also induces the YL band which is absent in PL spectra from as grown samples. The energy position of the implantation-induced YL band depends on the implantation dose. Finally, the features between 1.44–1.68 eV in Fig. 3 are the second order peaks.

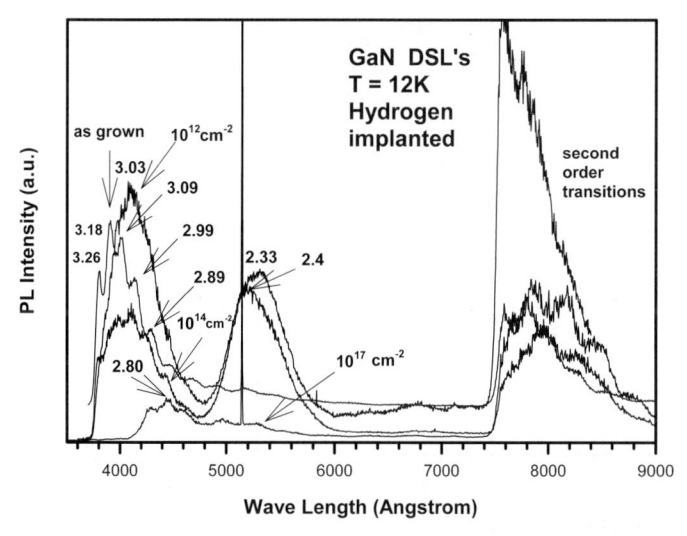

Fig. 3 Photoluminescence spectra from H- implanted GaN doping superlattices.

3   Raman Spectroscopy from GaN DSL's

Raman spectra from implanted samples are presented in Fig. 4. Only two Raman transitions at 572 cm^{-1} and 734 cm^{-1} are well resolved, which are usually observed in bulk GaN. Only a negligible low-energy shift of the modes is observed. The sapphire mode at 419 cm^{-1} is also affected by implantation, indicating that hydrogen ions reached the substrate. The second order Raman spectra are much more influenced by the implantation and doping, as shown in Fig. 5. For comparison we show in Fig. 6 the second order Raman spectra collected from bulk GaN/SA at different spots.

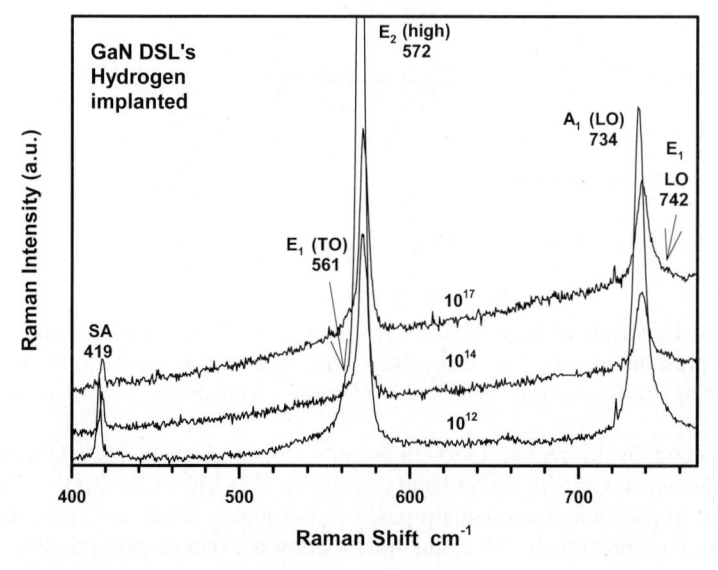

Fig. 4 Raman spectra from hydrogen implanted GaN doping superlattices.

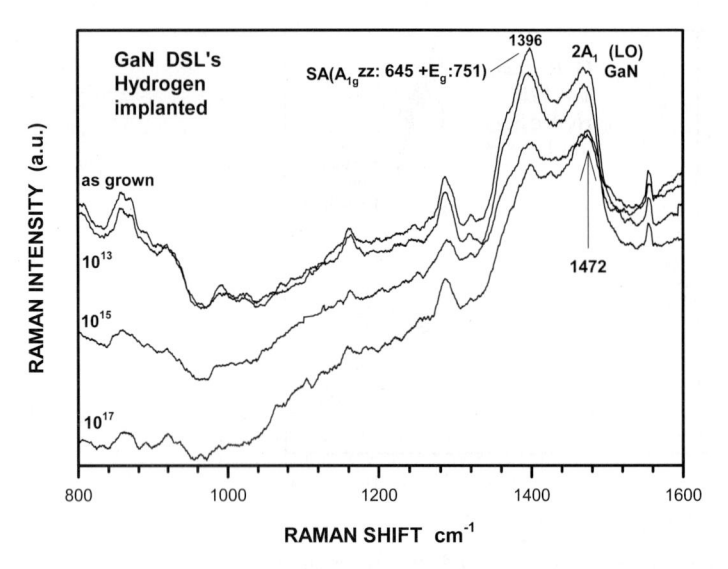

Fig. 5 Second order Raman spectra from hydrogen implanted GaN doping superlattices.

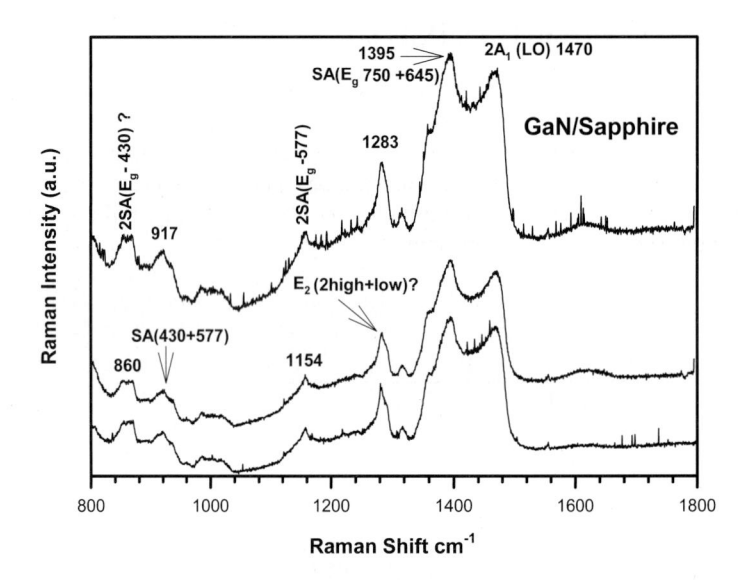

Fig. 6 Second order Raman spectra from GaN bulk.

Results of the group-theoretical analysis of two-phonon (overtones and combinations) Raman selection rules for GaN/SA [5] are presented in the Fig. 6 together with experimental spectra. Clearly, the two-phonon transitions originating from sapphire: 860, 917, 1154 cm^{-1}, vanish due to doping and implantation.

Figure 7 displays Raman spectra in the region 1550–2500 cm^{-1} from implanted GaN DSL's. The hydrogen dose of 10^{17} cm^{-2} induces an optically active band centred at 1618 cm^{-1} (200 meV), which is absent in as grown GaN DLS's. In addition, several sharp peaks in the region 2120–2045 cm^{-1} appear. These bands have also been observed previously [6]. In the next section we will discuss possible origin of the observed features.

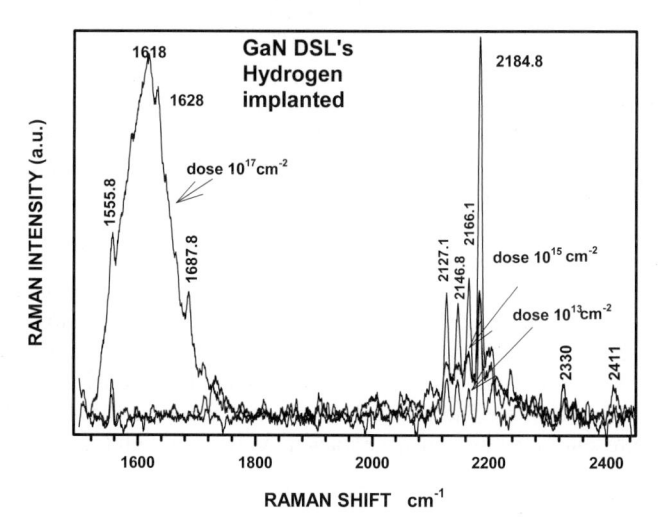

Fig.7 High-frequency spectra from H- implanted GaN doping superlattices

4 Modelling and discussion

4.1 PL Features

The design parameters, $d_n = d_p = 50$ nm, $N_d = N_a = 10^{18}$ cm^{-3} with eight periods of n-GaN/p-GaN, are very similar to those for GaAs DSL's studied in Ref. [4], where the main optically active PL band originates from the recombination between the spatially separated electrons (Te) in the conduction subbands and holes in the acceptor (Zn) impurity band, across the indirect gap in real space (see Fig.23 in Ref.[4]). This band is very strongly laser power and temperature dependent [6–9] because the carrier concentration is tunable and also due to the enhancement of the lifetime. In the effective mass (EFM) approach, the effective band gap and subband structure are also tunable. The fact that several PL optically active transitions from GaN DSL's present in Fig. 1 are independent of temperature and laser power may imply that we have new DAP band, induced by impurities in n (Si) and p (Mg) layers. The question is whether the Ploog and Doehler [4] EFM approach is applicable to the GaN DSL's. Applying the GaN DSL's design parameters to the periodic potential V_0, effective band gap $E_g^{eff,0}$ and the subband energies;

$$V_0 = (4\pi e^2 / \kappa_0) N_d d^2 / 8 , \qquad (1)$$

$$E_g^{eff,0} = E_g^0 - 2V_0 + \varepsilon_{c,0} + \varepsilon_{vh,0} , \qquad (2)$$

$$\varepsilon_{c,\mu} = (h/2\pi)(4\pi e^2 N_d / \kappa_0 m_c)^{1/2} (\mu + 1/2) , \qquad (3)$$

where all the quantities have their usual meaning [4], we obtain for the ground state: $V_0 = 0.6353$ eV, $\varepsilon_{c,0} = 13.58$ meV, $\varepsilon_{v,0} = 8.03$ meV, and $E_g^{eff,0} = 2.251$ eV. The energy separation of conduction subbands is 27.16 meV, whereas for the Mg–acceptor holes slightly above the valence band the separation is 1.14 meV. The peak energy separation between PL bands (Figs. 1, 2) in the region 2.8–2.09eV is about 110 meV. The effective band gap under illumination increases (see Fig. 19 in [4]) while the energy separation between subbands decreases. Taking into account our numerical values for the $E_g^{eff,0}$ in the ground state, it is still possible to associate the PL transitions in Figs. 1 and 2 with the recombination of electrons and holes across indirect gap in real space (between layers). However, the PL spectra do not show an appreciable shift of energy peak positions with the laser power. This could be due to the very narrow region (2.5–3.4 mW) of the available power of our facility.

In short, further experimental studies are needed to clarify the origin of the PL transitions. Particularly helpful would be "spin-density (single particle) excitation" experiment by means of Raman scattering spectroscopy, see Ref. [4, p. 351]. By this technique one measures the intersubband energies. Therefore, the experimental numerical values of ε under illumination would be available.

The main feature of Fig. 3 is the hydrogen implantation induced YL band centred at 2.33 eV (for the dose of 10^{12} cm^{-2}) and at 2.4eV (for the dose of 10^{17} cm^{-2}). This band is usually associated with deep levels caused by Ga or nitrogen vacancies. The YL band involves transitions from shallow donor states to the deep acceptor state. Implanted hydrogen impurities can be incorporated up to four different states at a gallium vacancy. The presence of H causes the vacancy energy level to split and shift towards the valence band. The shallow donor states are most likely the Si impurities in n-layers and implanted hydrogen, while the deep acceptors are likely Mg impurities and nitrogen vacancies. The latter could be responsible for unusual high energy peak positions (2.33 and 2.4 eV) of the YL band. However, one should note that in bulk GaN the YL band is due to transitions between shallow silicon donors and gallium or nitrogen vacancies. The n layers of the GaN DSL's are heavily doped by silicon, and despite of that the YL band is absent in the as grown samples. Therefore, there are no transitions across indirect gap in real space. Nevertheless, the DAP present in the as grown samples may involve such transitions.

4.2 Raman spectroscopy feature

Figures 4 to 7 display the Raman spectra of bulk GaN and as grown and hydrogen implanted GaN DSL's. The most striking results are presented in Figs. 5 and 7. The second order Raman spectra are the ones which are most influenced by implantation. Figure 5 (see Fig. 6 for comparison) shows that at the highest dose of implantation, 10^{17} cm^{-2}, several combinations and overtones of the first order vibrational modes disappear. The first order modes experience only a slight low-energy shift. The Raman spectra collected in the region 1550–2500 cm^{-1} exhibit a broad hydrogen implantation induced structured band of appreciable intensity centred at 1618 cm^{-1}. The origin of the band is unknown. It may originate from hydrogen–silicon–magnesium complexes or extended defects (of about 200 meV energy). Several sharp peaks appear at around 2120 to 2415 cm^{-1}. We tentatively assign these modes to H-decorated defects [10] such as N vacancy (V_N) and extended defects or possibly to Ga–H complexes generated by the high dose (10^{17} cm^{-2}) implantation. In conclusion, PL and RS reveal several interesting features of as grown and H-implanted GaN DSL's. Within the limit of our PL experimental facilities (narrow region of the laser power) it is not clear whether the GaN DSL's exhibit typical tunable doping superlattices properties such as GaAs - DSL's [4].

References

[1] M.A. Khan, R.A. Skogman, and J.M. van Hove, Appl. Phys. Lett. **56**, 257 (1990).
[2] K. Itoh, T. Kawamoto, H. Amano, K. Hoiramatsu, and I. Akasaki, Jpn. J. Appl. Phys. **30**, 1924 (1991).
[3] S.C. Jain, M. Willander, J. Narayan, and R. Van Overstraeten, J. Appl. Phys. **87**, 965 (2000).
[4] K. Ploog and G.H. Doehler, Adv. Phys. **32**, 285 (1983).
[5] H.W. Kunert, Appl. Surf. Sci. **212–213**, 890 (2003).
[6] H.W. Kunert, J.B.Malherbe, and D.J Brink, Appl. Suf. Sci. **135**, 29 (1998).
[7] H.W. Kunert, D.J. Brink, A. Donnadieu, K. Zeaiter, C. Llinares, J. Allegre, G. Leveque, and J. Camasel, phys. stat. sol. (b) **210**, 699 (1998).
[8] H.W. Kunert, J.B. Malherbe, D.J. Brink, R.Q. Odendaal, L.C. Prinslo, J. Camassel, J. Allegre, K. Zeaiter, C. Linares, Appl. Surf. Sci. **166**, 77 (2000).
[9] H.W. Kunert, D. Dale, M. Hayes, J. Malherbe, V. Kononenko, and J. Barnas, Cryst. Res. Technol. **38** 344 (2003).
[10] H. Harima, T. Inoue, Y. Sone, S. Nakashima, M. Ishida, and M. Taneya, phys. stat. sol. (b) **216**, 789 (1999).

Energy transfer to Er^{3+} ions in silicon-rich-silicon oxide: efficiency limitations

D. Kuritsyn[*,1,2], **A. Kozanecki**[1], **H. Przybylińska**[1], and **W. Jantsch**[3]

[1] Institute of Physics, Polish Academy of Sciences, Warsaw, Poland
[2] Institute for Physics of Microstructures, RAS, GSP-105, 603950 Nizhny Novgorod, Russia
[3] Institute of Semiconductor and Solid State Physics, Johannes Kepler University, Linz, Austria

Received 15 September 2003, accepted 22 October 2003
Published online 22 December 2003

PACS 61.72.Ww, 78.55.Hx

Results of studies of excitation mechanisms of Er ions in silicon-rich silicon oxide (SRSO) are presented. It is shown that in SRSO two mechanisms of excitation are in competition: a resonant one and via defects. The resonant channel of excitation of the Er^{3+} 4f-shell is suppressed by the silicon excess related defects. Decay time measurements of erbium luminescence confirm the existence of isolated Er ions, characterised by a long time constant of ~6 msec, and those strongly coupled, with the time constant of 0.5 msec, whose luminescence is quenched at 300 K. Limitations to energy transfer process due to distance dependence of the transfer rate and little spectral overlap of interacting states are discussed.

© 2004 WILEY-VCH Verlag GmbH & Co. KGaA, Weinheim

1 Introduction

Silicon-based semiconductors doped with erbium are still very attractive as materials for optical devices operating at 1.5 µm, because of the possibility of on-chip integration with microelectronic devices [1]. Silicon-rich silicon oxide (SRSO) is one of the very promising materials as temperature quenching of Er photoluminescence (PL) is strongly reduced in comparison with crystalline silicon [2, 3]. This is because of the large band gap of amorphous SRSO and the larger binding energy of defect states than in silicon.

A key question for inventing optoelectronic devices based on SRSO is good understanding of excitation mechanisms of Er ions. This is particularly important for materials containing different phases like SRSO:Er containing silicon nanocrystals (nc-Si), as the Er ions are most probably located outside nc-Si in silica environment. Nc-Si play the role of luminescence sensitizers which absorb the pump energy due to their large absorption cross section (of the order of 10^{-16} cm^{-2}) and then transfer it to Er ions which emit light at 1.54 µm [4, 5].

A few models describing energy transfer from nc-Si to Er ions were proposed. Some authors suggest that the dominant mechanism is nonradiative energy transfer from excitons in nc-Si [6, 7] to Er ions located in silica environment at a certain distance from nc-Si. Other authors point to the possibility of a more complex mechanism, in which energy is transferred first from excitons to centres at the Si/SiO_2 interface and/or to some defect states in SiO_2, and then transferred to Er ions [8].

Recently, Watanabe et al. [9] presented evidence that excitation energy is indeed transferred resonantly from excitons in nc-Si to Er ions. This observation clearly shows that the overall PL efficiency is on one hand enhanced by nc-Si, whereas on the other the condition of energy conservation in energy transfer to narrow excited states of Er^{3+} may reduce the efficiency by orders of magnitude and, as a consequence, the beneficial effect of the sensitising action of nc-Si may be reduced.

---

[*] Corresponding author: e-mail: kuritsyn@ifpan.edu.pl, Phone: +48 22 843 66 01 3321, Fax: +48 22 843 09 26

© 2004 WILEY-VCH Verlag GmbH & Co. KGaA, Weinheim

Another problem to be solved is the question about the number of Er ions which can be excited by one nanocrystal. There still exists controversy on it – some authors suggest that only one Er ion can be excited [6], other argue that the number of excitable Er ions can be as high as fifty [10]. This question is very important for the determination of the PL efficiency, because if only a fraction of Er ions can be excited, it would limit the possibility of obtaining laser action. The determination of the number of excitable Er ions should also bring some information about the distribution of nc-Si – Er atoms pairs (i.e. preferential pairing) and on the character of interaction of nc-Si and Er ions.

In this work we present results of studies of energy transfer form nc-Si to Er and discuss limitations to this process. We also refer to the problem of the number of excitable Er ions.

2 Experiment

Thin layers of SiO_2 thermally grown on (100) silicon wafers and reference samples of crystalline quartz were implanted at 300 K with a dose of 10^{15} cm^{-2} of Er^+ ions. The implantation energy was 800 keV, the projected range R_p and straggling ΔR_p - 288 and 66 nm, respectively, and the peak concentration of Er was 6×10^{19} cm^{-3}, as determined using secondary ion mass spectroscopy [11].

SiO_2:Er layers were implanted with three energies of 100, 150 and 200 keV of Si^+ ions to doses of 2.0, 3.0 and 4.5×10^{16} cm^{-2}, respectively. This implantation yields a nearly flat profile with ~7% excess Si concentration. Some other samples were implanted with higher Si doses to produce a 15% excess of Si. The implanted samples were annealed at temperatures up to 1100 °C in flowing nitrogen gas. The transmission electron microscopy images of the Si and Er implanted samples annealed at 1100 °C show a uniform distribution of Si nanocrystals of 1.4±0.5 nm diameter, at a concentration of ~2×10^{19} cm^{-3}.

PL and photoluminescence excitation (PLE) spectra were excited with an u nfocussed beam of an optical parametric oscillator tuned in the wavelength range of 430–680 nm, operating at an output energy density of 2±0.4 mJ/cm^2, a pulse duration of 6 ns, and a repetition rate ranging from 5 Hz for Er doped quartz up to 20 Hz for nc-Si. The Er emission was detected using a liquid nitrogen cooled Ge detector.

3 Results

In Fig. 1 the PL spectra of SRSO in the visible are presented. Broad bands located near 760 nm are due to nc-Si [3]. It is seen that for a 15% excess of Si the PL is red shifted in comparison with a 7% sample, confirming that sizes of nanocrystals are larger, as expected.

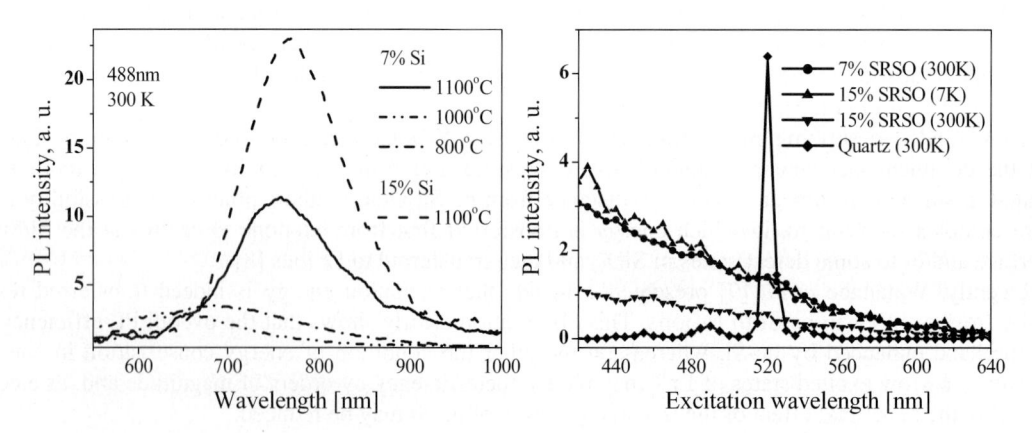

Fig. 1 Luminescence due to defects in SRSO. The sample annealing temperature is given in the figure.

Fig. 2 PLE spectra of Er^{3+} in in SRSO. The measurement temperature is indicated in the figure.

The Er^{3+} PLE spectra for SRSO with 7 and 15% excess Si annealed at 1100 C are presented in Fig. 2, in comparison to Er implanted quartz. In quartz, only resonant transitions to the excited states of Er are observed, with the dominant transition to the $^2H_{11/2}$ state at 520 nm. In the PLE spectra of SRSO the lines due to resonant absorption are not observed, however, the Er PL can be excited with a broad range of wavelengths. A broad short-wavelength absorption is observed in all Si implanted SiO$_2$:Er samples, independent of annealing temperature [12]. It is, apparently, associated with excess silicon and not with any specific type of defect due to ion implantation. At 7 K the PLE intensities for SRSO:Er samples with 7% and 15% excess Si are practically the same (therefore we show here the PLE spectrum only for one of them), however, quenching of the Er^{3+} PL is observed at 300 K for 15% excess Si. It is seen that in SRSO the Er^{3+} PL intensity for off-resonance excitation does not exceed that for resonant excitation of quartz neither at RT nor at low temperatures. In our opinion, there is competition between nc-Si and Er ions for pump energy. This competition is apparently won by nc-Si due to their large absorption cross-section, but only a small part of the absorbed energy seems to be transferred to erbium ions.

We also measured the decay time of Er^{3+} PL as a function of excitation wavelength. Such a procedure allows us to distinguish different types of Er-centres. Decay time for isolated Er ions in silica should be long, whereas Er centres strongly interacting with nc-Si should reveal much shorter decay times as a result, for example, of energy back transfer.

In Fig. 3 decay times of Er^{3+} PL at RT for different excitation wavelengths are presented. Excitation at 520 nm corresponds to resonant transition to the $^2H_{11/2}$ state. The excitation wavelength dependence is observed only for 7% excess Si, whereas for 15% SRSO it is very weak. This result can be explained if one assumes that two processes contribute to decay: a slow one, most probably due to isolated Er ions, and a fast one for centres strongly coupled to nc-Si.

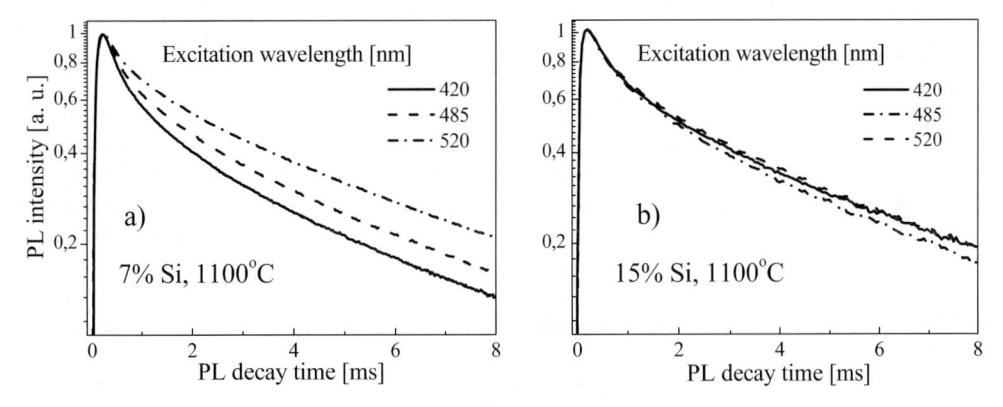

Fig. 3 Photoluminescence decay traces at 300 K of Er^{3+} in SRSO: (a) for 7% excess Si, and (b) for 15% excess Si.

4 Discussion

The experimental data presented here suggest that nc-Si are not as efficient sensitizers of Er^{3+} PL as frequently claimed. Apparently, there are limitations in the energy transfer process from nc-Si to Er ions. In our opinion two factors have to be taken into considerations: (i) spectral overlap of excitons in nc-Si and Er^{3+} ions, and (ii) distance dependence of the transfer.

The probability of energy transfer is proportional to the so called spectral overlap integral Ω:

$$\Omega = \int g_s(E) \, G_a(E) \, dE \quad (1)$$

where $g_s(E)$ and $G_a(E)$ are normalized spectral functions of the sensitizer and acceptor, respectively. As for a single nanocrystal the energy density of excitonic states can be approximated by a δ-Dirac function, the spectral integral will be non-zero in the rare cases when the exciton energy is equal to the energy of one of the excited Er^{3+} states. In SRSO there is a distribution of sizes and, consequently, energy gaps of nc-Si, so only quite a small fraction of nc-Si will have energies resonant with the excited states of Er^{3+}. As a result the PL intensity of Er will be in the first approximation proportional to the number of nc-Si with exciton energy resonant with the excited states of Er and to the fraction of the total concentration of Er ions, which may accept energy from nanocrystals. To give an impression about orders of magnitude - the distance between excited states of Er^{3+} is 2000–3000 cm^{-1} and the width of the Stark split excited manifold 100 cm^{-1}, it reduces the efficiency of energy transfer by two orders of magnitude. Of course phonon-assisted processes may improve the situation to some degree, but in general such processes are less probable.

Another factor influencing the transfer probability is the distance dependence of the transfer process. In general, the problem can be treated in two ways. In the strong coupling regime, the exchange interaction between closely spaced entities is important. Wave functions of electrons and holes within nanocrystals decay quickly outside nc-Si, as they must penetrate a high potential barier of SiO_2 surrounding nc-Si. In the absence of preferential pairing of nc-Si and Er ions it would restrict the number of excitable ions to a small percent of the total value, limiting the PL efficiency even further. In the weak-coupling regime nc-Si and Er ions interact by the transfer of energy from one to another through nonradiative processes such as multipolar electromagnetic interactions, which in principle allows to excite every Er ion with non-zero probability. According to Forster-Dexter theory [13] the probability of energy transfer for dipole–dipole interaction is proportional to R^{-6}, where R is distance between centres. As the average distance between nc-Si and Er ion for the concentrations used is ~1–2 nm many Er ions can be excited via electromagnetic interaction, provided energy is conserved. Therefore, experimental evidence for excitation of only one ion by one nc-Si [5] may suggest either preferential pairing of nc-Si and Er or it shows that a highly limited number of Er ions can be excited. In this sense the results of Kik and Polman [5] and Franzo et al. [4] do not contradict each other.

5 Summary

We showed that in SRSO:Er containing nc-Si there exist two types of Er centers – isolated ones and those strongly coupled to defects. The well known effect of enhancement of the Er emission for optical pumping in the blue-green range of wavelength is due to large absorption cross section of nc-Si, however, this effect does not compensate losses in resonant optical excitation of Er ions. This is because of the distance dependence of the transfer rate and little spectral overlap of the interacting excitonic states with the 4f-states of Er^{3+} ions.

Acknowledgments Work was partly supported by the KBN grant No. 1508/P03/2003/25 in Poland and in Austria by FWF, ÖeAD, and GMe (Vienna).

References

[1] S. Coffa, G. Franzo, and F. Priolo, Mater. Res. Soc. Bull. **23**, 2325 (1998).
[2] G. van den Hoven, J. H. Shin, A. Polman, S. Lombardo, and S. U. Campisano, J. Appl. Phys. **78**, 2642 (1995).
[3] M. Fujii, M. Yoshida, Y. Kanzawa, S. Hayashi, and K. Yamamoto, Appl. Phys. Lett. **71**, 1198 (1997).
[4] G. Franzo, D. Pacifici, V. Vinciguera, F. Priolo, and F. Iacona, Appl. Phys. Lett. **76**, 2167 (2000).
[5] P. G. Kik and A. Polman, J. Appl. Phys. **88**, 1992 (2000).
[6] P. G. Kik, M. L. Brongersma, and A. Polman, Appl. Phys. Lett. **76**, 2325 (2000).
[7] S-Y. Seo and J. Shin, Appl. Phys. Lett. **78**, 2709 (2000).
[8] C. E. Chryssou, A. J. Kenyon, T. S. Iwayama, C. W. Pitt, and D. E. Hole, Appl. Phys. Lett. **75**, 2011 (1999).
[9] K. Watanabe, M. Fujii, and S. Hayashi, J. Appl. Phys. **90**, 4761 (2001).
[10] D. Pacifici, G. Franzo, F. Priolo, F. Iacona, and L. Dal Negro, Phys. Rev. B **67**, 245301 (2003).
[11] L. Palmetshofer, M. Gritsch, and G. Hobler, Mat. Sci. & Eng. B **81**, 83 (2001).
[12] A. Kozanecki, B. J. Sealy, K. Homewood, S. Ledain, W. Jantsch, and D. Kuritsyn, Mat. Sci. & Eng. B **81**, 23 (2001).
[13] D. L. Dexter, J. Chem. Phys. **21**, 836 (1953).

Gap state absorption in AlGaN photoconductors and solar-blind photodetectors

V. Lebedev[*,1], **I. Cimalla**[1], **U. Kaiser**[2], and **O. Ambacher**[1]

[1] Zentrum für Mikro- und Nanotechnologien, Technical University of Ilmenau, 98684 Ilmenau, Germany
[2] University of Jena, Institut für Festkörperphysik, 07743 Jena, Germany

Received 15 September 2003, accepted 19 September 2003
Published online 15 January 2004

PACS 81.15.Hi, 71.35.Cc, 68.55.Jk

AlGaN photoconducting layers and solar blind UV-detectors with high responsivities in a narrow range of photon energies were grown by molecular beam epitaxy on c-plane sapphire substrates. The applicability of photothermal deflection spectroscopy and the spectral photocurrent measurements for absorption studies in III-nitride thin films is investigated. Process dependent variations in direct band-gap energies, Urbach's tail widths, bulk and surface defect densities in the films are given as a measure of composition and structural quality of epitaxial layers. The difference of more than one order of magnitude in the photocurrent above the band gap has been measured in back- and top-illuminated UV-photodetector heterostructure demonstrating a functionality of the integrated filter (cut-off energy ~ 4.85 eV). The UV/visible contrast of more than four orders of magnitude has been observed.

© 2004 WILEY-VCH Verlag GmbH & Co. KGaA, Weinheim

1 Introduction A large variation in the band gaps of wurtzite $Al_xGa_{1-x}N$ alloy (3.42 – 6.2 eV at 300 K) makes it an ideal candidate for light detecting in the 200–400 nm ultraviolet (UV) spectral range [1]. The superior performance of solar-blind AlGaN UV-photodetectors has been recently demonstrated [2–4]. AlGaN-based detectors are essential for applications such as flame sensors and UV biosensors. However, these applications require a photoresponse in a narrow spectral band with photon energies close to a high background radiation demanding a high selectivity devices.

The AlGaN-based photodetector with a high photoresponse in a narrow spectral range have been recently released [2] by means of a combination of three AlGaN epilayers with different alloy composition acting as optical filter, isolator, and detector layer, respectively. Peak responsivity for this device was achieved on the level of 35 A/W at 305 nm. In this paper we present the studies on a high performance AlGaN photodetector based on a similar heterostructure designed for light detection in the spectral range of 240–300 nm (peak sensitivity at 280 nm). Because a device performance of solar-blind AlGaN photodetectors is mainly limited by a high density of structural defects in the epilayers, the detailed knowledge of the defect structure is necessary for the performance improvements. The measurements of absorption in a sub-band gap energy range is the most direct way to obtain this information. In this work, spectral photocurrent measurements (SPC) and photothermal deflection spectroscopy (PDS) combined with optical transmission (OT) and spectral ellipsometry (SE) have been applied to obtain a sub-band and near band gap absorption in III-nitride structures [2, 5, 6]. The advantage of CPM and PDS, in comparison with optical techniques (particularly transmission) is their high sensitivity at low absorption spectral regions ($10^{-7} < \alpha d < 1$, α-absorption coefficient, d-film thickness) and low sensitivity to a scattering effect. Therefore, the combination of all these techniques allows the detection of nearly all absorption-related phenomena in the photoconductors as above as below the band gap.

---

[*] Corresponding author: e-mail: vadim.lebedev@tu-ilmenau.de, Phone: +49 3677 693410, Fax: +49 3677 6933

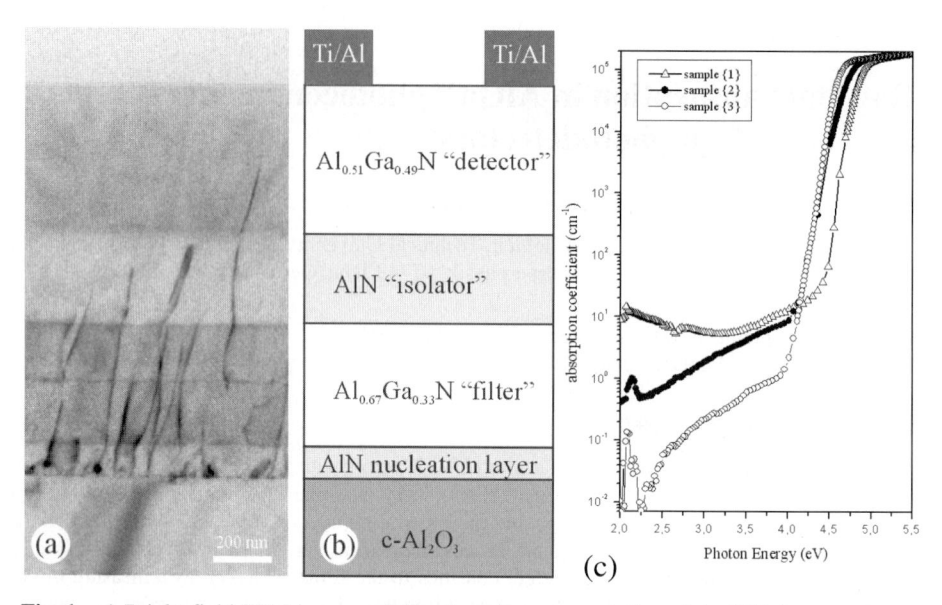

Fig. 1 a) Bright-field TEM image and b) schematic representation of the UV-detector structure with an integrated filter. c) The absorption coefficients of single-layer photoconductors {1} and {2}, and detector structure {3} summarised from SPC, OT and SE measurements. The SPC data are rescaled to fit the absolute values for $\alpha < 10^2$ cm^{-1}.

2 Results and discussion

Nominally undoped AlGaN layers were grown on c-Al$_2$O$_3$ substrates by plasma assisted molecular beam epitaxy (PAMBE). In the UV detector structure (Fig. 1b), Al$_{0.67}$Ga$_{0.33}$N optical filter layer ($E_g \propto 4.8$ eV, $d \sim 0.2$ μm) has been grown on sapphire coated with an AlN nucleation layer ($d \sim 40$ nm). The filter layer absorbs the photons with energies above 4.8 eV and acts as a cut-off filter. An AlN epilayer ($d \sim 100$ nm) then was grown as an insulating barrier ($E_g = 6.2$ eV). Finally, 0.3 μm thick Al$_{0.51}$Ga$_{0.49}$N layer ($E_g \sim 4.6$ eV) has been grown on top acting as an UV photoconductor. Thus, only the photons with energies in the range of 4.6–4.8 eV (240–300 nm) can contribute to the photocurrent.

A typical full width at half maximum of ~0.2° has been measured by XRD ((0002)$_{AlGaN}$, $\theta/2\theta$ scan). Undoped AlGaN was insulating. A residual n-type doping is $\sim 1 \times 10^{16}$ cm^{-3}, according to C–V measurements. A root mean square (RMS) roughness of 0.5–0.6 nm has been detected by AFM.

The detector devices were processed by evaporation of Ti/Al (20/80 nm) rectangular-shape contacts, with a contact spacing of 0.5 mm and an effective area of 50 mm^2. The spectral response has been measured by SPC and PDS with a wavelength resolution of 2 nm in the range of 200–600 nm using a 150 W xenon-arc lamp, L.O.T. Oriel monochromator and lock-in technique ($f_m = 10$ Hz) at ambient temperature. A He–Ne laser was used as a probe beam for PDS measurement. In our optical scheme, the sample was placed in a quartz cuvette filled with a high-purity perfluorhexan ($\partial n/\partial T \approx 4 \times 10^{-4}$ K^{-1}). An absorption in the film is resulting in a certain heat release, causing gradient changes in the refraction index of the liquid near the film surface. This change is proportional to the amount of absorption and can be measured by the deflection of the probe beam on a position sensitive detector (see [5] for details). The angle of probe beam deflection $\Delta\varphi$ can expressed as $\Delta\varphi(\alpha) = n^{-1}(\partial n/\partial T) P(\lambda)(1 - \exp(-\alpha d))$, where n and T is a refractive index and temperature of the liquid, respectively, α is an absorption coefficient, P is a power of incident beam at wavelength λ, and d is a film thickness (see for details [12]).

Spectra of the sub-band gap absorption are generally decomposed into "band to tail" and "band to defect" type transitions. The first type is responsible for the exponential increase at the absorption edge,

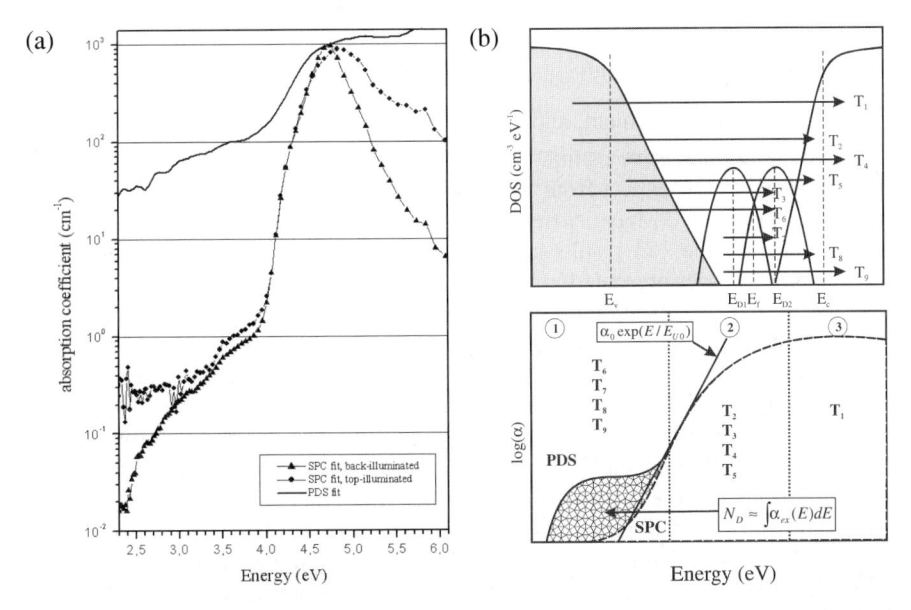

Fig. 2 a) PDS (solid lines) and SPC (line + symbol) curves on the absolute α scale for the back- and top-illuminated photodetector {3}. b) Common transition model for *n*-type semiconductor (transitions T_1–T_9) [12].

which is commonly described by the *Urbach's rule* and follows the relation [8, 9]:

$$\alpha(E) = \alpha_0 \exp\left\{(E - E_0)/E_u\right\} = \alpha_{U0} \exp\left\{E/E_{U0}\right\} \tag{1}$$

where α_0 is a material parameter, E_u the *Urbach energy* representing the width of the exponential absorption edge, and E_0 is an *Urbach parameter* that is a measure of the total disorder, static and thermal, present in the film.

In the particular case of III-nitrides, the *Urbach's* tale, which is mainly a consequence of the carrier interaction with lattice and lattice imperfections (defects, impurities etc.), is also influenced by a constant electric field embedded in the structure [1]. Shallow levels (like Si) seem also to modify the optical absorption edge abruptness and usually an empirical *Urbach* potential can be obtained experimentally [10]. The second type of transition shows up as a plateau (or close dependence) at low photon energy. In this region of low DOS α is usually orders of magnitude lower compared to that for energies at the band edge. The OT and SE methods are not sensitive enough to detect these transitions which are important for fitting the exponential absorption edge by the Urbach expression [9].

In Fig. 1c the absorption coefficients for the UV-detector structure {3} and AlGaN single-layer photoconductors with compositions close to the "filter" {1} and "detector" {2} layers of the UV-detector are presented. The curves were summarised on the basis of PDS and SPC for the sub-band gap energies, and OT and SE for a high-energy range near and above the band gap. The PDS and SPC data are rescaled to fit the absolute values in the vicinity of the band edge E_G (for $\alpha < 10^2$ cm^{-1}). The absorption curves have been fitted with the Urbach expression (1). Derived Urbach parameters are summarised in Table 1.

Table 1 Summary of *Urbach* parameters for UV-detector and AlGaN single-layer photoconductors.

sample	E_G, eV	$\alpha_0 \cdot 10^4$, cm^{-1}	E_0, eV	E_u, meV	$\int \alpha_{ex} dE$, cm^{-2}
{1}	4.85	12.16	4.85	51.5	0.0235
{2}	4.65	4.035	4.65	72.8	0.0239
{3}	4.60	8.531	4.60	55.5	0.0286

We can also assume that the defect density N_D in the epilayer is proportional to the integral over the sub-band gap energies $N_D \approx \int \alpha_{ex}(E)\,dE$, where $\alpha_{ex}(E) = \alpha_{PDS}(E) - \alpha_0 \exp(E/E_{U0})$, α_{PDS} is a *total* absorption derived from PDS measurement, and $\alpha_0 \exp(E/E_{U0})$ is an exponential fit of the *Urbach's* tail (see Fig. 2b) [11, 12]. The qualitative estimation of N_D in the sub-band gap region reveals very similar defect density values, as we can expect from the similar growth conditions (see Table 1). Also, we can conclude that samples {1} and {3} have close structural quality confirmed by the derived parameters in the *Urbach* region. The sample {2} results are reflecting the compositional fluctuation in the AlGaN layer leading to a higher disorder presented in the film.

A low energy absorption tail (<4 eV) has been fitted with SPC curves (Fig. 1c). The obvious responsivity has been observed in the region well below the band gap, which is mainly associated with deep centers responsible for non-linear trapping processes [10]. In Fig. 1c, the sample {1} shows the highest level of absorption at low energies indicating a high density of defects such as Al/Ga vacancies, incorporated O, C atoms and defects localized in lattice discontinuities. The difference of one order of magnitude in α for back- and top illuminated UV-detector demonstrates a functionality of the integrated filter (cut-off ~ 4.85 eV). The UV/visible α contrast of more than four orders of magnitude has been observed.

The integrated sub-band gap absorption obtained from SPC is more then one order of magnitude lower in contrast to PDS data. This is due to the difference in the detection mechanism provided by these two methods (see Fig. 2b). PDS are sensitive to all absorption events representing the *total* absorption in the film (α_{PDS}). According to Fig. 2b, all the pointed transition (T_1–T_9) contribute to the PDS signal. The SPC signal depends only on intraband (T_1), "band tail to band" (T_4, to delocalised states at and above the mobility edge E_c) and "defect to band" (T_9) transitions. Transitions T_2, T_5 and T_8 do not participate in the photocurrent due to the low mobility in the localised conduction band tail states [12]. Thus, α_{SPC} relates to the generation of photocarriers and is affected only by the defects located in the region with the highest conductance. Thus, α_{SPC} represents only the bulk absorption in the effective layer conduction section.

3 Conclusions

A device performance of modern solar-blind AlGaN photodetectors is mainly limited by a high density of defects responsible for reduced UV/visible contrast and slow photoresponse. In this work, SPC and PDS methods combined with transmission and ellipsometry measurements have been applied to obtain a sub-band and near band gap absorption in AlGaN photodetectors to provide detailed information about defect states in a wide energy range. The difference of more than one order of magnitude in absorption coefficients above the band gap has been measured in a back- and top-illuminated photodetector demonstrating a functionality of the integrated filter. The UV/visible contrast of more than four orders of magnitude has been detected despite of the significant photoreponse well below the band gap associated with deep center states in the epilayers. On the basis of derived *Urbach's* parameters, the compositional, crystalline and optical quality of the heterostructures has been estimated.

Acknowledgements The authors would like to thank Dr. R. Goldhahn for ellipsometry analysis and all our colleagues for many helpful discussions. This work was supported by the Thüringer Ministerium für Wissenschaft, Forschung und Kultur (project B609-02004). The authors also acknowledge an ONR NICOP Grant.

References

[1] O. Ambacher, J. Phys. D: Appl. Phys. **31**, 2653 (1998).
[2] U. Karrer, A. Dobner, O. Ambacher, and M. Stutzmann, J. Vac. Sci. Technol. B **18**, 757 (2000).
[3] M. Khan, J. Kuznia, J. Van Hove, D. Olson, M. Blasingame, and L. Reitz, Appl. Phys. Lett. **60**, 2917 (1992).
[4] D. Walker, X. Zhang, P. Kung, A. Saxler, S. Javadpour, J. Xu, and M. Razeghi, Appl. Phys. Lett. **68**, 2100 (1996).
[5] W. B. Jackson, N. M. Amer, A. C. Boccara, and D. Fournier, Appl. Opt. **20**, 1333 (1981).
[6] M. Vanecek, J. Kocka, J. Stuchlik, and A. Triska, Solid State Commun. **39**, 1199 (1981).
[7] Z. Hecht, Optics (Addison, New York, 1974).

[8] F. Urbach, Phys. Rev. **92**, 1324 (1953).
[9] A. Meeder, D. Fuertes Marron, A. Rumberg, M. Lux-Steiner, V. Chu, and J. Conde, J. Appl. Phys. **92**, 3016 (2002).
[10] E. Munoz, E. Monroy, J. Pau, F. Calle, F. Omnes, and P. Gibart, J. Phys., Condens. Matter **13**, 7115 (2001).
[11] W. B. Jackson and N. M. Amer, Phys. Rev. B **25**, 5559 (1982).
[12] H. Curtins and M. Favre, Amorphous silicon and related materials **1A**, 329 (1988).

GaN on Si substrates for LED and LD applications

Suzuka Nishimura[*], **Satoru Matsumoto**[1], and **Kazutaka Terashima**[2]

[1] Keio University, 3-14-1 Hiyosi Kouhokuku Yokohama, Kanagawa 223-8522, Japan
[2] Shonan Institute of Technology, 1- 1- 25 Tsujido-nishikaigan, Fujisawa, Kanagawa 251- 8511, Japan

Received 15 September 2003, revised 15 October 2003, accepted 15 October 2003
Published online 22 December 2003

PACS 42.70.Hj, 81.10.Bk

GaN epitaxial layer has been grown on 2 inch diameter Si substrates by using highly conductive BP buffer crystal. It has been found that the 2-inch GaN on BP/Si wafers are remarkably flat after epitaxial process. BP layer has grown flat and continuously on a large size silicon wafer. GaN has been found to grow continuously on such a substrate. BP crystals have been found to be highly conductive and controllable depending on the gas ratio. In addition, UV illumination enables us to study the chamber conditions such as contamination by impurity.

© 2004 WILEY-VCH Verlag GmbH & Co. KGaA, Weinheim

1 Introduction

Recently, blue LED and LD have attracted much attention of many researchers including application fields [1, 2]. The GaN crystals are usually grown on Al_2O_3 substrates, so the device structure is complicated and the cleavage face is not so apparent. We have reported the growth of GaN on Si by using BP thin layer crystal as a buffer crystal with growing in home made apparatus [3–7].

The objectives are 1. To make low cost devices, 2. To grow cubic type GaN or phase matched wurtzite GaN on BP/Si crystals, 3. To fabricate opto-electronic integrated circuit devices (OEIC) in the near future.

We have studied the growth of BP with large diameter Si substrates of 2 inch in diameter by using a mass-producing MOVPE apparatus. As a starting gas, TEB (Triethyl boron) and TBP (t-butyl phosphine) were used with hydrogen as a carrier gas. The substrates were markedly flat when we used a low temperature buffer layer.

This paper describes the growth of BP layer on 2 inch diameter Si(100) substrates. The electrical property variation and the possibility of light emitting devices will be discussed.

2 Experimental

Si(100) wafers of 2 inch in diameter were used as a substrate. Hydrogen gas was used as a carrier gas for growing BP crystals. The mixed gas was used to grow GaN and InGaN. As source gases, TBP and TEB were used for growing BP on Si substrates. TMG, TMI and NH_3 were used for growing GaN and InGaN. The crystals obtained were evaluated by SEM and surface X-ray diffraction. Some wafers were illuminated by a UV lamp.

3 Results and discussion

The growth of BP on Si(111) or Si(100) has been carried out at 1000–1100°C after low temperature layer of 300–500 Å has been grown. The substrates have been markedly flat if low temperature layer has been

---

[*] Corresponding author: e-mail: t985008@mate.shonan-it.ac.jp, Phone: +81 466 30 0226, Fax: +81 466 30 0226

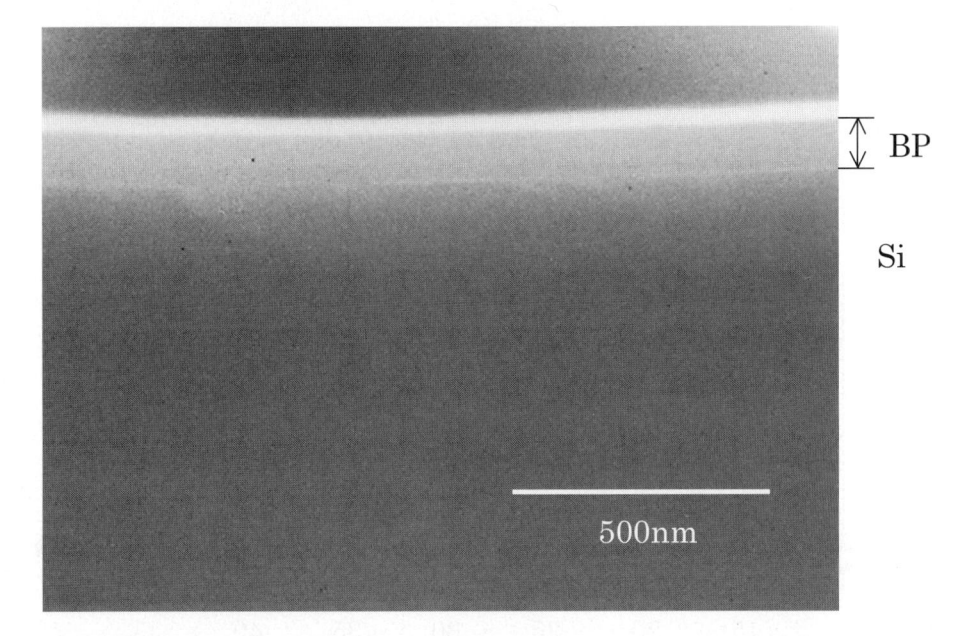

Fig. 1 Cross sectional SEM image of BP layer on Si(100) substrate.

successfully grown even by using face-down system (periphery area has not been grown). An example of BP on Si(100) is shown in Fig. 1 by observing SEM image. The typical thickness is about 0.1μm. It has been found that the void existence and large defect generation have not been detected at the interface of BP and Si. The continuous BP layer has been grown on Si surface as shown in Fig. 1. The X-ray diffrac-

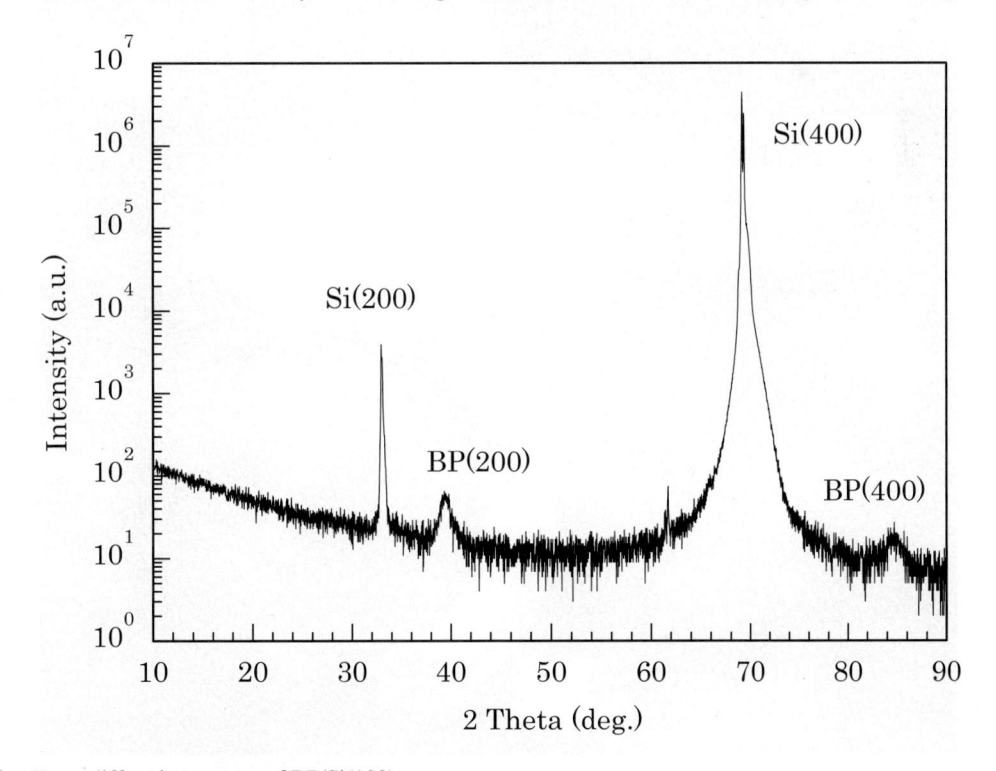

Fig. 2 X-ray diffraction pattern of BP/Si(100).

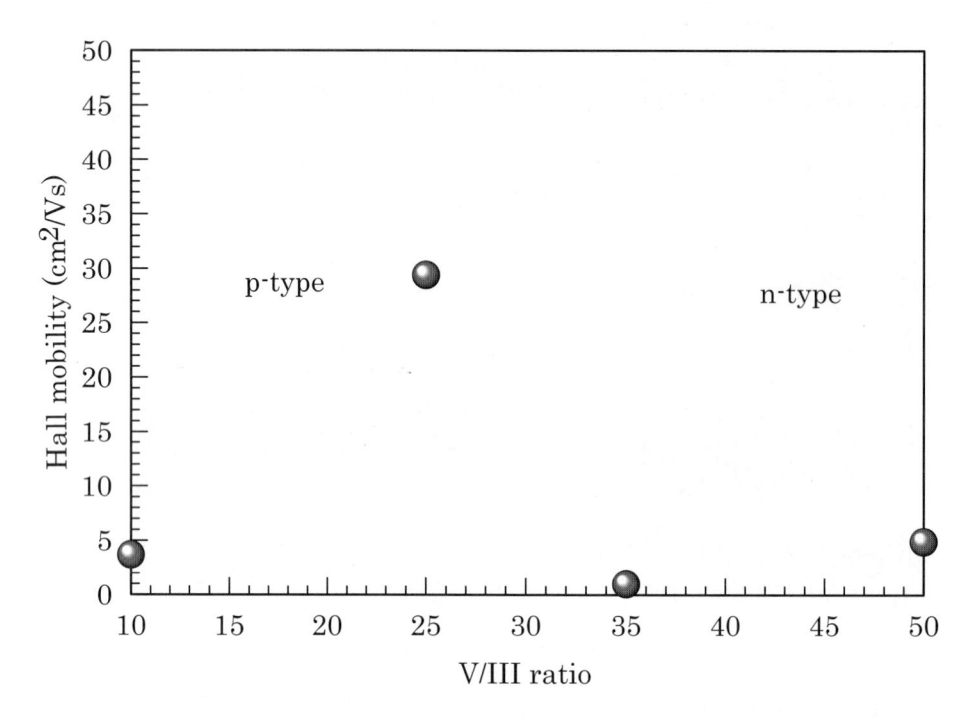

Fig. 3 Electrical conductive type variation as a function of V/III ratio.

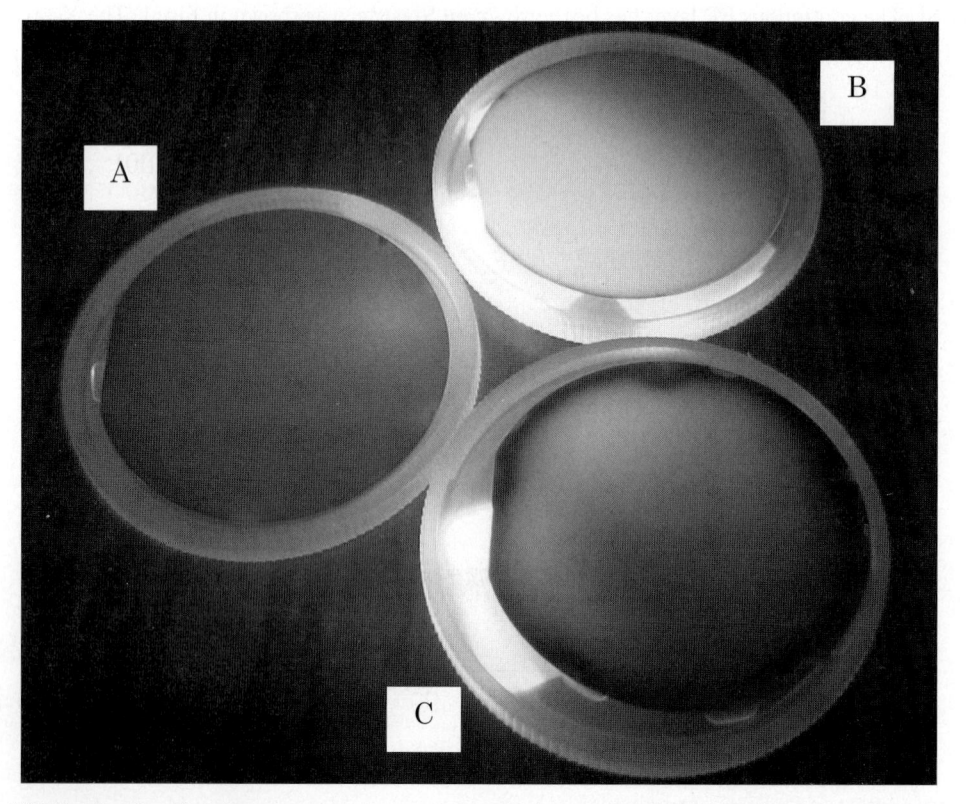

Fig. 4 UV illumination of wafers.

© 2004 WILEY-VCH Verlag GmbH & Co. KGaA, Weinheim

tion of the markedly thin BP epitaxial layer is shown in Fig. 2. BP(200) and (400) diffraction has been clearly observed with Si(400) and Si(200) diffraction. The Si(200) diffraction may be due to the surface atomic re-construction or residual strain of Si surface.

The electrical resistivity of as-grown BP layer grown at low temperature around 400°C showed semi-insulating. The electrical conductivity remarkably increased after the growth at relatively high temperature. The resistivity is around 0.2–0.6 Ω cm, it may be depend on the residual impurity. The electrical conductivity of initial region grown at low temperature also shows high conductivity. This variation has an advantage of the vertical type devices such as LED and LD. The electrical conductivity varies depending on the growth conditions. A typical approximate variation is shown in Fig. 3 as a function of V/III ratio. If the V/III ratio is small the conductive type has a tendency of p type. At the V/III ratio is large the conductive type changes to n type. The ionicity of BP crystal is small, because the electro negativity of boron atom is 2.0 and phosphorus atom is 2.1. So, the conductive type of BP layer will be controlled optimizing to the device structures by intentional doping of impurity.

To elucidate the presence of residual impurities in the growing chamber, GaN and InGaN layer was grown on Si substrates and UV light was illuminated to the wafers. One of the results is shown in Fig. 4. The wafer denoted A is one of the normal wafers of GaN on Si. The wafer denoted B is the GaN on Si, but the residual impurity is high in the GaN layer depending on the growing system difference. The yellow illumination has been observed from wafer B. One of the candidates of the impurities is the carbon incorporated during the growth. The yellow illumination corresponds to carbon existence and/or defect at the interface. This result means wafer B is unusable for optical devices of short wavelength. The wafer C has InGaN multi quantum layers with 5–10 Å. The content of In atom is high, so InN precipitated in the layers. We are studying the growth conditions for fabricating optical and electric devices on Si substrates. The results of test devices will be published in the near future.

4 Conclusion

GaN epitaxial layer has been grown on Si substrates by using BP buffer crystal. It has been found that the 2-inch GaN on BP/Si wafers are remarkably flat after epitaxial process. In addition, UV illumination enables us to study the chamber conditions such as contamination by impurity. The contaminated layer has been easily detected by this technique. The machine condition will be visible by this illumination.

Acknowledgements We are much indebted to VTERA Co. in Taiwan for growing GaN on Si substrates. We thank Dr. S. Yoshida of FURUKAWA electric Co. for his helping of SEM observation. We also thank Prof. N. Kieda of Shonan Institute of Technology for his help of X-ray diffraction and fruitful discussions though this study.

References

[1] S. Nakamura, M. Senoh, S. Nagahama, N. Iwasa, T. Yamada, T. Matsushita, H. Kiyoku, and Y. Sugimoto, Jpn. J. Appl. Phys. **35**, L74 (1996).
[2] G. P. Yablonskii, E. V. Lutsenko, V. N. Pavlovskii, I. P. Marko, B. Schineller, M. Heuken, and K. Heime, Mater. Sci. and Eng. B **80**, 322–326 (2001).
[3] S. Nishimura, H. Hanamoto, K. Terashima, and S. Matsumoto, Mater. Sci. and Eng. B **93**, 135–138 (2002).
[4] S. Nishimura and K. Terashima, Mater. Sci. and Eng. B **82**, 25–26 (2001).
[5] S. Nishimura and K. Terashima, Mater. Sci. and Eng. B **75**, 207–209 (2000).
[6] S. Nishimura and K. Terashima, Appl. Surf. Sci. **159–160**, 288–291 (2000).
[7] S. Nishimura, S. Matsumoto, and K. Terashima, Opt. Mater. **19**, 223–228 (2002).

Sensitive In$_{0.53}$Ga$_{0.47}$As/InP (SI) magnetic field sensors

T. Przesławski[*,1], **A. Wolkenberg**[1,2], **K. Regiński**[1], and **J. Kaniewski**[1]

[1] Institute of Electron Technology, Al. Lotników 32/46, 02-668 Warsaw, Poland
[2] Institute of Materials Engineering, Technical University of Częstochowa, Al. Armii Krajowej 19, 42-200 Częstochowa, Poland

Received 15 September 2003, revised 5 November 2003, accepted 5 November 2003
Published online 22 December 2003

PACS 72.80.Ey, 73.50.Td, 85.30.Fg

The usefulness of InGaAs/InP heterostructures for magnetic field sensor applications has been studied. Basic parameters of the Hall and magnetoresistive devices have been determined. Magnetic field sensitivities in a wide temperature range from 3.5 K to 300 K for layers with different carrier concentration from $2 \cdot 10^{20}$ m^{-3} to $8 \cdot 10^{23}$ m^{-3} have been measured. It is concluded that optimized lattice-matched InGaAs/InP heterostructures can be used as very sensitive magnetic field sensors.

© 2004 WILEY-VCH Verlag GmbH & Co. KGaA, Weinheim

1 Introduction Magnetic micro sensors using either the transverse Hall or the magnetoresistance effects, with practical applications in factory automation, home and automobile electronics, are gaining in commercial importance [1, 2]. A review of magnetometer systems used to measure magnetic fields is presented in [3, 4]. For ten years InGaAs material system has been proposed for such devices [5]. We report here on the use of In$_{0.53}$Ga$_{0.47}$As/InP heterostructures for the fabrication of Hall and magnetoresistance sensors with high sensitivity and high magnetic resolution. Sensor application of the layers results from better transport properties of In$_{0.53}$Ga$_{0.47}$As/InP comparing to GaAs and InGaAs/GaAs. This report is an appraisal of the aforementioned materials samples usage in Hall or magnetoresistor sensors construction as signal or measurement elements. Here, device structure and fabrication procedure of the sensor is proposed. The material characteristics together with the calculation methods are described.

2 Experimental The galvanomagnetic methods were used for characterization of the materials for sensors. Hall effect and magnetoresistance measurements were carried out using the van der Pauw technique on (5 × 5) mm or (10 × 10) mm samples. Studies have been performed in a closed-cycle He cryostat, in the dark in the temperature range from 3.5 K to 300 K. Carrier concentration n_H, carrier mobility μ_H, magnetoresistivity ρ_B, and relative resistance $\Delta\rho/\rho_0$ were determined at two magnetic fields: 0.02 T and 1.4 T. Basic properties of the samples studied are listed in Table 1.

Carrier concentration in the layers changed in the range from 2.2×10^{20} m^{-3} to 8.5×10^{23} m^{-3}. We assume that carriers concentration profile in the In$_{0.53}$Ga$_{0.47}$As layer as in all MBE layers is constant across the thickness with some increase at the interface with InP substrate [6]. The heterostructures were grown by means of Molecular Beam Epitaxy (MBE). Details of the growth of lattice-matched In$_{0.53}$Ga$_{0.47}$As/InP heterostructures have been given elsewhere [7–10]. Sensor parameters were calculated assuming length to width ratio (L/W) = 1 and the geometric correction factor equal to G ~ 0.77 (after Lippmann, and Kuhrt [11]).

---

[*] Corresponding author: e-mail: tprzes@ite.waw.pl, Phone: (48-22) 548 79 33, Fax: (48-22) 847 06 31

© 2004 WILEY-VCH Verlag GmbH & Co. KGaA, Weinheim

Table 1 Basic properties of the samples at 300 K.

Sample no. Magnetic field [T]	Metalurg. thickness t [μm]	Electron concentr. n_H [m^{-3}]	Electron mobility μ_H [m^2/Vs]	Resistivity ρ_B [Ω m]	Absolute sensitivity $\gamma_0 \left[\frac{V_H}{T}\right]$	Relative resistance $\Delta\rho/\rho_0$
MBE 273	4					
0.02		2.2×10^{20}	0.728	0.04	1.11	0.0015
1.42		2.3×10^{20}	0.696	0.052	1.06	0.31
MBE 225	1.0					
0.02		8.5×10^{23}	0.53	0.000014	0.00028	0.0002
1.42		8.2×10^{23}	0.55	0.000014	0.00029	0.034

3 Results and discussion

In order to discuss the sensor performance one should consider the sensitivity dependence on several physical and geometrical parameters. The supply-current related sensitivity, γ, of Hall sensor was calculated using following equation [12–14].

$$\gamma = \frac{V_H}{IB} = G\frac{R_H}{t} \quad \left[\frac{V}{AT}\right] \qquad (1)$$

where V_H -Hall voltage [V], B-the normal component of magnetic induction [T], I-bias current [A], G-geometrical correction factor, R_H -Hall coefficient [m^3 C^{-1}.], t-thickness of the active layer [m] (metallurgical). The current sensitivity of magnetoresistor, S_I, is resistance R dependent on magnetic field and according to [4] is given by

$$S_I = \frac{dV}{I\,dB} = \frac{dR(B)}{dB} \quad \left[\frac{\Omega}{T}\right] \qquad (2)$$

Simultaneously the voltage sensitivity of magnetoresistor, S_V, can be expressed as

$$S_V = \frac{dV}{V\,dB} = \frac{dR}{R\,dB} \quad \left[\frac{1}{T}\right] \qquad (3)$$

Of practical value in the design of magnetoresistors are the approximate analytical expressions made with L/W ≥ 0.4. At weak magnetic field ($\mu_H B < 1$) both sensitivities, can be described as follows:

$$S_I = \frac{\mu_H B L}{n_H e t W}\left(1 - 0{,}54\frac{L}{W}\right) \qquad (4)$$

and

$$S_V = \left(1 - 0{,}54\frac{L}{W}\right)\mu^2 B^2 \qquad (5)$$

Simultaneously, at strong magnetic filed ($\mu_H B > 1$) one can assume after Heremans [4] and Wick [15] that

$$S_I = \frac{N_{el}}{n_H e t} \qquad (6)$$

and

$$S_V = \frac{\mu_H}{\mu_H B + 5{,}9\dfrac{L}{W} - 3.9} \qquad (7)$$

To study the influence of carrier concentrations on γ, S_I and S_V parameters their temperature dependence was determined. Temperature changes of the current sensitivity for Hall effect devices are presented in

Fig. 1 and 2. Significant difference in $\gamma(T)$ characteristics of (MBE 273)-Fig. 1 and Si doped layers (MBE 225)-Fig. 2 was detected. The value of sensitivity γ is nearly constant at investigated temperature range for highly doped sample only (Fig. 2). In the case of (273 MBE) sample γ parameter drastically decreases at temperatures higher than 100 K. At lower temperatures current sensitivity of the structure is nearly fixed Therefore this kind of semiconductor can be used at lower temperatures as magnetic field sensor and at higher temperatures (>100 K) as temperature sensor.

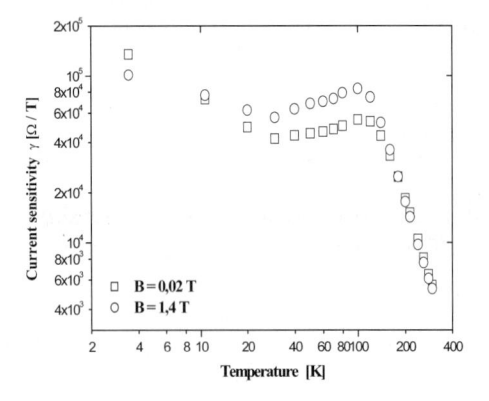

Fig. 1 Current-related sensitivity for Hall sensor vs.temperature for the sample MBE 273.

Fig. 2 Current-related sensitivity for Hall sensor vs.temperature for the sample MBE 225.

Similarly for magnetoresistance sensor strong differences in $S_I(T)$ characteristics were observed. Currant sensitivity of undoped sample is independent on temperature at T ≤ 100 K-Fig. 3. On the other hand S_I exhibits an almost linear decrease with increasing temperature. Identical behavior has been observed at weak magnetic field ($\mu_H B < 1$) as well as at strong magnetic filed ($\mu_H B > 1$). In contrast to that, S_I shows nearly temperature independent dependence in Si doped sample labeled MBE 225-Fig. 4. This indicates that $In_{0.53}Ga_{0.47}As/InP$ layer with electron concentration 8.2×10^{23} m^{-3} at T = 300 K can be treated as degenerated semiconductor.

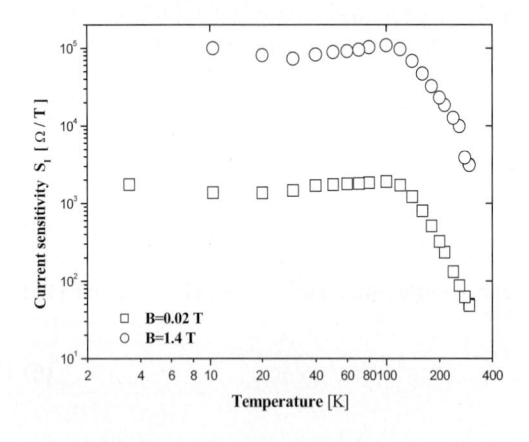

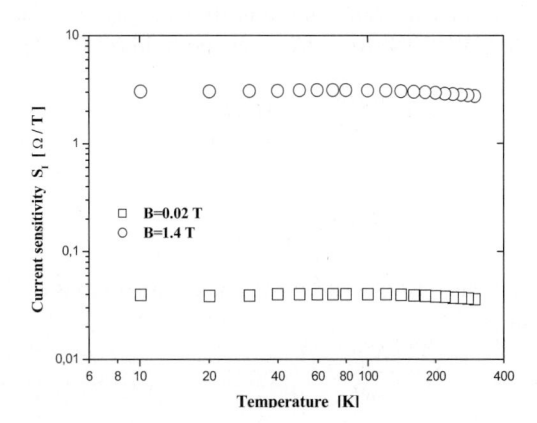

Fig. 3 Current sensitivity of magnetoresistor vs. temperature for the sample MBE 273.

Fig. 4 Current sensitivity of magnetoresistor vs. temperature for the sample MBE 225.

Voltage sensitivity of undoped sample labeled MBE 273 exhibits similar features in temperature dependence as it was observed for current sensitivity. However, decrease of $S_V(T)$ at T > 100 K have been found to be less pronounced at stronger magnetic field-Fig. 5. In the case of Si doped sample (MBE 225) voltage sensitivity have been found to be independent on temperature-Fig. 6. Additionally strong magnetic field dependence has been found.

From application point of view undoped sample labeled MBE 273 seems to be most suitable.

For this heterostructure sensitivity γ is almost independent on magnetic field.
On the basis of experimental data, presented in Fig. 1–6 one can conclude that strongly Si doped $In_{0.53}Ga_{0.47}As/InP$ layers can be used for the fabrication of measurement sensors operating in wide temperature range. They exhibit good thermal stability. More pure layers are preferable to be applied to produce measurement sensors, which can operate at low temperatures, below 100 K. On the other hand pure InGaAs layers are useful in signal sensor applications operating at T > 100 K due to high values of magnetic sensitivities γ, S_I and S_V. The layers are especially suitable for fabrication Hall devices.

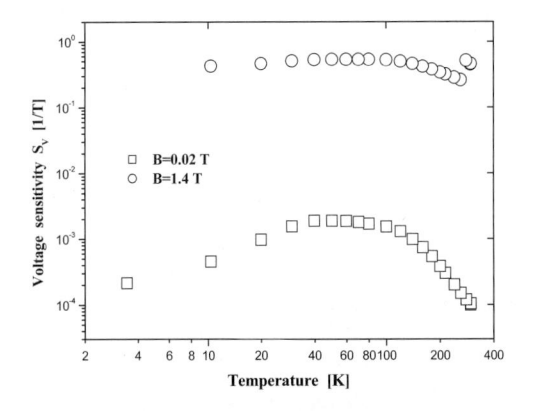

Fig. 5 Voltage sensitivity of magnetoresistor vs. temperature for the sample MBE 273.

Fig. 6 Voltage sensitivity of magnetoresistor vs. temperature for the sample MBE 225.

Absolute sensitivity for the pure material at T = 300 K equals to 1.1 V/T and it is more than 4×10^3 times higher than γ_0 detected for highly doped sample, this sensivity seem to be magnetic field independent for both samples (Table 1). Such γ_0 behaviour classifies InGaAs/InP material as useful for sensors construction.

4 Conclusions Galvanomagnetic properties of $In_{0.53}Ga_{0.47}As/InP$ (SI) layers in weak as well as strong magnetic fields have been studied. Undoped layers appears to be a promising material for magnetic filed sensors operating at temperatures lower than 100 K. On the other hand Si doped layers seems to be more suitable for higher temperature application. The layers exhibit large sensitivity and linearity of parameter changes on magnetic field applied. It has been shown that $In_{0.53}Ga_{0.47}As/InP$ (SI) heterostructures can be used for the fabrication of sensitive Hall sensors and magnetoresistive elements. Magnetic sensitivities of Hall sensor, γ, and magnetoresistor element, S_I and S_V, have been determined. The parameters are better than that obtained for similar GaAs based devices [5]. The best magnetic sensitivity was detected for low electron concentration $In_{0.53}Ga_{0.47}As/InP$ (SI) magnetoresistors.

Acknowledgements The authors want to thank Prof. M. Bugajski for stimulating discussions. This work was partially supported by the State Committee for Scientific Research (KBN, Poland) under Grant No. 4T11B 066 22.

References

[1] J. Heremans, D.L. Partin, C.M. Trush, and L. Green, Semicond. Sci. Technol. **8**, 424 (1993).
[2] Y. Sugiyama, Bull. Electrotech. Laboratory, no. **54**, 65 (1990).
[3] S. Foner, IEEE Trans. Magn. **17**, 3358 (1981).

[4] J. Heremans, J. Phys. D: Appl. Phys. **26**, 1149 (1993).
[5] R. Kyburz, J. Schmidt, R. S. Popovic, and H. Melchior, Sensors Materials **6**, 279 (1994).
[6] T. Przesławski, A. Wolkenberg, K Regiński, J. Kaniewski, and J. Bąk-Misiuk, Thin Solid Films **367**, 232 (2000).
[7] T. Przesławski, A. Wolkenberg, K. Regiński, and J. Kaniewski, Optica Applicata **XXXII**, 511 (2002).
[8] W. Strupiński, M. Czub, M. Wójcik, J. Gaca, J. Sass, and B. Surma, Electron Technology **29**, 162 (1996).
[9] Z. Dziuba, T. Przesławski, K. Dybko, M. Górska, and J. Marczewski, J. Appl. Phys. **85**, 6619 (1999).
[10] A. Wolkenberg, T. Przesławski, J. Kaniewski, and K. Regiński, J. Phys. Chem. Solids **64**, 7 (2003).
[11] H. J. Lippmann and F. Kuhrt, Z. Naturforsch. **13a**, 474 (1958).
[12] S. Del Medico, T. Benvalton, G. Guillot, M. Gendryt, M. Oustric, J. Tardy, G. Hollinger, A. Chovet, and N. Mathieu, Semicond. Sci. Technol. **11**, 576 (1996).
[13] R. S. Popović, Hall Effect Devices, Bristol, Adam Hilger, p. 206 (1991).
[14] R. S. Popović and B. Halg, Solid-State Electron. **31**, 1681 (1988).
[15] R. F. Wick, J. Appl. Phys. **25**, 741 (1954).

Kinetics of radiative recombination of the "pseudo-donor–pseudo-acceptor" pairs in InGaN/GaN

A. J. Zakrzewski[*,1]

[1] Institute of Physics, Polish Academy of Sciences, Al. Lotnikow 32/46, 02-668 Warsaw, Poland

Received 15 September 2003, revised 24 September 2003, accepted 24 September 2003
Published online 22 December 2003

PACS 78.47.+p, 78.55.–m, 78.67.–n

It was shown that the recombination kinetics of electrons and holes localized on potential fluctuations in quantum wells could be calculated exactly in terms of hypergeometric functions. Our algorithms are much easier for coding and much faster and more accurate then previously developed. Moreover, our approach enables real time processing of experimental data.

© 2004 WILEY-VCH Verlag GmbH & Co. KGaA, Weinheim

1 Introduction

Semiconductor heterostructures are now basic components of many modern devices. Due to rapidly growing needs and expectations of electronic and optoelectronic industries, the number of produced and investigated various low-dimensional systems is steadily increasing. It is not strange, that the time spent on characterization of such systems is a crucial factor for further development. In the following we will show, that the analysis one of the most important low temperature recombination mechanisms of photo-excited carrier i.e. the kinetics of radiative recombination of the donor–acceptor pairs can be substantially improved and speeded up.

Recently, Morel et al. [1] introduced a new method for characterization the smoothness of the interfaces in InGaN/GaN quantum wells. As it is well known, the potential fluctuations at the interfaces in such systems can act as traps for free carriers and the recombination of these carriers may yield a direct information about these traps and spatial extension of the wavefunctions of trapped carriers. The idea of Morel at al. is based on closed analogy between decay kinetics observed in their experiments and those observed earlier by other authors in bulk materials doped with shallow donors and shallow acceptors. In the latter case observed luminescence originates from radiative recombination of electrons trapped on donors and holes trapped on acceptors. Such processes were thoroughly investigated by Thomas et el. and the results were fully described in their classic paper [2]. Additional information can be also found in [3] and results of recent studies of donor–acceptor pair luminescence in ZnSe and GaN can be found in [4] and [5], respectively.

2 Calculations of the decay kinetics

Assuming the acceptor concentration to be much greater then the concentration of donors (the so-called uncompensated case) the decay of the luminescence I(t) is given by:

$$I(t) = -\frac{d\langle Q(t) \rangle}{dt} \tag{1}$$

---

[*] Corresponding author: e-mail: zakrz@ifpan.edu.pl

© 2003 WILEY-VCH Verlag GmbH & Co. KGaA, Weinheim

where:

$$\langle Q(t) \rangle = \exp(nv_n \rho \int_0^\infty [\exp[-W(r)t]-1]r^{n-1}dr) \qquad (2)$$

and ρ is the initial density of acceptors, $v_n = \pi^{n/2}\Gamma(n/2+1)$ is the volume of the unit n-dimensional sphere and W(r) denotes transition rate for a donor–acceptor pair with separation r. Thomas et al. proposed, that W(r) is proportional to the overlap of the electron and hole wavefunctions and for shallow, hydrogenic impurities is simply given by:

$$W(r) = W_0 \exp(-\frac{r}{a}) \qquad (3)$$

where parameter a is one half of the effective Bohr radius of the shallower impurity.
Equation 1 and 2 show that calculation of I(t) is reduced to the calculation of two integrals:

$$I_1 = \int_0^\infty W(r)\exp[-W(r)t]r^{n-1}dr \qquad (4)$$

$$I_2 = \int_0^\infty [\exp[-W(r)t]-1]r^{n-1}dr \qquad (5)$$

In the case of carriers localized on local imperfections at the interfaces the attractive potential is not long range Coulombic, but a short range one, similar in character to that produced by isoelectronic traps. Moreover, internal electric field separates holes and electrons along the growth axis, so electrons are trapped on potential fluctuations ("pseudo-donors") at one interface and holes on potential fluctuations ("pseudo-acceptors") at the second one. In this case recombination dynamics resembles recombination dynamics of two-dimensional pairs, with parameter W_0 setting the time scale of the decay kinetics and its shape determined by the concentration of traps and spatial extension of the wavefunctions of the captured carriers [1].

The exact form of the trapping potential is not known, but Morel at al. approximated it with two-dimensional harmonic well. In this case transition rate W(r) equals:

$$W(r) = W_0 \exp(-\frac{r^2}{a^2}) \qquad (6)$$

and parameter a can be named a characteristic distance for the extension of the wavefunction [1].

As Morel et al. pointed out, one cannot separate ρ from a. Only their product $\eta = \rho a^2$ can be effectively determined from the fit of experimental data to theoretical formulae. However, η has still a direct physical meaning: it characterizes the nanotexture and can be used to quantify "a degree of localization" of carriers among samples grown at different conditions [1].

It turns out that both integrals I_1 and I_2 with W(r) given by Eqn. (6) can be calculated exactly in terms of generalized hypergeometric functions $_pF_q$:

$$I_1 = \frac{W_0 a^2}{2} {}_1F_1[\{1\},\{2\},-W_0 t] \qquad (7)$$

$$I_2 = -W_0 t \, _2F_2[\{1,1\},\{2,2\},-W_0 t] \qquad (8)$$

or, in terms of "more elementary" functions:

$$I_1 = \frac{a^2}{2t}(1-\exp(-W_0 t)) \qquad (9)$$

$$I_2 = \frac{a^2}{2}[-\gamma - \Gamma(0, W_0 t) - \ln(W_0 t)] \tag{10}$$

where γ is Euler's constant ($\gamma = 0.577216...$) and $\Gamma(0,x)$ is incomplete gamma function. Calculations of I_2 can be further simplified. First, we splitted time domain into two subdomains: $0 \leq W_0 t \leq 10$ and $W_0 t > 10$, (we choose 10 as a splitting value bearing in mine experimental results of Morel et al.). In the first subdomain, we constructed mini-max rational approximation $R_{3,4}$ to I_2:

$$I_2(x) = \frac{a^2 \sum_{i=0}^{3} b_i x^i}{2 \sum_{i=0}^{4} c_i x^i} \tag{11}$$

which approximates I_2 with accuracy better than $5 \cdot 10^{-7}$, and the numerical values of coefficients $\{b_i\}$ and $\{c_i\}$ are listed in Table 1.

Table 1 Values of coefficients of rational approximation to I_2, Eqns. (10) and (11).

i	b_i	c_i
0	$1.729683133081238 \cdot 10^{-12}$	1
1	-1.000001733154165	0.400083955402468
2	-0.150071120939064	0.065216629076836
3	-0.020783094981582	0.004531985789551
4		-0.000015403024613

In the second subdomain we can simply neglect incomplete gamma function bearing in mind that $\Gamma(0,10) \approx 4.16 \cdot 10^{-6}$, and $\Gamma(0,x)$ is a monotically decreasing function of x.

3 Conclusion

We have calculated exactly decay kinetics of the "pseudo-donor–pseudo-acceptor" pairs luminescence in the case of unequal concentrations of "pseudo-donors" and "pseudo-acceptors". Our final formulae turn out to be very simple and give efficient way for analyzing experimental data. Moreover, they can be easily adapted for a description of decay kinetics in other systems, like ion–ion resonant energy transfer processes or "standard" bulk donor–acceptor pair luminescence.

Acknowledgment This research was supported by grant no. 5 P03B 007 20 of KBN for 2001-2003 years

References

[1] A. Morel, P. Lefebvre, T. Taliercio, T. Bretagnon, B. Gil, N. Grandjean, B. Damilano, and J. Massies, Physica **E17**, 64 (2003).
[2] D. G. Thomas, J. J. Hopfield, and W. M. Augustyniak, Phys. Rev. **140**, A202 (1965).
[3] P. J. Dean, in "Progress in Solid State Chemistry" vol. 8, J. O. McCaldin, and G. Somorjai (Eds.), Pergamon Press, Oxford (1973).
[4] P. Baume, S. Strauf, J. Gutowski, M. Behringer, and D. Hommel, J. Cryst. Growth **184/185**, 531 (1998).
[5] S. Strauf, S. M. Ulrich, P. Michler, J. Gutowski, T. Bottcher, S. Figge, S. Einfeldt, and D. Hommel, phys. stat. sol. (b) **238**, 379 (2001).

Optical and magnetic resonance investigations of ZnO crystals doped with TM ions

V.Yu. Ivanov[1], **M. Godlewski**[*,1,2], **S. Yatsunenko**[1], **A. Khachapuridze**[1], **Z. Gołacki**[1], **M. Sawicki**[1], **A. Omel'chuk**[3], **M. Bulany**[4], and **A. Gorban'**[4]

[1] Institute of Physics, Polish Acad. of Sciences, Al. Lotników 32/46, 02-668 Warsaw, Poland
[2] College of Science, Dep. of Mathematics and Natural Sciences, Cardinal S. Wyszyński University, Warsaw, Poland
[3] Institute of Semiconductors Physics, Kiev, Ukraine
[4] Dnepropetrovsk National University, Dnepropetrovsk, Ukraine

Received 15 September 2003, accepted 17 September 2003
Published online 22 December 2003

PACS 75.25.+z, 75.50.Pp, 76.30.Fc, 78.55.Et

We analyze optical and magnetic properties of ZnO crystals grown by chemical transport method using chlorine as the transporting agent. Doping during the growth process with several transition metal ions is applied in an attempt to create observable magnetic ordering in the crystals. Our magnetic resonance and magnetization measurements indicate a paramagnetic phase in all obtained crystals.

1 Introduction

ZnO is a promising material for applications in electronics and optoelectronics due to the large band gap value and also to the large value of excitons binding energy making this material suitable for blue – UV light sources emitting at the room temperature. Zinc oxide doped with transition metal (TM) ions is also expected to find applications in spintronics [1-7]. Moreover, ZnO is useful as a substrate material for GaN-based devices [8] which at present are the most important for short wavelength optoelectronics. Unfortunately, bulk ZnO crystals are grown at quite challenging conditions. Melting temperature of ZnO is about 2000 °C, which limits the selection of available growth methods [9].

We describe below the properties of several ZnO crystals grown by the chemical transport with doping during the process with different TM oxides. In particular, the magnetic and optical properties are studied in the search for indications of magnetic ordering in the samples.

2 Experimental

The TM doped ZnO crystals were obtained by the chemical vapor growth method using chlorine as the transporting agent. Commercially available ZnO and TM oxide (TMO) powders were mixed together and put into a quartz ampoule with the inner diameter of 20 mm and the length of 150 mm. The ampoule was filled with the chlorine gas up to 500 Torr, sealed off and placed in a vertical furnace with a temperature gradient. The source material was kept at 1050 °C. The crystals grew as platelets of about 5 mm long edges in the cooler end of the ampoule at about 1000 °C. The processes performed at lower source temperatures, e.g. at 900 °C or 850 °C, delivered needle-like crystals of 1-2 mm thick and 5-8 mm long. For these needles growth along the c axis dominated.

---

[*] Corresponding author: e-mail: godlew@ifpan.edu.pl, Phone +48 22 8436861, Fax + 48 22 8430926

Only partial reproduction of the TMO concentrations was observed in the crystals. For example, using the source material with 5 mol % of MnO, crystals with 3% of Mn cationic component were obtained. Samples containing Ti (less than 1 %), Mn (up to 3 %), Cu (less than 1 %), Co (up to 5 %) were obtained. All the samples were of the n-type, as tested with the thermoprobe method. As-grown crystals were investigated without any post-growth thermal treatment. For comparison, reference samples obtained by the chemical transport with hydrogen as the transporting agent were studied.

The PL measurements were performed in a wide temperature range from the liquid helium temperature up to the room temperature, using a conventional PL set up. UV line of Coherent Co. argon laser was applied for the photo-excitation. A conventional X-band ESR spectrometer was used for electron-spin resonance experiments. The low temperature ESR measurements were performed at either liquid helium or liquid nitrogen temperature. The ESR studies were also led for the Mn-doped samples at the room temperature. The low-temperature magnetization experiments were carried out using a home-built SQUID magnetometer.

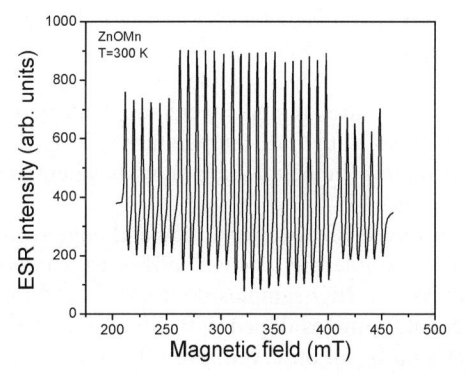

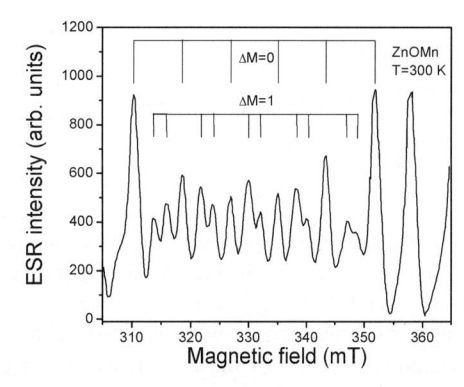

Fig. 1 (left) Room temperature ESR spectrum of ZnMnO sample with low Mn content.

Fig. 2 (right) Central part of the Mn^{2+} magnetic resonance showing well-resolved structure of the signal.

3 Results

Magnetic properties of the ZnTMO crystals were studied with ESR and SQUID. All analyzed samples show magnetic resonances typical for a paramagnetic phase. Characteristic ESR signals of Mn^{2+} ions (see Figs. 1 and 2) and Co^{2+} (Fig. 3) are observed. The Mn^{2+} signals can be detected up to the room temperature.

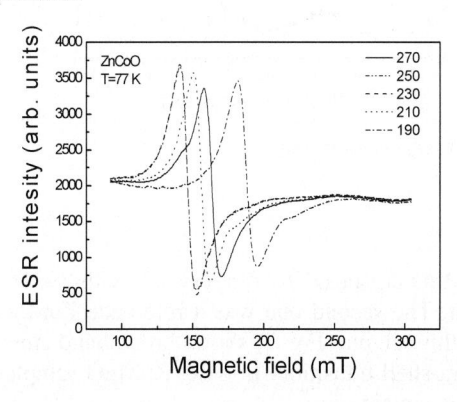

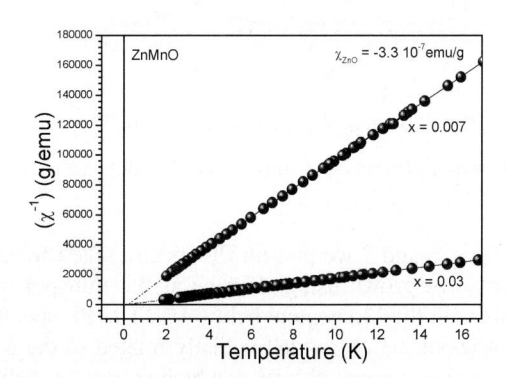

Fig. 3 (left) ESR spectrum of Co^{2+} ion in a ZnCoO sample taken at the liquid nitrogen temperature. Angular dependence of the signal is shown by selecting the ESR spectra taken for five different angles in the (110) plane.

Fig. 4 (right) Magnetization properties of two ZnMnO crystals with Mn contents of 0.7 and 3 %.

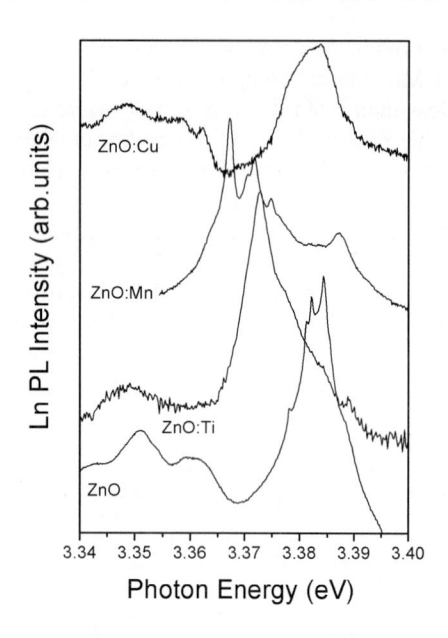

Fig. 5 Low temperature edge part of the PL spectrum for ZnO samples doped with various TM ions.

We could notice for the samples with lower Mn concentrations a resolved fine and hyperfine structure, seen in Fig. 1 as five sets of six hyperfine lines. Closer inspection of the pattern (Fig. 2) reveals further details of the ESR spectrum. So-called spin forbidden transitions are observed.

The high quality of the spectrum indicates good crystalline quality of our crystals and it allows to notice that the Mn^{2+} ions are in the paramagnetic phase. The existence of paramagnetic phase was confirmed by the SQUID data shown in Fig. 4 for two ZnMnO samples with different Mn concentrations. These measurements were supplemented by those performed down to the mK temperatures [7]. Even at the lowest temperatures (in the mK range), no hystheresis of magnetization was observed, as discussed in details in the reference [7].

Most of the samples doped with the TM ions show well-resolved PL spectra in the band edge region of the emission, as shown in Fig. 5. Shifts, peaks broadening and intensity changes of bound excitonic PLs are observed, but edge part of the spectrum is present for all ZnO samples doped with TM ions. Surprisingly different effects were noticed in the reference ZnO samples doped with Mn^{2+} ions during the hydrogen transport. For these samples, the edge PL emission is not observed, even though the samples were very homogeneous and transparent.

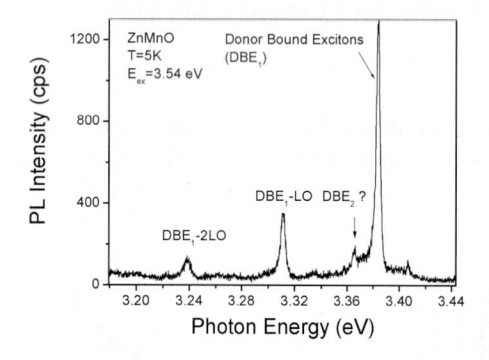

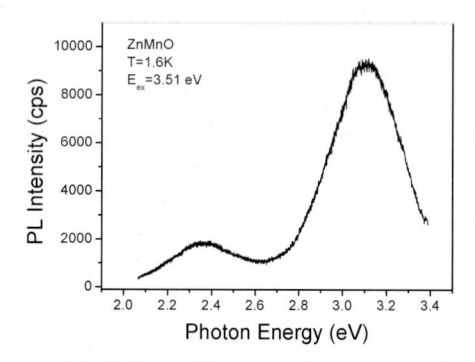

Fig. 6 (left) Low temperature edge part of the PL spectrum for ZnMnO with 3 % Mn.

Fig. 7 (right) Broad band emissions of Mn doped ZnO.

In Figs. 6 and 7 we present PL spectra taken for two ZnMnO crystals. The first crystal, with 3% Mn content, was grown using chlorine as the transporting agent. The second one was a reference ZnMnO sample with the Mn content below 1%. The PL spectrum of this sample (Fig. 7) shows broad band emission at about 2.3 eV, which is likely related to the Mn^{2+} intra-shell transition. In other ZnMnO samples this emission was not observed, which at present is difficult to explain.

In addition to the PL spectra, we measured their temperature dependencies (Fig. 8). Bound excitonic emissions are stable with the increasing temperature up to the liquid nitrogen range, but closely to the room temperature, they are replaced by the free excitonic emission.

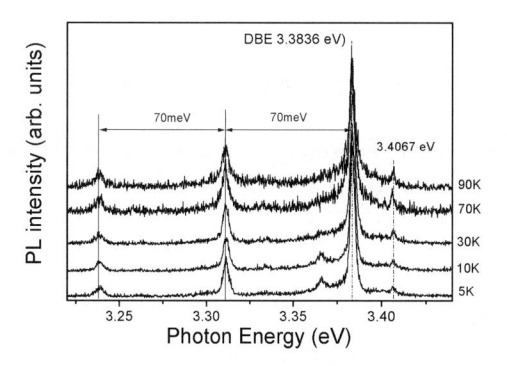

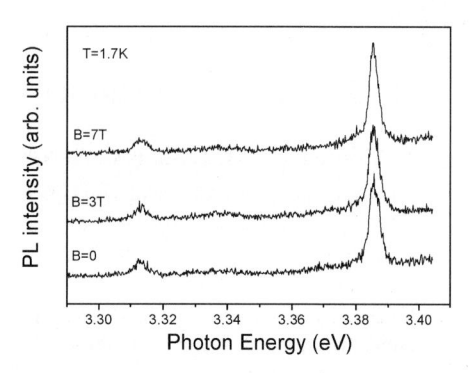

Fig. 8 (left) Temperature dependence of the PL spectrum for ZnMnO with 3 % Mn.

Fig. 9 (right) Magnetic field dependence of the PL spectrum of Mn doped ZnO with 3 % Mn.

We also investigated influence of the magnetic field on the edge emissions (Fig. 9). No shift of spectral positions of the PL lines was observed up to 7 T. Such shift is observed in the case of II-Mn-VI samples showing strong magnetic interactions between localized magnetic moments of Mn ions and spins/magnetic moments of free carriers [10]. Line splitting could not be resolved even at 7 T. The observed PL lines were too broad to achieve such splitting and thus to perform Zeeman effect analysis of the DBE origin, similar to that recently reported [11].

4 Conclusion

ZnO crystals doped with various transition metal ions were grown. The crystals show high structural and optical quality. We studied them in the search for indications of ferromagnetic ordering. We conclude from the ESR and magnetization measurements that all the samples show a paramagnetic phase. Even for the samples with a few percent of TM ions, no magnetic ordering was identified.

Acknowledgements This work was partly supported by grant No. PBZ-KBN-044/P03/2001 and 7 T08A 006 20 of KBN (Polish State Committee for Scientific Research).

References

[1] T. Dietl, H. Ohno, F. Matsukura, J. Cibert, and D. Ferrand, Science **287**, 1019 (2000).
[2] T. Dietl, J. Appl. Phys. **89**, 7437 (2001).
[3] H. Saeki, H. Tabata, and T. Kawai, Solid State Commun. **120**, 439 (2001).
[4] H.-J. Lee, S.-Y. Jeong, C. R. Cho, and C. H. Park, Appl. Phys. Lett. **81**, 4020 (2002).
[5] M. Kohls, M. Bonanni, L. Spanhei, D. Su, and M. Giersing, Appl. Phys. Lett. **81**, 3858 (2002).
[6] S-J. Han, J. W. Song, C.-H. Yang, S. H. Park, J.-H. Park, Y. H. Jeong, and K. W. Rhie, Appl. Phys. Lett. **81**, 4212 (2002).
[7] X. Gratens, V. Bindilatti, N.F. Oliveira, Jr., Y. Shapira, S. Foner, Z. Gołacki, and T.E. Haas, Phys. Rev. B submitted.
[8] M. Godlewski and E. M. Goldys, in: Optoelectronic Properties of Semiconductors and Superlattices: III-Nitride Semiconductors: Optical properties II, ed. M.O. Manasreh and H.X. Jiang, Taylor & Francis Books, Inc., New York, London, 2002, ch. 7, and references given in.
[9] R. Triboulet, Proc. SPIE **4412**, 1 (2001).
[10] M. Godlewski, V.Yu. Ivanov, A. Khachapuridze, and S. Yatsunenko, phys. stat. sol. (b) **229**, 533 (2002).
[11] M. Strassburg, A. Rodina, M. Dworzak, U. Haboeck, I.L. Krestnikov, O. Gelhausen, A. Hoffmann, M. Phillips, H.R. Alves, A. Zeuner, D.M. Hofmann, and B.K. Meyer, Proc. II-VI Conference, Niagara Falls, New York, USA, Sept. 22-26, 2003, phys. stat. sol. (c), in press.

Optical and RBS studies in Tm implanted ZnO samples

T. Monteiro[*,1], **M. J. Soares**[1], **A. Neves**[1], **M. Oliveira**[1], **E. Rita**[2,3], **U. Wahl**[2,3], and **E. Alves**[2,3]

[1] Departamento de Física, Universidade de Aveiro, 3810-193 Aveiro, Portugal
[2] ITN, Estrada Nacional 10, 2686-953 Sacavém, Portugal
[3] CFNUL, Av. Prof. Gama Pinto 2, 1699 Lisboa, Portugal

Received 15 September 2003, accepted 17 September 2003
Published online 22 December 2003

PACS 61.72 Dd, 61.72 Vv, 78.55 Et

We report on optical and structural analysis of Tm implanted ZnO [0001] single crystals. The samples were implanted at room temperature with 150 keV Tm$^+$ ions with a nominal fluence of 5×10^{16} Tm$^+$/cm^2 and subsequently air annealed for 30 min at 800 °C, 900 °C and 950 °C. The implantation damage and annealing effects were investigated with Rutherford Backscattering/Channelling Spectroscopy. We observe that following implantation the majority of Tm ions are incorporated on Zn sites. The optical properties of as-implanted and annealed samples have been studied by low temperature photoluminescence measurements. Well defined Tm-related near infrared emission were observed upon above band gap excitation and the data are consistent with the presence of multi Tm-related optical centers.

© 2004 WILEY-VCH Verlag GmbH & Co. KGaA, Weinheim

1 Introduction Due to its potential applications for electroluminescent devices the doping of wide band materials, such as GaN and ZnO, with rare earth (RE) ions is nowadays an interesting field of study. Trivalent RE ions such as Eu^{3+}, Er^{3+} and Tm^{3+} are known to be suitable dopants for red, green and blue emitters, respectively [e.g. 1–7]. For instance, *Steckl, et al* [1 and references therein] reported recently the achievement of electroluminescence devices on GaN films doped with these RE ions. Following the interest in applied devices, and in order to improve the use of these ions in the visible spectral region, the understanding of the RE ions fundamental properties in wide band gap semiconductors, such as its site location and surroundings, optical ion activation mechanisms and emission properties, are currently a motivation for several research groups [e.g. 1–7].

The aim of this work is, by means of the Rutherford Backscattering (RBS) and Channeling (RBS/C) techniques and Photoluminescence Spectroscopy (PL), to provide further information on the fundamental properties of thulium ions when incorporated by ion implantation in ZnO single-crystals. More precisely, the RE ion's lattice site location, the level of incorporated damage and a correlation of these two features with the optical ion activation, will be done.

2 Experimental details High-quality wurtzite [0001] ZnO single-crystals were implanted at room temperature (RT) with 150 keV Tm^{3+} ions at a fluence of 5×10^{16} cm^{-2}. Following implantation, the ZnO single-crystal was divided in four smaller samples, being three of them subjected to one single 30 min air annealing step, at temperatures of 800 °C, 900 °C and 950 °C. Crystalline quality, damage recovery and Tm lattice site location were evaluated by means of the Rutherford Backscattering Spectrometry and Channeling techniques, for which a 1 mm collimated 2 MeV ^{4}He$^+$ beam was used. The samples were mounted in a computer-controlled two-axis goniometer with an accuracy of 0.01°. The backscattered particles were detected with two Si surface barrier detectors placed at 160° and 170° with respect to the beam direction (IBM geometry) and with energy resolutions of 13 keV and 16 keV, respectively.

---

[*] Corresponding author: e-mail: tita@fis.ua.pt, Phone: +351 234 370 824, Fax: +351 234 424 965

Photoluminescence measurements were carried out with a 325 nm cw He–Cd laser using an excitation power density ~0.6 W.cm^{-2}. The PL was measured at 13 K with the sample mounted in the cold finger of a closed cycle helium cryostat. The emitted light was collected in a 90° geometry, dispersed by a Spex 1704 monochromator (1m, 1200 mm^{-1}) and detected by a cooled Hamamatsu R928 photomultiplier. The presented spectra are all uncorrected to spectral responses.

3 Results and discussion Figure 1 shows the random and <0001>-aligned RBS spectra for the as-implanted sample

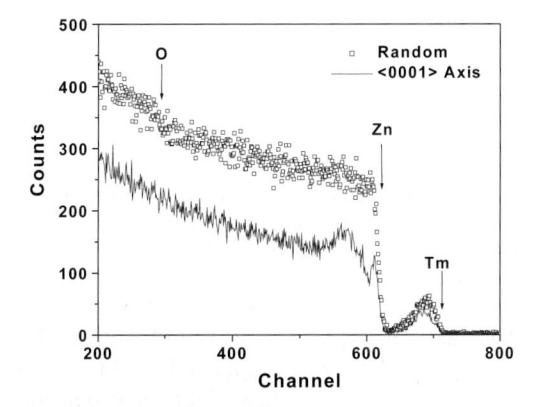

Fig. 1 Random and [0001]-aligned RBS/C spectra for the ZnO single-crystals following implantation with 5×10^{16} Tm^{3+}/cm^2.

The surface-edge for the different elements (O, Zn and Tm) can be clearly distinguished, as well as the damage region produced by ion implantation. The reduced Zn signal in the aligned spectrum, when compared with the random acquisition yield, indicates that the as-implanted sample is below the amorphisation threshold. For this sample a 60% minimum yield (χ_{min}) was measured, which is quite large comparing to the 2% value found in virgin single-crystalline ZnO. Both the minimum yield and the calculated defects density (1.12 x10^{17} at/cm^3) are larger than the ones found for lower dose (5x10^{15} cm^{-2}) Tm implanted ZnO crystals, under the same conditions [8]. Figure 2 shows the detailed angular scans performed along the [0001] direction, for the as-implanted sample.

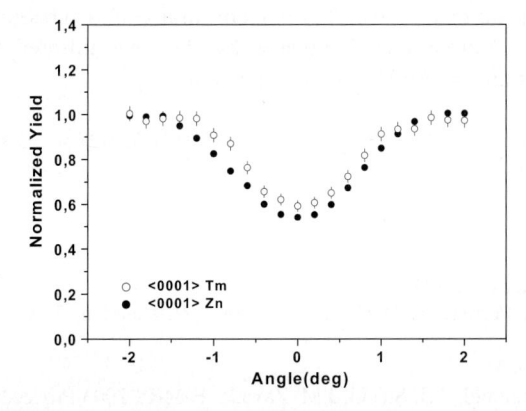

Fig. 2 Angular scans along the [0001] direction measured following implantation.

The Tm and Zn related curves almost overlap, suggesting that Tm and the Zn sub-lattice are subjected to the same disorder. The minimum yields found for both elements were χ^{Tm} = 59% and χ^{Zn} = 54%. the Using the equation $(1-\chi^{Tm})/(1-\chi^{Zn})$ we estimated that 89% of Tm ions are substitutional at the Zn site (S_{Zn}) along the c-axis.

Figure 3 shows the 13 K observed PL with above band gap excitation. This emission is due to intraionic transitions at the Tm ion.

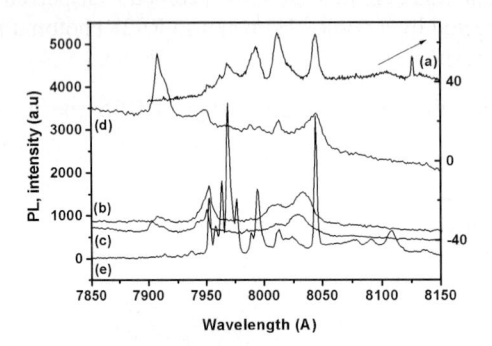

Fig. 3 Tm-related PL spectrum at 13 K, upon above band gap excitation, for the 5×10^{16} Tm^{3+}/cm^2 RT implanted ZnO crystals: (a) as-implanted; after a 30 min air annealing at (b) 800 °C, (c) 900 °C and (d) 950 °C; (e) Tm-related PL spectrum observed upon the same conditions for the 5×10^{15} Tm^{3+}/cm^2 RT implanted ZnO, after a 30 min 800 °C air annealing.

Ion optical activation is achieve in the as-implanted state (spectrum (a) from Figure 3). The spectrum is similar to the one observed after annealing for 5×10^{15} Tm^{3+}/cm^2 implanted ZnO (spectrum (e) from Figure 3) [8]. Upon heat treatment at three different temperatures new Tm-related optical centers are observed as shown by the spectra (b), (c) and (d) of Figure 3.

Within the ZnO's band gap the Tm^{3+} ion ($4f^{12}$ configuration) has a $^3H_6$ ground state manifold and $^3F_4$, $^3H_5$, $^3H_4$, $^3F_3$, $^3F_2$ and $^1G_4$ excited manifold states. With the used excitation condition and in this particular spectral region the intraionic luminescence is expected to be dominated by the $^3H_4 \rightarrow ^3H_6$ transition.

The appearance of different spectral features after sample annealings suggests that either Tm ions environment had changed or that the emission arises from new Tm related complexes.

To achieve a better understanding of these optical centers, further work is under way to establish the kinetics and excited energy levels from where luminescence originates. To evaluate the Tm lattice location and defect recovery for the annealed samples, RBS/C analysis is being carried out.

4 Conclusions 5×10^{16} Tm^{3+}/cm^2 ions were incorporated into ZnO [0001] single crystals by room temperature ion implantation. RBS/C results indicate that 90% of Tm is incorporated into substitutional Zn sites after the implantation. The Tm-related emission is dominated by the near infrared transition $^3H_4 \rightarrow ^3H_6$, and new Tm-related optical centers are observed after thermal annealing.

Acknowledgements This work was funded by the FCT, Portugal (project CERN/FIS/43725/2001). E.Rita and U. Wahl acknowledge their fellowships supported by the FCT.

References

[1] Y. Q. Wang, A. J. Steckll, Appl. Phys. Lett. **82**, 502 (2003).
[2] T.Monteiro, C. Boemare, M.J. Soares, R.A. Sá Ferreira, L. D. Carlos, K. Lorenz, R. Vianden, E. Alves, Physica B **308–310**, 22 (2001).
[3] M. Pan, A.J. Steckl, Appl. Phys. Lett. **83**, 9 (2003).
[4] U. Hömmerich, Ei Ei Nyein, D.S. Lee, J. Heikenfeld, A.J. Steckl, J.M. Zavada, E-MRS 2003 Proceedings, to be published.
[5] K. Lorenz, E. Alves, U. Wahl, T. Monteiro, S. Dalmasso, R.W. Martin, K.P. O'Donnell, R.Vianden & RENiBEL Network, E-MRS 2003 Proceedings, to be published.
[6] H.J. Lozykowski, W.M. Jadwisienczak, I. Brown, Appl. Phys. Lett. **74**, 1129. (1999).
[7] T. Monteiro, J. Soares, M.R. Correia, E. Alves, J. Appl. Phys. **89**, 6183 (2001).
[8] E. Rita, E. Alves, U. Wahl, J.G. Correia, A. J. Neves, M.J. Soares, T. Monteiro Physica B (to be published).

Defect dynamics in P⁺ implanted 6H-SiC studied by positron annihilation spectroscopy

G. P. Karwasz[*,1,2], **R. Rurali**[3], **G. Consolati**[4], and **P. Godignon**[3]

[1] INFM, Dipartimento di Fisica, Universitá di Trento, via Sommarive, 14, 38050 Trento, Italy
[2] Institute of Physics, Pomeranian Pedagogical Academy, Arciszewskiego 22B, 76-200 Słupsk, Poland
[3] Centre Nacional de Microelectrònica (CNM – CSIC), Campus de Bellaterra, 08193 Bellaterra, Spain
[4] INFM, Dipartimento di Fisica, Politecnico di Milano, Piazza Leonardo da Vinci, 32, 20133 Milano, Italy

Received 15 September 2003, revised 3 October 2003, accepted 3 October 2003
Published online 22 December 2003

PACS 61.72.Tt, 78.70.Bj

We study P⁺-implanted 6H-SiC samples by means of a variable energy slow positron beam. In as-grown samples we observe a slow fall of the annihilation S-parameter from the surface to the bulk value, indicating a long diffusion length, i.e. absence of positron-trapping defects. This result is also confirmed by positron-lifetime measurements yielding essentially a single-component lifetime of 148 ps. In the implanted samples we detect a wide flat region of slightly risen defectiveness. However, we note that, in order to reproduce reasonably well the depth of the damaged layer, one has to assume the presence of strong electric fields in the implanted region. With annealing, S-curves show a rising maximum, moving towards surface – probably, in as-implanted samples the defects are decorated by P⁺ ions and do not trap positrons in effective way. After the highest temperature annealing a highly defected region extends for the first 50 nm depth. We perform also atomic force microscopy to monitor the evolution of the surface defects.

© 2004 WILEY-VCH Verlag GmbH & Co. KGaA, Weinheim

1 Introduction Silicon carbide (SiC) is a wide band-gap material ideally suited for high-power and high-temperature applications. After being a long-standing promising material, nowadays it is used in realistic, industrial device fabrication. There is a growing interest in P as a n-type dopant for SiC, especially in all those applications in which a high doping dose is required. In such conditions, N – the other typical donor – suffers of serious problems from the viewpoint of the activation efficiency [1], induced by the self-passivation mechanisms that bedevil N aggregates [2]. Unfortunately, most of the doping species, including N and P, have very low diffusion coefficients in SiC, making ion implantation the only viable technique to achieve localized doping to form p–n junctions. This is a serious drawback for P, as the damage created in the host lattice will be comparatively higher than in the case of the lighter N ion. For these reasons, characterizing the formation and the evolution of the defect layer in P implanted SiC is very important, especially when relatively high concentration of dopants are needed and P should be preferred to N as a n-type impurity.

Positron annihilation spectroscopy (PAS) using low energy beams is an efficient experimental tool to probe the depth-dependence of the free-volume created by the implantation process. Injecting positrons with well-controlled energy and studying the Doppler broadening of the annihilation photopeak provides a valuable insight on the electronic structure of the material as a function of the depth.
In this work we have used PAS to study P-implanted 6H-SiC before and after the thermal annealing, comparing the results with the reference as-grown material. In order to obtain a more complete picture,

---

[*] Corresponding author: e-mail: karwasz@science.unitn.it

we have also performed positron annihilation lifetime spectroscopy (PALS) of the bulk material and atomic force microscopy (AFM) to monitor the changes which the surface undergoes.

2 Experimental methods Samples of 6H-SiC 0.6 mm thick (from Cree Inc.) were implanted with P[*] at 280 keV at room temperature with a fluency of 10^{14} cm^{-2}. TRIM modeling shows a maximum concentration of P atoms of about 7×10^{18} cm^{-3} positioned at 260–300 nm and a distribution width (FWHM) of 150 nm. Successively, samples were annealed at 1550 °C for 30 min.

Slow positron beam technique [3] and positron lifetime techniques [4] have been described previously. Briefly, the 511 keV γ-annihilation peak is broadened by the non-zero momentum of the electron with whom the positron annihilates. The shape "S" parameter describes the ratio between the area of the central part and the whole area of the annihilation peak: ±0.85 keV window is used in the present data. The rise of S-parameter indicates the lowering of electron momenta – the annihilation takes place with valence electrons, i.e. in vacancy-like sites.

3 Results and discussion Ohshima and co-workers [5] studied P-implanted SiC by PAS, reporting a complete removal of the positron detectable defects after a 1550 °C thermal annealing. In our work, however, we have focused on a higher-dose, higher-implantation energy process, so that it is interesting to analyze the differences in the evolution of the damaged layer.

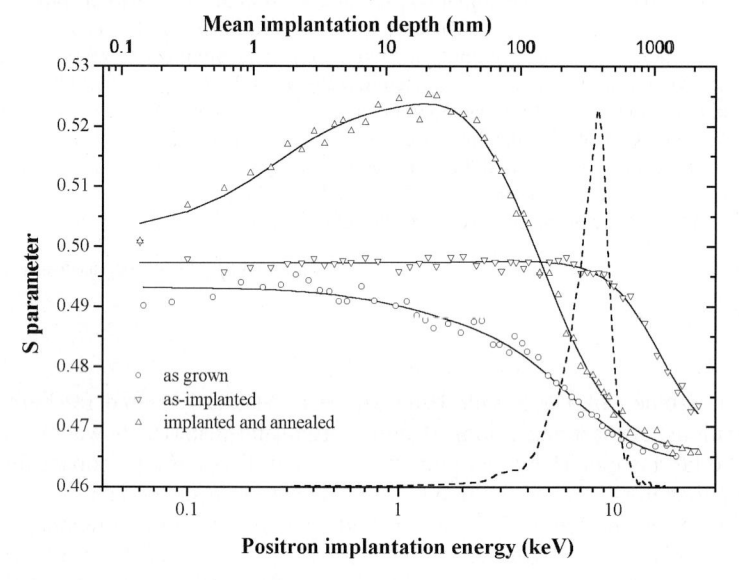

Fig. 1 „S"- positron annihilation parameter vs. depth: experimental points and VEPFIT model. Dashed line shows the simulated P[+] implanted profile.

In Fig. 1 we plot the S-parameter vs. depth of the as-grown, the as-implanted and the annealed sample. These curves have been fitted by means of the VEPFIT package [6], which solves the positron diffusion equation assuming a box-shaped defect layer structure. The results are summarized in Table 1.

Table 1 Results of the VEPFIT analysis.

	S_{surfac}	S_{layer}	Layer thickness [nm]	S_{bulk}
as-grown	0.493	–	–	0.465
as-implanted	0.505	0.497	250/700[*]	0.469
Annealed	0.497	0.524	40	0.466

[*] see text

In "as grown" sample the S-parameter changes smoothly from a value of 0.493 (typical for the surface) down to 0.465 typical for a defect free bulk material [5]. Also a long diffusion length (190 nm) indicates absence of positron-trapping defects in as grown SiC.

Initially, the S-curve after implantation has been approximated by a two-layer model, but this yields very poor chi-square values. Therefore, a three-layer model has been proved, under following assumptions:
1) the bulk value and bulk diffusion length have been fixed to "as-grown" values; 2) the surface S-value has been fixed close (0.505) to that in as-grown sample. A very good agreement between the fit and experimental values has been obtained assuming presence of strong (15000 V/cm) opposite electric fields, trapping positrons in the defected layer, see the scheme shown in Table 2. The border of the first layer corresponds to the "left-side" of the P-implantation peak, see Fig. 1. The border of the second layer (700 nm) is somewhat deeper than the range of P^+ ions (about 400 nm) – this can be explained by kind of ballistic effect – the presence of electric field slows down all positrons, also those very fast, traveling through the implanted layer, giving therefore an impression that this layer extends deeply into the material. The S-value in the implanted region practically coincides with values obtained by Ohshima et al. [5] and attributed to a pair (C, Si) vacancy. The diffusion length of positrons in the implanted region is shorter (65 nm) than in the bulk.

Table 2 VEPFIT layered model of implanted samples.

Layer 1	Layer 2	Bulk
Electric	Field	
13500 →	← 16500 V/cm	0

Borders of layers ↑ 250 nm ↑ 700 nm
P^+ *implantation profile* ↑ ↑↑↑ ↑

In annealed samples no defects are seen by positrons in the implanted region: the S-parameter shows a higher (0.524) value only in the very near-to-surface region and falls quickly with rising depth to the "no-defect" bulk value. The diffusion length in this near-to-surface region is rather long, 130 nm.

Ohshima et al. [5] observed a similar trend in samples thermally treated between 800 °C and 1200 °C: the defected layer moved to the surface and the S-parameter rose from 1.06 to 1.09 times the "no-defect" S-value. However, contrarily to what reported in Ref. [5], in the present measurements the thermal treatment has not proved to be sufficient to anneal all the vacancy-related centers out. The defects migrate toward the surface and accumulate there, leading to a coalescence, as can be clearly seen looking at the results of Table 1, where the fitted defect layer is much thinner than in the as-implanted sample, but it is characterized by a higher S_d, indicating larger voids. In a preliminary way, we note that the relative S-value in annealed samples is double, 1.12 (taking "no-defect" as 1.0) compared to the implanted sample value (i.e. 1.06 in relative units).

Bulk PALS measurements cannot resolve different depths and they only average on all the sample. However, lifetime data could provide useful complementary information to Doppler broadening beam measurements. Lifetime and intensity of the fitted components are collected in Table 3.

Table 3 Lifetime τ and intensity *I* of the two fitted component of the PALS measurements.

	τ_1[ps]	I_1	τ_2[ps]	I_2
as-grown	147	99.83	2267	0.17
as-implanted	148	99.91	3050	0.10
annealed	151	99.77	1305	0.23

As expected, there is only one lifetime component due to the positron annihilation in the material. The concentration of vacancy-type defects is not high enough to resolve the lifetime of positrons in the bulk and another (longer) lifetime component due to the positrons annihilating in the vacancies. However, we have been able to fit also a longer component whose intensity is very small, but appearing systematically. We speculate that this longer lifetime may be due to surface and sub-surface states, where nanovoids sufficiently large could possibly form, allowing the formation of positronium (Ps).

The AFM images seem to confirm such interpretation. As can be seen in Fig. 2, the implantation leaves the surface morphology essentially unaltered. It is with the annealing that the deep defects (accordingly also to our PAS results) emerge to the surface, creating the characteristic step structure (see Fig. 2c) that is probably due to vacancy coalescence in the sub-surface layers.

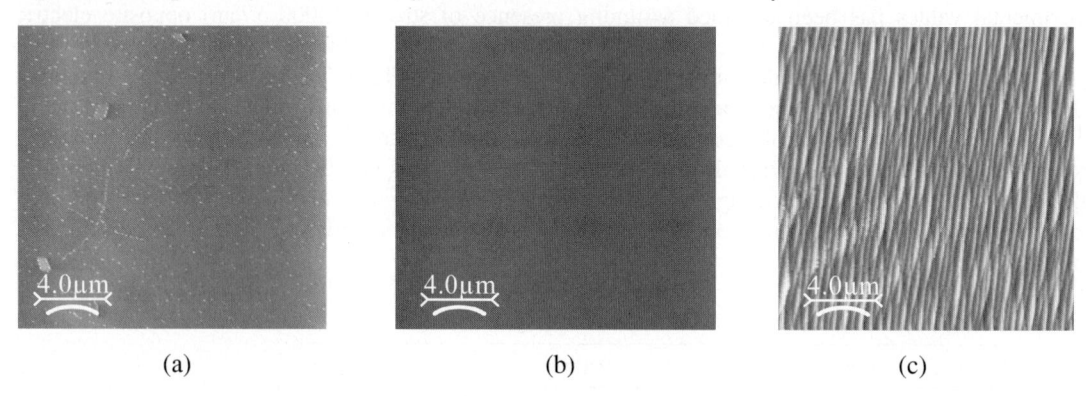

Fig. 2 AFM images of the (a) as-grown, (b) as-implanted and (c) annealed samples.

4 Conclusions

Positron annihilation measurements (low S-parameter, long diffusion length of 200 nm, short 148 ps lifetime) indicate no defects in as-grown SiC. Defects emerge in P^+ as-implanted samples but at apparently wrong depth. The discrepancy between TRIM and VEPFIT modeling can be explained assuming the existence of a strong electric field (>10 kV/cm) pulling positrons into the implanted region. This is the most surprising result of the present research – we are not aware of indications of such fields (in samples before annealing) obtained by other techniques. A tentative explanation is double-fold: negative space charge in some defect-related sites, or simply the enhanced drift of positrons into the highly defected region. The electrical field disappear with the annealing, so the physics underlying this effect is related to the implantation-induced defects and/or implanted atom distribution.

Acknowledgements R. Rurali is supported by Training and Mobility Network of EU (ATOMCAD).

References

[1] M. Laube, F. Schmid, G. Pensl and G. Wagner, Mater. Sci. Forum **389–303**, 791 (2002).
[2] R. Rurali, P. Godignon, J. Rebollo, E. Hernández and P. Ordejón, Appl. Phys. Lett. **82**, 4298 (2003).
[3] R.S.Brusa, G.P.Karwasz, M.Bettonte and A.Zecca, Appl. Surf. Sci. **116**, 59 (1997).
[4] G. Consolati and F. Quasso, J. Chem. Phys. **114**, 2825 (2001).
[5] T. Ohshima, A. Uedono, K. Abe, H. Itoh, Y. Aoki, M. Yoshikawa, S. Tanigawa and I. Nashiyama, Appl. Phys. A **67**, 407 (1998).
[6] A. van Veen, H. Schut, J. De Vries, R. A. Haakvoort and M. R. Ijpma, AIP Conf. Proc. **218**, 171 (1990).

Deep UV detection by CVD diamond position sensitive devices

G. Mazzeo[1], **G. Conte**[*,1], **M. C. Rossi**[1], **S. Salvatori**[1], and **V. Ralchenko**[2]

[1] INFM and Electronic Engineering Dept., Univ. "Roma Tre", Via Vasca Navale, 84 – 00146 Rome, Italy
[2] General Physics Institute, Russian Academy of Sciences, 38 Vavilov Str., 199991 Moscow, Russia

Received 15 September 2003, revised 2 October 2003, accepted 2 October 2003
Published online 22 December 2003

PACS 73.40.Ty, 73.50.Pz, 73.61.Ng, 85.60.Gz

The realization of position sensitive detectors for deep UV light monitoring is reported. Simple voltage division structures have been developed based on photolithography defined resistive element. Deviations from linearity of few percentage with the beam impinging position have been achieved.

© 2004 WILEY-VCH Verlag GmbH & Co. KGaA, Weinheim

1 Introduction

Diamond is considered the elective material for electronic devices operating in extreme environments. In particular, its wide band-gap (5.5 eV) and its high thermal conductivity (20 W cm^{-1} K^{-1}) make diamond the best candidate for the realisation of UV detectors able to work with the high energy photons and beam intensities typical of pulsed excimer lasers. The use of photoconductive CVD diamond devices for monitoring pulsed UV sources in the field of ULSI photolithography and micro-machining has been recently proposed [1]. Indeed, the need of decreasing the active dimension of the electronic devices has lead to the use of excimer laser radiation (ArF 193 nm, F2 157 nm) for the photoresist exposure. For such kind of applications, however, the principal request is for a fast and sensitive detector in order to control the photoresist dosimetry. Moreover, low damage to the high energy used is a requirement not completely satisfied by silicon based devices. Diamond, with its threshold for radiation damage around 100 mJ/cm^2, meets the specifications. On the other hand, the power levels normally used in current photolithography applications are well belowe of those sustained by CVD diamond. Such kind of application, however, requires not only the intensity but also the position of the beam. The field of position sensitive devices (PSDs) has recorded many contributions on 1D and 2D detectors based principally on amorphous [2] and crystalline silicon [3]. Untill now, very few contributions have been focused on the realization of PSDs aiming to the control of the beam position and intensity, principally owing to non-uniformity in the quality of CVD diamond. Huge variations in sensitivity as the position of the spot varies have been reported [4] in comparison with the performances of silicon UV devices. Notwithstanding, imaging deep-UV light detectors have been developed by using high quality CVD diamond [5]. With the aim of contributing in understanding potentiality and performances of CVD diamond detectors we started a research activity aimed to realize area sensitive detectors for deep-UV or extreme-UV light and soft X-ray sources. In this paper we report on the realization of linear PSDs based on resistive current division structures able to monitor the position and intensity of a 193 nm laser beam. Sensitivity and linearity have been studied as a function of the beam intensity.

---

[*] Corresponding author: e-mail: gconte@ele.uniroma3.it, Phone: +39 06 5517 7268, Fax: +39 06 5579 078

2 Materials and technology

High quality, 8x8 mm^2 large and 0.8 µm thick, polycrystalline diamond deposited by a microwave CVD process has been used in this work. After the growth, the diamond film was mechanically polished to obtain mirror-like surfaces. Chromium strips, 5 mm long, have been realised by physical vapour deposition together with coplanar Silver contacts on the diamond surface. Fine line photolithography has been then used to define Chromium structures with resistance values in the kΩ range. A Neweks PSX100 ArF laser (4.5 mJ per pulse, FWHM=3 ns, 10–100 Hz) has been used to shine light on the realized devices and to test their performance in respect to the beam point stability and intensity variations. The output voltage signals have been recorded by using a Le Croy Wavepro 960 digital sampling oscilloscope.

3 Results and discussion

The current versus voltage curves in the dark and under UV light illumination are reported in Fig. 1. The characteristics are ohmic and symmetric in the probed range. The sensitivity ratio of the diamond specimen, defined as photocurrent over current in the dark, I_{ph}/I_d, is in excess of 10^7 for medium illumination conditions. The photoconductive quantum yield measured in steady state conditions is shown in Fig. 2. The UV-VIS selectivity ratio is higher then 10^3. The onset of optical absorption is around 1.2 eV, corresponding to the Fermi level position, confirmed by the Arrhenius plot shown in the inset of the same figure. Around 5 eV extended states optical transitions are apparent with an exponential increase corresponding to an Urbach tail with slope equal to 70 meV. Over 5.5 eV, the photoconductive response reduces due to surface recombination.

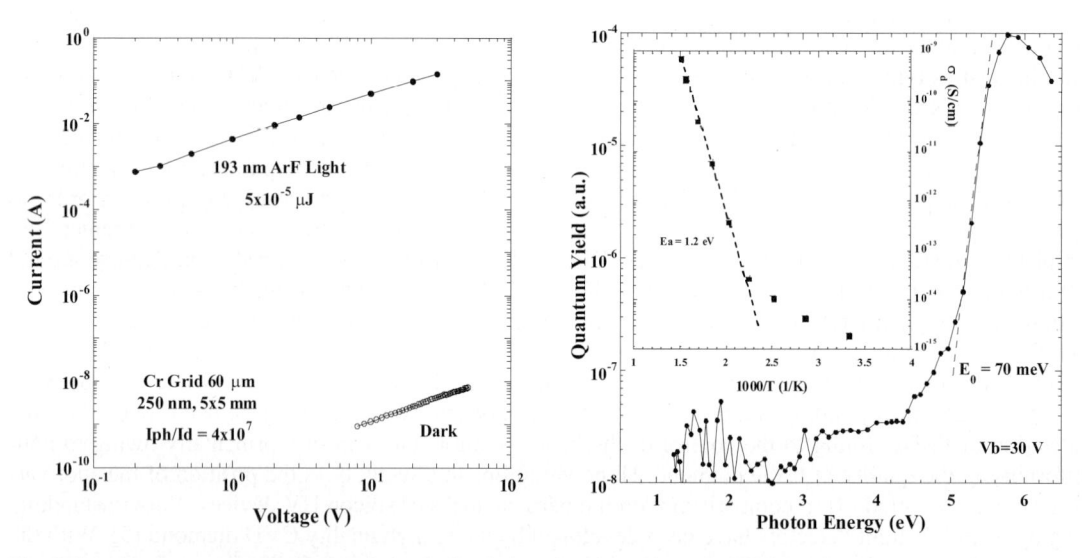

Fig. 1 Current versus voltage characteristics in the dark and under 193 nm irradiation.
Fig. 2 Photoconductive quantum yield of the same samples. Inset: Dark conductivity Arrhenius plot.

A schematic of the realised linear PSDs together with the impedance adapter is shown in Fig. 3. Due to the high ratio I_{ph}/I_d measured on this specimen, we can assume that, impinging into a position x inside the active area, the beam will cause a shorting contact between the resistive element, R_e, and the floating silver contact, in such a way producing a voltage drop on the sensing resistance, R_p, equal to V_0. Being the PSD active area L in length and applying a bias voltage, V_b, to the resistive element, the voltage division across the sensing resistance can be expressed by the relationship

$$V_0 = \frac{(1-x/L)}{1+R_{ON}/R_p + R_e/R_p(x/L)(1-x/L)} V_b, \quad (1)$$

where R_{ON} is the resistance value of the illuminated diamond zone. The 47±1Ω resistance at the exit of the voltage follower coupled to 50 Ω on the oscilloscope gives an output signal spanning over the $0-V_b/2$ range. The external resistance, R_p, provides the possibility to control the linearity of the response optimising the whole circuit coupling. Indeed, the non-linearity on the impinging position are determined by the R_e/R_p ratio whereas the sensitivity on the light beam intensity is determined by the R_{ON}/Rp ratio. Due to the high sensitivity ratio evidenced in Fig. 1 but also at lower illumination rate, $R_{ON}/Rp \approx 0$ can be assumed. Moreover, by choosing a small enough R_e/R_p ratio, a linear dependence on the impinging position is obtained.

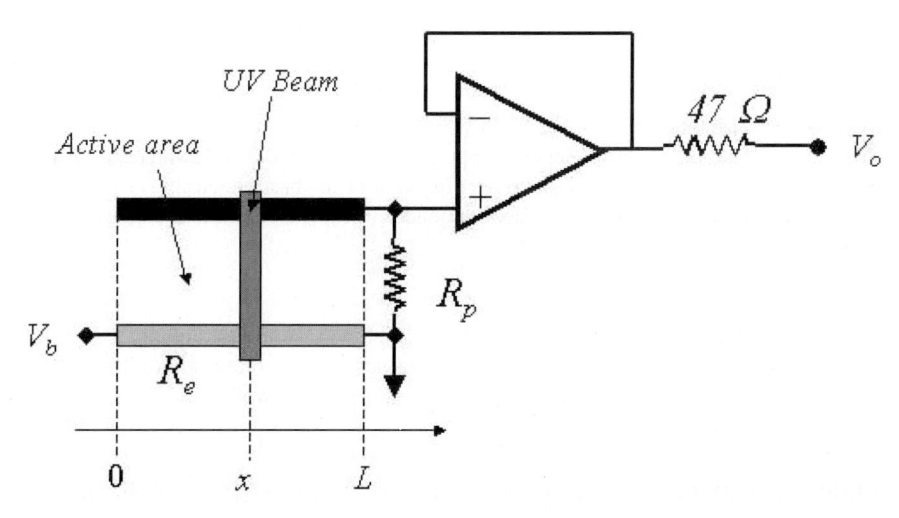

Fig. 3 Scheme of the linear PSD based on resistive voltage division. The coupling electronics is also illustrated. R_e is the chromium resistive element, whereas V_b represents the applied bias and R_p the external sensing resistance.

The normalized response of a 5 mm long active area PSD is reported in Fig. 4 versus the impinging beam abscissa in the case of a 3.3 kΩ resistance R_e. The expected response of a distributed resistive element is achieved. The influence of the variation in the R_p value are also observed. Continuous lines represent the best fitting with expression (1). As expected a strong non linearity is evidenced when small R_p/R_e ratios are used. Increasing this ratio improves the linearity as a function of the beam impinging position. Taking into account the beam spot width, deviations of few percentage from linearity are observed. These results are in good agreement and confirm the indications of prototypes realized with finite resistances [6].

The dependence on the beam intensity was also analysed and the results of the output voltage versus the beam impinging position are reported in Fig. 5 for three different UV beam intensities. To reduce the laser intensity, the cascade reflections on a couple of quartz beam splitters was used. Data in Fig. 5 refer to a R_p/R_e ratio equal to 2 and to two orders of magnitude of beam attenuation. Inside the experimental errors, very low dependence on the beam intensity is here observed. This result address the possibility to use this very simple PSD structure to monitor the position or the angular deflection of a laser beam independent of the light intensity in the µJ per pulse range.

Windings observed in Fig. 4 and 5 is believed to be associated to non-homogeneities in the distributed resistance (i.e. chromium thickness).

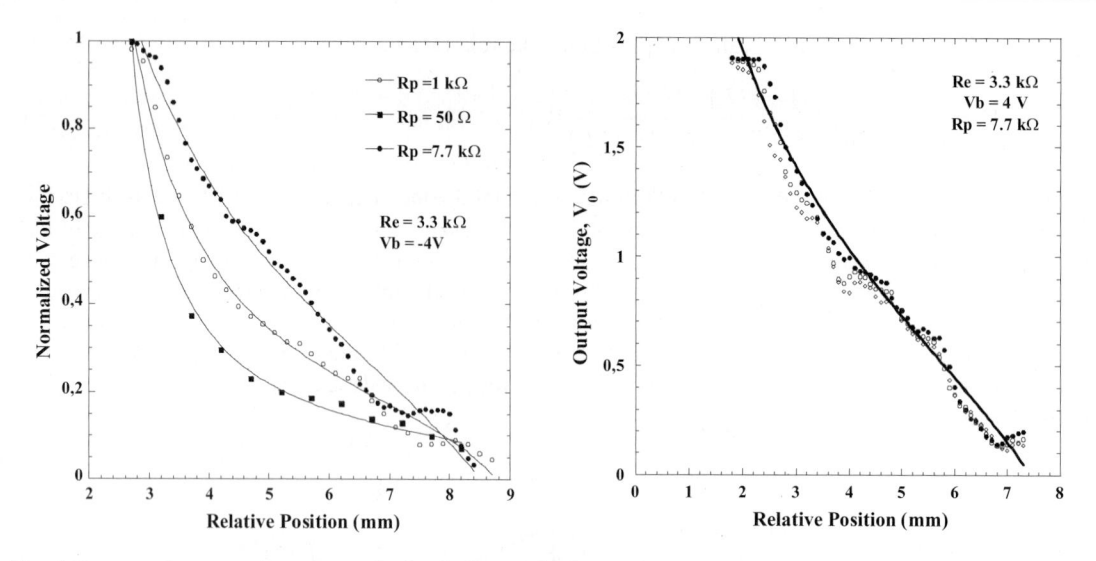

Fig. 4 Output voltage response versus the impinging position as a function of the external resistance R_p of a 5.5 mm long active area PSD with a 3.3 kΩ sensitive element. Continuous lines are the best fit with expression (1).

Fig. 5 Output voltage versus the impinging position as a function of the beam intensity (◊) 1/100, (○) 1/10, (●) 1. The beam spot has been estimated to be 400 µm wide. Continuous line is the best fit with expression (1).

4 Summary

A very simple contact structure has been realized and tested to demonstrate the feasibility of CVD-diamond based position sensitive detectors for UV excimer laser radiation monitoring. Non-linearity with the impinging position of few percentage have been achieved as well as insensitivity to beam intensity in the µJ per pulse range.

Acknowledgement This work has been carried out in the framework of NATO - Collaborative Linkage Grant #979594.

References

[1] M. D. Whitfield, S. P. Lansley, O. Gaudin, R. D. McKeag, N. Rizvi, and R. B. Jackman, Diamond Relat. Mater. **10**, 693 (2001). Diamond Relat. Mater. **10**, 715 (2001).
[2] A. Toneva and D. Sueva, Sensors and Actuators **73**, 210 (1999).
[3] J. Henry and J. Livingstone, IEEE Sensors J. **2**, 372 (2002).
[4] J. Hiscock and A. T. Collins, Diamond Relat. Mater. **8**, 1753 (1999).
[5] S. P. Lansley, O. Gaudin, H. Ye, N. Rizvi, M. D. Whitfield, R. D. McKeag, and R. B. Jackman, Diamond Relat. Mater. **11**, 433 (2002).
[6] G. Mazzeo, S. Salvatori, M. C. Rossi, A. Della Scala, and G. Conte, Intl. Conf. on Sensors and Microsystems, AISEM 2003, February 11–14, Trento, Italy.

The interaction of Pr$_2$O$_3$ with 4H-SiC(0001) surface

A. Goryachko[*], I. Paloumpa, G. Beuckert, Y. Burkov, and **D. Schmeißer**

Brandenburg Technical University Cottbus, Department of Applied Physics/Sensorics, Postfach 101344, 03013 Cottbus, Germany

Received 15 September 2003, accepted 17 September 2003
Published online 22 December 2003

PACS 77.55+f, 77.84.Bw, 79.60.Jv

Praseodymium oxide (Pr$_2$O$_3$) as a hetero-oxide on SiC(0001) represents a promising semiconductor / high-K dielectric material combination. We prepared thin films (<3 nm) of Pr$_2$O$_3$ on 4H-SiC(0001) surfaces by a wet chemical process involving aqueous solutions of Pr(NO$_3$)$_3$. Synchrotron radiation photoelectron spectroscopy (SR-PES) is used to study the chemical composition of these films after deposition and annealing in 300–900 °C range. The Si2p and C1s core level emissions are analysed in detail and reveal that praseodymium-silicate is formed at the interface. Our data are compared to corresponding studies of the Pr$_2$O$_3$ interaction with Si(001) surfaces.

© 2004 WILEY-VCH Verlag GmbH & Co. KGaA, Weinheim

1 Introduction

Silicon carbide (SiC) is a unique semiconducting material suited extremely well for high-power, high-frequency, and high-temperature applications. The 4H-SiC(0001) surface of its 4H polytype is chosen as it has a high potential to be used in field effect transistors (FETs) [1].

SiO$_2$ is a natural oxide not only for Si, but for SiC as well. The dielectrics with permeativity K higher than K$_{SiO_2}$ ~ 4, are needed to reduce the electric field in high-voltage MOS devices for higher electrical breakdown reliability. The oxides of rare-earth elements are good candidates to replace SiO$_2$, and indeed, HfO$_2$-based metal oxide semiconductor (MOS) structures are already reported with low interface state densities on the 4H-SiC(0001) surface [2]. Pr$_2$O$_3$ is a unique dielectric with K$_{Pr_2O_3}$ ~ 30 and extremely low leakage current [3]. Thus, 4H-SiC - Pr$_2$O$_3$ material combination is very interesting for possible MOS applications. The Pr$_2$O$_3$ dielectric was already deposited on Si(001) substrate by e-beam evaporation of Pr$_6$O$_{11}$ targets leading to high quality Pr$_2$O$_3$ films [4, 5]. A different approach is the preparation of Pr$_2$O$_3$ films by a wet-chemical method with Pr(NO$_3$)$_3$ precursor [6], which is attractive for low cost commercial applications. That is why, we have chosen the wet chemistry deposition of Pr$_2$O$_3$ on the 4H-SiC(0001) surface.

The chemical and thermal stability are the most important issues in evaluation of any high-K material to be used in combination with corresponding semiconductor. The technological processes of creating MOS devices include annealing steps, sometimes as high as 1000 °C which take place after the dielectric layer is already deposited. Therefore, it is always a matter of concern if dielectric will not react or intermix with semiconductor material, since the loss of the desired dielectric properties may occur. In this work we have investigated chemical and thermal stability of Pr$_2$O$_3$ on the 4H-SiC(0001) surface against annealing in vacuum up to 900 °C. The X-ray photoelectron spectroscopy (PES) can provide valuable information on the chemical state of Pr, O, C and Si species in the film and at the interface with SiC. Furthermore, if performed with a synchrotron radiation (SR-PES) of varying photon energy, it can show depth distributions of the relevant species in the near-surface region as smaller photon energies produce surface sensitive spectra.

---

[*] Corresponding author: e-mail: goryach@tu-cottbus.de, Phone: +49-(0)355-69-49-10, Fax: +49-(0)355-69-39-31

2 Experimental

The substrates used in this work were 4H-SiC(0001) epitaxial films grown on 4H-SiC(0001) substrates obtained from the Institute of Crystal Growth (Berlin, Germany) [7]. Pr_2O_3 films were deposited from an aqueous $Pr(NO_3)_3$ solution on the 4H-SiC(0001) surface in ambient environment. The 4H-SiC(0001) surface was pretreated with HF to remove the native Si oxide layer. Within the UHV system, by heating the SiC crystal through direct current we first applied a 300 °C annealing step for several minutes. Further one minute annealing steps at 600 °C and 900 °C were performed in order to study the thermal stability of so-formed oxide on the SiC surface. The SR-PES measurements were performed on the U49/2-PGM2 beam line at the Bessy-II (Berlin, Germany) synchrotron radiation facility [8].

3 Results

Figure 1 demonstrates the bulk specific SR-PES Si2p spectra obtained with photon energy of 1200 eV (a) and surface specific spectra with 240 eV (b). The spectra were taken after annealing the dried samples at 300 °C, 600 °C, and 900 °C. For comparison a spectrum of SiO_2 layer on SiC surface obtained with photon energy of 490 eV is displayed in (a) [9]. Two major components are present in the Si2p emission: a SiC-related component and the partially oxidised Si related component. The latter one grows with increasing of annealing temperature and belongs to Pr silicate. In the surface specific spectrum (Fig. 1b) the SiC component is extremely weak already after 600°C annealing and is practically invisible after 900 °C.

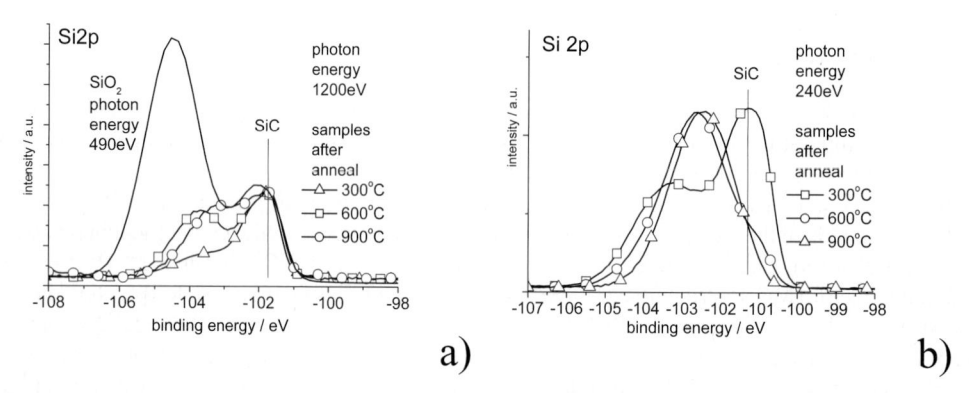

Fig. 1 SR-PES spectra of Si2p emission obtained with photon energies of 1200 eV (a) and 240 eV (b) of the 4H-SiC(0001) surface after covering with aqueous Pr-nitrate solution and annealing at 300 °C, 600 °C, and 900 °C. For comparison a Si2p emission of a native SiO_2 excited by 490 eV photons is given in (a).

In Fig. 2 we show the O1s emissions obtained after the same treatments as in Fig. 1. Both, a bulk-specific (a) and surface-specific (b) spectra are shown. In these spectra we observe three components of distinct binding energies. The leftmost component at ~530 eV is indicative of Pr_2O_3, thus showing that the bulk Pr_2O_3 oxide phase is created by Pr-nitrate decomposition already after 300 °C. This component disappears after annealing at higher temperature, which means that Pr_2O_3 reacts with a SiC surface. As a result two components of higher binding energies grow stronger, belonging to Pr-silicate and Pr-carbonate.

Next, in Fig. 3 we show the SR-PES spectra with 490 eV photon energy of C1s and valence band emissions of our sample after the same treatments. The dominating SiC substrate related component at ~283 eV binding energy is seen in the C1s spectrum (Fig. 3a) A rather strong graphite related component of higher binding energy is already pronounced after the 300 °C annealing procedure. After 600 °C and 900 °C annealing, however, that shape changes drastically by shifting the emission towards

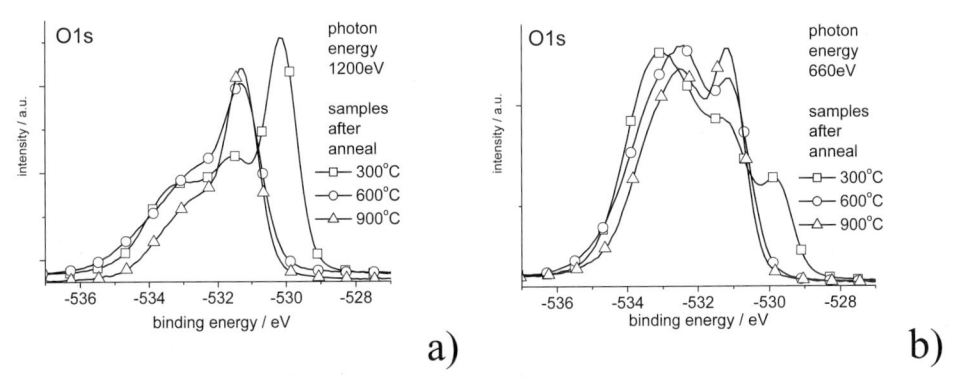

Fig. 2 SR-PES spectra of O1s emissions obtained with photon energies of 1200 eV (a) and 660 eV (b) from the 4H-SiC(0001) surface after covering with aqueous Pr-nitrate solution and annealing at 300 °C, 600 °C, and 900 °C.

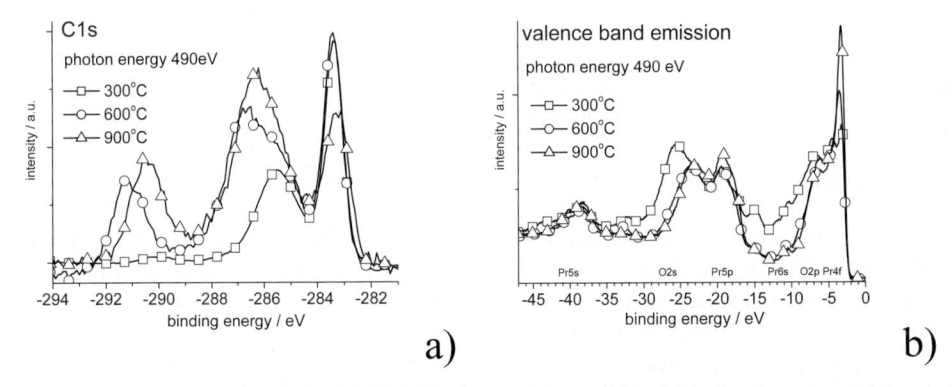

Fig. 3 SR-PES spectra obtained with 490 eV photon energy of C1s (a) and valence band (b) emissions of the 4H-SiC(0001) surface after covering with aqueous Pr-nitrate solution and annealing to 300 °C, 600 °C, and 900 °C.

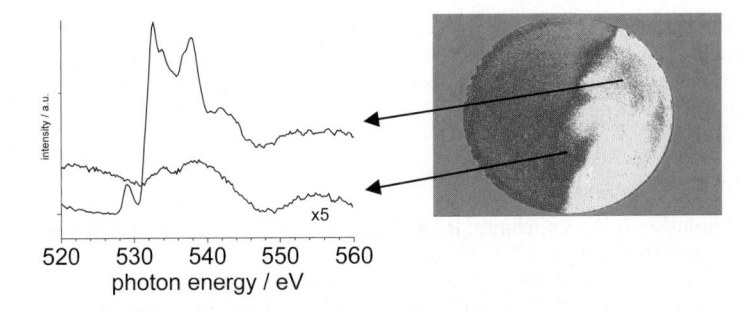

Fig. 4 Pr_2O_3 film on the 4H-SiC(0001) substrate imaged with photoelectrons emitted near the O1s absorption edge. Field of view is 140 μm. Two XAS spectra are obtained in two representative points of the image.

higher binding energies. In particular, we notice strong features characteristic of C–O bonds and surface carbonates (Fig. 3a). Also, there is a clear distinction in valence band structure of Pr_2O_3 on the one hand, and of the silicate–carbonate mixture on the other hand. This is visible in Fig. 3b, where the valence band spectra obtained with 490 eV photon energies are given for the same sample. The valence band structure changes substantially between 300 °C and 600 °C, but than stays unchanged up to 900°C.

Finally in Fig. 4 we demonstrate the image of the Pr_2O_3 film obtained with photoelectrons emitted near the O1s absorption edge, together with X-ray absorption structure (XAS) in two representative points of the image. The right-hand half of the image corresponds to the upper XAS spectrum, with Pr$_2$O$_3$ characteristic structure [5], but the left-hand half and lower spectrum do not have such characteristic peaks. Therefore, our substrate is not yet completely covered by wet chemistry deposited Pr$_2$O$_3$ film.

3 Discussion

The experimental results presented above indicate that Pr$_2$O$_3$ film can be obtained by the wet chemistry deposition on the 4H-SiC(0001) substrate, using Pr(NO$_3$)$_3$ precursor. The distinctive components in the Si2p and the O1s emissions (Figs. 1–2) demonstrate that Pr-silicate is formed at the interface, similarly to the case of the Si(001) substrate [5]. At the same time we observe strong graphite contribution (Fig. 3a). It is not yet clear whether graphite is formed by nitrate interaction or result from the HF etching. The silicate is stable at least up to 900 °C as indicated from the respective contributions in the Si2p and O1s (Figs. 1–2). A new feature in comparison with Si(001) case is Pr-carbonate formation (Figs. 2–3a), formed at high temperatures and appearing as a mixture of silicate and carbonate on top of the Pr-silicate covered SiC surface.

Since it appears that the praseodymium-silicate is thermally stable in contact with SiC, it will be interesting to explore its dielectric properties in the future. If the dielectric constant of this material is sufficiently higher than of SiO$_2$, it could be a valuable alternative as a gate dielectric of SiC-based MOS devices. The application of this silicate could thus offer a substantial thermal budget for the MOS production process involving SiC substrates in power electronics.

4 Conclusion

In conclusion we have investigated the thermal stability of thin Pr$_2$O$_3$ films deposited on the 4H-SiC(0001) epi-layer substrates by wet chemical deposition. Pr$_2$O$_3$ can be reliably formed after drying the aqueous solution of Pr(NO$_3$)$_3$ and annealing at 300 °C in vacuum environment. The oxide reacts with the SiC and forms praseodymium-silicate which is stable up to 900 °C in contact with the SiC substrate. Our current studies involve the improvement of the sample pretreatment and the homogeneity of the film growth. Both are prerequisite for an investigation of the dielectric properties for the possible application as a high-K gate dielectric material.

Acknowledgements We are especially thankful to Dr. D. Siche from the Institute of Crystal Growth (Berlin, Germany) for supplying the 4H-SiC samples and to Dr. P. Hoffmann from the Brandenburg Technical University Cottbus (Germany) for the help with experimental setup. Further, the technical help of the BESSY II (Berlin, Germany) synchrotron radiation facility staff is acknowledged.

References

[1] A. P. Zhang, L. B. Rowland, E. B. Kaminsky, J. W. Kretchmer, R. A. Beaupre, J. L. Garett, J. B. Tucker, B. J. Edward, J. Foppes, and A. F. Allen, Solid-State Electron. **47**, 821 (2003).
[2] V. V. Afanas'ev, A. Stesmans, F. Chen, S. A. Campbell, and R. Smith, Appl. Phys. Lett. **82**, 922 (2003).
[3] H. J. Osten, J. P. Liu, P. Gaworzewski, E. Bugiel, and P. Zaumseil, Technical Digest IEDM 653 (2000).
[4] D. Schmeißer and H.-J. Müssig, Solid-State Electron. **47**, 1607 (2003).
[5] D. Schmeißer, Materials Science in Semiconductor Processing **6**, 59 (2003).
[6] G. A. M. Hussein, B. A. A. Balboul, M. A. A-Warith, and A. G. M. Othman, Thermochim. Acta **369**, 59 (2001).
[7] D. Siche, H.-J. Rost, J. Doerschel, D. Schulz, and J. Wollweber, J. Cryst. Growth **237–239**, 1187 (2002).
[8] P. Hoffmann, D. Schmeißer, G. Roters, and Z. Nenyei, Thin Solid Films **428**, 216 (2003).
[9] D. Schmeißer, D. R. Batchelor, R. P. Mikalo, P. Hoffmann, and A. Lloyd-Spetz, Appl. Surf. Sci. **184**, 340 (2001).

Hydrogenated nanocrystalline silicon carbide: fabrication, properties and heterostructure device application

H. Colder[1], **P. Marie**[1], **L. Pichon**[*,2], and **R. Rizk**[1]

[1] LERMAT, CNRS 2149, 6 bd Maréchal Juin, 14050 Caen, France
[2] GREYC, UMR 6072, 6 bd Maréchal Juin, 14050 Caen, France

Received 15 September 2003, revised 14 October 2003, accepted 14 October 2003
Published online 22 December 2003

PACS 73.40.Lq, 73.50.–h, 73.61.Le

Silicon carbide thin layers have been obtained by reactive magnetron sputtering in a hydrogen-rich plasma at substrate temperatures, T_s, varying between 200°C and 600°C. The analysis of the infrared absorption spectra reveal an abrupt transition from amorphous phase to a crystalline fraction f_c exceeding 30% when T_s is increased to 300°C. The f_c value continues to increase for higher values of T_s. The evolutions of the optical parameters deduced from the modeling of the spectroscopic ellipsometry spectra are quite consistent with the observed structural change. The increase of the refractive index reflects the improvement of both crystallinity and compactness of the layers, while the crystallized SiC samples appear more absorbent than monocrystalline SiC. The heterojunction diodes fabricated from the deposition of our layers on c-Si wafers have shown good rectifying behavior, as well as a low leakage current.

© 2004 WILEY-VCH Verlag GmbH & Co. KGaA, Weinheim

1 Introduction

Silicon-based devices are nowadays widely predominating the microelectronic industry, apart from some severe conditions such as high temperature and harsh environments. Indeed, silicon loses its reliable properties at high temperatures and high voltages, as well as under radiative or corrosive ambient. These unfavorable conditions imply the use of silicon compatible wide bandgap semiconductor, and in particular silicon carbide. Many efforts have been devoted recently to the growth of crystalline SiC (c-SiC) on Si for the development of heterojunction devices such as bipolar transistors [1], solar cells [2], photodetectors [3] and electroluminescent components [4]. However, the high temperature processing required so far for the growth of SiC would generate high density of defects at the SiC/Si interface, owing mainly to the 20% lattice mismatch. To overcome this difficulty, the lowering of the temperature deposition of crystallized silicon carbide appears of paramount importance. Our group has recently succeeded in the achievement of significantly nanocrystallized silicon carbide (nc-SiC) layers for substrate temperature T_s, ranging between 300 °C and 600 °C, together with high performance n-type nc-SiC/ p-type c-Si heterojunction diode processed at 600°C [5, 6]. These layers and systems were fabricated by reactive magnetron sputtering under pure hydrogen plasma with a relatively low growth rate (0.5 – 0.8 nm/min). To increase this deposition rate, improve the layer compactness and also explore the diode performance processed at lower temperatures, we present in this paper a study dealing with samples obtained by reactive magnetron sputtering under a hydrogen-rich plasma containing also argon. The quality of the material is checked by the fabrication, characterization and also comparison of SiC/Si heterojunction diodes obtained from samples deposited at two different temperatures, 400 °C and 600 °C.

---

[*] Corresponding author: e-mail: L.Pichon@greyc.ismra.fr or lpichon@greyc.ismra.fr

2 Experimental details

The layers used for this study were deposited by reactive magnetron co-sputtering of carbon and silicon either on (100) silicon or quartz substrates. The silicon target was a highly doped (10^{19} Sb.cm^{-3}) target covered by an adjusted number of carbon chips, aiming at the achievement of quasi-stoichiometric SiC layers [7]. Before deposition, the chamber was evacuated down to a few 10^{-8} Torr, while the substrate temperature, T_s was brought to a value ranging between 200 °C and 600 °C. Subsequently, both argon and hydrogen gases (6N purity) were introduced until reaching an hydrogen partial pressure ratio P_{H2}/P_{tot} of about 80%. The total pressure, $P_{ot} = P_{H2} + P_{Ar}$ and the radio-frequency power density were kept constant at 50 mTorr and 200 W, respectively.

The SiC layers were investigated by infrared (IR) absorption measurements made under Brewster incidence between 400 and 4000 cm^{-1}, by means of a Fourier transform Nicolet 750-II spectrometer. The optical properties were determined from spectroscopic ellipsometry (SE) spectra recorded in the range 1.5 – 5 eV under an incidence angle of 66.2° and using a Jobin-Yvon ellipsometer (UVISEL). The SE spectra consist in the measured ψ and Δ ellipsometric angles defined from the fundamental equation of ellipsometry : $\dfrac{\overline{r_p}}{\overline{r_s}} = \tan\Psi \ \exp i\Delta$, where $\overline{r_p}$ and $\overline{r_s}$ are the complex reflection coefficients for parallel and perpendicular polarisation of the light, respectively.

The n-type nc-SiC/p-type c-Si heterojunctions are obtained as follows. At first SiC films are deposited at $T_s = 400$ °C or $T_s = 600$ °C on p-type (10^{15} B cm^{-3}) (100) Si wafer, following the above-described conditions. On the 250-nm thick nc-SiC:H layer, the plots and contacts were patterned by an appropriate lithography. Due to its chemical stability and hardness, SiC was etched by reactive ion plasma using a O_2 / SF_6 mixture and a power of 200 W. The SiC plots were cleaned by RCA method, before the deposition of the aluminum 1% silicon alloy contacts on both sides of the wafer. The 300-nm thick contacts were then lithographically patterned to define ohmic contacts whose quality was improved by an annealing at 390°C in forming gas (9:1 N_2/H_2).

3 Results and discussion

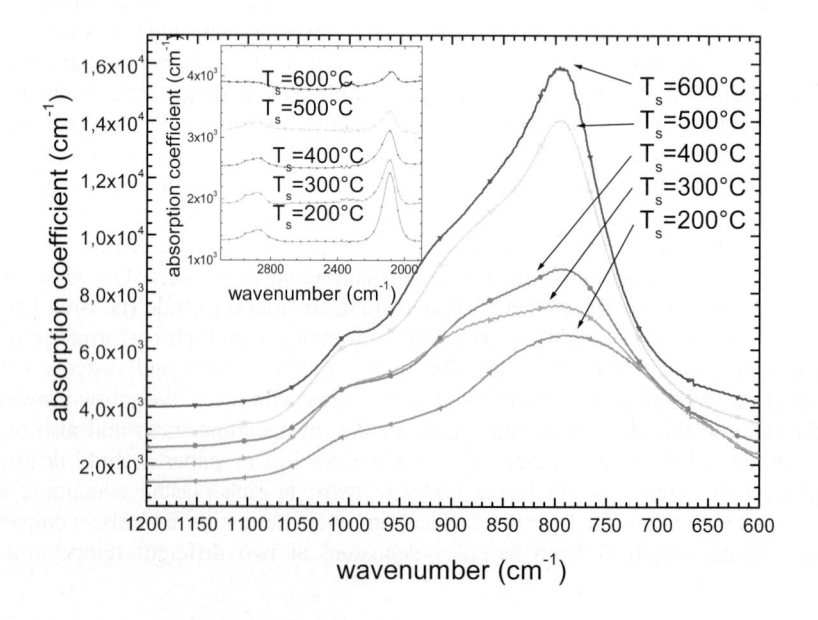

Fig. 1 IR spectra recorded in the 500-1200 cm^{-1} region for the SiC layer obtained with the indicated Ts values. The inset shows the corresponding spectra recorded in the 1800-3000 cm^{-1} region.

© 2004 WILEY-VCH Verlag GmbH & Co. KGaA, Weinheim

The films were first analyzed by means of IR measurements. Fig. 1 shows the of IR spectra recorded in the region between 500 and 1200 cm^{-1} for the films obtained at the indicated T_s values. The IR spectra show a predominant band peaking at around 790-800 cm^{-1}, which corresponds to the stretching mode vibrations of the Si-C bond [9]. This peak is shouldered by two structures located at about 900 and at 1000 cm^{-1} and which can be assigned to Si-H$_2$ bending and to C-H wagging or rocking modes [9], respectively. The presence of these bonded hydrogen is confirmed by the detection of their stretching counterparts at 2100 cm^{-1} and 2 900 cm^{-1}, respectively, as show in the inset of Fig. 1.

Although the Si-C mode is both present and dominant in all the spectra, the peak shape evolves with T_s, from a Gaussian form for T_s = 200 °C to a nearly Lorentzian for T_s = 600 °C, together with a gradual decrease of the full width at half maximum from 200 to 80 cm^{-1}. This behavior reflects some phase transition, since for amorphous SiC the distribution of Si-C bond lengths and angles follows a Gaussian law, whereas for crystalline SiC such a distribution is rather Lorentzian. In this connection, the spectra of Fig. 1 consist in a mixture of Gaussian and Lorentzian components that can be determined after deconvolution, aiming at providing an estimate of the crystalline fraction, f_c. According to an earlier reported approach [10], f_c is evaluated from f_c = L/(L+G), L and G being the area of the Lorentzian and the Gaussian components, respectively. Fig. 2 shows the variation of f_c versus T_s: for T_s below 200 °C, the film is quasi amorphous but exhibits a radical phase transition for $T_s \geq 300$ °C, with a f_c value of more than 60% for T_s = 500-600 °C.

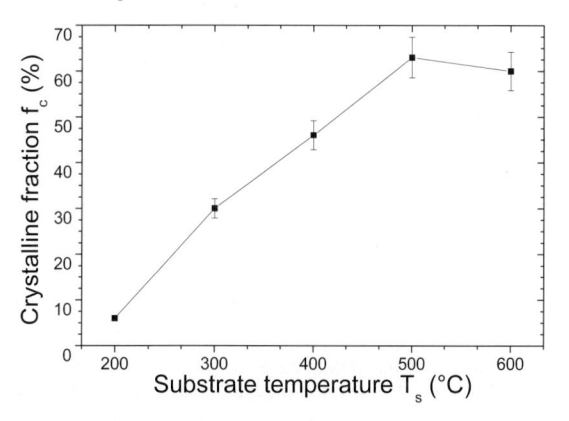

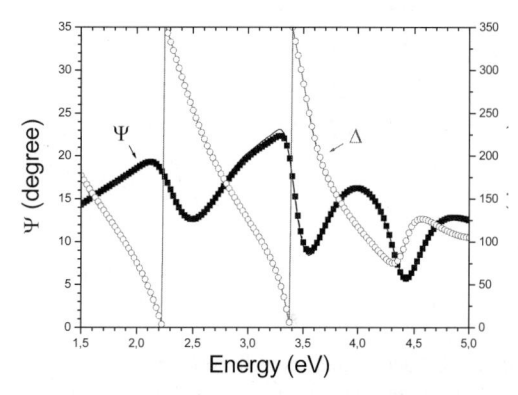

Fig. 2 Evolution of the crystalline fraction f_c as a function of T_s.

Fig. 3 Typical ES spectra. Experimental data are labeled by ■ for Ψ and ○ for Δ. The calculated spectra is represented by full lines.

This direct crystallization at $T_s \geq 300$ °C is assigned to the high reactivity of the silicon-based [11] and carbon-based [12, 13] radicals interacting with the growing surface. Such interactions result in the selective etching of weak and distorted near-surface bonds and hence leads to the relaxation of the network towards the crystalline state and the concomitant desorption of hydrogen through the volatile species formed during the interactions. This is supported by the continous decrease of both 2100 and 2900 cm^{-1} stretching modes in the inset of Fig. 1, which is quite compatible with the improvement of the SiC crystallization.

The above-described analyses of phase transition were complemented by SE measurements aiming at correlating the structural features to some optical properties. The detailed studies dealing with the modeling of the SE spectra are reported elsewhere [14, 6]. Using the Forouhi-Bloomer model [13], a nearly perfect agreement between both measured and calculated spectra was found and can be observed in Fig. 3 for a total thickness of the layer equivalent to that determined by optical transmission and for a 12 nm-thick overlayer having a void fraction of 23 %. This overlayer is assumed to reflect the surface roughness that could govern the optical response of the layer probed by ellipsometry.

Fig. 4 shows the refractive index spectra n(E) for the indicated values of T_s and reproduces also the n(E) plot for monocrystalline cubic β-SiC [10]. Two main features can be noticed from the evolution of n(E) against T_s: (i) n(E) increases with T_s all along the spectral range and approaches continously the n(E) spectrum of β-SiC, (ii) the n(E) peak shifts towards high energies when T_s is increased. Indeed, the n(E) maximum shifts gradually from about 4.6 eV for T_s = 300 °C to 5.2 eV for T_s = 500 °C. It shifts further for T_s = 600 °C, and approaches the first maximum of the refractive index at 6.0 eV that corresponds to the first direct transition between the valence band and the conduction band [17, 18]. Consequently, the gradual evolution of both value and maximum of n(E) towards their counterparts for monocrystalline β-SiC, clearly indicates that more and more SiC nanocrystals are formed within the layers and this is quite compatible with the concomitant increase of the crystalline fraction. The influence of T_s on the absorption coefficient α is shown in the inset of Fig. 4 which also compares the spectra obtained for different T_s values to that relating to 6H-SiC [19]. It appears clearly that the nc-SiC samples are more absorbent than monocrystalline Si, all along the spectra range. It is worth noticing that α(E) values tend towards that of 6H-SiC when T_s is increased, indicating an improvement of the layer crystallinity.

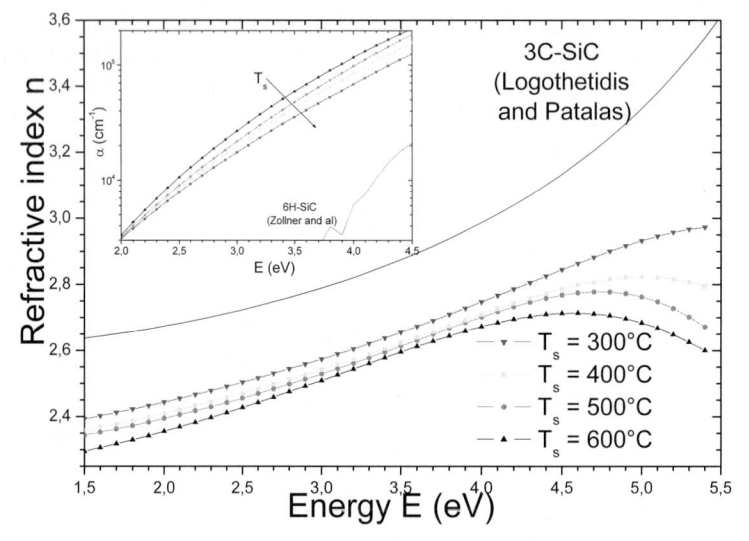

Fig. 4 Refractive index spectra n(E) of the nc-SiC:H layers obtained at the indicated T_s values, as deduced from the ES measurements. Also reported is the n(E) spectrum of 3C-SiC [16]. The inset shows the absorption coefficient spectra, α(E), of the SiC:H layers deposited at the indicated T_s values, as deduced from the SE measurements. For comparison, the α(E) spectra corresponding to 6H-SiC is also reported [19].

To take advantage of the crystalline quality of our nc-SiC:H layers, we proceeded to the fabrication, characterization and comparison of two n-type nc-SiC:H/p-type c-Si heterojunction diodes made from the layers deposited at 400 °C and 600 °C.

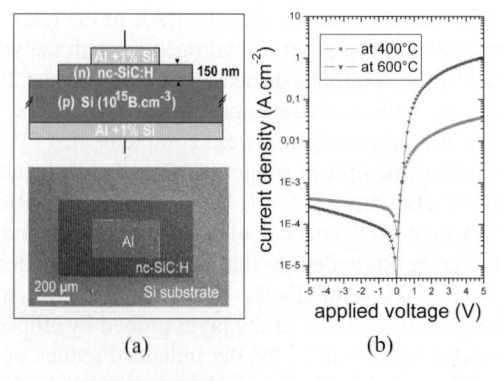

Fig. 5 (a) Schematic cross section (top side) and top view (bottom side) of the diode. (b) Typical current density-voltage characteristics of the nc-SiC:H/c-Si heterojunction diodes.

Fig. 5 compares typical room-temperature characteristic plots of current density versus voltage (I-V) of the two diodes processed at 400 °C and 600 °C. The rectifying character of both diodes is clearly indicated by the I-V curves. As expected, the rectification ratio is much higher for the diode processed at 600 °C, reaching about 10^4 against less than 10^2 at ±2 V for that processed at 400 °C. Concerning the leakage current, it is estimated to a few tens of $\mu A.cm^{-2}$ for voltages lower than −1 V and is therefore, competitive with those obtained from similar diodes processed at temperatures as high as 800-1000 °C [20, 21]. Nevertheless, the most revelant result lies in the promising features exhibited by the diode processed at a temperature as low as 400 °C. The rectifying ratio of about 10^2 at ±2 V and the leakage current of hundred of $\mu A.cm^{-2}$ are supposed to improve with a prolonged low temperature annealing (300 - 400 °C) that is expected to enhance the material crystallinity through the critical role of the reactive dihydride species present in the samples, as already observed for the case of silicon [22].

4 Conclusion

In summary, this work reports the possible achievement of crystallized silicon carbide layers at temperatures starting from 300 °C under a hydrogen-rich plasma that still allows a relatively high deposition rate. The modeling of the spectroscopic ellipsometry spectra by means of the Forouhi-Bloomer model has enabled the determination of the optical parameters such as the refractive index and the absorption coefficient. The refractive index improves with the substrate temperatures reflecting the behavior of the crystalline fraction of the layers. The two heterojunction diodes fabricated from the crystallized silicon carbide at two different temperatures, 400 °C and 600 °C, have shown promising performances. The diode processed at 600 °C has exhibited a rectifying ratio of 10^4 at ± 2 V which decreases, as expected to 10^2, for the diode processed at the lowest temperature (400 °C) never reported.

Acknowledgements The authors wish to acknowledge Prof. O. Bonnaud from the Institut d'Électronique et de Communication de Rennes (IETR) for his help in devices fabrication.

References

[1] R. Acubilla, D. Bardes, A. Orpella, J. Calderer, L.F. Marsal, J. Pallarès, and X. Correig, Mater. Scienc. Forum **264–268**, 1455 (1996).
[2] F. Smole and J. Furlan, J. Appl. Phys. **72**, 5964 (1992).
[3] D. Caputo, G. de Cesaré, F. Irrera, and F. Palma, IEEE Trans. Electron Dev. **43**, 1351 (1996).
[4] Y.A.Chen, C.F. Chiou, W.C. Tsay, J.W. Hong, and C.X Chang, IEEE Trans. Electron Dev. **44**, 1360 (1997).
[5] S. Kerdiles, A. Berthelot, F. Gourbilleau, and R. Rizk, Appl. Phys. Lett. **76**, 2373 (2000).
[6] S. Kerdiles, R. Madelon, and R. Rizk, Appl. Surf. Sci. **284**, 150 (2001).
[7] S. Kerdiles, R. Rizk, A. Perez-Rodriguez, B. Garrido, O.Calvo-Barrio, and J.R. Morante, Solid-State Electron. **42**, 2315 (1998).
[8] Y. Katayama, K. Usami, and T. Shimado, Phil. Mag. **43**, 283 (1981).
[9] P.I. Rovira and F. Alvarez, Phys. Rev. B **55**, 4426 (1997).
[10] P. Musumeci, R. Reitano, L. Calcagno, F. Roccaforte, A. Makhtari, and M.G. Grimaldi, Phil. Mag. B **76**, 323 (1997).
[11] D.G. Moon, B.H. Jung, J.N. Lee, B.T. Ahn, H.B. Im, and S.W. Kang, J. Mater. Sci : Mater in Electron. **5**, 364 (1994).
[12] T. Hatayama, T. Fuyukim, and H. Matsunami, Jpn. J. Appl. Phys. **35**, 5255 (1996).
[13] S. Desgupta, S. Gosh, S.T. Kshirsagarm, and Ray, Thin Solid Film **295**, 37 (1997).
[14] S. Charvet, R. Madelon, F. Gourbilleau, and R. Rizk, J. Lumin. **80**, 257 (1999).
[15] A.R. Forouhi and I. Bloomer, Phys. Rev. B **34**, 7018 (1986).
[16] S. Logothetidis and L. Palatas, J. Appl. Phys. **80**, 1768 (1996).
[17] W.R. Lambrecht, S. Limpijumnong, S.N. Rashkee, and B. Segall, phys stat. sol. (b) **202**, 5 (1997).
[18] G. Theodorou, G. Tsegas, and E. Kaxiras, J. Appl. Phys. **85**, 2179 (1999).
[19] S. Zollner, J.G. Chen, E. Duda, T. Wetteroth, S.R. Wilson, and J.N. Hilfiker, J. Appl. Phys. **85**, 8353 (1999).
[20] M.I. Chaudhry and R.L. Wright, Appl. Phys. Lett. **59**, 51 (1991).
[21] L.F. Marsal, J. Pallarès, X. Coreig, A. Orpella, D. Bardès, and R. Alcubilla, J. Appl. Phys. **85**, 1213 (1999).
[22] F. Gourbilleau, A. Achiq, P. Voivenel, and R. Rizk, Thin Solid Films **337**, 74 (1995).

Self-passivation mechanisms in clusters of N dopants in SiC

R. Rurali[*,1,2], **E. Hernández**[2], **P. Godignon**[1], **J. Rebollo**[1], and **P. Ordejòn**[2]

[1] Centre Nacional de Microelectrònica, CNM-CSIC, Campus de Bellaterra, 08193 Bellaterra, Barcelona, Spain
[2] Institut de Ciencia de Materials de Barcelona, ICMAB – CSIC, Campus de Bellaterra, 08193 Bellaterra, Barcelona, Spain

Received 15 September 2003, revised 3 October 2003, accepted 3 October 2003
Published online 22 December 2003

PACS 31.15.Ar, 71.15.Mb, 71.20.Nr

Although having proved to be one of the most successful n-type dopant in SiC for many applications, nitrogen exhibits a troublesome nature when a high-dose implantation is required. Experimental measurements showed that phosphorus is more suited in such conditions. We proposed previously that the reason may be the self-passivation of the N clusters that are likely to form in the high-concentration limit. In this paper we discuss in details the passivation mechanisms that are relevant for the different N aggregates that can form and that are the ultimate responsible of the low activation rate of high-dose N-implanted SiC. We have found that all the N complexes that are likely to form in SiC do self-passivate. However, the passivation mechanisms are not always the same and they are due in turn to the formation of a bound molecule or to a charge-trapping effect.

© 2004 WILEY-VCH Verlag GmbH & Co. KGaA, Weinheim

1 Introduction

A high breakdown field and a high thermal conductivity have concurred in recent years to make silicon carbide (SiC) definitely stand out among other wide band-gap semiconductors [1]. SiC brings together the physical properties suited for high power operations with an excellent mechanical and physical stability up to very high temperatures. The possibility to achieve high-doping doses is extremely interesting to obtain high quality ohmic contacts and low resistive sources and drains in FET based devices. However, this aim cannot always be trivially fulfilled because some problems may arise when the solubility limit of the chosen dopant starts to be approached. In this work we discuss the effects of N high-dose doping in SiC, proposing an explanation of the experimental results [2] based on the local electronic structure of the defects and of the defect-complexes. Besides our previous report of self-passivation of high-dose n-type doping [3], N clustering was studied theoretically by Malhan et al. [4], although they restricted to second-neighbour complexes.

2 Methodology

All the calculations presented in this work consist of full structural relaxations and have been performed with the Density Functional Theory (DFT) [5, 6] package Siesta [7]. We have used the Local Density Approximation (LDA) [8, 9] and non-local pseudopotentials of the Troullier-Martins type [10]. The one-electron wave functions have been represented using a double-ζ polarised basis set. The effect of spin-polarisation has been checked in some test structures, but it turned out to be negligible. The calculations have been carried out in a 64-atom supercell of 3C-SiC, with a grid of $2 \times 2 \times 2$ k-points following the Monkhorst-Pack sampling scheme [11]. Structure geometries were optimised until all the forces on the atoms were reduced under the limit of 0.04 eV/Å.

---

[*] Corresponding author: e-mail: riccardo.rurali@cnm.es, Phone: +34 93 594 7700, Fax: +34 93 580 1496

3 Results and discussion

3.1 Isolated N dopants
We have first studied isolated N in SiC, finding that substitution at C site, N_C, is favoured over substitution at Si site, N_{Si} [3, 12]. However, while N_C retains the tetragonal symmetry, N_{Si} relaxes to an off-centre position, leading to a three-fold coordination with three C atoms. The off-centre relaxation has a limited impact with respect to the formation energy of the substitutional, N_{Si}^{off} is only ~ 0.2 eV more stable than N_{Si}^{tetra}. On the other hand, the change in the electronic structure is much more relevant. As it can be seen in Fig. 1, the off-centre geometry results in an intrinsic passivation of the originally shallow level that would correspond to the unrelaxed N_{Si}.

When the concentration of N dopants is not especially high, these considerations are not very important, because most of the N atoms will be found as N_C, as the difference in stability with N_{Si}^{off} amounts to few eV. N_C is characterised by a shallow donor level [12], which can therefore be easily thermally activated. As we will discuss further on, the role played by N_{Si}^{off} is more important when a high doping dose might force the formation of N aggregates. In that condition substitution at a Si site will become more likely, because many C site will already be occupied, and the deep state has important consequences on the electrical characteristics of the material.

3.2 Aggregates of N dopants
In order to focus our attention on the problems that may arise when the concentration of N dopants is very high, we have analysed the stability and the electronic structure of different N aggregates. We have considered the following cases: (a) first neighbour substitutional, (b) split interstitial and (c) substitutional plus interstitial. These complexes are intended to model some typically interesting situations from the viewpoint of dopant passivation.

Firstly, we have calculated the aggregation probability of these compounds. The aggregation energy can be defined as follows:

$$E_{agg} = E_{complex} - nE[N_C] \tag{1}$$

where n is the number of N atoms involved in the complex (so $n = 2$ in the cases that we have considered). Eq. (1) indicates the energy gain (or loss) for gathering n dopants with respect to their most stable configuration, thus N_C in this case. None of the complexes that we have taken into account exhibits a spontaneous tendency to aggregation, as can be seen in Table 1 where all the aggregation energies are positive. However, all the aggregates that we have considered are metastable, therefore, if they form as a results of the non-thermal equilibrium conditions that the material experiences during the implantation cascade, they must overcome a barrier to dissolve. In other words, although our simple model cannot provide information about the kinetic of the complexes formation, we think it is accurate enough to infer on their stability and their electronic structure.

The relative stabilities of the aggregates considered have been already discussed elsewhere [3]. Each one of these complexes is relevant under different conditions of doping and stoichiometry, thus studying all of them is necessary to provide a thorough picture of the passivation mechanisms. Therefore, from now on we will focus on the analysis of their electronic structures.

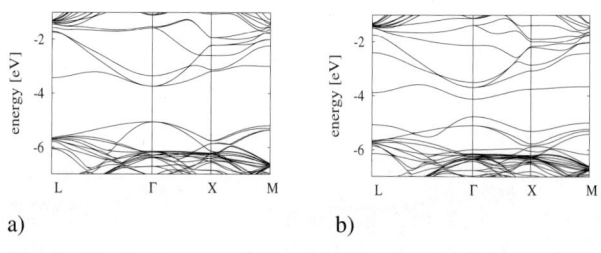

Fig. 1 Band structure of N_{Si} a) in the tetragonal geometry and b) in the more stable off-centre geometry. As an effect of the off-centre relaxation the originally shallow doping level *falls* deeper into the gap. Our calculations are not capable of giving an accurate estimate of the position of the shallow states within the band gap, due to the small size of the supercells compared to the spread of the doping states. However, the passivation due to the off-centre relaxation can be clearly appreciated.

Table 1 Aggregation energy of the complexes studied. $\Delta\mu$ account for the stoichiometric condition of the material and varies between $-\Delta H_f$ for C-rich condition and ΔH_f for Si-rich condition. $\Delta H_f \sim 0.4$ eV is SiC formation heat [13].

Complex	Formation energy [eV]
first neighbour substitutional	$3.74 + \Delta\mu$
split interstitial	$3.08 + \frac{1}{2}\Delta\mu$
substitutional plus interstitial	$3.71 + \frac{1}{2}\Delta\mu$

3.2.1 First neighbour substitutional This configuration models one of the most typical condition to which high-dose doping may lead: the efficient occupation of neighbouring lattice sites. We have already discussed that N favours substitution at C sites, but in the high-dose limit there is a growing probability that substitution at Si sites may also occur. Although the most stable configuration for an isolated N substituting at a Si site is N_{Si}^{off}, the plain N_{Si} is also a local minimum. In our structural relaxations, to find the off-centre geometry the symmetry must be *artificially* broken to observe a transition, which would spontaneously occur at $T > 0$. In the case of a first neighbour substitutional dimer, the N_{Si} relaxes automatically to the off-centre geometry, because, being the N_C one of its neighbours, the perfect tetragonal symmetry is perturbed. As we have already discussed above, the N_{Si}^{off} is an intrinsically passivated configuration, as its asymmetric geometry leads to a deep level inside the band-gap. However, in this aggregated configuration it has the significant side effect of passivating also the otherwise shallow level of the N_C. The deep N_{Si}^{off} state is only half occupied and can therefore be further populated by the loosely bound electron of N_C, leading to a full passivation of the two N dopants, as shown in Fig. 2a.

3.2.2 Split interstitial The split interstitial configuration is intended to model the competition for the same site of two N atoms, which end up sharing it. An isolated N atom favours substitution at a C site, thus it was natural to study a split interstitial at a C site too. In such a configuration, the N atoms were found to relax to an inter-atomic distance of around 1.5 Å. The N_2 molecule in vacuum was calculated, finding a N-to-N length of 1.11 Å, therefore the effect of the crystal host consists mainly in an elongation of the molecule. Also the band diagram of this kind of N complex showed a deep, passivated state inside the band-gap [see Fig. 2b], however the reduced distance between the dopants and the resemblance with the N_2 molecule suggests a different interaction. We are in presence of an ordinary, plain N_2 molecule, slightly stretched by the interaction with the crystal. The increased separation between the two N will lead to more loosely bound molecule, but the passivated state will essentially generate from the π states of the molecular orbitals.

3.2.3 Substitutional plus interstitial The last representative configuration that we have taken into account derives from considering the perturbation effect that a N interstitial may exert on a nearby

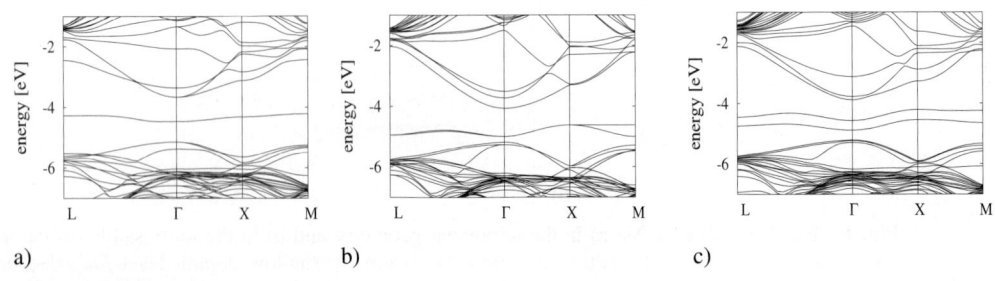

Fig. 2 Band structure of a) substitutional N first neighbours, b) N split interstitial at a C site and c) substitutional N plus interstitial hexagonal N. In all the cases the presence of a deep level inside the band-gap indicates the self-passivation.

substitutional. In the high-dose limit this could be a relatively frequent case, given the rather unlikely occupation of a Si site. The passivation mechanism follows the same pattern described for the first neighbour substitutional. It is the N interstitial here that provides the deep, half-populated state that in case of the first neighbour substitutional was due to the asymmetric relaxation of the N_{Si} and that, like here, passivates the shallow N_C state [see Fig. 2c].

4 Conclusions N high-doping in SiC is highly critical due to the tendency to self-passivation of the aggregate that may form during the implantation cascade. We have discussed the electronic structure of some representative N complexes, showing that the passivation mechanism can follow different patterns, according to the aggregate involved, and is thus far to be unique.

In the split interstitial arrangement, the N atoms form a bound state similar to the N_2 molecule, except for a longer bond length induced by the host lattice. On the other hand, in the remaining configuration analysed, the N dopants have an inter-atomic separation sufficiently long as to prevent such a strong interaction. However, the off-centre N_{Si} in one case and the N interstitial in the other, have an associated, delocalised deep state that, being half- populated, passivates the shallow N_C state.

Acknowledgements The authors are extremely grateful for enlightening discussions to the late J. L. Mozos whose memory will live on with us. R. Rurali is supported by a Training and Mobility EU Grant (ATOMCAD). This work is supported in part by the Spanish Ministry of Science and Technology (BFM2000-1312-C02 and BFM2002-03278).

References

[1] Silicon Carbide: A Review of Fundamental Questions and Applications to Current Device Technology, edited by E. J. Choyke, H. Matsunami, and G. Pensl, Akademie Verlag, phys. stat. sol. (b) **202**, No. 1, (1997).
[2] M. Laube, F. Schmid, G. Pensl, and G. Wagner, Mater. Sci. Forum **389–303**, 791 (2002).
[3] R. Rurali, P. Godignon, J. Rebollo, E. Hernández, and P. Ordejón, Appl. Phys. Lett. **82**, 4298 (2003).
[4] R. K. Malhan, J. Kozima, T. Yamamoto, and A. Fukumoto, Mater. Sci. Forum **389–303**, 541 (2002).
[5] P. Hohenberg and W. Kohn, Phys. Rev. B **136**, 864 (1964).
[6] W. Kohn and L. J. Sham, Phys. Rev. **140**, A1133 (1965).
[7] J. Soler, E. Artacho, J. D. Gale, A. García, J. Junquera, P. Ordejón, and D. Sánchez-Portal, J. Phys.: Condens. Matter **14**, 2745 (2002).
[8] D. M. Ceperley and B. J. Alder, Phys. Rev. Lett. **45**, 566 (1980).
[9] J. P. Perdew and A. Zunger, Phys. Rev. B **23**, 5048 (1981).
[10] N. Troullier and J. L. Martins, Phys. Rev. B **43**, 1993 (1991).
[11] H. J. Monkhorst and J. D. Pack, Phys. Rev. B **8**, 5747 (1973).
[12] H. H. Woodbury and G. W. Ludwig, Phys. Rev. **124**, 1083 (1961).
[13] J. E. Northrup and S. B. Zhang, Phys. Rev. B **47**, 6791 (1992).

Micro-Raman study of laser damage in CdTe

M. J. Soares [*], J. C. Lopes, M. C. Carmo, and A. Neves

Departamento de Física, Universidade da Aveiro, 3810-193 Aveiro, Portugal

Received 15 September 2003, accepted 17 September 2003
Published online 22 December 2003

PACS 71.55.Gs, 78.30.Fs

The development of electronic devices involves a variety of technologies including laser-assisted doping of materials. However, laser irradiation can induce severe damage of the electronic materials, changing the structure by phase segregation and promoting defect migration and aggregation. So it is very important to know the extent and relevance of such side effects. CdTe is a wide band gap semiconductor largely used in photonic applications such as radiation detectors and solar cells. Laser irradiation of CdTe induces significant structural modifications localized in a thin surface layer. The relationship between laser power density, local temperature and structural changes is discussed by monitoring the relative intensity of the Te modes in Raman spectra. Local temperature is measured through the anti-Stokes/Stokes ratio of the same modes. We also show that the surface decomposition occurs at temperatures well below the melting point.

© 2004 WILEY-VCH Verlag GmbH & Co. KGaA, Weinheim

1 Introduction

The observation of Te optical modes on the surface of CdTe using micro-Raman methods has been reported by several authors [1–4]. However the origin of these modes is a subject of controversy. Shin et al. [1] have attributed them to Te precipitates present as a part of growth process of CdTe. Picos-Vega et al. [3] studied CdTe samples grown either with Cd or Te excess and observed in Te rich material Te related optical modes at 123 cm^{-1} and 142 cm^{-1}, whereas in Cd rich material modes at 167 cm^{-1} and 334 cm^{-1} were measured. The authors concluded that the Te modes were due to small Te aggregates related with poor stoichiometry and disappeared gradually with the increase of cadmium content. In a previous work [4] we showed by micro-Raman that visible radiation leads to the formation of Te aggregates on the irradiated surface, compromising any conclusion about Te inclusions on CdTe surface, and speculated that heating by laser light could break chemical bonds and produce Cd evaporation, thus leaving an excess of Te on the surface.

In this work we report on measurements of temperature through micro-Raman spectra in CdTe samples excited either below or above the gap (1.4 eV, 840 nm at room temperature). In the latter case, the experimental parameters were similar to those where optical damage (i.e., formation of Te aggregates on the sample surface) is observed.

2 Experimental details

The samples were prepared from commercially available CdTe single crystals grown by a vertical Bridgman process. The Raman spectra were recorded in backscattering geometry at room temperature, using either a 64000 Jovin Yvon spectrometer with the 515 nm argon laser line or a Bruker RFS 100/S spectrometer with the 1064 nm Nd:YAG laser line. The power densities were at least ~10^3 W/cm^2, which is the minimum required to have a detectable Raman signal.

---

[*] Corresponding author: e-mail: jorge@fis.ua.pt, Phone: +351 234 370 284, Fax: +351 234 424 965

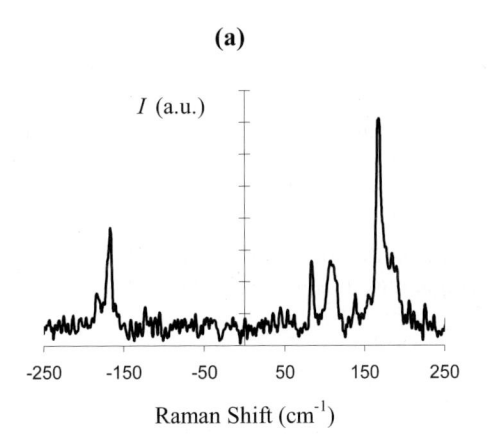

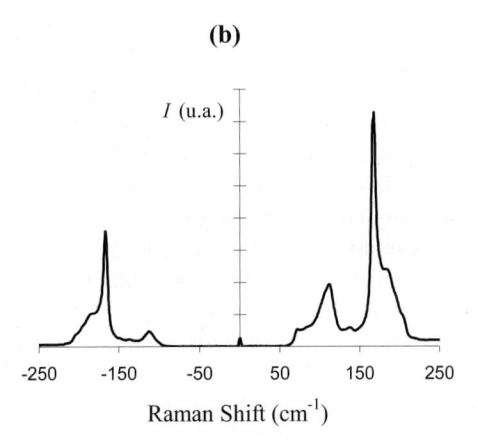

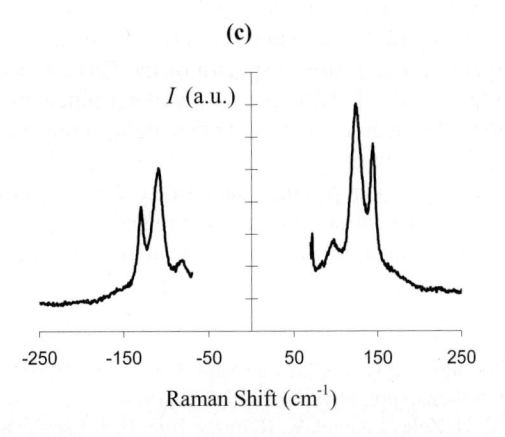

Fig. 1 Stokes and anti-Stokes spectra of CdTe at room temperature. The exciting wavelengths and power densities are: (a) 1064 nm, ~10^3 W/cm^2; (b) 1064 nm, ~10^4 W/cm^2 ~10^4 W/cm^2; (c) 515 nm, ~10^6 W/cm^2. In (c) previous to recording the spectra the sample was exposed to the same radiation for 30 min.

3 Results and discussion

In Fig. 1(a) and (b) we show spectra recorded with excitation under the gap at two power densities. The Te modes are not observed. The relevant features are the CdTe LO mode (167 cm^{-1}) and the plasmon (~190 cm^{-1}); the bands at lower wave numbers (70 and 111 cm^{-1}) are not related to Te [2]. Increasing the power density by one order of magnitude does not change the mode pattern nor the ratio of the areas under the CdTe modes in the Stokes and anti-Stokes regions. All this remains unchanged no matter how long the sample is irradiated.

That ratio can be used to estimate the local temperature. In fact, the intensity of a Stokes line (I_S) is only weakly dependent on temperature (T) while the anti-Stokes line (I_{AS}) shows a strong relation to temperature, and their ratio is given by [5]:

$$\frac{I_{AS}}{I_S} = \left(\frac{v_L + \Delta v}{v_L - \Delta v}\right)^4 \exp\left(-\frac{hc\Delta v}{kT}\right) \qquad (1)$$

Here v_L is the laser wave number and Δv is the Raman shift; the other symbols have the usual meaning. In the case of Fig. 1a and b, that ratio is ≈ 0,5, which means that the local temperature is just about room temperature.

The spectra in Fig. 1(c) were obtained by exciting above the gap. We can see that the CdTe mode is now a weak feature, and the Te modes prevail. This means there is a significant formation of Te aggregates at the surface (optical damage). As we have shown in [4], the damage at these optical densities increases with exposure time for a while and then stabilizes. The spectra shown were recorded after a 30 min. exposure time, when the intensity no longer changes. The ratio of the areas under Te modes is ≈ 0.7, which by Eq. (1) corresponds to a temperature of about 200 degrees above room temperature. For 515 nm irradiation the absorption coefficient is four orders of magnitude higher than for 1064 nm excitation and a heating is expected. However the temperature reached is much lower than the melting point of CdTe, so that evaporation of Cd from the surface cannot occur.

4 Conclusions

For excitation power densities required to record a Raman spectra, formation of Te aggregates on the CdTe sample surface only occurs if the excitation is above the gap. We showed that, for excitation below the gap, the local temperature – determined using similar experimental conditions, that is, through Raman scattering, from the Stokes and anti-Stokes spectra of the CdTe modes – is close to room temperature; for excitation above the gap, the local temperature (determined from the Stokes and anti-Stokes spectra of Te modes) is about 200 degrees higher. These higher temperatures are much lower than the CdTe melting point and therefore the optical damage cannot be caused by evaporation of Cd.

Having ruled out that heating by itself can produce the surface damage, another explanation is thus called for. Some authors speculated that the photo-induced aggregation of Te involves hole migration [6] or vacancy migration [7]. These ideas must now be explored and tested in the lab.

References

[1] S. H. Shin, J. Bajaj, L. A. Moudy, and D. T. Cheung, Appl. Phys. Lett. **43**, 68 (1983).
[2] P. M. Amirtharaj and F. H. Pollak, Appl. Phys. Lett. **45**, 789 (1984).
[3] A. Picos-Vega, M. Becerril, O. Zelaya Angel, R. Ramirez Bon, F. J. Espinoza Beltran, J. Gonzalez Hernandez, S. Jienez Sandoval, and B. Chao, J. Appl. Phys. **83**, 760 (1998).
[4] M. J. Soares and M. C. Carmo, in: David L. Andrews (ed.), Raman Spectroscopy and Light Scattering Technologies in Materials Science, Proc. SPIE **4469**, 57 (2001).
[5] R. J. Nemanich, D. K. Biegelsen, R. A. Street, and L. E. Fennell, Phys. Rev. B **29**, 6005 (1984).
[6] S. Sugai, Jpn. J. Appl. Phys. **30**, L1083 (1991).
[7] C. Uzan, R. Legros, Y. Marfaing, and R. Triboulet, Appl. Phys. Lett. **45**, 879 (1984).

© 2004 WILEY-VCH Verlag GmbH & Co. KGaA, Weinheim

Photo-ESR and optical studies of Cr photoionization transition in CdZnSe:Cr crystals

K. Swiatek[*,1], **M. Godlewski**[1,2], and **T.P. Surkova**[3]

[1] Institute of Physics, Polish Academy of Sciences, Al. Lotnikow 32/46, 02-668 Warsaw, Poland
[2] College of Science, Cardinal S. Wyszynski University, Warsaw, Poland
[3] Institute of Metal Physics, Ural Division of RAS, 620219 Ekaterinburg, Russia

Received 15 September 2003, revised 15 October 2003, accepted 15 October 2003
Published online 22 December 2003

PACS 71.55.Gs, 76.30.Fc, 78.55.Et

Photoionization transitions of chromium ions in $Cd_xZn_{1-x}Se$:Cr bulk crystals ($0 \leq x \leq 0.3$) were studied in wide temperature range (4–300 K). By monitoring changes of the Cr^{1+} electron spin resonance signal under external illumination, we are able to determine energies of $Cr^{2+} \Rightarrow Cr^{1+} + h_{VB}$ transition in different $Cd_xZn_{1-x}Se$ host crystals. At low temperature we observe metastable population of photo-excited Cr^{1+} centres.

© 2003 WILEY-VCH Verlag GmbH & Co. KGaA, Weinheim

1 Introduction In recent 15 years, transition metal – doped II–VI compounds received considerable interest due to possible application in mid infrared semiconductor lasers, blue – green lasers and other optoelectronic devices [1, 2]. Chromium – doped chalcogenides, especially ZnSe:Cr^{2+} [2, 3] and ZnS:Cr^{2+} [2], have been demonstrated to be attractive for room – temperature laser operation. A typical feature of transition metal dopands in semiconductors is an existence of several possible charge states of a given impurity ion in a host lattice. Neutral charge state of chromium in cation position of wide band gap II–VI compounds is Cr^{2+}, but Cr^{1+} may also be present. Thus, charge transfer processes between the valence band (VB) and the $Cr^{2+/1+}$ ion energy level are possible. As a consequence of such proces, an absorption band below the band-to-band edge may be observed. Recent studies have shown that the $Cr^{2+} + h\nu \Rightarrow Cr^{1+} + h_{VB}$ photoionization transitions ($h\nu > 2$ eV) is an efficient channel for excitation of the Cr^{2+} intra-ion infrared emission and laser action in ZnSe:Cr [4–7]. Since common used semiconductor – based laser diodes emit light in energy range of 1.9–2.0 eV, it is interesting to look for a host crystal in which energy of chromium $2+ \Rightarrow 1+$ photoionization transition coincides with energy of light emited by the laser diodes.

In this work we study $2+ \Rightarrow 1+$ photoionization transition of chromium ions in $Cd_xZn_{1-x}Se$:Cr bulk crystals ($0 \leq x \leq 0.3$) in wide temperature range (2–300 K). Monitoring changes of the Cr^{1+} ESR (electron spin resonance) signal induced by external illumination we can determine energy of $Cr^{2+} \Rightarrow Cr^{1+} + h_{VB}$ transition in different $Cd_xZn_{1-x}Se$ host crystals.

2 Experiment $Cd_xZn_{1-x}Se$:Cr samples were bulk crystals grown by chemical transport method. The crystals were chromium doped, with the impurity concentration of 10^{18} cm^{-3}. The ESR experiments were performed using a Bruker 300 X-band spectrometer equipped with a continuous He gas-flow Oxford Instruments cryostat. Illumination of samples, mounted in a microwave cavity was done with laser (Ar or He–Cd) sources or with a high-pressure mercury lamp and set of interference filters or a single-grating monochromator. Absorption spectra were measured using a Perkin-Elmer-Hitachi 350 spectrophotometer.

---

[*] Corresponding author: e-mail: swiat@ifpan.edu.pl, Phone: +48 22 8436861, Fax: +48 22 8430926

© 2004 WILEY-VCH Verlag GmbH & Co. KGaA, Weinheim

3 Results and discussion

The ESR spectra of $Cd_{0.10}Zn_{0.90}Se:Cr$ measured at 4 K are presented in Fig. 1. Spectrum of the sample cooled from room to helium temperature in dark (lower part of Fig. 1) shows well resolved patterns related to chromium ions in 2+ charge state and Mn^{2+} ions. ESR spectrum of the sample illuminated by 2.5 eV light at 4 K (upper part of Fig. 1) exhibits a sharp line with g-factor close to 2.0, related to Cr^{1+} ions. Comparison of these two situations suggests that illumination of the sample by external light at low temperature generates recharging of part of the Cr ions from 2+ to 1+ charge state.

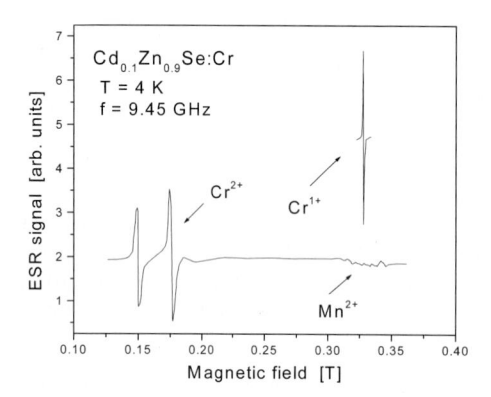

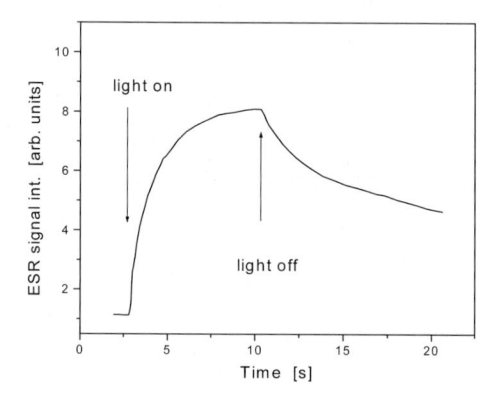

Fig. 1 (left) ESR spectra of $Cd_{0.10}Zn_{0.90}Se:Cr$ crystal measured at 4 K in dark (lower part) and after illumination by external light of energy 2.5 eV (upper part – only Cr^{1+} ESR signal is shown).

Fig. 2 (right) Sequence of steps in photo-excitation experiment. Cr^{1+} ESR signal in $Cd_{0.10}Zn_{0.90}Se:Cr$ crystal is monitored (T = 4 K).

Fig. 2 presents changes of the Cr^{1+} ESR signal in $Cd_{0.10}Zn_{0.90}Se:Cr$ measured at 4 K after illumination of the sample by light of energy $h\nu = 2.5$ eV. We observe at first a relatively fast increase of the Cr^{1+} ESR signal to some saturation level and then a slow decrease of the signal when the illumination was turn off.

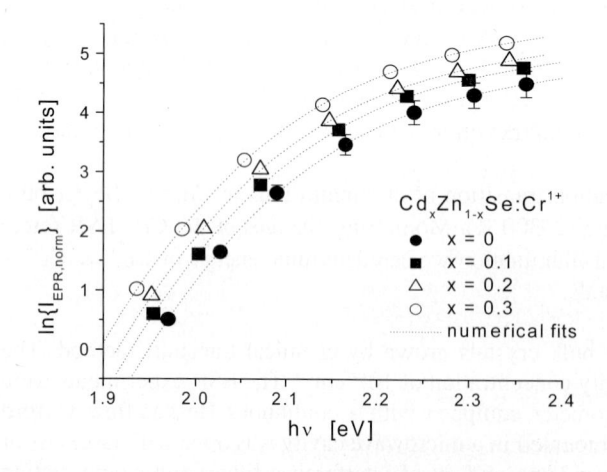

Fig. 3 Spectral dependences of the Cr^{1+} excitation process in four $Cd_xZn_{1-x}Se:Cr$ crystals (x = 0, 0.1, 0.2, 0.3). The experimental data are corrected for an effect of decrease of light penetration through the sample in the range of large absorption coefficients.

This kind of investigations of ESR-active impurities in semiconductors was developed in past. There was detailed discussion on types of temporal evolution of increasing or decreasing intensity of ESR signals induced by switching light on/off [8]. For further study we select the Cr^{1+} ESR signal as a "detector" of recharging processes in $Cd_xZn_{1-x}Se:Cr$ because of following reasons:

(i) ESR signal of the Cr^{1+} impurity ion ($d^5(^6S)$ ground state configuration) was observed in wide teperature range (4 – 250 K) in all studied crystals (x = 0, 0.1, 0.2, 0.3). Integrated intensity of the Cr^{1+} signal exhibited typical 1/T temperature dependence.

(ii) ESR signal of the Cr^{2+} impurity ion ($d^4(^5D)$ ground state configuration) was observed only at low temperatures (T < 20 K) and only in crystals with low cadmium content (x = 0, 0.1).

In our case it is useful to analyse total increase of the Cr^{1+} ESR signal from "dark" to "saturation" values after switching the light on. To evaluate the spectral dependence of photo-excitation process of the Cr^{1+} ESR signal in $Cd_xZn_{1-x}Se:Cr$ we performed a sequence of light on/off measurements using different energy of external light at each step (Fig.3). After each measurement with a given energy of light we heated up the sample to temperature of 250 K and cooled it down again to 4 K in dark, to achieve the same initial conditions (the same Cr^{1+} ESR signal intensity).

During the Cr photoionization transition (PT), energy hv of light excites an electron from the VB to the $Cr^{2+/1+}$ energy level in the forbidden gap of the host crystal: $Cr^{2+} + h\nu \Rightarrow Cr^{1+} + h_{VB}$. This kind of PT's was studied theoretically in past by many authors. One of the approaches was developed by Kopylov and Pikhtin [9] and then extended by Langer et al. [10] to account for the electron–phonon interaction in the configuration coordinate model. The photoionization cross section $\sigma_0(h\nu)$ has following form:

$$\sigma_0(h\nu) = \frac{1}{\sqrt{\pi}} \int_{-\beta}^{\infty} dz (1 + \frac{\Gamma z}{h\nu}) \exp(-z^2) \sigma_{el}(E_{opt}, h\nu + \Gamma z); \quad (1)$$

$$\beta = (h\nu - E_{opt})/\Gamma, \quad (2)$$

$$\Gamma = \frac{\omega_{ex}}{\omega_0} \left[2(E_{opt} - E_{th}) \hbar \omega_0 cth(\hbar \omega_0 / 2kT) \right]^{1/2}, \quad (3)$$

$$\sigma_{el}(E_{opt}, h\nu) \sim \frac{(h\nu - E_{opt})^{1/2}}{(h\nu)^3}, \quad (4)$$

where E_{opt} and E_{th} are optical and thermal ionization energies, respectively. ω_0 and ω_{ex} are the frequencies of phonons coupled to the ground and excited states, respectively. Parameter Γ describes the effect of broadening of the PT absorption band at non-zero temperatures. Results of numerical fit of the theoretical formula $\sigma_0(h\nu)$ to experimental data are shown in Fig. 3 (dotted curves). Values of fitting parameters E_{opt} and Γ are summarized in Table 1.

Table 1 Values of fitting parameters E_{opt} and Γ for theoretical $\sigma_0(h\nu)$ dependence.

x composition	E_{opt} (eV)	$\Gamma(eV)$
0.0	2.223 ± 0.004	0.22 ± 0.03
0.1	2.215 ± 0.005	0.22 ± 0.04
0.2	2.206 ± 0.005	0.23 ± 0.04
0.3	2.196 ± 0.005	0.24 ± 0.05

We obtained almost linear dependence of chromium ionization energy E_{opt} on change of x composition in the $Cd_xZn_{1-x}Se:Cr$ crystal. It is worth mentioning that appropriate Cr photoionization bands we observed also in low temperature absorption spectra. Moreover, at low temperature we observed metastable population of photo-excited Cr^{1+} centres. Free holes created in the VB under 2+ to 1+ ionization of chromium, are efficiently trapped by shallow acceptor centres.

4 Conclusions By use of the photo-ESR technique we have determined the position of $Cr^{2+/1+}$ energy level with respect to the valence band edges of the $Cd_xZn_{1-x}Se:Cr$ ($0 \leq x \leq 0.3$). Due to a change of the energy of Cr photoionization transition in host crystals with different cation composition, it is possible to tune in a precise way the pumping energy of potential lasing system based on $Cd_xZn_{1-x}Se:Cr$ crystals.

Acknowledgment This work was partly supported by grant no. 5P03B00720 of KBN (Poland).

References

[1] M.A. Haase, J. Qiu, J.M. Depuydt, and H. Cheng, Appl. Phys. Lett. **59**, 1272 (1991).
[2] L.D. Deloach, R.H. Page, G.D. Wilke, S.A. Payne, and W.F. Krupke, IEEE J. Quantum Electron. **32**, 885 (1996).
[3] R.H. Page, K.I. Schaffers, L.D. Deloach, G.D. Wilke, F.D. Patel, J.B. Tassano Jr., S.A. Payne, W.F. Krupke, K.-T. Chen, and A. Burger, IEEE J. Quantum Electron. **33**, 609 (1997).
[4] V. Yu. Ivanov, Yu. G. Semenov, M. Surma, and M. Godlewski, Phys. Rev. B **54**, 4696 (1996).
[5] S. Bhaskar, P.S. Dobal, B.K. Rai, R.S. Katiyar, H.D. Bist, J.-O. Ndap, and A. Burger, J. Appl. Phys. **85**, 439 (1999).
[6] A.V. Podlipensky, V.G. Shcherbitsky, N.V. Kuleshov, V.I. Levchenko, V.N. Yakimovich, M. Mond, E. Heumann, G. Huber, H. Kretschmann, and S. Kuck, Appl. Phys. B **72**, 253 (2001).
[7] E. Sorokin and I.T. Sorokina, Appl. Phys. Lett. **80**, 3289 (2002).
[8] M. Godlewski, J. Appl. Phys. **56**, 2901 (1984).
[9] A.A. Kopylov and A.N. Pikhtin, Fiz. Tverd. Tela **16**, 1837 (1974) (Sov. Phys. Solid State **16**, 1200 (1975)).
[10] J.M. Langer, Lect. Notes Phys. **122**, 123 (1980).

1.54 µm luminescence quenching of erbium-doped hydrogeated amorphous silicon deposited by D.C. magnetron sputtering

M. Kechouane[*,1], **D. Biggemen**[2], and **L. R. Tessler**[2]

[1] Thin Films and Semiconductors Laboratory, USTHB, Physical Faculty, PO Box 33, El Alia – Bab Ezzouar, 16111 Algiers, Algeria
[2] Instituto de Fisica "GlebWataghin", UNICAMP, C. P. 6165, 13083-970 Campinas, SP, Brazil

Received 15 September 2003, revised 18 November 2003, accepted 18 November 2003
Published online 22 December 2003

PACS 71.23.Cq, 71.55.Jv, 78.55.Qr, 78.66.Jg

Erbium was doped onto hydrogenated amorphous silicon using the D.C. magnetron sputtering technique. After annealing under oxygen atmosphere at 350 °C, a-Si:H<Er> films with variable optical gaps exhibits efficient room temperature at 1.538 µm. The intensities of the 1538 and 1549.6 nm Er^{3+} lines at room temperature were 15 and 12%, respectively, of those at 15K. This result suggests that DC sputtered a-Si:H is a suitable host for efficient Er^{3+} emission at room temperature. Amorphous silicon has an erbium solubility higher than 10^{20} cm^{-3} and a greatly reduced thermal quenching of the erbium PL due to a less efficient excitation energy transfer. The observation of efficient room temperature luminescence from erbium-doped a-Si:H films involves impurities such as oxygen. We found that the increase in the optical gap and the electrical resistivity of the a-Si:H<Er> films produced at low substrate temperature (high defect density, low gap) is more important than that in material deposited at 240 °C.

© 2004 WILEY-VCH Verlag GmbH & Co. KGaA, Weinheim

1 Introduction

Erbium doped silicon is of particular interest for optoelectronic systems due to the characteristic emission from erbium. Er^{3+} ions are characterized by an unfilled $4f$ shell surrounded by an external close shell. Their intra-$4f$ transitions are screened by the exterior $5s$ and $5p$ shells thus their energy is not sensitive to the details of the host matrix. Erbium shows a luminescence peak at 1.54 µm, which corresponds to the transmission maximum of conventional silica, based optical fibers.

Rare earth ions-doped semiconductors are very attractive materials for light emitting devices, but in order to develop devices of technological relevance the understanding of the factors which influence their light emitting yield is necessary.

Er^{3+} has been introduced into several host materials, particularly in those compatible with conventional Si technology. Er doping of crystalline silicon however presents several limitations, e. g. a limited solubility ($< 10^{19}$ cm^{-3}) or the strong thermal quenching of the photoluminescence at room temperature (RT). The use of an amorphous matrix allows to surmount these problems. This is the reason why several research teams are studying the erbium doping of a-Si:H [1–3]. Hydrogenated amorphous silicon is a promising host material to obtain intense Er^{3+} emission due to its large modulable bandgap [4]. Since amorphous silicon is a defective, glassy phase, it has an erbium solubility higher than 10^{20} cm^{-3} and a greatly reduced thermal quenching of the Er-PL due to unefficient excitation energy back transfer. On the other hand, a-Si:H is known to contain high amounts of oxygen due to the preparation technique

---
[*] Corresponding author: e-mail: kechouanem@yahoo.fr, Phone: (213) 21 24 79 50 ext 062

itself. The oxygen concentration is in excess of 10^{19} cm^{-3}, which is larger than that optimum concentration obtained by implantation on c-Si [5].

In a previous work [6] we have report a successful process of preparation of erbium doped a-Si:H films by DC sputtering technique which allows the incorporation of high erbium concentration (1%).

In this study, we present a photoluminescence characterization of thin hydrogenated amorphous silicon layers doped with erbium by the D.C. magnetron sputtering technique. The temperature dependence (quenching) and the effect of the thermal annealing used in order to activate the Er^{3+} ions in a-Si:H are studied.

2 Experimental

Highly erbium doped hydrogenated amorphous silicon (a-Si:H <Er>) thin films were deposited using a D.C. magnetron sputtering system. The undoped silicon target was partially covered by small metallic erbium platelets and was sputtered in a DC discharge using an argon-hydrogen gas mixture. Details of the deposition parameters and experimental procedure have been described elsewhere [7, 8]. In this paper we studied several samples with different Er concentrations and prepared at substrates temperatures of 100 and 240 °C.

The optical film parameters (optical gap and refractive index) were determined from optical transmission measurements in the 0.4–2.5 μm range. The electrical resistivity was determined using the four-probe technique.

Near infrared photoluminescence measurements were performed using a variable temperature closed-cycle helium cryostat (15 to 300 K). The 488 nm line of an Ar$^+$ ion laser (P = 70 mW) was used as excitation source. Light emitted from the samples was collected by f/1 optics and dispersed through a 30 cm spectrograph. A cooled 512-element InGaAs diode array detector was used to detect the signal, which was integrated during 10 sec.

Table 1 summarises the results secondary ion mass spectrometry and transmission measurements made on various a-Si:H and a-Si:H<Er> samples in order to determine the erbium content, the film thickness and the optical Tauc gap. The Er/Si column expresses the relative target surface ratio and the values of the bandgap are related to the as-deposited films.

Table 1 The main characteristic parameters of various a-Si:H and a-Si:H<Er> substrates.

Sample	Deposition temperature (°C)	Hydrogen partial pressure (Pa)	Thickness (μm)	Optical gap (eV)	$\frac{Er}{Si}$ (%)
X4	240	0.2	0.90	1.86	0
X16	240	0.015	0.85	1.5	1
X12	240	0.2	0.68	1.75	0.84
X2	100	0.2	0.54	1.96	0
Y7	100	0.01	0.94	1.21	1.7
X15	100	0.035	0.94	1.53	1.52

3 Results and discussions

Hydrogen is crucial for improving the electrical and optical properties of a-Si:H. It passivates dangling bonds that acts as (non radiative recombinaison centers) and decreases density of defects states in the gap. The incorporation of hydrogen also has other effects, in particular the reduction of the structural disorder and the widening of the gap. The first of these effects is associated to the relaxation of the silicon matrix; the second is an alloying effect due to the substitution of Si–H bonds for Si–Si bonds. In figure 1 the Tauc optical bandgap E_T is plotted as a function of the hydrogen partial pressure for a-Si:H films deposited at 100 °C and 240 °C with and without erbium. Figure 1 shows an increase in the optical gap when the hydrogen pressure is increases in the deposition chamber. The optical gap of DC sputtered a-Si:H was previously found to increase linearly with increasing hydrogen content in a-Si:H films [4]. We can also observe that the optical gap of a-Si:H<Er> is lower than that undoped films, probably due to the metallic character of the erbium atoms.

The 1.54 μm photoluminescence (PL) of Er^{3+} was measured on a-Si:H and a-Si:H<Er> films at temperatures between 15 and 300 K. The luminescence at 1.54 μm wavelength is undetectable in as-deposited films. In this case most of the erbium atoms remain in a metallic or silicide form and therefore are optically inactive. These films were annealed at 350 °C for a period of 30 minutes in a flowing oxygen stream. Figure 2 shows the PL spectra of a-Si:H<Er> films (X12, X15, X16 and Y7) measured at 15 K under identical conditions and plotted on the same scale such the PL intensities can be compared quantitatively. The crystal field splitting of the surrounding matrix leads to a characteristic double peak structure at 1549.6 and 1538 nm. The narrow peak at 1529.7nm is due to the third harmonic of an Ar laser plasma line. A dominant peak emission at 1538 nm is observed for all samples, which is at the same position with that found for crystalline Si:Er. The relatively large width of the main peak (10 nm) is a result of Stark splitting of the excited (4I13/2) and the ground state (4I15/2) in the host field. This figure shows clearly that the PL intensity increases with the increase of hydrogen content of the a-Si:H<Er> films (see Table 1).

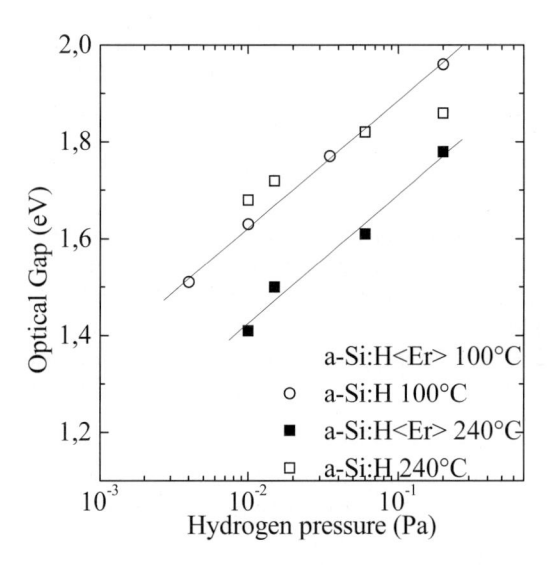

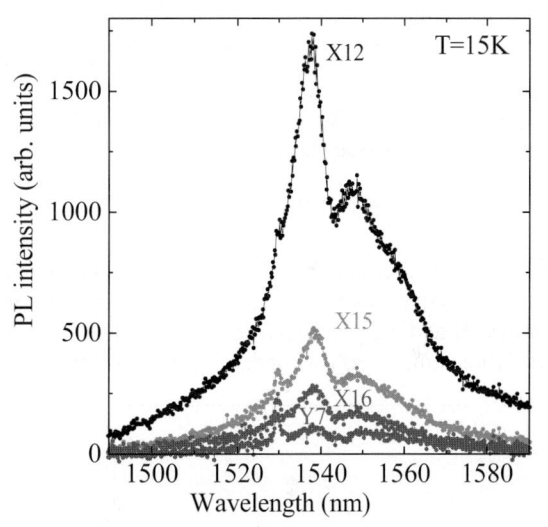

Fig. 1 The optical bandgap Et versus the hydrogen partial pressure for a-Si:H and a-Si:H<Er> Si:H<Er> films deposited at 100 °C and 240 °C.

Fig. 2 PL spectra of a-Si:H<Er> films (samples X12, X15, X16 and Y7) at 15 K under identical conditions and plotted on the same intensity scale.

One of the major problems in the development of erbium-doped silicon related materials is the strong quenching behaviour of both electro-and photoluminescence when the temperature increases from 77 K to room temperature. The PL intensity decreases, in the case of crystalline silicon, by two orders of magnitude. Figure 3 shows the infrared PL spectra, for temperature between 15 and 300 K, measured in an erbium-doped a-Si:H film, deposited at 240 °C and annealed at 350 °C for 30 minutes. The wavelengths of these PL lines, do not depend on the doping level of the a-Si:H<Er> films and on the temperature (between 15 and RT), indicating that these peaks are due to the intra-4f shell transitions in the Er^{3+} ions.

Figure 4 represents the temperature dependence of the integrated area of the dominant peak at 1538nm. The Er^{3+} emission at 1.538 μm exhibits a low thermal quenching and became very weak at room temperature, especially for the Y7 sample. The intensities of the 1538 and 1549.6 nm Er^{3+} lines at room temperature were 15 and 12%, respectively, of those at 15K. This result suggests that a-Si:H is a suitable host for efficient Er^{3+} emission at room temperature. It has been suggested that the strong thermal quenching of photoluminescence in c-Si(Er) arises from the thermally induced depopulation of the Er-induced impurity donor state, which presumably is located 15–170 meV below the conduction band edge. Within this interpretation the relative weak temperature dependence in a-Si:H<Er> suggest that the Er-related level is positioned deeper in the forbidden gap in a-Si:H that in c-Si(Er) [1].

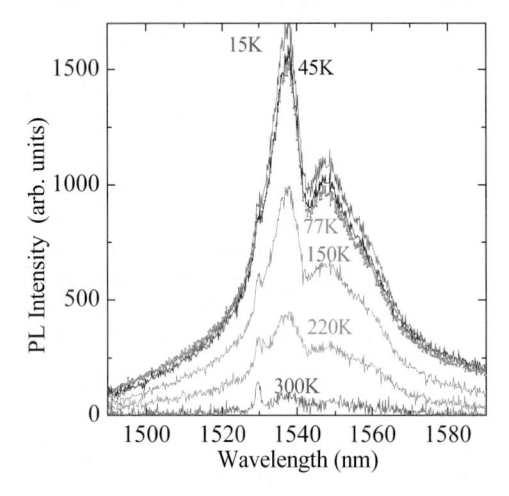

Fig. 3 PL spectra at different temperatures for a-Si:H<Er>, deposited at 240 °C and annealed at in various a-Si:H<Er> samples.

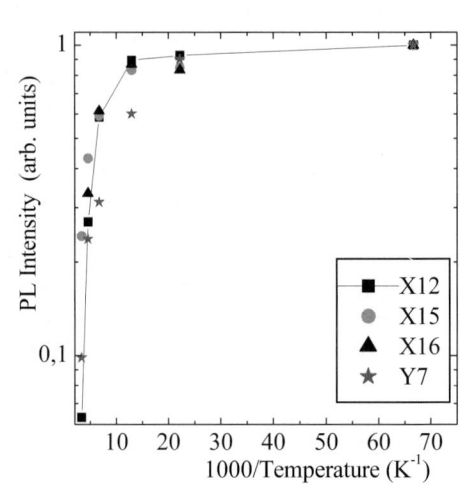

Fig. 4 Temperature dependence of the normalized intensity of the dominant peak emission at 1538 nm 350 °C for 30 minutes.

In our previous work [7] we have clearly shown that the structure of the a-Si:H film is affected when erbium is incorporated onto the host matrix. We are thus interested in the structural modifications induced by the thermal annealing used in order to activate the Er^{3+} optical centers. In figures 5 and 6 the static refractive index and the electrical resistivity are plotted as a function of the optical gap energy E_T for a-Si:H and a-Si:H<Er> films deposited at 100 and 240 °C before (open symbols) and after (full symbols) annealing at 300 °C under oxygen atmosphere.

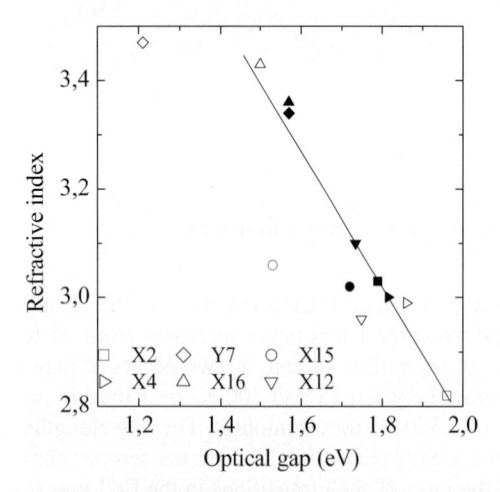

Fig. 5 Dependence of the static refractive index with the optical gap energy E_T of a-Si:H and a-Si:H<Er> films deposited at 100 and 240 °C before (open symbols) and after (full symbols) annealing at 300 °C under oxygen atmosphere.

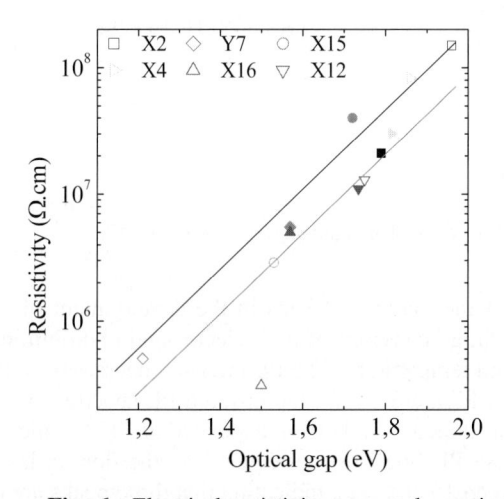

Fig. 6 Electrical resistivity versus the optical gap energy E_T of a-Si:H and a-Si:H<Er> films deposited at 100 and 240 °C before (open symbols) and after (full symbols) annealing at 300 °C under oxygen atmosphere.

At a given temperature of growth, figure 5 shows the well known decrease of the static refractive index when the optical gap and the hydrogen content increase. After annealing at 300 °C under oxygen atmosphere, excepted for the sample X12, the increase in the optical gap and the electrical resistivity of the a-Si:H<Er> films deposited at 100 °C are more important than those in material deposited at 240 °C. In addition to reducing the quenching of the erbium PL, the presence of impurities can increase the solubility of erbium in the matrix [9].

For the undoped a-Si:H films, we observe a decrease of the electrical resistivity and an increase of the refractive index correlated to a decrease of the optical gap during annealing.

3 Conclusion

We have observed 1.538 µm room temperature photoluminescence of erbium-doped hydrogenated amorphous silicon, a-Si:H<Er> prepared by the D.C. magnetron co-sputtering technique.

After annealing in a flowing oxygen stream at 350 °C, a sharp luminescence peak at 1.54 µm was observed from 15 K to room temperature. In a previous work, we have reported that the oxygen contents in our a-Si:H<Er>, determined from the SIMS profile were estimated at around 0.4–2%. Oxygen is known to enhance the PL by increasing the number of radiative Er^{3+} optical centers and to produce an increase of the band gap. The Er^{3+} emission around 1.54 µm exhibits a low thermal quenching when the temperature increases from 77 K to room temperature.

These preliminary results are encouraging for possible applications of a-Si:H<Er> in optoelectronics devices (waveguides, light emitting diodes etc.). In order to increase the Er^{3+} emission, an investigation of the details of the annealing temperature effects are necessary. Work is in progress to address theses questions.

References

[1] M. S. Bresler, O. B. Gusev, V. Kh. Kudoyarova, A. N. Kuznetsov, P. E. Pak, E. I. Terukov, I. N. Yassievich, B. P. Zakharchenya, W. Fuhs, and A. Sturms, Appl. Phys. Lett. **67**, 3599 (1995).
[2] J. H. Shin, R. Serna, G. van den Hoven, and A. Polman, Appl. Phys. Lett. **68**, 997 (1996).
[3] R. Zanatta, L. A. O. Nunes, and L. R. Tessler, Appl. Phys. Lett. **70**, 511 (1997).
[4] N. Beldi, A. Rahal, D. Hamouli, M. Aoucher, T. Mohammed-Brahim, D. Mencaraglia, Z. Djebbour, O. Glodt, J. Sib, C. Longeaud, J. P. Kleider, L. Chahed, and Y. Bouizem, Proceedings of the 11th European Photovoltaic Solar Energy Conference, Montreux, edited by L. Guimaraes, W. Pak, C. De Reyff, H. Kiessand P. Helm, 533 (1992).
[5] Moutonnet, H. L'Haridon, P. N. Favennec, M. Salvi, M. Gauneau, F. Arnaud d'Avitaya, and J. Chroboczek, Mater. Sci. Eng. B **4**, 428 (1989).
[6] M. Kechouane, N. Beldi, O. Mouheb, T. Mohammed-Brahim, A. S. Barrière, H. L'Haridon, and M. Gauneau, Phil. Mag. B **79**, 1205 (1999).
[7] M. Kechouane, N. Beldi, T. Mohammed-Brahim, H. L'Haridon, M. Salvi, M. Gauneau, and P. N. Favennec, Philos. Mag. B **77**, 137 (1998).
[8] P. N. Favennec, H. L'Haridon, D. Moutonnet, M. Salvi, and M. Gauneau, Jpn. J. Appl. Phys. **29**, L524 (1990).
[9] S. Coffa, G. Franzo, F. Priolo, A. Polman, and R. Serna, Phys. Rev. B **49**, 16313 (1994).

Lasers and medicine

Przemysław J. Deren[*,1]

[1] Institute of Low Temperatures and Structure Researches, Polish Academy of Sciences, P.O. Box 1410, 50-950 Wrocław, Poland

Received 15 September 2003, accepted 22 October 2003
Published online 22 December 2003

PACS 42.62.Be, 79.20.Ds., 87.15.Mi, 87.50.Hj, 87.63.Lk, 87.64.Ni

This paper presents physical bases of laser light interaction with biological tissue. Invasive and non-invasive laser applications like laser knife (scalpel), biostymulation – Low Level Laser Therapy, laser diagnosis and therapy especially Photodynamic Diagnosis and Therapy are reviewed.

© 2004 WILEY-VCH Verlag GmbH & Co. KGaA, Weinheim

1 Introduction

Light has always been a symbol of goodness and purity. Ancient Egyptians disinfected surgeon tools in flame; cured leukoderma using rays of the Sun. Herodotus in his papers also mentioned positive influence of Sun light on human health.

The old good methods have been forgotten for ages. Beginning of the XXth century brought certain interest in light therapy when sunbath was introduced to cure some dermatosis. However the situation changed drastically when physicists constructed first LASER. At that moment new possibilities for diagnosis and therapy in medicine appeared, because laser-light had properties that had never been observed before. Neither natural source of light like Sun nor artificial one like electric bulb produces light so monochromatic, directed and coherent.

Water, hemoglobin and melanin – three main tissue components – have their specific absorption. Since laser-light is monochromatic selective impact on tissue components is possible. Because laser light is parallel it is possible to introduce all power of light into a fiberoptic and deliver it to the desired place in the body. Coherentness and polarization of laser-light may be significant in wound treatment by weak laser radiation.

Physical bases of laser-light impact on human tissue are presented. Laser diagnosis and laser therapy, in particular:
– proper selection of laser for cutting of tissue,
– Low Level Laser Therapy (bio-stimulation) to improve wounds healing,
– Photodynamic Therapy (PDT) a new method for diagnosis and treatment of tumours and,
– emission properties of sick and health tissue
are reviewed in this paper.

2 Selection of laser light for medical purpose

Fig. 1 presents absorption spectra of three main components of human body: melanin, hemoglobin and water [1]. As biological tissue contains mainly water (for example a baby body contains about 70 % of water) it is evident that wavelength of a laser scalpel should match maxima of water absorption. Accord-

---

[*] Corresponding author: e-mail: deren@int.pan.wroc.pl, Phone: +48 71 34 350 21, Fax: +48 71 3 44 10 29

ing to the picture, the best laser scalpel should work at 2,94 µm (Er: YAG). Commonly used CO_2 laser would be a good choice too, because at far IR water absorption is high. In the 70-ties a continuous wave (cw) Nd:YAG lasers were used for tissue cutting. The 1064 nm wavelength of Nd:YAG laser matches a window in the water absorption band. At in wavelength absorption of light is five orders of magnitude weaker than at 2.94 µm. Because of weaker absorption first laser light will coagulate tissue, than tissue will be carbonised, and eventually evaporated. Tissue will be cut bloodlessly (because blood vessels will be coagulated) but the round's wound will be burned and scorched. This, for the first sight, unwilling side effect is blessed however in tumor surgery, because 1064 nm laser light penetrates deeply and destroy tumor's bed.

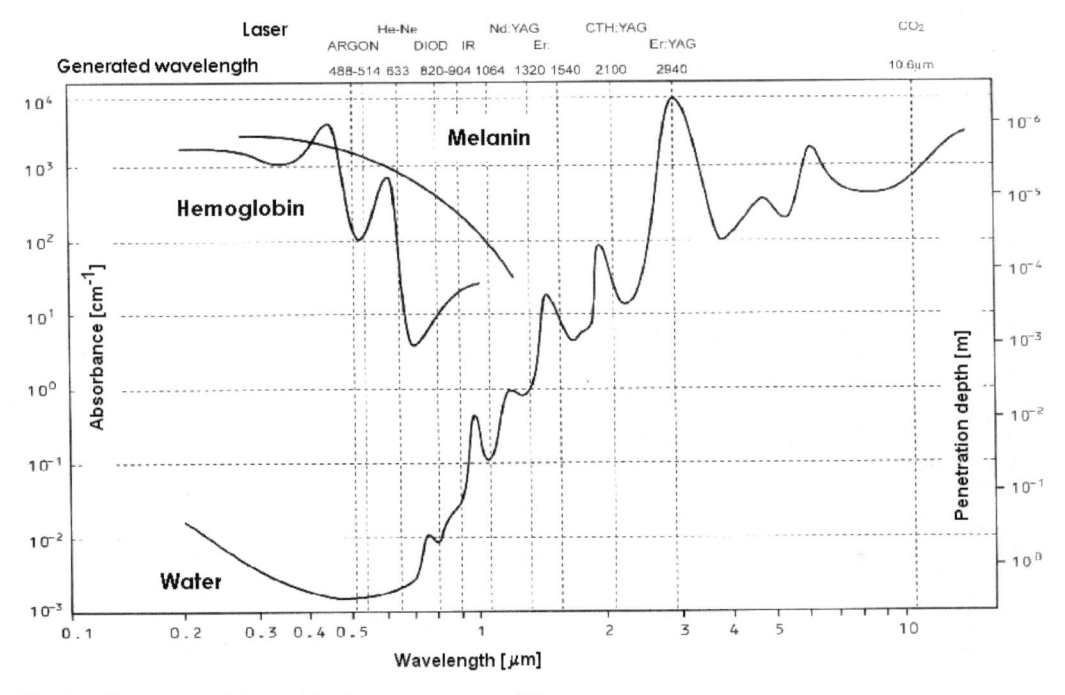

Fig. 1 Absorption of three main tissue components [1].

It is also possible to act selectively on blood vessels. Second harmonic of the Nd:YAG (532 nm) or Argon lasers (514,5 nm) continues to play a role in the treatment of vascular lesions that have been successfully treated by the argon laser including several angiomas, and venous lakes. Other application of these lasers is treatment of mature, hypertrophic port wine stains in adults. Pulse Argon or Nd:YAG lasers are used in ophthalmology . It is possible to help a patient with unstuck retina. This very important part of the eye is vasculatured. Therefore it is possible to weld the retina to the bottom of the eye with an argon laser.

However, 514.5 nm laser light is also slightly absorbed by macula, therefore the best solution for retina welding is Krypton laser working at 568 nm. At this wavelength macula does not absorb the light, Hemoglobin has two absorption maxima at 548 nm and 577 nm. Therefore it is possible to weld the retina very close to the macula using Krypton laser.

It is also possible to have an impact on melanin, because green, blue and violet lasers match very well melanin absorption. Application of tunable laser enables to adjust laser wavelengths to the maximum of a dye absorption used in tattoos.

The region from 670 nm to 900 nm is called Optical Window in Tissue because there is very low absorption of water, hemoglobin, and melanin. Light from this range deeply penetrates human body (up to 6 cm). Therefore red light is used in biostymulation and laser diagnosis, for example in photon tomography.

Proper choice of laser wavelength according to the medical application is essential. However, there is another very important factor, which must be considered when action of laser light on biological tissue is planned. It is energy density (or power density) of laser light versus interaction time with the tissue (see Fig. 2). Fraction of Joule or single Joules per cm^2 interacting with tissue even long time may provoke only photochemical reactions, similar to reaction of skin after exposition to the Sun radiation. Single Jules of laser light interacting with tissue during one tens of seconds will enhance temperature of the tissue. At the beginning it may occur reversible coagulation, then denaturisation and homogenous volume coagulation. Increasing the energy densities will result in evaporation and cutting. These considerations are true for continuous wave (cw) and pulse lasers, however we have to emphasise that for cw or relatively long pulses (up to tens microseconds) all processes are heat assisted. Therefore such process cannot be used for revasculisation to remove thrombotic sclerotic material because such laser pulse will provoke thermal spasm of the blood vessel.

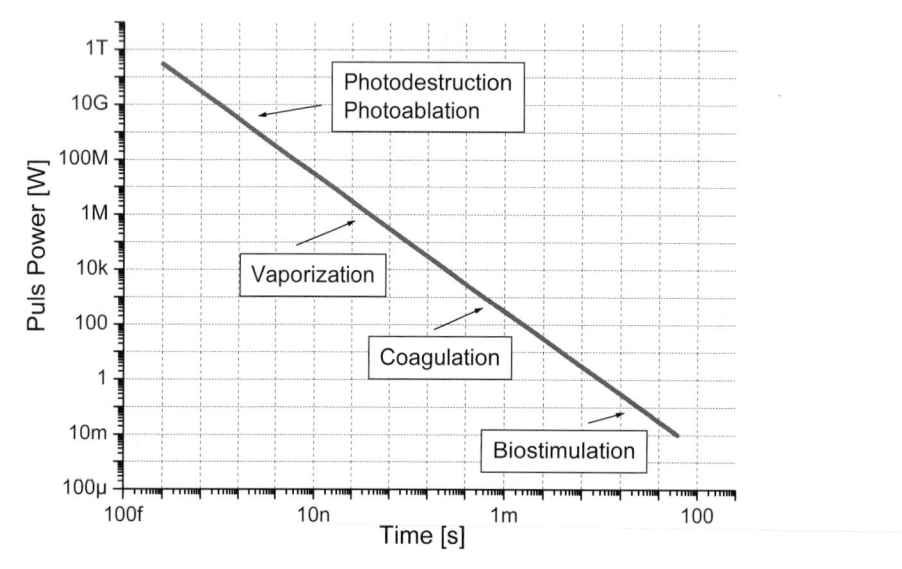

Fig. 2 Power of a 0.3 J laser pulse versus pulse time. A laser with such tunability of pulse' length does not exists; similar behaviour could be observed when Q-switch is turn on in for example Nd:YAG laser.

At high energies with short (nano or pico seconds) pulses obtained power density is also very high. Produced by laser radiation electrical field is so large that ionises the molecules and photoablation, photodiscruption is observed. This process is free of carbonisation of the tissue and therefore could be very useful in angioplasty – for recanalization of coronary vessels. Recently this feature of short laser pulses has found an application in plastic surgery. Instead of cutting the skin and stretching it on the cranium a pulse CO_2 laser is used. This new technique is not so harmful, painful and sometimes burden with risk of bad esthetical effects as classical surgery is. The laser works like a smoothing plane. 10,64 μm of CO_2 laser emission is absorbed very well by tissue (see Fig 1), almost 100 % of it is absorbed in a very thin layer. Because of short pulse working regime, there are no harmful thermal effects. Absorbed laser power removes (vaporises) thin layer (several μm) and slightly hits collagen beneath. Then reversible coagulated collagen rebuilds in a new, much elastic form. The results are very interesting revealing healthier looking skin beneath. The patient should avoid the Sun exposure during a weak and it is the only side effect.

Another application of intense pulse technique is again ophthalmology. It is possible to correct even up to 6 Dioptres of Myopia, Hyperopia or Astigmatism by removing thin layer from the cornea. Inert pressure on the eyeball will change the curvature of the eye changing focus of the whole optical system. Such an operation is performed using excimer lasers with energy about 100 mJ at 308 nm and pulse length 10 ns.

3 Biostimulation

In opposition to rather violent applications of laser light biostimulation concerns very gentle use of a laser. Biostimulation is known also as Low Level Laser Therapy (LLLT). Usually a laser works at the region of the skin optical window with very low power radiation –about 200 mW. It was showed that laser radiation may improve healing of wounds, broken bones and additionally has an anaesthetic action.

Many studies show that the LLLT acts by:
– improving in vitro fibroblast proliferation,
– enhancing cutaneous wound tensile strength,
– affecting the mitotic rate
– minimising inflammation
– better reduction of edema
– improving of skin regeneration and enhancement of collagen synthesis.
– causing densing epidermis and collagen fibber, which were thicker and arranged better.

It could be not excluded that biostimulation is a good example of placebo effect. However, many reports show that this method cure animals and has an impact on biological cells in vitro.

Up to now the LLLT mechanism(s) are not known. Several hypotheses have been proposed one for example stresses on stimulation of cell respiration by either the endogenous porphyrins in the cell or by cytochromes.

Our hypothesis is based on observation that collagen grows in a very regular manner. Tropocollagen - the protein substance from which collagen fibbers are formed has exactly 260 nm, and it is arranged in very regular way. Because laser light is highly coherent and polarised, it introduces an order in illuminated tissue. Therefore we believe that experiments with tuneable lasers may show the best results of biostimulation at wavelength equal 3 or 4 multiple of tropocollagen length i.e. at 780 nm or 1040 nm, respectively.

4 Laser diagnosis and therapy

One of the most promising methods in tumour therapy is photodynamic diagnosis (PDD) and photodynamic therapy (PDT). The idea of PDD is based on observation that fast growing tumour cells absorb and hold a dye much better than healthy cells. Fig. 3 presents normalised luminescence intensity of a dye present in healthy and in tumour cells [2]. We can learn from this picture, that after 5, 6 days dye is virtually absent in healthy cells and reside in tumour. Now if the body is excited by light, which wavelengths matches absorption of the dye, tumour cells will shine indicating their presence. Dye should be "compatible" with our body. Therefore it is prepared on the base of heme. Iron is removed from heme and one obtains hematoporfiryn. Nowadays Porfimer Sodium (Photofrin) is the only dye approved by Food and Drug Administration (FDA). This drug may be applied by injection, per or (orally) or directly as a gel to the skin, mucous membrane. The only drawback is that the patient should avoid direct Sun or strong light exposure during and after the treatment.

After diagnosis a doctor may very easily cure the patient. It is enough to change illumination wavelength to red one. About 630 nm (it depends on a dye) excites very well singlet states of porphyrin. Then excitation energy is transferred to the long-living triplet state, which decay time is about 10 ms. This time is long enough to excite oxygen being delivered continuously by blood vessels. So-called Singlet Oxygen is very toxic. In water solution its decay time is 2–5 µs, but it is enough to destroy tumour cells. To be precise it provokes apoptosis, a programmed cell death [3]. This fact distinguishes very much the all-traditional methods from PDT. Because PDT is a oriented method; it destroys only sick, malignant cells, secondly using the PDT we do not observe acute necrosis. Necrosis is caused by any external reason like gamma irradiation – very often used to combat cancer, chemotherapy, temperature etc. In necrosis the tissue is swollen, finally the membrane of a cell is broken and immune system should also combat all products of such inflammation process.

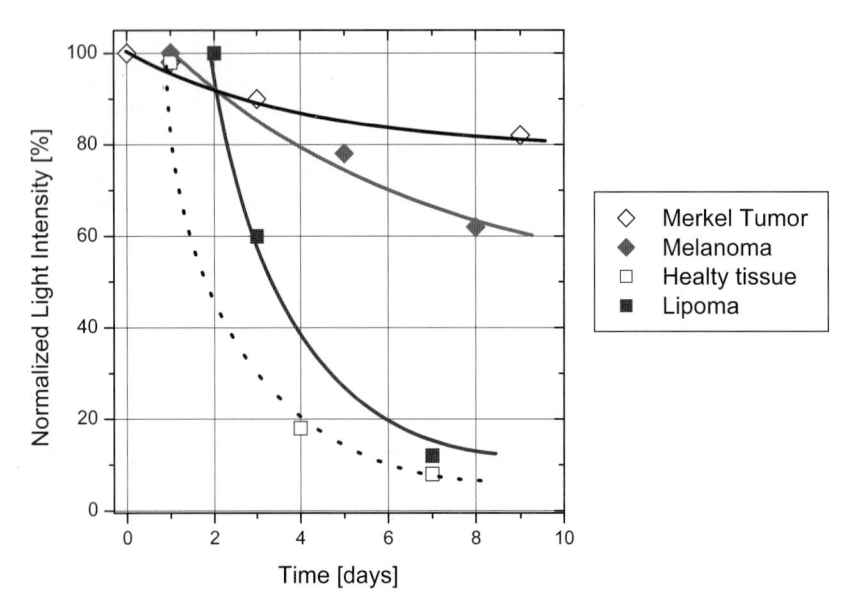

Fig. 3 Normalized light intensity of a dye in different tissues [2].

5 Summary

This short article does not fulfil the broad subject concerning lasers in medicine. The author does not, for example, describe Raman scattering used for medical diagnosis or very promising results obtained in photon imaging, which one day will replace X-ray diagnosis and also development of work on new dyes, like for example chlorin, used in PDT.

Today this unique tool, as laser is, allows for bloodless surgery, high precision surgery *(angioplasty)*, - surgery of „impossible cases" like saving eyesight, better, healthier and versatile diagnosis and thanks to biostimulation shorter time of convalescence. However we are still waiting for propitious ending of investigations on PDT and for LLLT mechanisms explanation.

References

[1] Z. Jankiewicz, "Lasery stosowane w medycynie", VII Krajowa Szkola Optoelektroniki, Zegrze 1993.
[2] Fotodynamiczna Metoda Rozpoznawania i Leczenia Nowotworów (Photodynamic Method of Diagnosis and Therapy of Tumours), ed. Alfreda Graczyk, Dom Wydawniczy Bellona, Warsaw 1999.
[3] Sydney Brenner, H. Robert Horvitz, John E. Sulston, The Nobel Prize in Physiology or Medicine 2002 for their discoveries concerning "genetic regulation of organ development and programmed cell death"., http://www.nobel.se/medicine/laureates/index.html

UV-written channel waveguides in Er^{3+}-doped Bi$_2$O$_3$-based glass

A. Favre[*,1], **E. Lee**[1], **V. Apostolopoulos**[1], **C.B.E. Gawith**[1], **C.Y. Tai**[1], **E. Taylor**[1], **Y. Kondo**[2], and **F. Koizumi**[2]

[1] Optoelectronics Research Centre–University of Southampton, Southampton, SO17-1BJ, UK
[2] Asahi Glass Co. Ltd., Kanagawa-ku, Yokohama, 221-8755, Japan

Received 15 September 2003, revised 15 October 2003, accepted 15 October 2003
Published online 22 December 2003

PACS 42.70.Ce, 42.70.Gi, 42.82.Et

The present paper describes the fabrication of optical channel waveguides in an Er^{3+}-doped Bi$_2$O$_3$-based glass using direct laser writing with a UV beam at 244 nm. A buried positive refractive index change was achieved through this method and a description of the observed phenomena is given, based on optical and electron microscopy.

© 2004 WILEY-VCH Verlag GmbH & Co. KGaA, Weinheim

1 Introduction

There is an urgent need for optical amplifiers with a wide and flat gain spectrum in the telecommunication window in order to increase the capacity in wavelength division multiplexing network (WDM) systems. For metro use, a compact amplifier is required to meet low spatial and cost effective demand. Compact amplifiers such as erbium-doped planar waveguide amplifiers have been proposed. Compared to silicate glasses, Bi$_2$O$_3$-based glass can accept a high concentration of Er^{3+} (up to 13,000 ppm) with broadband emission and negligible concentration quenching [1–3]. This represents a potential for making compact amplifiers such as erbium-doped planar waveguide amplifiers.

Also, the property of photosensitivity in glasses has recently caught attention and been exploited for producing in a simple manner optical structures and devices under light irradiation [4, 5]. For example, direct writing into glasses using a UV laser beam is an attractive way of producing optical channel waveguides via a fast single-step process [6–8]. The aim of the present study is the fabrication of low-loss channel waveguides for rare-earths doped devices using UV writing. This paper presents the effects of the exposure of the Bi$_2$O$_3$-based glass to a UV beam and some preliminary optical characterisations.

2 Experimental

The direct UV-writing apparatus consists of a frequency doubled Ar ion laser (Coherent FRED Sabre 500) with continuous wave output at 244 nm. The samples were positioned on a vacuum chuck connected to a computer controlled translation stage, which shifted perpendicularly to the incident UV laser beam at different speeds (Fig. 1). The sample is an Er^{3+}-doped bulk Bi$_2$O$_3$-based glass containing 10,000 ppm erbium. Various fluences were used by varying the laser power and the writing speed to evaluate the behaviour of the glass for different conditions. Three series of fifteen exposures each were tested with a power of 10, 20 and 30 mW, and scan rates between 0.001 and 0.5 m/s. The spacing between the lines was 100 µm and between each set of fifteen was 200 µm. The UV spot size was around 6 µm diameter and was focused on the sample surface. The resulting sample was observed with an optical microscope

---

[*] Corresponding author: e-mail: af1@orc.soton.ac.uk, Phone: +00 44 2380593172, Fax: +00 44 2380593149

and a scanning electron microscope (SEM) (LEO 430) on the surface and the cross-section. Laser light was launched into the channels in order to check the mode profile at 633, 980 and 1550 nm. The refractive index change was deduced from numerical aperture measurement through the trigonometric method, after imaging the far-field intensity pattern at 633 nm on a CCD camera at various distances from the output of the waveguide.

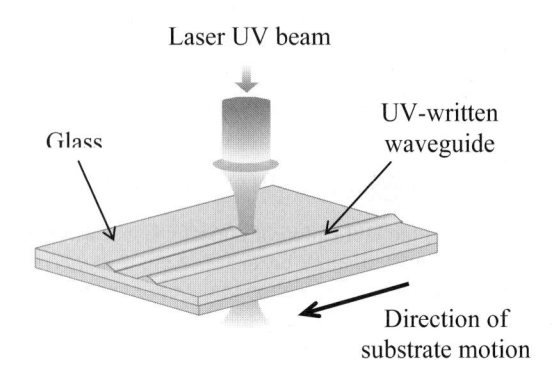

Fig. 1 Scheme showing the direct UV-writing of channel structures in glass samples.

3 Results

3.1 Material aspects

Under the UV beam, visible tracks at the surface of the sample appeared. For high laser fluences, physical damage and ablation of the glass occurred. A dark, uneven region can be seen under the exposed area (Fig. 2), which is not guiding any light. The dimensions of this area increased with decreasing scan speed for a given laser power. This was confirmed by SEM and surface profile measurements. Figure 3 gives the topography and corresponding compositional contrast of the sample. The area exposed to the laser appears darker than the bulk glass around it, which suggests that this region should be of lower density, hence lower refractive index.

The mechanism for photosensitivity varies depending on the type of glass. It is suggested that in the Bi_2O_3-based glass all the UV light is absorbed within the first few microns of the sample, which leads to the formation of a lower refractive index region at the glass surface, and a guiding smaller denser area under this region (Fig. 4).

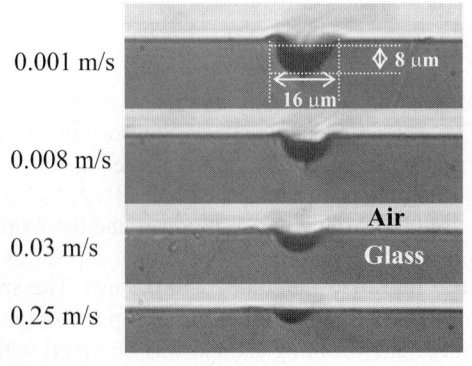

Fig. 2 Cross-sectional optical microscope picture of four UV-written channels with associated scan velocity (with a laser power of 30 mW).

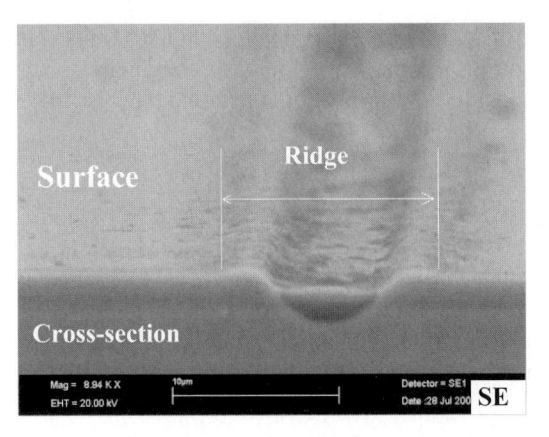

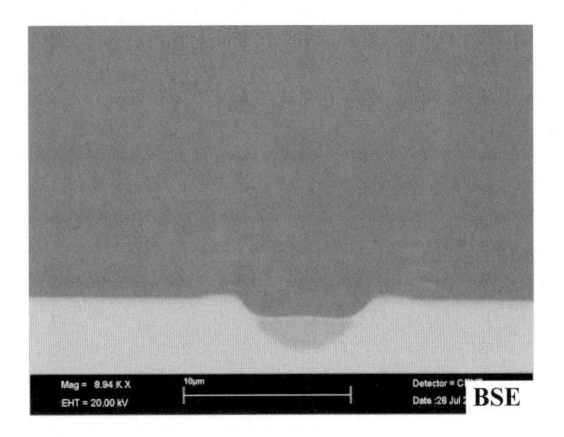

Fig. 3 Secondary electron (SE) and corresponding backscattered electron (BSE) SEM pictures of a waveguide written with a laser power of 20 mW and a scan speed of 0.08 m/s.

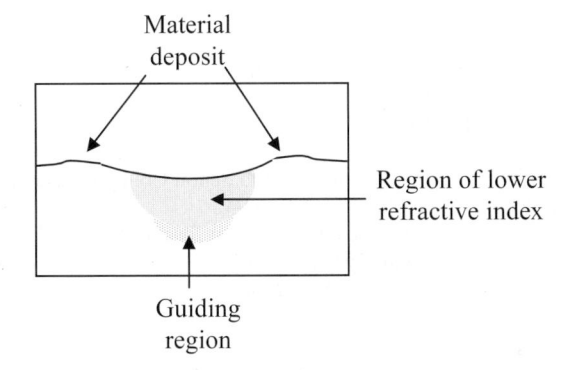

Fig. 4 Representation of the cross-section of a UV-written channel in the Bi_2O_3-based glass.

3.2 Optical characterisations

All the channels guide single mode at 633 nm and 980 nm, but not at 1550 nm. Figure 5 shows the near field image of the modal output at 980 nm from a channel waveguide written with a laser power of 30 mW and a scan velocity of 0.03 m/s. A Gaussian profile can be fitted.

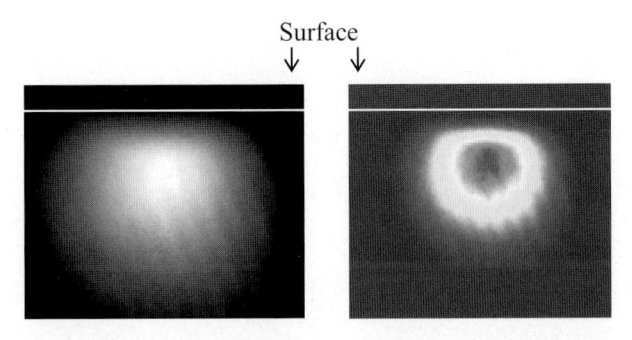

Fig. 5 Modal output at 980 nm of a channel written with a laser power of 30 mW and a scan velocity of 0.03 m/s.

A refractive index change of 4×10^{-4} at 633 nm was assessed for a channel written with a laser power of 30 mW and a scan velocity of 0.5 m/s.

The total loss in a channel written with a laser power of 30 mW and a scan velocity of 0.03 m/s was measured to be 4 dB/cm. This value includes propagation loss of the waveguide and coupling loss between the planar end faces and the fibres at the input and output interfaces. For comparing, the loss measured in a Bi_2O_3-based glass fibre is 0.7 dB/m.

4 Conclusion

In this study, the technique of direct laser writing using UV light at 244 nm was employed for engineering optical channel waveguides in a novel oxide glass composition based on Bi_2O_3. Light confinement was observed at 633 and 980 nm. Despite different experimental writing conditions with a wide range of laser power and scan velocity, no guiding mode could be detected at 1550 nm. Essential parameters were identified, which can allow future processing of the required channel waveguides, by using for example a bigger laser spot size.

Acknowledgement This work is funded by Asahi Glass Co.

References

[1] N. Sugimoto, J. Am. Ceram. Soc. **85**(5), 1083 (2002).
[2] S. Tanabe et al., J. Lumin. **87–89**, 670 (2000).
[3] Y. Kondo et al., Proc. of the Optical Fiber Communication conference, Anaheim, California, USA, 17–22 March 2002.
[4] H. Ebendorff-Heidepriem et al., Proc. of the 7th Otto Schott Colloquium, Jena, Germany, 7–10 June 2002.
[5] M. Svalgaard et al., Electron. Lett. **30**(17), 1401 (1994).
[6] A. Favre et al., Proc. of Photonics West conference, San Jose, California, USA, 25–31 July 2003.
[7] C. Contardi et al., J. Non-Cryst. Solids **291**, 113 (2001).
[8] C.B.E. Gawith, Appl. Phys. Lett. **81**(19), 3522 (2002).

Electrical properties of the five-component chalkogenides of silver

O. L. Kheifets-Kobeleva[*,1], **V. B. Zlokazov**[1], **N. V. Melnikova**[1], **L. L. Nugaeva**[1], **L. Ya. Kobelev**[1], and **Ya. L. Kobelev**[2]

[1] Ural State University, Physical Department, Lenina ave. 51, 620083 Ekaterinburg, Russia
[2] Institute of Metal Physics, S. Kovalevskaya st., 18, 620219 Ekaterinburg, Russia

Received 15 September 2003, revised 26 October 2003, accepted 26 October 2003
Published online 22 December 2003

PACS 72.80.Jc

The electrical properties of $AgGeSbS_{3x}Se_{3(1-x)}$ ($x = 0.4$-0.7) were investigated by means of impedance measurements in the frequency range between 10 Hz and 800 kHz and at temperatures between 78 K and 500 K. In all investigated chalcogenides ionic conductivity (Ag^+) was found. The onset of ionic transport was found at 250 K-300 K, depending on the composition. The complex impedance and admittance plots, the electrical properties are given.

© 2004 WILEY-VCH Verlag GmbH & Co. KGaA, Weinheim

1 Introduction

For the development of cryogenic microelectronics a new materials with purely ionic or mixed electronic-ionic conductivity (including materials with high resistance) which should work at low (down to 77 K) temperatures are needed, also requires the new the semiconductors materials, which combine the superionic properties at low temperatures with other useful properties (optical, magnetic et al). The electrical properties of some quaternary Ag-containing chalcogenides have been investigated by E.R. Baranova et al. [1, 2]. The present work deals with the synthesis and investigation of the electrical properties of a number of new five-component compounds of the form $AgGeSbS_{3x}Se_{3(1-x)}$ ($x = 0.4$-0.7). Our aim was to study whether the synthesised compounds are ionic conductors, and if so, to specify the temperature regions were ionic transport becomes significant. We present here the impedance diagrams and the temperature dependences of the total electric conductivity and the dielectric permittivity of the synthesised compounds in temperature range between 78 K and 500 K at the frequency of 1592 Hz. Also the methods of synthesising are described.

2 Experimental

2.1 Synthesis of the materials

All the compounds were prepared by sintering stoichiometric quantities of the elements in quartz ampoules, which were evacuated to 10^{-4} Pa and then filled with deoxygenated helium. The sintering was carried out in a regime of slow heating with long standing at several fixed temperatures. These temperatures corresponded to the melting and vaporing (sulphur) points of all the initial components. For each compound was found its own regime producing a homogenous material. The maximum sintering temperature depended on the composition of the samples and did not exceed 1200 K. The tablet-shaped specimens to be investigated were cut out of the bars and polished.

---

[*] Corresponding author: e-mail: Olga.Kobeleva@usu.ru; Phone.: +7 3432 531981

© 2004 WILEY-VCH Verlag GmbH & Co. KGaA, Weinheim

2.2 Electrochemical cell

All the measurements were carried out in a specially constructed cell. For the impedance measurements the ionically blocking copper or ionically reversible silver electrodes were used. The value of the ionic conductivity was measured using a Wagner nonsymmetrical cell [3].

2.3 Impedance measurements

The measurements were carried out with a Solartron 1170 Frequency Analyser in combination with an electrochemical interface Solartron 1186, and with an ac bridge R5021 (Russia). The impedance diagrams were measured within 10^{-2}-10^5 Hz at the temperatures given.

2.4 Conductivity and dielectric permittivity measurements

The conductivity and dielectric permittivity single frequency (1592 Hz) investigations were carried out with an ac bridge TESLA BM484 (Czech Republic) which measures conductivity and geometrical capacitance using the model equivalent circuit containing a parallel connection of resistance and capacitor. The error did not exceed 0.05%.

3 Results and discussions

3.1 Impedance measurements

Impedance measurements were carried out to separate the bulk sample properties from the properties at electrode/sample interface. The dispersion spectra of the $AgGePbS_{3x}Se_{3(1-x)}$ compounds were analysed at 300 K. The impedance plots of $AgGePbS_{3x}Se_{3(1-x)}$ (x = 0.4, 0.6, 0.7) are shown in Fig. 1. All the spectra are characterised by the presence of two clearly distinguished regions - HF and LF. The equivalent circuit of the cell can be presented as two straightforwardly connected parts, one of which describes the bulk response and the other deals with the electrode processes.

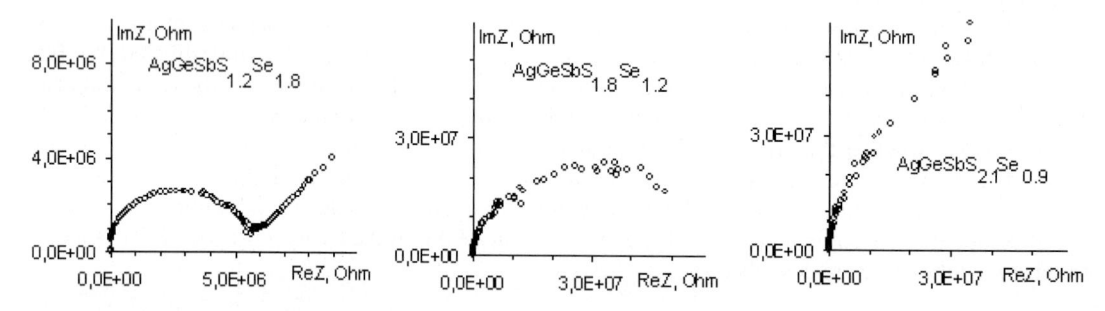

Fig. 1 Hodographs of impedance $AgGeSbS_{3x}Se_{3(1-x)}$ at temperature 300 K.

At high frequencies the form of the impedance plots may be well approximated by semi-circles which either go through the origin (0,0), with the centers being lower then the abscissa axis in all cases. The corresponding part of equivalent circuit can be approximated by a parallel resistance, and geometrical capacitance and a constant phase element (CPE). This, in its turn, may be caused by fractal-like structures, producing specific charge distributions and giving this effect [4, 5]. Since in our case CPE is a feature of bulk response, these must be an inner structures in polycrystals - e.g., grain boundaries, dislocations and other lattice defects with fractal distribution. With increase x centers of a semicircle displace.

Exponent CPE is equal 0.95, 0.78 and 0.92 for x = 0.4, 0.6 and 0.7 accordingly.

During the measurements the voltage at the electrodes was chosen to be in the range of 10-30 mV and the current in the sample was very small (nanoamperes). Therefore one can neglect the difference in stoichiometry of the sample near reversible and blocking electrodes and difference between bulk and near-electrode stoichiometry. Nevertheless, the properties of near-electrode regions, of course, would differ. Since our aim was not to investigate all the electrochemical processes in the cell, but to define the temperatures of the onset of ionic transport (if any). So, a detailed analysis of the low frequency parts of dispersion spectra and construction of equivalent circuits was not worked out.

3.2 Conductivity measurements and determination of the dielectric constant

Electrical conductivity of the compounds AgGeSbS$_{3x}$Se$_{3(1-x)}$ was investigated at a fixed frequency of 1592 Hz which belongs to the HF region of the impedance measurements. The values of the conductivity and geometrical capacitance of the empty cell are σ<2 pS, C = 0.3 pF at T = 78 K and σ = 70 pS, C = 0.4 pF at T = 300 K. The temperature dependence of the conductivity of all the compounds are of semiconductor type (σ=σ$_0$exp(-E$_a$/kT)) with different values of the activation energy at different temperatures (see Fig. 2.). To separate ionic and electronic contributions to the current we carried out measurements in a cell with ionic electrodes (Ag$_4$RbI$_5$).

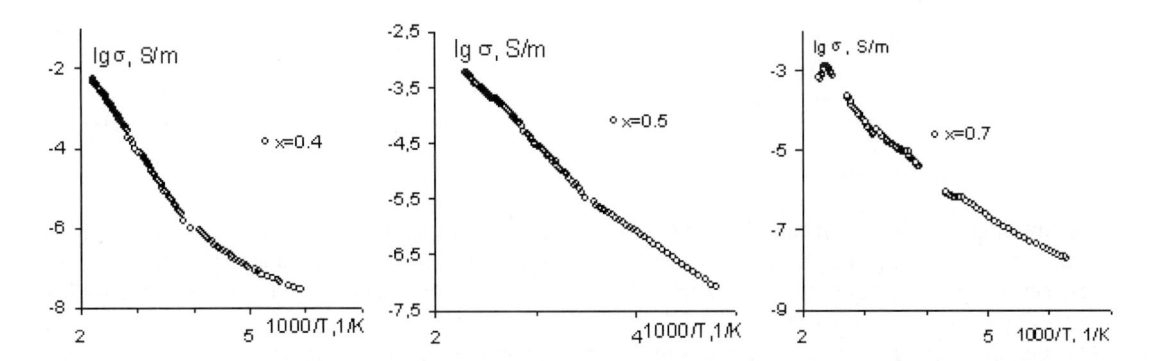

Fig. 2 Temperature dependence of the conductivity.

The values of dielectric permittivity ε were calculated from the magnitudes of the capacitance which had been measured at the frequency of 1592 Hz. The temperature dependencies of the dielectric permittivity AgGeSbS$_{3x}$Se$_{3(1-x)}$ are given in Fig. 3. Again, the temperature regions where the slopes of the ε(T)-curves increase rapidly correspond to the onset of ionic conductivity, since the latter leads to appearance of free charges in the bulk and hence possibility of large polarization.

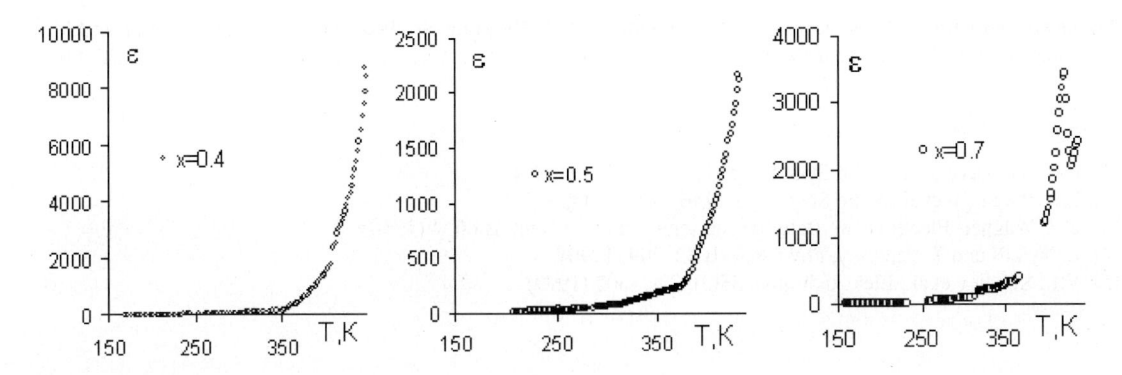

Fig. 3 Temperature dependencies of dielectric permittivity.

Investigations of the ε(T) temperature dependences, together with the results from the conductivity measurements enabled us to establish the temperature regions for the onset of ionic transport more precisely (Table 1).

Table 1 The regions of begining the ionic conductivity, energy of activation, part of ionic conductivity and specific conductivity.

Compound	T, K	E_a, eV	σ, S/m	$σ_i/σ$, T = 300 K, 1592 Hz	regions of begining the ionic conductivity
$AgGeSbS_{1.2}Se_{1.8}$	<250 K	0.09	16.4×10^{-6}	98%	270-290K
	250 K-300 K	0.17			
$AgGeSbS_{1.5}Se_{1.5}$	<280 K	0.02	7.6×10^{-6}	82%	270 K-280 K
	280 K-380 K	0.03			
$AgGeSbS_{1.8}Se_{1.2}$	250 K-285 K	0,62	1.98×10^{-6}	67%	330 K-340 K
	285 K-450 K	0.19			
$AgGeSbS_{2.1}Se_{0.9}$	<170 K	0.08	8.2×10^{-6}	64%	300 K-310 K
	170 K-400 K	0.34			
	400 K-440 K	0.14			

4 Conclusions

Our study indicates that synthesised compounds possess mixed electronic-ionic conductivity. In compounds with x = 0.7 at T~420 K is observed ferroelectric phase transition. All compounds are the mixed ionic-electronic conductors with temperature of beginning of the ionic current 270 K-340 K, the part of ionic conductivity depends from x (64%-98%).

The area of a beginning of ionic current in the investigated compound is higher, than for compounds which containing only sulfur or a selenium, however the part of the ionic current is higher, than for compounds with an only selenium. The increase of a part of a selenium results in increase of ionic conductivity. The conductivity at T = 300 K for all compounds is lower on the order, than for $AgGeSbS_3$ and on some orders is lower, than for $AgGeSbSe_3$. As comparison of the compounds containing an arsenic instead of antimony, part of ionic current and the area of temperatures of a beginning of the ionic current and the values of conductivity are comparable.

It is possible, that other ratio of sulphur and selenium in compounds will result in the further growth of conductivity and downturn of temperatures of a beginning the ionic of current. Also it will be interesting to investigate copper-containing chalkogenides.

Acknowledgements This work in part was supported by CRDF (grant No. EK-005-X1) and grant CRDF and Min. of Education of the Russian Federation (Post Doctoral Fellowship, award EK-005-X1, annex 07, No Y1-E-05-09)".

References

[1] E.R. Baranova et.al., Solid State Ionics **124**, 255–261 (1999).
[2] E.R. Baranova et.al., Solid State Ionics **146**, 415 (2002).
[3] C.Z. Wagner, Electrochem. Berichte Bunsenges. Phys. Chem. B **60**, 4 (1956).
[4] L. Nyikos and T. Pajkossy, Phys. Rev. B **42**, 709 (1990).
[5] V.L. Kobelev et al., Elektrokhimiya **35**(3), 294–302 (1999).

Light scattering topography of excimer grade CaF$_2$ crystal

Yasunao Oyama [1], **Suzuka Nishimura** [2], and **Kazutaka Terashima**[*,1]

[1] Materials Science and Technology, Shonan Institute of Fujisawa, Kanagawa, Japan
[2] Science and Engineering, Keio University, Hiyoshi, Yokohama, Kanagawa, Japan

Received 15 September 2003, revised 17 October 2003, accepted 17 October 2003
Published online 22 December 2003

PACS 42.60.By, 42.81Dp, 78.35.+c

We have measured the scattering substance of calcium fluoride used for the excimer lithography grade materials by the new type of light scattering topography method. This method is thought to be effective for the observation method, which is no need of cutting samples. Some scattered particles have been observed in the matrix of calcium fluoride. The fog-like hazy calcium fluoride, which are occurred by annealing at high temperature, have been observed by the method. The small scattering substances, which coordinate along the structure of the crystal, and the large substances, which are disordered, form the particles. The small one is thought to be related to the structure of crystal of calcium fluoride.

© 2003 WILEY-VCH Verlag GmbH & Co. KGaA, Weinheim

1 Introduction Calcium fluoride (CaF$_2$) is a representative alkaline earth fluoride that has been the subject of experimental and theoretical studies for years. It has been one of the important materials used in the design of optical components. Recently, many efforts have been focused on the growth of large-scale crystals for the use in UV lens of IC lithography. It is called excimer grade CaF$_2$. We have published the fundamental melt properties such as melt density and surface tension [1, 2]. It is important to investigate the optical properties of CaF$_2$ for the purpose of using for IC lithography. Light scattering method is a very valuable way to investigate imperfections in crystals. Topographic observation is preferable to precisely characterize the inside of specimens. The light scattering topography method is used to the angle of 90 degrees light scattering [3, 4] and effective for the observation of small impurities of crystal. Dislocations in a LEC-GaP have been observed by this method [5]. And the observation of imperfections in LiF, MgO and Ruby has been investigated [4]. On the other hand X-ray topographic investigation of growth defects in natural CaF2 crystals was reported [6]. In this paper we have performed an improvement of light scattering method and observed the excimer grade CaF$_2$ by the improved method.

2 Experimental A basic constitution of the light scattering topography apparatus is shown in Fig. 1. A green semiconductor laser is used as an incident light. The wavelength of the laser is 532 nm. Scattering light is observed by a microscope, which consists of 5 times as much objective and detected by CCD (charge-coupled device). Data of microphotograph have been observed by a monitor and captured into PC and the stage of the sample can be moved with a motor and has been controlled by a stage controller and PC. Conventional optical geometry is in Fig. 2(a). The surface across the laser and the surface across the scattering light are different. Then it is necessary to polish the three surface of crystal by the geometry to avoid the effect of light scattering caused by the loughness of the surface. An improved optical geometry of the topography in this study is shown in Fig. 2(b). It is possible to observe by polishing only one surface of crystal. The improved method has an advantage of process of cutting and polishing crystals. Crystals of CaF$_2$ have been grown by the Bridgman Stockbarger technique under vacuum pressure (ca.~10^{-3} Pa). A graphite crucible is used for the growth. For purification of the crystals from oxygen

---

[*] Corresponding author: e-mail: terasima@mate.shonan-it.ac.jp, Phone: +81 466 30 0226, Fax: +81 466 30 0226

impurity some amount of PbF$_2$ was added. Oxygen impurity was introduced when crystals were grown without addition of PbF$_2$. Starting material of the crystal was the powder of CaF$_2$, which grade is 4N purity. The grown crystals have been annealed between 1200 K and 1500 K. Crystals have been sliced like a disk and polished one surface.

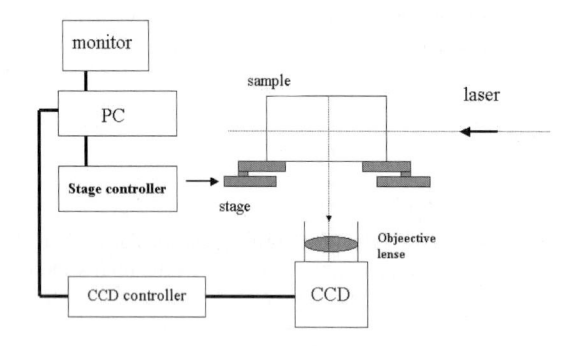

Fig. 1 Constitution of the light scattering topography apparatus.

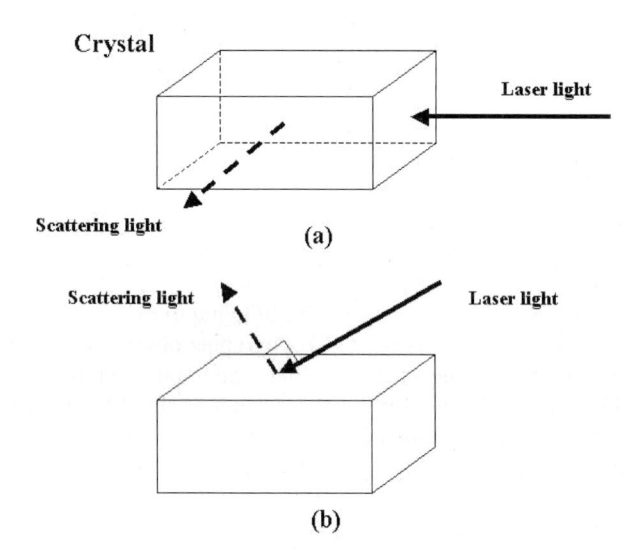

Fig. 2 Optical geometry of the scattering apparatus ((a) conventional (b) this study).

3 Results and discussion Figure 3 shows microphotographs of the scattering light patterns from a CaF$_2$ crystal. Crystal was annealed about 1450 K and fog-like hazy area was able to be observed by no use of microscope. The microphotograph is focuced to fog-like hazy area. Two types of particles have been observed in the microphotograph. One is the large scattering particle which size is about $1*10^{-6} \sim 5*10^{-6}$ m and the other is curtain-like scattering particle which is less than $5*10^{-6}$ m. Figure 4 shows the magnification of curtain-like scattering particles. The former particles exist independently and are thought to be the inclusions which consist impurities like rare earth element or other metal element. The latter particles are very small and their arrangement is thought to be related to the direction of crystal structure. These particles are thought to occur by means of the vacancies of CaF$_2$ crystal. These vacancies occur by the order–disorder phase transition of CaF$_2$ about 1424 K [7, 8] because the position of F$^-$ ion is disordered and F$^-$ interstitial and F$^-$ vacancy occurs. Then Frenkel defect occurs above 1424 K [7].

The vacancies of CaF$_2$ are substituted by the O^{2-} ions or another anions of impurities. Another microphotograph is shown in Figure 5. Scattering particles are the former type and arranged in a line. These particles are thought to be collected along dislocations of crystal. These arrangements are occured because the spatial gaps exist near dislocations. Then the inclusions are thought to be gathered along dislocations. To investigate component of scattering particles and the mechanism of ordering scattering particles are the next subject in the future.

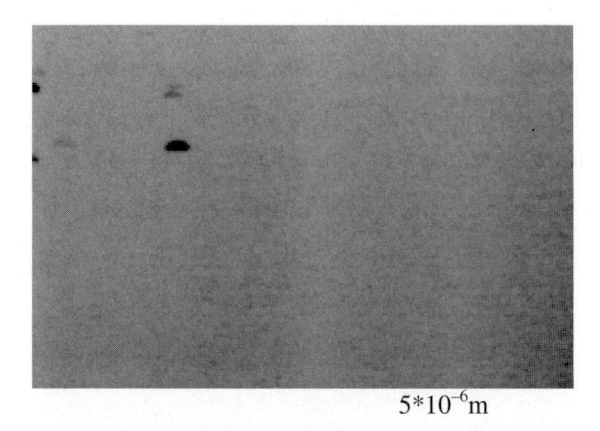

Fig. 3 Light scattering topography of CaF$_2$ crystal

Fig. 4 Light scattering topography of CaF$_2$ crystal.

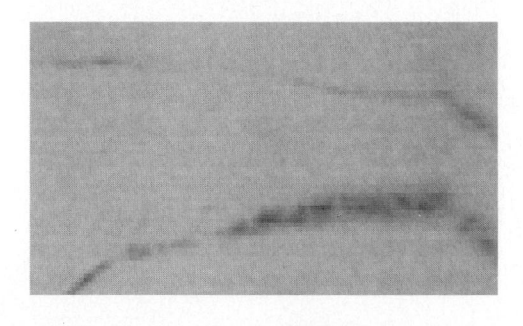

Fig. 5 Light scattering topography of CaF$_2$ crystal.

4 Summary The improvement of a light scattering topography have been performed and the light scattering of excimer grade CaF_2 have been observed. Two types of particles are shown. The former particles are about $1*10^{-6}$–$5*10^{-6}$ m and exist independently or arranged in a line because they are gathered along the dislocation of crystal and the latters are less than $5*10^{-6}$ m and curtain-like particles. The direction of their arrangements are thought to be related to the direction of the crystal structure.

References

[1] S. Jingu, X. Chen, S. Nishimura, Y. Oyama, and K. Terashima, J. Cryst. Growth **237–239**, 1797 (2002).
[2] X. Chen, S. Jingu, S. Nishimura, Y. Oyama, and K. Terashima, J. Cryst. Growth **240**, 445 (2002).
[3] I. N. Guseva, J. Cryst. Growth **3/4**, 723 (2002).
[4] V. Vedam, R. Stein, J. Appl. Phys. **37**, 2551 (1966).
[5] M. Tajima and T. Iizuka, Jpn. J. Appl. Phys. **15**, 651 (1976).
[6] C. C. Desai, Cryst. Res. Technol. **16**, 437 (1981).
[7] W. Hayes, Contemp. Phys. **27**, 519 (1986).
[8] B. F. Naylor, Contemp. Phys. **27**, 519 (1986).

Effect of deposition conditions and annealing on residual stress of ITO films magnetron sputtered on silica

A. Užupis[*,1], **S. Tamulevičius**[2], **R. Butkutė**[3], **B. Vengalis**[3], and **V. Lisauskas**[3]

[1] Lithuanian University of Agriculture, Faculty of Water and Land Management, Department of Physics, Universiteto 10, Akademija, LT-4324 Kaunas, Lithuania
[2] Institute of Physical Electronics, Kaunas University of Technology, Savanorių 271, LT-3009 Kaunas, Lithuania
[3] Semiconductor Physics Institute, Goštauto 11, LT-2600 Vilnius, Lithuania

Received 15 September 2003, revised 28 October 2003, accepted 28 October 2003
Published online 22 December 2003

PACS 61.10.Nz, 68.35.Gy, 68.37.Ps, 68.60.Bs, 72.80.Sk

Thin films of indium-tin oxide (ITO) (d ~ 1000 nm) were deposited on heated (T_d = 20–500 °C) amorphous quartz substrates by reactive DC-magnetron sputtering followed by post-deposition annealing at T ≤ 750 °C in oxygen and vacuum. The films were deposited under Ar:O_2 (1:1) pressure of about 1–3 Pa. Crystalline structure and surface quality of the films were studied by X-ray diffraction, scanning electron microscopy and atomic force microscopy. Optical interferometry was used to investigate residual stress in the films prepared under various deposition and annealing conditions. Influence of oxygen content and structural quality of the films on the residual stress have been defined.

© 2004 WILEY-VCH Verlag GmbH & Co. KGaA, Weinheim

1 Introduction Indium-tin oxide films exhibiting high electric conductivity and high transparency in the visible spectra region are widely used as transparent electrodes in flat panel displays and various other electroluminescent devices. ITO thin films of various crystalline quality are routinely prepared on various crystalline and amorphous substrates by a number of deposition techniques [1–4]. There are indications that most of the films grown at high temperatures (above 300 °C) exhibit residual stress depending on both substrate material and deposition conditions [5].

It is known that residual stress arising from thermal expansion coefficient mismatch and variety of other effects such as contamination, defects, solid state transformation etc. can influence electrical, optical, or magnetic properties of thin films. In addition, residual stress may be manifested by cracking or peeling of the film and bending the substrate. Geometrical factor becomes of high importance in the large area applications: because of the curvature of formed thin film-substrate structure the following technological processing steps like patterning or etching can be impossible.

This paper concentrates on the optimization of both deposition conditions and post-deposition annealing of the ITO films grown by DC magnetron sputtering onto silica substrates. We were focussing on the film properties demonstrating low resistivity, high optical transmittance and minimal residual stress. The influence of the different process parameters such as deposition temperature, oxygen flow rate, annealing temperature and oxygen pressure on electrical, mechanical and optical properties of the prepared ITO films was studied.

2 Experimental The ITO films were deposited on polished silica substrates with typical dimensions 60 × 40 × 0,27 mm using a reactive DC magnetron sputtering. The target was a disk of 100 mm in diameter prepared from the In–Sn (0,91:0,09) alloy. In order to remove possible contaminants, the target

---

[*] Corresponding author: e-mail: arnuzu@ktu.lt, Tel.: + 370 37 752363, Fax: +370 37 397492

was pre-sputtered for 10 min before deposition. The sputtering was performed at a total Ar:O_2 (1:1) pressure of about 1–3 Pa. The target-to-substrate distance was 40 mm. The substrate temperature, measured by a thermocouple, was set in the range 100 ÷ 750 °C. Thickness of all prepared ITO films was ~1000 nm. The films grown at elevated temperatures were cooled down to room temperature at the same oxygen pressure conditions. Part of the films was annealed additionally at different temperatures (T ≤ 750 °C) for 1 h, in vacuum (p(O_2) ~ 10^{-3} Pa), air or oxygen ambient (p(O_2) ~ 10^5 Pa).

Crystalline structure of the prepared ITO films was studied by X-ray diffraction (XRD) using a diffractometer operating with CuK_α radiation in a conventional $\Theta - 2\Theta$ set-up. The mean grain size for the films was estimated from the XRD data applying Scherrer's formula [6]. Scanning electron microscope (SEM) [Jeol JSM-IC25S] and Quesant atomic force microscopy (AFM) [Quesant Q-Scope 250] were employed to evaluate the surface morphology of the films. A triangular Si cantilever with 10° conical tip and a force constant of ~0,16 N/m was used, in contact mode.

The resistance of the grown films was measured by applying a conventional four-point probe method. Carrier density was determined from the Hall effect measurements.

Optical transmittance, $T(\lambda)$, and reflectance, $R(\lambda)$, spectra were measured at room temperature in the spectral range of 300–3000 nm. The plasma frequency of carriers $\omega_p = (4\pi e^2 N/m_\sigma \varepsilon_\infty)^{1/2}$ (here N and m_σ are carrier density and effective mass, respectively and ε_∞ is the high frequency dielectric constant of the material) as well as film thickness t_f were estimated by numerical fitting of $T(\lambda)$ and $R(\lambda)$ spectra to the corresponding experimental curves.

The in-plane residual stress for the films was calculated according to Stoney's formula [7]. The radii of the substrate were defined using a laser interferometer, which has been described earlier [8].

3 Results and discussion XRD measurements revealed various crystalline qualities of the ITO films deposited and annealed at various temperatures. The films deposited at room temperature were amorphous although they showed preferential (100) texture after post-deposition annealing. At the same time, the films grown on heated substrates showed existence of both (110) and (100) textures. The relative intensity of the (110) reflection was found to increase gradually with the annealing temperature. This is in agreement with similar results reported earlier by Vossen [9] and Lampert et al. [10].

The average crystallite size as found from the XRD data increased from 13 nm to 31 nm with the deposition temperature increase from room temperature up to 550° $\Psi >$ XRD data of ITO films annealed in vacuum and oxygen are displayed in Table 1. Fig. 1 shows AFM images (1 μm × 1 μm of two different ITO films. Following the figure, it can be seen that ITO films deposited at higher temperature demonstrate lower surface roughness compared to those sputtered at room temperature.

Fig. 2 shows temperature-dependent resistivity of ITO films deposited at various temperatures.

The decrease of film resistivity from 8×10^{-2} Ωcm to about 5.5×10^{-3} Ωcm with T_d increase up to 750 °C may, probably, be understood assuming improved crystallinity leading to lower point defect density as well as lower density of intergrain boundaries in the films synthesized at higher temperatures.

Table 1 Dominating XRD patterns and grain size of ITO films deposited at T_d = 20 and 550 °C and annealed additionally at various temperatures (T_a).

1-hour annealing at T_a (°C)	T_d = 20 °C, annealed in O_2 (p ~ 10^5 Pa)		T_d = 550 °C, annealed in vacuum (p ~ 10^{-3} Pa)	
	Grain size (nm)	Reflection hkl	Grain size (nm)	Reflection hkl
Not annealed	13	440	30	400
T_a = 300 °C	14	440	31	400
T_a = 550 °C	15	440	32	400
T_a = 750 °C	17	440	35	400

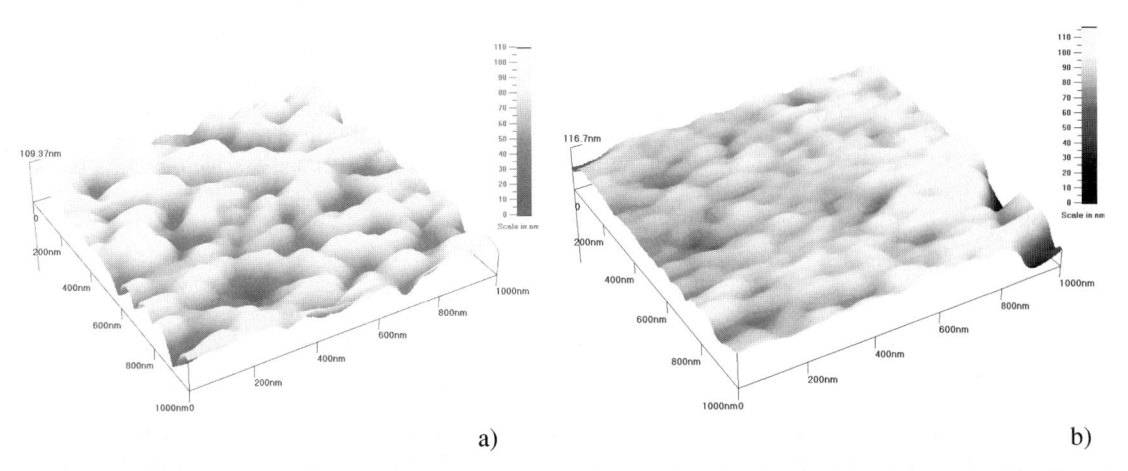

Fig. 1 AFM images (1 μm × 1 μm of ITO films (d = 1000 nm) deposited by dc sputtering at: a) –550 °C and annealed in vacuum at 550 °C; b) – room temperature and annealed in vacuum at 625 °C.

Fig. 3 shows the effect of annealing in vacuum on the resistivity of the ITO films. Significant resistivity decrease indicated for the films after their annealing in vacuum may be understood taking into account increased concentration of oxygen vacancies and hence density of carriers in the films investigated. It should be noted, however, that oxygen diffusion into ITO film from either oxygen atmosphere or air may only occur at higher temperatures (T ≥ 200 °C [11]).

The measured T(λ) and R(λ) spectra revealed clearly defined plasma anomalies in the near IR region for all the films investigated. Typical values of the characteristic plasma wavelength $\lambda_p = 2\pi c/\omega_p$ and optical gap of 2–3 μm and about 3.5 eV, respectively, have been indicated for the films annealed in oxygen. Significantly lower λ_p values (0.8–1.5 μm) and wider optical gap (up to about 4.2 eV) indicated for the films annealed in vacuum are caused mainly by carrier density increase. Due to increased free carrier absorption, all vacuum-annealed ITO films demonstrated slightly lower optical transmittance in the visible spectral region compared to similar films annealed in oxygen. It is also worth noting that the films deposited at room temperature showed slightly lower transmittance and significantly wider absorption edge compared to similar films grown at higher temperatures.

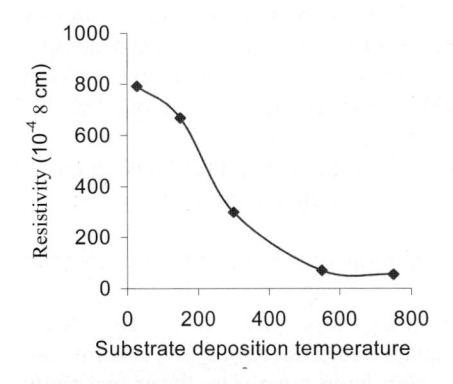

Fig. 2 Dependence of resistivity of the ITO films as a function of substrate deposition temperature deposited in a total Ar:O$_2$ (1:1) pressure of about 1–3 Pa. The film thickness was ~1000 nm for all films.

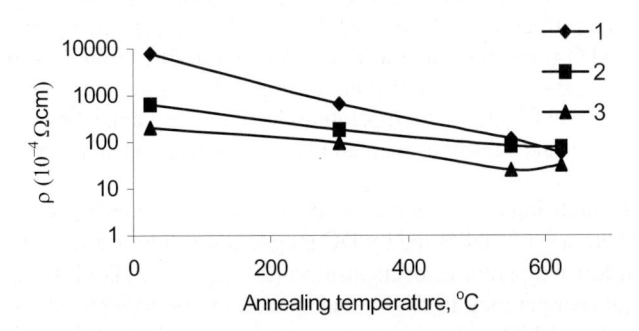

Fig. 3 Dependence of resistivity as a function of annealing temperature (in vacuum (p(O$_2$) ~ 10^{-3} Pa) for the ITO films deposited (in a total Ar:O$_2$ (1:1) pressure of about 1-3 Pa) at temperatures: 1 – Room temperature, 2 – 300 °C, 3 – 550 °C.

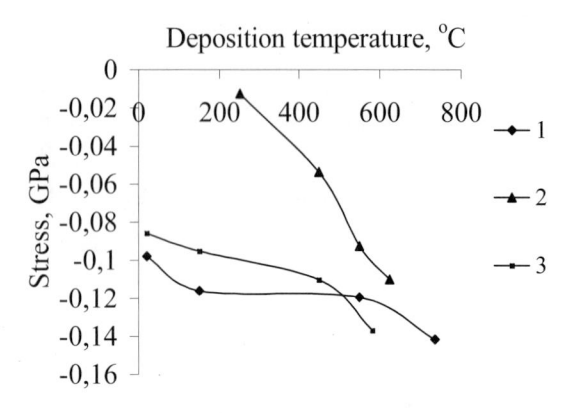

Fig. 4 The effect of substrate temperature on the residual stress measured for the as deposited ITO films (1), and those annealed for 1 h at 300 °C in O_2, $p(O_2) \sim 5 \cdot 10^4$ Pa (2) and vacuum, $p(O_2) \sim 10^{-3}$ Pa (3).

The residual stress of the ITO films is displayed in Fig. 4. Following the figure it can be seen that the residual stress in the ITO film-quartz structure is compressive. Furthermore it increases with the deposition temperature and is dependent on the annealing conditions (curves 1–3).

The stress doesn't depend strongly on deposition temperature up to 250 °C. However, there is a large increase in compressive stress in the ITO films between the deposition temperatures of 250 and 600 °C. It can be seen that one can minimize these stresses by subjecting the film-substrate structure to a post-deposition annealing treatment (Fig. 4). The effect of annealing on residual stress is more effective for low deposition temperatures. As long as the coefficients of thermal expansion of the substrate and the film and the elastic constants of the film do not change significantly in the temperature range of interest, the thermal stress should vary almost linearly with deposition temperature. The thermal stress in the room temperature deposited ITO thin films is obviously zero [12]. One can expect that variation of compressive stresses in ITO films at high temperatures may occur due to incorporation of oxygen atoms into the lattice. According to our results, both the components, i.e. the grain size and oxygen incorporation contribute to the creation of tensile stress.

Dependence of stress in the prepared ITO films on grain size can be explained using the grain boundary model suggesting that tensile stresses are built up due to the constrained volume contraction. The volume contraction may occur when two grains coalesce to form a single grain, thereby eliminating a grain boundary [12] (the difference between the grain boundary thickness and the interatomic spacing is equal to the amount of shrinkage).

The magnitude of stress in the as-deposited films correlated strongly to the microstructure of the films. Certainly, the films deposited at high temperatures (~300 °C), were found to be crystalline with a relatively large grain size. Thus, one can conclude that large grain size supports high intrinsic stresses (~ –0.12 GPa). Meanwhile the films deposited at low temperature with fine-grained and less densely packed structure showed smaller intrinsic stresses. We point out large increase in compressive stress in the ITO films deposited at temperatures ranging from 250 to 600 °C although it has been found in this work that one can minimize these stresses by post-deposition annealing of the films in air. Thus, combining deposition at different temperatures with varying annealing conditions one can obtain ITO films on amorphous quartz substrates with good electrical and optical properties as well as minimized residual stress.

4 Conclusions Correlation between electrical, mechanical, optical properties and thermal treatment of ITO thin films deposited by DC magnetron sputtering was defined.

At low temperatures of deposition (20–200 °C), ITO films showed high resistivity (~6–8 × 10^{-2} Ωcm), high transparency in the visible spectra region (≥90%) and dominant compressive stress. The films deposited at higher temperatures were found to be polycrystalline with large grain size, lower resistivity (~1–3 × 10^{-2} Ωcm) and transmission in the visible region up to 85%.

Kinetics of stress versus annealing in different ambient are caused either by oxygen content variation and growth of the grain size. ITO thin films with optimized optical, electrical properties and minimized stresses can be fabricated for use as transparent electrodes by optimizing either deposition temperature or post-deposition annealing conditions.

References

[1] M. Mizuhashi, Thin Solid Films **76**, 97 (1981).
[2] J. C. Jin, I. Hamberg, and C. G. Grangvist, Appl. Phys. **57**, 149 (1987).
[3] J. Kane, H. P. Schweizer, and W. Kern, J. Electrochem. Soc. **123**, 273 (1976).
[4] E. Shanthi, A. Bannerjee, V. Dutta, and K. L. Chopra, Thin Solid Films **71**, 237 (1980).
[5] S. Bhagwat and R. P. Howson, Surface and Coatings Technol. **111**, 163 (1999).
[6] D. Mardare, M. Tasca, M. Delibas, and G. I. Rusu, Appl. Surf. Sci. **156**, 200 (2000).
[7] G. G. Stoney, Proc. R. Soc. Lond. A **82**, 172 (1909).
[8] S. Tamulevičius, Vacuum **51**, No. 2, 127 (1998).
[9] J. L.Vossen, RCA Rev. **32**, 289 (1971).
[10] C. M. Lampert, Optical materials technology for energy efficiency and solar energy conversion XII, SPIE, 2017, 1993.
[11] S. K.Park, J. I. Han, W. K. Kim, and M. G. Kwak, Thin Solid Films **397**, 49 (2001).
[12] S. B. Desu, D. P. Vijay, S. Ramanathan, H. D. Bhatt, and S. Tirumala, Thin Solid Films **350**, 21 (1999).

Thermoluminescence of doped YAlO₃ crystals

Ya. Zhydachevskii[*,1,2], **A. Durygin**[2], **A. Suchocki**[2], **A. Matkovskii**[1,3], **D. Sugak**[1], and **Z. Frukacz**[4]

[1] Lviv Polytechnic National University, Bandera 12, 79646 Lviv, Ukraine
[2] Institute of Physics, Polish Academy of Sciences, Al. Lotnikow 32/46, 02668 Warsaw, Poland
[3] Institute of Physics, University of Rzeszow, Rejtana 16A, 35310 Rzeszow, Poland
[4] Institute of Electronic Materials Technology, Wolczynska 133, 01919 Warsaw, Poland

Received 15 September 2003, revised 10 October 2003, accepted 10 October 2003
Published online 22 December 2003

PACS 61.72.Ji, 61.80.Ba, 78.60.Kn

The paper presents results of thermally stimulated luminescence (TSL) studies of YAlO₃ crystals both pure and doped with rare-earth (Ho^{3+}, Tm^{3+}, Er^{3+}) and Mn ions performed in the 20–600 K temperature range. The activation energy and frequency factor of traps have been determined. Nature of the traps and role of dopant ions are discussed.

© 2004 WILEY-VCH Verlag GmbH & Co. KGaA, Weinheim

1 Introduction

YAlO₃ (YAP) single crystals doped with rare-earth ions (REI) are attractive materials for active elements for solid-state lasers [1]. Recently it was shown that Mn-doped YAP crystals have a high application potential for optical data storage [2] and laser light optical limiters [3].

The presence of lattice point defects, including color centers, strongly influences the properties of YAP crystals. The color centers created in YAP crystals during growth process as well as under various external influences (UV and γ-irradiation, thermal treatments) have been studied in details by means of optical spectroscopy [4–7]. As far as we know, substantial thermally stimulated luminescence studies of YAP crystals are absent except the studies of Ce-doped YAP crystals (see e.g. Ref. [8–10]). These TSL studies have been performed from the point of view of influence of intrinsic point defects on scintillation properties of YAP:Ce crystals.

In the present work we report results of TSL measurements of YAP crystals both nominally pure and doped with various dopants (Ho^{3+}, Tm^{3+}, Er^{3+}, Mn). The measurements were performed in the 20–600 K temperature range after UV light irradiation of the crystals.

2 Experimental samples and techniques

The following crystals were studied in the present work: nominally pure YAP, YAP:Er (Y$_{0.5}$Er$_{0.5}$AlO$_3$), YAP:Ho (0.1 at.%, 0.3 at.%, 5 at.%) grown by the Czochralski technique in iridium crucibles under nitrogen atmosphere in Institute of Electronic Materials Technology, as well as YAP:Tm (4 at.%) and YAP:Mn (0.05 at.%) grown by the Czochralski technique under conditions described in [11,12] and [2] respectively.

The TSL studies of the crystals were performed separately in the 20–320 K and 300–600 K ranges with the 0.1 K/s heating rate. For this purpose crystals were preliminary exposed with Ar$^+$-laser light (351 nm) or Xe-lamp (UV part of spectrum) at temperature of 20 K (for the 20–320 K experiment) or

---

[*] Corresponding author: e-mail: crystal@polynet.lviv.ua, Phone: +38 0322 398153, Fax: +38 0322 742164

© 2004 WILEY-VCH Verlag GmbH & Co. KGaA, Weinheim

300 K (for the 300–600 K experiment). In the first case (20–320 K) a Leybold cryogenator with LTC60 temperature controller was used. In the second case (300–600 K) a specially designed compact furnace with *T*-type thermocouple connected with Keithley 2000 multimeter was used.

The TSL glow intensity was measured using one of the two apparatus. The first one was used for detecting of relatively low intensity of TSL emission. It was a Hamamatsu cooled photomultiplier with S-20 photocathode. In the second case Jobin Yvon-Spex Triax 320 monochromator with a CCD camera was used for registration of relatively high intensive TSL emission. Application of the monochromator with CCD camera allowed also to obtain spectra of TSL emission in the 350–900 nm region during the TSL experiments as well as photoluminescence spectra under continuous excitation of samples.

The temperature controller and the monochromator with CCD camera were interfaced by IEEE 488 (GPIB) to a PC computer where the experimental data were processed and recorded.

3 Experimental results and discussion

Our measurements of TSL glow curves in the 20–320 K temperature range show the presence of three commensurable and well separated glow peaks with maxima near 175 K, 210 K and 255 K that dominate for all the crystals studied (Fig.1). In the case of pure YAP and YAP:Ho crystals the peak with maximum near 210 K consists of two neighboring peaks at 195 K and 205 K. In general, four glow peaks with maxima near 175 K, 195 K, 205 K and 255 K (marked as C, D, E and F respectively) were observed. In some crystals two (or maybe more) feebly pronounced glow peaks (marked as A and B) were observed at temperature lower than 140 K. Intensity of these peaks in pure YAP and YAP:Ho crystals was two orders of magnitude lower than intensity of C, D, E and F peaks (see Fig. 1). In YAP:Tm crystal, for which the overall TSL intensity was very low, the A and B peaks were commensurable with C, D/E and F peaks. Besides, the glow peak at 320 K marked as G was observed in some crystals. Intensity of G peak in pure YAP and YAP:Ho crystals was one order of magnitude lower than intensity of C, D, E and F peaks, whereas in YAP:Mn crystal G peak was dominant.

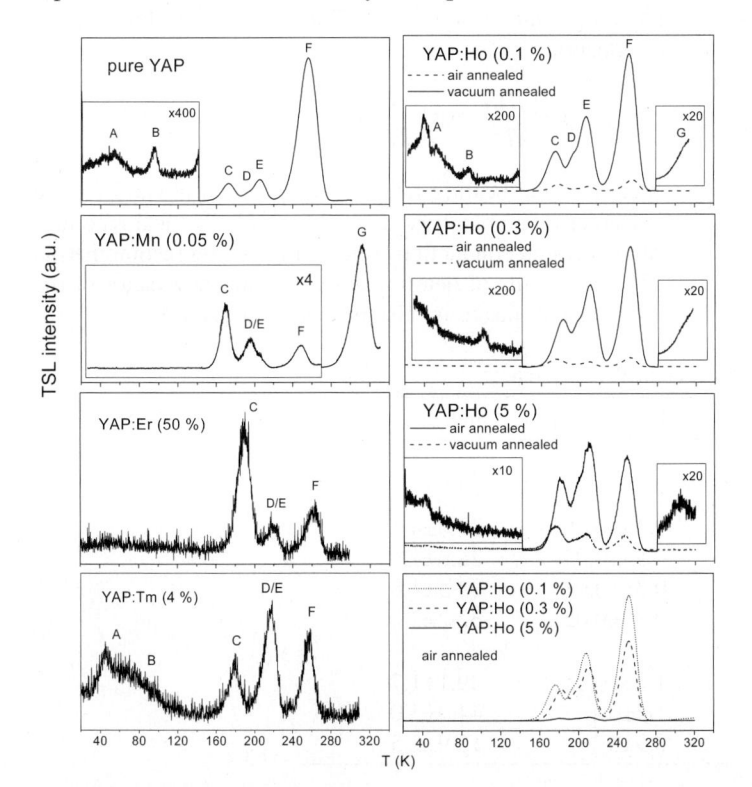

Intensities of TSL emission strongly differ one from another for various crystals studied. The highest TSL intensity was observed for YAP:Mn, pure YAP and YAP:Ho crystals. The TSL intensities of YAP:Tm and YAP:Er crystals were much lower.

The TSL intensity registered at the same experimental conditions for the crystals preliminary annealed in reducing atmosphere (hydrogen or vacuum, T=1300 K) is about one order of magnitude lower than the TSL intensity of as-

Fig. 1 The TSL glow curves of various YAP crystals after UV light irradiation at 20 K.

grown crystals or crystals annealed in air (T=1300 K) (see Fig. 1). This observation we explain by the mean that the reducing annealing decreases absorption of the crystals in UV region (see e.g. Ref. [8]). In other words, such annealing decreases a concentration of point defects, which can release charge carriers under excitation.

The influence of content of rare-earth dopant on TSL glow has been studied on YAP:Ho crystals with different concentrations of Ho ions. Comparison of TSL glow curves of these crystals recorded at the same experimental conditions reveals the following regularities. The total intensity of TSL decreases with increasing of concentration of the dopant ions. Besides, the correlation of intensities of different TSL peaks changes, namely, the intensity of F peak decreases with respect to intensities of other (C, D and E) peaks (see Fig. 1).

Recording of TSL glow curves in the in the 300–600 K temperature range was successful for nominally pure YAP and YAP:Mn crystals (Fig. 2), which have the highest TSL intensity in visible in the low temperature (20–320 K) range. Besides G peak at 320 K, the TSL glow peaks with maxima near 360 K and 400 K (marked as K and L respectively) were observed in the pure YAP and YAP:Mn crystals. Intensity of G peak in the pure YAP crystal was about five times lower than the intensity of L peak. Intensity of G peak in the YAP:Mn crystal was about three orders of magnitude lower than intensity of K and L peaks, therefore G peak is unnoticeable in the 300–600 K experiment conditions for this crystal. Besides, the glow peak at 500 K marked as M was observed in the YAP:Mn crystal.

In order to determine the activation energy of traps (E_T), we used the initial rise method. To separate the TSL glow peaks the partial heating procedure was used [13]. This procedure was performed for YAP:Ho (0.1%) crystal in the 20–320 K range and for YAP:Mn crystal (for details see Ref. [14]) in the 300–600 K range. The summarized results are presented in Table 1.

Analysis of shape of TSL peaks testifies the first-order recombination kinetics for all the TSL peaks observed. The frequency factor s of a certain TSL peak was evaluated from the equation:

$$s = \frac{\beta E_T}{kT_{max}^2} \exp\left(\frac{E_T}{kT_{max}}\right), \quad (1)$$

where β is the heating rate. This expression is valid for the first-order recombination kinetics for both classical recombination through conduction band and tunneling recombination [15]. The results of determination of frequency factor of the observed TSL peaks are also presented in Table 1.

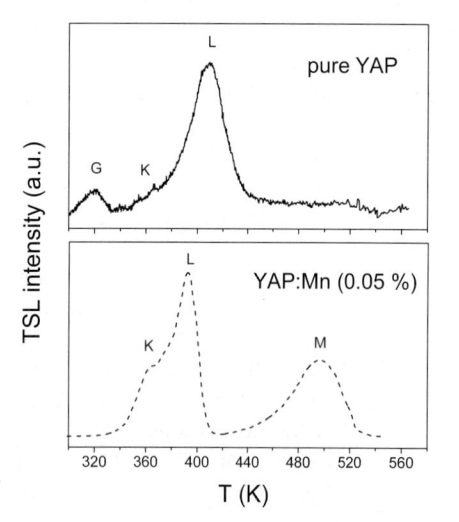

Fig. 2 The TSL glow curves of nominally pure YAP and YAP:Mn crystals after UV light irradiation at room temperature.

Table 1 Parameters of traps observed in YAP crystals.

TSL peak	T_{max}, K	E_T, eV	ln s
A	~50	–	–
B	~80	–	–
C	175	0.48±0.06	27.8±4.1
D	195	0.55±0.03	28.6±1.9
E	205	0.68±0.03	34.5±1.8
F	255	0.80±0.05	32.2±2.4
G	320	–	–
K	360	1.35±0.05	39.1±1.7
L	400	1.7±0.1	44.9±3.0
M	500	0.9±0.1	15.9±2.5

Fig. 3 Spectrum of TSL emission of YAP:Ho (0.1%) crystal recorded at 250 K after UV light irradiation at 20 K.

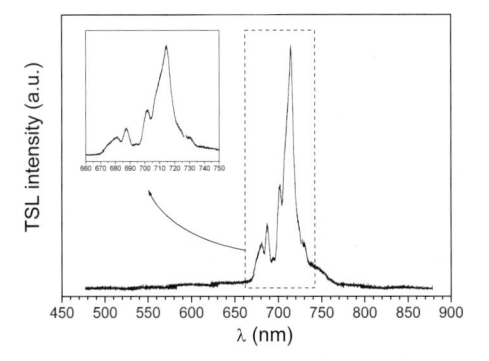

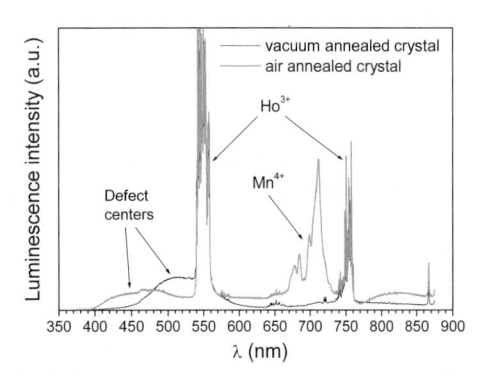

Fig. 4 Photoluminescence spectra of YAP:Ho (0.1%) crystal under continuous Ar$^+$-laser (λ=351 nm) excitation at 300 K.

Here it should be noted, that a number of traps with the depth of 0.12–0.40 eV that correspond to TSL glow peaks at T≤150 K as well as a trap with the depth of 0.5 eV (T_{max}~154 K) were identified in YAP:Ce crystals in Ref. 10. The value of 0.5 eV is very close to the activation energy determined by us for C peak. Two dominant TSL peaks at 400 K and 470 K (E_T=1.04 eV and 1.36 eV respectively) were observed at elevated temperature in YAP:Ce at 1 K/s heating rate [8]. The activation energy of 1.36 eV is very close to the energy determined by us for K peak.

As it was shown earlier [7], the transient optical absorption with maximum near 15000 cm^{-1} induced in YAP crystals by UV excitation is caused by one type centers (E_T=0.75 eV) in pure YAP crystals and two types centers (E_T=0.11 eV and E_T=0.75 eV) in Nd-doped YAP crystals. The value of E_T=0.75 eV agrees with the trap depth determined for F peak, which is present in all the crystals studied and dominates in pure YAP crystal. This fact allows us to correspond this TSL peak with the transient optical absorption at 15000 cm^{-1} described in [7].

Measuring of spectra of TSL emission in the 350–900 nm region performed for C, D/E and F glow peaks for pure YAP, YAP:Ho and YAP:Mn crystals as well as for K, L and M glow peaks for YAP:Mn crystal showed that the TSL emission of these crystals corresponds to the characteristic luminescence of Mn^{4+} ions (transition $^2E \rightarrow ^4A_2$) (see Fig. 3). Obviously manganese ions are present in the nominally pure YAP and YAP:Ho crystals as uncontrolled impurity and play role of glow centers and most probably of recombination centers during TSL experiments.

For comparison, the photoluminescence measurements at continuous excitation by UV light were performed on these crystals. In such a way the presence of Mn^{4+} impurity ions in pure YAP and YAP:Ho crystals was confirmed (Fig. 4). Moreover, the luminescence of Mn^{4+} ions in these crystals is much lower or fully absent for the crystals preliminary annealed in reducing atmosphere. This testifies the decreasing of concentration of Mn^{4+} ions in the reduced crystals, possibly by the Mn^{4+}+$e^- \rightarrow$Mn^{3+} recharging scheme.

The presence of manganese impurity ions in pure YAP and YAP:Ho crystals most probably is a reason of relatively high TSL intensity of these crystals in comparison with other crystals studied. Note, that our photoluminescence studies did not revealed the presence of manganese ions in YAP:Tm and YAP:Er crystals. In the crystals that do not contain manganese ions, the TSL emission can take place via rare-earth ions due to energy transfer processes between point defects and rare-earth ions that was observed in YAP:Tm crystals [12].

The fact that concentration of Mn^{4+} ions is strongly reduced in pure YAP and YAP:Ho crystals preliminary annealed in reducing atmosphere can be an additional explanation of the much lower TSL intensity of these crystals in comparison with the same crystals annealed in oxidizing atmosphere.

4 Conclusions

The performed study allowed us to distinguish several TSL glow peaks in the 20–600 K temperature range that are characteristic for YAP crystals after UV light irradiation. The activation energy and frequency factor of the traps were determined.

The correlation between traps, which are apparent in TSL, and color centers have been established. In particular, it was shown that the transient optical absorption of YAP crystals near 15000 cm^{-1} correlates with the low-temperature TSL glow peak at 255 K (E_T=0.80±0.05 eV).

It was revealed that Mn^{4+} ions present in YAP crystals both as intentional dopant and uncontrolled impurity are effective glow centers in TSL experiments.

Acknowledgements The work was supported by the Ukrainian Ministry of Education and Science (project No. 0100U000498, Acronym: Ion). The support by G.B. Loutts in providing the YAP:Mn crystals is gratefully acknowledged. Ya. Zhydachevskii gratefully acknowledge the personal studentship of Polish Ministry of Education.

References

[1] A.A. Kaminskii, Crystalline Lasers: Physical Processes and Operating Scheme (Boca Raton, 1996).
[2] G.B. Loutts, M. Warren, L. Taylor, R.R. Rakhimov, H.R. Ries, G. Miller, M.A. Noginov, M. Curley, N. Noginova, N. Kukhtarev, H.J. Caulfield, and P. Venkateswarlu, Phys. Rev. B **57**, 3706 (1998).
[3] R.L. Hutcheson, and R.W. Equall, Laser light optical limiter (US Patent 6,243,219 B1. Jun. 5, 2001).
[4] J. Kvapil, B. Perner, M. Koselja, and Jos. Kvapil, Czech. J. Phys. B **40**, 99 (1990).
[5] A.I. Riabov, G.N. Pirogova, V.E. Kritskaja, N.S. Stelmakh, V.M. Sorokin, G.A. Yermakov, and V.A. Akkerman, Neorganicheskie Materialy **28**, 178 (1992) (in Russian).
[6] V. Akkerman, G. Ermakov, and V. Ljubchenko, Proceedings of XII International Conference on Defects in Insulating Materials **1**, 489 (1992).
[7] A. Matkovski, A. Durygin, A. Suchocki, D. Sugak, G. Neuroth, F. Wallrafen, V. Grabovski, and I. Solski, Optical Mater. **12**, 75 (1999).
[8] A.J. Wojtowicz, J. Glodo, W. Drozdowski, and K.R. Przegietka, J. Lumin. **79**, 275 (1998).
[9] J. Glodo, and A.J. Wojtowicz, J. Alloys Comp. **300–301**, 289 (2000).
[10] A. Vedda, M. Martini, F. Meinardi, J. Chval, M. Dusek, J.A. Mares, E. Mihokova, and M. Nikl, Phys. Rev. B **61**, 8081 (2000).
[11] D.I. Savytskii, L.O. Vasylechko, A.O. Matkovskii, I.M. Solskii, A. Suchocki, D.Yu. Sugak, and F. Wallrafen, J. Cryst. Growth **209**, 874 (2000).
[12] D. Sugak, A. Durygin, A. Matkovskii, A. Suchocki, I. Solskii, D. Savitskii, Y. Zhydachevskii, F. Wallrafen, and K. Kopczynski, Cryst. Res. Technol. **36**, 1223 (2001).
[13] S.W.S McKeever, Thermoluminescence of Solids, Cambridge Solid State Science (Cambridge University Press, Cambridge, U.K., 1985).
[14] Ya. Zhydachevskii, A. Durygin, A. Suchocki, A. Matkovskii, D. Sugak, G.B. Loutts, and M.A. Noginov, J. Lumin. (to be published).
[15] I.F. Chang, and P. Thioulouse, J. Appl. Phys. **53**, 5873 (1982).

Schottky barrier and ageing effect studies in Au(Cu)/p-CdTe contacts

Ye. O. Bilevych[*]**, A. V. Sukach, and V. V. Tetyorkin**

Institute of Semiconductor Physics of NAS of Ukraine, 45, pr. Nauki, 03028 Kyiv, Ukraine

Received 15 September 2003, accepted 2 October 2003
Published online 15 January 2004

PACS 73.30.+y, 73.40.Sx, 78.66.Hf, 81.65.–b

The Schottky barrier formation are investigated in Au(Cu)/p-CdTe contacts prepared by electroless deposition of metals on chemically etched surfaces. The Schottky barrier height are studied in freshly prepared contacts as well as stored under normal laboratory condition during one-year period. The potential barrier height is determined from the photoemission current spectra measurements. The secondary ion-mass spectroscopy (SIMS) profiling has been carried out to study the compositional structure in the contacts. The effect of thermal annealing on electrical and photoelectrical properties of the contacts was studied.

© 2004 WILEY-VCH Verlag GmbH & Co. KGaA, Weinheim

1 Introduction

Preparation of both ohmic and blocking contacts is of great importance for manufacture of X-ray and γ-ray detectors on CdTe single crystals. In order to utilize detectors with high performance reliable contacts with predictable characteristics are desired. However, preparation of such contacts on high-resistance crystals of CdTe is still a problem under investigation. It is complicated by intrinsic properties of CdTe such as low temperature conductivity, poor mechanical characteristics and unstable surface. In this study we report the results of an investigation of ageing effects in Au and Cu contacts made on chemically etched surfaces of p-CdTe single crystals. Ageing effects in metal contacts to n-CdTe were investigated in [1, 2]. Similar investigations in contacts to p-CdTe has not been performed yet. The formation of the Schottky barriers in p-CdTe is not known exactly. So far the contacts prepared on vacuum-cleaved and chemically etched in Br$_2$-methanol solution surfaces were mainly investigated [4–9].

2 Experimental results and discussion

The starting materials were single crystals of (100) CdTe of p-type conductivity grown by Bridgman technique. The hole concentration $p = (8–10) \cdot 10^{15} \cdot cm^{-3}$ and mobility of the order of 100 cm^2/V · s were measured at room temperatures. In order to study the role of chemical etching several etchants with different composition were prepared, Table 1. Au and Cu layers were electroless deposited into the chemically etched surfaces of the CdTe. Their thicknesses ranged from 500 to 600 nm for electrical and SIMS profile measurements. Photoelectrical measurements were performed on contacts with semitransparent metallic layers with thickness of approximately 50 nm. Experimental measurements, including SIMS profiles, were performed within 48 hours of the samples being chemically etched. The SIMS profile measurements were taken with IMS-4F ("Cameca"). All measurements were also repeated in the one-year-old contacts in order to investigate ageing effects. The contacts were stored under normal laboratory conditions.

---

[*] Corresponding author: e-mail: bilevych@usp.kiev.ua, Phone: +38 044 265 5755, Fax: +38 044 265 8342

Table 1 Some characteristics of as-deposited Au/*p*-CdTe contacts.

Etchant number	Etchant composition	Barrier height, φ_b^p, eV	$E_g - \varphi_b^p$, eV
1	HNO$_3$-HCl-citric acid	(0.06–0.20) ± 0.025	(1.30–1.44)±0.025
2	HNO$_3$-HCl-tartaric acid	0.30 ± 0.025	1.20±0.025
3	HNO$_3$-HJ-tartaric acid	0.45 ± 0.025	1.05±0.025
4	Br$_2$-methanol	0.50 ± 0.025	1.00±0.025
5	HNO$_3$-HBr-tartaric acid-washing	0.60 ± 0.025	0.90±0.025
6	HNO$_3$-HBr-tartaric acid	(0.70–0.90)±0.025	(0.60–0.80)±0.025

The barrier height values measured from the valence band maximum (VBM) in as-deposited contacts are listed in Table 1. It was found that each etchant introduces distinct change in the barrier height φ_B. The barrier heights obtained for Au and Cu contacts were approximately the same. It means that barrier height does not depend on the metal identity for the metals used. The contact prepared on the surface etched in HNO$_3$-HCl-citric acid has the lowest values of the barrier height. It has been used as an ohmic contact in electrical and photoelectrical measurements. Ageing of the contacts stored under normal laboratory conditions results in pronounced changes in the barrier height. The barrier height in the one-year-old contacts was found to range from ~0.55 eV to ~0.8 eV. It is increased in contacts with low values of φ_B (~0.4 eV) and remained practically unchanged in other contacts with high values of the barrier height.

The SIMS profiles of the constituents in the starting wafer before the metal deposition is shown in Fig. 1. The surfaces enriched with Te were obtained for all etchants. The higher barrier heights are observed for the surfaces more depleted with cadmium. The SIMS profiles of Au in as-deposited and one-year-old one contacts are shown in Fig. 2. The abrupt interfaces were not observed for both metals despite the fact that they were chemically deposited at room temperatures. Also, no interaction between the metal used and CdTe constituents has been observed in as-deposited contact. The distribution of the metals throughout the interface was found to be homogeneous and reaction zones were not observed. In the one-year-old contact the concentration of Cd and Te atoms at the interface is markedly increased in comparison with as-deposited contact and reaction zone is clearly seen, Fig. 2.

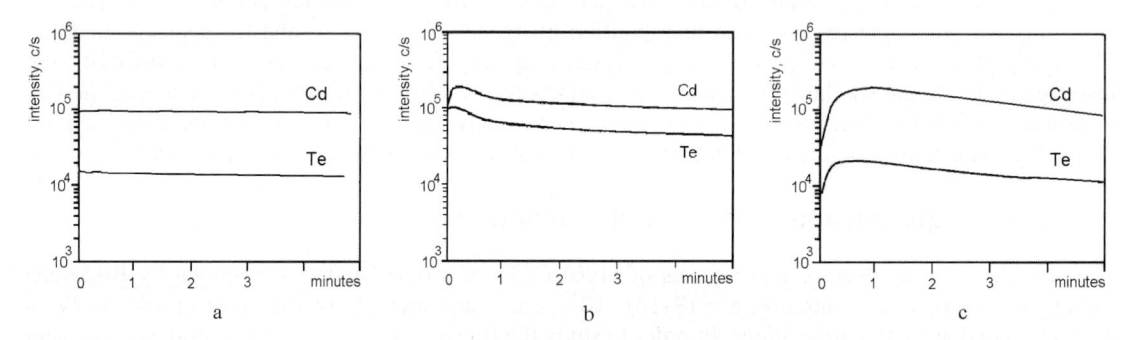

Fig. 1 SIMS profiles of cleaved a) and etched in etchant No. 1 b) and No. 6 c) surfaces.

The thermal annealing of contacts was performed in hydrogen atmosphere at 300 °C for 1 h. The contacts deposited on the surfaces etched in two solutions (1 and 6, Table 1) were annealed. The annealing has different effect on electrical and photoelectrical properties of contacts prepared on different surfaces. The contacts with the highest value of the barrier height were found to exhibit the nonlinear behavior of the current–voltage characteristics before and after annealing. The contacts with the lowest value of the barrier height had nonlinear characteristic before annealing and linear one after that. Also, the long wavelength Fowler's "tails" were not observed in annealed contacts for both etchants, Fig. 3. This experimental fact indicates that potential barriers at the interface may have different nature in unannealed and annealed contacts.

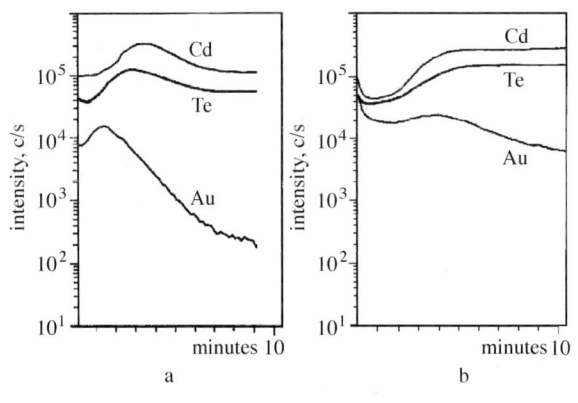

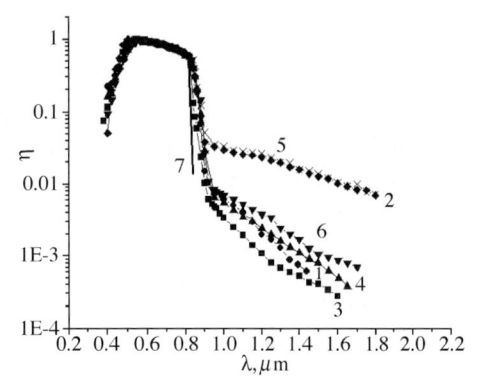

Fig. 2 SIMS profiles of as-deposited a) and one-year-old b) contacts for the surface etched in etchant No. 6.

Fig. 3 Photoemission current spectra in as-deposited (1–6, see Table 1) and annealed (7) Au/p-CdTe contacts.

The Fermi level pinning at the interface may be caused by i) native defects introduced during chemical etching [9], ii) interdiffusion and chemical reactions at the interface and iii) disorder induced gap states. Native defects may be introduced before and after the metal deposition procedure. No evidences have been obtained for chemical reactions between the metal used and semiconductor constituents in as-deposited contacts. The distribution of Au and CdTe constituents at the interface is homogeneous, Fig. 3. It is likely that in as-deposited contacts the Fermi level can be pinned by native defects (or their complexes) localized in the interfacial region. It is known that Cd vacancies and Te interstitial are regarded as the dominant native defects in CdTe. Unfortunately, results experimentally measured as well as theoretically calculated energies of native defects in CdTe have contradictory character [10]. Therefore, additional experimental and theoretical investigations of deep defect states in the gap are required to understand the Fermi level pinning in contacts on *p*-type CdTe.

The interdiffusion and chemical reactions can also influence physical properties of metal-semiconductor interfaces. The outdiffusion of Cd and Te leaves behind vacancies and probably interstitials. The indiffusion of Au or Cu can lead to i) introducing of additional defect states in the gap, ii) doping of interface region or iii) formation of new chemical composition different from the host semiconductor. Besides this, the new alloyed metal with work function different from the original metal can be formed. It is known that Au serves as acceptor dopant in CdTe and introduces deep acceptor levels at 0.3–0.4 eV above the VBM. Doping with Cu introduces defect levels at approximately the same energies above the VBM. However, simple doping effect can't explain the multiple pinning of Fermi level at the interface.

Unstable behavior of the barrier height during ageing as well as the broadening of the interface can in principle be explained by existence of chemical reactions in investigated systems. In this case experimental data are to be treated in terms of formation of new chemical species at the interface. Several possible reactions have to be examined in investigated contacts. The exchange reaction caused by interaction between Au and Te was analyzed by Brillson [11]. One stable compound $AuTe_2$ is known in Au–Te system [12]. There are also some metastable and high-pressure phases. Different $AuTe_2$ modifications were observed from Mossbauer spectra studies [13]. Also, stable compound Cu_2Te is known in Cu–Te system. Formation of both compounds at elevated temperatures were experimentally observed in Au(Cu)/CdTe contacts by Cordes and Schmid-Fetzer [3]. Other possible reactions include interaction between the contact metal and native oxide as well as cadmium [1]. The last one characterize possible alloying between the metal and Cd. Generally, the interface region may be a mixture of different chemical compounds. There has been suggested that in such case the barrier height can be estimated within the simple Schottky model with the modified work function [14]. Previously the contacts prepared on Br_2-methanol-etched surfaces were investigated for several metals [3]. The barrier heights were obtained for Cu (0.65 eV), Ag (0.73 eV), Au (0.64 eV) and Ni (0.65 eV). However, it is hard to believe that the work function prepared in different systems and at conditions has approximately the same value.

Absence of Fowler's spectra in annealed contacts may be explained by displacement of potential barrier far away from the initial metals and formation of heterojuncton structures. Experimental measurements of high-frequency capacitance–voltage characteristics in investigated contacts indicate that the interface region has higher resistivity than the bulk crystal. This experimental fact is in favor of formation of semiconductor compound rather than the metallic one. In this case investigated contacts may be treated as heterojunctions. Similar results were obtained earlier in [5] at much higher temperatures (T ≅ 400 °C). However, the formation of heterojunctions even at lower temperatures should be investigated in the future.

Another possible model of the Schottky barrier formation in investigated contacts stems from the fact that wet chemical etching produces not only non-stoichiometric but also highly disodered surfaces. Naturally, the most disordered surface is obtained for the etchant which cause the largest deviation from the stoichiometry. After the metal deposition the disorder interface region is formed. Chemical composition and physical properties of this region can be related with chemical reactions at the interface. For example, interface region can be doped with the contact metal. The width of the interface region is appreciable so it can be treated in the same way as bulk disordered semiconductor. In particular, the theory of highly doped and highly compensated semiconductors can be used for explanation of the Fermi level pinning at the band gap. The fluctuations in concentration of native defects and foreign impurities can lead to formation of large-scale fluctuation potential and subsequent shift of Fermi level towards the middle of the gap.

3 Conclusions

Experimental data obtained in the present investigation result in the following conclusions:
– it is shown that in aged (stored under normal laboratory condition during one-year period) and thermally annealed contacts the mechanisms of the Schottky barrier formation are different;
– it is very likely that in thermally annealed contacts heterostructures are formed at rather low temperatures (T = 300 °C);
– in as-deposited Au(Cu)/*p*-CdTe contacts the Schottky barrier height depends strongly on the composition of etchant and can be varied in a wide range of energies.

References

[1] I. M. Dharmadasa, Prog. Cryst. Growth Charact. **36**, 249 (1998).
[2] I. M. Dharmadasa, C. J. Blomfield, G. E. Gregory, and J. Haigh, Semicond. Sci. Technol. **9**, 185 (1994).
[3] H. Cordes and R. Schmid-Fetzer, Semicond. Sci. Technol. **10**, 77 (1995).
[4] A. K. Wahi, G. P Carey, K. Miyano et. al., J. Vac. Sci. Technol. A **8**, 1152 (1990).
[5] J. P. Ponpon, M. Saraphy, E. Buttung, and P. Siefert, phys. stat. sol. (a) **59**, 259 (1980).
[6] T. L. Chu and Shirley S. Chu, J. Appl. Phys. **58**, 4296 (1985).
[7] D. J. Friedman, I. Lindau, and W. E. Spicer, Phys. Rev. B **37**, 731 (1988).
[8] A. Musa, J. P. Ponpon, J. J. Grob et al., J. Appl. Phys. **54**, 3260 (1983).
[9] W. E. Spicer, I. Lindau, P. R. Skeath et. al., Phys. Rev. Lett. **44**, 420 (1980).
[10] D. M. Hoffman, W. Stadler, P. Christmann, B. K. Meyer, Nucl. Instrum. Methods Phys. Res. A **380**, 117 (1996).
[12] L. J. Brillson, R. Z. Bachrach, R. S. Bauer, and J. C. McMenamin, Phys. Rev. Lett. **42**, 397 (1979).
[13] G. van Tendeloo, P. Gregoriadis, and S. Amelinckx, J. Solid State Chem. **50**, 321 (1983).
[14] C. C. Tsuei and E. Kankeleit, Phys. Rev. **162**, 312 (1967).
[15] J. L. Freeouf and J. M. Woodall, Appl. Phys. Lett. **39**, 727 (1981).

Comparative analysis of (0001)GaN and (001)GaAs growth kinetics under Ga-rich conditions

V. G. Mansurov*, Yu. G. Galitsyn, and **K. S. Zhuravlev**

Institute of Semiconductor Physics, Siberian Branch of the Russian Academy of Sciences, Lavrentiev avenue, 13, 630090 Novosibirsk, Russia

Received 19 September 2003, accepted 3 November 2003
Published online 22 December 2003

PACS 81.05.Ea, 81.15.Hi

The comparative study of the GaAs and GaN growth by molecular beam epitaxy under Ga rich conditions has been performed. The growth rate is measured by the reflection high energy electron diffraction. Different behavior of the growth rate versus substrate temperature for both materials is revealed: the growth rate increases with temperature for GaN (normal kinetics) and decreases for GaAs (anomalous kinetics). Kinetic models for both materials in the frame of the mean field kinetics equations are developed.

© 2004 WILEY-VCH Verlag GmbH & Co. KGaA, Weinheim

1 Introduction It is well known that the bimolecular reaction of dissociation of As_4 molecules on the surface ($As_4 + As_4 \rightarrow 2As_2 + As_4^{gas}$) plays an important role in the molecular beam epitaxy (MBE) of GaAs using Ga and As_4 fluxes. This reaction provides the formation of As_2 active intermediate complexes. This is the necessary stage in the chain of successive steps of surface reactions for incorporation of the V-group atoms into lattice sites.

On the other hand, the analysis of the growth kinetics of GaN by MBE using Ga and NH_3 fluxes under Ga rich conditions reveals the importance of the recombination of NH_2 radicals: $NH_2 + NH_2 \rightarrow N_2^{gas} + 2H_2^{gas}$. But in contrast to GaAs in this case the bimolecular reaction decreases the concentration of the active species (NH_2 radicals) on the surface.

In the present work the comparative study of the GaAs and GaN growth kinetics under Ga-rich conditions is performed by reflection high energy electron diffraction (RHEED) in order to clarify the role of these bimolecular reactions in epitaxial growth of GaAs and GaN.

2 Experimental The growth of gallium arsenide on (001)GaAs substrates was carried out in B E-32 system (Riber) using Ga and As_4 fluxes under Ga-rich conditions in the temperature range 500-600 °C. The growth of GaN was carried out in CBE-32 system (Riber) using Ga and NH_3 fluxes. Sapphire (0001) wafers were used as substrates for growth. The procedure of the smooth (0001) GaN surface preparation for experiments includes: a nitridation step ($T_s \sim 850-900$ °C), a growth of thin (~200 Å) AlN nucleation layer and a GaN (~0.5μm) buffer layer growth under stoichiometry or slightly Ga-rich conditions. In order to measure the growth rate under the Ga rich conditions RHEED specular beam (SB) intensity evolution with time during successive supplying of the growth components onto surface was monitored. The strictly determined amount of Ga was deposited onto GaAs(GaN) surface without V-group component fluxes, then the surface was exposed to $As_4(NH_3)$ fluxes. The time (τ) needed to transform excess Ga into GaAs(or GaN) was measured as a function of the V-group components pressure (P) and substrate temperature (T_s). The growth rate (V_{growth}) is proportional to $1/\tau$.

---

* Corresponding author: e-mail: mansurov@thermo.isp.nsc.ru, Phone: +007 3832 341 945, Fax: +007 3832 341 945

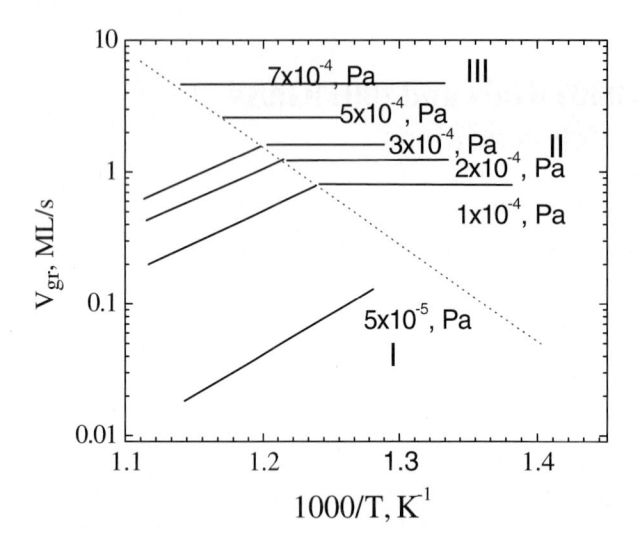

Fig. 1 GaAs growth rate as a function of sample temperature for different As$_4$ pressures.

3 Results and discussion

3.1 GaAs growth kinetics
The growth rate of GaAs under Ga-rich conditions versus reciprocal temperature for different As$_4$ pressure is shown on Fig. 1. Several research groups experimental results are presented in the Fig. 1: I – our measurements [1]; II – Garcia, J. C. et al. [2]; III – Tok, E. S. et al. [3].

In order to explain the experimental data a kinetic model was developed in the framework of mean field kinetic equations:

$$As_4^{gas} \underset{k_2}{\overset{k_1}{\rightleftarrows}} As_4^{ads} \quad \text{adsorption and desorption} \tag{1}$$

$$As_4^{ads} + As_4^{ads} \xrightarrow{k_3} 2As_2^{chem} + As_4^{gas} \quad \text{bimolecular dissociation} \tag{2}$$

$$As_2^{chem} + 2Ga \xrightarrow{k_4} 2GaAs \quad \text{incorporation} \tag{3}$$

$$As_2^{chem} \xrightarrow{k_5} As_2^{gas} + Ga \quad \text{desorption of As}_2 \tag{4}$$

where k_1, k_2, k_3, k_4, k_5 – are the kinetic constants. Of course, any of these kinetic (k_i) constants depends on the temperature $k_i(T) = k_0 \cdot \exp\left(-\dfrac{E_i}{k_B T}\right)$.

In the framework of mean field equations and assuming steady-state conditions for intermediate surface species [As$_4^{ads}$] and [As$_2^{chem}$] analytical expressions for the growth rate can be derived:

$$V_{growth} = \frac{k_1 P \cdot k_4}{k_4 + k_5} \cdot g(\lambda) \tag{5}$$

where $k_1 P$ – is a As$_4$ incident flux onto surface,

$$g(\lambda) = \frac{(\sqrt{1+\lambda} - 1)^2}{\lambda} \quad \text{and} \quad \lambda = 4 \cdot \left(\frac{k_1 P \cdot k_3}{k_2^2}\right) = 4 \cdot \frac{C_1}{C_2}.$$

© 2004 WILEY-VCH Verlag GmbH & Co. KGaA, Weinheim

$C_1 = k_1P/k_2$ – is a stationary concentration of As_4^{ads} at equality of adsorption and desorption fluxes, $C_2 = k_2/k_3$ As_4^{ads} stationary concentration of As_4^{ads} at equality of desorption flux in Eq. (1) and reaction flux in Eq. (2).

There are two limits for the growth rate:

$$V_{growth} = \frac{k_1 P \cdot k_4}{k_4 + k_5} \quad \lambda < 1 \quad \text{high temperature (>570 °C)} \tag{6}$$

$$V_{growth} = k_3 \left(\frac{k_1 P}{k_2}\right)^2 \quad \lambda > 1 \text{ low temperature (<530 °C)} \tag{7}$$

3.2 GaN growth kinetics The GaN growth rate as a function of pressure is shown in Fig. 2 for different temperatures. It was revealed also that the growth rate depended on NH_3 pressure as $V \sim P^n$, where n increased with increasing substrate temperature from 0.5 (below 760 °C) to 1 (above 800 °C). It should be stressed that in contrast to GaAs growth rate for GaN increases with temperature increasing.

For GaN the following kinetic model was developed:

$$NH_3^{gas} \underset{k_2}{\overset{k_1}{\rightleftarrows}} NH_3^{ads} \quad \text{adsorption and desorption} \tag{8}$$

$$NH_3^{ads} \xrightarrow{k_3} NH_2^{ads} + H^{ads} \quad \text{dissociation} \tag{9}$$

$$NH_2^{ads} + NH_2^{ads} \xrightarrow{k_4} N_2 \uparrow + 2H_2 \uparrow \quad \text{recombination} \tag{10}$$

$$NH_2^{ads} + Ga^{ads} \xrightarrow{k_5} Ga-N + H_2 \uparrow \quad \text{incorporation into lattice site} \tag{11}$$

where k_1, k_2, k_3, k_4, k_5 – are the set of GaN growth kinetic constants.

In the same assumptions as for GaAs we can derive for GaN:

$$V_{growth} = k_3 \cdot \frac{k_1 P}{k_2 + k_3} \cdot f(\lambda)$$

where $f(\lambda) = 2\left\{\frac{\sqrt{1+\lambda}-1}{\lambda}\right\}$, and $\lambda = \frac{4 \cdot k_1 P \cdot k_3 \cdot k_4}{k_5^2 \cdot (k_2 + k_3)}$ – is similar to parameter λ for GaAs.

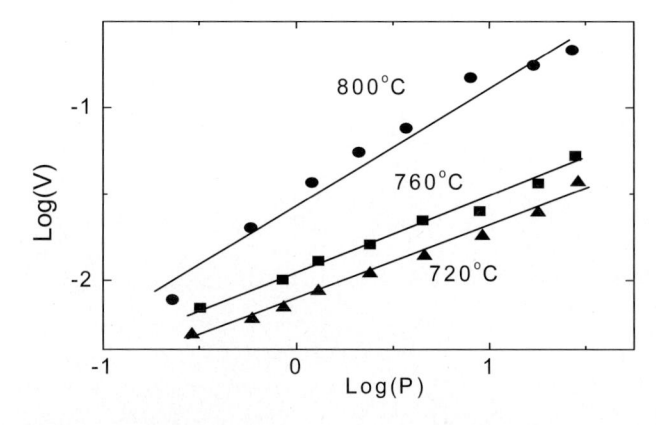

Fig. 2 GaN growth rate as a function of NH_3 pressure for different sample temperatures.

© 2004 WILEY-VCH Verlag GmbH & Co. KGaA, Weinheim

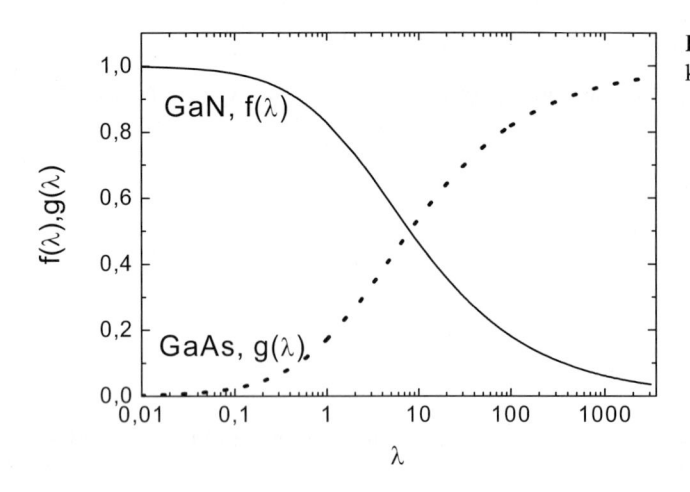

Fig. 3 Characteristic functions of the growth kinetics for GaAs $g(\lambda)$ and for GaN $f(\lambda)$.

$$V_{growth} = \frac{k_1 P \cdot k_4}{k_4 + k_5} \quad \lambda \gg 1 \text{ high temperature } (>570 \text{ °C}) \tag{6}$$

$$V_{growth} = k_3 \left(\frac{k_1 P}{k_2}\right)^2 \quad \lambda \ll 1 \text{ low temperature } (<530 \text{ °C}) \tag{7}$$

Now it is interesting to compare the characteristic functions for GaAs $g(\lambda)$ and for GaN $f(\lambda)$. These characteristic functions are shown in Fig. 3. For both materials the parameter λ depends on the substrate temperature and pressure (of NH_3 or As_4) in a similar way. Namely, it decreases with increasing substrate temperature and increases with increasing NH_3 (As_4) pressure. The values of both characteristic functions vary significantly in the same parameter range $1 < \lambda < 100$.

4 Conclusion Thus the comparative study of the GaAs and GaN growth kinetics under Ga-rich conditions is performed, the kinetics models are developed in agreement with experimental data.

Acknowledgment The authors are grateful to the RFBR financial support (grant 02-03-32307).

References

[1] Yu. G. Galitsyn, V. G. Mansurov, and I. I. Marakhovka, Phys. Low-Dim. Struct. **5–6**, 75 (1997).
[2] J. C. Garcia, C. Neri, and J. Massies, J. Cryst. Growth **98**, 511 (1989).
[3] E. S. Tok, J. H. Neave, J. Zhang et al., Surf. Sci. **371**, 277 (1997); Surf. Sci. **374**, 397 (1997).

Growth kinetics of (0001)GaN from Ga and NH₃ fluxes

V. G. Mansurov*, Yu. G. Galitsyn, V. V. Preobrazhenskii, and K. S. Zhuravlev

Institute of Semiconductor Physics, Siberian Branch of the Russian Academy of Sciences, Lavrentiev avenue 13, 630090 Novosibirsk, Russia

Received 19 September 2003, accepted 24 November 2003
Published online 22 December 2003

PACS 68.43.Mn, 68.55.Ac, 81.15.Hi

Ga desorption and GaN growth kinetics by molecular beam epitaxy using Ga and NH₃ fluxes under Ga-rich conditions were investigated. Ga desorption activation energy of 2.4 eV from (0001)GaN was measured. It was experimentally determined that the growth rate depends on the NH₃ pressure as P^n, where n increased with increasing substrate temperature from 0.5 (below 760 °C) to 1 (above 800 °C). A kinetic model of growth in the framework of the mean field kinetic equations has been developed.

© 2003 WILEY-VCH Verlag GmbH & Co. KGaA, Weinheim

1 Introduction Recently the growth of high quality III-nitrides attracted a great deal of interest because of its usefulness for modern optoelectronics as well as high temperature and power devices. The growth of GaN by molecular beam epitaxy (MBE) using Ga and NH₃ is a sequence of elementary steps: adsorption-desorption, dissociation of NH₃, recombination of NH₂ radicals, incorporation of Ga and N atoms into lattice sites. This means that without the knowledge of the parameters of the microscopic surface processes it is not clear how to choose the optimum growth conditions.

In the present work the kinetics of GaN growth under Ga-rich conditions has been investigated by reflection high energy electron diffraction (RHEED).

2 Experimental The growth experiments were carried out in a CBE-32 system (Riber) using Ga and NH₃ fluxes. Sapphire (0001) wafers were used as substrates for growth. The procedure to prepare smooth (0001) GaN surfaces for experiments includes: a nitridation step (T_s ~ 850–900 °C), growth of a thin

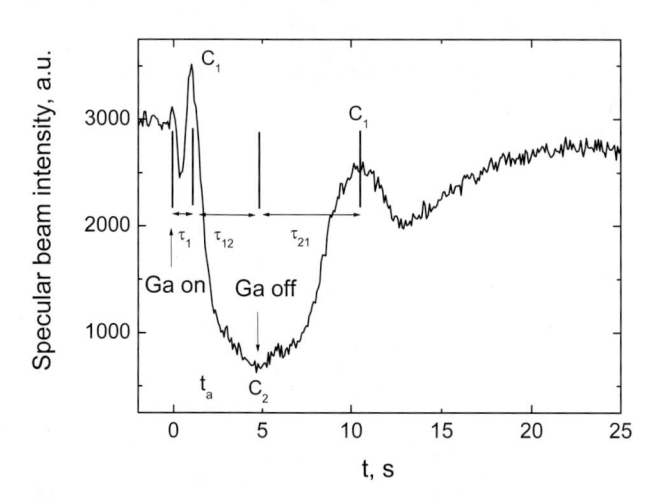

Fig. 1 RHEED specular beam (SB) intensity evolution with time during Ga adsorption and desorption onto a GaN surface.

* Corresponding author: e-mail: mansurov@thermo.isp.nsc.ru, Phone: +007 3832 34 19 45, Fax: +007 3832 34 19 45

(~200 Å) AlN nucleation layer and a GaN (~0.5μm) buffer layer growth under stoichiometric or slightly Ga-rich conditions. RHEED specular beam (SB) intensity evolution with time during successive supply of the growth components onto a GaN surface was monitored. The typical SB intensity evolution curve is shown in figure 1.

A strictly determined amount of Ga was deposited onto the GaN surface (during the time t_a) without NH$_3$ flux first, then the surface was exposed to vacuum (Ga desorption) or NH$_3$ flux (GaN growth). The time needed to desorb the Ga or to transform it into GaN was measured as a function of the Ga surface concentration, NH$_3$ flux and substrate temperature Ts.

3 Results and discussion

3.1 Ga desorption The dependence of desorption time (τ_{21}) versus deposition time ($t_a = \tau_1 + \tau_{12}$, see Fig. 1) for different substrate temperatures is shown in figure 2.

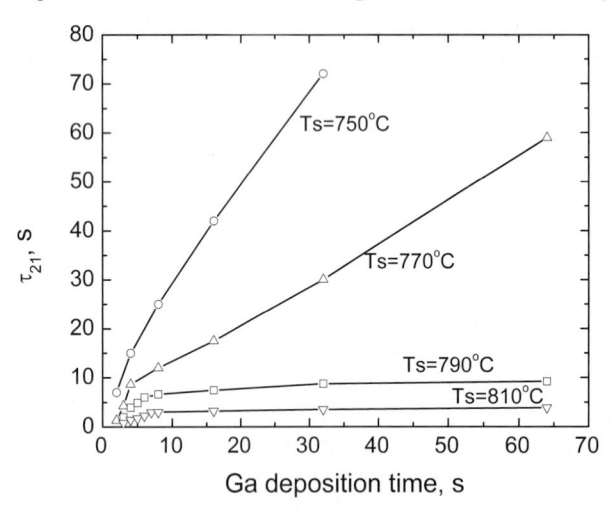

Fig. 2 The dependence of Ga desorption time (τ_{21}) versus Ga deposition time (t_a) for different substrate temperatures.

The Ga desorption can be described by a simple kinetic model. Let C denote the surface coverage of excess Ga deposited onto the GaN surface. The time dependence of C (during Ga deposition) is defined by equation:

$$\frac{dC(t)}{dt} = J_{Ga} - q_{des} \cdot C(t) \qquad (1)$$

where J_{Ga} – is the incident Ga flux, q_{des} – is the kinetic constant of Ga desorption. Let $C_1 = C(\tau_1)$ – and $C_2 = C(t_a)$, where $t_a = \tau_1 + \tau_{21}$.

The solution Eq. (1) is:

$$C(t) = \frac{J_{Ga}}{q_{des}} \cdot (1 - \exp(-q_{des} \cdot t)) \qquad (2)$$

$$C_1 = \frac{J_{Ga}}{q_{des}} \cdot (1 - \exp(-q_{des} \cdot \tau_1)) \qquad (3)$$

$$C_2 = \frac{J_{Ga}}{q_{des}} \cdot (1 - \exp(-q_{des} \cdot t_a)) \qquad (4)$$

When the Ga flux is switched off we have

$$\frac{dC(t)}{dt} = -q_{des} \cdot C(t) \qquad (5)$$

The solution of Eq. (2) is:

$$C(t) = C_2 \cdot \exp(-q_{des} \cdot t) \qquad (6)$$

$$C_1 = C_2 \cdot \exp(-q_{des} \cdot \tau_{21}) \qquad (7)$$

We assume that peaks in intensity evolution curve at moments τ_1 and $t_a + \tau_{21}$ correspond to the same excess Ga concentration. Equating Eq. (3) to Eq. (7) and using experimental values of t_a, τ_1, τ_{21} we found q_{des}. We determined the q_{des} at different temperatures. By assuming the following dependence of q_{des} on temperature: $q_{des}(T) = q_0 \cdot \exp(-E_a/k_B T)$, we found the values of $q_0 = 10^{10}$ 1/s and $E_a = 2.4$ eV. The determined value of E_a is in accordance with the literature data [1].

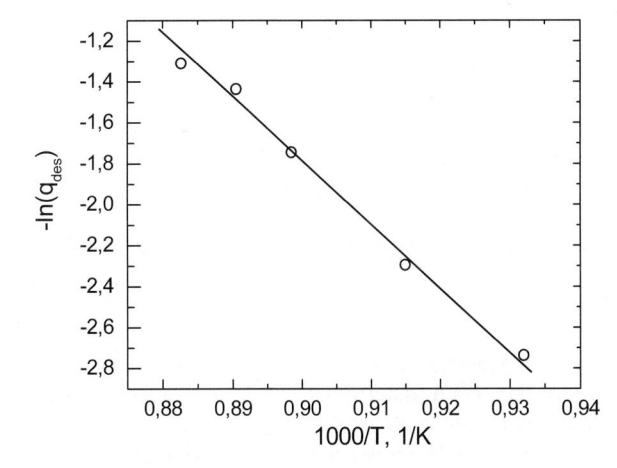

Fig. 3 The Ga desorption kinetic constant as function of temperature

3.2 GaN growth The GaN growth rate as a function of pressure is shown in figure 4 for different temperatures. It was revealed also that the growth rate depended on NH_3 pressure as $V \sim P^n$, where n increased with increasing substrate temperature from 0.5 (below 760 °C) to 1 (above 800 °C). It should be noted that in Ga-rich conditions the growth rate increases with temperature increasing.

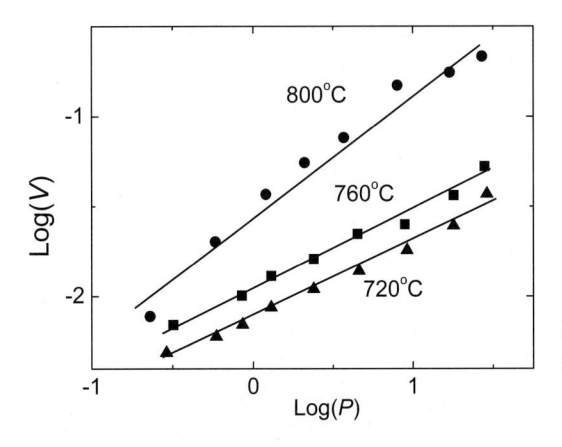

Fig. 4 GaN growth rate as a function of NH_3 pressure for different sample temperatures.

In order to explain the experimental data a kinetic model was developed in the framework of mean field kinetic equations:

$$NH_3^{gas} \underset{k2}{\overset{k1}{\rightleftarrows}} NH_3^{ads} \quad \text{(adsorption and desorption)} \qquad (8)$$

$$\text{NH}_3^{ads} \xrightarrow{k3} \text{NH}_2^{ads} + \text{H}^{ads} \quad \text{(dissociation)} \tag{9}$$

$$\text{NH}_2^{ads} + \text{NH}_2^{ads} \xrightarrow{k4} \text{N}_2 \uparrow + 2\text{H}_2 \uparrow \quad \text{(recombination)} \tag{10}$$

$$\text{NH}_2^{ads} + \text{Ga}^{ads} \xrightarrow{k5} \text{Ga-N} + \text{H}_2 \uparrow \quad \text{(incorporation into lattice site)} \tag{11}$$

k_1, k_2, k_3, k_4, k_5 – are the kinetic constants of the corresponding elementary reactions.

In the framework of mean field equations and assuming steady-state conditions for intermediate surface species [NH_3^{ads}] and [NH_2^{ads}] analytical expressions for the growth rate can be derived:

$$V = k_3 \cdot \frac{k_1 P}{k_2 + k_3} \cdot 2\left\{\frac{(1+y)^{1/2}-1}{y}\right\}, \quad \text{where} \quad y = \frac{4 \cdot k_1 P \cdot k_3 \cdot k_4}{k_5^2 \cdot (k_2 + k_3)} \tag{12}$$

V – have the unit of flux (cm^{-2} s^{-1}) and the growth rate can be transformed to unit of ML/s, P – ammonia pressure, $k_1 P$ – is the incident NH_3 flux.

From the above expression it is clear that the dependence of the growth rate from ammonia pressure and substrate temperature is complex. In the two limits we can deduce that $V \sim P^{0.5}$ when $y \gg 1$ and $V \sim P$ when $y \ll 1$. As a rule an activation barrier of incorporation (k_5) is higher than a barrier of radicals' recombination (k_4), the expression $\frac{4 \cdot k_1 P \cdot k_3}{(k_2 + k_3)}$ weakly depends on Ts, so it is naturally to suppose that decreases when T_s increases.

4 Conclusion The Ga desorption rate and GaN growth kinetics under Ga-rich conditions have been investigated. The kinetic constant of Ga desorption has been determined with activation energy $E = 2.4$ eV. It is shown that growth rate ~ Pn where n increased with increasing substrate temperature from 0.5 (below 760 °C) to 1 (above 800 °C). A kinetic growth model has been developed in good agreement with experimental data.

References

[1] P. Hacke, G. Feuillet, H. Okumura, and S. Yoshida, Appl. Phys. Lett. **69**, 2507 (1996).

Influence of TBAs flow in MOCVD growth on nitrogen incorporation in GaAsN alloy and its optical quality detected by photoreflectance

W. Rudno-Rudzinski[*,1], **R. Kudrawiec**[1], **J. Misiewicz**[1], **J. Derluyn**[2], and **I. Moerman**[2]

[1] Institute of Physics, Wrocław University of Technology, Wybrzeże Wyspiańskiego 27, 50-370 Wroclaw, Poland
[2] INTEC, Department of Information Technology, Ghent University, St. Pietersnieuwstraat 41, B9000 Gent, Belgium

Received 19 September 2003, accepted 24 November 2003
Published online 22 December 2003

PACS 61.10.Nz, 81.15.Gh, 78.20.Jq, 78.40.Fy

In this work there has been studied GaNAs bulk layers grown by metal organic chemical vapour deposition (MOCVD) with different tributylarsine (TBAs) flow and the same flow of triethylgallium (TEGa) and unsymmetrical dimethyl-hydrazine (DMHy). High resolution X-ray diffraction (HR-XRD) and room temperature photoreflectance (PR) have been carried out in order to evaluate the nitrogen content and bandgap energy and quality of the GaAsN layer. It has been observed that the TBAs flow influences of the nitrogen incorporation. The variation of the TBAs from 37 to 1182 μmol/min causes variation of nitrogen content in GaAsN layer in the range of 1.5–4 %. The best optical quality of GaAsN layer and the highest nitrogen incorporation have been found for the TBAs flow in the range of 296 and 591 μmol/min.

© 2004 WILEY-VCH Verlag GmbH & Co. KGaA, Weinheim

1 Introduction

GaAsN is an important semiconductor alloy both for its potential device applications [1, 2] and for being a new type of alloy with interesting properties. The incorporation of small amounts of N leads to giant band structure changes in the host semiconductor GaAs. A reduction of the band gap by more than 100 meV per atomic percent of nitrogen has been first observed in GaAsN alloys grown by plasma-assisted MOCVD [3, 4]. So far, many groups have experimentally demonstrated the reduction of the bandgap of the GaAsN and other III-V-N alloys with a small amount of N incorporation (e.g. GaInNAs, GaAsSbN). Most recently, it has been shown that the reduction of the band gap can be explained by an interaction between the localised nitrogen levels resonant with the conduction-band states of the host semiconductor matrix materials [5].

The large difference in lattice constants and the structural mismatch between the GaN and GaAs create major difficulties for material fabrication and are the origin of poor quality of this system. An increase in N composition in the alloy required to realize desired material properties can only be achieved by using non-equilibrium growth method, e.g. molecular beam epitaxy (MBE) or MOCVD. It is expected that in this case the growth conditions should more strongly affect the alloy quality in comparison with conventional alloys grown closer to thermal equilibrium conditions. It is obvious that the growth temperature strongly influences the layer quality. However, an investigation of different growth parameters, other than temperature, seems to be also very interesting. This paper shows the systematic study of correlation between TBAs flow in MOCVD process and optical quality of bulk GaNAs.

---

[*] Corresponding author: e-mail: wojciech.rudno-rudzinski@pwr.wroc.pl, Phone: +48 71 320 23 58, Fax: +48 71 328 36 96

2 Experiment

All layers were grown on a Thomas Swan vertical rotating disk close-coupled showerhead reactor at pressure 76 Torr under hydrogen carrier gas. Group III and group V source materials were switched in separate manifolds and were mixed only after the gasses exit the showerhead. Source materials were triethylgallium (TEGa), tributylarsine (TBAs) and unsymmetrical dimethyl-hydrazine (DMHy). The growth temperature of all layers was 530 °C. Quantum wells were post-growth annealed for 10 minutes at 700 °C. Growth rates were 1.08 μm/h for bulk GaAs and GaNAs layers. Samples were grown on semi-insulating <100> GaAs substrates, with a 150 nm GaAs buffer layer, a 300 nm GaNAs layer and a 30 nm GaAs top layer. During growth of bulk samples the TEGa flow was held constant at 50.9 μmol/min, DMHy at 4283.59 μmol/min, whereas TBAs flow was varied from 36.9 μmol/min to 1181.68 μmol/min.

Optical measurements were carried out in standard configurations, using 650 nm semiconductor laser as an excitation source in PL and pump beam in PR experiments. Signal was detected with thermoelectrically cooled InGaAs photodiode. More details of PR measurement may be found elsewhere [6].

3 Results

The structural quality of GaNAs layers has been studied by high resolution x-ray diffraction (HR-XRD) using a Philips X'Pert Pro diffractometer. Figure 1 shows HR-XRD rocking curves of GaAsN layers with TBAs flow between 37 and 1182 μmol/min. In this case, GaAsN layers on GaAs substrates are under tensile strain. The narrow peaks and Pendelloesung fringes indicate good quality and sharp interfaces of these structures. The shift of GaAsN-related peak means that the nitrogen content depends on TBAs flow. In order to determine the layer content we have simulated HR-XRD data by dynamical simulation theory. The obtained nitrogen content is presented in Fig.1 at the GaAsN-related peak.

In order to determine bandgap energy and optical quality of GaAsN layers room temperature (RT) PR spectra have been measured. These spectra are shown in Fig. 2. Optical transition from GaAs layer is visible on all spectra at energy of 1.42 eV. Apart from that strong resonance, another resonance appears below 1.4 eV. It is much weaker and broader than dominating optical transition from GaAs layer, but nevertheless its energy and broadening may be well determined from fitting procedure [7] (third derivative gaussian line-shape (TDGL)). The transition below GaAs originates from the GaNAs layer and its energy defines the energy gap of the alloy. In addition, an oscillation is seen on the spectrum of sample with 37 μmol/min flow below the energy bandgap of GaAs. This feature is attributed to interference phenomena [6, 8] which do not influence the energy and broadening of the bandgap-related resonance. An analysis he GaAsN bandgap energy in the framework of band anticrossing model (BAC) model [5] allows to determine

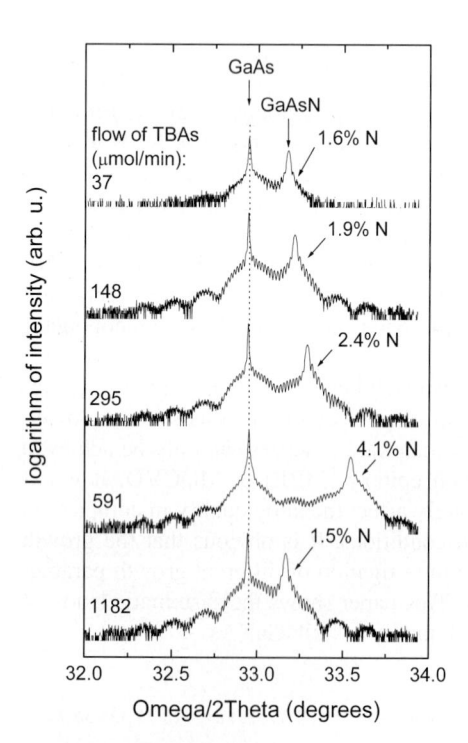

Fig. 1 HR-XRD data of a 300 nm thick GaAsN layers obtained under different TBAs flow.

the nitrogen content. In this model the bandgap energy of GaAsN alloy is described by following equation:

$$E_{GaAsN} = \frac{1}{2}\left[E_N + E_{GaAs} \pm \sqrt{\left[E_N - E_{GaAs}\right]^2 + 4C_{MN}^2 x}\right], \qquad (1)$$

where E_N, E_{GaAs}, C_{MN}, and x are energy of resonant nitrogen level, energy of host semiconductor (in this case GaAs), matrix element, and nitrogen content, respectively.

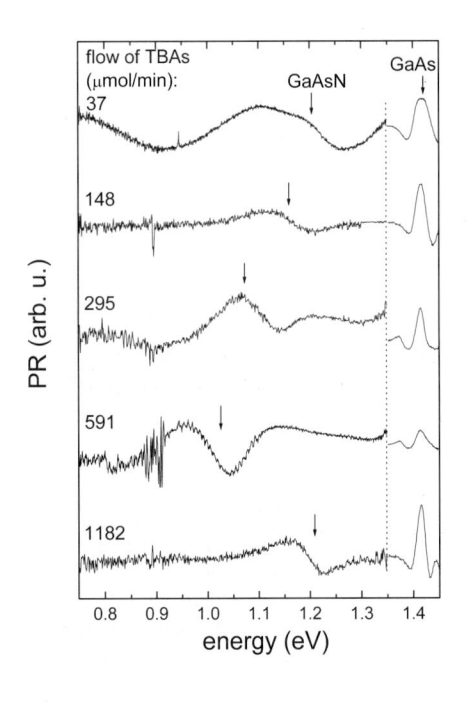

Fig. 2 Room temperature PR spectra of GaNAs bulk layers grown under different TBAs flow. Arrows point out the optical transitions.

After Shan et al. [5] it was assumed that E_N and C_{MN} equal 1.65 and 2.7 eV, respectively. The bandgap energy of GaAsN (E_{GaAsN}) was taken from the fitting analysis of PR data, and the nitrogen content was extracted from Eq. (1). The nitrogen content obtained from both HR-XRD and PR measurements versus TBAs flow is presented in Fig. 3. The agreement between the two approach is satisfactory. It means that the BAC model within considered range of nitrogen content quite precisely describes the bandgap energy of GaAsN alloy.

Since the incorporation of nitrogen changes the lattice constant and introduces stress into structure, the splitting of light and heavy hole energy levels should appear. However, its energy is much smaller than broadening parameter, which is rather large for all samples (~100 meV) and thus cannot be observed. The magnitude of broadening indicates that the optical quality of this compound is poor (alloy inhomogeneity) in comparison with a GaAs layer (e.g. buffer layer). In this case, the broadening parameter is a good indicator of GaAsN quality and will be consider bellow.

In Fig. 1 it is well seen that GaAsN layers with different content and quality can be obtained at fixed TEGa and DMHy flows and various TBAs flow. With the increase in the TBAs flow from 37 to 591 μmol/min an increase in nitrogen content is observed. Above flow of 591 μmol/min the nitrogen content decreases dramatically. This anomalous incorporation efficiency of nitrogen is certainly not due to gas phase prereactions and, therefore, must be surface related [9]. One possible explanation is that an arsenic covered surface actually promotes the adsorption and reaction of the DMHy molecules [9].

The optical quality of samples, indicated by the broadening of PR resonanses, also shows non-monotonic behaviour as a function of TBAs flow (see in Fig. 3). The broadening parameter decreases for small values of TBAs flow, but after reaching minimum it increases, which means that the optical quality of sample drops down. In our case the best optical quality can be obtained for the TBAs flow in the range of 296–591 μmol/min. The non-monotonic behaviour of broadening parameter is directly associated with the defect formation in GaAsN layer. We suppose that As-related defects could be responsible for the layer quality. Probably at low TBAs flow (<148 μmol/min) the number of As-related defects is high and causes significant broadening of PR line. With the increase in TBAs flow the number of the defects decreases and it improves the layer quality (i.e. the broadening of PR line decreases). Above TBAs flow of 591 μmol/min broadening parameter starts to increase probably due to other kind of defects whose number increases for higher TBAs flows. So far, an As$_{Ga}$ antisite [10, 11], an N interstitial [12-14] and a Ga vacancy [14] have been identified in the arsenide-nitrides.

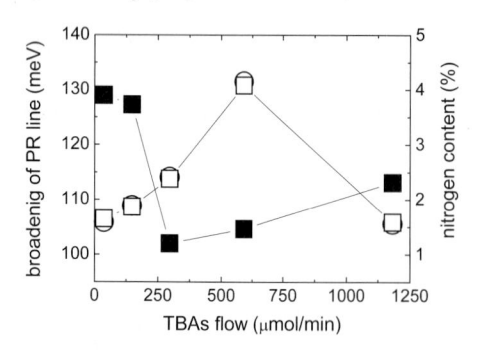

Fig. 3 Open circles and squares – nitrogen incorporation in GaNAs as a function of the TBAs-flow, taken from HR-XRD and PR, respectively. Full squares – broadening of PR resonanses. The solid lines are a guide to the eye.

An increase in the layer quality despite the increase of nitrogen incorporation has been noticed in this paper. Generally, it was observed that an increase of nitrogen content leads to the degradation of GaAsN quality. In this case, at fixed TEGa and DMHy flows, optimal quality of GaAsN layer was found to be for some GaAsN content which could be identified by the change in TBAs flow. It is expected that for other TEGa and DMHy flows optimal quality of GaAsN layer could be found for other TBAs flow. The adequate TBAs flow may be identified on the basis of comparison of PR line broadening, as was shown in this paper. In order to control both nitrogen content and layer quality the flow of every source has to be optimised [9].

4 Conclusions

In this paper we have shown that the TBAs flow is an essential parameter for nitrogen incorporation in MOCVD process. It is possible to obtain the nitrogen content as high as 4% only increasing its value. Above certain TBAs flow nitrogen incorporation efficiency drastically drops. Optical quality of samples is a non-monotonous function of TBAs flow. This change in the quality has been attributed to TBAs flow dependent formation of As vacancies in GaAsN layer. For some TBAs flow the material shows good optical properties but the increase in arsenic concentrations above certain level strongly degrades the structure.

Acknowledgements This research was supported in part by the Committee for Scientific Research in Poland under Grant No. T11B 00823, Grant No. 2P03B 10825.

References

[1] M. Kondow, K. Uomi, A. Niwa, T. Kikatani, S. Watahiki, and Y. Yazawa, Jpn. J. Appl. Phys. **35**(2B), 1273 (1996).
[2] J. F. Geisz and D. J. Friedman, Semicond. Sci. Technol. **17**, 769 (2002).
[3] M. Weyers, M. Sato, and H. Ando, Jpn. J. Appl. Phys. **31**, 853 (1992).
[4] S. Sakai. Y. Ueta, and Y. Terauchi, Jpn. J. Appl. Phys. **32**, 4413 (1993).
[5] W. Shan, W. Walukiewicz, J. W. Ager III, E. E. Haller, J. F. Geisz, D. J. Friedman, J. M. Olson, and S. R. Krutz, Phys. Rev. Lett. **82**, 1221 (1999).
[6] J. Misiewicz, P. Sitarek, G. Sek, and R. Kudrawiec, Mater. Sci. **21**, (2003).
[7] O. J. Glembocki, Proc. SPIE **1286**, 2 (1990).
[8] O. J. Glembocki and B. V. Shanabrook, Semiconductors and Semimetals 36, Chapter 4.
[9] J. Derluyn, I. Moerman, M. R. Leys, G. Patriarche, G. Sek, R. Kudrawiec, W. Rudno-Rudzinski, K. Ryczko, and J. Misiewicz, J. Appl. Phys. **94**, 2752 (2003).
[10] N. Q. Thinh, I. A. Buyanova, P. N. Hai, W. M. Chen, H. P. Xin, and C. W. Tu, Phys. Rev. B **63**, 033203 (2001).
[11] N. Q. Thinh, I. A. Buyanova, W. M. Chen, H. P. Xin, and C. W. Tu, Appl. Phys. Lett. **79**, 3089 (2001).
[12] S. G. Spruytte, C. W. Coldren, J. S. Harris, W. Wampler, P. Krisprin, K. Ploog, and M. C. Larson, J. Appl. Phys. **89**, 4401 (2001).
[13] T. Ahlgren, E. Vainonen-Ahlgren, J. Likonen, W. Li, and M. Pessa, Appl. Phys. Lett. **80**, 2314 (2002).
[14] W. Li, M. Pessa, T. Ahlgren, and J. Dekker, Appl. Phys. Lett. **79**, 1094 (2001).

Investigation of Si–Ge whisker growth by CVD

A. A. Druzhinin and **I. P. Ostrovskii**[*]

National University "Lvivska polytechnika", 1 Kotlyarevskii st., Lviv, Ukraine

Received 19 September 2003, accepted 3 November 2003
Published online 22 December 2003

PACS 61.50.–f, 72.80.Cw

The present paper deals with study of $Si_{1-x}Ge_x$ ($x = 0 \div 0.11$) whiskers growth in sealed Si–Ge–Br systems by CVD method. An influence of multiple doping (Zn, Hf, Ni, Mn) on morphology of the whiskers was investigated. A model based on evaluation of impurity activity coefficients in Si crystals was proposed. High impurity activity coefficients were established to correspond to submicron whisker growth. As the impurity activity coefficient decreases, the whisker morphology changes from twinned ribbon to needle-like crystals with perfect surface of facets.

© 2004 WILEY-VCH Verlag GmbH & Co. KGaA, Weinheim

1 Introduction A study of whiskers is of great interest because of their peculiar properties appearing due to 1D growth. Impurities are known to be important at whisker growth [1]. They serve as initiators of whisker growth due to VLS mechanism [2]. Therefore, growth of whiskers doped with single impurity was mainly under consideration. Our previous study [3, 4] was devoted to an experimental investigation of growth mechanisms of Si and Si-Ge whiskers doped with single impurities Au, B, Pt. The growth of whiskers with multiple doping was insufficiently studied.

The present paper deals with studies of $Si_{1-x}Ge_x$ ($x = 0 \div 0.11$) whisker growth in sealed Si–Ge–Br systems by CVD method with aim to provide the crystals with multiple doping and morphology predicted beforehand.

2 Results and discussion $Si_{1-x}Ge_x$ ($x = 0 \div 0.11$) whiskers were grown by CVD method in a sealed tube. Ge composition of grown whiskers was controlled by micro probe analysis. Bromine was used as transport agent, its concentration falls within the range 0.5–1 mg/cm^3. Temperature of evaporation zone consists of 1200 °C, while temperature of crystallization zone is equal to 800–1000 °C. Time processing was nearly 1.5 hours. The following combinations of doping impurities were used: <Ni+Pt>, <Ni+B>, <Ni+Mn>, <Ni+Pt+Mn>, <Au+B+Pt>, <La+Au>, <Zn+Hf+Au>. A choice of impurity types was determined as by fundamental study of low dimensional structure peculiarities [3] and by a possibility of whisker's applications [5]. For instance, behaviour of Ni as initiator of Si–Ge whisker growth is not as yet studied, while the whisker doping with rarely earth elements leads to increase of their photoconductance what is prospective for photoconverters design.

Crystals of various diameters and shapes were obtained depending on their doping. So, growth of submicron whiskers ($d = 0.1–0.9$ μm) prevails at <Ni+Pt> doping (Fig. 1a), twinned crystals (1–2 μm in diameter) create at <Ni+B> doping (Fig. 1b); needle-like crystals ($d = 5–150$ μm) grow at <Ni+Pt+Mn>, <Au+B+Pt>, <La+Au>, <Zn+Hf+Au> multiple doping (Fig. 1c).

To study an influence of multiple doping on crystalline structure and morphology of the whisker we use a model based on evaluation of impurity activity coefficients in Si–Ge crystals.

---

[*] Corresponding author: e-mail: iostrov@polynet.lviv.ua; phon.: 380322721632

© 2004 WILEY-VCH Verlag GmbH & Co. KGaA, Weinheim

We propose that doping of the crystals during the growth process is determined by some coefficient of impurity activity. This parameter shows how easy an impurity enters into the crystal. At multiple doping some effective coefficient of impurity activity serves as characteristic parameter. It can be expressed as multiplication of the impurity activity coefficients of separate dopants:

$$Ka^{eff} = Ka_1 \cdot Ka_2 \cdot Ka_3 \ldots Ka_n \tag{1}$$

The partial coefficient of impurity activity in silicon can be drawn from the following thermodynamical equation [6]:

$$kT \ln(1/Ka) = (1-x)E' + E_D \tag{2}$$

where x is molar fraction of impurity content in Si, E' is an energy necessary to replace the atoms of Si lattice for correspondent atoms of the impurity, E_D is an energy of lattice deformation, that can be written as

$$E_D = 8\pi \, Gr_{Si}(r_D - r_{Si})^2 \tag{3}$$

where G is Young's modul, r_D and r_{Si} are thetraedrical radiuses of silicon and impurity, respectively. The value E' can be expressed by use of specific warmth of sublimation ΔH_{si} and ΔH_D in such way

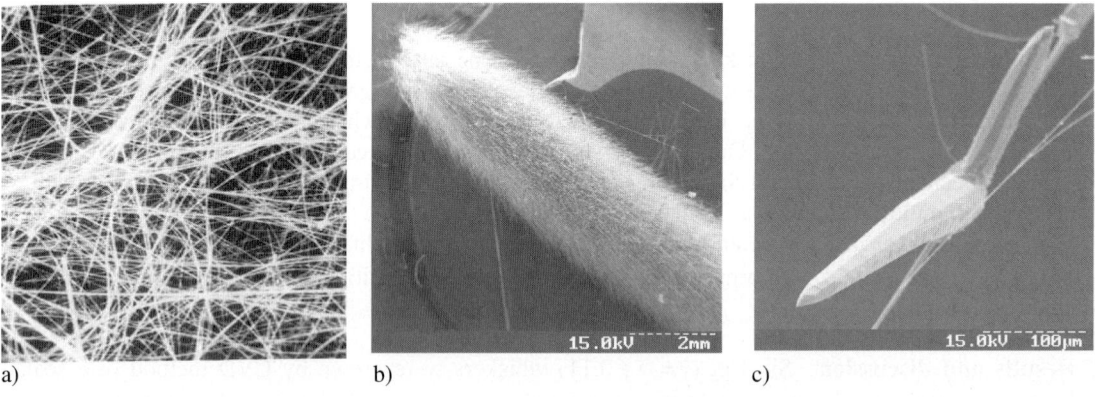

$$E' = (\Delta H_{si} - \Delta H_D)^2/(\Delta H_{si} + \Delta H_D) \tag{4}$$

Fig. 1 Growth of various morphological types of Si-Ge whiskers depending on their doping: a) submicron whiskers ($d = 0,1-0,9$ μm) created at Ni–Pt doping impurities; b) twinned crystals with 1–2 μm in diameter obtained in Ni–B impurity system; c) needle-like crystals with diameters 5–150 μm grown at <Ni+Pt+Mn> doping impurities.

Taking into account equation (2), the coefficients of impurity activity for some impurities in Si have been estimated. An analysis of equation (2) has shown, that if the thetraedrical radius of impurity atom strongly differs from Si one, the coefficient of activity for such impurity is very small. But if the radius of impurity and silicon atom are almost identical, then the coefficient of activity depends on relation of thermodynamical parameters of these materials and increases at impurity concentration increase. The obtained results are present in Table 1. The Table 1 contains specification and initial concentrations of dopants, estimated effective coefficients of impurity activity as well as the specific resistance and a short description of morphological peculiarities of the grown whiskers. As you can see from the Table 1, a certain correlation exists between the estimated impurity activity coefficients and the whisker morphology. If the coefficient of impurity activity decreases from 0.025 to $10^{-6} - 10^{-10}$ the following changes takes place:
a) the whisker size increases from 0,2–10 μm to 50–100 μm;
b) the crystal morphology changes from cylindrical submicrometer whiskers (at $K_a = 10^{-2}$) to ribbon and twines (at $K_a = 10^{-3}-10^{-6}$) and further to wires of large diameter (at $K_a = 10^{-6} - 10^{-10}$).

© 2004 WILEY-VCH Verlag GmbH & Co. KGaA, Weinheim

Therefore, it was shown that the coefficient of impurity activity can be used as characteristic parameter of the growth of the doped whiskers with certain diameter and morphology. For instance, doping by Ni and Pt leads to thin whiskers growth (Table 1) due to high coefficient of impurity activity. An additional entering of Mn, that itself is not an initiator of whisker growth but has small coefficient of activity, decreases an effective coefficient of activity to 10^{-6} (Table 1) that provides the broadening of growth zone: the whiskers of large diameter and twinned crystals are created.

Table 1 Some technological parameters of Si-Ge whiskers with complex doping (K_a is an effective coefficient of impurity activity, ρ is specific resistance of the whiskers).

Type of impurity n, mg/cm^3	K_a	ρ, $\Omega \cdot$ cm	Size and shape of the crystals
Ni, Pt $2.7 \cdot 10^{-1}$ $1.3 \cdot 10^{-1}$	0.025	–	Submicron whiskers
Ni, Pt $5 \cdot 10^{-2}$ $8 \cdot 10^{-2}$	0.025	0.2	thin and submicron whiskers
Ni, B $4 \cdot 10^{-2}$ $2.4 \cdot 10^{-2}$	0.005	0.007	Whiskers
Ni, B $1.5 \cdot 10^{-2}$ $5 \cdot 10^{-2}$	0.005	0.001–0.0005	twinned crystals and whiskers
Se, B 10^{-1} $2.4 \cdot 10^{-2}$	0.001	0.01	twinned crystals and whiskers
Ni, Mn $2 \cdot 10^{-1}$ $7 \cdot 10^{-2}$	0.0005	700	thin whiskers (<15 μm)
Ni, Pt, Mn $2 \cdot 10^{-1}$ $8 \cdot 10^{-2}$ $7 \cdot 10^{-2}$	$5 \cdot 10^{-6}$	1.0	thin and thick whiskers
Au, B, Pt $2 \cdot 10^{-2}$ $4 \cdot 10^{-2}$ $5 \cdot 10^{-2}$	$5 \cdot 10^{-6}$	0.04	needle-like crystals (<50 μm)
La, Au $6 \cdot 10^{-2}$ $2 \cdot 10^{-2}$	10^{-10}	130	needle-like crystals (50 μm)
Zn, Hf, Au $4 \cdot 10^{-2}$ $4 \cdot 10^{-2}$ $2 \cdot 10^{-2}$	10^{-10}	0.01–0.3	needle-like crystals (100 μm)

A change of the whisker morphology is caused by an influence of multiple doping on growth velocity of different facets of the crystals. Therefore, we measure velocity of axial growth of the whiskers obtained at various doping. According with the method described in [7, 8] the growth rate V was determined as a ratio of a whisker length (l) to a growth time. Since various doping leads to growth of the whiskers with different diameters, we obtain size dependence of growth velocity $V(d)$ (Fig. 2).

Taking into account a model proposed we can explain the above results. It should be noted that the whiskers of different diameters are created at various homogeneous conditions. So, entering of dopants with high activity initiated the growth of the whiskers of the least diameters, that takes place at inhomogeneous conditions. It leads to great velocities of sub micron whisker growth (Fig. 2) and can cause disturbance of their crystal structure [4]. Low coefficient of impurity activity provides the growth of the whiskers of large diameters that occurs in homogeneous conditions. Therefore they should have perfect crystal structure and rather small growth velocities.

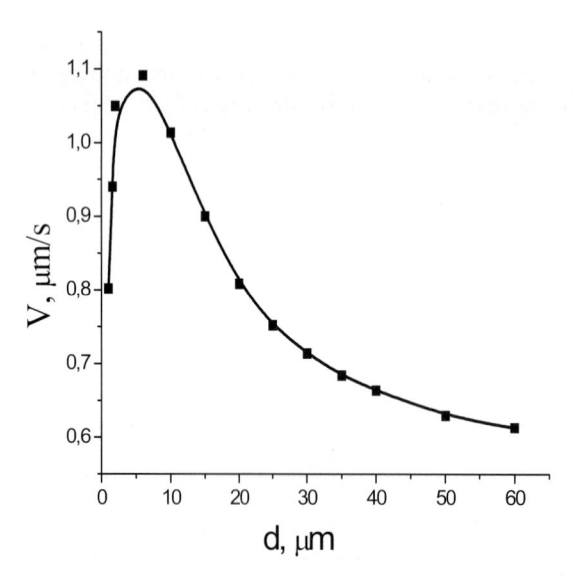

Fig. 2 Size dependence of growth velocity of Si-Ge whiskers.

3 Conclusions

Si–Ge whisker growth was investigated by the CVD-method in the Si–Ge–Br system. The whisker growth may occur under non-equilibrium conditions. The estimated parameters such as coefficient of impurity activity characterizes the degree of non-equilibrium. It was shown that Si–Ge whiskers of smaller diameters grow under non-equilibrium conditions, while growth of the crystals of larger diameters is almost homogeneous. This fact should have an effect on morphology and physical properties of crystals of different diameters, and therefore, on operating characteristics of microsensors designed on their base.

References

[1] V. A.Voronin, I. L. Maryamova, and A. S. Ostrovskaya, Cryst. Prop. and Prepar. **36–38**, 340 (1991).
[2] E. I. Givargizov, Growth of needle-like crystals and plates from vapour, Moskow, 1977 (in Russian).
[3] A. I. Klimovskaya, I. P. Ostrovskii, and A. S. Ostrovskaya, phys. stat. sol. (a) **153**, 465 (1996).
[4] A. I. Klimovskaya, I. P. Ostrovskii, R. I. Baitsar, and A. S. Ostrovskaya, J. Phys.: Condens. Matter. **7**, 1229 (1995).
[5] S. S. Varshava and I. P. Ostrovskii, Sensors and Actuators A **99**, 134–136 (2002).
[6] R. I. Baitsar and A. S. Ostrovskaya, Bul. of Lviv politech univ. **280**, 3–5 (1994) (in Ukrainian).
[7] E. Givargizov, J. Cryst. Growth **31**, 20 (1975).
[8] A. Schetinin, A. Dunajev, and O. Kozenkov, Izv. Vuzov. Fizika **3**, 111 (1982) (in Russian).

Lateral alignment of SiC dots on Si

V. Cimalla[*,1], **J. Pezoldt**[1], **Th. Stauden**[1], **A.A. Schmidt**[2], **K. Zekentes**[3], and **O. Ambacher**[1]

[1] Center for Micro- and Nanotechnologies, Technical University Ilmenau, 98693 Ilmenau, Germany
[2] A.F. Ioffe Physico-Technical Institute of RAS, 26 Polytechnicheskaya Str., 19402 St. Petersburg, Russia
[3] Foundation for Research and Technology-Hellas, P.O. Box 1527, Heraklion, Crete, 71110, Greece

Received 19 September 2003, accepted 3 November 2003
Published online 22 December 2003

PACS 68.65.Hb, 81.07.Ta, 81.16.Dn

An alignment of self-assembled SiC dots grown by molecular beam epitaxy on Si substrates is demonstrated. Atomic force microscopy was applied showing the possibility to control the lateral ordering in linear chains and in dense dot arrays. The large lattice mismatch between Si and SiC of 20% stimulates a three-dimensional nucleation on the substrate. The formation of well ordered monoatomic, and biatomic steps as well as step bands on (100) and (111) Si was used, offering the advantage to not need additional processing steps to define the alignment. However, during the SiC nucleation on Si the substrate participates in the reaction to SiC by lateral Si diffusion resulting in an unstable surface during the growth. Thus, the reproducible control of the nucleation sites independent on the movement of the steps during the growth is a critical issue to achieve the lateral alignment.

© 2004 WILEY-VCH Verlag GmbH & Co. KGaA, Weinheim

1 Introduction

Self-organization phenomena in order to produce quantum dot structures represents a new research area in epitaxial growth techniques. In contrary, antidot systems increased much less attention. They are acting as non-linear scattering centers and influence drastically the lateral conductivity, the photoconductivity or the magnetotransport in vertically confined layer structures like two dimension electron gases (2DEG). Regular arrays of antidots offers the additional possibility to modulate the lateral conductivity or to introduce an artificial anisotropy into the system which has prospective applications for narrow band photodetectors and magnetic frequency dependent switches.

Previous studies concentrated on the formation of antidots in a 2DEG in III-V semiconductors using self-assembled semiconducting islands with a wider bandgap [1] or using etched holes [2, 3]. Much less work is done for the integration of antidots into the Si technology, e.g. in MOS structures [4]. The wide band gap semiconductor SiC is a candidate for the creation of antidots in Si. It can be grown epitaxially on Si [5], however, only a few investigations of SiC nanocrystals on Si exist [6–8] and no demonstration of an alignment can be found in the literature. The large lattice mismatch of 20% stimulates a three-dimensional nucleation on the substrate. However, nonuniformity in the size and the distribution of the islands strongly limit the practical application. Thus, the control of the nucleation sites of SiC is a precondition for a lateral alignment of the SiC nuclei. In the Ge cluster formation technology on Si several ways to localize the nucleation sites are known, e.g., ordering on lithographically pre-patterned substrates [9], on dislocation networks [10], and along step arrays on vicinal substrates [11]. The latter method of two-dimensional ordering has the advantage to not need additional processing steps. However, during the SiC nucleation on Si the substrate participates in the reaction to SiC by lateral Si diffusion. Consequently the surface is unstable during the growth which complicates an alignment control. In this

---

[*] Corresponding author: e-mail: volker.cimalla@tu-ilmenau.de, Phone: +49 3677 69 3408, Fax: +49 3677 69 3355

work we describe a method to form regular arrays of SiC dots on Si based on the control of the nucleation sites by step arrays. We demonstrate how the growth temperature and time can be used to tune the size and distribution of the SiC nanostructures in order to align them regularly on step arrays.

2 Experimental

SiC dots were grown supplying only carbon in a solid source molecular beam epitaxy [12, 13]. The grain size and the morphology were determined by tapping mode AFM (DI Nanoscope) with an etched single crystal Si tip. Data were collected from area scans of (1x1) to (5x5)µm^2 with a 512x512-pixel resolution. The surface was characterized by power spectrum density (PSD) analysis [13], which deconvolutes the roughness as a function of surface lateral length. The off-cut angle of the used substrates was estimated from AFM measurements and confirmed by x-ray diffraction with an uncertainty of 0.05°.

3 Results and discussion

The nucleation and the growth of the SiC islands require the lateral migration of both Si and carbon and the diffusion length L is in the order of the mean separation distance between the nuclei on the surface [14]. However, on vicinal substrates the migration of adatoms from one terrace to another across the step for incorporation into dots may be prohibited due to an energy barrier (Schwoebel barrier[1] formed at the step edge. Thus, the steps effectively reduce the reservoir of atoms available for dot growth. Consequently the terrace length on the Si substrate determines the maximum diffusion length for the adatoms. If the terrace width w matches the temperature dependent nucleus distance, only one island nucleation per terrace appear (Fig. 1a). If the nucleus distance is larger than the terrace width, a random nucleation on the terraces occur (Fig. 1b), and if the nuclei distance is smaller, not on every terrace islands nucleate (Fig. 1c). In the latter two non-matching cases irregular distributed islands are formed [13].

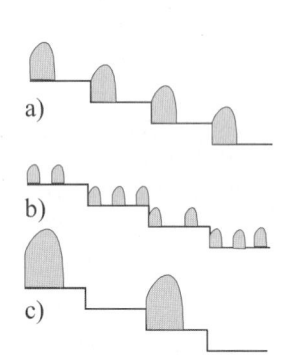

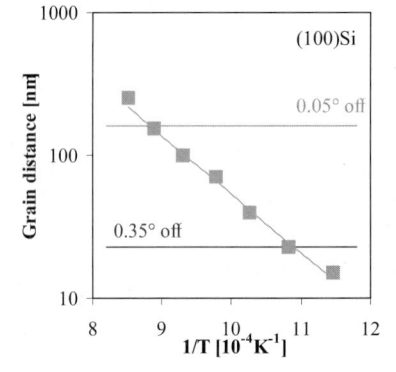

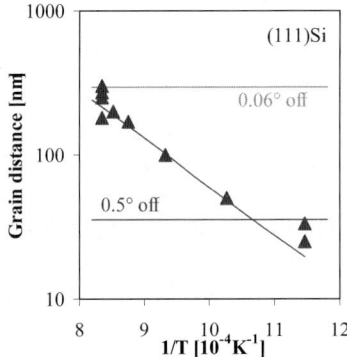

Fig. 1 Alignment of SiC nuclei on a stepped Si surface if a) $w \sim L$, b) $w > L$ and c) $w < L$.

Fig. 2 Equilibrium distance of SiC nucleation sites as a function of the growth temperature at a carbon flux of 10^{13} cm^{-2}s^{-1} on (100)Si [13] and (111)Si. Horizontal lines indicate terrace widths for the vicinal substrates used in this study. Crossing points indicate the optimum temperature for SiC dot alignment along step edges.

Previous studies [7, 13] have shown that if the growth temperature is increased the SiC island density decreases, i.e., the island separation increases (Fig. 2). The nucleation density can also be increased at increased carbon fluxes. The terrace width mainly depends on the off-cut angle of the substrate. For nominal surfaces with off-cut angles below 0.5° monoatomic and biatomic steps, which are defining the terrace width, can be prepared on (100)Si and (111)Si, respectively. Examples of terrace width values for substrates, used in this study, are represented as lines in Fig. 2. Moreover, for vicinal surfaces step bunching can increase the terrace width. Thus, if the growth and surface preparation conditions are tuned in a way that the terrace width matches the island separation distance, a linear ordering should be achieved.

© 2004 WILEY-VCH Verlag GmbH & Co. KGaA, Weinheim

However, this simplified picture does not hold for the nucleation of SiC islands on Si. The main difficulty to grow regular island arrays is the participation of the Si substrate in the reaction with the incoming carbon flux to SiC. It requires a lateral transport of Si to the growing islands resulting in a movement of the step edges during the growth. SiC islands are known to pin steps [15] and the nucleation density increases in the very early stage of growth up to an effective SiC coverage of 10% [16]. Thus, even if regular step or step band arrays can be prepared on the Si surface and the initial SiC islands nucleate regularly, the Si surface transport and the step pinning on islands create irregular step bands. Further nucleation cannot occur anymore on straight step band edges and the distribution of the nuclei will be irregular. This behavior was observed at high temperatures (around 1000 °C, Fig. 3), since at temperatures above 950 °C (i) a remarkable sublimation of Si occurs, and (ii) the growth mode transforms from a quasi two-dimensional nucleation to a pure three-dimensional mode with a higher need of Si. Both effects increase the velocity of the step wandering and consequently favor irregularities.

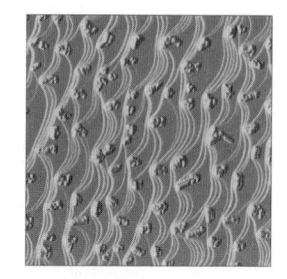

Fig. 3 Shaded AFM image of SiC islands on a 3° vicinal (111)Si surface (5x5 µm²).

A simple geometric model (Fig. 4) can evaluate this step movement and the consequences for an alignment of the SiC islands. From the nucleation rate the amount of Si needed for the growth of the islands can be calculated. This number N of Si atoms causes a step movement Δw:

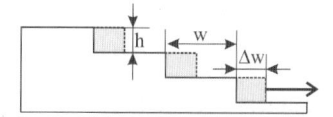

Fig. 4 Schematic representation of the step movement on a vicinal surface

$$N = n\Delta w L^2 \tan\varphi, \quad (1)$$

where n, ϕ, and L^2 are Si density, off-cut angle, and area unit, respectively. On the selected (111)Si substrates with an off-cut angle of 0.06° steps with an average terrace width of 300 nm are formed. According to the matching condition in Fig. 2 the nucleation temperature has to be adjusted to a value as high as 925 °C in order to match terrace width and nucleus separation. The equilibrium island density is reached at an effective surface coverage of 8% resulting in a step movement around 5 nm. The evolution of the distribution and size was presented in a previous paper [16]. After reaching the saturation island density at the selected growth conditions a regular decoration of the steps occur (Fig. 5a). The step edges are still running parallel and through the islands indicating a movement below 10 nm in agreement with calculation. This linear alignment of the SiC islands remains stable at longer growth times even if now a partial step pinning could be observed at the irregularities along the step [16]. To obtain the positional correlation properties of the dots, the PSD of the AFM image is calculated, which yields information on the preferred spatial frequencies (Fig. 5a). It shows a broad peak perpendicular to the steps and a constant value at low wave numbers k along the steps. These differences indicate the periodicity of the steps and the lack of correlation of the islands along the steps. In addition, the correlation length differs for the two directions from 250 nm to 300 nm for the mean island separation along the steps (corresponding to Fig. 2b) and the terrace width perpendicular to the steps, respectively. Thus, even if an exact matching was not obtained, the SiC islands tend to form chains along the steps instead of being periodically spaced.

This one-dimensional surface anisotropy is not efficient enough to induce a periodic two-dimensional lateral ordering of the dots. One way is to decrease the average terrace width on the Si surface to a value comparable to the typical SiC islands dimensions in order to obtain a perfect 1D alignment of the islands along the steps. This was achieved by increasing the substrate misorientation. For the two-dimensional ordering still an exact matching of the terrace width and the equilibrium island density is necessary. Thus, lower growth temperatures were necessary (Fig. 2). An example on (100)Si with the relatively high off-cut angle is presented in Fig. 5b. At the off-cut angle of 0.35° monoatomic steps were formed after the precleaning and the removal of a sacrificial oxide layer. These step bands are very regular and parallel with a mean terrace width around 25 nm as measured by AFM. The image in Fig. 5b was taken after the deposition of effectively 2.5 nm SiC and represents an almost coalescent surface. The dot

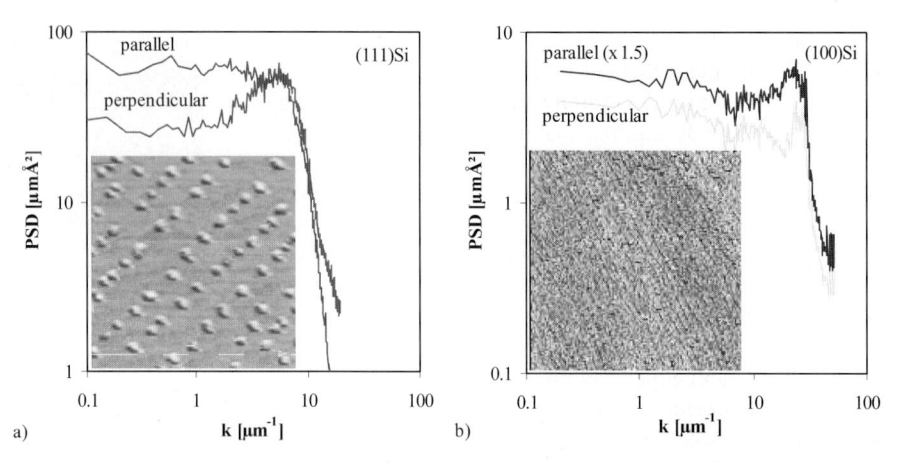

Fig. 5 PSD parallel and perpendicular to the step bands for SiC chains on (a) (111)Si, 0.06° off at 925 °C and (b) (100)Si, 0.35° off at 700 °C. Inlets are the corresponding AFM images {2x2μm^2, z-scale (a) 20 nm, (b) 2 nm}

alignment in both directions is clearly visible in the PSD (Fig. 5b), even if the peak for the PSD along the steps is still broader. These results proof the possibility of controlling the islands spatial distribution by the nucleation density and the step periodicity.

4 Summary

Lateral aligned self-assembled SiC dots have been grown for the first time on Si substrates by molecular beam epitaxy. The alignment was defined by the formation of well-ordered monoatomic and biatomic steps as well as step bands on (100) and (111) Si. These terraces promote an alignment along their step edges. However, the adjustment of the nucleation density according to the terrace width depends critically on the Si transport and the stability of the steps on the surface. The arrangement was controlled by three key parameter: (i) the temperature and the flux control the nucleation density, (ii) the processing time defines the size, and (iii) the terrace width has to be adjusted close to the average distance of the nuclei. As a result an ordering in linear chains and in dense dot arrays can be achieved.

Acknowledgements This work was partially supported by the Research Grant FMRX-CT97-0141 and the DAAD program "Internationale Qualitätsnetzwerke (IQN)", project "Ilmenau Coordination center".

References

[1] Yu. Vasiyev, S. Suchalkin, M. Zundel, et al., Appl. Phys. Lett. **75**, 2942 (1999).
[2] B.G.L. Jager, S. Wimmer, A. Lorke, et al., Phys. Rev. B **63**, 045315 (2001).
[3] Lorke, S. Wimmer, B. Jager, J.P. Kotthaus, W. Wegscheider, and M. Bichler, Physica B **249**, 312 (1998).
[4] A.Huber, I. Jejina, H. Lorenz, et al., Semicond. Sci. Technol. **10**, 365 (1995).
[5] S. Nishino, J.A. Powell, and H.A. Will, Appl. Phys. Lett. **42**, 460 (1983).
[6] Y. Sun, T. Ayabe, and T. Miyasato, Jpn. J. Appl. Phys. **38**, L1166 (1999).
[7] F. Scharmann, P. Maslarski, W. Attenberger, et al., Thin Solid Films **380**, 92 (2000).
[8] J. Yang, X. Wang, G. Zhai, N. Cue, and X. Wang, J. Cryst. Growth **224**, 83 (2001).
[9] G. Jin, J.L. Liu, S.G. Thomas, Y.H. Luo, K.L. Wang, and B.-Y. Nguyen, Appl. Phys. Lett. **75**, 2752 (1999).
[10] F. Leroy, J. Eymery, P. Gentile, and F. Fournel, Appl. Phys. Lett. **80**, 3078 (2002).
[11] K. Sakamoto, H. Matsuhata, M.O. Tanner, D. Wang, and K.L. Wang, Thin Solid Films **321**, 55 (1998).
[12] J. Pezoldt, V. Cimalla, Th. Stauden, et al., Diam. Rel. Mater. **6**, 1311 (1997).
[13] V. Cimalla and K. Zekentes, Mater. Sci. Forum **389–393** (2002) 747.
[14] A. Pimpinelli and J. Villain, Physics of Crystal Growth, Cambridge University Press, 1998, p. 181ff.
[15] Y. Ishikawa, N. Ikeda, M. Kenmochi, and T. Ichinokawa, Surf. Sci. **159**, 256 (1985).
[16] V. Cimalla, J. Pezoldt, Th. Stauden, Ch. Förster, O. Ambacher, A. A. Schmidt, and K. Zekentes, submitted to Appl. Phys. Lett.

The role of Ge predeposition temperature in the MBE epitaxy of SiC on silicon

F. M. Morales[*,1]**, Ch. Zgheib**[2,3]**, S. I. Molina**[1]**, D. Araújo**[1]**, R. García**[1]**, C. Fernández**[4]**, A. Sanz-Hervás**[5]**, P. Masri**[3]**, P. Weih**[2]**, Th. Stauden**[2]**, V. Cimalla**[2]**, O. Ambacher**[2]**, and J. Pezoldt**[2]

[1] Dpto. de Ciencia de los Materiales e Ingeniería Metalúrgica y Química Inorgánica, Universidad de Cádiz, Apdo. 40 Puerto Real, 11510-Cádiz, Spain
[2] FG Nanotechnologie, Zentrum für Mikro- und Nanotechnologien, Postfach 100565, 98684 Ilmenau, Germany
[3] Groupe d'Etude des Semiconducteurs, CNRS-UMR 5650, Université de Montpellier 2, cc074, 12 Place Eugene Bataillon, 24095 Montpellier, France
[4] Dpto. de Química Física, Universidad de Cádiz, Apdo. 40 Puerto Real, 11510-Cádiz, Spain
[5] Dpto. de Tecnología Electrónica, ETSI Telecomunicación, Universidad Politécnica de Madrid, 28040 Madrid, Spain

Received 19 September 2003, accepted 3 November 2003
Published online 22 December 2003

PACS 61.10.Nz, 61.14.Lj, 68.35.Fx, 68.37.Lp, 68.49.Sf, 68.55.Jk

An alternative method for stress relaxation in the SiC/Si heteroepitaxial system based on the incorporation of a group-IV element (germanium) into the interface between SiC and Si is presented. We have investigated the effect of the temperature during the predeposition of 1 ML of Ge on a silicon (111) surface, prior to the MBE carbonization. The resulting structures were investigated by transmission electron microscopy (TEM), x-ray diffraction (XRD), secondary ion mass spectrometry (SIMS) and Fourier transform infrared spectroscopy (FTIR). The crystalline quality of the 3C-SiC layer was improved by the Ge predeposition, as shown by XRD measurements and by the smaller damping constants of the TO and LO phonons in the FTIR analysis. As the temperature of Ge predeposition increases, the SiC layer exhibits lower residual strain and larger grain size. Furthermore, the incorporation of Ge at the interface suppresses the outdiffusion of Si from the substrate to the surface of the growing SiC layer and, therefore, impedes the formation of voids at the SiC/Si interface. TEM and SIMS results revealed a strong segregation of Ge at the interface leading to an increased stress relaxation, in agreement with theoretical predictions.

© 2004 WILEY-VCH Verlag GmbH & Co. KGaA, Weinheim

1 Introduction There have been several attempts to increase the range of materials compatible with Si technologies. Heterostructures of silicon carbide on silicon (SiC/Si) have proved useful in micro- and opto-electronic applications and also as templates for the overgrowth of SiC or III-group nitrides [1]. Due to the high difference in lattice parameter and thermal expansion coefficient between SiC and Si, carbonization of Si is the most common and extended method for improving the quality of SiC/Si interfaces before a subsequent epitaxial growth. However, in these "carbon-modified" surfaces, the misfit strain relieved by the formed misfit dislocation network is not able to achieve a fully relaxed 3C-SiC thin layer (normally in tension [2]). This residual strain added to a significant component of strain generated near the interface region of the Si substrates (usually compressive) often causes extended defects such as planar defects in the SiC growing layer, voids at Si behind the heterointerface and bending [3].

In order to improve the electrical and structural quality of these SiC/Si heteroepitaxies, the combination of carbonization and growth of layered systems based on group-IV elements (i.e., adding Ge in the

---

[*] Corresponding author: e-mail: fmiguel.morales@uca.es, Phone: ++ 34 956 01 6335, Fax: ++ 34 956 01 6288

initial steps of growth) is starting to be considered [4]. In these approaches, volume effects originated by Ge incorporation play an important role since Ge atoms have a larger size than Si and C. Additionally, the use of MBE techniques allow a layer by layer "in-situ" control of the deposition process and carbonization or growth at relatively low temperatures (<1000 °C), with associated advantages such as lower diffusion coefficients, suppression in the deterioration (structural and morphological defects) of the SiC films and feasible implementation in microelectronic industries.

In this report, we present the growth and characterization of SiC layers grown on Si(111) substrates with Ge-modified interfaces and the influence of varying the temperature of Ge atoms predeposition before SiC growth.

2 Experimental

120 nm-thick SiC layers were grown by solid-source molecular beam epitaxy (SSMBE) on on-axis boron doped p-type Si(111) wafers (resistivity of 1 to 10 $\Omega \cdot$ cm) in a UMS 500 Balzers MBE system. The growth procedure consists of the following process steps: (1) hydrogen plasma cleaning of the silicon substrates, (2) annealing at 750 °C for 1 h in the deposition chamber, (3) 1 ML Ge deposition on the (7×7)-Si reconstructed Si surface at temperatures ranging from room temperature to 660 °C by electron beam evaporation (ML with respect to the Si(111) surface), (4) deposition of C at 325 °C, (5) gradual increase of the substrate temperature in steps of 50 degrees up to the final growth temperature for SiC epitaxy, (7) the SiC deposition started at 850 °C with a growth rate of 1 nm/min under Si rich conditions and continuously operating Si and C sources. The epitaxial growth was carried out at 1000 °C. Other two samples previously carbonized on its Si surfaces, one by SSMBE and the other by RTCVD (for details see [5]), were also subject to step (7) and 120 nm SiC epitaxial growth in order to study the difference when Ge is not added. Table 1 shows the main differences among the samples. The evolution of the growth process was controlled "in-situ" by reflection high energy electron diffraction (RHEED), spectroscopic ellipsometry (SE) and "ex situ" by Auger electrons spectroscopy (AES) and atomic force microscopy (AFM).

Further "ex-situ" investigations of the epitaxial layers were carried out to study the thickness and crystalline structure by the following techniques: secondary ion mass spectrometry (SIMS), transmission electron microscopy (TEM), selected-area electron diffraction (SAED), high-resolution electron microscopy (HREM), X-ray diffraction (XRD) and Fourier-transform infrared spectroscopy (FTIR). Depth profiling by SIMS was carried out with an IMS 4F CAMECA. An O_2^+ primary beam with impact energy of 14.96 kV was used. The incidence angle was 60°. Specimens were prepared for cross-section TEM (XTEM) and planar-view TEM (PVTEM) using mechanical thinning and Ar^+ milling at 4.5 kV in a Gatan Dual Ion Mill system. Conventional TEM was carried out in a JEOL JEM-1200EX electron microscope; a JEOL JEM-2000EX/THR was used for HRTEM. FTIR spectra with a resolution of 4 cm^{-1} in reflectivity mode were collected with a μ-FTIR Nicolet Impact 410 equipment attached to a FTIR microscope IR Plan Advantage. XRD patterns were measured with a Bede D^3 diffractometer. θ/2θ scans were recorded around the 111 and 333 reflections of Si and 111, 222 and 333 of SiC. Pseudo-Voigt functions were fitted to the SiC peaks to deduce the grain size and microstrain of the samples [6]. Rocking-curves (r-c) around the 111, 222 and 333 reflections of SiC were also measured.

Table 1 Temperature of Ge predeposition (TP) and quantity of Ge predeposited [Ge] before the subsequent growth of a 120 nm SiC layer.

Sample	A	B	C	D	E	F
T P (°C)	20	325	500	660	20	20
[Ge] (ML)	1	1	1	1	0	0

3 Results and discussion

3.1 FTIR measurements

Reflectance FTIR allows the direct identification of the SiC layer. Due to the relative good crystalline quality and small thickness of the SiC layers [7], FTIR spectra did not show typical SiC/Si Reststrahlen bands. Anyway, the observed peaks proved the good quality of the SiC material. An isolated peak with a maximum located around 790.7 cm^{-1}, corresponding to the (w_{TO}) "stretching" mode of almost stoichometric SiC, is observed for all the SiC layers. The measured values are close to the wavenumber observed by other authors for similar structures or bulk SiC (from 794 to 796 cm^{-1}) [8, 9]. The minimum located around 966 cm^{-1} is associated with the LO phonon mode of 3C-SiC (ideally near 971 cm^{-1}). The observed small shifts are related to residual stress in the non-fully relaxed SiC layers. The stress could indicate the presence of defects, disorders, doping effects or the influence of the Si substrate on the interface region [10]. Figure 1(a) shows an FTIR spectrum measured from sample A. The rest of the samples provided almost identical spectra. We have not found a clear relation between the SiC peak positions and the temperature of Ge predeposition.

3.2 XRD measurements θ/2θ patterns give information on interplanar spacings along the selected orientation. In our case, from the Bragg angle of the 111, 222 and 333 SiC peaks, we have accurately deduced the interplanar spacing along the [111] direction (d_{111}). Fig. 1(b) shows a representative θ/2θ pattern. The spacing d_{111} changes from 2.5124 Å to 2.5132 Å as the temperature of Ge predeposition (TP) goes from 20° to 660°. All these values are smaller than the bulk SiC d_{111} value (2.5170 Å). Therefore, the SiC layers are not fully relaxed, but are still subject to a residual in-plane tensile strain probably due to the lattice mismatch with the Si substrate. The in-plane strain can be estimated using conventional elasticity theory, yielding 0.86% for TP = 20° and 0.72% for TP = 660°. This is shown in Table 2, along with the values of the SiC/Si mismatch for the different samples. Thus, Ge predeposition reduces the residual mismatch stress of the SiC layer.

By analyzing the shape of the SiC 111 peak it has been possible to quantify the grain size in each layer. The grain size increases with the temperature of Ge predeposition, ranging from 480 Å for TP = 20° to more than 1000 Å for TP = 660°. The full-width-at-half-maximum (FWHM) of the rocking-curves decreases slightly as the temperature of Ge predeposition increases (see Table 2). Fig 1(c) shows an example. It must be noted that rocking-curves provide a measurement of the dispersion of the crystal orientation along a certain axis (the [111] axis in our case). Therefore, the increase of the Ge predeposition temperature has positive effects for the crystalline quality of the SiC layers. It seems that the characteristics of the Ge-covered interface become more adequate for the epitaxy of SiC as the temperature is increased. The reason for this improvement will be discussed in the TEM section.

Table 2 summarizes the most important numeric results. Here, d_{111} is the measured {111} interplanar spacing of the 3C-SiC; $\varepsilon_\parallel$ is the calculated in-plain strain; m is the mismatch of the SiC layer with the Si substrate; r-c FWHM is the average FWHM of the rocking-curves around the 111, 222 and 333 SiC reflections; and D is the grain size.

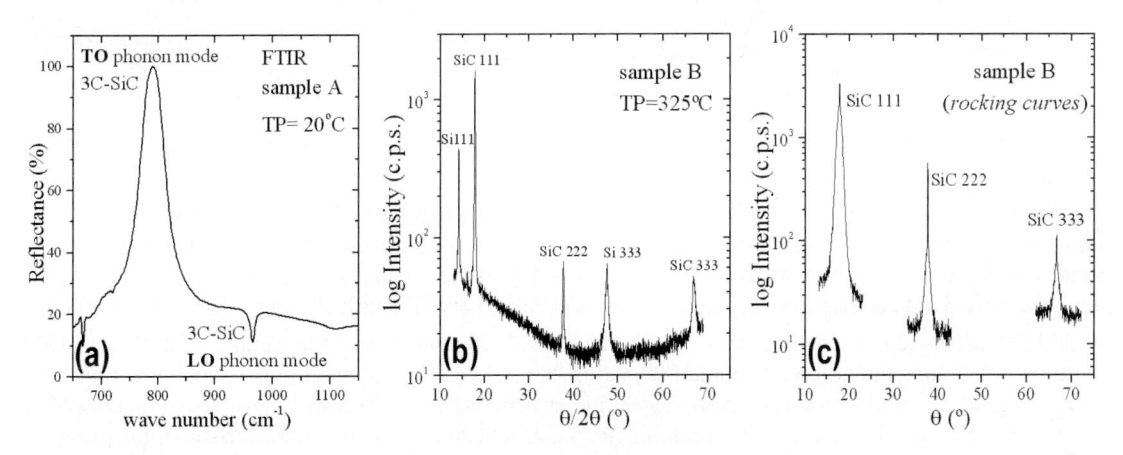

Fig. 1 FTIR spectra for sample A with Ge predeposited at 20 °C (a); XRD study: θ/2θ pattern (b) and rocking-curves (c) of sample B with Ge predeposited at 325 °C.

Table 2 Summary of XRD data for samples A–D.

SAMPLE	d_{111} (Å)	in-plane strain $\varepsilon_\parallel$ (%)	mismatch (m) (%)	r-c FWHM (°)	D (Å)
A	2.5124	0.86	19.04	1.062	480
B	2.5132	0.72	19.15	1.059	590
C	2.5130	0.74	19.14	1.044	990
D	2.5132	0.72	19.15	1.028	1070

3.3 SIMS measurements The experimental depth profiles show a similar aspect for samples A–D, with three different regions. Figure 2(a) shows a SIMS depth profile for sample D from which the SiC layer and the Si substrate can be identified clearly. The x-axis scale was calibrated by assigning the SiC thickness (120 nm) to the point where the Si signal varies by 50%. The interface region is defined by using the silicon signal variation (16%–84%); its borders are marked by dashed lines in Fig. 2(a). Germanium atoms are located at this interface layer; they diffuse over 30 nm into the Si substrate, having a maximum near the initial Si surface. The C distribution inside the Si substrate, in terms of $^{28}Si^{12}C^+$ counts, shows a deep tail with a marked decay that vanishes. It is confined from the interface region to the limit of the Ge signal indicating a suppressed diffusion by the SiC forming elements. Moreover, the detected amount of Ge at the interface was independent on the predeposition temperature and no signal of Ge is detected in the SiC layer and at the SiC surface, indicating the absence of Ge out-diffusion in the growth experiments. Figure 2(b) compares the Ge signals of all the samples. The interface width and the FWHM of the Ge distribution are very similar, having a minimum for samples B and C. Nevertheless, the interface thickness is a few nanometres for all the samples.

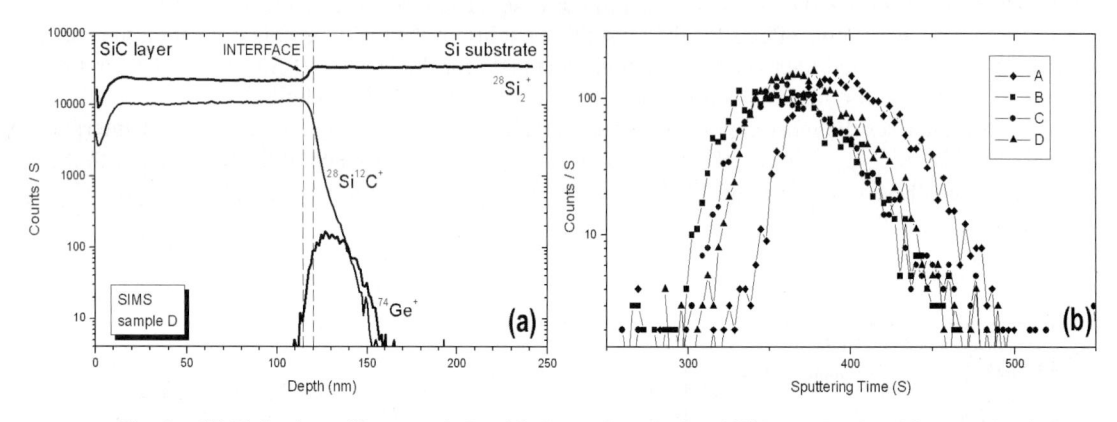

Fig. 2 SIMS depth profiles: sample D with Ge predeposited at 660 °C (a); Ge signal for samples A–D.

3.4 TEM measurements The β-SiC(111)//Si(111) structures exhibit SiC with good crystalline orientation with respect to the Si substrate. Extended defects were found in all the samples studied by TEM (with and without Ge predeposition). The SiC layer is almost fully relaxed, as concluded from the SAED analysis carried out for samples prepared in both XTEM and PVTEM orientations. Figure 3 corresponds to SAED patterns registered along the [111] (a) and the [1$\bar{1}$0] zone axes (b) in sample A. In these SAED patterns, reflections due to typical cubic SiC and Si lattices perfectly aligned in the same direction indicate the good orientation between both crystalline structures. The XTEM/SAED pattern for sample A shows elongated diffuse contrasts between the SiC main reflections that are characteristic of the presence of planar defects. This elongated contrast is not observed between PVTEM/SAED patterns, where the main feature is the presence of multiple diffraction produced by the interaction of SiC/Si. Figure 3(c) is a

PVTEM/DF two-beam image using the $(02\bar{2})$ reflection near the [111] zone axis of sample A. The contrasts due to the high density of defects in the SiC layer make difficult the detailed identification of each of them, even by two-beam techniques. The inserted SAED pattern in figure 3(c) corresponds to the diffraction of the same free-standing SiC layer. The intense dots are the main reflections of 3C-SiC, while the set of less bright diffraction dots (some of them encircled) correspond to {4/3 2/3 2/3} forbidden reflections of SiC that appear when a high density of planar defects (twins, stacking faults…) parallel to the (111) plane exists. There are some other SiC diffraction dots (far from the transmitted-central axis) with a triangular distortion also related to the presence of defects.

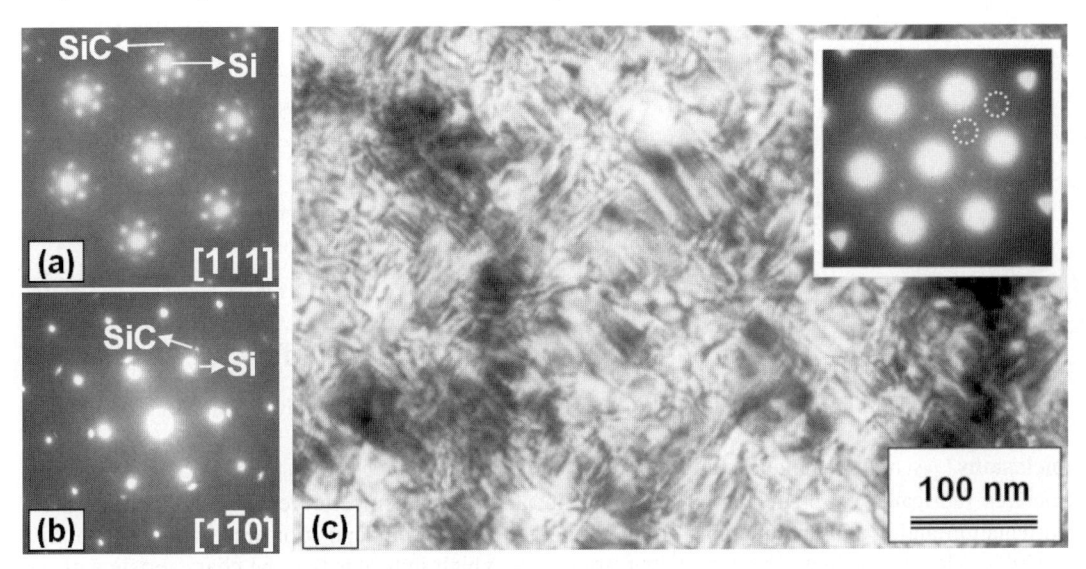

Fig. 3 Sample A: SAED patterns registered in PVTEM (a) and XTEM (b) orientations of 3C-SiC/ /Si heterostructure. PVTEM micrograph and associated SAED pattern for the 120 nm isolated SiC layer (c).

The most important feature, which distinguishes samples with or without predeposited Ge, is the presence of voids in their structure. These defects are associated with outdifussion of Si induced by the stress inherent to the high lattice mismatched SiC/Si system. In contrast, typical voids are not present or only occasionally present in the Ge-modified interfaces. In order to study the void suppression a detailed PVTEM study was carried out. Parts (a), (b) and (c) of figure 4 correspond to PVTEM/BF images of samples F (RTCVD-carbonized without Ge), E (SSMBE-carbonized without Ge) and A (1ML of Ge predeposition and SSMBE-carbonized), respectively. Samples without Ge show a high density of voids, whereas the sample with Ge is not affected by these defects. The voids size distribution over the Si surfaces in samples E and F show an exponential decay similar to that previously observed in SSMBE-carbonized (001) and (111) Si surfaces (without epitaxial overgrowth of SiC) [5]. In this case, equilateral triangles (corresponding to inverted hollow pyramids) are more defined and sharper than those observed in the SSMBE-carbonized case. These voids are one order of magnitude larger than those previously observed (tens of thousand nm^2 in comparison with the current units of μm^2). Indeed, the latter difference is proportional to the thickness of the SiC layer. In the case of RTCVD carbonization plus SiC growth, for which the best sealing properties were predicted, voids change from being occasional and very small (in the carbonized state) to having similar average sizes than in the SSMBE carbonization plus SiC epitaxy case. However, these RTCVD voids grow in a narrower range of sizes. Histograms shown in figure 4(d) represent the void density for samples E (SSMBE-carbonized) and F (RTCVD-carbonized).

HREM micrographs at the SiC/Si heterointerface show two clear crystalline structures related to the relaxed Si substrate and the SiC lattices. Besides, a distinct contrast (structure) is observed at the heterointerface. This new structure is not uniform; its thickness varies between 1 and 3 nm depending on the region studied. It is pseudomorfically grown on the Si substrate because the lattice planes maintain their continuity, and a misfit dislocation network is placed just on top of it. This interlayer stops the formation of voids, thus producing a combination of: (a) increase of crystalline density, (b) sealing of Si out-diffusion channels, and (c) stress reduction. Some roughness associated to atomic steps that originates threading planar defects towards the SiC overlayer are also visible. HREM observations indicate that C and Ge atoms inside the Si substrate at the SiC/Si interface are interstitially located, since Si lattice planes show similar spacings to those measured in the relaxed bulk Si substrate. Other complementary theoretical [4] and experimental [11] studies propose that the new phase corresponds to a $Si_{1-x-y}C_xGe_y$ solid solution.

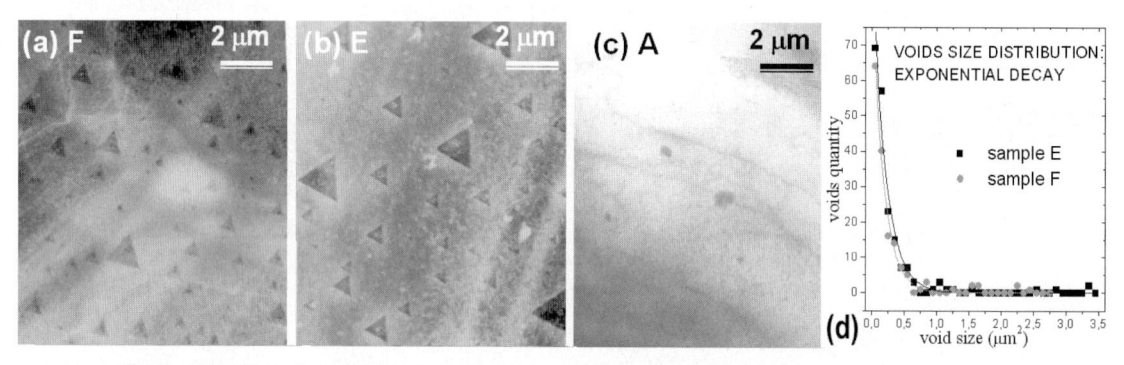

Fig. 4 PVTEM micrographs of samples F (a), E (b) and A (c) and voids study for samples E and F (d).

4 Conclusions
A multitechnique study has been carried out to investigate β-SiC(111)//Si(111) heterostructures fabricated by predepositing 1ML of Ge at different temperatures. The SiC layers show good orientation and crystalline quality. As the temperature of Ge predeposition was increased, the SiC layer exhibited lower residual strain and larger grain size, which indicates that the characteristics of the Ge-covered interface become more adequate for the epitaxy of SiC as the predeposition temperature increases. Ge segregation is limited to the surface of the Si substrate and does not extend into the SiC layer. The formation of a $Si_{1-x-y}C_xGe_y$ layer at the interface region prevents the Si outdifussion and, consequently, the formation of voids

Acknowledgements Support by Centre of Excellence CELDIS, contract ICA1-CT-2000-70018 and by MCYT, MAT2000-0478-P4-02 is gratefully acknowledged. The authors thank Dr. E. Bustarret for the fruitful discussions on FTIR results.

References

[1] V. Lebedev, J. Pezoldt, V. Cimalla, J. Jinschek, F. M. Morales, and O. Ambacher, phys. stat. sol. (c) **0**, 183 (2002).
[2] F. M. Morales, S. I. Molina, D. Araújo, R. García, V. Cimalla, and J. Pezoldt, Diam. Rel. Mater. **12**, 1227 (2003).
[3] J. Camassel, S. Contreras, and J. L. Robert, C. R. Acad. Sci. Paris, t. 1, S. IV, 5 (2000).
[4] P. Masri, N. Moreaud, N. Rouhani Laridjani, J. Calas, M. Averous, G. Chaix, A. Dollet, R. Berjoan, and C. Dupuy, Mater. Sci. Eng. B **61–62**, 535 (1999).
[5] F. M. Morales, S. I. Molina, D. Araújo, V. Cimalla, and J. Pezoldt, Mater. Sci. Forum **433–436**, 285 (2003).
[6] Th. H. de Keijser, E. J. Mittemeijer, and H. C. F. Rozendaal, J. Appl. Cryst. **16**, 309 (1983).
[7] R. T. Hola, P. H. Klein, and P. E. R. Nordquist, Jr., J. Appl. Phys. **60**, 1479 (1986).
[8] D. Olego, M. Cardona, and P. Vogl, Phys. Rev. B **24**, 3878 (1982).
[9] H. Ohyama, T. Suzuki, K. Nishi, T. Mitsuyu, and T. Tomimasu, Appl. Phys. Lett. **71**, 823 (1997).
[10] M. B. el Mekki, J. Pascual, M. Androulidaki, K. Zekentes, J. Camassel, and J. Stoemenos, Diam. Rel. Mater. **6**, 1772 (1997).
[11] J. Pezoldt, Ch. Zgheib, P. Masri, M. Averous, F. M. Morales, R. Kosiba, G. Ecke, P. Weih, and O. Ambacher, Surf. Inter. Anal., (submitted).

3C-SiC:Ge alloys grown on Si (111) substrates by SSMBE

P. Weih[1], V. Cimalla[1], Th. Stauden[1], R. Kosiba[1], G. Ecke[1], L. Spiess[1], H. Romanus[1], M. Gubisch[1], W. Bock[2], Th. Freitag[3], P. Fricke[3], O. Ambacher[1], and J. Pezoldt[*, 1]

[1] Zentrum für Mikro- und Nanotechnologien, TU Ilmenau, PF100565, 98693 Ilmenau, Germany
[2] Institut für Oberflächen und Schichtanalytik, Universität Kaiserslautern, Erwin-Schrödinger Str., Gebäude 56, 67663 Kaiserslautern, Germany
[3] SIMS-Labor, Institut Fresenius, Zur Wetterwarte 10, Haus 107, 01109 Dresden, Germany

Received 19 September 2003, accepted 3 November 2003
Published online 22 December 2003

PACS 68.35.Dv, 68.37.Lp, 68.55.Nq, 81.05.Je, 81.15.Hi, 81.70.Jb

In this work for the first time a comprehensive research of $(Si_{1-x}C_{1-y})Ge_{x+y}$ thin films epitaxially grown on Si (111) substrates by solid source molecular beam epitaxy are presented. The layers were grown at substrate temperatures ranging from 900 °C to 1040 °C with a growth rate of 0.6 nm/min. They were analysed by atomic force microscopy, scanning electron microscopy, X-ray diffraction, Auger electron spectroscopy, secondary ion mass spectroscopy and transmission electron microscopy. The structural analysis revealed that the grown epitaxial layer consists of the cubic polytype. A maximum Ge incorporation of 0.16% was achieved in epitaxial layers grown at 900 °C. It was obtained that the Ge concentration decreases with increasing growth temperature.

© 2004 WILEY-VCH Verlag GmbH & Co. KGaA, Weinheim

1 Introduction

Silicon carbide exhibits higher saturation electron drift velocity, break down field and thermal conductivity in comparison to the conventional semiconductor material Si. For this reason SiC is an important wide band gap semiconductor material that provides a great potential for applications in electronic devices, operating at high temperature, high frequency and harsh environment conditions. SiC exhibits also some disadvantages that lead to limitations in electronic applications such as: (1) relative low carrier mobilities, (2) absence of well developed techniques for the band-gap engineering and heterostructure formation. Furthermore, if the 3C-SiC polytype with the highest carrier mobilities is desired, no bulk material substrates are available. One way to smooth out some of this disadvantages is the use of Ge during the growth of 3C-SiC on silicon. The incorporation of Ge into the interface between SiC and Si leads to stress reduction in the heteroepitaxial layer [1]. Furthermore, if SiC will be alloyed with Ge, an increase of the carrier mobility can be expected, because of the higher carrier mobilities in Ge. This provides the opportunity to improve the properties of SiC based devices. Additionally, a modification of the band gap is possible. This can be used for band gap engineering applications [2, 3]. The disadvantage of this alloy is the predicted thermodynamical metastability of the binary GeC system with respect to decomposition into its elemental constituents (enthalpy of formation of GeC is positive with a value of $\Delta H = 0.27$ eV/atom). For this reason the GeC alloy does not occur under equilibrium conditions [4–6]. Although the group IV elements such as Ge are immiscible in crystalline SiC under equilibrium conditions (max. solid solubility 3×10^{20} atoms/cm^3 at 2500 °C [7]). Therefore nonequilibrium methods of material synthesis must be chosen to form the required composition. Such methods are (1) ion beam

---
[*] Corresponding author: e-mail: joerg.pezoldt@tu-ilmenau.de, Phone: +49 3677 693408, Fax: +49 3677 693355

© 2004 WILEY-VCH Verlag GmbH & Co. KGaA, Weinheim

synthesis [3, 8, 9], (2) plasma assisted deposition and (3) MBE at low temperatures [10]. In the present study our effort was to obtain Ge incorporation into the 3C-SiC lattice by using solid source molecular beam epitaxial (SSMBE) growth on Si substrates in order to create a cubic SiC:Ge alloy. The dependence of the Ge concentration from the growth temperature and the distribution of the Ge in the epitaxial layers are investigated.

2 Experimental

The thin films (thickness ~120 nm) were grown by solid source molecular beam epitaxy (MBE) on *on-axis* (111) Si wafers at different substrate temperatures in the range from 900 °C to 1040 °C with a growth rate of 0.6 nm/min. Prior to epitaxial growth the Si (111) substrates were cleaned in low energy argon-hydrogen plasma without any wet cleaning steps. This treatment results in a hydrogenated surface with an rms roughness below 1 nm. After an annealing step at 780 °C it transforms into a smooth surface with a (7 × 7) reconstruction. The carbonisation of the substrates was carried out by depositing 6 ML carbon at 325 °C followed by a gradual increase of the substrate temperature up to the final SiC growth temperature. Prior to the alloy growth a 10 nm thick 3C-SiC buffer layer was grown. The deposition of the $(Si_{1-x}C_{1-y})Ge_{x+y}$ epitaxial layer was carried out under Si rich conditions and continuously operating Si, C and Ge sources. The Ge flux was varied according to the growth temperature in such a manner that the Ge supply to the surface was kept constant. The stability of the deposition conditions was controlled by monitoring the appearance or the disappearance of 3D-Si nucleation or the intensity of (3x3)-Si related surface reconstruction streaks on the 3C-SiC (111) surface with *in situ* high energy electron diffraction (RHEED). The structural and morphological investigations of the samples were carried out by using atomic force microscopy (AFM), scanning electron microscopy (SEM), X-ray diffraction (XRD) and transmission electron diffraction (TEM). AFM, SEM, XRD and TEM were carried out on TopoMetrix TMX 2000 Scanning Probe Microscope in contact mode, XL 30 LaB6 ESEM (FEI), Siemens D5000 Diffractometers, and TEM Tecnai 20 S-TWIN, respectively. The chemical composition was determined by energy dispersive X-ray analysis (EDX) with SEM and TEM, Auger electron spectroscopy (AES) and secondary ion mass spectrometry (SIMS). The sputter depth profiling in conjunction with AES was carried out with a Riber ASC2000 using 1 keV Ar^+ ions impinging onto the surface under 60° with respect to the surface normal. The SIMS spectra were measured with a CAMECA IMS 4F system using O^+ primary ions and energy of 12.5 keV. For the calibration of the Ge concentration within the SiC layers a 6H-SiC sample implanted with $5H10^{14}$ cm^{-2} Ge at 200 keV and annealed at 1050 °C was used.

3 Results and discussion

The morphology of the sample surface was investigated by means of AFM and SEM. Large area (70 × 70 μm^2) and small area (5x5 μm^2) AFM scans were carried out. The large area scans show hillocks on the surface with a size of 1 to 5 µm in diameter and a height of 100 to 450 nm (Fig. 1).

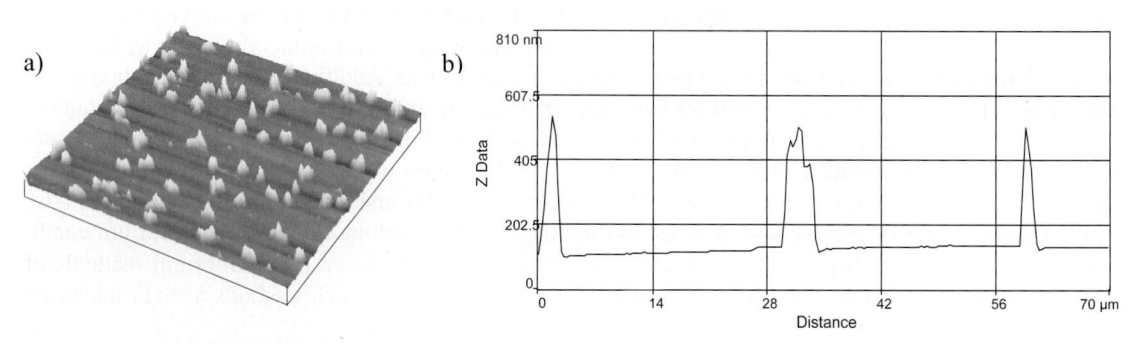

Fig. 1 (a) AFM surface Topography image (70 × 70 μm^2 scan); (b) line scan across the hillocks

© 2004 WILEY-VCH Verlag GmbH & Co. KGaA, Weinheim

In comparison to the reference sample where SiC was grown without Ge an increase of the hillock size and a decrease of the hillock density was observed. The small area scans revealed a relative smooth surface with root mean square (rms) value between 0.97 and 3.4 nm in the regions without hillocks. Furthermore, in this region the terrace width increases with increasing growth temperature.

Further information about the morphology and the chemical composition of the layers gave SEM investigations. They were obtained with accompanying EDX analysis. The EDX measurements were carried out with the minimal possible acceleration voltage (6 keV) in order to minimize the electron penetration in the depth of the Si substrate. The local EDX analysis revealed a lateral inhomogeneous distribution of Ge across the layer (Fig. 2). The Ge signal was detected only in regions where hillocks can be found. Thus if Ge is incorporated into the epitaxial grown $(Si_{1-x}C_{1-y})Ge_{x+y}$ layers the Ge concentration is lower than the sensitivity of the EDX method.

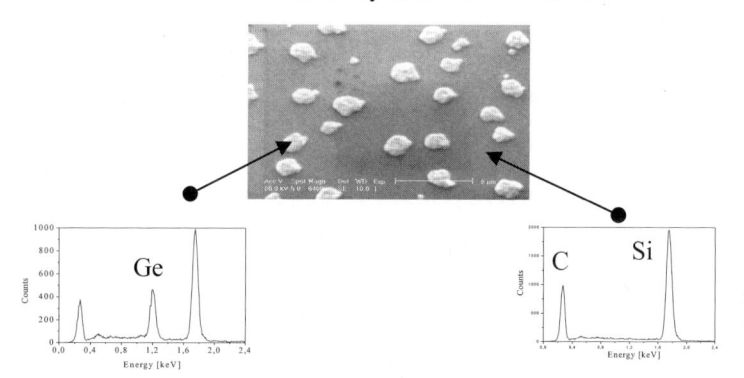

Fig. 2 SEM image of the layer surface and EDX spectra from the hillock and from the layer

Fig. 3 TEM cross-section micrograph of the SiC:Ge layer grown on Si (111) substrate

In order to obtain additional information about the origin, the growth mode and the composition of the hillocks cross-sectional TEM analysis was carried out. As can be seen in Fig. 3 the growth of the hillocks is started not directly on the Si surface at the beginning of the heteroepitaxial growth. There is a clear boundary between the epitaxial layer and the hillock. TEM accompanying EDX investigations were carried out in order to determine the local composition of the structures on the samples. In Fig 4 the spectras of the local EDX analysis measured on the hillock (region (a)) and in the epitaxial layer (region (b)) are presented. It is evident that in both regions Ge was detected. The intensity of the Ge signal obtained on the hillock is higher then the Ge signal measured in the epitaxial layer. The intensity of carbon peak is lower in the hillock region compared to the signal detected in the epitaxial layer. This results indicate on a $Si_{1-x}Ge_x$ composition of the hillocks.

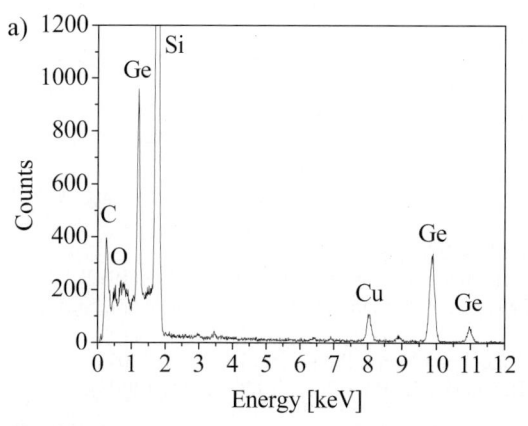

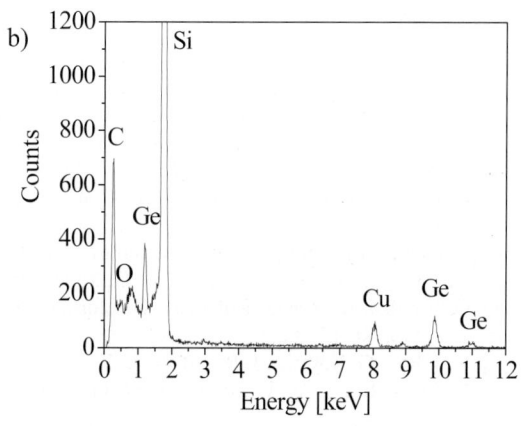

Fig. 4 EDX point analysis a) from the $Si_{1-x}Ge_x$ hillock and b) from the $(Si_{1-x}C_{1-y})Ge_{x+y}$ grown layer

The possible reason of the hillock formation is three fold. Firstly, they can be formed during the cooling down of the sample after the epitaxial growth by nucleating from the accumulated Si and Ge excess on the $(Si_{1-x}C_{1-y})Ge_{x+y}$ surface. Secondly, they can appear during the growth when flux fluctuation lead to a slightly higher Si and Ge excess leading to three dimensional nucleation in regions which cannot be controlled by RHEED. Thirdly, Ge deposition on SiC surfaces promote the formation of three dimensional Si nuclei. This was tested by depositing Ge on (3H3)-Si reconstructed 3C-SiC surfaces at the growth temperature. In this case the reconstruction vanished and three dimensional Si diffraction spots appeared even if the Ge deposition was below one monolayer with respect to the silicon carbide surface.

The carried out AES and SIMS meaurements revealed a detailed picture of the Ge distribution in the depth of the layers. The AES depth profiling detected an inhomogeneous Ge distribution in the depth. With AES the Ge signal was detected only at the $(Si_{1-x}C_{1-y})Ge_{x+y}$/Si interface region. This means that Ge segregates at the interface between the two materials. The carried out more sensitive SIMS measurements allowed to detect dependence of the Ge concentration on the growth temperature (Fig. 5). The highest average Ge content over the layer was found to be 0.16%. With increasing growth temperature the Ge content in the epitaxial layer decreases indicating on a kinetic limitation of the Ge incorporation during the chosen growth conditions.

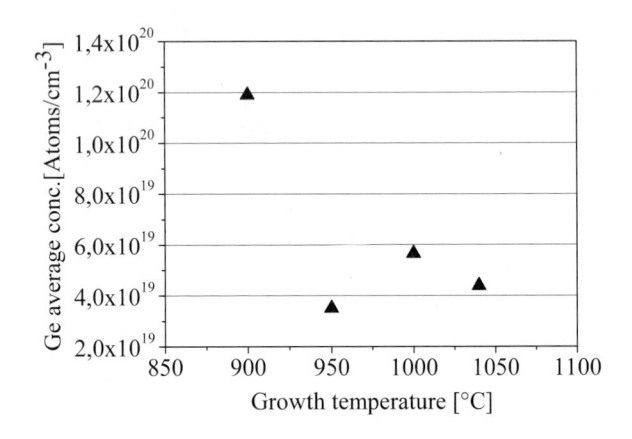

Fig. 5 Dependence of the Ge average concentration in the epitaxial layer versus growth temperature

4 Summary

For the first time a 3C-$(Si_{1-x}C_{1-y})Ge_{x+y}$ alloy was grown by solid source molecular beam epitaxy. The maximal average Ge concentration over the layers in order of 0.16% was obtained at 900 °C. At higher temperatures the Ge content decreases. An optimisation of the growth process conditions will be necessary to suppress the hillock formation and to increase the Ge content in the SiC layer

Acknowledgements This work is supported by the TMWFK under contract number B 607-02006. The authors are grateful to E. Remdt for the cautiously TEM sample preparation.

References

[1] P. Weih, V. Cimalla, C. Förster, J. Pezoldt, Th. Stauden, L. Spies, H. Romanus, M. Hermann, M. Eickhoff, P. Masri, and O. Ambacher, Mater. Sci. Forum **433–436**, 233 (2003).
[2] R. Khenta, H. Batache, M. Sagnoun, M. Driz, M. Rerat, and B. Abbar, Physica B **336**, 321 (2003).
[3] G. Katulka, C. Guedj, R.G. Wilson, C. Swann, M.W. Tsao, and J. Rabolt, Appl. Phys Lett. **74**, 540 (1999).
[4] W. Sekkal and A. Zaoui, New J. Phys. **4**, 9 (2002).
[5] O. F. Sankey, A. A. Demov, W. T. Petuskey, and P. F. McMillan, Modell. Simul. Mater. Sci. Eng. **1**, 741 (1993).
[6] R. Pandey, M. Rerat, C. Darrigan, and M. Causa, J. Appl. Phys. **88**, 6462 (2000).
[7] Yu. A. Vodakov, E. N. Mokhov, M. G. Ramm, and A. D. Roenkov, Springer Proc. Phys. **56**, 329 (1992).
[8] G. Katulka, K. Roe, J. Kolodzey, G. Eldridge, R. C. Clarke, C. P. Swann, and R. G. Wilson, Appl. Surf. Sci. **175–176**, 505 (2001).
[9] K. J. Roe, G. Katulka, J. Kolodzey, S. E. Saddow, and D. Jacobson, Appl. Phys. Lett. **78**, 2073 (2001).
[10] M. Diani, L. Kubler, L. Simon, D. Aubel, I. Matko, and B. Chenevier, Phys. Rev. B **67**, 125316 (2003).

Growth of InSb thin films on GaAs(100) substrates by flash evaporation epitaxy

M. Oszwaldowski[*], **T. Berus, A. Borowska, R. Czajka**, and **M. Zimniak**

Instytut Fizyki, Politechnika Poznanska, ul. Nieszawska 13a, 60-965 Poznan, Poland

Received 19 September 2003, accepted 3 November 2003
Published online 22 December 2003

PACS 73.50.Dn, 73.61.Ey, 81.15.Ef

A two-stage flash-evaporation epitaxy of InSb thin films on GaAs(100) substrates is developed. In the first, low temperature stage, a buffer layer of thickness of about 20 nm is deposited. In the second, high temperature stage the bulk of the film is deposited. Thus obtained InSb films have good structural and electrical properties, comparable with those obtained by MBE. They can be used for practical applications, e.g. for Hall sensors.

1 Introduction

The main practical applications of indium antimonide are the Hall sensors and the IR detectors. In both the applications, the material in the form of an epitaxial film is an optimum. Unfortunately, in the case of InSb, there is no substrate material that is cheap, chemically stable, insulating and has the lattice parameter matching that of InSb. In that unfortunate situation, semi-insulating GaAs (i-GaAs) is commonly taken to be the best choice for the substrate material. However, the mismatch in the lattice parameter between GaAs and InSb is very large, about 14%, and this makes the epitaxial growth of InSb a problem. Such a large misfit in lattice parameters leads to a Stransky – Krastanov mode of growth, and thereby the growth of smooth and structurally perfect films is not a trivial matter. In the initial stage of the growth, regular islands (called "quantum boxes") of InSb are formed on the surface of GaAs. The quantum boxes and their potential applications are an important reason of the renewal of the interest in the study of the InSb/GaAs system observed in recent years.

A large number of papers has been devoted to the epitaxial growth of InSb on GaAs. In most of them the MBE was used for growing the films. It has become clear from those investigations that there is no single model of the growth, and, in general, the growth mechanism is complicated. It depends both on the substrate orientation and its surface preparation and on the growth conditions, such as substrate temperature and associated surface reconstructions, the Sb/In ratio, deposition rate and so on.

It is relevant to note that recently Kanisawa and co. [1] has proposed a two – stage growth, which seems to eliminate the initial Stransky – Krastanov mode of growth for the case the GaAs oriented in the (111)A plane. They deposited the first 3 nm InSb layer at 310–325 °C and, in the second stage, the rest of the film was deposited at 380–395 °C. Due to the quasi – two – dimensional growth the InSb films were smooth and had very good electrical properties. That two – stage growth did not improve the film properties deposited on the substrate oriented in the (100) plane.

In the present contribution we present the results of a two – stage deposition of InSb epitaxial films onto GaAs(100) substrates by the flash evaporation method. The film growth occurs in a standard evaporation plant, and thereby the method is of use for commercial applications. An important role in the flash

---

[*] Corresponding author: e-mail: maciej.oszwaldowski@put.poznan.pl

evaporation epitaxy plays a thin low temperature InSb buffer layer. We show that the quality of the films obtained at optimal technological conditions is comparable to those obtained by MBE and MOCVD.

2 Experimental results and discussion

The apparatus for the flash evaporation of InSb films used in the present experiments was described in previous papers [2, 3]. Hence we give details on the film preparation only.

The films were evaporated on epi-ready i-GaAs(100) substrates obtained from The Institute for Electronic Material Technology, Warsaw. As in the previous experiments with the use of the GaAs substrates oriented in the (110) and (111) planes [3], we have found that the films deposited at substrate temperatures $T_s < 670$ K show smooth, mirror–like surfaces, but have low electron mobility, whereas those obtained at $T_s > 670$ K show a considerably higher mobility, but they have rough surfaces. The latter evidences that the film growth is far from the layer-by-layer mode. We have also confirmed that to obtain single phase InSb films, it is necessary to perform the evaporation process with the mixture of InSb and Sb powders, which gives the Sb/In ratio higher that unity.

In further experiments we have found that to obtain high electron mobility films with smooth surfaces, it is necessary to adopt a two-stage deposition process. In the first stage, a thin InSb buffer layer is formed at $T_s < 570$ K. In the following we present result essentially for $T_s = 520$ K. The buffer thickness <t> used in the following is an averaged value, determined from the amount of the InSb evaporated. In the second stage, the bulk of the film is deposited at $T_s > 690$ K. In the following we present results for $T_s = (670 \pm 5)$ K and the total film thickness of about 2 μm.

The dependences of the electron concentration at 77 K and the electron mobility at 300 K and 400 K on the buffer layer thickness, as obtained from the Hall measurements, are shown in Fig. 1 a and b.

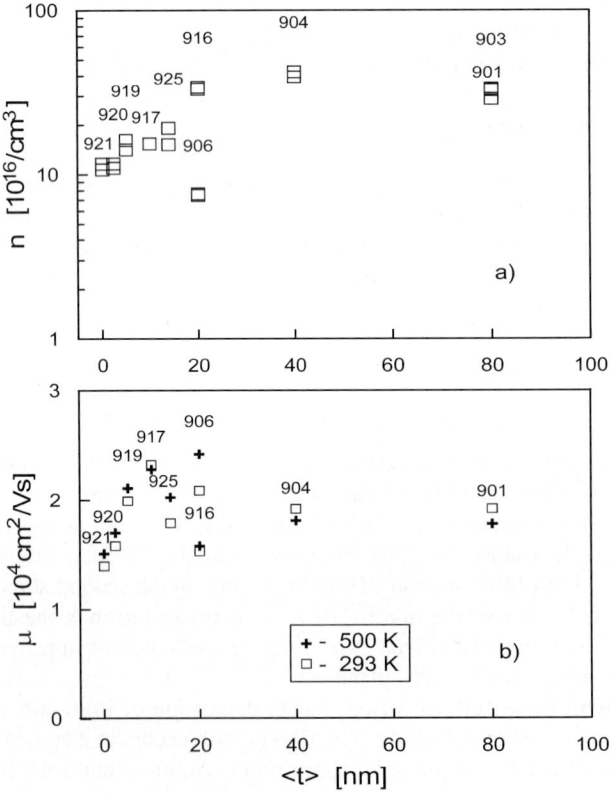

Fig. 1 Dependence of the electron concentration, a) and the electron mobility, b) on the buffer layer thickness <t>.

The concentration is measured at 77 K and the mobility at 293 K and 500 K. The series number of the films are given above the experimental points.

The concentration is measured on two samples of each series. The series number is given above the experimental points. It is seen that the concentration difference within the series is rather small. On the other hand, the difference between the series can be considerable. The reason of that is not known yet.

For the sake of clarity, in the case of the mobility we present results for one sample of each series, but for two temperatures: 293 K and 500 K. As seen in the figure, most samples show an anomaly in the temperature dependence of the mobility: the mobility at 500 K is higher than that at 293 K. This anomaly is typical for InSb films, and it was also observed in the epitaxial films obtained by MOCVD [4].

The temperature dependences of the electron concentration and mobility are shown in Fig. 2a and b.

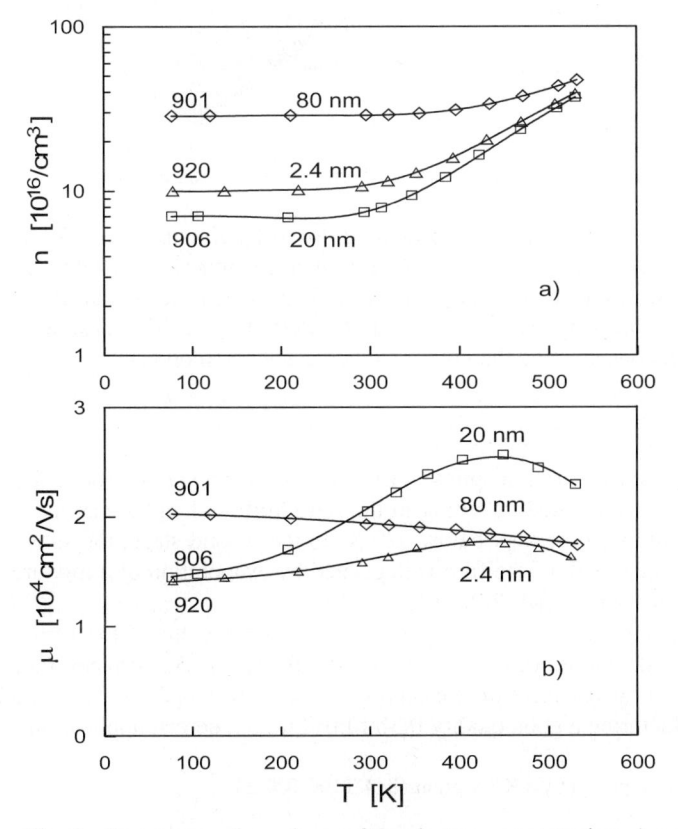

Fig. 2 Temperature dependence of the electron concentration, a) and the mobility, b) in three InSb films with the buffer layers 2.4 nm, 20 nm and 80 nm. The respective serial numbers are: 920, 901 and 906.

It is seen that the anomalous dependence µ(T) is rather typical for samples with the electron (the effective donor) concentration smaller than 10^{17} cm^{-3}, and can disappear in the films with higher electron concentration. Therefore, it is rather the high temperature mobility that characterize the quality of InSb films for applications in electronic devices working at room or elevated temperatures. Taking this into account, we can conclude from Fig. 1b that the best InSb films are those deposited on the buffer layer of thickness of about 20 nm.

X-ray studies reveal that the films obtained are well oriented in the (100) plane. No lines from other orientations are revealed. In comparison with the films with no buffer layer, the films grown on the buffer layer are smoother. The most smooth and mirror–like are the films grown on the buffer layers of about 20 nm thick. This observation is confirmed by studies with a Nomarsky optical microscope.

In order to elucidate the role of the buffer layer in the film growth, atomic force microscopy (AFM) studies have been performed in the contrast mode. Fig. 3 shows the AFM images of two buffer layers with <t> = 10 nm and <t> = 80 nm.

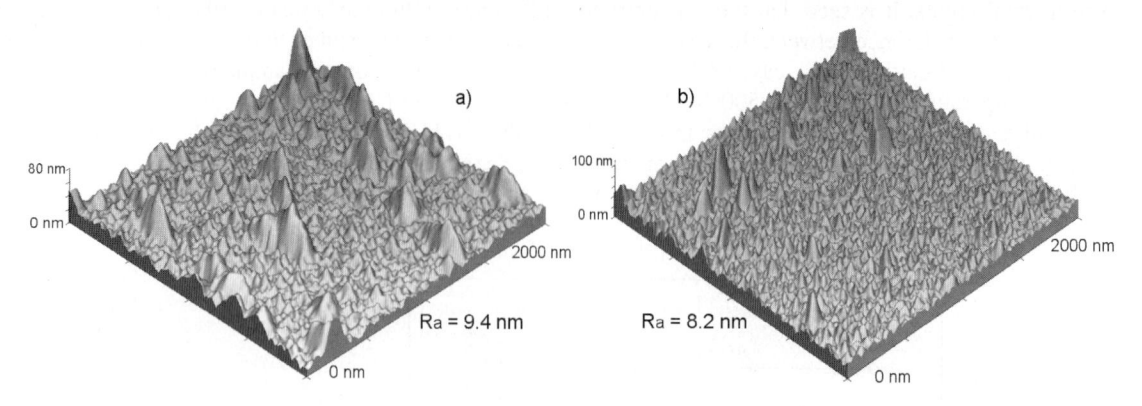

Fig. 3 AFM images (contact mode) of two buffer layers. a) <t> = 10 nm and b) <t> = 80 nm.

It is seen that the layers have an island structure, evidencing Stransky-Krastanov or Volmer-Weber mode of growth. The roughness of both films, Ra < 10 nm. This small roughness, associated with a high density of the islands, is preserved in the second stage of the growth, which means that in the second stage the growth has more or less a layer-by-layer (Frank-van der Merve) mode. That desirable mode of growth results in a decreased defect density, and thereby in an increase in the mobility.

3 Conclusions

We have developed a two-stage flash-evaporation epitaxy of InSb thin films on GaAs(100) substrates. In the first stage, pure InSb powder is evaporated, resulting in the ratio Sb/In = 1 in the vapour stream. In that stage the substrate temperature has to be kept below 570 K. In the second stage, the substrate temperature is raised above 720 K and the bulk of the film is deposited by evaporation of a mixture of InSb and Sb powders, in proportion ensuring the ratio Sb/In $\geq$ 1.5.

The obtained in the first stage smooth buffer layer allows the further growth of the film bulk in a layer-by-layer like mode. As a result, smooth InSb epitaxial films with mirror-like surfaces are obtained. They show high electron mobility at given electron concentrations. The developed method is relatively simple and can be used for manufacturing a good quality InSb films for commercial applications.

Acknowledgements This work was supported by a KBN grant No 4 T10C 037 24.

References

[1] K. Kanisawa, H. Yamaguchi, and Y. Huirayama, Appl. Phys. Lett. **76**, 589 (2000).
[2] M. Oszwaldowski and M. Slany, Vacuum **43**, 617 (1992).
[3] C. Besikci, Y. H. Choi, R. Sudharsanan, and M. Raseghi, J. Appl. Phys. **73**, 5009 (1993).

Properties of MBE Cd$_x$Hg$_{1-x}$Te/GaAs structures modified by ion-beam milling

V. V. Boboyashchyy[1], S. A. Dvoretsky[2], I. I. Izhnin[*,3], N. N. Mikhailov[2], Yu. G. Sidorov[2], F. F. Sizov[4], V. S. Varavin[2], and V. A. Yudenkov[3]

[1] Kremenchuk State Polytechnical University, 20 Pershotravneva St., 39614 Kremenchuk, Ukraine
[2] Institute of Semiconductor Physics, SB RAS, 13 Lavrentieva Av., 630090 Novosibirsk, Russia
[3] R & D Institute for Materials SRC "Carat", 202 Stryjska St., 79031 Lviv, Ukraine
[4] Institute of Semiconductor Physics, NASU, 45 Nauki Av., 03028 Kiev, Ukraine

Received 19 September 2003, accepted 3 November 2003
Published online 15 January 2004

PACS 61.72.Vv, 61.80.Jh, 66.30.Jt, 72.80.Ey, 73.61.Ga

Modification of electrical properties of *n*- and *p*-type MBE mercury–cadmium-telluride (MCT) structures with passivation gradient band gap layers under ion milling was investigated. Main features of the Cd$_x$Hg$_{1-x}$Te electrical properties modification with such treatment also hold for MBE MCT structures in comparison with bulk samples. It is shown that a significant decrease of the *p–n* conversion rate for such structures in comparison with homogeneous bulk samples at other equal conditions is caused by the presence of the wide band gap passivation layer. The diffusion-like character of the *p–n* conversion front movement is confirmed. The nature of additional "heavy" electrons in MBE structures is proposed.

© 2004 WILEY-VCH Verlag GmbH & Co. KGaA, Weinheim

1 Introduction Following the proposition of Wotherspoon [1], an ion milling (IM) method for creation of *p–n* junctions in p-type MCT wafers was developed. The list of observed effects concerning *p*-to-*n*-type conductivity conversion under IM of p-Cd$_x$Hg$_{1-x}$Te was given in [2]. The main results were obtained either on bulk single-crystal samples or LPE epitaxial layers [1, 3–5]. Up to now only a small number of investigations have been made for IM of molecular beam epitaxy (MBE)-grown Cd$_x$Hg$_{1-x}$Te layers [6–8], although such layers are the most promising for large-format diode array manufacture, especially compound structures containing some passivation layers with high Cd content. However, according to results of [7, 8], in MBE MCT layers with CdTe passivation layers, the converted depth changes linearly with milling time and is inversely proportional with vacancy concentration in layers thinner than 10 µm. This shows that the filling of Hg vacancies in this region during conversion is limited by the rate of supply of extra Hg due to the milling process. It corresponds to the results [1, 9] but it is opposite to the conclusions of [3, 4, 10]. Besides, in [8] any dependence of conversion depth on Cd content in active MCT layers was not observed and this does not match experimental results [11] and theoretical predictions [10].

Preliminary results of electrical properties modification in MBE Cd$_x$Hg$_{1-x}$Te layers under IM were presented in our previous publication [6]. The main aim of this paper is the subsequent investigation of the electrical properties modification of MBE Cd$_x$Hg$_{1-x}$Te structures with gradient band gap passivation layers and the study of possibilities of stable *p–n* junction manufacturing in such structures by IM.

2 Experiment MBE structures were grown on (013) GaAs substrates with 4.5-µm CdTe buffer layers. In these structures the Cd$_x$Hg$_{1-x}$Te (*x* = 0.22) active layers of thickness *d* were protected by ≈ 1.5-µm

---

[*] Corresponding author: e-mail: granat@ipm.lviv.ua, Phone: +380 322 63 22 19, Fax: +380 322 63 50 69

gradient band gap passivation layers with increasing Cd content up to $x \approx 0.4$ near both interfaces. The composition and the depth of growing layers were in-situ continuously controlled by means of an ellipsometer [12]. As-grown structures were of n-type conductivity and then they were converted to p-type by thermal annealing in Ar or He atmosphere at 185 °C (see Table 1). It is assumed that type conductivity conversion at such annealing is due to Hg vacancies formation in the layers.

Samples of initial n-type and converted p-type from two structures grown in different technological processes were studied with or without a preliminary chemical etched passivation top layer. The IB-3 (EIKO, Japan) etching system with Ar$^+$ ions was used to perform IBM under the following conditions: ion energy $E = 500$ eV, current density $j = 0.25$ or 0.3 mA/cm^2, and milling time $t = 5-12$ min. The operation table of the etching system with samples was cooled during milling by water.

The electrical properties of initial and ion-milled samples were measured at 77 K in magnetic fields between 0.01 and 1.5 T in the Van der Pauw configuration. The analysis of Hall data was performed in the framework of a two-carrier model (see Table 1). Vacancy concentration N_A was estimated from these data taking into account the average degree of the vacancies ionization that depends on their concentration and x-value [13]. The depth of the $p–n$ junction and the carrier concentration distribution over sample depth were studied using the "step by step etch" technique. In this technique, semiconductor layers are sequentially stripped through chemical etching and measurements of the differential Hall effect and the conductivity are performed. The thickness of material removed at etching degree was controlled at every stage of the process.

Table 1 Properties of MBE Cd$_x$Hg$_{1-x}$Te structures.

# Structure	# sample	d, μm	Annealing	N_A, cm^{-3}	Type	$p, (n)^*$, cm^{-3}	$\mu, (\mu_n)^*$, cm^2/V s	$p_l, (n_h)^*$, cm^{-3}	$\mu_l, (\mu_{nh})^*$, cm^2/V s
#1	#1-1	8.5	75 min	–	n	1.2×10^{13}*	129000*	3.6×10^{14}*	18600*
#1	#1-2	8.5	34 h	1.4×10^{16}	p	1.4×10^{16}	480	1.2×10^{13}	9880
#2	#2-2	6	15 min	1.0×10^{17}	p	8.5×10^{16}	260	1.7×10^{14}	5600

3 Results and discussion – n-type samples

Before IM the surface variable band gap passivation layer (~1-μm thick) was chemically etched from this sample. Measurements and analysis of the Hall coefficient magnetic field dependences before and after removing the surface passivation layer indicated the presence of two types of electrons in the initial n-Cd$_x$Hg$_{1-x}$Te MBE structure with high (usual electrons) and low ("heavy" electrons) mobility (see Table 1), which was not typical for bulk n-Cd$_x$Hg$_{1-x}$Te sam-

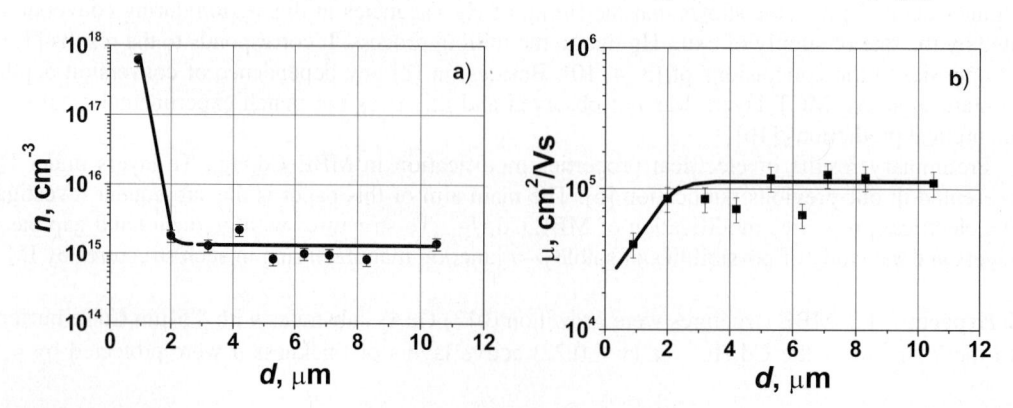

Fig. 1 Electron-concentration a) and mobility b) distributions at 77 K over #1-1 sample depth after IM.

ples. Up to now the nature of heavy electrons (electrons in layers with low electron mobility) is not completely clear.

The most probable reason for their existence may be the availability of an interface layer between active and buffer layers in such structures, and the influence of structural and other defects [14]. The sample was milled with current density $j = 0.25$ mA/cm^2 and milling time $t = 300$ s that corresponded to an ion dose $\Phi = 5 \times 10^{17}$ cm^{-2}.

Eleven successive steps of chemical etching (CE) with the measurement of conductivity and Hall coefficient magnetic field dependences $R_H(B)$ at 77 K after each step were carried out to estimate the electron concentration and mobility distributions over the depth of this sample after IM. The appearance of $R_H(B)$ for the first ten steps of CE was similar to that for the initial sample before IM and indicated the presence of two types of electrons. But, after the last (11th) step of CE, the Hall coefficient did not depend any more on magnetic field, which suggests the presence only of one type of electron. At the same time the Hall electron mobility, which does not depend on the sample depth, was equal to 18000 cm^2/V s, which agrees well with the heavy-electron mobility in the initial sample. Taking into account that the total thickness of the etched layer during the first 10 CE steps is ~10.5 µm, after the 11th one the position of the layer of the residual sample coincides with the region of the gradient interface passivation layer between the active and the CdTe buffer layers; it can therefore be concluded that heavy electrons in the investigated MBE structure are situated in a thin transition layer between the narrow-gap active layer and the CdTe buffer layer.

Determined profiles of electron-concentration and mobility distributions are shown in Fig. 1. It can be seen that the electron-concentration and mobility distributions over the investigated MBE Cd$_x$Hg$_{1-x}$Te sample are similar to those for bulk homogeneous n-Cd$_x$Hg$_{1-x}$Te layers modified by IM [3]. Namely, in a thin subsurface (damaged) layer (~2 µm) the electron concentration falls from surface to sample depth and further, in the main part of the n-type layer, has a constant value. At the same time the electron mobility, quite on the contrary, in the damaged layer rises from surface to sample depth and further, in the main part of the n-type layer, has a constant value. The subsurface damaged layer is characterised by high electron concentration with low mobility that are conditioned by the creation of additional structural defects at IM. The electron concentration in the main part of the n-layer for a structure modified by IM equals $(1-1.2) \times 10^{15}$ cm^{-3} and agrees well with estimations of donor defect concentrations (that were assumed [15] to be the antisite Te ions) for MBE Cd$_x$Hg$_{1-x}$Te layers, grown at the same conditions [15].

p-type samples For each structure the carrier concentration distribution over the sample depth was investigated both for samples #1-2-1 and #2-2-1 with a surface passivation layer and for similar samples #1-2-2 and #2-2-2 with a previously chemically etched surface gradient band gap passivation layer of thickness ~ 1 µm. All samples were milled with the same ion dose $\Phi = 1.4 \times 10^{18}$ cm^{-2}. Carrier concentration distributions for these samples are presented in Fig. 2a, b.

It was revealed that IM of samples with passivation layers results in the creation of n-type layers with thicknesses approximately 5.5 µm and 2.2 µm for samples #1-2-1 and #2-2-1 respectively (Fig. 2a). The correlation between the thickness of converted layers d_n and the experimental conditions agrees well with the results of theoretical analysis [10] and is determined by initial acceptor concentration as: $d_1/d_2 = 2.5$; $(N_{A2}/N_{A1})^{0.5} = 2.7$. Thus $d_1/d_2 \approx (N_{A2}/N_{A1})^{0.5}$. This definitely indicates a diffusion-like process of conversion front movement in samples (see [3, 4, 10]). But the thickness of the converted layers was much smaller than had been predicted by theory [10] and that experimentally observed in [3, 4] for uniform bulk Cd$_x$Hg$_{1-x}$Te samples with the same acceptor concentration and IM conditions. It is presumed that these distinctions are due to the presence of the wide band gap gradient surface layer, which results in decreasing of the conversion rate because of the internal electric field action [10]. This conclusion confirms also the data of Fig. 2b. For samples with previously chemically etched passivation layers the p–n type conductivity conversion was observed over the whole active-layer depth.

The character of carrier concentration distribution over the depth for both groups of MBE samples was the same as for the uniform bulk MCT samples (Fig. 2a, b) [3, 5]. Namely, the electron concentration in the thin subsurface damaged layer (~2–3 µm) with high electron concentration and low mobility decreases from surface to sample depth and then, in the main part of the n-type layer, has a constant value

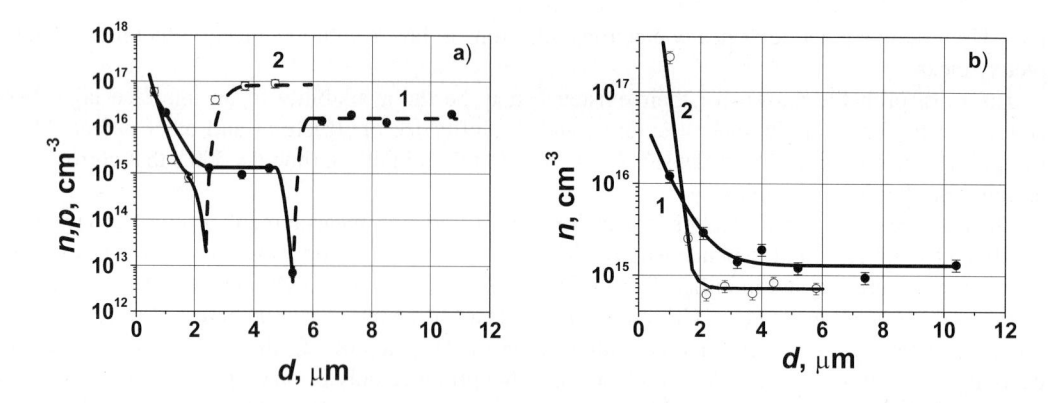

Fig. 2 Electron (solid) and hole (dashed) concentration distributions at 77 K over the samples' depth after IM: a) for samples #1-2-1 (1) and #2-2-1 (2) with passivation layer; b) for samples #1-2-2 (1) and #2-2-2 (2) with passivation layer etched.

that corresponds to the residual donor concentration. That is why the electron concentrations in the main part of the *n*-type layer after IM were equal for all samples prepared from structure #1. Hole concentration and mobility in the unconverted region of the formed *p–n* structures are the same as in the samples before IBM.

4 Conclusions In general, the main regularities of electrical properties modification in uniform bulk MCT layers at IM remain also for MBE $Cd_xHg_{1-x}Te$ structures. A significant decrease of the *p–n* conversion rate was observed here for MBE structures with surface passivation gradient wide gap layers in comparison with the conversion rate for uniform bulk MCT layers. It confirms the conclusions [10, 11] about the dependence of the conversion depth in *p*-MCT layers on the *x*-value. Also observed is the square-root inversely proportional dependence of conversion depth on acceptor concentration; this is evidence of the diffusion-like character of *p–n* conversion front movement. On the basis of the obtained data on "step-by-step" chemical etching of the MBE *n*-$Cd_xHg_{1-x}Te$ structure there was concluded the existence of "heavy" electrons with low mobility at the interface between the active and buffer layers in such types of structures.

References

[1] J. T. M. Wotherspoon, UK Patent No. GB 2095898, 1981.
[2] E. Belas, R. Grill, J. Franc, P. Moravec, R. Varghova, P. Höschl, H. Sitter, and A. L. Toth, J. Cryst. Growth **224**, 52 (2001).
[3] V. I. Ivanov-Omskii, K. E. Mironov, and K. D. Mynbaev, Semicond. Sci. Technol. **8**, 634 (1993).
[4] E. Belas, P. Höschl, R. Grill, J. Franc, P. Moravec, K. Lischka, H. Sitter, and A. Toth, Semicond. Sci. Technol. **8**, 1695 (1993).
[5] I. I. Izhnin, A. I. Izhnin, K. R. Kurbanov, and B. B. Prytuljak, Proc. SPIE **3182**, 383 (1996).
[6] V. V. Bogoboyashchiy, A. P. Vlasov, S. A. Dvoretsky, I. I. Izhnin, D. Y. Protasov, L. N. Romashko, and Y. G. Sidorov, in: Proc. 2nd Russian–Ukraine Workshop "Nanophysics and Nanoelectronics", Kiev, Ukraine, 22–24 November 2000, p. 63 (in Russian).
[7] R. Haakenaasen, T. Colin, H. Steen, and L. Trosdahl-Iversen, J. Electron. Mater. **29**, 849 (2000).
[8] R. Haakenaasen, T. Moen, T. Colin, H. Steen, and L. Trosdahl-Iversen, J. Appl. Phys. **91**, 427 (2002).
[9] M. V. Blackman, D. E. Charlton, M. D. Jenner, D. R. Purdy, J. T. M. Wotherspoon, C. T. Elliott, and A. M. White, Electron. Lett. **23**, 978 (1987).
[10] V. V. Bogoboyashchiy and I. I. Izhnin, Russ. Phys. J. **43**, 627 (2000).
[11] K. D. Mynbaev, N. L. Bazhenov, V. A. Smirnov, and V. I. Ivanov-Omskii, Pisma Zh. Tekh. Fiz. **28**, 64 (2002) (in Russian).
[12] V. S. Varavin, S. A. Dvoretsky, V. I. Liberman, N. N. Mikhailov, and Yu. G. Sidorov, J. Cryst. Growth **159**, 1161 (1996).

[13] V. V. Bogoboyashchyy, Condens. Media Interfaces **3**, 28 (2001) (in Russian).
[14] V. S. Varavin, A. F. Kravchenko, and Yu. G. Sidorov, Fiz. Tekh. Poluprov. **35**, 1036 (2001) (in Russian).
[15] Yu. G. Sidorov, S. A. Dvoretsky, V. S. Varavin, N. N. Mikhailov, I. V. Sabinina, A. P. Shotov, Yu. G. Selivanov, A. V. Babushkin, and V. V. Kopylov, in: Proc. 2nd Russian–Ukraine Workshop "Nanophysics and Nanoelectronics", Kiev, Ukraine, 22–24 November 2000, p. 109 (in Russian).

Formation of Ge nanoislands on pure and oxidized Si surfaces by MBE

A. I. Nikiforov[*,1], **V. V. Ulyanov**[1], **A. G. Milekhin**[1], **O. P. Pchelyakov**[1], **S. A. Teys**[1], **S. Schulze**[2], and **D. R. T. Zahn**[2]

[1] Institute of Semiconductor Physics SB RAS, Lavrentjeva 13, Novosibirsk 630090, Russia
[2] Institut für Physik, Technische Universität Chemnitz, 09107 Chemnitz, Germany

Received 19 September 2003, accepted 3 November 2003
Published online 15 January 2004

PACS 68.55.Ac, 68.65.Hb

The discussion deals with experimental data on the process of formation of self-organized Ge islands on an oxidized atomically pure Si(100) surface. Unlike the Stranski–Krastanow mechanism, which is characteristic of Ge growth on a pure silicon surface, the Volmer–Weber growth mechanism is observed on the oxidized silicon surface. The growth process is accompanied by a considerable change (up to 7%) in the surface unit cell of Ge relative to the parameters of Si. The generated nanoislands are less than 10 nm in base size and have a density of more than 2×10^{12} cm^{-2} for a Ge film not thicker than five monolayers.

© 2004 WILEY-VCH Verlag GmbH & Co. KGaA, Weinheim

1 Introduction The phenomenon of self-organization of nanoislands is of interest to several areas of solid-state physics. For physics of surface and condensed media, the actual importance are studies of the mechanisms of nanostructure growth and atomic processes occurring on the surface during the growth. On the other hand, these objects are also of interest to semiconductor physics due to their potential application as quantum-sized nanostructures. Among these objects is the "Ge on Si" heterosystem with Ge nanoislands behaving as quantum dots. Dislocation-free germanium islands of 10 to 100 nm in size are detected on the Si(100) surface after formation of a continuous Ge film [1]. The islands can be minimized in size to provide the appearance of the effects of size quantization even at temperatures as low as ambient [2].

The smallest germanium islands growing on the pure silicon surface are 15 nm in size. The size can be lowered by growing germanium on an oxidized atomically pure surface prepared *in situ* in the MBE installation. It has been for known for a long time that an oxide layer can be generated on the silicon surface under ultra-vacuum conditions. The possibility of choosing regimes of etching and growth of the oxide film depending on oxygen pressure and temperature has been demonstrated for the first time elsewhere [3]. Growth of germanium islands on the pre-oxidized silicon surface allows the islands to be decreased considerably in size at a higher density. The islands grown on the oxidized Si(111) surface are shown [4, 5] to be 10 nm in lateral size at the density higher than 10^{12} cm^{-2}. Shklyaev et al. [4] supposed that this is the case of local disoxidation of silicon by germanium (disproportionation) accompanied by desorption of germanium monoxide. Germanium nanoislands conjugated coherently with silicon are nucleated in these sites.

2 Experimental A MBE installation Katun-C equipped with two electron-beam evaporators for Si and Ge was used for synthesis. Analytical equipment of the chamber included a quadrupole mass spectrometer, a quartz thickness monitor and a reflection high energy (20 kV) electron diffraction (RHEED) appa-

---

[*] Corresponding author: e-mail: nikif@isp.nsc.ru, Phone: +(3832)333-286, Fax: +(3832)333-502

ratus. Diffraction patterns were monitored during the growth using a CCD camera online with a PC. The software allowed both the whole images and chosen fragments of the diffraction patterns to be monitored at the rate of 10 frames/s. Ge grew at the rate of 10 ML/min and the temperature was varied up to 700 °C. Silicon wafers misoriented by less than 0.5° were used as substrates. Before a Ge film started growing, the substrate was annealed and a buffer Si layer was grown to reconstruct the original surface. A MBE installation was used for oxidation at an oxygen supply up to 10^{-4} Pa and the substrate temperature of 400–500 °C. Then germanium was deposited on the oxidized surface.

3 Results and discussion The process of Ge film growth was controlled using RHEED patterns by registering both qualitative changes in the structure and morphology of the growing film and quantitative information about elastic deformation of the surface unit cell [6]. The initial stage of germanium film growth on the oxidized Si surface was analyzed by registering variations in the intensities of specular reflection and three-dimensional diffraction reflection (3D reflection). These values were very sensitive to variations in the surface roughness, while the appearance of 3D reflection indicated the presence of 3D objects on the surface under study. The occurrence of intensity oscillations of the specular reflection during growth on the pure surface, extinction of the reflection and the appearance of 3D reflection at the germanium film thickness of more than four monolayers argue for 2D growth of the wetting layer followed by formation of three-dimensional islands. As to the Ge film growth on the oxidized surface, the reflection intensities are only varied after deposition of one monolayer and the specular reflection intensity does not oscillate. These observations indicate the absence of the stage of wetting layer formation. While the first monolayer is deposited on the SiO_2 surface, an adsorbed Ge layer is formed, which is transformed into three-dimensional islands during the growth of the second and next monolayers. Therefore, germanium film growth on the oxidized silicon surface occurs by the Volmer–Weber mechanism and not by the Stranski–Krastanow mechanism, which is characteristic of the growth of a pure Si surface.

Deformations of the germanium lattice were studied by measuring the in-plane lattice constant $a_\parallel$ of the surface 2D unit cell (Fig. 1). For this purpose variations in the distance between reflections corresponding to the $a_\parallel$ parameter were registered in the diffraction pattern.

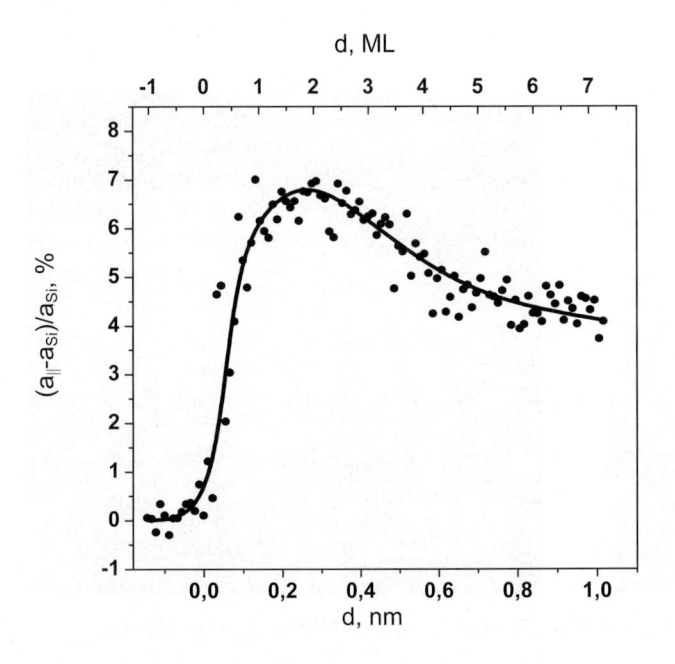

Fig. 1 In-plane lattice constant changes during growth of the Ge film on the oxidized Si(100) surface at 600 °C.

The film growth results in changes in the surface-cell constant of the Ge lattice against the constant of the Si lattice, which remains constant. The difference may reach 7% as reported for the growth on a pure Si(100) surface [6]. Elastically strained islands grow first, then $a_\parallel$ decreases to the value characteristic of bulk germanium, which indicates complete plastic relaxation of the islands. The mode of parameter $a_\parallel$ variations is similar to that observed during germanium heteroepitaxy on the pure Si(100) surface except for a considerably smaller region of existence of strained Ge islands; as a result the maximal constant of a surface 2D unit cell is characteristic of 3D islands formed after deposition of one germanium layer. Hence, the maximal elastic deformation occurs already in the 3D island nuclei.

Depending on the thickness of the deposited germanium layer, the islands are different in size and density. Figure 2 shows a STM image of an array of Ge islands on the silica surface obtained after deposition of 0.5- and 2-nm germanium at the substrate temperature of 650 °C. An increase in the effective thickness of deposited germanium results in the formation, along with small-sized islands, of islands with the size larger by an order of magnitude at a considerably lower density. The small-sized germanium islands are not altered considerably in size and density upon the appearance of coarse islands. Hence, the amount of germanium fed to the growth surface is consumed for augmentation of the dimensions of the coarse islands and not to increase the dimensions of the small islands. It follows from Fig. 1 that the large islands are fully relaxed and the small ones are elastically strained. The constant density and size of small islands upon the appearance of large islands indicate that the former possess an energetically favorable structure and shape, as well as being highly resistant to variations in thermodynamic and kinetic conditions on the surface. More evidence is the high temperature of the growth in comparison with that of the formation of "hut" clusters on a pure Si(100) surface. Therefore, a bimodal size and density distribution of islands is observed on the oxidized silicon (100) surface at the germanium film thickness of more than five monolayers.

These samples were also used for HREM studies in plan-view and cross-sectional conditions. Figure 3 shows pertinent TEM images. They also argue for the bimodal size distribution of germanium islands on the atomically clean silica surface. At a higher plan-view resolution, there appear Moiré bands, indicating relaxation of the islands. For the film not thicker than 1 nm, the islands are less than 10 nm in base size at the density higher than 2×10^{12} cm^{-2}.

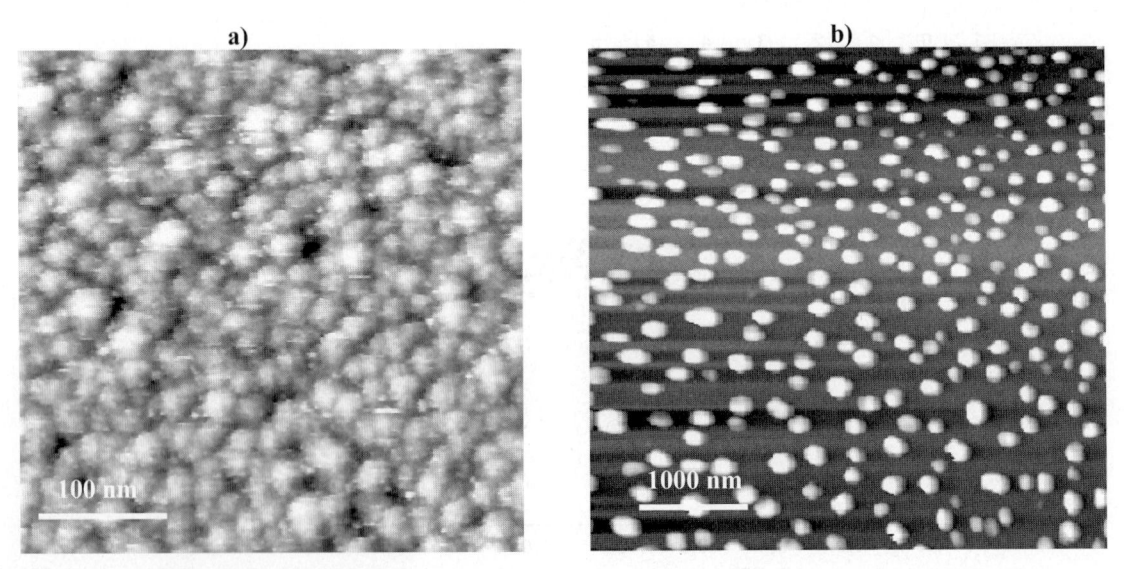

Fig. 2 STM image of an array of Ge islands on silica surface $T_s = 650$ °C, $d_{Ge} = 3$ ML (a), $d_{Ge} = 14$ ML (b).

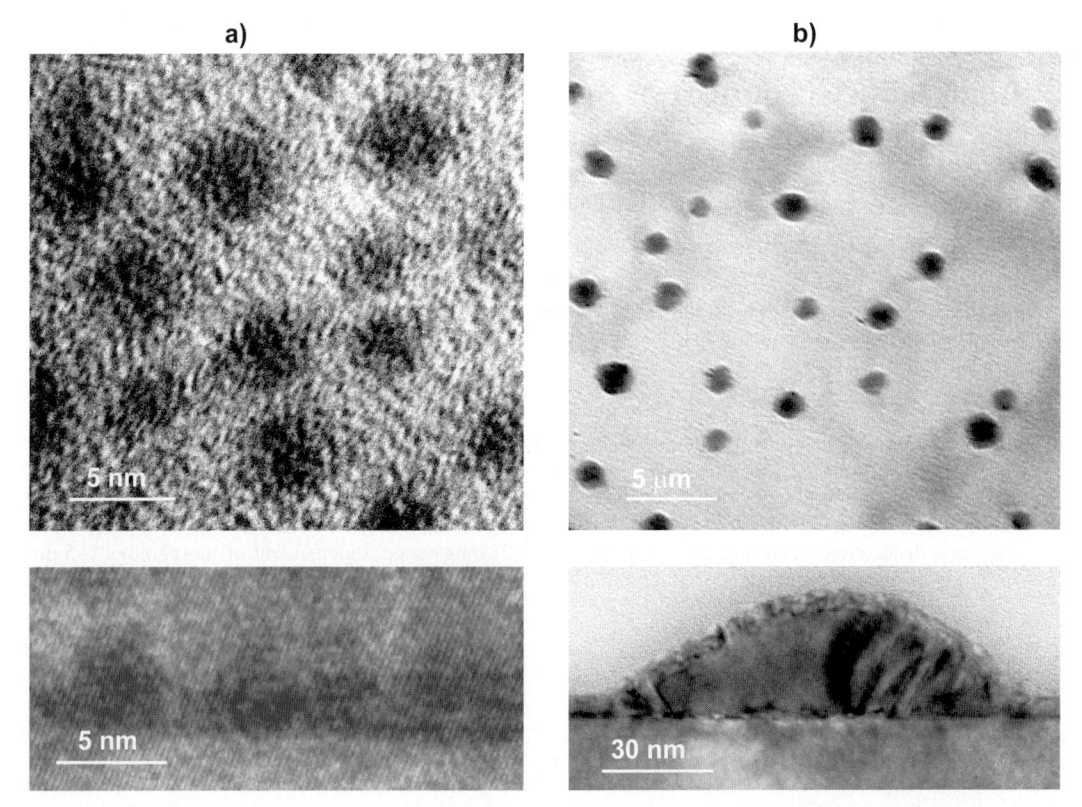

Fig. 3 TEM image of an array of Ge islands on silica surface $T_s = 650$ °C, $d_{Ge} = 3$ ML (a), $d_{Ge} = 14$ ML (b).

4 Conclusions Variations in the lattice constant of Ge films were determined in the course of the MBE film growth on SiO$_2$ surfaces. The germanium film growth on silica, unlike the growth on the pure silicon surface, proceeds without formation of the wetting layer and the changes in the surface lattice of Ge relative to the surface lattice of Si, as in the case of the growth on the pure silicon surface. It is demonstrated that Ge islands smaller than 10 nm in size are formed at the density of 2×10^{12} cm^{-2} on the silica surface. The bimodal island size distribution is observed at thicknesses of more than five monolayers.

Acknowledgements The work is supported by the Russian Foundation for Basic Research (Grants 03-02-16468, 03-02-16506 and 02-02-17746) and INTAS (Grant 01-0444).

References

[1] D. J. Eaglesham and M. Cerullo, Phys. Rev. Lett. **64**, 1943 (1990).
[2] A. I. Yakimov, A. V. Dvurechenskii, Yu. Yu. Proskuryakov, A. I. Nikiforov, and O. P. Pchelyakov, Thin Solid Films **336**, 332 (1998).
[3] J. J. Lander and L. Morrison, J. Appl. Phys. **33**, 2098 (1962).
[4] A. A. Shklyaev, M. Shibata, and M. Ichikawa, Phys. Rev. B **62**, 1540 (2000).
[5] A. Barski, M. Derivaz, J. L. Rouviere, and D. Buttard, Appl. Phys. Lett. **77**, 3541 (2000).
[6] A. I. Nikiforov, V. A. Cherepanov, O. P. Pchelyakov, A. V. Dvurechenskii, and A. I. Yakimov, Thin Solid Films **380**, 158 (2000).

Light induced contrast in Kelvin Force Microscopy of GaN epilayers

R. Bozek[*,1], **K. Pakula**[1], and **J. M. Baranowski**[1]

[1] Institute of Experimental Physcis, Hoza 69, 00-681 Warsaw, Poland

Received 19 September 2003, accepted 3 November 2003
Published online 22 December 2003

PACS 07.79.–v, 61.72Hh, 68.37.–d

The results of investigations of undoped GaN heteroepitaxial layers with Atomic Force Microscopy and Kelvin Force Microscopy are presented. The measurements of surface potential were performed in dark and under illumination below and above the energy gap of GaN. In dark no fluctuations of potential were observed. Moderate illumination revealed fluctuations of (1) short range (~100 nm and ~0.1 V), related to threading dislocations with a screw component and (2) long range, independent of morphology (~5 μm and ~0.3 V). Spectral dependence and possible sources of those fluctuations are discussed.

© 2004 WILEY-VCH Verlag GmbH & Co. KGaA, Weinheim

1 Principle of measurements

Kelvin Force Microscopy (KFM) [1] combines the idea of tapping mode Atomic Force Microscopy (AFM) with a Kelvin probe to measure the distribution of the surface potential for conducting samples. The principle of its operation is based on a capacitor consisting of a conducting AFM tip and the sample, formed when the tip is lifted at small height above the surface. In the KFM mode, the tip is not excited to oscillations by a piezo, as in the tapping AFM. Instead external voltage $V_{DC} + V_0 \sin \omega t$ is applied between the tip and the sample. The frequency ω of the AC component corresponds to the mechanical resonance of the cantilever with the tip. The force acting between the plates of a capacitor connected to a power supply is equal to the derivative of energy stored in it and hence proportional to the square of potential difference between its plates. This is the sum of the external voltage and the "built in" contact potential difference V_{cpd} and hence is equal $V_{cpd} + V_{DC} + V_0 \sin \omega t$.

Squaring that sum leads to a conclusion that the component of the force at the resonant frequency ω is proportional to $(V_{cpd} + V_{DC})V_0 \sin \omega t$. Because of the resonance, the tip is very sensitive to the excitation at that frequency. So the condition $V_{cpd} + V_{DC} = 0$, is used to find the value of V_{cpd}, which in turn reflects the potential. The precision of such measurements (both the value and resolution) is limited mainly by the strayed capacitance. Because the force is dependent on the derivative of the capacitance in respect of distance, the tip should be kept as close to the sample as possible. Topography and potential are measured simultaneously "line by line". At first the profile of the surface is found. Next tip tracks the surface at the constant height. It enables to correlate the potential and topography, especially with threading dislocations represented by pits on the surface [2].

Some results of such investigations for GaN have been already published. Koley and Spencer [3] showed correlation of peaks in potential and surface pits, suggesting negative charging of threading dislocations in n-type GaN. On the contrary Krtschill et al. [4] reported peaks in potential only in the case of p-type GaN, while they failed to observe any contrast in n-type material. We report on the dependence of the potential distribution on illumination with photons with energy close to band gap of GaN.

---

[*] Corresponding author: e-mail: Rafal.Bozek@fuw.edu.pl, Phone: +48 22 5546861, Fax: +48 22 6219712

2 Experiment

The measurements were performed using Nanoscope IIIa with an extender electronics module (Digital Instruments). The source of light was a 150 W halogen lamp with a monochromator and focusing lens.

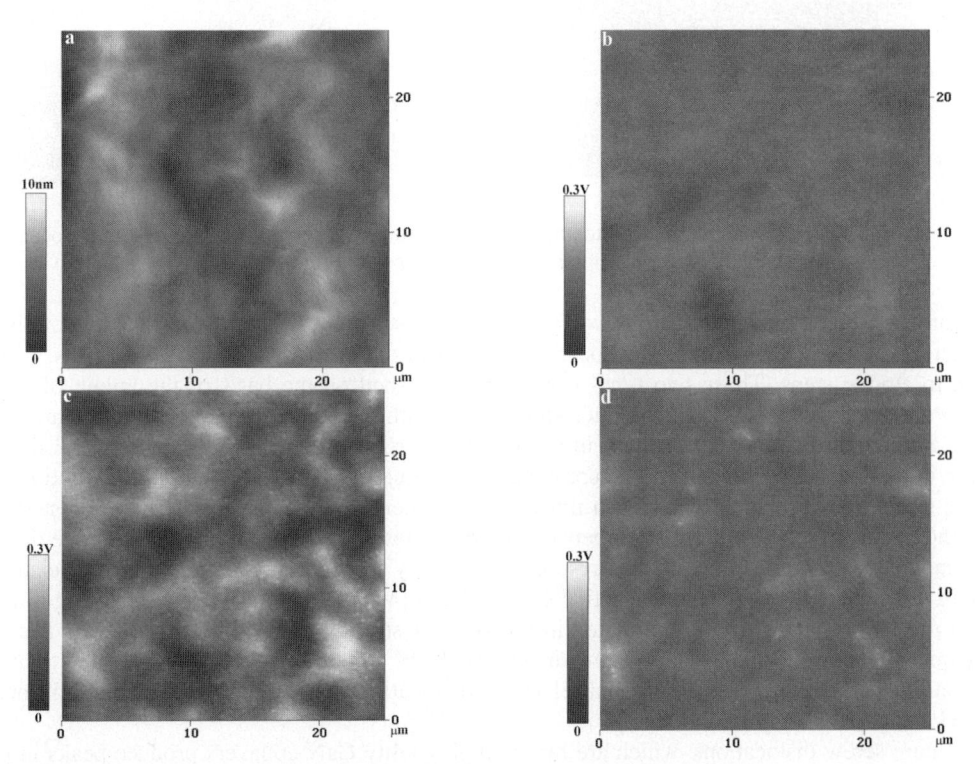

Fig. 1 Large scale maps of topography (a) and surface potential recorded (b) in dark, (c) with moderate (~$0.1 I_{max}$), and (d) strong (I_{max}) illumination. Photon energy was equal 3.30 eV.

The investigated samples were undoped GaN epilayers grown on sapphire by Metal-Organic Vapour Phase Epitaxy with the total density of threading dislocations 3×10^8 cm^{-2}. That value, reduced by one order of magnitude in comparison with reference samples of the same thickness, was achieved by incorporating *in-situ* a Si$_x$N$_y$ mask. The growth details has been described elsewhere [5]. Reduced density of dislocations makes easier correlation of topography and potential images.

The basic structure contained GaN layer grown at 1100 °C followed by a 1 μm GaN layer grown at 1025 °C. Those layers differed substantially in their electrical parameters. GaN grown at 1100 °C exhibited high resistivity, while that grown at 1025 °C was n-type with electron concentration of the order 1×10^{16} cm^{-2}.

The investigated samples are characterised with flat morphology in the scale of tens of micrometers (Fig. 1a). Potential scans without additional illumination are almost featureless (Fig. 1b). Moderate (~$0.1 I_{max}$) illumination below energy gap of GaN (3.30 eV) strongly enhances the image (Fig. 1c). The fluctuations with the amplitude of 0.3 V and lateral size of a few micrometers are pronounced. Further increment of the intensity of illumination causes the amplitude of fluctuations to decrease below 0.05V for I_{max} (Fig. 1d). Long range fluctuations were distinct only in the case of below band gap illumination. They are almost invisible in the case of light above the band gap, independently of its intensity. The possible explanation of those fluctuations are local variations of energy gap due to residual strain or not uniform distribution of defects, that changes their charge state under illumination.

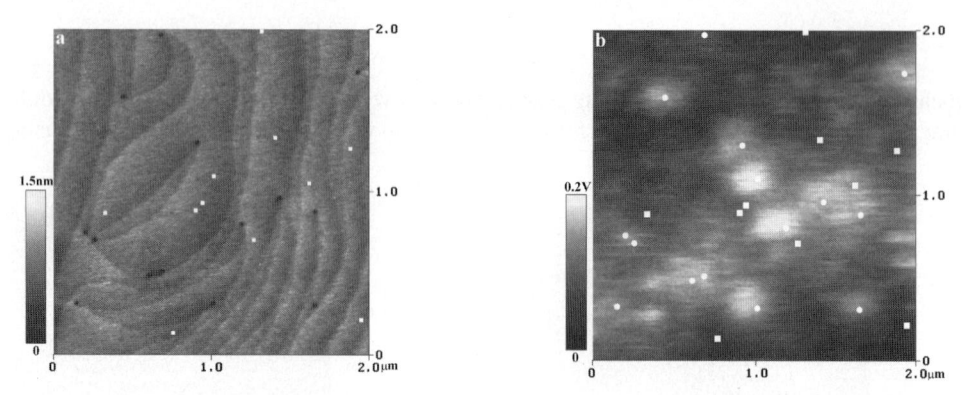

Fig. 2 High resolution scans of topography (a) and surface potential (b). Positions of dislocation pits are marked with squares (edge type) or circles – (mixed type only right image).

Figure 2a presents topography of a 2 × 2 µm area. That scale is sufficient to resolve pits of all types of dislocations. Pure screw or mixed dislocations can be easily recognised due to large pits, which give origin to atomic steps. Those two types can be distinguished on the basis of the layout of the atomic steps. Pure edge type dislocations produce smaller pits with diameter less than 20 nm and are independent of atomic steps. To visualise them in print, they have been marked with squares. All dislocations visible in the Fig. 2a are of mixed or pure edge type. Figure 2b presents potential distribution obtained for the same area. The sample has been illuminated at energy 3.30 eV with intensity adjusted to maximise the contrast. To show correlation between both images, the pits of the mixed and edge dislocations were marked with circles and squares respectively. Every mixed dislocation can be associated with a peak in potential and so an area of negative spatial charge, which may arise due to charging of the dislocation itself or other defects accumulated in the field of strain generated by dislocation. The opposite assignment cannot be done, because some distinct peaks of potential have not been related to any mixed dislocation. The fluctuations in potential related to dislocations can be visualised using light below and above band gap with adjusted intensity.

The pure screw dislocations, which are rare in high quality GaN epilayers produce peaks in potential and hence are subject to negative charging as mixed type dislocations. The pure edge dislocations are not correlated with distinct features in the potential image, however most of them can be found in the areas of lower potential. It is not sufficient to draw a conclusion on their charge.

Edge dislocations in the samples of worse quality very often form long strings corresponding to low angle boundaries among crystal grains (not visible in the presented image). In such a case they can be more clearly related to low potential.

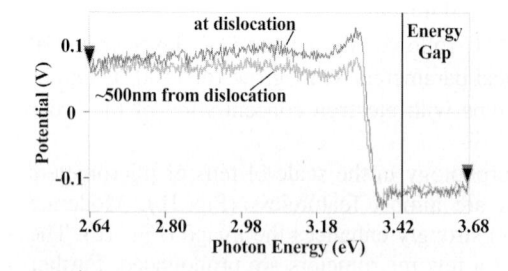

Fig. 3 Spectral dependence of the potential. The presented curves have been recorded directly at a mixed type dislocation (upper one) and at a point separated by ~0.5 µm (lower one) for intense (I_{max}) illumination.

The spectral dependence of surface potential have been found by turning off the slow axis movement of the microscope scanner and repeating measurements for the same line while varying the wavelength of light. The curve measured at dislocation (Fig. 3) starts to separate at energy about 2.8 eV. It is interpreted as the beginning of the accumulation of negative charge at dislocation and hence reveals the position of the energy level of the acceptor state related to dislocation. The maximum vertical distance between both curves ~60 mV (highest contrast) is reached for photon energy 3.30 eV, corresponding to the onset of downward slope of potential. The lowering of potential is caused by accumulation of electrons close to the surface obtained due to the structure of the sample.

3 Discussion

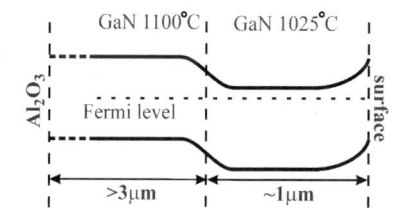

Fig. 4 Structure and band line-up of investigated samples.

As it has been explained, the idea of KFM is based on a capacitor consisting of the tip and the sample. However surface of semiconductor layers often happens to be depleted due to surface states which cause upward bending of the bands (Fig. 4). It means, that one "plate" is buried at some depth and does not match the very surface of the sample. It has been also pointed, that the relative separation of both plates is crucial for the resolution.

In our experiment, the key point is the structure of the samples. GaN layers grown at 1100 °C and 1025 °C characterised with different positions of Fermi level form a homojunction. The barrier created in the conduction band prevents electrons generated with light from escaping into deeper regions. The surface potential measured for a single GaN layer increases under illumination, while in the case the investigated double structure its decrement has been obtained. It is consistent with the fact, that illumination has not produced contrast for homogenous layers grown at single temperature.

The advantage of the proposed structure is the possibility of tuning the density of electrons in the conduction band with light. If the density of electrons is low (illumination with $0.1 I_{max}$), the electrons visualise fluctuations because they are distributed according to the shape of the potential generated by spatial charge. High density of electrons (illumination with I_{max}) screens fluctuations so the images look like those obtained for n-type samples.

Another role of light is changing charge states of defects. The light below energy gap penetrates the sample, while that above gap affects only layer adjacent to the surface. It explains why fluctuations in the large scale are observed only in the first case.

4 Conclusions

Layered structure for investigations of the distribution of potential in GaN has been proposed. The performed measurements have shown strong dependence of the fluctuations of the potential on both the photon energy and intensity of illumination. Threading dislocations with a screw component have been related to negative charges in the sample. No assignment could be done in case of edge type dislocations.

Acknowledgements The paper was partially supported by the Polish Government Strategic Program "Blue optoelectronics" 8T11C 001 2000C/5013.

References

[1] H.O. Jacobs et al., Ultramicroscopy **69**, 39 (1997).
[2] B. Heying et al., J. Appl. Phys. **85**, 6470 (1999).
[3] G. Koley and M.G. Spencer, Appl. Phys. Lett. **78**, 2873 (2001).
[4] A. Krtschil, A. Dadgar, and A. Krost, J. Cryst. Growth **248**, 542 (2003).
[5] K. Pakula et al., in 10th European Workshop on Metalorganic Vapour Phase Epitaxy, Lecce, 2003, Booklet of Extended Abstracts, The University of Lecce (Italy), 2003, pp. 335–339.

Optical quality improvement of InGaAs/AlAs/AlAsSb coupled double quantum wells grown by molecular beam epitaxy

J. Kasai[*, 1], **T. Mozume**[1], **H. Yoshida**[1], **T. Simoyama**[1], **A. V. Gopal**[**, 1], and **H. Ishikawa**[1]

[1] FESTA Laboratories, Femtosecond Technology Research Association, 5-5 Tokodai, Tsukuba 300-2635, Japan

Received 19 September 2003, accepted 3 November 2003
Published online 22 December 2003

PACS 42.70.Km, 42.70.Nq, 73.61.Ey, 78.30.Fs

We have grown InGaAs/AlAs/AlAsSb coupled double quantum wells (C-DQWs) with AlAs diffusion-stopping layers by molecular beam epitaxy. An obtained sample had many cross-hatched lines, suggesting relatively poor structural quality. Optical measurements, however, revealed that the optical quality of the C-DQWs was greatly improved compared to earlier C-DQWs without AlAs diffusion-stopping layers. The intersubband absorption saturation intensity in the present C-DQW sample was extremely low, measuring 34 fJ/μm^2 at the optical communication wavelength of 1.62 μm, while ultrafast response times of about 600 fs were maintained.

© 2004 WILEY-VCH Verlag GmbH & Co. KGaA, Weinheim

1 Introduction Intersubband transitions (ISBTs) in quantum wells (QWs) offer significant potential for device applications [1]. One promising application of ISBTs is a high-speed all-optical switch that can be used in ultrafast optical communication systems with bit rates exceeding 1 Tb/s. The ultrafast relaxation times and large optical nonlinearities of ISBTs give rise to ultrafast response times and low absorption saturation intensities (Is) of optical switches, respectively. An InGaAs/AlAsSb system on InP is one of the most suitable systems for this application. Although the response time of this system is slower than that of a GaN/AlGaN system [2], Is, which is related to switching energy, is much less than in a GaN system.

We previously reported on near-infrared intersubband transitions [3] and low Is [4, 5] using a In-GaAs/AlAsSb single quantum well. The Is was finally reduced to 3 fJ/μm^2 when AlAs diffusion-stopping layers were inserted into the well-barrier interfaces [6]. Our most recent study indicates that the absorption response time is greatly reduced to 690 fs when we use InGaAs/AlAs/AlAsSb coupled double quantum wells (C-DQWs) [7]. This achievement led to the expectation that the application of diffusion-stopping layers to C-DQWs will improve the ISBT properties with respect to both Is and response time. We therefore tried to grow C-DQWs epitaxially with AlAs diffusion-stopping layers. This paper reports on the effects of AlAs diffusion-stopping layers on the structural and optical properties of C-DQWs.

2 Sample growth InGaAs/AlAs/AlAsSb C-DQW structures were grown on Fe-doped (001) InP substrates by molecular beam epitaxy. Elemental Ga, In, and Al were used for group III growth species, while Sb$_2$ and As$_2$ were used for group V. The Sb$_2$ and As$_2$ fluxes were supplied using cracker cells. The n-type doping was performed with Si. The substrate temperature during growth was 460 °C.

Our goal is the fabrication of all-optical switches with ridge waveguide structures. One problem faced in pursuing this goal is that due to the small refractive index of the AlAs layer, the ridge waveguides tend to become leaky. In fact, computer simulations of ridge waveguide structures with InAlAs cladding lay-

---

[*] Corresponding author: e-mail: kasai@festa.or.jp, Phone: +81 29 847 5181, Fax: +81 29 847 4417
[**] Present address: The Imai quantum computation and information project, Japan Science and Technology Corporation.

© 2004 WILEY-VCH Verlag GmbH & Co. KGaA, Weinheim

ers reveal that the waveguides with AlAs diffusion-stopping layers in earlier C-DQW structures [7] were leaky. Thus to enhance optical confinement in the waveguide core, we modified the C-DQW structure as follows: (1) the thickness of the AlAsSb barrier layers was reduced, (2) the In composition and layer thickness in the InGaAs wells were increased, and (3) the stack period of the C-DQWs was increased.

The sample structure realized in the present study is shown in Figs. 1a and b. The C-DQW sample that is doped in the well regions to about 1×10^{19} cm^{-3} has 93 periods of 4 nm AlAs$_{0.56}$Sb$_{0.44}$/6 monolayers (ML) AlAs/2.8 nm In$_{0.80}$Ga$_{0.20}$As/4 ML AlAs/2.8 nm In$_{0.80}$Ga$_{0.20}$As/5 ML AlAs/4 nm AlAs$_{0.56}$Sb$_{0.44}$. The InAlAs cladding layer was lattice-matched to InP with an In composition of 0.52. We were able to confirm that 1-mm-long ridge waveguides fabricated from this sample are capable of transmitting near-infrared light adequately [8].

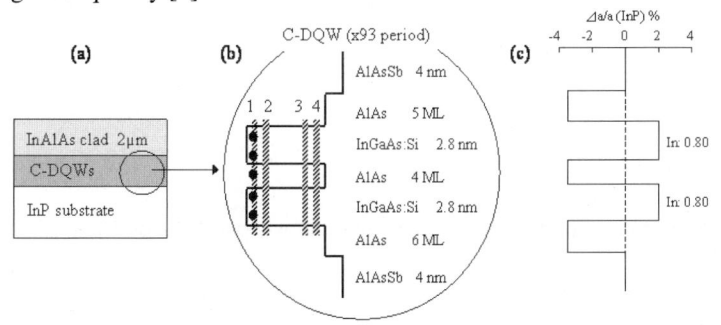

Fig. 1 (a) Schematic of sample layer structure, (b) conduction-band alignment of C-DQW with AlAs diffusion-stopping layers, (c) lattice mismatch between the InP substrate and constituent layers.

3 Structural properties Figure 2a shows an optical micrograph of the sample. The sample has many cross-hatched lines. We examined the structural quality of the sample by 4-crystal x-ray diffraction rocking curve measurements with a Philips MRD system. As shown in Fig. 2b, the obtained rocking curve has broad satellite peaks. These results suggest that the present sample may not be of the highest structural quality. A trial to improve the structural quality will be discussed later. Although the structural quality of the present sample may not be good, optical measurements of the sample reveal good optical quality, as described below.

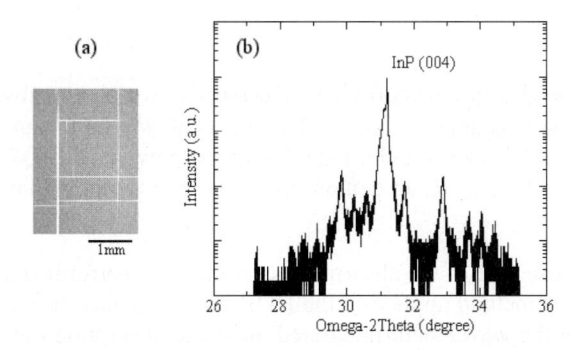

Fig. 2 (a) Optical micrograph of surface morphology for C-DQWs with AlAs diffusion-stopping layers. (b) X-ray diffraction spectrum of the sample.

4 Absorption spectra We measured the absorption spectra using a Fourier-transform infrared (FTIR) spectrometer (Bruker 66V/S FTIR). To enhance light absorption along the growth axis, we used multiple reflection geometry with 45° polished mirror facets (see the inset of Fig. 3). Figure 3 shows the absorption spectrum of the present sample and the previous C-DQWs without AlAs stopping layers [7]. Our C-DQWs are designed such that the QWs have four subbands in the conduction band (see Fig. 1b) and the wavelength of the 1–4 transition becomes the 1.55-µm optical communication wavelength. As shown in the spectrum of the present sample, an absorption peak caused by the 1–4 transition is clearly visible around the communication wavelength. The absorption coefficient at the peak was estimated to be 1,200 cm^{-1}, indicating significant enhancement of the absorption intensity compared to that of earlier C-DQWs without AlAs layers. In general, absorption coefficients of QWs are dependent on their doping density. However, the present and previous samples have the same doping density. Thus enhancement of the absorption intensity indicates that AlAs stopping layers significantly improve the optical quality of C-DQWs.

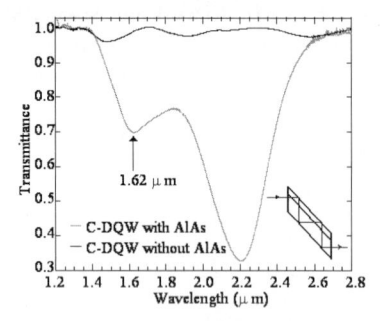

Fig. 3 Intersubband absorption spectra of C-DQWs with and without AlAs diffusion-stopping layers. The inset shows multiple reflection geometry.

5 Absorption saturation

5.1 Saturation intensity We next measured the Is by the same geometric method mentioned before, this time implementing an optical parametric amplifier (100-kHz repetition rate, 150-fs pulse-width). Figure 4 (a) shows the transmittance at the absorption peak of 1.62 μm as a function of the input pulse energy density. The transmittance spectrum indicates clear absorption saturation and absorption recovery caused by two-photon absorption in the InP substrate. The data can be analyzed by pulse transmission modelling of the thickness-dependent change in the incident pulse intensity and two-photon absorption:

$$\frac{dI}{dz} = -\alpha I ,$$

where

$$\alpha = \frac{\alpha_0}{1 + \frac{I}{I_s}} + \beta I .$$

Here, I is the incident light intensity, α_0 is the absorption coefficient, and β is the two-photon absorption coefficient in the InP substrate. Using the given parameters, $\alpha_0 = 1,200$ cm^{-1} and $\beta = 0.06$ cm/GW [5], Is was estimated to be 34 fJ/μm^2. The Is value is much smaller than that of the previous C-DQW structure without AlAs stopping layers (1 pJ/μm^2) [7]. This result demonstrates that AlAs stopping layers can be effectively used with C-DQWs.

5.2 Response time To determine the response time of the absorption saturation, we performed a pump-probe measurement using the same geometric method applied in the FTIR measurements. In this measurement, we measured the time evolution of the wavelength-integrated interband absorption induced by the intersubband excitation in the C-DQW sample (see the inset of Fig. 4b). The pump pulse excites carriers from a lower conduction subband into an upper conduction subband. Then the excited carriers relax into the lower subband. The intersubband relaxation involving intrasubband relaxation is related to the absorption recovery of the lower conduction subband, and can be observed by monitoring the interband absorption. Thus the measured decay time of interband absorption corresponds to the response time of the absorption saturation.

In our measurements, the probe pulse of white-light was generated by irradiating a light pulse from the optical parametric amplifier onto a sapphire plate. The pulse light transmitted through the sample was detected by InGaAs photodiode arrays coupled to a monochromator. Figure 4b shows the temporal evolution of interband-absorption integrated over the range from 1.10 to 1.25 μm when intersubband excitation consisted of a 1.62-μm pump pulse. As shown in Fig. 4b, an ultrafast decay of the interband absorption was clearly observed. The solid line in the figure shows a single exponential decay with a decay time of 600 fs. Thus the response time of the absorption saturation can be estimated to be 600 fs. This indicates that inserting AlAs stopping layers into C-DQWs doesn't impair ultrafast response.

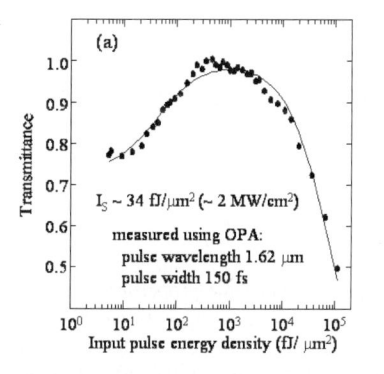

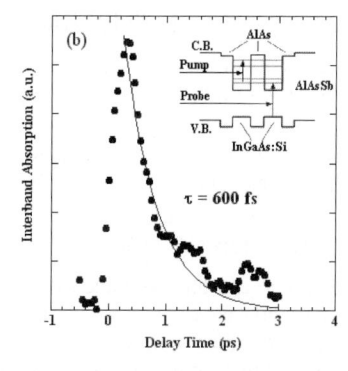

Fig. 4 (a) Transmittance at the wavelength of 1.62 μm as a function of the input pulse energy density measured with multiple reflection geometry. The solid line shows a fit with the results from pulse transmission modelling. (b) The time evolution of the wavelength-integrated interband absorption induced by the intersubband excitation in the C-DQW sample. The solid line indicates an exponential decay curve with a decay time of 600 fs. The inset shows band alignment of a C-DQW structure.

6 Optimization of C-DQW structures Although the results of the optical measurements clearly demonstrate the potential of C-DQWs with AlAs diffusion-stopping layers for ultrafast ISBT switches, certain problems remain to be resolved. As mentioned before, the sample has relatively poor structural quality. This poor structural quality may be caused by a large residual strain in the sample. Figure 1 (c) shows a lattice mismatch between the InP substrate and constituent layers. Since an AlAs layer features a large lattice mismatch with the InP, the insertion of AlAs layers into C-DQWs introduces a large strain into the sample. We are now trying to grow C-DQWs with thinner AlAs layers to accommodate this strain. Preliminary results of the trial to optimize the thickness of AlAs layers indicate that cross-hatched lines can be reduced by reducing AlAs thickness. We can also improve x-ray diffraction rocking curves and FTIR spectra. We confirmed that further optimization of growth parameters improves the structural quality of C-DQW structures.

Acknowledgements One of authors (J. K.) would like to thank Dr. Asakawa for his continuous encouragement. This work was performed under the management of Femtosecond Technology Research Association supported by the New Energy and Industrial Technology Development Organization.

References

[1] E. Rosencher, B. Vinter, and B. Levine (eds.), Intersubband Transitions in Quantum Wells (Plenum Press, New York, 1992).
[2] N. Iizuka, K. Kaneko, N. Suzuki, T. Asano, S. Noda, and O. Wada, Appl. Phys. Lett. **77**, 648 (2000).
[3] T. Mozume, H. Yoshida, A. Neogi, and M. Kudo, Jpn. J. Appl. Phys. **38**, 1286 (1999).
[4] A. V. Gopal, H. Yoshida, A. Neogi, T. Mozume, N. Georgiev, T. Simoyama, O. Wada, and H. Ishikawa, Jpn. J. Appl. Phys. **40**, L1015 (2001).
[5] A. V. Gopal, H. Yoshida, A. Neogi, T. Mozume, N. Georgiev, T. Simoyama, O. Wada, and H. Ishikawa, Electron. Lett. **37**, 1265 (2001).
[6] T. Mozume, N. Georgiev, T. Simoyama, A. V. Gopal, and H. Ishikawa, Post Deadline Papers of 14th International Conference of Indium Phosphide and Related Materials, Stockholm, Sweden, 2002.
[7] T. Akiyama, N. Georgiev, T. Mozume, H. Yoshida, A. V. Gopal, and O. Wada, Electron. Lett. **37**, 129 (2001).
[8] T. Simoyama H. Yoshida, J. Kasai, T. Mozume, A. V. Gopal, and H. Ishikawa, IEEE Photon. Technol. Lett. (in press).

X-ray photoemission study of manganese thin films deposited on a layered semiconductor

F. Mirabella[*,1], **B. A. Parkinson**[2], and **J. Ghijsen**[1]

[1] Lise Laboratory, University of Namur (FUNDP), 61 rue de Bruxelles, B-5000 Namur, Belgium
[2] Department of Chemistry, Colorado State University, Fort Collins, Colorado 80523, USA

Received 19 September 2003, accepted 3 November 2003
Published online 22 December 2003

PACS 68.47.Fg; 73.20.At, 81.05.Hd, 79.60.Bm

Germanium sulphide (GeS) and tin diselenide ($SnSe_2$) present anisotropic properties due to their layered crystal structures. These crystals are composed of atomic layers interacting with each other by van der Waals forces only. Recently these materials have been used to prepare a new diluted magnetic semiconductor (DMS) by substituting some Ge or Sn atoms by manganese in the crystal lattice, forming $Ge_{1-x}Mn_xS$ and $Sn_{1-x}Mn_xSe_2$. To compare Mn behaviour within the layered crystals, Mn/GeS and Mn/$SnSe_2$ thin films have been grown and investigated in situ by X-ray induced photoemission. Here is reported the XPS analysis of in situ grown Mn/$SnSe_2$ for coverage ranging from sub-monolayer to thin films. The properties of these films were investigated from the structural and reactivity point of view. XPS measurements suggest that an Mn-Sn alloy is formed at the interface and that Mn atoms first deposit as an atomic netting pattern dictated by the typical structure of the $SnSe_2$ (001) surface.

© 2004 WILEY-VCH Verlag GmbH & Co. KGaA, Weinheim

1 Introduction

Tin diselenide ($SnSe_2$) crystals have a layered structure (space group P_{3m1}) in which Sn planes are sandwiched by Se planes (Fig. 1). The Se-Sn-Se unit layers repeat along the c crystal axis interacting by van der Waals forces only giving rise to the anisotropic electro-optical properties of this material. Unit cell parameters are $a = 3.81$ Å and $c = 6.14$ Å [1]. From an electrical point of view, $SnSe_2$ is an indirect gap semiconductor ($E_g^i = 1.0 \pm 0.2$ eV) [2, 3]. From 1999, this compound has been used as the precursor to grow a diluted magnetic semiconductor (DMS) by partial substitution of tin by manganese, forming $Sn_{1-x}Mn_xSe_2$ [4, 5]. In the present work we report the growth and X-ray photoemission spectroscopy (XPS) study of manganese thin films on $SnSe_2$ (001) surface. This is thought to be a useful comparison material to improve our knowledge of $Sn_{1-x}Mn_xSe_2$ layered DMS and is also, in the light of the present analysis, an interesting system in itself. To our knowledge, it is the first time $SnSe_2$ is used as a substrate for thin film deposition. As a consequence of its structural properties, fresh $SnSe_2$ (001) faces are easily obtained by cleaving the crystal under ultra-high vacuum (UHV). The resulting top crystal layer, shown in Fig. 1, has a hexagonal symmetry and is made of completely bound atoms. It is therefore interesting to see how metallic films will interact with such a "non-reactive" surface.

---

[*] Corresponding author: e-mail: frederic.mirabella@fundp.ac.be, Phone: +32 81 725 237, Fax: +32 81 724 595

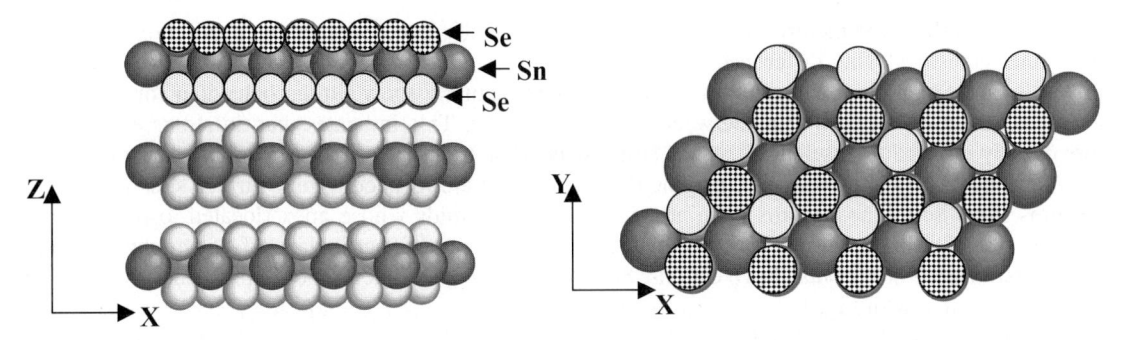

Fig. 1 SnSe$_2$ crystal structure. For convenience, selenium atoms in the top layer have been depicted differently depending on their position above or below the cation (Sn) layer. Right: SnSe$_2$ (001) surface.

2 Experimental

The tin diselenide single crystal used in this work was made by heating stoichiometric amounts of highest purity Sn and Se in a sealed glass ampoule for 12 h at 800 °C. Fresh (001) surfaces were obtained by cleaving the crystal under UHV and were checked by low energy electron diffraction (LEED). Manganese films were deposited on this surface at room temperature by evaporation from a Knudsen cell. A quartz thickness monitor monitored films thickness. XPS analyses were performed in situ with a Scienta ESCA-300 spectrometer running with 20 eV pass energy and 0.5 mm entrance slit that results in an experimental resolution of 0.27 eV measured at the silver Fermi level. Binding energies are referred to the Ag 3d$_{5/2}$ line at 368.2 eV. XPS line shape and position were analysed with a home-developed software.

3 Results and discussion

3.1 Growth mode study

Venables *et al.* demonstrated how XPS (as well as Auger spectroscopy) line quantification is suitable to analyse the growth mode of a solid film on a solid surface [6]. Integrating areas of core levels peaks from the substrate and the adsorbate and normalising them to the proper photoemission cross-section gives atomic concentrations present in the analysed sample. Plotting these data vs. the film thickness results in a curve whose shape depends on the growth mode of the film: layer-by-layer, island or layer plus island mode. Figure 2 shows this result for Mn/SnSe$_2$ (dotted lines) as well as a simulation of XPS intensities obtained with a home-developed software (full lines). The curve behaviour indicates that manganese films grow on SnSe$_2$ in the Stransky-Krastanov mode (layer and islands). The calculated curve indicates that 5×10^{11} islands per square centimetre are formed. The type of compounds constituting the interface layer between SnSe$_2$ and Mn islands will be revealed in the next section after discussion of chemical states analysis.

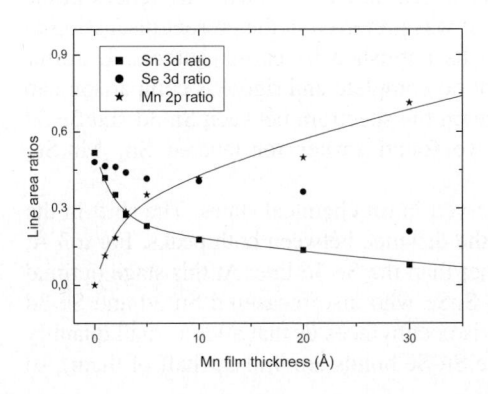

Fig. 2 Evolution of Mn, Sn, and Se concentration ratios with the film thickness.

Moreover, in the same picture, one can observe a strong difference between attenuation of Sn 3d and Se 3d line intensities. This difference extends beyond these particular core level peaks; also observed with Sn 3p and Se 3p photoemission lines. Measurements actually show similar behaviour for Se 3d and Se 3p peak attenuation, as well as between Sn 3d and Sn 3p. This indicates that tin atoms emission is more quickly reduced than selenium as Mn atoms deposit on the SnSe$_2$ surface. Looking carefully at the substrate surface geometry, one can imagine that manganese atoms occupy first surface sites just above tin atoms since Se topmost atoms form atomic inverted pyramids whose apex (located lower than the basis) are occupied by tin atoms. It is thus natural to assume that this particular surface structure favours the location of adsorbed atoms above cations, Sn atoms in our case. The proposed picture, shown in Fig. 3, can be compared with Fig. 1.

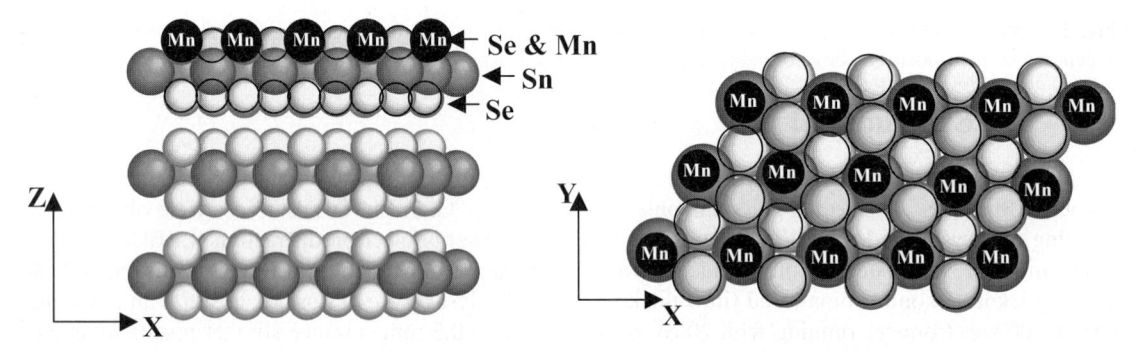

Fig. 3 Proposed picture of the first step of Mn/SnSe$_2$ film growth. Manganese atoms deposit first above Sn atoms resulting in stronger attenuation of tin photoemission lines with respect to Se signals.

3.2 Chemical states and reactivity

Sn 3d, Mn 2p and Se 3d photoemission lines are reported in Fig. 4 for increasing Mn coverage, up to 50 Å. Concerning Sn, one can observe one single chemical state on the clean substrate with the Sn 3d$_{5/2}$ component located at 486.2 eV, which shifts toward lower binding energy as Mn begins to be deposited on the surface. This chemical shift seems to be due to a chemical state reduction of tin atoms from Sn^{4+} to Sn^{2+}. From a 5 Å film, a second chemical state appears at 484.9 eV that corresponds to Sn^0 chemical state [7]. As the Mn quantity increases, the intensity of this metallic tin peak increases with respect to the Sn^{2+} contribution. The Se 3d line recorded at 53.7 eV for clean SnSe$_2$ shows weaker reactivity since only a small shift of the peak toward higher binding energies is detected as Mn is adsorbed.

Though SnSe$_2$ has been investigated by photoemission since the 70's, no information is available regarding core level binding energies. From the results of Schlaf et al. who studied the deposition of SnSe$_2$ on mica and highly oriented pyrolytic graphite (HOPG), one can estimate the Sn 3d peak position at 486.0 and 485.8 eV respectively [8, 9]. It must be emphasised that in the cited references, binding energies were referred to the C 1s line of graphite at 284 eV, which may cause small differences in the comparison of absolute value peak positions. For that reason, it was preferred in the present discussion to use distances between XPS lines for comparisons with results published by earlier authors. Unfortunately, Schlaf et al. have not investigated any Se lines so that no complete and rigorous comparison can be made. In the case of clean SnSe$_2$ we report here a distance on the spectrum between Sn 3d and Se 3d lines of 432.5 eV. This value is very close to the one we found earlier for layered Sn$_{1-x}$Mn$_x$Se$_2$ (432.6 eV) [5].

As Mn is deposited on SnSe$_2$, noticeable changes are observed in tin chemical states. The shift in the Sn 3d line while the Se 3d line remains at 53.5 eV reduced the distance between both peaks. For a 3 Å, the Sn 3d line is located at 485.7 eV, which is 432.0 eV higher than the Se 3d line. At this stage optimal agreement is found with data reported by Shalvoy et al. on SnSe who also measured Sn 3d and Se 3d lines at 485.7 eV and 53.7 eV, respectively [10]. This comparison convinces us that after a small quantity of manganese has been deposited on the SnSe$_2$ surface, some Sn-Se bonds are broken (half of them), so that SnSe is detected instead of SnSe$_2$. Due to the surface structure, already mentioned in the previous section, one can imagine that tin atoms are still bound to selenium locating in the lower layer while manganese reactivity breaks bonds between tin and the topmost selenium atoms. These Se anions will now be able to form a compound with adsorbed manganese.

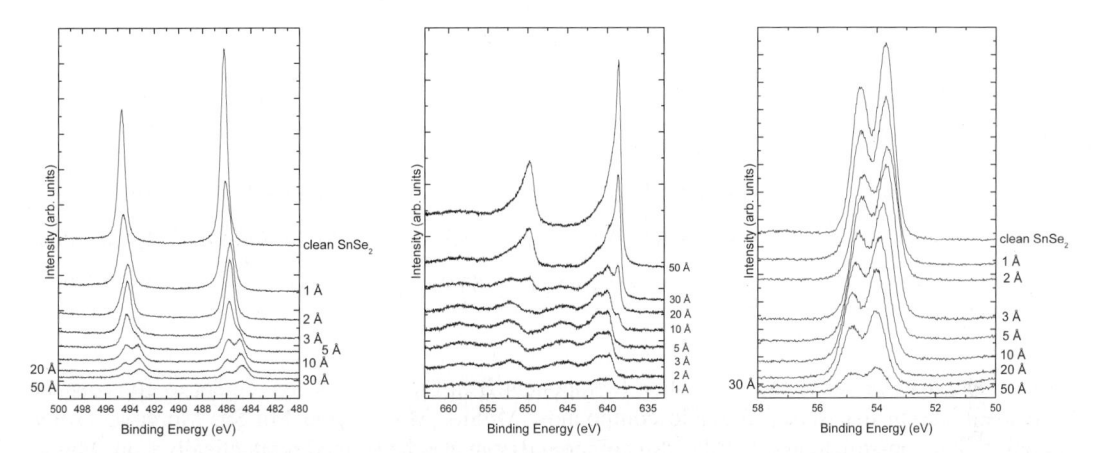

Fig. 4 Investigated core levels for various film thickness. Sn 3d line (left), Mn 2p line (middle) and Se 3p (right).

The evolution of the Mn 2p line presented in Fig. 4 shows that two chemical states of Mn are present from the earliest growth stage since two peaks, lying at 639.9 and 641.3 eV, are visible. A well-known strong satellite structure is also present at 644.9 eV. Though one of the peaks can be assigned to Mn^{2+} by comparison with literature data, no Mn 2p line is reported for binding energies between 639.1 and 640.1 eV. This binding energy range is intermediate between values attached to Mn^0 and Mn^{2+} species [7]. It can thus be thought that the line detected at 639.9 eV belongs to an unusual chemical state of manganese atoms. As the film thickness increases, another contribution appears in the Mn 2p spectrum, corresponding to Mn^0 (639.1 eV). This new component shifts toward lower binding energies as the substrate is covered by the transition metal. The Mn^0 line from a 20 Å Mn film is located at 638.6 eV. Since this peak appears only after 5 Å deposition, we attribute it to Mn islands that nucleate at the interface layer. All the discussed chemical states of manganese are clearly visible in Fig. 5, which was recorded for a coverage of 10 Å.

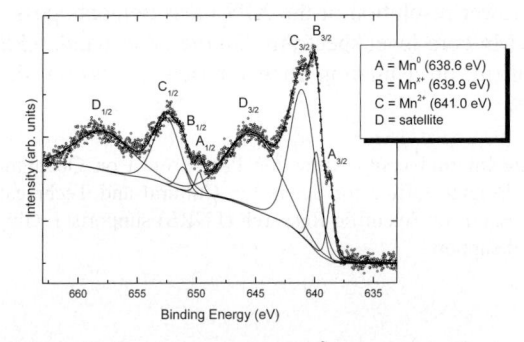

Fig. 5 Mn 2p spectrum for 10 Å Mn/SnSe$_2$. Three chemical states are clearly visible as well as the satellite structure due to charge transfer from ligand to metal.

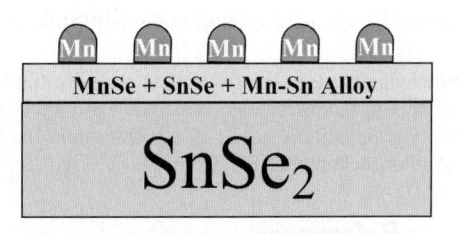

Fig. 6 Stranski-Krastanov growth model proposed for the Mn/SnSe$_2$ system.

Considering the Se 3d line, we observe a distance of 587.2 eV to Mn^{2+} contribution in the Mn 2p spectrum. Now a good agreement is found with the work on MnSe by Mandale et al. [11]. Though absolute binding energies reported by the author (641.8 eV for Mn 2p and 54.6 eV for Se 3d) are different from our values, the separation between both lines is (587.2 eV) the same value we report here. Table 1 presents a summary of binding energies measured in this study, as well as chemical species assignements.

Table 1 Binding energy of core levels investigated for various growth stages of $Mn/SnSe_2$.

Film thickness	Sn 3d (eV)	Mn 2p (eV)	Se 3d (eV)	Assignment
Clean $SnSe_2$	486.2	–	53.7	$SnSe_2$
1 Å	486.0	639.9 + 641.3	53.7	SnSe + MnSe + Mn^0
5 Å	484.9 + 485.8	639.1 + 639.9 + 641.0	53.8	SnSe + MnSe + Mn-Sn alloy
10 Å	484.8 + 485.8	638.6 + 639.9 + 641.0	53.9	SnSe + MnSe + Mn-Sn alloy

To deal with what we called the Mn^{x+} state (component B in Fig. 5), it is convenient to consider the Mn-Sn phase diagram that shows 3 stable compounds: $MnSn_2$, Mn_3Sn_2 and Mn_3Sn. Moreover, some authors report that non-stoichiometric $Mn_{2-x}Sn$ compounds can also be formed peritectically from Mn_3Sn [12]. Analysing the possible interaction of atomic species present in our system, it seems that one of these Mn-Sn alloys can actually be formed at the $Mn/SnSe_2$ interface. In that material, manganese behaves like a Mn^{2+} species and tin atoms are in a pure metallic state. It would be interesting to investigate Mn-Sn alloys by photoemission spectroscopy to confirm this hypothesis.

The variation in relative intensity between the Sn^{2+} and Sn^0 components of the Sn 3d line pointed out earlier, can now be recalled to make a last observation concerning the 3 to 10 Å coverage range. When more Mn is deposited, more Sn-Se bonds are broken with the consequence that the quantity of formed Mn-Sn bonds at the $Mn/SnSe_2$ interface increases with respect to SnSe.

4 Conclusions

Manganese films have been deposited on $SnSe_2$, a layered semiconductor, and analysed in situ by XPS from submonolayer range to quasi-complete attenuation of the substrate signals. Putting together the discussion of growth mode and chemical state analysis of the $Mn/SnSe_2$ system studied, we present in Fig. 6 a model in which Mn islands grow on $SnSe_2$ after the formation of an interface monolayer in which evidence has been made of the presence of SnSe, MnSe as well as an Mn-Sn alloy. This lead us to reconsider our previous work on Mn/GeS [13], where lower resolution of the XPS measurements possibly prevented us from detecting all components of the Mn core-level spectrum. On the other hand, XPS investigation of Mn-Sn alloy is recommended to confirm, on a binding energy based discussion, the presence of such a compound at the $Mn/SnSe_2$ interface.

Acknowledgements This work was partially funded by the Interuniversity Attraction Pole Project on Quantum size effects in nanostructured materials (PAI P5/01) of the Belgian Office for Scientific, Cultural and Technical Affairs. F. Mirabella is holder of a FRIA grant; the Belgian Fund for Scientific Research (FNRS) supports J. Ghijsen. Authors acknowledge F. Come and V. Oguz for technical support.

References

[1] C. Domingo, R. S. Itoga, and C. R. Kannewurf, Phys. Rev. **143**, 536 (1966).
[2] B. L. Evans, R. A. Hazelwood, Brit. J. Appl. Phys. **2**, 1507 (1969).
[3] C. Julien, S. I. Saikh, and G. A. Nazri, Mat. Sci. Eng. B **15**, 73 (1992).

[4] F. Mirabella, J. Ghijsen, R. L. Johnson, Z. Golacki, and B. A. Orlowski, J. Alloy Comp. **328**, 166 (2001).
[5] F. Mirabella, G. Schmerber, Z. Golacki, R. L. Johnson, and J. Ghijsen, submitted for publication.
[6] J. A. Venables, G. D. T. Spiller, and M. Handbücken, Rep. Prog. Phys. **47**, 399 (1984).
[7] J. F. Moulder, W. F. Stickle, P. E. Sobol, and K. D. Bomben, Handbook of X-ray Photoelectron Spectroscopy, Perkin-Elmer Co., USA (1992).
[8] R. Schlaf, D. Louder, O. Lang, C. Pettenkofer, W. Jaegermann, K. W. Nebesny, P. A. Lee, B. A. Parkinson, and N. R. Armstrong, J. Vac. Sci. Technol. A **13**, 1761 (1995).
[9] R. Schlaf, N. R. Armstrong, B. A. Parkinson, C. Pettenkofer, and W. Jaegermann, Surf. Sci. **345**, 1 (1997).
[10] R. B. Shalvoy, G. B. Fisher, and P. J. Stiles, Phys. Rev. B **15**, 1680 (1977).
[11] A. B. Mandale and S. Badrinarayanan, J. Elect. Spect. Rel. Phenom. **53**, 87 (1990).
[12] M. Stange, H. Fjellvag, S. Furuseth, and B. C. Hauback, J. Alloy Comp. **259**, 140 (1997).
[13] F. Mirabella, R. L. Johnson, and J. Ghijsen, Surf. Sci. **506**, 172 (2002).

Optical investigations of two dimensional electron gas in the AlGaN/GaN heterostructures

R. Kudrawiec[*,1], **M. Syperek**[1], **J. Misiewicz**[1], **R. Paszkiewicz**[2], **B. Paszkiewicz**[2], **M. Tłaczała**[2], and **W. Strupinski**[3]

[1] Institute of Physics, Wrocław University of Technology, Wybrzeże Wyspiańskiego 27, 50-370 Wrocław, Poland
[2] The Faculty of Microsystem Electronics and Photonic, Wrocław University of Technology, Janiszewskiego 11/17, 50-372 Wrocław, Poland
[2] Institute of Electronic Materials Technology, Wólczyńska 118, 01-919 Warszawa, Poland

Received 19 September, accepted 17 November 2003
Published online 22 December 2003

PACS 77.65.Ly, 78.67.De

Undoped and Si doped AlGaN/GaN heterostructures have been investigated in reflectance (R) and photoreflectance (PR) spectroscopies. In the case of R spectra, features associated with GaN and AlGaN layers are observed for these two heterostructures. The GaN- and AlGaN-related transitions are better visible in PR spectra. In addition, a feature associated with two dimensional electron gas (2DEG) has been observed in PR spectrum of Si doped AlGaN/GaN heterostructure. This feature is rather not observed for the undoped AlGaN/GaN heterostructure. For the undoped heterostructure a huge internal electric field (284 kV/cm) has been found.

AlGaN/GaN heterojunction field-effect transistors (HFETs) emerged as attractive transistors suitable for high-power and high-temperature electronics [1–3]. The heart of the device is the 2DEG, which is formed at the AlGaN/GaN interface. Large piezoelectric and spontanieous polarization fields occuring in AlGaN/GaN heterostructures can generate a 2DEG without doping [4–6]. The polarization depends on the crystal direction as well as the growth sequence of different layers along the growth direction. Consequently, the crystal direction [0001] of a GaN film can either be parallel or antiparallel to the growth direction, leading to epilayers with two different polarities, Ga- and N-polar films. This phenomenon and other factors like e.g. AlGaN layer thickness should influence on the 2DEG formation at AlGaN/GaN interface. In consequence, AlGaN/GaN heterostructures with different electrical and optical properties can be obtained. In our opinion this subject needs intensive investigations because many features of AlGaN/GaN heterostructures are uncertain. Previous study in this area has confirmed the existence of 2DEG at AlGaN/GaN interface [7–14]. However, usually AlGaN/GaN HFETs are characterised with electrical measurements which need contacts and little work is devoted to contactless investigations of 2DEG in AlGaN/GaN. It is well known that contacts change the band bending at surface of semiconductor heterostructures. It is expected that the electric contact can strongly influence on 2DEG formation in AlGaN/GaN heterostructures especially if the AlGaN layer is thin. In addition, it is know that GaN and AlGaN surface properties are very sensitive to the ambient medium. Therefore, a contactless experimental technique seems to be the excellent method to investigate the 2DEG formation in AlGaN/GaN heterostructures. Such method is the PR spectroscopy, which is a contactless variation of electroreflectance (ER) spectroscopy [15, 16]. In this paper we have applied this technique to detect 2DEG formation in AlGaN/GaN heterostructures.

---

[*] Corresponding author: e-mail: robert.kudrawiec@pwr.wroc.pl, Phone: +48 071 320 23 58, Fax: +48 071 328 36 96

The studied heterostructures were grown on c-plane sapphire in a low pressure, vertical flow MOVPE system. Trimethylgallium (TMGa), trimethylaluminum (TMAl) and ammonia (NH$_3$) were used along with H$_2$ carrier gas. The epitaxial process parameters were optimised [17]. The 30 nm thick Al$_x$Ga$_{1-x}$N undoped and Si doped Al$_x$Ga$_{1-x}$N ($N_D = 10^{19}$ cm^{-2}) layers (with a nominal Al content of x = 0.18) were grown on 2 μm thick undoped high temperature GaN layers deposited on GaN buffer layer. Electrical properties of the Al$_x$Ga$_{1-x}$N/GaN heterostructures were determined by impedance spectroscopy method performed in the range 80 Hz ÷ 10 MHz with HP 4192A Impedance Meter using a mercury probe [18]. In the case of our MOCVD growth process Ga-polar GaN films are grown on sapphire substrates using GaN buffer layer. The electric and photoluminescence properties of the structures also support Ga-polar layer.

The PR measurement was performed in the so called bright configuration where the sample was illuminated by white light from a halogen lamp (150 W) serving as a probe beam source, at near normal incidence. The reflected light was dispersed through a 0.55 m focal length single grating monochromator and detected by R647P Hamamatsu photomultiplier. For photomodulation 275 nm line of an Ar$^+$ laser was used as a pump beam, that was mechanically chopped at frequency of 224 Hz. The signal was recorded by a model 3R830 DSP lock-in amplifier. The reflectance (R), measurement was performed using the same apparatus. The spectral resolution was kept on the level of several tenths of meV.

Figure 1 (a) and (b) shows room temperature PR and R spectra of undoped and Si doped AlGaN/GaN heterostructures, respectively.

In the case of R spectra, typical oscillations associated with an interference phenomenon are visible below the energy band gap of GaN. Above the GaN energy gap a peak associated with a transition in AlGaN layer is observed. The energy band gap of AlGaN layers estimated on the basis of the peak position is 3.84 and 3.89 eV for undoped and doped AlGaN layers, respectively. It corresponds to the Al content of 15 and 17 % for undoped and Si doped AlGaN layers, respectively [19]. Unfortunately, it has been found that the AlGaN-related transition is observed at different energy. However, the energy difference is rather small (less than 50 meV it corresponds to a 2% difference in Al content). Therefore, this undoped heterostructure can be considered as a reference sample for the Si doped AlGaN/GaN heterostructure. PR spectra recorded for these two heterostructures exhibit more details than R spectra. The increase in the noise of PR signal with the increase in energy is associated with a weak efficiency of the tungsten lamp in the UV spectral

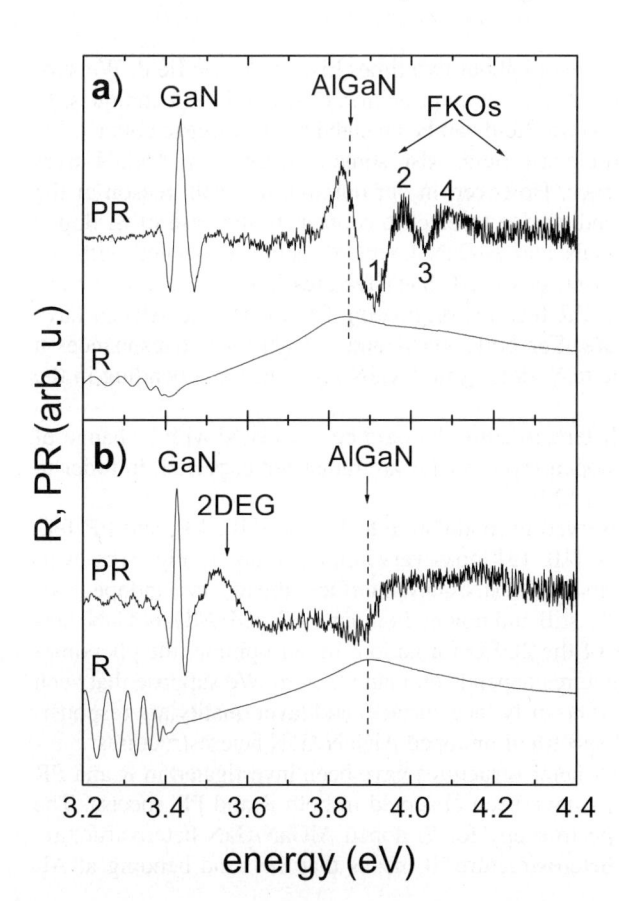

Fig. 1 Room temperature R and PR spectra of undoped (a) and Si doped (b) AlGaN/GaN heterostructures.

region. In the case of PR of the undoped AlGaN/GaN heterostructure transitions originating from both GaN and AlGaN layers are very well visible. The AlGaN-related resonance possesses Franz-Keldysh oscillations (FKOs) [20] which are associated with an internal electric field in the AlGaN layer. The AlGaN-related FKOs were analyzed using the asymptotic expression for ER [20, 21]

$$\frac{\Delta R}{R} \propto \exp\left[\frac{-2\Gamma\sqrt{E-E_g}}{(\hbar\theta)^{3/2}}\right] \cdot \cos\left[\frac{4}{3}\left(\frac{E-E_g}{\hbar\theta}\right)^{3/2} + \phi\right] \cdot \frac{1}{E^2(E-E_g)}, \quad F = \frac{(\hbar\theta)^{3/2}\sqrt{2\mu}}{e\hbar}, \quad (1)$$

where $\hbar\theta$ is the electro-optic energy, Γ is the linewidth, ϕ is a phase factor, F is the electric field, and μ is the electron hole reduced mass ($\approx 0.2 m_e$). The field estimated from the period of FKOs is 284 kV/cm (see Fig.2). Such huge value of this field is typical for GaN-based structures [13, 23, 24, 25].

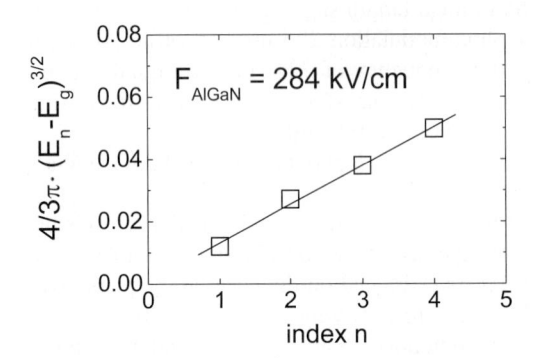

Fig. 2 Analysis of AlGaN-related FKOs observed for undoped AlGaN/GaN heterostructure.

Kurtz et al. [13] in very similar structures has observed about two times higher electric field. We suppose that the difference could be associated with the difference in principles of ER and PR techniques. In the case of used by Krutz contact ER the built-in electric field can be changed by the electric contact. PR is contactless technique and probes really existed electric field. Also some differences in AlGaN layer thickness and content can influence on the difference. However, in our opinion, the main reason of the difference is associated with a change in band bending due to electric contact. In the case of Si doped AlGaN/GaN heterostructure beside signal from GaN and AlGaN layers a feature associated with the presence of 2DEG at AlGaN/GaN interface has been observed. Such features is not observed for undoped heterostructure. In addition, it is noted that PR features originating from GaN and AlGaN have changed in comparison to undoped heterostructure. For both, GaN- and AlGaN-related resonances a change of PR phase is observed. It means that due to Si doping in AlGaN layer the band bending in the heterostructure has changed [24].

AlGaN-related FKOs disappear for the doped heterostructure. It could be associated with a change of the internal electric field, however in the case of doped layers FKOs are rather not expected. In order to observe FKOs a pure and undoped layer is necessary [20].

The presence of 2DEG has been previously observed in modulation techniques [10, 13], and PR feature which has been attributed to the 2DEG in Ref. [10, 13] looks very similar to this in Fig. 1(b). Ambacher et al. [4, 5] suggest that the 2DEG appears at AlGaN/GaN interface already for undoped AlGaN/GaN heterostructures. However, its origin is still unknown. For our undoped AlGaN/GaN heterostructure we have not found a strong evidence of the 2DEG formation. In our opinion, the phenomenon of 2DEG formation in AlGaN/GaN heterostructures depends on many factors. We suppose that such factors like the inhomogeneity in surface polarity (Ga- or N-face surface) and layer quality are responsible for the absence of 2DEG-related feature in PR spectra of undoped AlGaN/GaN heterostructures.

In summary, undoped and Si doped AlGaN/GaN heterostructures have been investigated in R and PR spectroscopies. GaN and AlGaN related transitions have been observed in both R and PR spectra. The 2DEG-related feature has been observed in PR spectroscopy for Si doped AlGaN/GaN heterostructure. This feature is rather not observed for undoped heterostructure. It means that the band bending at Al-

GaN/GaN interface is not exactly the same for these two heterostructures. In addition, quite different phase of PR resonances confirms that the band bending in these two samples is different. The band bending is responsible for formation of 2DEG at the AlGaN/GaN interface. For undoped AlGaN/GaN heterostructure, the AlGaN-related transition possesses FKOs which are associated with the internal electric field. This field estimated from the period of FKOs equals 284 kV/cm.

Acknowledgements The research was supported in part by the Committee for Scientific Research in Poland under Grant No. T11B 00823, Grant No. 4T11B06124 and by the Centre for Advanced Materials and Nanotechnology, Wroclaw University of Technology, Wroclaw Poland.

References

[1] S. N. Mohammad, A. A Salvador, and H. Morkoc Proc. IEEE **83**, 1306 (1995).
[2] G. J. Sullivan, M. Y. Chen, J. A. Higgins, J. W. Yang, Q. Chen, R. L. Pierson, and B. T. McDermott, IEEE Electron Device Lett. **19**, 198 (1998).
[3] J. Burm, W. J. Schaff, L. F. Eastman, H. Amano, and I. Akasaki, Appl. Phys. Lett. **68**, 2849 (1996).
[4] O. Ambacher, J. Smart, J. R. Shealy, N. G. Weimann, K. Chu, M. Murphy, W. J. Schaff, and L. F. Eastman, J. Appl. Phys. **85**, 3222 (1999).
[5] O. Ambacher, B. Foutz, J. Smart, J. R. Shealy, N. G. Weimann, K. Chu, M. Murphy, A. J. Sierakowski, W. J. Schaff, L. F. Eastman, R. Dimitrov, A. Mitchell, and M. Stutzmann, J. Appl. Phys. **87**, 334 (2000).
[6] R. Dimitrov, M. Murphy, J. Smart, W. Schaff, J. R. Shealy, L. F. Eastman, O. Ambacher, and M. Stutzmann, J. Appl. Phys. **87**, 3375 (2000).
[7] M. Ashif Khan, J. N. Kuzina, J. M. Van Hoffe, N. Pan, and J. Carter, Appl. Phys. Lett. **60**, 3027 (1992).
[8] J. M. Redwing, M. A. Tschler, J. S. Flynn, S. Elhamri, M. Ahoujja, R. S. Newrock, and W.C. Mitchel, Appl. Phys. Lett. **69**, 963 (1996).
[9] H. Alause, C. Skierbiszewski, M. Dyakonov, W. Knap, M. L. Sadowski, S. Huant, J. Young, M. Ashif Khan, and Q. Chen, Diam. Relat. Mater. **6**, 1536 (1997).
[10] D. Y. Lin, Y. S. Huang, Y. F. Chen, and K. K. Tiong, Solid State Commun. **107**, 533 (1998).
[11] L. K. Li, B. Turk, W. I. Wang, S. Syed, D. Simonian, and H. J. Stormer, Appl. Phys. Lett. **76**, 742 (2000).
[12] H. K. Kwon, C. J. Eiting, D. J. H. Lambert, B. S. Shelton, M. M. Wong, T. G. Zhu, and R. D. Dupuis, J. Cryst. Growth **221**, 362 (2000).
[13] S. R. Kurtz, A. A. Allerman, D. D. Koleske, and G. M. Peake, Appl. Phys. Lett. **80**, 4549 (2002).
[14] N. Maeda, K. Tsubaki, T. Saitoh, T. Tawara, and N. Kobayashi, Optical Materials **23**, 211 (2003).
[15] F. H. Pollak, in: Handbook on Semiconductors, edited by T. S. Moss, vol. 2, (Elsevier Science, Amsterdam, 1994), pp. 527–635.
[16] J. Misiewicz, P. Sitarek, G. Sek, and R. Kudrawiec, Materials Science **21**, (2003).
[17] R. Paszkiewicz, R. Korbutowicz, D. Radziewicz, M. Panek, B. Paszkiewicz, J. Kozłowski, B. Boratyński, and M. Tłaczała, SPIE 3725, **21** (1999).
[18] R. Paszkiewicz, B. Paszkiewicz, J. Kozlowski, and M. Tlaczala, J. Cryst. Growth. **248**, 487 (2003).
[19] I. Vurgaftman and J. R. Meyer, J. Appl. Phys. **94**, 3675 (2003).
[20] H. Shen and M. Dutta, J. Appl. Phys. **78**, 2151 (1995).
[21] D. E. Aspnes and A. A. Studna, Phys. Rev. B **7**, 4605 (1973).
[22] C. Wetzel, T. Takeuchi, H. Amano, and I. Akasaki, J. Appl. Phys. **85**, 3786 (1999).
[23] Y. T. Hou, K. L. Teo, M. F. Li, K. Uchida, H. Tokunaga, N. Akutsu, and K. Matsumoto, Appl. Phys. Lett. **76**, 1033 (2000).
[24] W. Krystek, F. H. Pollak, Z. C. Feng, M. Schurman, and A. Stall, Appl. Phys. Lett. **72**, 1353 (1998).
[25] A. Drabińska, Acta Physica Polonica A **104**, 149 (2003).

Resistivity and mobility in ordered InGaP grown by MOVPE

S. Hasenöhrl*, J. Betko, M. Morvic, J. Novák, and J. Fedor

Institute of Electrical Engineering, Slovak Academy of Sciences, Dubravska cesta 9, 841 04 Bratislava, Slovak Republic

Received 19 November 2003, accepted 24 November 2003
Published online 22 December 2003

PACS 72.20.My, 72.80.Ey, 73.61.Ey, 81.15.Gh

A series of five ordered $In_xGa_{1-x}P$ epitaxial layers with varied composition were prepared using MOVPE technique. Their electrical properties were evaluated from temperature dependent conductivity and Hall measurements using the van der Pauw method as well as from measurements of the physical and geometrical magnetoresistances in [0–11] and [011] crystallographic directions. The magnetoresistance coefficient C was defined and was used as a parameter for the assessment of the sample inhomogenity. The ratio $C_{[0-11]}$ and $C_{[011]}$ was the criterion for the sample anisotropy. We found that ordered layers exhibit the strong dependence of transport parameters on layer composition. Samples with the lattice mismatch larger than $\pm 2 \times 10^{-3}$ are inhomogeneous and layers slightly mismatched from the substrate at the growth temperature exhibit anisotropy of transport parameters.

© 2004 WILEY-VCH Verlag GmbH & Co. KGaA, Weinheim

1 Introduction

The MOVPE grown $In_xGa_{1-x}P$ layers are often used in device structures, like heterojunction bipolar transistors, high electron mobility transistors, solar cells, light-emitting and laser diodes. Typical for this ternary prepared by MOVPE or MBE is that the group III atoms are not arranged randomly but they tend to create an ordered CuPt-B type structure. The extent of group III atoms rearrangement is significantly influenced by the growth conditions and layers with various degree of order are usually prepared [1]. Ordering lowers the crystal symmetry what leads to anisotropic behaviour of material parameters [2]. Parameters of complete device structures, particularly of those where the carrier transport proceeds parallel to substrate-structure interface, are significantly influenced by transport properties in InGaP layer. Anisotropy of device parameters was reported e.g. for laser diodes [3], but it is expected to occur also in HEMTs.

Transport properties of $In_xGa_{1-x}P$ have been studied in several papers [4–6]. The value of mobility derived from Hall measurements in van der Pauw configuration varied from 400 to 2 000 $cm^2V^{-1}s^{-1}$ at 300 K. The dependence of mobility on the degree of order was declared, but the influence on lattice mismatch was not investigated. Layers with $\Delta a/a < 10^{-3}$ were declared as lattice matched [6]. Effect of misfit strain on electrical properties was studied in [7]. Samples were not measurable at 77 K due to the high resistivity. Hall measurements showed negligible effect of misfit strain on electrical properties at room temperature. It was remarked that possible effect can be screened by large compensation.

Transport anisotropy in ordered $In_xGa_{1-x}P$ was reported firstly in [8]. It was shown, that anisotropy was not observable in disordered layers and in ordered ones it was well-defined in undoped and strained ($\Delta a/a \sim 7 \times 10^{-3}$) samples. The conductivity was calculated from I–V curves measured at room temperature in various crystallographic directions.

The temperature dependence and impurity concentration dependence of the Hall mobility in $In_{0.49}Ga_{0.51}P$ was calculated in [9]. The calculations were done only for $x_{In} = 0.49$, but we will use this source for the comparison of our results.

---

* Corresponding author: e-mail: Stanislav.Hasenohrl@savba.sk, Phone: +421 2 5477 5826, Fax: : +421 2 5477 5816

The aim of this paper is to contribute to the knowledge of transport properties in ordered $In_xGa_{1-x}P$. We will present experimental results from temperature dependent measurements of transport properties in [0–11] and [011] crystallographic directions. Magnetoresistance measurements are used for the evaluation of mobility. The magnetoresistance coefficient C will be defined as the measure of sample inhomogenity. It will be shown that the transport properties are anisotropic and they depend strongly on layer composition.

2 Experimental The undoped ordered $In_xGa_{1-x}P$ epitaxial layers were prepared in low pressure MOVPE AIX 200 reactor. A 300 nm thick highly resistive ($\rho \geq 7.5 \times 10^4$ Ωcm) [10] GaAs buffer layer was deposited on semiinsulating exactly (100) oriented GaAs substrate prior the growth of ternary layer. The growth conditions and layer properties are presented in [11]. Five layers from a series of six with varied composition x_{In} are the subject of this study. The first one with x_{In} of 0.388 was unmeasurable because of the high resistance. The composition of sample 2 is $x_{In} = 0.441$, sample 3 is exactly lattice matched to GaAs ($x_{In} = 0.486$), composition x_{In} of sample 4 is 0.498, of sample 5 it is 0.507 and sample 6 has composition $x_{In} = 0.552$.

Electrical properties were evaluated from conductivity and Hall measurements using the van der Pauw method (Hall vdP measurements) as well as from measurements of the physical (PMR) and geometrical (GMR) magnetoresistances (magnetoresistance measurements). Measurements were performed in continuous flow cryostat on samples in the dark with temperature ranging from 5 to 300 K.

Hall measurements were realized on square 5×5 mm^2 samples with applied current of 50 µA and magnetic field of 0.36 T. Magnetoresistance measurements were done on 4-point probes. Two stripes 1×8 mm^2 oriented in [0–11] and [011] directions were prepared from each sample. Outer contacts are numbered 4 and 3, inner contacts 1 and 2. Contacts in pairs 4–1 and 2–3 were placed close each to other.

2.1 Hall measurements in van der Pauw configuration Results from temperature dependent Hall and conductivity measurements were presented in [10]. Parameters were denoted as apparent ones, because the ratio of resistances R_1 to R_2 measured in the vdP configuration indicated inhomogeneity of the samples. The mobility measured in the van der Pauw configuration is in this paper designated as the Hall vdP mobility $\mu_{H\,vdP}$.

2.2 Magnetoresistance measurements on four-point probes The magnetoresistance in magnetic field varied from –540 to +540 mT was measured with stripes in two configurations: i) the physical magnetoresistance (PMR) configuration and ii) the geometrical magnetoresistance (GMR) configuration described below.

In the PMR configuration the current I_{43} was passed through contacts 4–3 and the voltage was measured between contacts 1–2. This is a standard configuration for the resistivity measurement in homogeneous samples. Here, the resistivity is given as

$$\rho = \frac{U_{12}}{I_{43}} \frac{w \cdot d}{L_{12}}, \tag{1}$$

where w is the sample width, d is the layer thickness and L_{12} is the distance between contacts 1 and 2.

Dependence of $\Delta\rho/\rho_0$ measured in the PMR configuration represents the physical magnetoresistance. After Look [12] it is defined as

$$\left(\frac{\Delta\rho}{\rho_0}\right)_{PMR} \cong \xi_{PMR} \cdot \mu_H^2 \cdot B^2, \tag{2}$$

where ξ_{PMR} is the physical magnetoresistance coefficient (usually about 0.1), μ_H is the Hall mobility related to the conductivity μ as $\mu_H = r\,\mu$ (r is the Hall factor, usually close to 1) and B is the magnetic field.

As ingomogeneities are expected in the stripes, the resistivity calculated from Eq. (1) is further denoted "apparent" (ρ_{app}) and the magnetoresistance even in PMR configuration is not pure physical magnetoresistance but contains also a geometrical magnetoresistance component. Therefore, Eq. (2) can be rewritten as

$$\left(\frac{\Delta\rho_{app}}{\rho_{0,app}}\right) \cong C \cdot \mu_H^2 \cdot B^2 , \qquad (3)$$

where C is a magnetoresistance coefficient which is equal to ξ_{PMR} for homogeneous sample and it is equal to about 1 for strongly inhomogeneous sample (in this case also μ_H represents an apparent value and can much differ from the conductivity mobility). So, C can be used as a parameter for the assessment of the sample inhomogenity and from the ratio C_1 (for stripes in [0–11] direction) and C_2 (for stripes in [011] direction) we can judge the sample anisotropy.

For an estimation of the Hall mobility (μ_H) we have used the GMR configuration – the current was passed through contacts 4–1 or 2–3 and the voltage was measured on the same contacts. The resistance R was calculated simply as the ratio of the voltage and the current neglecting the contacts resistances. The distances between contacts 4–1 and 2–3 are short and the geometrical magnetoresistance is expected to be measured in this case:

$$\left(\frac{\Delta R}{R_0}\right)_{GMR} \cong \mu_{GMR}^2 \cdot B^2 , \qquad (4)$$

where μ_{GMR} is the geometrical magnetoresistance mobility which can be considered to be close to the Hall mobility. Note, for each of the individual sample (stripe) the pair of contacts (4–1 or 2–3) with larger magnetoresistance effect was accepted for estimation of μ_{GMR}.

3 Results and discussion

Results obtained from electrical measurements can be discussed in two subgroups. The first one concerns samples 2 (x_{In} = 0.441) and 6 (x_{In} = 0.552) with mismatch larger than $\pm 2 \times 10^{-3}$.

Tensile strained sample 2 shows different temperature dependence of resistivity in [0–11] and [011] directions. At 300K both values are comparable, but at 5 K the $\rho_{[0-11],app}$ is more than 4 times higher than $\rho_{[011],app}$. The μ_{GMR} is lower in the [0–11] direction than in [011] direction (Table 1). In the vdP configuration this anisotropy in transport parameters is reflected in the $(R_{[0-11]}/R_{[011]})_{vdP}$ ratio decrease with decreasing temperature (Fig. 2a). The vdP method is, due to the sample inhomogeneity, charged by the large error what results in large value of $\mu_{H\ vdP}$ at 5 K. Physical magnetoresistance measured on the [0–11] oriented stripe contains strong geometrical magnetoresistance component and both PMR and GMR magnetoresistance curves are nearly identical. Magnetoresistance coefficient $C_{[011]}$ lies in the interval from 0.1 at 300 K to 0.03 at 5 K (usual value for ξ_{PMR} [12]), while $C_{[0-11]}$ is much higher with the utmost value of 0.97 at 120 K. Sample is strongly inhomogeneous in the [0–11] direction. As we have shown in [11], this sample includes strained areas in form of [0–11] oriented narrow stripes. Sample exhibits transport anisotropy along this direction what results in higher $\rho_{[0-11],app}$ and lower $\mu_{GMR[0-11]}$.

Compressive strained sample 6 exhibit larger inhomogenity in the [011] direction. The resistivity in this direction was higher by one to two orders of magnitude than in all other samples. Due to its very high value the sample was unmeasurable under 80 K. The strong GMR component at PMR configuration was present in both directions; for temperatures over 175 K it was stronger in sample oriented in [011] direction (Fig. 2e). Synchrotron topography [11] revealed strained stripe regions oriented in both [0–11] and [011] directions with higher density of [0–11] oriented stripes. Surface structure was typically crosshatched with misfit dislocations probably created. Despite the highest In content this sample exhibit the lowest mobility. Transport properties are evidently influenced by the presence of crystallographic defects and the difference of $\mu_{GMR\ [011]}$, $\mu_{GMR\ [0-11]}$ and $\mu_{H\ vdP}$ from the conductivity mobility μ can be significant.

© 2004 WILEY-VCH Verlag GmbH & Co. KGaA, Weinheim

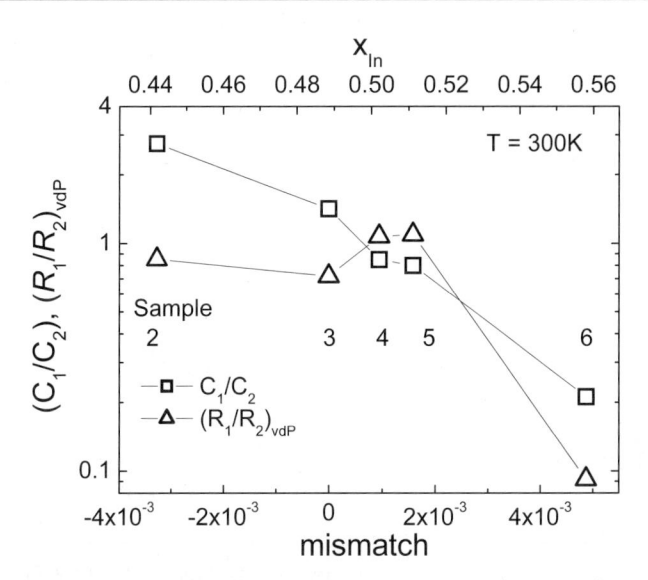

Fig. 1 Ratio of magnetoresistance coefficients C_1 to C_2 measured at 300 K. Index 1 denotes direction [0–11], index 2 [011]. For comparison with Hall measurements the ratio of $(R_1/R_2)_{vdP}$ is plotted. All parameters are dimensionless and are depicted as a dependence of layer mismatch and composition, respectively, for the set of five samples.

The second group of samples include those nearly lattice matched to the substrate. Samples 3, 4 and 5, as can be seen from Fig. 1, keep values of $C_{[0-11]}$ and $C_{[011]}$ close to 0.12 at 300 K and ratios $C_{[0-11]}/C_{[011]}$, $\rho_{[0-11],app}/\rho_{[011],app}$, and $(R_{[0-11]}/R_{[011]})_{vdP}$ are close to 1. The more detailed analysis of results shows that there are some differences.

Surprisingly, not the sample 3 with unmeasurable misfit, but the sample 4, with misfit $\Delta a/a = 9.4 \times 10^{-4}$, exhibit the best equality of parameters in the both orientations. Values of ρ_{app} and μ_{GMR} in the [0–11] and [011] directions are nearly identical in the whole temperature interval from 5 to 300 K. The temperature dependence of magnetoresistance coefficient C is similar for both orientations, only with rather higher decrease towards low temperature in the [0–11] direction than in the [011] direction (Fig. 2c). As we stated in [11], this sample was probably exactly lattice matched to the substrate at the growth temperature. The Hall data of this layer, Hall concentration and Hall mobility, comply very well with those calculated in [9]. As can be seen in the Table 1, the Hall mobility $\mu_{H\,vdP}$ is in a very good agreement with mobility μ_{GMR} derived from magnetoresistance.

As we presented in [10] the sample 3 with unmeasurable misfit at room temperature and the sample 5 with $\Delta a/a = 1.58 \times 10^{-3}$ showed significantly higher Hall vdP mobilities than the sample 4. This trend was confirmed by magnetoresistance measurements, however the GMR mobility is lower than the Hall vdP mobility (Table 1). Sample 3 shows different temperature dependence of ρ_{app} in [0–11] and [011] directions with $\rho_{[011],app} > \rho_{[0-11],app}$. The temperature dependence of $\rho_{[011],app}$ has a minimum at 120 K, while $\rho_{[0-11],app}$ monotonously decreases towards low temperature. Magnetoresistance coefficient C is in the [0–11] direction nearly constant (0.15–0.12) and in [011] direction it decreases from 0.12 at 300 K to 0.01 at 5 K. This indicates that the transport properties are orientation dependent (Fig. 2b.). Sample 5 shows similar temperature dependence of ρ_{app} in [0–11] and [011] directions with ρ_{app} monotonously decreasing towards low temperature. Magnetoresistance coefficient C is nearly constant in the [011] direction (0.15–0.11), but in the [0–11] direction it decreases from 0.12 at 300 K to 0.01 at 5 K (Fig. 2d).

Physical and geometrical magnetoresistance curves of samples 3, 4 and 5 are well distinguished and the magnetoresistance coefficient does not indicate that there is a strong inhomogenity in the samples. It is important that the temperature dependence of ρ and C depend on the crystallographic orientation in samples 3 and 5. This can indicate the anisotropy in transport properties.

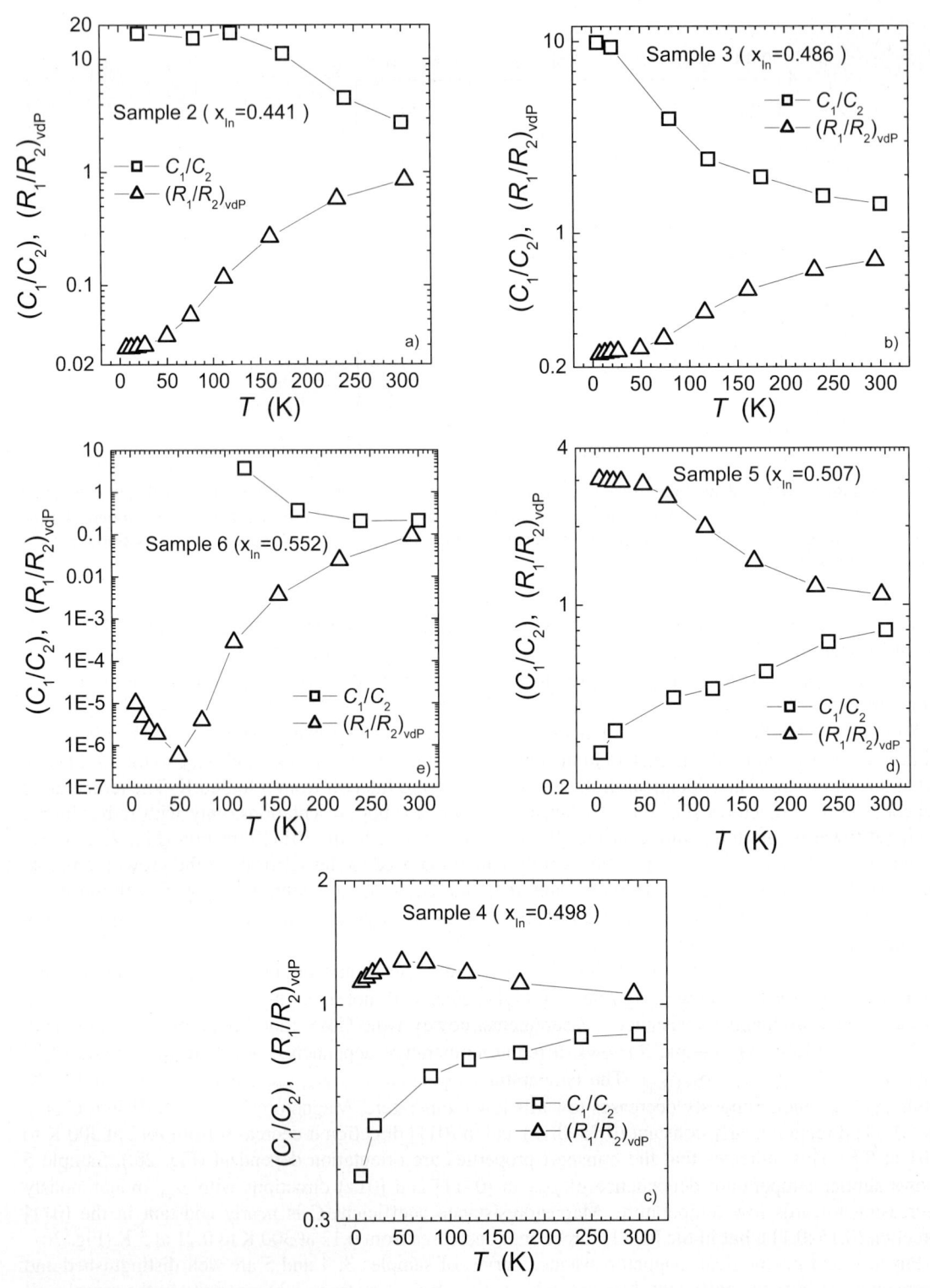

Fig. 2 Magnetoresistance coefficients C_1 to C_2 ratio indicating the dependence of transport properties on crystallographic direction as a function of temperature. Index 1 denotes direction [0–11], index 2 [011]. For comparison with Hall measurements the ratio of $(R_1/R_2)_{vdP}$ is plotted.

Table 1 Mobility in cm^2 V^{-1} s^{-1} measured by Hall and magnetoresistance measurements.

Sample No	x_{In}	μ GMR [011] 300 K	μ GMR [0–11] 300 K	μ H vdP 300 K	μ GMR [011] 175 K	μ GMR [0–11] 175 K	μ H vdP 175 K	μ GMR [011] 80 K	μ GMR [0–11] 80 K	μ H vdP 80 K	μ GMR [011] 5 K	μ GMR [0–11] 5 K	μ H vdP 5 K
2	0.441	3 610	2 370	3 640	6 190	2450	8 610	9 030	2 930	16 390	8 690	1 740	17 530
3	0.486	3 700	3 600	4 190	7 800	7 080	8 950	13 120	11 530	15 770	13 670	10 660	16 360
4	0.498	3 250	3 520	3 220	5 710	6 270	5 520	8 310	9 050	8 300	7 040	8 470	7 630
5	0.507	3 500	3 780	3 730	6 590	7 360	7 900	10 600	12 120	13 360	10 390	12 090	13 401
6	0.552	2 850	2 260	1 900	4 050	2 780	1 840	4 070	2 740	1 150	-	-	24

4 Conclusion Magnetoresistance measurements in [0–11] and [011] directions were useful for determination of transport properties in In$_x$Ga$_{1-x}$P epitaxial layers. It was revealed, that the transport properties are very sensitive to the layer composition and that the layer lattice matched to the GaAs substrate at the growth temperature is electrically homogeneous. The Hall mobility and resistivity measured in this sample by Hall and magnetoresistance measurements seem to be very close to the true values of μ_H and ρ in this material. Layers slightly mismatched from above mentioned composition show different transport properties in [0–11] and [011] directions indicated by the different temperature dependence of resistivity and mobility in [0–11] and [011] directions. Higher low temperature mobility measured in these samples is in a correlation with the difference of $C_{[0-11]}/C_{[011]}$ and $\rho_{[0-11],app}/\rho_{[011],app}$ ratios from 1. At this moment we are not able to conclude if this higher mobility is a result of inaccuracy in evaluating mobility from Hall and magnetoresistance measurements or if it is based on the change of physical parameters in In$_x$Ga$_{1-x}$P due to the strain. Mobility values calculated from Hall and magnetoresistance measurements for samples with the lattice mismatch larger than ±2 × 10^{-3} seems to be incorrect because of the large sample inhomogenity.

Acknowledgements This research was supported by VEGA Agency under Grant No. 2/7195/20 and The European Community IST Programme No. 2001-32793.

References

[1] G. B. Stringfellow, MRS Bull. **22**, 27 (1997).
[2] A. Zunger, MRS Bull. **22**, 20 (1997).
[3] H. Fuji, Y. Ueno, A. Gomyo, K. Endo, and T. Suzuki, Appl. Phys. Lett. **61**, 737 (1992).
[4] D. J. Friedman, A. E. Kibbler, and M. J. Olson, Appl. Phys. Lett. **59**, 2998 (1991).
[5] F. A. J. M. Driessen, G. J. Bauhuis, S. M. Olsthoorn, and L. J. Giling, Phys. Rev. B **48**, 7889 (1993).
[6] B. Zhang, S. Lan, L.-Q. Li, W.-J. Xu, C.-Q. Yang, and H.-D. Liu, Solid State Commun. **92**, 419 (1994).
[7] K. Ozasa, M. Yuri, S. Tanaka and H. Matsunami, J. Appl. Phys. **68**, 107 (1990).
[8] K. L. Chernyak, A. Osinsky, H. Temkin, A. Mintairov, I. G. Malkina, B. N. Zvonkov, and Yu. N. Safanov, Appl. Phys. Lett. **70**, 2425 (1997).
[9] C. Besikci and M. Razeghi, IEEE Trans. Electron Devices **41**, 1066 (1994).
[10] S. Hasenöhrl, J. Novák, R. Kúdela, J. Betko, M. Morvic, and J. Fedor, J. Cryst. Growth **248**, 369 (2003).
[11] S. Hasenöhrl, R. Kúdela, J. Novák, T. O. Tuomi, and L. Knuuttila, Mater. Sci. Eng. B **88**, 134 (2002).
[12] D. C. Look, Electrical Characterization of GaAs Materials and Devices (John Wiley & Sons, 1989), chap. 1.

Structural characterization of original 3D gallium structures grown by LP-MOCVD

L. Imhoff[*,1], **M. Sacilotti**[2], **C. Josse Courty**[1], **M. Mesnier**[1], **M C. Marco de Lucas**[1], and **S. Bourgeois**[1]

[1] Laboratoire de Recherches sur la Réactivité des Solides, UMR 5613 CNRS-Université de Bourgogne,
[2] Couches Minces et Nanostructures, FR 2604 CNRS-Université de Bourgogne,
 9 avenue Alain Savary, BP 47870, 21078 Dijon Cedex, France

Received 19 September 2003 accepted 3 November 2003
Published online 22 December 2003

PACS 61.10.Nz, 64.60.My, 68.37.Hk, 68.90+g, 81.05.Bx, 81.15.Gh

This study is concerned with the growth and characterization of metallic gallium 3D structures, obtained with a single growth step, by the LP-MOCVD technique on various substrates. Commercial organometallic is used as gallium precursor and nitrogen as carrier gas. The growth temperature and the reactor pressure are ranking between 500 and 700 °C, and between 150 and 700 torr, respectively. Depending on the elaboration conditions, different 3D structures are obtained such as droplets, cauliflowers, aggregates or thin stems, with micrometer sizes. The morphology, substrate surface density and thermal stability are studied by optical and scanning electron microscopy. At last, X-ray microanalysis and X-ray diffraction are carried out in order to determine the elemental composition and the crystallographic structure of these 3D objects.

© 2004 WILEY-VCH Verlag GmbH & Co. KGaA, Weinheim

1 Introduction

The elaboration of metallic three dimensional (3D) structures, composed of tin [1], gallium [2] or indium [3], with dimensions in the micro or nanometer range presents substantial interest in numerous research fields. From fundamental point of view, these objects open the way to a new class of materials with remarkable effects [4]; for instance, some modifications of the optical properties can be observed [5]. From device applications point of view, the fabrication of patterned 3D structures with quantum properties is an attractive technological challenge for electronic or optoelectronic for example [6, 7]. Unfortunately, it remains difficult to obtain such devices without several fabrication processing steps [8]. In this work, we present results concerning metallic gallium 3D structures, and in particular micrometric gallium droplets. In the literature, some gallium droplets obtained intentionally [9, 10] or not [11–14] have already been reported. But, to our knowledge, none of them has been morphologically and structurally described. The aim of this paper is to report the growth and characterization of metallic gallium 3D structures, obtained with a single growth step, by the LP-MOCVD technique on various substrates.

2 Experimental

The 3D gallium structures growths were carried out in a low pressure metal-organic chemical vapor deposition system (LP-MOCVD) [15], with a deposition temperature varying between 500 and 700 °C, and a reactor pressure between 150 and 700 torr. Trimethyl-gallium (TMGa) was used as the gallium source, maintained in a cylinder at −10 °C. Nitrogen, used as carrier gas, was bubbled through TMGa, so

---

[*] Corresponding author: e-mail: luc.imhoff@u-bourgogne.fr, Phone: +33 (0)3 80 39 61 61, Fax: +33 (0)3 80 39 38 19

the obtained TMGa flow rate was 1 sccm. An additional N_2 flow was used in order to control the partial pressure of TMGa. Different types of materials were used as substrates: Si (100), GaAs (100), TiO_2 (110) and amorphous SiO_2. Prior to the deposition process, all the substrates were chemically cleaned with a 6:1:1 solution made of sulphuric acid, hydrogen peroxide and deionised water, and then N_2 blow-dried. Scanning electron microscopy (SEM) was used to study the 3D structures morphology (JEOL JSM 6400F typically working at 20 keV). X-ray analysis was done by energy dispersive spectrometry (EDS - OXFORD – Inca energy software). The surface density and thermal stability of the droplets were observed by Nomarski optical microscopy (OM). The crystallographic structure of the deposits was analyzed by small incidence angle X-ray diffraction (XRD) (Cu K_α tube; INEL CPS 120 curved detector).

3 Results and discussion

Different 3D structures were obtained such as thin stems, cauliflowers, droplets or aggregates, depending on the elaboration conditions, that is to say the deposition temperature, the deposition duration and the total and TMGa partial pressures.

First, the effect of deposition temperature on the structures morphology was observed, in the case of a 15 minutes growth, 700 torr and 140 mtorr for the total and TMGa partial pressures, respectively. For deposition performed below 500 °C, no growth is obtained. The first deposits grow at 500 °C as packs of thin stems (Fig. 1a) with an unique nucleation point for a pack, and with stems lengths of a few microns. For temperatures between 500 °C and 600 °C, some cauliflowers like structures are observed (fig. 1b) with sizes between 5 and 20 µm. For temperatures higher than 600 °C, droplets with a montgolfier shape are obtained (Fig. 1c), with a nucleation point corresponding to the montgolfier base, and with a diameter ranking between 0.5 and 5 µm. At this stage, no differences in the shapes are observed for depositions on different substrates: whatever the substrate is, the same shapes are observed for a given temperature. However, a variation of the deposits size and the surface density is linked to the substrate nature.

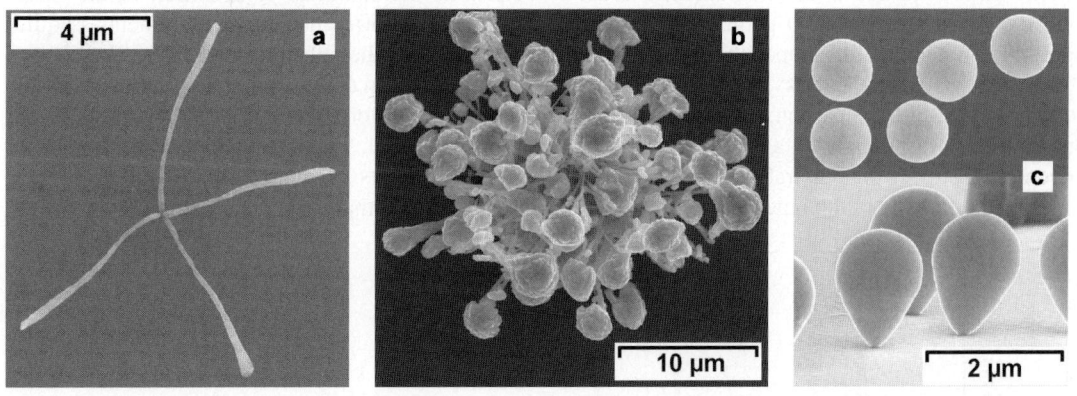

Fig. 1 SEM pictures of (a) thin stems obtained at 500 °C, (b) cauliflowers like structures obtained at 550 °C, and (c) droplets obtained at 650 °C (top view and side view). These structures are obtained on silicon substrates, for a 15′ growth, and 700 torr and 140 mtorr for the total and TMGa partial pressures, respectively.

Afterwards, the variation of the diameter and surface density of the droplets elaborated on Si (100) and GaAs (100) was studied as a function of the deposition temperature and the deposition duration. The total and TMGa partial pressures were kept to 700 torr and 140 mtorr, respectively. The diameter was measured by SEM and the density by OM observations. For 15 minutes growth and temperatures from 600 to 650 °C, the diameter of the droplets deposited on Si (100) increases from 2 to 3.5 µm, and the density seems to be stable around $4.6 \; 10^5 \; cm^{-2}$. On the opposite, in the case of GaAs (100), the diameter remains stable around 1.5 µm while the density increases from $2.5 \; 10^6 \; cm^{-2}$ to $3.2 \; 10^6 \; cm^{-2}$. From these results, an activation energy was calculated with an Arrhenius law, by calculating the volume of matter deposited per unit time, i.e. the ball volume multiplied by the surface density and divided by the deposition time. Values of 133 and 115 kJ are obtained for Si and GaAs substrates, respectively. So, although the diameter and surface density evolutions seem to be linked to the substrate, the activation energy is quite similar in both cases. Then, the variation of these two geometrical parameters was studied as a function of the deposition duration at 650 °C. An incubation time of about one minute is observed, followed by a constant increase of the diameter and the density for both substrates. Moreover, after a given time, depending on the substrate, a coalescence of the structures appears and leads to the formation of aggregates with dimensions of about 30 to 40 µm. At last, variations of the total pressure between 150 and 700 torr, and of the TMGa partial pressure between 15 and 140 mtorr haven't underlining any coherent tendency. More experiments are in progress in order to analyze the influence of these parameters.

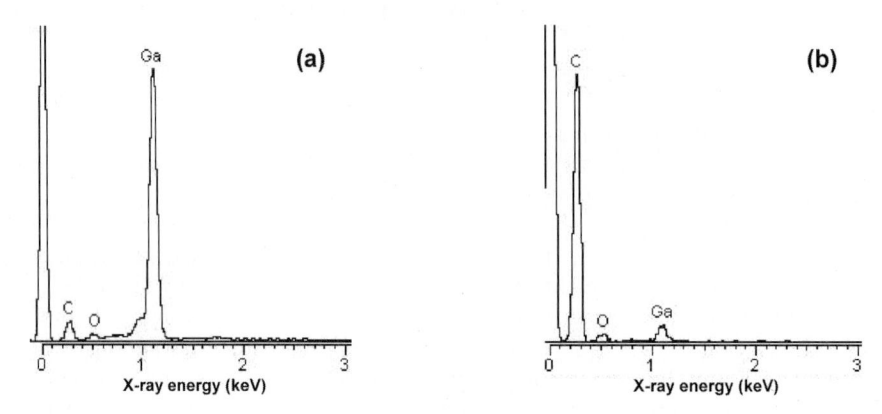

Fig. 2 EDX analyses of a 3D gallium structure, with (a) 20 keV and (b) 5 keV incident electron beams energy.

Some EDX analyses were undertaken in order to determine the chemical composition of the structures. By the fact of their 3D geometry, the quantification of the different chemical elements is quite impossible. Whatever the shape (montgolfier, cauliflower, ...) and the substrate are, the probe always detects gallium, carbon and oxygen (Fig. 2a). The carbon presence can come from the gallium precursor, and the oxygen, like the carbon, eventually from air exposition pollution. Moreover, some analyses were done on the montgolfiers with a smaller incident electron beam energy, *5* keV instead of *20* keV, in order to decrease the electron penetration depth (Fig. 2b). These experiments show that carbon and oxygen are the main components of the outer part of the objects, while gallium is mainly present in the core.

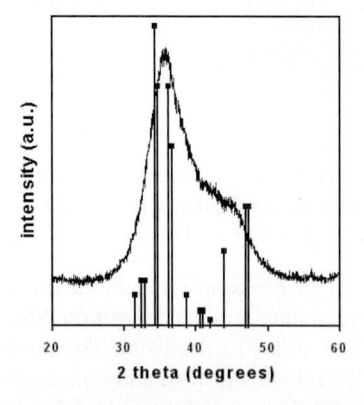

Fig. 3 Small incidence angle XRD of 3D gallium structure. Vertical lines correspond to Bragg reflections for a mixture of gallium metastable phases [16].

X-ray diffraction experiments were carried out in order to determine the crystallographic structure of the cauliflowers, droplets and aggregates. No diffraction is observed for the thin stems, but this result can be linked to the low quantity of matter of these objects. For the others objets, the same diffractogram is always obtained (Fig. 3). This diffractogram is not the one of the more stable gallium phase, i.e. the α phase, but the one corresponding to a mixture of metastables β, γ, δ, η and χ phases [16]. This kind of mixture has already been observed by electron microdiffraction in the case of metallic gallium nanoparticles embedded in an amorphous SiO_x matrix [17]. In this way, the α phase is not found, since its density is lower than the liquid one, but some of the metastable phases appear and are stable [18]. Moreover, some droplets were heated up to 200 °C during an OM observation, and no modification of the droplets shape is observed. This result seems to show that these structures are not only composed of gallium, because in this case, the droplets melting should be observed at such temperatures. On the one hand, from the XRD results, the inner part is composed of metastable metallic gallium. On the other hand, it is shown from EDX and thermal stability results, that this gallium core is covered by a skin probably composed of gallium, carbon and/or oxygen. It should be noticed that this hypothesis explains the metastable phases presence previously obtained in the case of gallium embedded in a matrix.

4 Conclusion

In this study, the growth of metallic gallium 3D structures obtained with a single LP-MOCVD growth step is reported. Different 3D structures are obtained such as thin stems, cauliflowers, droplets or aggregates, depending on the elaboration conditions. The thin stems are obtained for deposition temperatures around 500 °C, the cauliflowers for temperatures ranking between 500 °C and 600 °C, and the droplets with montgolfier shape for temperatures higher than 600 °C. The use of different kinds of substrates influences the size and the surface density of the structures, but not their shapes. The calculated activation energy seems to be similar for droplets grown on silicon and gallium arsenide. Aggregates are obtained for long deposit times after coalescence of smallest structures. XRD, EDX and thermal stability results show that the inner part of the droplets is composed of metastable metallic gallium, and that the droplets have an outer skin probably composed of gallium, carbon and/or oxygen. Moreover, according to the microprobe analysis, all the objects are composed by the same chemical elements, and according to XRD, they present the same metastable crystallographic structure. The potentialities of these 3D objects in particular for applications in optoelectronics will be the argument of a forthcoming paper.

References

[1] E. Søndergård, R. Kofman, P. Cheyssac, and A. Stella, Surf. Sci. **364**, 467 (1996).
[2] A. M. Malvezzi, M. Patrini, A. Stella, P. Tognini, P. Cheyssac, and R. Kofman, Mater. Sci. Eng. C **15**, 33 (2001).
[3] K. Deppert, M. H. Magnusson, L. Samuelson, J.-O. Malm, C. Svensson, and J.-O. Bovin, J. Aerosol Sci. **29** (5/6), 737 (1998).
[4] S. Stagira, M. Nisoli, S. De Silvestri, A. Stella, P. Tognini, P. Cheyssac, and R. Kofman, Chem. Phys. **251**, 259 (2000).
[5] A. Stella, P. Tognini, A. Cavalleri, A. M. Malvezzi, M. Nisoli, S. De Silvestri, P. Cheyssac, and R. Kofman, Thin Solid Films **318**, 73 (1998).
[6] T. Arai, H. Tobita, Y. Miyamoto, and K. Furuya, J. Cryst. Growth **221**, 212 (2000).
[7] S. C. Lee, L. R. Dawson, and S. R. J. Brueck, J. Cryst. Growth **240**, 333 (2002).
[8] L. Haderbache, R. Garrigos, R. Kofman, E. Søndergård, and P. Cheyssac, Surf. Sci. **410**, L748 (1998).
[9] K. Tsutsui, K. Uejima, and K. Kawasaki, Microelectron. Eng. **35**, 245 (1997).
[10] Z. W. Pan, Z. R. Dai, C. Ma, and Z. L. Wang, J. Am. Chem. Soc. **124** (8), 1817 (2002).
[11] Z. Liu, R. T. Lee, and G. B. Springfellow, J. Cryst. Growth **191**, 1 (1998).
[12] A. V. Blant, S. V. Novikov, T. S. Cheng, L. B. Flannery, I. Harrison, R. P. Campion, D. Korakakis, E. C. Larkins, Y. Kribes, and C. T. Foxon, J. Cryst. Growth **203**, 349 (1999).
[13] S. Y. Karpov, O. V. Bord, R. A. Talalaev, and Y. N. Makarov, Mater. Sci. Eng. B **82**, 22 (2001).
[14] H. Shin, D. B. Thomson, R. Schlesser, R. F. Davis, and Z. Sitar, J. Cryst. Growth **241**, 404 (2002).
[15] J. Decobert, and G. Patriarche, J. Appl. Phys. **92** (10), 5749 (2002).
[16] Heyding et al., J. Phys. Chem. Solids **34**, 133 (1973).
[17] P. Tognini, A. Stella, P. Cheyssac, and R. Kofman, J. Non-Cryst. Solids **249**, 117 (1999).
[18] P. Tognini, G. B. Parravicini, A. Stella, L. Fornari, R. Kofman, P. Cheyssac, and M. E. Giardini, Thin Solid Films **380**, 230 (2000).

MBE growth and characterization of InAs/GaAs for infrared detectors

K. Regiński[*,1], **J. Kaniewski**[1], **K. Kosiel**[1], **T. Przesławski**[1], and **J. Bąk-Misiuk**[2]

[1] Department of Physics and Technology of Low Dimensional Structures,
 Institute of Electron Technology, Al. Lotników 32/46, 02-668 Warsaw, Poland
[2] Laboratory of X-ray and Electron Microscopy Research, Institute of Physics,
 Polish Academy of Sciences, Al. Lotników 32/46, 02-668 Warsaw, Poland

Received 19 September 2003, accepted 3 November 2003
Published online 22 December 2003

PACS 68.55.–a, 73.61.Ey, 81.05.Ea, 81.15.Hi, 85.60.Gz

We discuss several problems of optimisation of the MBE growth of thick (3.5 ÷ 9 μm) InAs epilayers on GaAs(100) substrates. Three types of layers were grown: undoped, Si-doped, and Be-doped. The whole growth process has been divided into two parts: the initial stage and the main stage. For each stage, the optimum growth conditions comprising the substrate temperature, the growth rate, and the flux ratio have been found. The crystal quality and the transport properties of the layers have been analysed by the transmission electron microscopy, the double-crystal X-ray diffraction, and the Hall effect measurements. It has been demonstrated that epilayers grown under the optimum conditions have good structural and electrical properties.

1 Introduction The growth of InAs epitaxial layers on GaAs substrates has been a subject of intense studies over the past few years, due to important device application of such layers in optoelectronics. In the field of high-speed infrared detectors, InAs is especially attractive as an absorption layer, due to its narrow band gap (0.354 eV at 295 K) and high electron mobility. On the other hand, a semi-insulating GaAs is very convenient as a substrate, due to its transparency. However, the large lattice mismatch (7.3%) between InAs and GaAs causes major problems in fabrication of high quality epilayers.

Despite many efforts, the problem of optimum growth conditions for InAs/GaAs system is still open. Now, it is clear that two main factors play a crucial role: the proper choice of a technique used to reduce the misfit dislocation generation at the start of the growth and the optimum growth conditions during the continuation of the process. Since the classical paper of Kubiak et al. [1] several techniques have been developed to diminish the effect of lattice mismatch: growth of step-graded buffer layer, growth of continuously graded buffer layer, continuous grading, or growth of superlattices. However, the quality of the epilayer only slightly depends on the applied buffer layer. Much more important is a good control of the parameters of the MBE process at the start of InAs growth. This control is difficult due to instability of the system at the initial stage of the process: First, the optical parameters of the optical system (consisting of the substrate and the thin InAs layer) vary rapidly. The increased absorption causes the increase of temperature of the layer. At the same time large changes in emissivity and reflectivity make the optical pirometry very difficult. Growth conditions formulated in the next section allow for such difficulties. We also take into account that the optimum growth conditions for the nucleation differ from those for the main stage of process (cf. [2]).

The purpose of this paper is to find optimum growth conditions for growing thick (3.5 ÷ 9 μm) InAs epilayers on GaAs(100) substrates. These conditions should be proper for growing both the undoped and

---

[*] Corresponding author: e-mail: reginski@ite.waw.pl, Phone: +48 22 5487 920, Fax: +48 22 8470 631

the doped layers. To test the growth conditions, we will demonstrate that the epilayers grown under the optimum conditions have simultaneously a good crystal quality and good electrical properties.

2 Experimental The growth processes have been performed by Elemental Source MBE technique on RIBER 32P machine equipped with ABN 135L evaporation cells. The molecular fluxes were measured by a Bayard-Alpert gauge mounted on the sample manipulator. For monitoring the state of the crystal surface during the growth, the reflection high energy electron diffraction (RHEED) system with a 10 keV electron gun was used. The substrate temperature was precisely measured with a thermocouple and simultaneously with an IRCON Modline Plus pyrometer. More details about this control system one can find in [3].

The calibration of the growth rate of the layers was performed in two steps. In the first step, the necessary values of the temperatures of effusion cells were found by using temperature-flux calibration curves and well-known formulae for converting the beam equivalent pressure into the absolute value of flux [4]. In the next step, RHEED intensity oscillations were measured several times during different stages of the growth process and necessary corrections for In cell temperature were introduced. The flux of As_4 was also measured several times during the process and the temperature of As_4 cell was corrected. The growth processes were performed on the GaAs SI substrates of (100) orientation. The thickness of the substrates was either 450 µm or 1150 µm. No influence of the layer thickness on the quality of grown layers was observed.

After desorption of the oxide, the 1 µm GaAs buffer layer was grown. Before the end of the buffer, the temperatures of the substrate, of the In cell, and of the As_4 cell were set at the values relevant for starting the growth of InAs. The growth of InAs was started just after finishing the buffer layer without any growth interruption. Just after starting the InAs growth, the substrate temperature and the temperatures of In and As_4 cells were slowly ramped to the values proper for the main stage of growth. The process of ramping took about 20 min.

Fig. 1 Cross-section TEM image of an InAs layer grown on GaAs(100) substrate. Dislocations are concentrated at the interface GaAs/InAs.

After approximately 20 min of growth, the 2 × 4 surface reconstruction was well visible and RHEED intensity oscillations could be measured.

Structural characterisation of the samples was carried out by using the transmission electron microscopy (TEM) and the double-crystal X-ray diffraction (DXRD). Measurements of the rocking curve width (Full Width at Half Maximum – FWHM) for epitaxial layers were performed by using the symmetrical 400 reflection. Hall effect measurements of the layers were performed in the magnetic field 0.3 T at 300 K and at 77 K.

The method of growth presented above results in a large concentration of dislocations near the interface. TEM images show that the dislocations are mainly concentrated within the distance less than 2 µm from the InAs/GaAs interface (Fig. 1). In the layers grown at optimised conditions (formulated below), the dislocation density observed far from the interface was as low as $2 \times 10^6 \, cm^{-2}$.

3 Optimum growth conditions and properties of layers In a number of processes we have established the optimum growth conditions for the initial stage and for the main stage of the process. It is known that in the case of InAs/GaAs system, good structural quality of layers correlates with good electrical properties [1, 5, and 6]. Nevertheless, we have tested this fact on a series of samples grown at different temperatures under As-rich conditions (the As_4 flux was minimised). The results of the double-crystal X-ray diffraction and the Hall effect measurements at the room temperature are shown in Figs. 2 and 3, respectively.

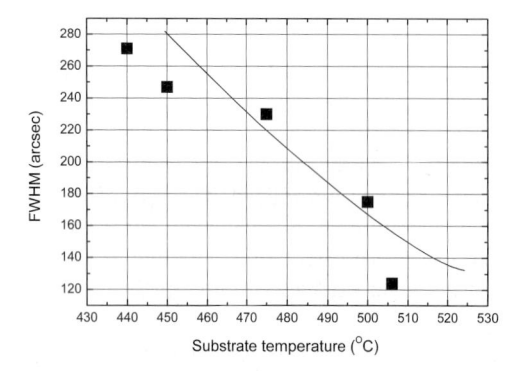

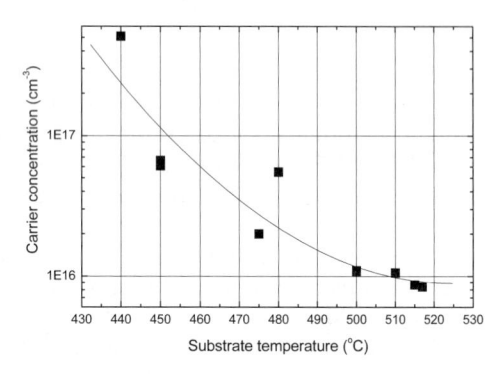

Fig. 2 FWHM of DXRD as a function of growth temperature for undoped InAs/GaAs(100).

Fig. 3 Electron concentration at 300 K as a function of growth temperature for undoped InAs/GaAs(100).

Combining the above results with those of growth rate and flux intensity measurements, we formulate the following optimum growth conditions: At the beginning of the initial stage of process, the substrate temperature is 490 °C, the growth rate is 1 µm/h, and As_4/In flux ratio is 20. For the main stage of the process, the substrate temperature is 515 °C, the growth rate is 1.5 µm/h, and As_4/In flux ratio is 16. During the main part of the process, the surface reconstruction 2 × 4 (not very strong but well visible and clear) has to be observed. The optimum growth conditions are critical: after a small increase of substrate temperature, or after a decrease of As_4 flux, we observed the 4 × 2 reconstruction. We have never seen the 1 × 1 reconstruction, as it was stated in [5].

Table 1 Properties of undoped InAs layers grown under optimum growth conditions.

thickness (µm)	300 K mobility (10^4 cm^2/V s)	300 K electron concentration (10^{15} cm^{-3})	77 K mobility (10^4 cm^2/V s)	77 K electron concentration (10^{15} cm^{-3})	FWHM (arcsec)
7	1.4	4.6	4.7	2.6	134
9	1.6	4.6	5.0	2.4	106

The physical properties of the epilayers grown at the optimum conditions are presented in Table 1. It is worth noticing that only small changes of the parameters are observed when the layers are thicker than 7 µm.

Doped layers were grown at the same optimum conditions. The Si-doping did not demonstrate any peculiarity and the electron concentration in the range from 5.0×10^{16} cm^{-3} to 2.5×10^{18} cm^{-3} could be controlled with high accuracy. In the case of Be-doping used to obtain p-type material, calibration procedure seems to be more complicated. Results obtained from Hall effect measurements are presented in Figs. 4 and 5.

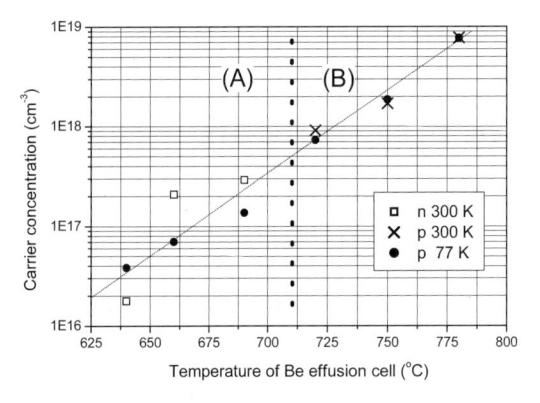

Fig. 4 Carrier concentration in Be doped InAs layers determined from Hall effect measurements.

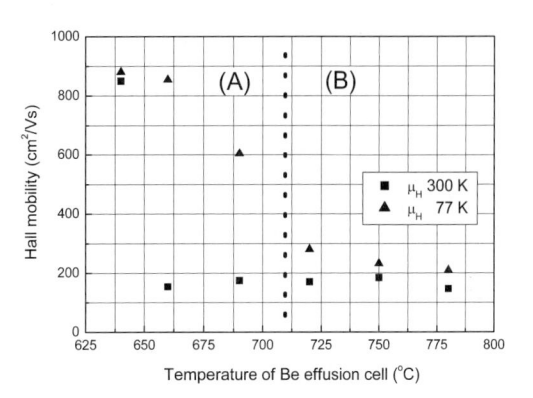

Fig. 5 Hall mobility of carriers in Be doped InAs layers.

It should be noticed that the hole concentration, p is determined properly from Hall effect measurements performed at 300 K only when $p > 3 \times 10^{17}$ cm^{-3} – region labelled (B) in Figs. 4 and 5. At the lower level of p-type doping, results obtained at 77 K are more suitable for doping calibration – region labelled (A) in Figs. 4 and 5. It is clearly seen that Hall effect measurements of slightly Be doped InAs layers exhibit n-type like behaviour at 300 K and p-type like behaviour at 77 K (Fig. 4). On the other hand, the Hall mobility of carriers in the same group of layers exhibits a strong increase at 77 K and different temperature dependence (Fig. 5). All the detected effects are associated with the two-carrier transport and are related to a big difference between electron and hole mobility. In order to determine the accurate value of p, the values of hole and electron mobility have to be known and, therefore, more extended transport measurements are needed. Similar effects are observed in many narrow gap semiconductors.

4 Conclusions On the basis of many MBE processes we have formulated the optimum growth conditions for the InAs/GaAs system. These conditions are proper for growing both the undoped and the doped layers. Applying the optimum conditions we have grown several epilayers of good structural and electrical properties. The range of thickness of the layers corresponds to the typical values needed for fabrication of optimized infrared detectors.

References

[1] R. A. A. Kubiak, E. H. C. Parker, S. Newstead, and J. J. Harris, Appl. Phys. A **35**, 61 (1984).
[2] A. Ohtake and M. Ozeki, Appl. Phys. Lett. **78**, 431 (2001).
[3] J. Muszalski, Thin Solid Films **367**, 299 (2000).
[4] G. J. Davies and D. Williams, in: E. H. C. Parker (Ed.), The Technology and Physics of Molecular Beam Epitaxy, (Plenum, New York, 1985) p. 38.
[5] S. M. Newstead, R. A. A. Kubiak, and E. H. C. Parker, J. Cryst. Growth **81**, 49 (1987).
[6] S. Kalem, J. Appl. Phys. **66**, 3097 (1989).

Investigation of Indium Tin Oxide (ITO) films for the VCSEL laser with dielectric Bragg reflectors

H. Wrzesińska[*,1], **L. Ilka**[1], **D. Wawer**[1], **K. Hejduk**[1], **A. Kudła**[1], **M. Bugajski**[1], and **E. Łusakowska**[2]

[1] Institute of Electron Technology, Al. Lotników 32/46, 02-668 Warszawa, Poland
[2] Institute of Physics Polish Academy of Sciences, Al. Lotników 32/46, 02-668 Warszawa, Poland

Received 19 September 2003, accepted 3 November 2003
Published online 22 December 2003

PACS 42.70. Km, 68.37.Ps, 68.47.Fg, 78.20.Ci, 81.15.Cd, 81.40.Ef

This paper investigates ITO films for the VCSEL (vertical cavity surface emitting laser) lasers emitting at 1000 nm band. Looking for low-resistance VCSEL structures a new laser with dielectric distributed Bragg reflectors (DBR) and ITO films was designed. Here we examine the influence of deposition process parameters on the optical and electrical parameters of ITO films. These films were deposited using the dc magnetron sputtering system with In_2O_3/SnO_2 (90/10) source. After annealing at 500 °C, the transmittance of tested ITO films was in the range of 84%–98% for the 1000 nm wavelength, whereas the resistivity of all samples of the order of tens of $\mu\Omega$cm. The refractive index of ITO films, measured for 1000 nm, does not depend on the partial pressure of argon and oxygen, but depends on the deposition rate. For 180 nm/min deposition rate the refractive index is about 2, whereas for 70 nm/min – only 1.75. Surface of ITO films was smooth (roughness R_a is of the order of several tens of Å).

© 2004 WILEY-VCH Verlag GmbH & Co. KGaA, Weinheim

1 Introduction

Idium tin oxide (ITO) is a wide bandgap (3.5–4.05 eV) [1, 2] material exhibiting high transparency (higher than 80%) in the visible and near-infrared region of spectrum. The ITO refractive index is about 1.96 [3]. Furthermore, ITO is the n-type, highly degenerated semiconductor which has relatively low electrical resistivity (about 2×10^{-4} Ωcm). Due to these properties, it is used for a wide range of optoelectronic applications and is becoming increasingly important in the production of electronic devices. One of the potential applications of ITO films is a new construction of vertical cavity surface emitting laser (VCSEL). Looking for low-resistance structure, VCSEL with dielectric distributed Bragg reflectors (DBR) and ITO films has been designed [4]. The ITO films should have high transmission (above 85%), low resistivity (about 10^{-4} Ωcm) and smooth surface to allow for deposition of dielectric Bragg reflectors.

The investigations of ITO films presented in this paper are devoted to VCSEL lasers emitting at 1000 nm band. The aim of this research was to develop technology of ITO films with high transmittance, which strongly depends on the film structure, modified by the deposition [5] and annealing conditions [6–9]. In this work we have concentrated on the optimization of the following of parameters deposition process: the partial argon pressure, and oxygen, the deposition rate and annealing temperature – aiming on achieving required optical and electrical parameters of ITO films and the improved surface smoothness.

---

[*] Corresponding autor: e-mail: wrzes@ite.waw.pl, Phone: +48 22 54 87 877, Fax: +48 22 84 70 631

2 Experimental

The 200 nm thick ITO films were deposited on the quartz substrates at ambient temperature using dc magnetron sputtering system with In_2O_3/SnO_2 (90/10) source. The deposition processes were performed with different partial pressures of argon and oxygen in processing chamber, ranging from 1×10^{-3} – 2.5×10^{-4} mbar and 2×10^{-2} – 2.5×10^{-3} mbar respectively. Deposition processes of ITO films in pure argon atmosphere were also done. Annealing processes were made at open system in the air atmosphere at 350 °C and 500 °C, for 1 hour.

Spectral characteristics of ITO films were measured for wavelenght range 400–1100 nm by two-beam LAMBDA spectrometer system. The resistivity of films was calculated on the resistance measurement by a standard four-probe technique. Optical constants n, k were determined with the VASE spectroscopic ellipsometer and surface smoothness was tested with the use of atomic force microscope (AFM).

3 Results and discussion

3.1 Transmittance

For the 1000 nm wavelength transmittance of ITO films after deposition was in the 60%–93% range, depending on the partial pressure of argon and oxygen. After annealing at 500 °C, transmittance increased to 84–100%. Fig. 1 shows the influence of annealing temperature on the transmittance of ITO films.

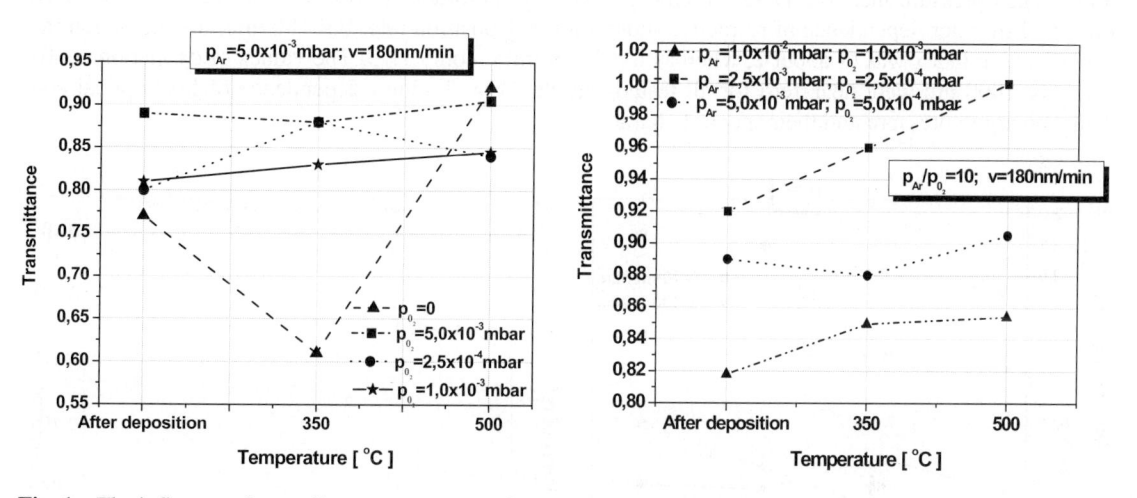

Fig. 1 The influence of annealing temperature on the transmittance of ITO films at 1000 nm.

3.2 Resistivity

The strong influence of annealing temperature and the pressure of reactive gases on the ITO films resistivity was observed. Resistivity measurements done before annealing of the ITO films, deposited in the pure argon atmosphere, show results ranging from several tens to several hundred of $\mu\Omega$cm. ITO films deposited in the same conditions but in the argon and oxygen mixed atmosphere had resistivity of a few Ωcm. The deposition rate did not affect ITO films resistivity. The lowest resistivities (several $\mu\Omega$cm) were obtained for ITO films deposited in pure argon atmosphere, after annealing at 350 °C and for films deposited in argon and oxygen mixture after annealing at 500 °C. The influence of annealing temperature on ITO film resistivity is shown in Fig. 2.

We have obtained ITO films with resistivities much lower than the once reported in the literature [10–12].

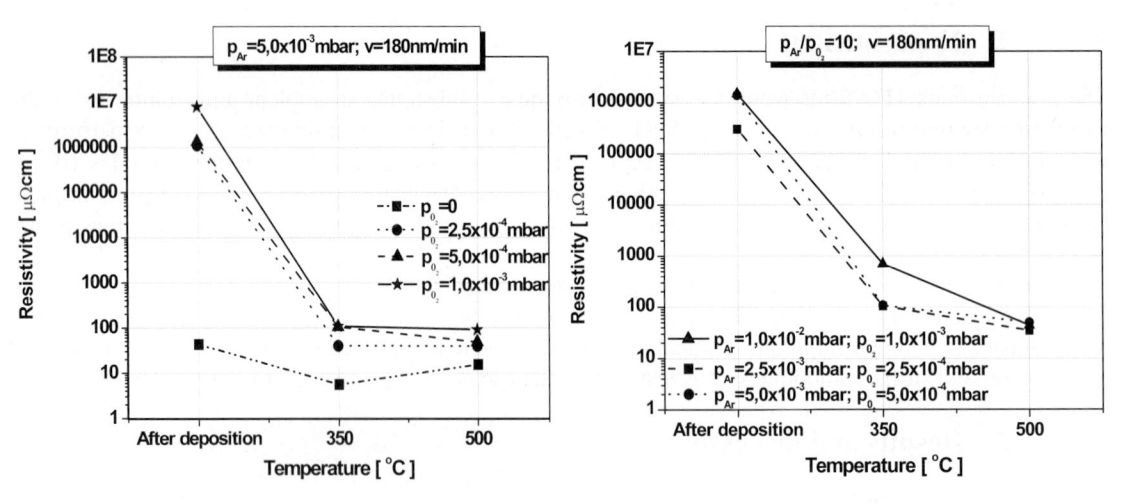

Fig. 2 The influence of annealing temperature on ITO films resistivity.

3.3 Optical constants n, k

The refractive index of ITO films measurements at 1000 nm showed only slight increase with the argon and oxygen pressure increase. Detected changes were of the order of 10%. On the other hand we have observed stronger dependence of refractive index on the deposition rate. For 180 nm/min deposition rate the refractive index was of about 2, whereas for 70 nm/min – only 1.75. The values of extinction coefficient for 1000 nm were near zero for all tested samples. Fig. 3 shows dependence of ITO optical constants on argon pressure and their spectral characteristics.

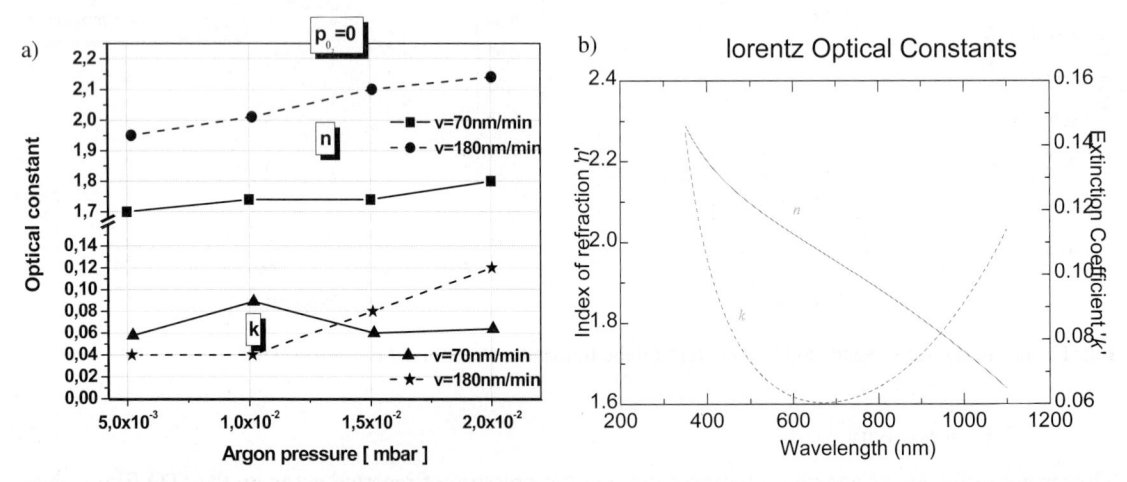

Fig. 3 Dependence of ITO optical constants for 1000 nm on the argon pressure (a); spectral dependence of optical constants (b)

3.4 Topography

3D AFM images of ITO films were investigated to confirm high surface smoothness. Roughness R_a was equal to 0.174 nm for as-deposited samples and 0.210 for annealed ones. Annealing temperature and deposition rate of influence the size of grains. The ITO films have structure with grain size of about 300 nm and 120 nm for deposition rate of 70 nm/min and 180 nm/min respectively. Fig. 4, 5 show the influence of annealing temperature and deposition rate on the structure of ITO films.

© 2004 WILEY-VCH Verlag GmbH & Co. KGaA, Weinheim

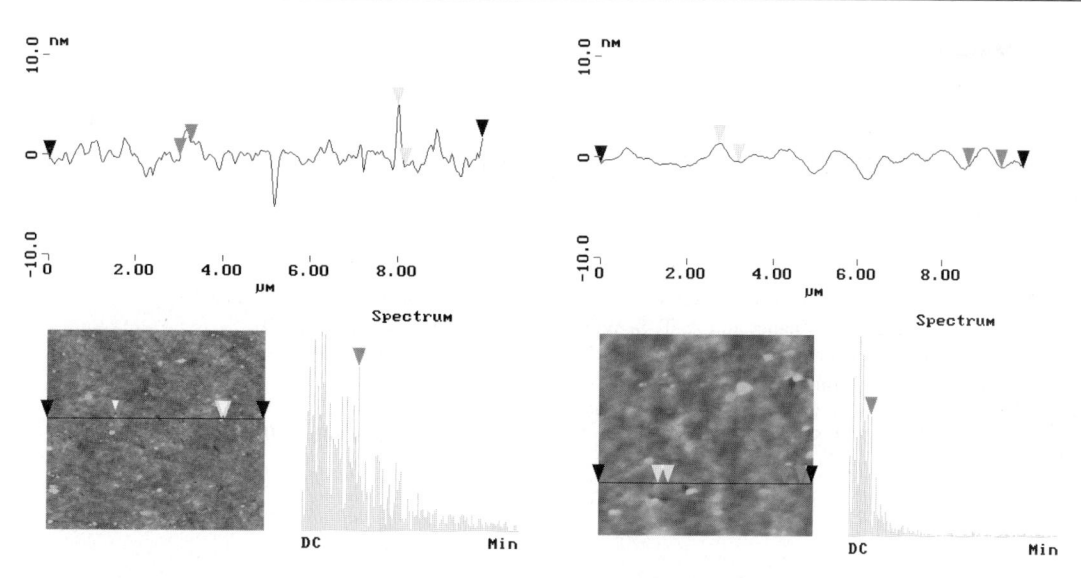

Fig. 4 Surface topography of ITO film deposited with the rate of 70 nm/min; $p_{Ar} = 1.0 \times 10^{-2}$ mbar, $p_{O2} = 0$. Left image – before annealing, right image – after annealing

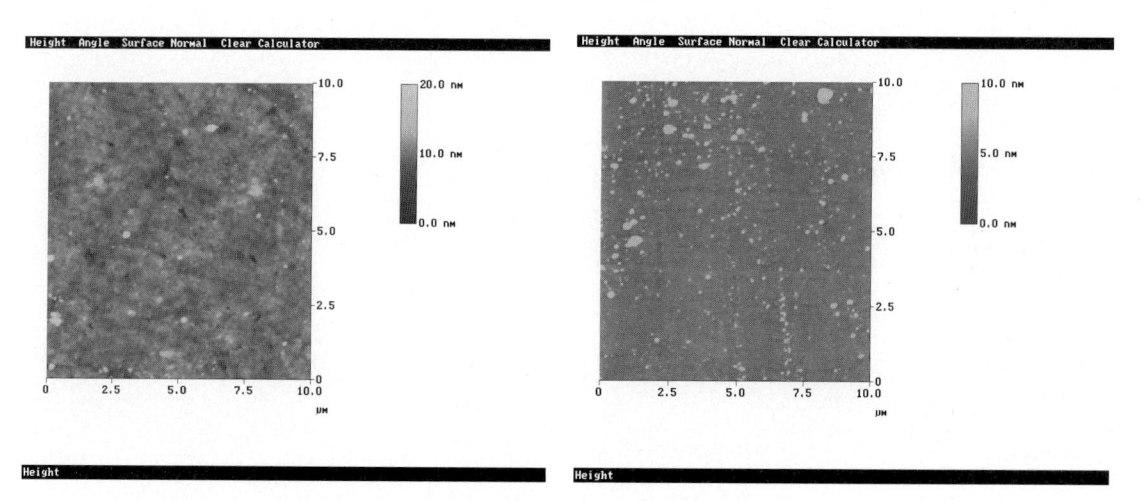

Fig. 5 Surface topography and essential parameters of ITO films deposited at $p_{Ar} = 1.0 \times 10^{-2}$ mbar, $p_{O2} = 0$. Left image – deposition rate – 70nm/min, right image – deposition rate – 180 nm/min

4 Conclusions

We have elaborated reproducible technological conditions of the ITO films deposition and annealing. Their optical and electrical parameters, as well as the surface structure allow for the application of ITO films in the new technology of VCSEL lasers. Optimum parameters of ITO (transmittance – 0.93 and resistivity – 15.4 μm) were obtained for films deposited at the rate 180 nm/min in pure argon atmosphere and annealed at 500 °C in air atmosphere for 1 hr. During deposition of this ITO films the argon pressure was kept at 5×10^{-3} mbar.

References

[1] N. Balasubramanian and A. Subrahmanyam, J. Phys. D: Appl. Phys. **22**, 206 (1998).
[2] J. C. C. Fan and J. B. Goodenough, J. Appl. Phys. **48**, 3524 (1997).
[3] J. Szczyrbowski, A. Dietrich, and H. Hoffmann, phys. stat. sol. (a) **78**, 243 (1983).
[4] A. Szerling, D. Wawer, K. Hejduk, T. Piwoński, A. Wójcik, B. Mroziewicz, and M. Bugajski, Optica Applicata, **XXXII** (3), 523 (2002).
[5] M. Quaas, H. Steffen, R. Hippler, and H. Wulff, Surf. Sci. **454–456**, 790 (2000).
[6] S. Shimado, I. Sato, K. Kodaira, and T. Matsushita, J. Electrochem. Soc. **134** (8), 2056 (1987).
[7] I. Baia, M. Quntinela, L. Mendes, P. Nunes, and R. Martins, Thin Solid Films **337**, 171 (1999).
[8] F. Zhu, C. H. A. Huan, K. Zhang, and A. T. S. Wee, Thin Solid Films **359**, 244 (2000).
[9] M. J. Alam and D. C. Cameron, Surface Coatings Technol. **142–144**, 776 (2001).
[10] A. Salehi, Thin Solid Films **324**, 214 (1998).
[11] K. Utsumi, O. Matsunga, and T. Takahata, Thin Solid Films **334**, 30 (1998).
[12] D. J. You, S. K. Choi, H. S. Han, J. S. Lee, and C. B. Lim, Thin Solid Films **401**, 229 (2001).

Revealing of threading and misfit dislocations in partially relaxed InGaAs/GaAs heterostructures

Oksana Yastrubchak[*,1], **Elżbieta Łusakowska**[1], **Andrzej Morawski**[1], **Oleg Demchuk**[2], and **Tadeusz Wosiński**[1]

[1] Institute of Physics, Polish Academy of Sciences, Al. Lotników 32/46, 02-668 Warsaw, Poland
[2] Institute of Organic Chemistry, Polish Academy of Sciences, Kasprzaka 44/52, 01-224 Warsaw, Poland

Received 19 September 2003, accepted 17 November 2003
Published online 22 December 2003

PACS 61.72.Ff, 61.72.Lk, 68.37.Ps, 68.55.Jk, 81.05.Ea, 81.15.Hi

Both threading and misfit dislocations in partially relaxed InGaAs/GaAs heterostructures with a small lattice-mismatch have been investigated by means of chemical etching and atomic force microscopy (AFM). An anisotropic etching in $HF-H_2SO_4-H_2O_2$ based aqueous solution was successfully used to determine the polarity of the crystallographic surfaces and an ultrasonic-vibration aided etching in CrO_3-HF-H_2O based solution was employed to reveal threading dislocations on the heterostructure surfaces. AFM measurements of surface morphology of the structures revealed a well-resolved cross-hatch pattern, which reproduced the network of underlying misfit dislocations, and, in addition, the outcrops of threading dislocations on the surface in a form of characteristic craters. Analysis of the results allowed drawing a conclusion on the mechanism of misfit strain relaxation in the investigated heterostructures.

© 2004 WILEY-VCH Verlag GmbH & Co. KGaA, Weinheim

1 Introduction Lattice-mismatched GaAs-based heterostructures are of current interest because of their application in high-speed and optoelectronic devices. Epitaxial growth of those heterostructures is accompanied by a strain in the epitaxial layer that results from a difference in lattice parameters between the substrate and the layer. If the epitaxial layer is thin enough, the lattice misfit can be accommodated by elastic strain of the layer. However, when the layer thickness exceeds a critical value, misfit dislocations are generated at the interface to relieve some of the strain. In heteroepitaxial semiconductor systems with zinc-blende structure and a small lattice mismatch, grown on (001)-oriented substrates, orthogonal arrays of 60° misfit dislocations lying along two different $\langle 110 \rangle$ crystallographic directions are formed at the interface. The misfit dislocations are associated with threading dislocations which propagate through the epitaxial layer up to the surface.

Dislocations play an important role in the technology of semiconductor materials and devices owing to their influence, mostly detrimental, on various electronic and mechanical processes. Current requirements for device miniaturization make the influence of dislocations on their performance and reliability increasingly important. Moreover, ideas of taking an advantage of specific electronic properties of dislocations in designing new electronic devices have been proposed; cf. [1]. To those ideas belongs the use of single misfit dislocations as a source and drain in a very short-channel field-effect transistor. More recently, dislocations have been utilized for the fabrication of silicon-based light-emitting diodes compatible with ultra-large-scale integration (ULSI) technology [2]. In those diodes a local strain field of intentionally introduced dislocation loops, which modifies the semiconductor band structure, provides spatial confinement of charge carriers and drastically improves the room-temperature efficiency of electroluminescence of the diodes.

---

[*] Corresponding author: e-mail: yastru@ifpan.edu.pl, Fax: +48 22 843 0926

© 2004 WILEY-VCH Verlag GmbH & Co. KGaA, Weinheim

In the present paper we report our results on investigation of InGaAs/GaAs heterostructures with a small lattice mismatch by means of chemical etching and atomic force microscopy aimed at revealing of both threading and misfit dislocations in the structures and concluding on the mechanism of misfit strain relaxation.

2 Heterostructure growth We investigated two types of InGaAs/GaAs heterostructures, which were grown by molecular beam epitaxy (MBE) on (001)-oriented n-type GaAs substrates with the etch pit density (EPD) of about 2×10^4 cm^{-2}. Each heterostructure contained 1 μm thick, Be-doped, p-type In-GaAs layer grown at a temperature of 530 °C on Si-doped n-type GaAs buffer layer, grown at 600 °C, to form a p-n junction near the interface. Two investigated structures differed by an In content in the In-GaAs layer, which amounted to 2.2% in the structure called A and 2.7% in the structure called B, as determined by energy dispersive spectrometry (EDS) in a scanning electron microscope (SEM). The InGaAs layers were grown under compressive misfit stress resulting from the lattice-mismatch between GaAs and the ternary compound, which was below 0.2% in the both structures. The layer thickness of 1 μm exceeded a little the critical value for misfit dislocation formation in the both structures, so they were only partially relaxed as confirmed by X-ray diffraction measurements [3].

3 Chemical etching The investigated samples have been subjected to chemical etching in order to (i) distinguish between the nonequivalent ⟨110⟩ crystallographic directions on the (001) face of heterostructures and (ii) reveal the terminations of threading dislocations at the surfaces of the structures. We used anisotropic etching in HF:H$_2$SO$_4$:H$_2$O$_2$ (2:2:1) mixture [4] diluted in 1:1 proportion in water at room temperature. This etching gives rise to a microrelief of quasiparallel grooves on the (001) GaAs surface [5], as shown in Fig. 1a. As demonstrated below, by comparison of this relief with the shape of dislocation etch pits we have shown that the grooves are aligned with the [$\bar{1}$10] crystallographic direction.

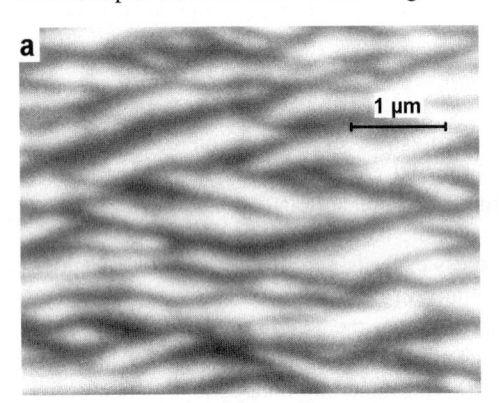

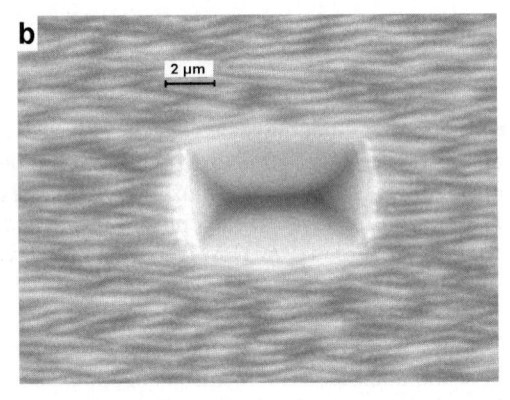

Fig. 1 Secondary electron images (SEI) in a scanning electron microscope (SEM) of (a) microrelief produced on epi-ready polished (001) surface of semi-insulating GaAs wafer by anisotropic chemical etching, described in the text, for 45 s at room temperature and (b) etch pattern obtained on (001) surface of the In$_{0.027}$Ga$_{0.973}$As/GaAs heterostructure by ultrasonic-vibration aided D$_{1:1}$S$_{1/2}$ etching for 30 s at room temperature, followed by the anisotropic chemical etching, same as in (a).

Conventional etching technique for revealing dislocations in GaAs by using molten KOH is rather inconvenient because it requires heating to a temperature of 350-400°C. In the present investigation we have adopted a method of ultrasonic-vibration aided etching proposed by Chen [6] who utilized, as a selective etchant, the CrO$_3$-HF-H$_2$O system, called DS, extensively studied by Weyher and van de Ven [7]. We performed etching in D$_{1:1}$S$_{1/2}$ solution (where, according to the notation used in [7], D$_{1:x}$ means dilution of 1 volume part of basic mixture with x volume parts of water and S$_{a/b}$ means a basic mixture consisting of a and b volume parts of HF (48 wt%) and CrO$_3$ (33 wt%) aqueous solutions, respectively), at room temperature. As a source of ultrasonic vibration we used a commercial ultrasonic scrubber operating at a frequency of 25 kHz and output power of 30 W. This etching produces rectangular etch pits with distinct {111} facets, as shown in Fig. 1b. The long axis of the pits on the (001) face is aligned with the [$\bar{1}$10] crystallographic direction because the {111}As faces are more rapidly dissolved by the etchant than the {111}Ga ones [8].

4 AFM results and discussion A systematic study of surface morphology of the both as-grown InGaAs/GaAs heterostructures has been performed by means of atomic force microscopy (AFM), which provides a powerful technique for microscopic surface roughness measurements. Typical surface morphology of the structure B, prior to chemical etching, is shown in Fig. 2. It presents a well-defined, so-called, cross-hatch pattern with ridges and trenches, of about 2 nm peak-to-valley amplitude, running along two perpendicular ⟨110⟩ directions on the (001) surface of the structure. Such a cross-hatched morphology occurs in many lattice-mismatched semiconductor systems, including SiGe/Si and various III-V compound heterostructures, grown on (001)-oriented substrates and subjected to strain relaxation by misfit-dislocation generation. Understanding of such a surface-relief formation is important for fabrication of low-dimensional devices based on those systems, which exhibit atomically smooth but mesoscopically rough surfaces and are not compatible with planar integrated-circuit technologies. Despite several mechanisms have been recently proposed to account for the cross-hatch development its origin remains controversial and unresolved [9].

In our recent paper [3] we have studied, by means of transmission electron microscopy (TEM) and electron-beam induced current (EBIC) mode of SEM, the structure of misfit dislocations generated at the interfaces of the same InGaAs/GaAs heterostructures investigated here. A direct comparison of the misfit-dislocation arrangement with the results of AFM measurements clearly demonstrated that the cross-hatched surface morphology of the heterostructures reproduced the structure of interfacial misfit dislocations. These findings demonstrate that the cross-hatch development results primarily from the misfit-dislocation generation. The linear density of misfit dislocations estimated from the AFM results for the structure B, shown in Fig. 2, was about 10^4 cm^{-1}. The density of misfit dislocations estimated from AFM results obtained for the structure A was lower by a factor of about 25%.

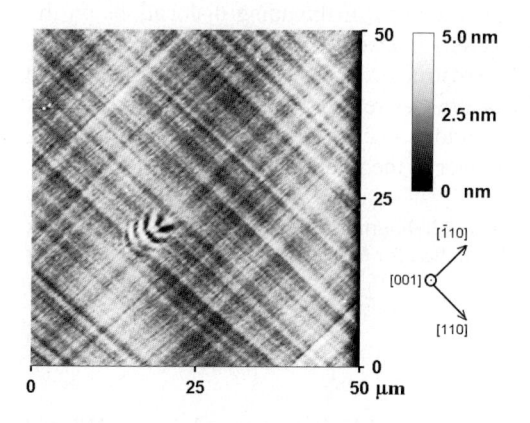

Fig. 2 Surface morphology of the In$_{0.027}$Ga$_{0.973}$As/GaAs heterostructure revealed by AFM showing a well-defined cross-hatch pattern. A characteristic feature visible on the structure surface reveals an outcrop of a threading dislocation.

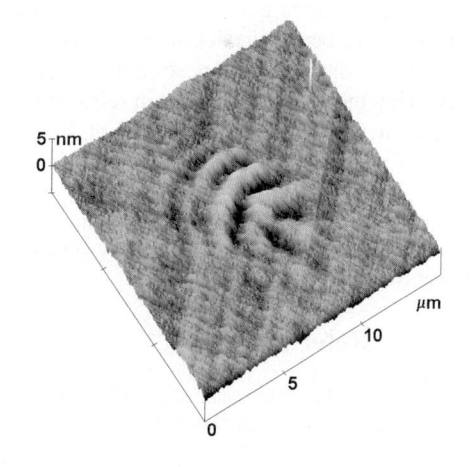

Fig. 3 Three-dimensional AFM image of surface morphology of the same heterostructure as in Fig. 2 showing an outcrop of a threading dislocation.

In addition to the cross-hatch pattern, our AFM investigations of the surface morphology of the heterostructures revealed characteristic features, one of which is visible in Fig. 2. A three-dimensional image of such a feature, under higher magnification, is shown in Fig. 3. It presents a crater, of about 5 nm in depth, surrounded by an asymmetric undulating structure. Selective chemical etching, which revealed a dislocation etch pits in the sites of the craters, allowed for concluding that the craters represent outcrops of threading dislocations lying on the {111} planes inclined to the (001) surface by the angle of

55°. However, the question remains at present what is the detailed origin of the undulating structure surrounding the craters. The density of the terminations of threading dislocations on the surface of the structures was estimated to be of the order of 10^4 cm^{-2}. It was distinctly larger for the structure A with respect to that for the structure B.

A larger density of misfit dislocations accompanied by a smaller density of threading dislocations in the structure B, with the higher In content, in comparison to those densities in the structure A is consistent with the mechanism of misfit strain relaxation proposed by Matthews et al. [10]. According to this mechanism, which is dominating in low-misfit systems [11], the formation of interfacial misfit dislocations occurs from pre-existing threading dislocations, originating from the substrate, by bending them and gliding on inclined slip planes, driven by the misfit stress, to the interface. The above mechanism makes use of the assumption that the substrate provides sufficient threading dislocations to generate all the misfit dislocations necessary to complete the strain relaxation. This assumption is fulfilled in our case of a small lattice mismatch. In cases where it is not fulfilled, then misfit-dislocation multiplication processes and nucleation processes are needed to complete the strain relaxation.

5 Summary and conclusions The structure of both threading and misfit dislocations in MBE-grown InGaAs/GaAs heterostructures with a small lattice-mismatch has been investigated by means of chemical etching and AFM. Two techniques of chemical etching have been successfully used to determine the polarity of the crystallographic surfaces and to reveal threading dislocations. We have shown that anisotropic etching in HF-H$_2$SO$_4$-H$_2$O$_2$ based aqueous solution allows to distinguish between the nonequivalent $\langle 110 \rangle$ crystallographic directions on the (001) face of GaAs crystals and InGaAs epitaxial layers. On the other hand, an ultrasonic-vibration aided etching in CrO$_3$-HF-H$_2$O based solution provides a very convenient method to reveal the terminations of threading dislocations on the (001) surface in a form of rectangular pits aligned with the $[\bar{1}10]$ crystallographic direction.

Systematic investigation of surface morphology of the InGaAs/GaAs heterostructures with the AFM technique revealed a well-resolved cross-hatch pattern, which reproduces the network of underlying misfit dislocations generated at the interface. In addition, the outcrops of threading dislocations on the surface have been revealed in a form of characteristic craters surrounded by an undulating structure. By comparing the dislocation structure in two investigated heterostructures with different In content, we have concluded that misfit strain relaxation in the epitaxial layers was realized in the model of Matthews et al., in which the development of interfacial misfit dislocations occurs from pre-existing threading dislocations, originating from the substrate, by bending them over to the interface.

Acknowledgements The authors would like to thank B. F. Usher (Melbourne) for growing the heterostructures investigated. This work has been partly supported by the State Committee for Scientific Research of Poland under Grant No. 2 P03B 077 24.

References

[1] S. Mil'shtein, phys. stat. sol. (a) **171**, 371 (1999).
[2] Wai Lek Ng, M. A. Lourenço, R. M. Gwilliam, S. Ledain, G. Shao, and K. P. Homewood, Nature **410**, 192 (2001).
[3] O. Yastrubchak, T. Wosiński, J. Z. Domagała, E. Łusakowska, T. Figielski, B. Pécz, and A. L. Tóth, J. Phys.: Condens. Matter (2003) in print.
[4] N. L. Dmitruk, O. Yu. Borkovskaya, O. I. Mayeva, and O. V. Fursenko, Microelectron. J. **27**, 37 (1996).
[5] N. L. Dmitruk, O. I. Mayeva, O. B. Yastrubchak, and G. V. Beketov, Acta Phys. Polon. A **94**, 285 (1998).
[6] N. Chen, J. Cryst. Growth **129**, 777 (1993).
[7] J. Weyher and J. van de Ven, J. Cryst. Growth **63**, 285 (1983).
[8] D. B. Holt, J. Mater. Sci. **23**, 1131 (1988).
[9] A. M. Andrews, J. S. Speck, A. E. Romanov, M. Bobeth, and W. Pompe, J. Appl. Phys. **91**, 1933 (2002).
[10] J. W. Matthews, S. Mader, and T. B. Light, J. Appl. Phys. **41**, 3800 (1970).
[11] J. Zou, D. J. H. Cockayne, and B. F. Usher, J. Appl. Phys. **73**, 619 (1993).

Magnetic properties of ultra-thin epitaxial V/Gd bilayers

P. Pankowski[*,1], **L. T. Baczewski**[1], **T. Story**[1] **A. Wawro**[1] **K. Mergia**[2], and **S. Messoloras**[2]

[1] Institute of Physics, Polish Academy of Sciences, Al. Lotnikow 32/46 02-668 Warsaw, Poland
[2] Institute of Nuclear Technology & Radiation Protection,
National Centre for Scientific Research "Demokritos", 15310 Aghia Paraskevi Attikis, Greece

Received 19 September 2003, accepted 17 November 2003
Published online 22 December 2003

PACS 75.60.Ej, 75.70.Ak, 75.70.Cn

Magnetic properties of a new system of epitaxial vanadium/gadolinium bilayers with different V thickness have been studied. The structure and quality of bilayers was determined by X-ray reflectivity measurements using synchrotron radiation. Hysteresis loops measurements suggest in-plane anisotropy for all samples (M_r/M_s close to 1). Basic magnetic properties: saturation magnetisation, coercive field and remanence were obtained. Average magnetic moment analysis gives a possibility of an induced magnetic moment on vanadium atoms.

© 2004 WILEY-VCH Verlag GmbH & Co. KGaA, Weinheim

1 Introduction

Magnetic materials in thin films have wide applications in the electronic industry during the last few years. The magnetic properties of vanadium (V) were so far a subject of many studies, theoretical as well as experimental ones. The V isolated atom possesses according to Hund's rules a magnetic moment of about 3 μ_B, but in the bulk it is paramagnetic. Theoretical investigations have shown that in thin film structures V may acquire a strong magnetic moment in the vicinity of a magnetic layer. Moreover, a magnetic order of (001) V surface was predicted at 3.8 K [1], and in subsequent experimental work [2] the existence of ferromagnetic order of (001) V surface in the temperature range 300 to 540 K was reported.

In the theoretical studies of V thin films significant discrepancies still exist between individual predictions concerning the alignment of the magnetic moments as well as their values. For ultra thin epitaxial V films on various substrates e.g. Ag, Au or Fe [3–5] the calculated magnetic moment per V atom was found to be more than 1.5 μ_B/atom. However, the theoretical calculations performed for different substrates significantly differ from experimental results. There are a few experimental papers devoted to V films deposited on 3d substrates, and even less concerning magnetic interactions between transition metals like V, Mn, Cr and magnetic rare earth (RE) elements.

The magnetic behaviour of the system where V atoms are in proximity of strongly magnetic Gd atoms is an interesting issue. V films of different thickness of the order of several monolayers (ML) were grown by molecular beam epitaxy (MBE) and structurally characterised by X-ray reflectivity. The bulk magnetic properties of the films were investigated by VSM measurements performed in an external magnetic field strong enough to saturate the samples. The attempt to estimate the magnetic moment of V atoms was done using values of Gd magnetic moments found in the literature for Gd/TM systems.

---

[*] Corresponding author: e-mail: panko@ifpan.edu.pl, Phone: +48 (22) 843-66-01, Fax: +48 (22) 843-13-31

2 Experimental details

The epitaxial V/Gd bilayers were grown using MBE at a vacuum level of the order of 10^{-10} Torr. In order to achieve epitaxial growth a sapphire substrate with orientation (11–20) covered by Mo buffer layer of 20 nm thickness was used and V layer was grown on Mo buffer. Samples with V thickness of 6, 8, 11 and 12 monolayers (ML) were fabricated. Directly on V layer a 20 ML Gd film was deposited. An Al cover layer of about 100 Å was used for protection against oxidation. All the materials were evaporated from electron guns. Deposition rate during growth process was kept at a level of 0.5 Å/s and was controlled by Sentinel III system based on Electron Impact Emission Spectroscopy (EIES). The Al_2O_3 substrate was chemically cleaned and degassed in vacuum at high temperature in order to avoid contamination. The Mo buffer layer was deposited at a temperature above 1000°C in order to obtain optimal crystalline structure and minimum surface roughness. The V layers deposition process was preformed at 700 °C and the Gd was deposited at room temperature. The resulting sample structure was Al_2O_3/Mo/V/Gd/Al. The quality of the interfaces and the crystallographic orientation were investigated in-situ by 12 kV Reflected High Energy Electron Diffraction (RHEED) and ex-situ by Atomic Force Microscope AFM. Auger spectroscopy (AES) was carried out also in-situ to check the chemical composition and surface contamination of the deposited metallic layers. The X-ray reflectivity measurements using synchrotron radiation were performed at room temperature at W1.1 station at HASYLAB in Hamburg, using a wavelength of 1.24 Å and Q-range from 3.5×10^{-3} to 1.15 Å^{-1}. The total magnetic moment of the sample was measured using standard VSM magnetometry in the in-plane applied external magnetic field ranging from –2 T to 2 T and at the temperature of 5 K.

3 Results and discussion

The optimisation of growth conditions (evaporation, deposition rate and temperature) allowed the growth of atomically smooth, flat V/Gd bilayers. The quality of the interfaces is crucial for the understanding of the magnetic behaviour since rough and diffuse interfaces induce additional effects. RHEED measurements showed that the spacing between the (110) Mo planes was equal to that of the bulk. The following relations between in-plane directions of the constituent layers are found: Mo(110)/V(110): [001]||[001], V(110)/Gd(0001): [001]||[11–20]. Such epitaxial relationships of V/Gd correspond to the Nishiyama-Wasserman orientation. Detailed discussion of the structural characterization will be published elsewhere [6]. From the X-ray reflectivity measurements the non-magnetic sample structure was determined. The wide Q-range, which is accessible with synchrotron X-ray, offers the possibility of an almost unique model for the nuclear structure even for ultra-thin bilayers. The structural parameters i.e. thickness, density and roughness of each layer, for each of the four samples were determined on the basis of the best fitting of calculated reflectivity values to the experimental data using the SimulReflec software [7]. The experimental X-ray reflectivity curves and their fittings for each V/Gd sample are presented in Fig. 1. Interface roughness of 2 to 5 Å for vanadium and 6 to 8 Å for gadolinium was found.

In Fig. 2 the hysteresis loops of the V/Gd samples measured by VSM are presented. Coercive field values are in the range from 0.044 to 0.052 Tesla. Saturation magnetisation determined for each sample was used to estimate the possible magnetic moments for V at T= 5 K assuming a constant Gd moment.

The saturation magnetisation M_s of each sample determined from M(H) measurements, after conversion into μ_B/atom was used to estimate a possible induced magnetic moment of V. Considering Gd magnetic moment at 5 K equal to its bulk value (7.44 μ_B/atom) and no magnetic moment for V, the estimated total magnetisations of the samples for V and Gd thickness determined from X-ray reflectivity measurements were too large (about ten times larger) in comparison to VSM results. Definitely Gd average moment in V/Gd system is much lower than the bulk value at 5 K contrarily to Gd moment of 7.3 μ_B/atom reported for Fe/Gd [8] and recently for Gd/Au [9]. The same group [9] measured Gd moment of 3.6 μ_B/atom at 5 K for Gd (60 Å)/V (20 Å) multilayer [9]. Assuming the reduced Gd magnetic moment of 3.6 μ_B/atom, recently found for Gd/V multilayers [9], an induced V magnetic moment can be expected, as it is shown in Table 1.

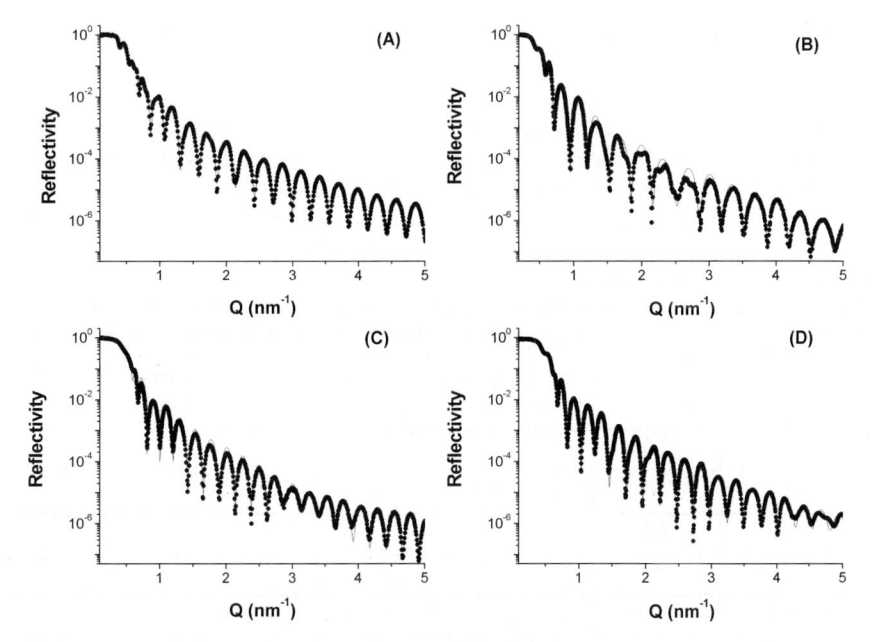

Fig. 1 X-ray reflectivity data from V/Gd bilayers. The solid lines are least squares model fittings. (A) 12 ML, (B) 11 ML, (C) 8 ML and (D) 6 ML of V.

Gd moment in Gd/V generally does not differ much from Gd/Cr system where a value of 3.3 μ_B/at was observed [10] but it seems that it varies as a function of V thickness. This can explain why the Gd moment values lower than 3.6 μ_B/atom have to be taken for samples A and C, (2.8 and 3.5 respectively) even assuming zero moment for vanadium. Such analysis of the average magnetic moment obtained from total saturation magnetisation does not give a definite answer whether there is an induced magnetic moment on V layer in V/Gd bilayers but provides a hint that it can be expected and it is worth to perform more sophisticated experiments. In order to clarify the situation different element sensitive experimental

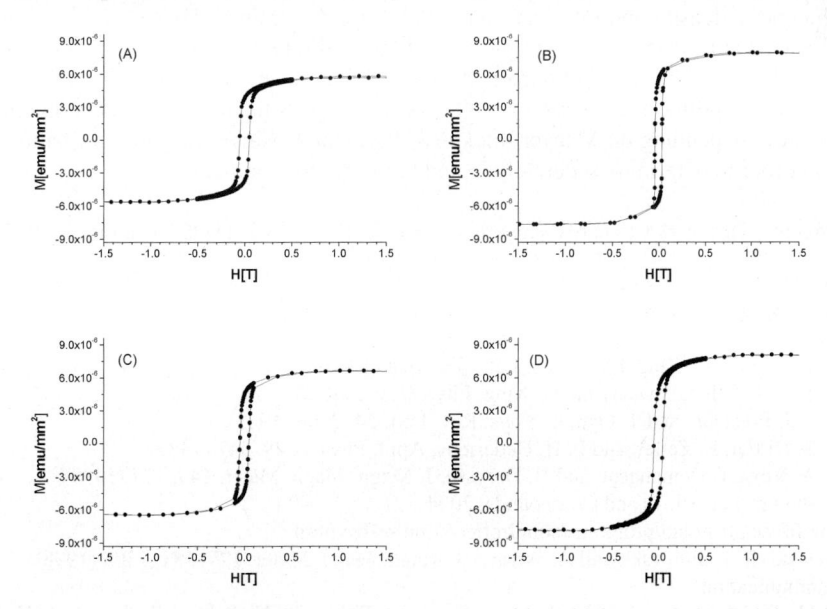

Fig. 2 The M(H) measurement of V/Gd bilayers for the samples with V thickness of 12 ML (A), 11 ML (B), 8 ML (C) and 6 ML (D) at T = 5 K. Magnetic field was applied in-plane of the samples.

methods should be applied i.e. XMCD (X-ray magnetic circular dichroism) or PNR (Polarized Neutron Reflectivity).

Even from the preliminary results obtained up to now it is clear that Gd moment in bi- multilayer system strongly depends on nature of adjacent element. Two different situations are observed: first, for Gd/Fe and Gd/Au where Gd moment is close to its bulk value and second: for Gd/V and Gd/Cr where it is reduced more than by half. As one can see this effect is not related to the magnetic nature of adjacent element rather to its density and/or crystallographic structure. Further studies are definitely necessary.

Table 1 Estimation of V magnetic moments from the VSM measured saturation magnetisations assuming reduced Gd moment The V and Gd layers thickness were determined from X-ray reflectivity fittings.

V [ML]	Element	Thickness [nm]	Magnetic moment [μ_B/atom]	M_{VSM} [emu/m^2]	Remanence magnetisation [emu/m^2]	Coercive field [T]
12 (A)	V	2.6	0	5.8	3.7	0.52
	Gd	7.4	2.8			
11 (B)	V	2.4	0.7	7.9	3.5	0.44
	Gd	6.7	3.6			
8 (C)	V	1.7	0	6.6	4.4	0.48
	Gd	6.8	3.5			
6 (D)	V	1.2	1.2	7.9	4.9	0.50
	Gd	6.9	3.6			

4 Conclusions

New system of V/Gd epitaxial bilayers has been studied with the V layer thickness varying from 6 to 12 ML keeping Gd thickness constant at 20 ML. The epitaxial relationships of V/Gd corresponded to Nishiyama-Wasserman orientation. From X-ray reflectivity analysis the structural parameters like individual layer thickness, density and interface roughness were determined. Hysteresis loops measurements suggested easy-plane anisotropy for all samples. Basic magnetic properties like saturation magnetization, coercive field and remanence were obtained. Average magnetic moment analysis, taking Gd moment equal to 3.6 μ_B/atom reported for system of V/Gd, gives the possibility of induced V magnetic moment in up to 1.2 μ_B/atom depending on V layer thickness. Element sensitive magnetic measurements have to be performed in order to determine separately V and Gd magnetic moments.

Acknowledgements This work was partly supported by EU RTD project SPINOSA contract no. IST-2001-33334.

References

[1] D. R. Grempel and S. C. Ying, Phys. Rev. Lett. **45**, 1018 (1980).
[2] C. Rau, C. Liu, A. Schmalzbauer, and G. Xing, Phys. Rev. Lett. **57**, 2311 (1986).
[3] C. L. Fu, A. J. Freeman, and T. Oguchi, Phys. Rev. Lett. **54**, 2700 (1983).
[4] S. Blugel, B. Dittler, R. Zeller, and P. H. Dederichs, Appl. Phys. A **49**, 547 (1989).
[5] P. Martin, A. Vega, C.Demangeat, and H. Dreysse, J. Magn. Magn. Mater. **148**, 177 (1995).
[6] to be published in J of Alloys and Compounds, 2004.
[7] http://www-llb.cea.fr/prism/programs/simulreflec/simulreflec.html
[8] L.T. Baczewski, R. Kalinowski, and A. Wawro, J. Magn. Magn. Mater. **177–181**, 1305 (1998).
[9] private communication.
[10] K. Mergia, L.T. Baczewski, S. Messoloras, S. Hamada, T. Shinjo, H. Gamari-Seale, and J. Hauschild, Appl. Phys. A **74**, 1520 (2002).

Author Index

Alves, E. . . . . (c) 254
Ambacher, O.   (a) 333, (c) 233, (c) 337, (c) 341, (c) 347
Anceau, S. . . . . (a) 190
Aoyagi, Y. . . . . (a) 190
Apostolopoulos, V.. . . . . (c) 295
Araújo, D. . . . . (c) 341
Ardaravičius, L. . . . . (a) 203
Arizmendi, L. . . . . (a) 253, (a) 298
Arnoult, A. . . . . (a) 357
Ascarelli, P. . . . . (a) 249
Avigal, Y. . . . . (a) 249

Baczewski, L. T. . . . . (c) 405
Bąk-Misiuk, J. (c) 392
Baranowski, J. M. . . . . (c) 193, (c) 364
Barcz, A. . . . . (a) 298, (c) 219
Barnas, J. . . . . (c) 223
Bauer, G. . . . . (a) 324
Bausá, L. E. . . . . (a) 289
Bedel-Pereira, E.. . . . . (a) 357
Bercha, D. M. (a) 304
Bergman, J. P. . . . . (a) 216
Berus, T. . . . . (c) 351
Betko, J. . . . . (c) 382
Beuckert, G. . . . . (c) 265
Biggemen, D. (c) 285
Bilevych, Ye. O. . . . . (c) 317
Bimberg, D. . . . . (a) 345
Bock, W. . . . . (c) 347
Bockowski, M. . . . . (c) 198
Bogoboyashchyy, V. V. . . . . (c) 355
Borowska, A. (c) 351
Böttcher, T. . . . . (a) 207, (a) 212
Bouchier, D. . . . . (a) 329, (a) 353
Bourgeois, S. (c) 388
Bożek, R. . . . . (a) 212, (a) 216, (c) 364
Brink, D.J. . . . . (c) 223
Bugajski, M. . . . . (c) 396
Bulany, M. . . . . (c) 250
Burkov, Y. . . . . (c) 265
Butkutė, R. . . . . (c) 307

Cabié, M. . . . . (a) 357
Carmo, M. C. (c) 278
Cheng, Yung-Chen. . . . . (a) 221
Choi, H. W. . . . . (c) 202
Chua, S. J. . . . . (c) 202
Chwalis, B. . . . . (c) 193
Cimalla, I. . . . . (c) 233
Cimalla, V. . . . . (a) 333, (c) 337, (c) 341, (c) 347

Clerc, C. . . . . (a) 329
Colder, H. . . . . (c) 269
Consolati, G. (c) 257
Conte, G.. . . . . (a) 249, (c) 261
Courty, C. Josse. . . . . (c) 388
Czajka, R. . . . . (c) 351
Czernecki, R. (a) 207

Dawson, M. D. . . . . (c) 202
Débarre, D. . . . . (a) 329, (a) 353
Demchuk, Oleg . . . . (c) 401
Deren, Przemysław J. . . . . (c) 290
Derluyn, J. . . . . (c) 329
Diale, M.W. . . . . (c) 223
Dmowski, L. H. . . . . (a) 190
Druzhinin, A. A. . . . . (c) 333
Durygin, A. . . . . (c) 312
Dvoretsky, S. A. . . . . (c) 355

Eastman, L. F. . . . . (a) 203
Ecke, G. . . . . (c) 347
Edwards, P. R. . . . . (c) 202

Favre, A. . . . . (c) 295
Fedor, J. . . . . (c) 382
Feduniewicz, A. . . . . (a) 320
Feng, Shih-Wei . . . . (a) 221
Fernández, C. . . . . (c) 341
Ferriol, M. . . . . (a) 284
Figge, S. . . . . (a) 207, (a) 212
Fiore, A. . . . . (a) 338
Fisher, M. . . . . (a) 364
Forchel, A. . . . . (a) 364
Freitag, Th. . . . . (c) 347
Fricke, P. . . . . (c) 347
Frukacz, Z. . . . . (c) 312

Galitsyn, Yu. G. . . . . (c) 321, (c) 325
García, R. . . . . (c) 341
Gawith, C.B.E. . . . . (c) 295
Gebicki, W. . . . . (c) 198
Gelhausen, O.. . . . . (a) 229
Georgiev, N. . . . . (a) 368
Ghijsen, J. . . . . (c) 372
Glukhanyuk, V. . . . . (a) 195
Godignon, P. . . . . (c) 257, (c) 274
Godlewski, M. . . . . (a) 207, (a) 212, (a) 216, (c) 213,
. . . . (c) 250, (c) 281
Gołacki, Z. . . . . (c) 250
Golaszewska, K. . . . . (c) 219

Goldys, E. M. . . . (a) 207, (a) 212, (a) 216, (a) 229,
. . . . (c) 213
Gopal, A. V. (c) 368
Gorban', A. . . . . (c) 250
Goryachko, A. . . . . (a) 245, (c) 265
Gotszalk, T. . . . . (c) 219
Grandjean, N. . . . . (c) 193
Grzegory, I. . . . . . (a) 207, (a) 216, (a) 320, (c) 193
Gubisch, M. . . . . (c) 347
Gwilliam, R. M. . . . . (a) 239

Hageman, P. R. . . . . (a) 199
Halbwax, M. . . . . (a) 329, (a) 353
Hardtdegen, H. . . . . (a) 312
Hasenöhrl, S. (c) 382
Hayes, M. . . . . (c) 223
Hegarty, S. P. . . . . (a) 345
Hejduk, K. . . . . (c) 396
Henkel, K. . . . . (a) 245
Hernández, E. . . . . (c) 274
Hinzer, K. . . . . (a) 345
Hirayama, H. (a) 190
Homewood, K. P. . . . . (a) 239
Hommel, D. . . . . (a) 207, (a) 212
Huyet, G. . . . . (a) 345

Ilka, L. . . . . (c) 396
Imhoff, L. . . . . (c) 388
Ishikawa, H. . . . . (c) 368
Ivanov, V.Yu. (c) 250
Iwamura, Yasuo . . . . (a) 225
Izaki, Ryohei . . . . (a) 225
Izhnin, I. I. . . . . (c) 355

Jantsch, W. . . . . (a) 195, (c) 229
Janus, P. . . . . (c) 219
Jastrzebski, C. . . . . (c) 198
Jeon, C. W. . . . . (c) 202
Juršėnas, S. . . . . (a) 199, (a) 221

Kaiser, U. . . . . (c) 233
Kalish, R. . . . . (a) 249
Kaluza, N. . . . . (a) 312
Kamińska, A. . . . . (a) 190, (a) 298
Kaminska, E. (c) 219
Kaminski, M. (c) 198
Kaniewski, J. . . . . (c) 242, (c) 392
Karwasz, G. P. . . . . (c) 257
Kasai, J. . . . . (a) 368, (c) 368
Kechouane, M. . . . . (c) 285
Khachapuridze, A. . . . . (c) 250
Kheifets-Kobeleva, O. L. . . . . (c) 299
Kim, Hyoun Woo . . . . (a) 235
Kim, Nam Ho . . . . (a) 235
Knap, W. . . . . (c) 193
Kobelev, Ya. L. . . . . (c) 299
Koizumi, F. . . . . (c) 295

Konczewicz, L. . . . . (a) 190
Kondo, Y. . . . . (c) 295
Konovalov, I. (a) 308
Kosiba, R. . . . . (c) 347
Kosiel, K. . . . . (c) 392
Kovsh, A. R. (a) 345
Kozanecki, A. . . . . (a) 195, (c) 229
Kruszka, R. . . . . (c) 219
Kuchuk, A. . . . . (c) 219
Kudła, A. . . . . (c) 396
Kudrawiec, R. . . . . (a) 364, (c) 329, (c) 378
Kunert, H. W. . . . . (c) 206, (c) 223
Kuo, Cheng-Ta . . . . (a) 221
Kurilčik, G. . . . . (a) 199
Kuritsyn, D. . . . . (c) 229

Le Thanh, V. (a) 353
Lebedev, V. . . . . (c) 233
Lee, E. . . . . (c) 295
Lefebvre, P. . . . . (a) 190
Łepkowski, S. P. . . . . (a) 190
Leszczynski, M. . . . . (a) 207, (a) 320
Liberis, J. . . . . (a) 203
Lichtenberger, H. . . . . (a) 324
Lisauskas, V. (c) 307
Liu, C. . . . . (c) 202
Lopes, J. C. . . . . (c) 278
Lourenço, M. A. . . . . (a) 239
Łusakowska, E. . . . . (a) 212, (a) 298, (c) 213, (c) 396,
. . . . (c) 401

Machatine, A.G.I. . . . . (c) 223
Makarov, Yu. N. . . . . (a) 312
Makhova, Liudmila V. . . . . (a) 308
Malherbe, J. . . . . (c) 223
Mansurov, V. G. . . . . (c) 321, (c) 325
Marco de Lucas, M. C. . . . . (c) 388
Marie, P. . . . . (c) 269
Markus, A. . . . . (a) 338
Martin, R. W. (c) 202
Masri, P. . . . . (c) 341
Massies, J. . . . . (c) 193
Matkovskii, A. . . . . (c) 312
Matsumoto, Satoru . . . . (c) 238
Matulionis, A. (a) 203
Mazzeo, G. . . . . (c) 261
McInerney, J. G. . . . . (a) 345
Melnikova, N. V. . . . . (c) 299
Mergia, K. . . . . (c) 405
Mesnier, M. . . . . (c) 388
Messoloras, S. . . . . (c) 405
Miasojedovas, S. . . . . (a) 199
Mikhailov, N. N. . . . . (c) 355
Mikhrin, S. S. (a) 345
Milekhin, A. G. . . . . (c) 360
Mirabella, F. (c) 372
Misiewicz, J. . . . . (a) 364, (c) 329, (c) 378

Author Index

Moerman, I. . . . . (c) 329
Molina, S. I. . . . . (c) 341
Monemar, B. . . . . (a) 216
Monteiro, T. . . . . (c) 254
Montoya, E. . . . . (a) 289
Morales, F. M. . . . . (c) 341
Morawski, Andrzej. . . . . (c) 401
Morvic, M. . . . . (c) 382
Mozume, T. . . . . (a) 368, (c) 368
Mühlberger, M. . . . . (a) 324
Myslivecek, J. . . . . (a) 324

Neves, A. . . . . (c) 254, (c) 278
Nguyen, Lam H. . . . . (a) 329, (a) 353
Nikiforov, A. I. . . . . (c) 360
Nishimura, Suzuka . . . . (c) 238, (c) 303
Novák, J. . . . . (c) 382
Nugaeva, L. L. . . . . (c) 299

O'Brien, D. . . . . (a) 345
Oliveira, M. . . . . (c) 254
Olszyna, A. . . . . (c) 213
Omel'chuk, A. . . . . (c) 250
Ordejòn, P. . . . . (c) 274
Ostrovskii, I. P. . . . . (c) 333
Oszwaldowski, M. . . . . (c) 351
Oyama, Yasunao. . . . . (c) 303

Pakula, K. . . . . (c) 364
Paloumpa, I. . . . . (c) 265
Pankowski, P. . . . . (c) 405
Papis, E. . . . . (c) 219
Parkinson, B. A. . . . . (c) 372
Pastuszka, B. . . . . (a) 320
Paszkiewicz, B. . . . . (c) 378
Paszkiewicz, R. . . . . (c) 378
Pchelyakov, O. P. . . . . (c) 360
Perlin, P. . . . . (a) 207
Péter, Á. . . . . (a) 284
Pezoldt, J. . . . . . (a) 333, (c) 337, (c) 341, (c) 347
Phillips, M. R. . . . (a) 207, (a) 212, (a) 229, (c) 213
Pichon, L. . . . . (c) 269
Piotrowska, A. . . . . (c) 219
Podsiadlo, S. . . . . (c) 198
Polgár, K. . . . . (a) 284
Ponchet, A. . . . . (a) 357
Porowski, S. . . . . . . (a) 207, (a) 216, (a) 320
Potemski, M. . . . . (a) 181, (c) 193
Preobrazhenskii, V. V. . . . . (c) 325
Prinsloo, L. . . . . (c) 223
Prystawko, P. . . . . (a) 207, (c) 193
Przesławski, T. . . . . (c) 242, (c) 392
Przybylińska, H. . . . . . (a) 195, (a) 216, (c) 229

Ralchenko, V. . . . . (c) 261
Ramírez, M.O . . . . (a) 289
Rebollo, J. . . . . (c) 274

Regiński, K. . . . . (c) 242, (c) 392
Ribbat, C. . . . . (a) 345
Rita, E. . . . . (c) 254
Rizk, R. . . . . (c) 269
Rocher, A. . . . . (a) 357
Romanus, H. . . . . (c) 347
Rossi, M. C. . . . . (a) 249, (c) 261
Rudno-Rudzinski, W. . . . . (c) 329
Rurali, R. . . . . (c) 257, (c) 274
Rushchanskii, K. Z. . . . . (a) 304

Sacilotti, M. . . . . (c) 388
Safonov, K. L. . . . . (a) 333
Salvatori, S. . . . . (a) 249, (c) 261
Sanz-Hervás, A. . . . . (c) 341
Sawicki, M. . . . . (c) 250
Schäffler, F. . . . . (a) 324
Schelling, C. . . . . (a) 324
Schmeißer, D. . . . . (a) 245, (c) 265
Schmidt, A. A. . . . . (a) 333, (c) 337
Schmidt, R. . . . . (a) 312
Schulze, S. . . . . (c) 360
Shao, G. . . . . (a) 239
Shealy, J. R. . . . . (a) 203
Siddiqui, M. S. A. . . . . (a) 239
Sidorov, Yu. G. . . . . (c) 355
Siekacz, M. . . . . (a) 320
Simoyama, T. . . . . (c) 368
Sizov, F. F. . . . . (c) 355
Skierbiszewski, C. . . . . (a) 320
Soares, M. J. . . . . (c) 254, (c) 278
Sokołowska, A. . . . . (c) 213
Spaziani, F. . . . . (a) 249
Spiess, L. . . . . (c) 347
SpringThorpe, A. J. . . . . (a) 345
Stauden, Th. . . . . (c) 337, (c) 341, (c) 347
Steins, R. . . . . (a) 312
Stępniewski, R. . . . . (a) 181, (c) 193
Story, T. . . . . (c) 405
Strojek, B. . . . . (c) 198
Strupinski, W. . . . . (c) 378
Suchocki, A. . . . . (a) 190, (a) 298, (c) 312
Sugak, D. . . . . (c) 312
Sukach, A. V. . . . . (c) 317
Surkova, T.P. . . . . (c) 281
Suski, T. . . . . (a) 190
Swiatek, K. . . . . (c) 281
Syperek, M. . . . . (c) 378
Szargan, R. . . . . (a) 308
Szeloch, R. . . . . (c) 219
Szmidt, J. . . . . (c) 213
Sznajder, M. . . . . (a) 304
Szyszko, T. . . . . (c) 198

Tai, C.Y. . . . . (c) 295
Talalaev, R. A. . . . . (a) 312

Tamulevičius, S. . . . . (c) 307
Taylor, E. . . . . (c) 295
Teisseyre, H. . . . . (a) 190
Terashima, Kazutaka . . . . (c) 238, (c) 303
Tessler, L. R. . . . . (c) 285
Tetyorkin, V. V. . . . . (c) 317
Teys, S. A. . . . . (c) 360
Tłaczała, M. . . . . (c) 378
Tripathy, S. . . . . (c) 202
Trucchi, D. . . . . (a) 249
Trushin, Yu. V. . . . . (a) 333
Tsang, Jian-Shihn . . . . (a) 221

Užupis, A. . . . . (c) 307
Ulyanov, V. V. . . . . (c) 360
Uskov, A. V. . . . . (a) 345
Ustinov, V. M. . . . . (a) 345

Varavin, V. S. . . . . (c) 355
Vengalis, B. . . . . (c) 307
Vertiatchikh, A. . . . . (a) 203
Voigtländer, B. . . . . (a) 324

Wahl, U. . . . . (c) 254
Wasilewski, Z. . . . . (a) 320
Watson, I. M. . . . . (c) 202
Wawer, D. . . . . (c) 396
Wawro, A. . . . . (c) 405
Weih, P. . . . . (c) 341, (c) 347
Werbowy, A. . . . . (c) 213
White, J. K. . . . . (a) 345

Wolkenberg, A. . . . . (c) 242
Wollweber, J. . . . . (a) 245
Wosiński, Tadeusz. . . . . (c) 401
Wrzesińska, H. . . . . (c) 396
Wysmołek, A. . . . . (a) 181, (c) 193

Yakovlev, E. V. . . . . (a) 312
Yam, V. . . . . (a) 329, (a) 353
Yamaguchi, Shigeo . . . . (a) 225
Yamamoto, Atsushi . . . . (a) 225
Yang, C. C. . . . . (a) 221
Yastrubchak, Oksana . . . . (c) 401
Yatsunenko, S. . . . . (c) 250
Yeromenko, Y. . . . . (a) 245
Yoshida, H. . . . . (c) 368
Yudenkov, V. A. . . . . (c) 355

Zahn, D. R. T. . . . . (c) 360
Zakrzewski, A. J. . . . . (c) 247
Zdrojek, M. . . . . (c) 198
Zekentes, K. . . . . (c) 337
Zettler, J.-T. . . . . (a) 312
Zgheib, Ch. . . . . (c) 341
Zheng, Y. . . . . (a) 329
Zhong, Z. . . . . (a) 324
Zhukov, A. E. . . . . (a) 345
Zhuravlev, K. S. . . . . (c) 321, (c) 325
Zhydachevskii, Ya. . . . . (c) 312
Zimniak, M. . . . . (c) 351
Zlokazov, V. B. . . . . (c) 299
Žukauskas, A. . . . . (a) 199

Information for conference organizers and guest editors

The third journal section *physica status solidi (c) – conferences and critical reviews* is devoted to the publication of proceedings, ranging from large international meetings to specialized workshops, as well as collections of topical reviews on various areas of current solid state physics research. The new series has been launched in December 2002 with volume **0** (2002/03). It is available both as a regular journal both online and in hardcover print volumes, to be delivered to subscribers, conference contributors and participants (upon arrangement with the organizers). Single copies of pss (c) may be ordered as a book using its ISBN number. Regular subscriptions to pss (c) are offered in combination with pss (a) and/or pss (b) .

Essential details concerning layout and organization of the new journal series are:

- pss (c) is published as a full hardcover-bound series, carrying a standard green-coloured cover design, individually adapted according to the organizers' request which includes conference designation, logo, names of Guest Editors etc.

- Proceedings issues contain all conference contributions which have been peer-reviewed and accepted by the Guest Editors. Upon special agreement between the pss journal editors and the Guest Editors, part of the conference papers may also be published simultaneously in an issue of pss (a) or (b). For all papers, strict criteria for journal publications, i.e. positive peer-review by independent referees, are obligatory. All papers are unambiguously citable as phys. stat. sol. (a), (b), or (c) journal articles and will be covered by standard reference databases.

- All articles are published online in PDF format at Wiley InterScience. Access for registered users (e. g. conference participants with special password) may be installed. The online version contains colour figures at no additional cost, regardless of their colour or black/white representation in print.

- The Editorial Office provides document templates and style files for Word and LaTeX, respectively, to be used by all authors, allowing an easy manuscript preparation and length estimate of their paper with respect to the page limits given by the organizers.

- The issue is completed by a table of contents in topical order, an author index, a preface, listings of conference committee members, organizers and sponsors, and any additional material, if desired.

- The usual service of the Editorial Office is available and includes support in the refereeing process, acceptance messages, PDF proofs (for typesetted papers), free PDF reprints (hardcopy reprints may be ordered) as well as individual communication with authors and organizers. The use of a Web-based software system for online submission and refereeing of papers is offered to Guest Editors.

- The editors of pss (c) aim at a timely, professional, and high-quality print and online publication of proceedings, typically within only four to six months after a conference.

- Various service packages for production are available, including either full typesetting of papers using electronic manuscript data or publication-ready delivery of manuscript files (prepared using the template/style files) by the organizers.

For further details as well as an individual offer for the publication of the proceedings of your forthcoming conference or of a special issue containing topical reviews, please contact the Editorial Office at pss@wiley-vch.de (for other contact information see the title page).